Membrane Transport in Biology – Volume I

Membrane Transport in Biology

Edited by

G. Giebisch · D. C. Tosteson · H. H. Ussing

Associate Editor

M. T. Tosteson

Volume I

Concepts and Models

Contributors

O. S. Andersen J. E. Hall D. J. Hanahan U. V. Lassen
P. K. Lauf R. J. Lefkowitz E. Racker B. E. Rasmussen
S. A. Rudolph F. A. Sauer C. W. Slayman G. Stark
O. Sten-Knudsen H. H. Ussing

Editor
D. C. Tosteson

With 108 Figures and 36 Tables

Springer-Verlag Berlin · Heidelberg · New York 1978

Professor Dr. Gerhard Giebisch
Yale University, School of Medicine, Department of Physiology
333 Cedar Street, New Haven, Conn. 06510 / USA

Professor Dr. Daniel C. Tosteson, Dean
Harvard Medical School
25 Shattuck Street, Boston, Mass. 02115 / USA

Professor Dr. Hans H. Ussing
University of Copenhagen, Institute of Biological Chemistry A
13 Universitetsparken, DK – 2100 Copenhagen

Dr. Magdalena T. Tosteson
Harvard Medical School, Department of Physiology
25 Shattuck Street, Boston, Mass. 02115 / USA

ISBN-13: 978-3-642-46372-3 e-ISBN-13: 978-3-642-46370-9
DOI: 10.1007/978-3-642-46370-9

Library of Congress Cataloging in Publication Data: Main entry under title: Membrane transport in biology. 1. Biological transport. 2. Membranes (Biology). I. Giebisch, G., 1927–; II. Tosteson, D. C., 1925–; III. Ussing, Hans H., 1911–. [DNLM: 1. Biological transport. 2. Cell membrane – Physiology. QH 509 M 533] QH 509-M44. 574.8'75. 78-17669.

Softcover reprint of the hardcover 1st edition 1978

Typesetting, printing and bookbinding: Druckerei G. Appl, Wemding
2122/3020–543 21

Preface

This Volume forms the cornerstone of this series of four books on Membrane Transport in Biology. It includes chapters that address i) the theoretical basis of investigations of transport processes across biological membranes, ii) some of the experimental operations often used by scientists in this field, iii) chemical and biological properties common to most biological membranes, and iv) planar thin lipid bilayers as models for biological membranes. The themes developed in these chapters recur frequently throughout the entire series.

Transport of molecules across biological membranes is a special case of diffusion and convection in liquids. The conceptual frame of reference used by investigators in this field derives, in large part, from theories of such processes in homogeneous phases. Examples of the application of such theories to transport across biological membranes are found in Chapters 2 and 4 of this Volume. In Chapter 2, Sten-Knudsen emphasizes a statistical and molecular approach while, in Chapter 4 Sauer makes heavy use of the thermodynamics of irreversible processes. Taken together, these contributions introduce the reader to the two sets of ideas which have dominated the thinking of scientists working in this field. Theoretical consideration of a more special character are also included in several other Chapters in Volume I. For example, Ussing (Chapter 3) re-works the flux ratio equation which he introduced into the field of transport across biological membranes in 1949. Andersen (Chapter 11) discusses some of the physico-chemical properties of bilayers which place constraints on the applicability of the theory of transport processes in homogeneous systems to these extremely thin and highly organized structures. Stark (Chapter 12) treats quite thoroughly the theory of carrier mediated ion transport across bilayers. These essays describe many of the concepts now available to and used by investigators of membrane transport in biology.

Measurements of the rates of transport of substances across membranes involve certain common experimental operations. In Chapter 3, Ussing reviews conceptual and technical aspects of isotopic tracers as non-perturbing probes of transport processes. He was one of the first scientists to use these techniques when radioactive isotopes became available in the 1940's and has continued to observe and participate in the gathering of information with these tools since that time. Through the use of isotopic tracers, the operational description of net transport processes as the resultant of oppositely directed unidirectional fluxes became possible for the first time, thus deepening our grasp of the kinetic mechanisms underlying these phenomena. For example, measurements of fluxes with tracers permitted Ussing to recognize exchange diffusion as an important and hitherto unsuspected pathway for the mixing of molecules separated by membranes. Electrical methods have also been usefully employed to characterize the transport of ions across membranes. Lassen and Rasmussen review the advantages and pitfalls of these techniques when applied to cell membranes in Chapter 5. They point out that the introduction of microelectrodes into small cells may seriously perturb the system under investigation. Several applications of electrical methods to detect ion movements within and across membranes are included in the three last chapters on bilayers.

Biological membranes share certain molecular properties which influence all transport processes. These shared characteristics form the substance of several papers in this Volume. Hanahan introduces these with a thoughtful review of the chemical composition of red cell membranes. Racker contributes a brilliant analysis of membrane enzymes with special emphasis on ATPases (Chapter 8). Slayman reviews the genetic determination of membrane transport systems in Chapter 7. Lauf explores the relation between immunological reactions and transport phenomena. Rudolph and Lefkowitz provide an introduction to the rapidly growing contemporary literature on membrane receptors which modulate and regulate transport. These Chapters are both distinctive expressions of important dimensions of membrane transport in biology and also useful summaries of material which will be helpful to readers of subsequent Volumes in this series.

The ingenious work of Mueller and Rudin in the early 60's began a new stage of research on transport across biological membranes. Before that time, the minds of most scientists in the field were animated by concepts derived from the analysis of bulk systems of membranes with dimension and physical properties that are markedly different from the structures which surround cells and organelles. Their development of relatively stable thin lipid bilayers made available for the first time a model with chemical composition and thickness comparable to the membranes found in living cells. The planar character of Mueller-Rudin membranes permitted simultaneous electrical and chemical transport measurements. The large number of productive insights into membrane transport which have resulted from studies of these structures are summarized in the last three chapters of this Volume. Andersen describes the features of unmodified bilayers, Stark considers carriers and Hall discusses channels. The relatively simple and defined character of these systems permits quantitative analysis of many concepts which are valuable in thinking about transport across more complicated biological membranes

New Haven, Boston, Copenhagen

G. Giebisch
D. C. Tosteson
H. H. Ussing

Contents

List of Contributors XIX

Chapter 1 – Membrane Transport in Biology
(H. H. Ussing) 1

Chapter 2 – Passive Transport Processes
(O. Sten-Knudsen) 5

A. Introduction 5

B. Fundamental Definitions 5
I. Flux 5
II. Types of Passive Transport 6
1. Diffusion 6
2. Migration 7
3. Convection 7
III. Flux Equations 7
1. Migration Flux 7
2. Convection Flux 9
3. Diffusion Flux 9
a) Fick's Law 10
b) The Driving Force behind the Diffusion Process 11
4. Diffusion and Migration Proceeding Concurrently 12
5. Convection with Superimposed Diffusion 13

C. Diffusion Processes: Macroscopic Treatment 13
I. The Diffusion Equation 14
1. Classification of Diffusion Processes 15
II. Stationary Processes in One Dimension 15
1. Steady-State Diffusion in a Plate 16
2. The Permeability Coefficient 17
3. Stationary Diffusion through Two Different Media 19
III. Time-Dependent Processes 20
1. Kinetics of Exchange between Two Phases Separated by a Membrane . 21
a) One of the Phases is Infinitely Large 21
(i) Outer Concentration Zero 21
(ii) Outer Concentration Finite, Inner Concentration Initially Zero 22

b) Both Phases Comparable in Size . . . 23
c) Unidirectional Fluxes . . . 26
2. Instantaneous Point Source (Green's Function) . . . 27
a) Solutions of Diffusion Problems by Means of Green's Function . . . 29
(i) Initial Uniform Distribution in the One Half-Space . . . 30
(ii) The One Half-Space is Separated by an Impermeable Wall . . 30
(iii) The Presence of an Absorbing Barrier . . . 31
(iv) Variable Flux into the Half-Space . . . 31
3. Diffusion Out of a Plate . . . 32
a) Concentration Profiles . . . 32
b) The Time Constant for the Exchange of the Mean Concentration . . 35
4. Establishing the Stationary Concentration Profile . . . 37

D. Diffusion Processes: Microscopic Aspects . . . 40
I. Brownian Movements . . . 40
II. Smoluchowski's Treatment . . . 41
1. Statistical Interpretation of the Diffusion Equation . . . 41
2. Random Walk in One Dimension . . . 42
3. The Einstein-Smoluchowski Equation . . . 45
III. Random Walk and Fick's Law (Einstein) . . . 46
IV. The Smoluchowski Equation . . . 48
V. Kramers' Equation . . . 51
VI. Diffusion Coefficient and Mobility . . . 51
1. Einstein's Relation . . . 52
2. Einstein-Stokes' Relation . . . 53

E. Diffusion and Superimposed Convection . . . 53
I. The Equation of Motion . . . 54
II. Steady-State Concentration Profile . . . 55
1. Stationary Transport through a Membrane . . . 55
a) Determination of the Flux . . . 55
b) Unidirectional Fluxes and Flux Ratio . . . 56
c) The Concentration Profile . . . 57

F. Electrodiffusion . . . 58
I. Conductance . . . 59
II. The Nernst-Planck Equations . . . 61
1. Various Equivalent Forms . . . 61
2. The Poisson Equation . . . 63
a) Electroneutrality . . . 64
b) The Constant Field . . . 65
III. Membrane Equilibrium . . . 65
1. Nonosmotic Equilibrium . . . 65
a) The Nernst Equation . . . 65
b) Equivalent Electrical Circuit for the Ion-Selective Membrane . . . 67
2. Donnan Equilibrium . . . 68
a) Thermodynamic Treatment . . . 68
b) Concentration and Potential Profiles between the Phases (The Poisson-Boltzmann Equation) . . . 73
IV. Diffusion Potentials . . . 80
1. Charging Time and Redistribution Time . . . 82
2. The Henderson Regime . . . 84

3. The Planck Regime 85
a) Planck's General Relations 86
b) The Electrical Equivalent Circuit for the Planck Regime 87
c) Planck's Expression for the Diffusional Potential 90
V. Electrodiffusion through Membranes 91
1. Single Salt 91
a) Diffusion Potential 91
b) Membrane Resistance 92
c) Equivalent Electrical Circuit 94
d) Electroneutrality 95
2. Ion-Selective Membrane 96
3. Membrane Separating Electrolytes Having a Common Ion 98
a) Flux Ratio 98
b) The Goldman Regime 99
(i) The Separate Ionic Currents and the Diffusion Potential 100
(ii) Total Membrane Current and Membrane Potential 102
(iii) Concentration Profiles and Membrane Potential 104
(iv) Ionic Conductances and Membrane Potential 106
c) Equivalent Electrical Circuits 109
Acknowledgements 110
List of Symbols 110
References 112

Chapter 3 – Interpretation of Tracer Fluxes
(H. H. Ussing) 115
A. Introduction 115
B. Fundamental Concepts 115
C. Tracer Permeability Coefficients 116
I. Measurement of Tracer Permeability Coefficients 116
II. Multicompartment Systems 117
D. The Concept of Unidirectional Flux 118
I. Unidirectional Fluxes 118
II. Isotope Effects 118
III. Associated Unidirectional Fluxes 119
IV. The Relation of Tracer Fluxes to Active and Passive Transport 119
V. Effects of Membrane Potentials on Ionic Fluxes 120
VI. Exchange Diffusion 120
VII. Limitations for Integration of Flux Equations 121
E. Flux Ratio Analysis 122
I. The Flux Ratio Equation 122
II. Derivation of the Flux Ratio Equation 123
III. Estimation of Electrochemical Potential Differences 126
IV. The Short-Circuiting Method 127
V. Flux Ratio with Solvent Drag 127
VI. Solvent Drag on Non-Electrolytes and Water 130
VII. Solute-Solute Interactions 131
VIII. Meaning of the Term "Interaction" 131
IX. Interpretation of Deviations from the Flux Ratio Equation 132

F. Examples . . . 133
I. The Short-Circuited Frog Skin . . . 133
II. Single-File Diffusion . . . 137
III. Solvent Drag Effects . . . 138

G. Concluding Remarks . . . 139

References . . . 139

Chapter 4 – Nonequilibrium Thermodynamics of Isotope Flow through Membranes
(F. A. Sauer) . . . 141

A. Introduction . . . 141

B. The System: Definitions and Mathematical Techniques . . . 141

C. Nonthermodynamic Considerations of Isotope Flow . . . 148

D. The Nonequilibrium Thermodynamic Approach . . . 156

E. Applications to Model Systems . . . 162

F. Summary . . . 166

Acknowledgements . . . 167

List of Symbols . . . 167

References . . . 168

Chapter 5 – Use of Microelectrodes for Measurement of Membrane Potentials
(U. V. Lassen and B. E. Rasmussen) . . . 169

A. Introduction . . . 169

B. Principles of Bioelectric Recording . . . 170
I. Electrode Chains and Junction Potentials . . . 171
II. Comments on Electronic Equipment . . . 175

C. The Glass Capillary Microelectrode . . . 178
I. The Suspension Effect . . . 181
II. Diffusion Regime of the Microelectrode Tip . . . 183

D. Potential Recording with Microelectrodes . . . 190
I. Penetration of the Cell Membrane . . . 190
II. Microelectrodes and Leaks in the Membrane . . . 194

E. Epilogue . . . 201

Acknowledgements . . . 202

References . . . 202

Chapter 6 – Chemical Composition of Membranes
(D. J. Hanahan) . . . 205

A. Introduction . . . 205

B. Some General Observations on Erythrocyte Composition . . . 206
I. General Composition . . . 206
II. Ion Composition . . . 206
III. Age-Related Patterns . . . 208
1. Intact Mixed-Age Erythrocytes . . . 208
2. Age (density)-Separated Erythrocytes . . . 210

C. Comments on Major Components of the Erythrocyte Membrane . . . 214
I. Lipid: an Appraisal of Composition and Orientation or Localization . . . 215
1. Some Structural Features of the Erythrocyte Lipids . . . 216
a) Neutral Lipid . . . 216
b) Phospholipids . . . 216
c) Sphingoglycolipids . . . 217
2. Observations on Types of Phospholipids Present in Human, Cow, and Pig Erythrocyte . . . 217
a) Fatty Acid Composition . . . 218
b) Positioning of Fatty Acids . . . 219
c) Importance of Fatty Acid Composition . . . 220
3. Localization of Lipids in Membranes: A Compositional Study of a Different Type . . . 221
a) Some General Observations . . . 222
b) Use of Enzymes as Probes for Location of Lipids in Membranes . . . 223
4. Summary of Observations on Phospholipase Action on Erythrocytes . . . 224
a) Phospholipase A_2 . . . 224
b) Phospholipase C . . . 225
c) Sphingomyelinase . . . 225
d) Combined Activity of Phospholipase C and Sphingomyelinase . . . 225
5. Development of the Concept of Asymmetric Location of Phospholipids in Membranes . . . 225
6. On the Validity of the Lipid Asymmetry Proposal . . . 226
7. Summary Statement . . . 231

II. Protein Composition . . . 231
1. General Comments . . . 232
2. Nature of Polypeptide Patterns on SDS-PAGE . . . 233
a) Specific Protein Components Revealed by SDS Gel Electrophoresis . . . 233
b) Studies on "Spectrin" of the Human Erythrocyte Membrane . . . 234
c) Observations on the Ox (Bovine) Erythrocyte Polypeptide Heterogeneity . . . 235
3. Summary Statement . . . 236

References . . . 236

Chapter 7 – Genetic Determination of Membrane Transport
(C. W. Slayman) . . . 239

A. Introduction . . . 239

B. Microorganisms . . . 240
I. Isolation of Transport Mutants . . . 240
II. Kinds of Genetic Analysis . . . 241
III. Examples of the Use of Genetic Analysis . . . 243

C. Higher Organisms . . . 246
I. Cystinuria . . . 246
II. HK/LK Erythrocytes . . . 247
D. Cultured Somatic Cells . . . 250
I. Methods for Selecting Transport Mutants . . . 251
II. The Kinds of Genetic Information that can be Obtained . . . 252
III. Ouabain-Resistant Mutants . . . 252
E. Conclusions . . . 254
References . . . 254

Chapter 8 – Mechanisms of Ion Transport and ATP Formation
(E. Racker) . . . 259
A. Translocation of Protons by the Oxidation Chain of Mitochondria and Chloroplasts . . . 259
I. The Chemiosmotic Mechanism of Mitchell . . . 259
II. Asymmetry of Oxidation Chain of Mitochondria and Chloroplasts . . . 261
III. The Coenzyme Q Cycle . . . 261
B. The Translocation of Protons by Bacteriorhodopsin . . . 263
I. Proton Movements and ATP Formation . . . 263
II. Mechanism of Proton Translacotion . . . 263
C. The Translocation of Protons by the Oligomycin- or Dicyclohexylcarbodiimide-Sensitive ATPase of Mitochondria, Chloroplasts and Bacteria . . . 265
I. Proton Movements and ATP Formation . . . 265
II. Properties of the Isolated Oligomycin-Sensitive ATPase Complex . . . 266
1. The Water-Soluble ATPase . . . 267
2. The Oligomycin-Sensitivity Conferral Protein (OSCP) . . . 269
3. The Heat-Stable Coupling Factor F_6 (F_{c2}) . . . 270
4. Coupling Factor 2 (F_2) . . . 270
III. Model of the Proton Pump and its Mode of Action . . . 271
1. The Mitchell Hypothesis . . . 272
2. The Phosphoenzyme Intermediate Hypothesis . . . 273
3. The Boyer-Slater Hypothesis . . . 273
D. Translocation of Calcium by the ATPase Complex of Sarcoplasmic Reticulum . 274
I. Properties of the Pump . . . 274
II. Properties of the Ca^{++}-ATPase Complex . . . 276
1. Latency of the Ca^{++}-ATPase . . . 276
2. Structural Properties of the Ca^{++}-ATPase Complex . . . 276
3. Catalytic Properties of the Ca^{++}-ATPase Complex . . . 277
II. The Reconstituted Pump and its Mechanism of Action . . . 277
E. Translocation of Sodium and Potassium Ions by the ATPase Complex of the Plasma Membrane . . . 281
I. Properties of the Pump . . . 281
II. Properties of the Na^{+}-K^{+}-ATPase Complex . . . 282
1. Latency of the ATPase . . . 282
2. Structural Properties of the Enzyme Complex . . . 282
3. Catalytic Properties of the Enzyme Complex . . . 283
III. The Reconstituted Pump . . . 284
F. Concluding Remarks . . . 286

Acknowledgement 287
Abbreviations 287
Addendum 287
References 287

Chapter 9 – Membrane Immunological Reactions and Transport
(P. K. Lauf) 291

A. Introduction: The Concept 291

B. Immunological Reactions and Membrane Transport Proteins 292
I. Introduction 292
II. Antibodies Against the Na^+-K^+-ATPase 293
1. Properties of Antigens and Antibodies 293
2. Sidedness of Binding and Immunological Effects on the Na^+-K^+-ATPase and its Partial Reactions 295
a) Sidedness of Binding 295
b) Immunological Effects on the Na^+-K^+-ATPase Activity 295
c) Effects of Partial Reactions of the Na^+-K^+-ATPase 296
3. Immunological Alteration of Cation Fluxes in Resealed Ghosts 297
4. Species and Organ Specificities of Immunological Reactions Involving the Na^+-K^+-ATPase 297
III. Antibodies Against Ca^{++}-ATPase of Sarcoplasmic Reticulum 298
IV. Conclusion 299

C. Immunological Reactions at the Outer Membrane Surface and Cation Transport in Erythrocytes 299
I. Introduction 299
II. Sheep Red Cells 301
1. Cation Transport, Genetics and Immunological Parameters 301
a) Cellular Cations and Genetics 301
b) Membrane Antigens and Genetics 302
c) Active and Passive Cation Transport 303
d) Ouabain Binding 304
e) Na^+-K^+-ATPase 304
2. The Effect of Antibodies on Cation Transport 305
a) Modification of Cation Pump and Leak Fluxes 305
b) Activation of the Na^+-K^+-ATPase 307
c) Correlation between Antigenic Sites and Na^+-K^+ Pumps 308
3. Properties of the ML Surface Antigens and Antibodies 310
a) Antigens 310
b) Antibodies 311
4. Developmental Aspects of Transport and Antigens 312
a) Red Cells of Newborn Sheep 312
b) Stress-Induced Erythrocyte Regeneration 313
III. Cation Transport Polymorphism and Antigenic Parameters in Red Cells of Ruminants Other than Sheep 315
1. Goat Red Cells 315
a) Cations and Antigens 315
b) Cation Transport and its Modification by Antibody 315
2. Cattle Red Cells 317

IV. Human Red Cells 318
1. The Rhesus Antigen Complex and Cation Transport 318
2. The En(a)-Negative Red Cell as Physiological Model 319
V. Conclusion 320

D. Membrane Immunological Reactions and Cation Transport in Lymphocytes and Other Cells 321
I. Lymphocytes 321
1. Introduction 321
2. Cellular Differentiation and Membrane Surface Receptors of Lymphocytes 321
3. Cellular and Membrane Morphological and Biochemical Changes Induced by Immunological Reactions in Lymphocytes 324
4. Modification of Monovalent Cation Transport 326
a) General Aspect of the Effect of Immunological Reactions 326
b) Cation Transport in the Absence of Immunological Reactions . . . 327
c) Cation Transport Changes Induced by Immunological Reactions . . 330
5. Requirement of Bivalent Cations for Lymphocyte Stimulation by Immunological Reactions 338
II. Tumor Cells 338
1. Introduction 338
2. Passive Permeability Changes Induced by Lectins 339
III. Conclusion 340

E. Summary and Prospectus 341

Acknowledgement 342

References 342

Chapter 10 – Membrane Receptors, Cyclic Nucleotides, and Transport

(S. A. Rudolph and R. J. Lefkowitz) 349

A. Introduction 349

B. Beta-Adrenergic-Receptor Binding in Avian and Amphibian Erythrocytes . . . 350

C. Beta-Adrenergic-Mediated Transport Processes in Avian and Amphibian Erythrocytes 352

D. The Amphibian Bladder 360

E. Cholera Enterotoxin 362

F. The Superior Cervical Ganglion 363

G. Nicotinic Cholinergic Receptors 363

H. The Heart 364

J. General Comments and Conclusions 365

References 366

Chapter 11 – Permeability Properties of Unmodified Lipid Bilayer Membranes

(O. S. Andersen) 369

A. Introduction 369

B. Lipid Bilayer Membranes 370
I. Capacitance 370
II. Composition 372
C. Transport Model and the Potential Energy Barrier 373
I. The Transport Model 373
II. Unstirred Layers 375
1. Stationary Fluxes 376
2. Transient Fluxes 376
3. The Membrane-Solution Interface 378
4. Chemical Reaction in Unstirred Layers 378
III. Potential Energy of Ions Within Lipid Bilayers 379
1. The Born Energy 379
2. The "Image" Force 380
3. Diffusion or Distortion? 384
IV. Potential Energy of Dipolar Molecules Within Lipid Bilayers 385
V. Interfacial Potentials 385
1. Diffuse Double-Layer Potentials 387
2. Dipole Potentials 388
VI. Hydrophobic Interactions 389
VII. The Potential Energy Barrier 390
D. Permeability to Neutral Solutes 392
I. Partition Coefficients 393
1. Nonpolar Solutes 393
2. Polar Solutes 394
II. Mobility 395
1. Indirect Measurements 395
a) Microviscosity 395
b) Walden's Rule 395
2. Direct Estimates 396
3. Variation through the Membrane 396
III. Permeability 397
1. H_2O 397
2. Organic Solutes 399
IV. The Rate-Limiting Barrier for Solute Movement 400
E. Ion Permeability 402
I. Tracer Flux Measurements 403
II. Anion Permeability 404
1. Stationary Conductance Changes 404
2. Translocation through the Membrane Interior 406
a) The Transport Model 408
b) Kinetics of Charge Translocation 415
c) Temperature-Dependence 419
d) Ion Translocation as a Function of Membrane Composition 419
III. Positive Ions 425
IV. Interactions Among Ions Absorbed into Lipid Membranes 426
1. Space Charge-Limited Conductance 427
2. Blocking Phenomena 428
3. The Three-Capacitor Model 430
a) Charge Adsorption 431
b) Charge Translocation 432
4. Discrete Charge Effects? 435

Acknowledgements 439
References 439

Chapter 12 – Carrier-Mediated Ion Transport Across Thin Lipid Membranes
(G. Stark) 447

A. Introduction 447
I. Carriers and Pores 448
II. A Survey of Suggested Ion Carriers 449

B. Carriers of Hydrogen Ions 450

C. Macrocyclic Carriers 456
I. Neutral Carriers 456
II. Charged Carriers 461

D. The Iodide-Iodine System 462

E. The Carrier-Transport Model 463
I. Kinetic Analysis of the Carrier Model 466
II. Valinomycin and Trinactin 470

F. Biological Implications 471

References 472

Chapter 13 – Channels in Black Lipid Films
(J. E. Hall) 475

A. Introduction 475

B. Basic Experiments 477
I. Demonstration of Conductance by Pore 478
II. Basic Conductance Characteristics 483
1. Steady-State Current-Voltage Curves 483
2. Conductance and Antibiotic Concentration 488
3. Kinetics of Conductance Development: Response to a Voltage Pulse . 491

C. Advanced Experiments 496
I. Introduction 496
II. Single-Step Experiments 498
1. The Probability Distribution 498
2. EIM and Hemocyanin: The Unit Event Explains High-Level Conductance 503
3. Noise Measurements and the Unit Conductance 506
4. Noise Measurements on Alamethicin 510
5. Noise Measurements on Monazomycin 512
6. Compounds with Unknown Unit Events: Summary 513
III. Conductance and Ion Selectivity of Unit Channels 513
1. Introduction 513
2. Conductance and Selectivity of the Gramicidin Unit Event 514
3. Conductance and Selectivity of the Unit Events of EIM and Hemocyanin 515
4. Conductance and Selectivity of the Alamethicin Unit Event Levels . . . 516

IV. Time Course of the Unit Event 517
V. Alteration of the Molecule and the Membrane 519
1. Effects of Membrane Composition 520
2. Alteration of the Pore Forming Molecule 523
D. Possible Molecular Mechanisms of Pore Formation 525
Acknowledgements 529
References 529

Subject Index 533

List of Contributors

Olaf S. Andersen
Cornell University, Medical College, Department of Physiology,
1300 York Avenue, New York, N.Y. 10021 / USA

James E. Hall
Dept. of Physiology, California College of Medicine, University of California,
Irvine, California 92717 / USA

Donald J. Hanahan
The University of Texas, Health Science Center at San Antonio, Department of Biochemistry, 7703 Floyd Curl Drive, San Antonio, Texas 78284 / USA

Ulrik V. Lassen
University of Copenhagen, August Krogh Institute, Zoophysiological Laboratory B, 13 Universitetsparken, DK-2100 Copenhagen

Peter K. Lauf
Department of Physiology, Duke University Medical Center,
Durham, North Carolina 27710 / USA

Robert J. Lefkowitz
Duke University Medical Center, M 3325,
Durham, North Carolina 27710 / USA

Efraim Racker
Cornell University, Section of Biochemistry, Molecular and Cell Biology,
Wing Hall, Ithaca, New York 14853 / USA

B. E. Rasmussen
University of Copenhagen, August Krogh Institute, Zoophysiological Laboratory B, 13 Universitetsparken, DK-2100 Copenhagen

Stephen A. Rudolph
Case Western Reserve University, School of Medicine, Dept. of Pharmacology,
Cleveland, Ohio 44106 / USA

Friedrich A. Sauer
Max-Planck-Institut für Biophysik, Kennedyallee 70, D-6000 Frankfurt/Main 70

C. W. Slayman
Yale University, School of Medicine, Department of Human Genetics,
333 Cedar Street, New Haven, Conn. 06510 / USA

Günther Stark
Universität Konstanz, Fachbereich Biologie, D-7750 Konstanz

Ove Sten-Knudsen
University of Copenhagen, Panuminstituttet, Department of Biophysics,
Blegdamsvej 3 C, DK-2200 Copenhagen N

Hans H. Ussing
University of Copenhagen, Institute of Biological Chemistry A,
13 Universitetsparken, DK-2100 Copenhagen

Chapter 1

Membrane Transport in Biology

H. H. Ussing

Recent years have seen an enormous increase in interest in the function of biological membranes. It has turned out that not only transport phenomena but also such important processes as oxidative metabolism, protein synthesis and several other synthetic processes are intimately connected with – and apparently dependent on – membrane processes. At the same time, the development of electron-microscopic methods has given reality to the originally hypothetical concept of plasma membranes and revealed a whole host of unknown membrane-covered organelles. This again means that arguments based on membrane properties have become integral parts of biological and medical thinking: one needs only to mention the role attributed to membrane processes in modern pharmacology and nerve physiology. But this prolific growth of interest in membranes has started out more or less independently in many areas of physiology, biochemistry, biophysics, anatomy, medicine, pathology, and pharmacology, all these areas developing their own nomenclature, their own experimental approaches and their own pet ideas and phobias. There has not been time for interdisciplinary exchange of information and ideas. There is a good chance that information or ideas which each of us is groping for in the dark are already available in another discipline engaged in membrane research.

At this juncture it would have been very nice indeed to be able to offer a single unifying idea to the reader, an idea which could make all other approaches superfluous. This, however, is not possible. What we can offer is a collection of chapters, characteristic samples of different approaches to the study of transport through biological membranes. Without being dogmatic about it we have arranged the samples in order of increasing complexity, encompassing as many types of cells and tissues as space permitted. Many important subjects had to be left out in order to allow for sufficient coverage of the examples chosen. The aim has been not to cater for the specialists, but rather to offer workers in any one region of membrane transport an opportunity to acquaint themselves with the ideas and approaches of people working on membrane transport in remote fields.

Unfortunately, in passing from one chapter to another, the reader will experience a "language problem". Due to its polyphyletic nature, membrane transport science has not yet agreed on a common set of symbols and concepts. The editors have endeavoured to "streamline" the symbols as much as possible, but even in their own chapters they have not been able to adhere faithfully to their own recommendations, and other authors have been allowed an even freer hand. The use of different symbols for the same physical quantity is just a nuisance we have to live with. It should be remembered, however, that different sets of symbols often reflect different ways of thinking. The physical description of an intricate biological object like a membrane is only made possible by the choice of parameters assumed to be the important ones. In other words, one has to choose a proper membrane model. Over the years, many models have been used, and each of them has left its mark in the form of equations and important parameters to be measured. Despite recent advances in the understanding of membrane structure and membrane function, it is often difficult to decide which formalism one should prefer in describing a given membrane phenomenon. The choice may vary from one object to the next. Nevertheless, some general rules as to the proper choice of model can be extracted from our present knowledge. The symbols used to describe membrane transport processes usually come from one of three physicochemical disciplines: kinetics, equilibrium thermodynamics, and irreversible thermodynamics. The kinetic treatment undoubtedly comes closest to the ultimate aim of a molecular description of the events involved in membrane transport. On the other hand, it is usually impossible to measure all the parameters necessary for the molecular kinetic treatment of the transport process. Even when we are dealing with "simple" diffusion, we usually lack precise information about the membrane thickness and the variations of solubility, diffusion resistance, and charge distribution along the diffusion path. But much worse: modern theories about membrane structure, as represented for instance by the fluid mosaic membrane theory of Singer and Nicolson, indicate that although plasma membranes may consist mainly of double leaflets of lipids, the penetration – notably of hydrophilic molecules – may depend on special protein molecules implanted in, and partly traversing, the lipid phase. These special proteins define the major transport pathways and also drastically modify the transport kinetics. Indeed, in living membranes we meet kinetic patterns that are foreign to simple diffusion theory, such as exchange diffusion and single file diffusion, not to speak of active transport. Nevertheless, kinetic approaches have turned out to work beautifully in many systems. When a full kinetic treatment is impossible for lack of detailed information, we have to fall back on thermodynamics and flux ratio analysis. We try to extract useful information from incompletely described physical systems. The application of equilibrium thermodynamics to transport processes seems to be a *contradictio in adjecto*. Nevertheless, classical thermodynamic considerations have served well in many instances, especially for systems in quasi-equilibrium.

In recent years irreversible thermodynamics has become widely used by workers in the transport field. Irreversible thermodynamics is designed to handle cases where there is interaction between different flows. In order to handle these interactions, the theory makes the simplifying assumption that a flow is always proportional to the force acting on the species in question. The theory is therefore applicable mainly under conditions where the interaction between certain flows is pronounced, but where the forces are so modest that the proportionality between force and flow still holds. Now we can see clearly that irreversible thermodynamics must be well suited to cases where several species of molecules and ions are moving through a water-filled pore where the water phase itself is moving under the influence of osmotic or hydrostatic force.

The other extreme is represented by, for instance, sodium ions entering a nerve fibre by way of a strictly sodium-selective channel under the influence of an electrochemical potential difference of, say, 200 mV. In this case there is no obvious interaction between flows, since only one species is moving, and the force is so violent that proportionality between force and flow cannot be assumed. Here an irreversible thermodynamic treatment would hardly be acceptable. This last-mentioned example is by no means unique. On the contrary, it is becoming increasingly apparent that the passage of most ions and hydrophilic molecules through cell membranes depends on highly selective pathways open only to one or a few species. This necessarily limits the interaction with other moving species to that mediated by changes in the electric potential of the bathing media. Furthermore, much of the passage through specific pathways exhibits strongly non-linear properties in relation to the forces involved. In such cases irreversible thermodynamics should be used with caution.

There can be no doubt that in years to come chemical and biochemical methods are going to play an increasingly important role in studies of membrane transport. As the molecular structures of the entities responsible for active transport, exchange diffusion, electro-diffusion etc. become known, many of the present-day formalisms are likely to become superfluous or obsolete. Although this stage has not been reached as yet, it is important to remember that transport studies can circumscribe the typical processes responsible for the transfer of substances, but that the refinement of the kinetic approach beyond a certain point may be less rewarding than a blunt biochemical attack. Then the transport worker has done his duty. The transport worker can go.

When this is said it should be remembered, however, that transport studies, besides being concerned with the organizational level of membrane structure and function, are also concerned with biological processes at the higher organizational levels of physiology, pharmacology and pathology. At these levels, the molecular events are usually unimportant and a description on the basis of active and passive transport rates, inhibitions, stimulations, conductances, capacitances, etc. are the useful parameters. Thus in the forseeable future transport studies will remain an important part of biology and an indispensable tool for medical science.

Chapter 2

Passive Transport Processes

O. STEN-KNUDSEN

This essay is dedicated to the memory of Albert Cass, *Jr. in recollection of a valuable friendship and stimulating discussions about many of the topics discussed here.*

A. Introduction

This chapter will provide an account of some of the physical mechanisms and concepts relating to the description of the passive transport of substances through homogeneous media (simple passive transport theory). Even though in many cases the passive transport of substances through biological membranes appears to take place by more complicated mechanisms, this does not mean that the theory of simple passive transport is of no value in describing passive transport through cell membranes. It is in fact difficult to understand and describe the behaviour of a complicated physical system such as that of a cell membrane, if we are not familiar with the mechanisms operating in the case of a simple system and with the laws governing such mechanisms. Further, in describing the transport of substances through cell membranes, we employ all those concepts which have been found helpful in the case of simple passive transport theory. The aim of these pages, therefore, is to provide an account of the most important elements of simple passive transport theory that can serve as a necessary background for understanding passive transport through cell membranes, and that can serve as a starting point for understanding those modifications it has been found necessary to introduce.

B. Fundamental Definitions

I. Flux

In any system, there will be passive transport of a substance provided the distribution of the substance in the system (for example between the intracellular and the extracellular phase) does not correspond to the thermodynamic equilibrium distribution of the substance. The passive transport of the substance thus reflects the tendency of the system to move towards the equilibrium position. In our discussion it will be assumed that the system is *isothermal,* and we will limit the discussion to *mass transport* in a solution, *i. e.,* transport of atoms, molecules and submicroscopic particles that can occur in an electrically neutral or an ionized state. Mass transport can be considered from two points of view: (a) In the *individual* mode of consideration, an attempt is made to describe how the individual particle moves about in the space available. (b) In the *collective* mode of consideration, a collection of particles is considered as a unit, and an attempt is made to describe the behaviour of the unit on the basis of our knowledge of the movements of the individual particles. If in the collection of particles the number moving in a given direction exceeds the number moving in the opposite direction, there will be a *net transport* in that direction. To characterize the intensity of the net transport of a given particle component in the system, it is convenient to introduce the *transport-flow density* or *flux*, J, of the transported component. The *flux* is defined as the amount of substance which per unit of time passes a unit of area placed at right angles to the direction of the transport flow. Flux has therefore the following dimensions:

$$J \equiv (\text{amount})\, \text{m}^{-2}\, \text{s}^{-1}\,, \tag{1}$$

where the amount is given in the units most suited to the given situation, *e. g.* number of particles, kg, cm^3, mol, etc.

II. Types of Passive Transport

Mass transport in an isothermal system takes place by the following mechanisms:

1. Diffusion

Because of the irregular thermal movements (molecular chaos) of both the dissolved particles and the molecules of the solvent, all the molecules in the system will continually change places and move in the available space in a completely random manner. If the concentration is uniform throughout the system, the thermal movements will not result in any net transport of dissolved particles in any given direction. On the other hand, if the dissolved particles are distributed nonuniformly throughout the system, there will be a net transport of particles in the direction of a lower concentration. This transport arises because the number of particles that wander *out* of a region of high concentration in a given interval of time is greater than the number of particles that wander *in* from a neighbouring region with lower concentration in the same interval. Transport of a substance arising from the thermal self-movements of the dissolved particles combined with the presence of a concentration gradient in the substance is called *diffusion*.

2. Migration

If each dissolved particle is under the influence of an external force, the effect of this will be that all the particles will experience an extra velocity component in the direction of the force, this extra component being superimposed on the thermal movements. The result is that each particle, and thereby also the collection of particles as a whole, moves with a given mean velocity in the direction of the force. In this situation, therefore, there will also be a net transport of particles, even though the concentration is uniform throughout the system. This type of transport, due to the presence of an external field of force acting on each dissolved particle, is called *migration.* The forces that can be of interest are *gravity* and other *g-forces,* and also *electrical forces* if the particles carry a net electric charge. Transport by migration will take place even though the system contains only a single dissolved particle. In contrast to this, it is only meaningful to talk of the transport of matter by diffusion, which is a collective phenomenon, when the number of dissolved particles is sufficiently high for the concentration of particles to constitute a continuum in time and space, *i. e.* provided it is possible to ignore the influence of thermal fluctuations on the particle concentration.

3. Convection

Finally, the transport of material can occur because the system as a whole is not at rest, but is flowing in a given direction as a result of a pressure fall throughout the system. This type of transport is called hydrodynamic flow or sometimes *convection.*

The transport of material often occurs by a combination of several of the mechanisms mentioned above. A combination which occurs frequently is diffusion superimposed on migration. If the transported substances are ions and the external driving force is an electric field, the transport process is called *electrodiffusion.*

III. Flux Equations

In this section, the explicit expressions will be established for the fluxes corresponding to the transport mechanisms mentioned above.

1. Migration Flux

We consider a particle of mass, m, which is suspended in a fluid. The particle is under the influence of an external force, X (*e. g.* gravity), acting in the positive direction of the axis. Let the velocity at time t be $v\,(\mathrm{ms}^{-1})$. During its movement, the particle must displace molecules of the solvent, and is thus subjected to a resisting force, X_f. As a first approximation, this is assumed to be proportional to the instantaneous velocity v, so that

$$X_f = -fv, \tag{2}$$

where the minus sign arises because the frictional force X_f and the velocity v are in opposite directions. The constant f is called the *friction coefficient* ($N \cdot s \cdot m^{-1}$) and depends on the shape and size of the particles and on the properties of the surrounding medium. The force acting on the particle is therefore $X + X_f = X - fv$, so that the equation of motion of the particle becomes

$$m \cdot \frac{dv}{dt} = X - fv.$$

When the movement of particle has become *stationary* ($dv/dt = 0$), $X - fv_s = 0$, where v_s is the stationary velocity of the particle. We therefore have

$$v_s = \frac{1}{f} X = BX, \tag{3}$$

where the new constant

$$B = 1/f = v_s/X \tag{4}$$

is called the *mechanical mobility* of the particle. It is thus equal to the stationary velocity of the particle under the influence of a unit force. In the SI system the dimensions of B are therefore

$$B \equiv m \cdot s^{-1} \cdot N^{-1}.$$

We now consider a collection of particles uniformly distributed throughout the solution, and all moving with the same stationary velocity v under the influence of the external driving force X. An element of surface with area A is then considered to be placed at right angles to the direction of movement of the particles. During time dt each separate particle will have moved a distance $v \cdot dt$. All the particles that are moving towards the surface element and at time t are present within the volume element defined by A and $v \cdot dt$ will therefore by time $t + dt$ have just passed the surface area A. This number of particles, dn, is equal to

$$dn = ANvdt,$$

where N is the number of particles per m^3. The flux is the number of particles passing a unit area per unit time. We therefore have $J = (dn/dt)/A$, or, invoking the expression above,

$$J = Nv, \tag{5}$$

which is the expression for the *migration flux*. This can also be expressed in terms of the driving force, X, which acts on each individual particle. If Eq. (3) is inserted into Eq. (5), we obtain

$$J = BNX, \tag{6}$$

i. e.

Migration flux = mobility · concentration · driving force per particle.

In Eq. (6) the concentration N is the number of particles per unit of volume and the flux is the number of particles passing a unit area in a unit time. Equation (6), of course, also holds if we use other measures of concentration. If both sides of the equation are divided by *Avogadro's number* $N_A = 6.023 \times 10^{23}$ molecules per mol we get

$$J = BCX, \tag{7}$$

where the unit of concentration is now mol · m^{-3}, and the flux is mol · m^{-2} · s^{-1}, whereas the mobility B and the driving force X still refer to the individual particles in the solution.

2. Convection Flux

This type of transport differs from the preceding type only in that both the dissolved particles and the solvent move with a velocity v. The tactics used to count the number of particles passing a unit area in a unit time are therefore the same as those used in the previous section. The expression for the *convection flux* therefore becomes

$$J = Nv, \tag{8}$$

or $J = Cv$ if the unit of concentration is mol · m^{-3}.

3. Diffusion Flux

Diffusion is a process leading spontaneously to equalization of the differences in concentration in a single phase. The mechanism by which the substance is transported from one region of the phase to another originates, as already mentioned, in random, irregular molecular movements which proceed continually (thermal movements). The diffusion laws link together the rate of transport of the diffusing substance and the concentration gradient responsible for this movement of the substance. In a diffusion process there will almost always be at least two substances involved, namely the dissolved substance and the solvent. It will therefore always be necessary to operate with at least two diffusion equations, usually one for each diffusing substance. In the case where only two components are involved (dissolved substance and solvent), however, there will only be two equations, which for physical reasons are subject to the constraint that the transport flow of the second component must equal but be of the opposite sign to that of the first component. In this case, therefore, it is enough to consider the transport flow for one component, usually the transport of the dissolved substance. Nevertheless it is always useful to bear in mind that (apart from some special cases, such as the diffusion of labelled water in water) we operate in principle with a transport equation for each component.

a) Fick's Law

The characteristic features of a diffusion process are: (a) the transport of matter by diffusion takes place only provided there is a nonuniform distribution of the molecules of the substance spatially; (b) the transport of the substance takes place in the direction in which the concentration of the substance decreases; (c) the rate of transport is greatest in the region with the steepest concentration profile. If the x-axis is oriented in the direction of transport, and if $C(x, t)$ is the concentration profile at time t and position x, then the above empirical facts can be combined in the following expression for the diffusion flux J:

$$J = -D\frac{\partial C}{\partial x} \quad \text{(Fick's law)}. \tag{9}$$

This expression was established by Adolf FICK (1855) in analogy to FOURIER's (1822) expression for the conduction of heat. The magnitude D, which is the *diffusion coefficient*, is characteristic for the type of molecule diffusing under the given conditions. This magnitude involves not only those factors determining the rate of transport, such as the size and shape of the molecule, but also characteristics of the surrounding medium (*e.g.* viscosity) through which the molecule is moving. D is not an absolute constant, as there is most often some degree of dependence of the magnitude D on the concentration of the diffusing substance. If we choose $J \equiv$ (amount of substance) $m^{-2}s^{-1}$; $C \equiv$ (amount of substance) m^{-3} and $x \equiv m$, we get

$$D \equiv m^2\, s^{-1}.$$

It is clear that D is independent of the units used to describe the amount of substance in J and C. In aqueous solution the majority of low-molecular substances have a diffusion coefficient of magnitude $10^{-9} - 10^{-10}$ $m^2\,s^{-1}$, or $10^{-5} - 10^{-6}$ $cm^2\,s^{-1}$.

Fick's law as written in Eq. (9) describes a one-dimensional diffusion process. In general Fick's law is written in the vector form

$$\mathbf{J} = -D\,\text{grad}\, C, \tag{10}$$

whereby it becomes independent of the reference system involved.

b) The Driving Force behind the Diffusion Process

As already mentioned, it is the combination of a nonuniform distribution of the dissolved molecules and the "random walk" of these that leads to the transport of a substance by diffusion. Thus, when a concentration gradient has been established, there is no question of the presence of a special "force of diffusion" driving each individual molecule in the direction of the fall in concentration. In certain formal calculations, however, it may nevertheless be convenient to ignore those processes that at the molecular level underlie the transport of material by diffusion, and instead regard the diffusion process as having arisen as a result of each separate particle being under the influence of a force driving the particle in the direction of the fall in concentration. The magnitude of this equivalent or fictive diffusion force can be determined as follows: we consider a point in space where the concentration is C and the concentration gradient is $\partial C/\partial x$. According to Fick's law, the diffusion flux is

$$J = -D\frac{\partial C}{\partial x}.$$

The right-hand side of this equation is now multiplied by $1 = (B/B) \cdot (C/C)$ and the terms arranged as follows:

$$J = BC\left\{-\frac{D}{B}\frac{1}{C}\frac{\partial C}{\partial x}\right\}.$$

This expression for Fick's law has exactly the same form as the expression for the migration flux $J = BCX$ in Eq. (7). It follows from this that the diffusion flux found in a region with a concentration gradient $\partial C/\partial x$ and a concentration C can be expressed formally as a migration process, driven by the equivalent diffusion force

$$X_{\text{dif}} = -\frac{D}{B}\frac{1}{C}\frac{\partial C}{\partial x} = -\frac{D}{B}\frac{\partial \ln C}{\partial x}. \tag{11}$$

This purely formal force X_{dif} represents the driving force per dissolved particle, which in a region with a uniform concentration C gives a transport of substance equal to the transport of substance taking place as a result of the presence of a concentration gradient of magnitude $\partial C/\partial x$. Equation (11) can be manipulated further by means of Einstein's relation $D = kTB$, where $k = 1.3804 \times 10^{-23}$ $J \cdot K^{-1}$ per molecule is *Boltzmann's constant* and T is the absolute temperature (see section D.VI.1). If this relation is inserted into Eq. (11), we then have, after multiplication by N_A/N_A, where N_A = Avogadro's number

$$X_{\text{dif}} = -\frac{1}{N_A}\frac{\partial}{\partial x}\{N_A kT \ln C\} = -\frac{1}{N_A}\frac{\partial}{\partial x}\{RT \ln c\},$$

since $N_A \cdot k = R$, where $R = 8.314\ J \cdot \text{mol}^{-1} \cdot K^{-1}$, is the *gas constant.* The right-hand side remains unchanged if the factor in brackets is replaced by the *molar chemical potential* of the diffusion substance

$$\mu = \mu_o + RT \ln C,$$

since μ_o is constant. Equation (11) can therefore also be written as

$$X_{dif} = \frac{1}{N_A}\left(-\frac{\partial \mu}{\partial x}\right). \tag{12}$$

The negative gradient of the chemical potential of the substance can thus be regarded as the driving force per mol for the diffusion process. The expression in brackets is often designated "the thermodynamic force" of the diffusion process. Nevertheless, giving this force a name does not alter the fact that the "diffusion force" is an equivalent force or a pseudoforce just like, for example, centrifugal force, even though it can be convenient at times to use it in formal calculations. As shown in section D, there is no continuously unidirectional force X_{dif} acting in the diffusion process, but on the contrary, a fluctuating force that continuously changes direction, the mean value of which is zero, even when taken over very brief intervals. If Eq. (12) is inserted into Eq. (7), we finally obtain

$$J = BC\frac{1}{N_A}\left(-\frac{\partial \mu}{\partial x}\right) = bC\left(-\frac{\partial \mu}{\partial x}\right), \tag{13}$$

where $B/N_A = b$ is called the *molar mechanical mobility* of the substance.

4. Diffusion and Migration Proceeding Concurrently

As mentioned earlier, the situation often arises that the transport process takes place both as a result of a concentration gradient being present in the region in question, $\partial C/\partial x$, and as a result of each particle being influenced by an outer driving force X. Formally, the flux at a point in space with concentration C can be written as

$$J = BCX_{tot}, \tag{14}$$

where force X_{tot} includes the equivalent force contribution originating from the diffusion process. The magnitude of this "diffusion force" is indicated, for example, by Eq. (11). We have therefore $X_{tot} = X - (D/B) \cdot (1/C) \cdot (\partial C/\partial x)$, which when inserted into Eq. (14) gives

$$J = -D\frac{\partial C}{\partial x} + BCX, \tag{15}$$

or in vector form

$$\mathbf{J} = -D \,\mathrm{grad}\, C + BC \tag{16}$$

This equation, the *Smoluchowski equation*, constitutes the basis for the later treatment of electrodiffusion in section F.

5. Convection with Superimposed Diffusion

The diffusion flux through a fixed plane of reference, where the concentration gradient is $\partial C/\partial x$ and the concentration is C, is

$$J_{\text{dif}} = -D\frac{\partial C}{\partial x}.$$

If a system with uniform concentration C moves as a whole with a velocity v with respect to the plane of reference, then, according to Eq. (8), the convection flux is $J_{\text{con}} = vC$. To make this flux equal to the diffusion flux above requires that the system moves with a convection velocity v_{dif} which satisfies the relation: $v_{\text{dif}} \cdot C = -D(\partial C/\partial x)$. Therefore, formally a convection process can be regarded as the equivalent of a diffusion process, provided the system moves with the equivalent convection velocity

$$v_{\text{dif}} = -D\frac{1}{C}\frac{\partial C}{\partial x}.$$

If the system with concentration gradient $\partial C/\partial x$ now moves with a superimposed convection velocity v at right angles to the plane of reference, the flux through this plane can then be expressed as

$$J = Cv_{\text{tot}},$$

where $v_{\text{tot}} = v + v_{\text{dif}}$ includes the equivalent convection velocity v_{dif} originating from the presence of the concentration gradient $\partial C/\partial x$. We therefore have $v_{\text{tot}} = v - (D/C)(\partial C/\partial x)$, which when inserted into the above equation gives

$$J = -D\frac{\partial C}{\partial x} + Cv \tag{17}$$

as an expression for the flux arising when a convection flux has a diffusion process superimposed. This equation is used for example in section E.I.1 in calculating the flux and concentration profile across a membrane through which the movements of water and solute occur simultaneously.

The derivations of Eqs. (15) and (17) are naturally purely heuristic. A more rigorous derivation must be based on molecular statistical considerations as first shown by SMOLUCHOWSKI (1915) and PLANCK (1917). An elementary introductory treatment of these topics is given in section D.

C. Diffusion Processes: Macroscopic Treatment

If we wish to determine the diffusion coefficient by means of Eq. (9), an arrangement must be found in which the flux J can be measured, while at the same time the magnitude of the concentration gradient is known. This is only possible in special cases. In addition, it would be desirable if we could calculate the distribution of a substance in space and time from one or other initial position. During the diffusion process, the steepness of a concentration profile as initially established will become reduced in step with the equalization of differences in concentration. The flux will therefore change with time as changes in concentration occur. We shall now establish the relationship between these two changes.

I. The Diffusion Equation

Let the diffusion process occur in the direction of the x-axis only. We calculate the increase of the amount of substance within a volume element bounded by two parallel planes of unit area situated at the positions x and $x + h$. This increase in time dt is

$$[J(x) - J(x+h)]\,dt = -\frac{\partial J}{\partial x} h dt = [C(x_o, t+dt) - C(x_o, t)]\,h,$$

where $C(x_o, t)$ is the concentration at the position $x \leqslant x_o \leqslant x + h$. Dividing by $h \cdot dt$ and letting both h and dt go to zero we obtain

$$\frac{\partial C}{\partial t} = -\frac{\partial J}{\partial x}, \tag{18}$$

which is the mathematical expression in one dimension for the principle of *mass conservation*. Combining Eq. (18) with Fick's law [Eq. (9)] gives the *diffusion equation*

$$\frac{\partial C}{\partial t} = \frac{\partial}{\partial x}\left\{D\frac{\partial C}{\partial x}\right\} = D\frac{\partial^2 C}{\partial x^2} + \frac{\partial D}{\partial x}\frac{\partial C}{\partial x},$$

which can also be written as

$$\frac{\partial C}{\partial t} = D\frac{\partial^2 C}{\partial x^2} + \left(\frac{\partial D}{\partial C}\right)\left(\frac{\partial C}{\partial x}\right)^2.$$

If the diffusion coefficient can be considered as independent of the concentration ($\partial D/\partial C = 0$) we obtain

$$\frac{\partial C}{\partial t} = D\frac{\partial^2 C}{\partial x^2}, \tag{19}$$

sometimes also called *Fick's second law,* which governs the distribution of the substance in one dimension by diffusion.

1. Classification of Diffusion Processes

A number of examples will be given of solutions of Eq. (19). This is a partial differential equation (of parabolic type) and its independent variables are positions in space and time. The solutions involving both variables are called *time-dependent* or *nonstationary solutions.* For large values of t the solutions will describe one of the following two situations:

(i) As a result of the diffusion process, the substance has become distributed uniformly throughout the accessible region, and all concentration gradients have therefore become eliminated. There is thus no further net flux in the system, which is now in its *position of equilibrium.* The condition for this can thus be described as

$$J = 0, \text{ Equilibrium state.} \tag{20}$$

(ii) By means of external constraints, for example a continuing supply and removal of the same amounts of substance at two different points in the system, it will be possible to maintain the concentration gradient, and thereby the flux through the system, without the concentration of substance changing with time at any point in the system. This is now in a *stationary state,* and solutions of the diffusion equation corresponding to this state are called *stationary* or *time-independent* solutions. Mathematically, the stationary solution is characterized by the condition

$$\left.\begin{aligned} \frac{\partial C}{\partial t} &= 0 \\ J &= \text{const.} \neq 0 \end{aligned}\right\} \text{ Stationary state} \tag{21}$$

being satisfied throughout the region under consideration.

Finally, there can be a transitional situation between the time-dependent and the stationary state, where the concentration changes with time, but at such a slow rate that at any time the concentration profile corresponds approximately to a stationary state. Such an exchange process is called a *quasistationary process.* This is the situation aimed at in experiments investigating exchange through membranes, since the stationary solution is always far simpler than the corresponding time-dependent solution. The following sections contain examples of solutions of the diffusion equation which are selected because of their general interest. Those who wish to go deeper into the subject are referred to the texts by Jacobs (1967), Crank (1956) and Carslaw and Jaeger (1959).

II. Stationary Processes in One Dimension

We will now consider the stationary one-dimensional diffusion process through a homogeneous plate (membrane), which develops when the concentration of the diffusing substance is maintained at fixed values on both sides of the membrane. The questions to be elucidated are: (i) which factors determine the magnitude of the flux, and (ii) the concentration profile in the plate.

1. Steady-State Diffusion in a Plate

A plate with thickness h is situated between phases i (inside) and o (outside), where the concentrations of substance are *maintained* at values $C^{(i)}$ and $C^{(o)}$. The coordinate system is arranged as shown in Fig. 1 A. In principle, the concentration profile through the membrane is determined by solving the diffusion equation, Eq. (19). This is done in section B. III. 4. in order to elucidate those factors determining the time required to achieve the stationary state. But as the diffusion process through the membrane is assumed here to have become stationary, $\partial C/\partial t = 0$ throughout. The concentration profile in the membrane is thus determined by invoking this condition into Eq. (19), *i. e.*

$$\frac{d^2C}{dx^2} = 0\,, \tag{22}$$

the general solution of which is

$$C(x) = Ax + B\,, \tag{23}$$

where A and B are constants. It follows that the concentration profile is *linear* in the one-dimensional stationary diffusion process. Constants A and B are found by adjusting Eq. (23) so that it satisfies the *boundary conditions* $C(x) = C^{(i)}$ for $x = 0$ and $C(x) = C^{(o)}$ for $x = h$. This gives

$$C(x) = \frac{C^{(o)} - C^{(i)}}{h}\,x + C^{(i)}, \tag{24}$$

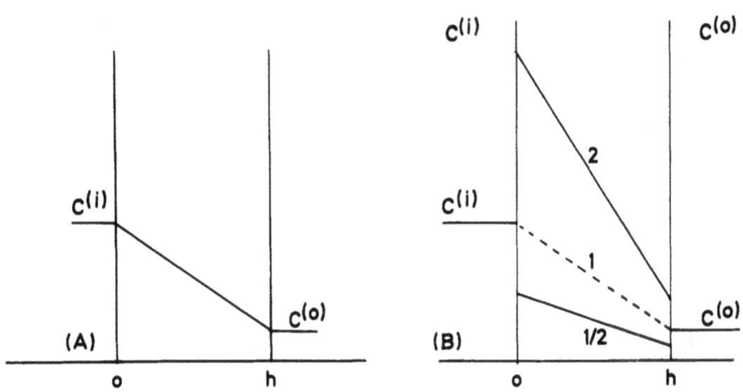

Fig. 1. Steady-state concentrations through a homogeneous membrane of thickness h. (A) Substance has same solubility in the membrane and surrounding media. (B) Different solubility in membrane and media, numbers indicating values of distribution coefficient

which is the expression for the concentration profile in the membrane during stationary diffusion. The flux (in the direction of the x-axis) is given by Fick's law, $J = -D(dC/dx)$, so that

$$J = -D\frac{C^{(o)} - C^{(i)}}{h} = \frac{D}{h}(C^{(i)} - C^{(o)}), \tag{25}$$

which is the solution to the problem. It should be noted that this solution is based on the simplest assumption that the substance has the *same solubility* both in the two phases (i) and (o) and in the membrane.

2. The Permeability Coefficient

We now consider the situation where the solubility of the diffusing substance in the membrane differs from the solubility in the phases on both sides of the membrane. The concentration profile is still determined by the general solution of Eq. (23), but the boundary conditions for $C(x)$ are now

$$C(x) = \alpha C^{(i)} \text{ for } x = 0; \text{ and } C(x) = \alpha C^{(o)} \text{ for } x = h,$$

where α is the *distribution coefficient* for the substance. If Eq. (23) is solved with these boundary conditions, we obtain

$$C(x) = \frac{\alpha}{h}[C^{(o)} - C^{(i)}]\,x + \alpha C^{(i)}. \tag{26}$$

Fig. 1B shows a sketch of the two concentration profiles corresponding to $\alpha_1 > 1$ and $\alpha_2 < 1$.
The flux now becomes

$$J = -\frac{\alpha D}{h}[C^{(o)} - C^{(i)}] = \frac{\alpha D}{h}[C^{(i)} - C^{(o)}]. \tag{27}$$

For a given set of values for $C^{(i)}$ and $C^{(o)}$, the magnitude of the flux is thus determined by the factor

$$P = \frac{\alpha D}{h} = \frac{\alpha kTB}{h}, \tag{28}$$

where the last expression is obtained by invoking the Einstein relation $D = kTB$ [see Eq. (93)]. Factor P is called the *permeability coefficient* of the membrane, or simply its *permeability* for the substance in question. Since the dimensions of the diffusion coefficient are $m^2 \cdot s^{-1}$ and α is a pure number, the dimensions of the permeability must be $m \cdot s^{-1}$.

In many cases the thickness h of the membrane is only known approximately, just as in many cases the solubility α cannot be determined directly. In those cases it is therefore not possible to determine the magnitude of D from flux experiments with any great precision. In such a situation, therefore, we are restricted to writing the flux equation, Eq. (27), in the form

$$J = -P[C^{(o)} - C^{(i)}] = P[C^{(i)} - C^{(o)}], \tag{29}$$

where our incomplete knowledge of the parameters determining the magnitude of the flux is symbolized by the permeability coefficient P. It is also this magnitude that can be directly determined if $C^{(i)}$ and $C^{(o)}$ are known and if the corresponding magnitude of the flux J can be measured.

It is nevertheless worth bearing in mind the influence of the solubility of the substance in the membrane and thereby of the distribution coefficient on the magnitude of the transport of the substance. Even though the diffusion coefficient D of the substance in the membrane is very much smaller than the diffusion coefficient D' of the substance in aqueous solution, the effect of this difference on the transport of matter through the membrane can be counteracted by a higher solubility of the substance in the membrane than in the solution. From Eq. (28), it is seen that the membrane will offer no transport barrier whatever in relation to the surrounding medium, provided

$$\alpha D = D'. \tag{30}$$

Conversely, the transport of a substance with the same diffusion coefficient in aqueous solution and in the membrane can nevertheless be restricted by the presence of the membrane. For this purpose it is only necessary that the solubility of the substance in the membrane is many times less than in the aqueous solution. A membrane thus presents the *greatest transport barrier* to those substances whose *mobility and solubility in the membrane* are much *less* than in surrounding medium.

It should be emphasized that the stationary transport by diffusion through a membrane covering a cylindrical or a spherical body reduces to an equation of the same form as Eq. (29) provided that the thickness of the membrane is very small compared with the radius.

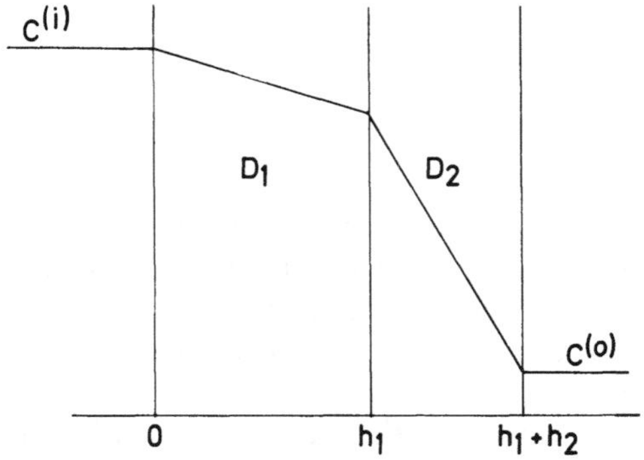

Fig. 2. Diffusion through a composite medium

3. Stationary Diffusion through Two Different Media

We consider the situation illustrated in Figure 2. The stratified plate (membrane) of thickness $h_1 + h_2$ consists of two different media of thickness h_1 and h_2, in which the diffusing substance has the diffusion coefficient D_1 and D_2. Let the concentrations in the surrounding media be maintained at $C^{(i)}$ and $C^{(o)}$ respectively. In the steady state the concentration profile in each region has to satisfy Eq. (22), *i. e.* $d^2C/dx^2 = 0$. Thus the concentration profiles are linear in both regions, viz. $C_1(x) = A_1x + B_1$ for $0 \leqq x \leqq h_1$ and $C_2(x) = A_2x + B_2$ for $h_1 \leqq x \leqq h_1 + h_2$. Furthermore, the concentration $C(x)$ and the flux have to be continuous across the boundary between the two media. This provides the two boundary conditions $D_1 (dC_1/dx) = D_2 (dC_2/dx)$ and $C_1(x) = C_2(x)$ for $x = h_1$, which together with the conditions $C_1 = C^{(i)}$ for $x = 0$ and $C_2 = C^{(o)}$ for $x = h_1 + h_2$ allow the determination of the constants A_1, A_2, B_1, and B_2. The final result is

$$C_1(\mathrm{x}) = C^{(i)} - \frac{D_2(C^{(i)} - C^{(o)})}{h_1D_2 + h_2D_1}\, x, \text{ for } 0 \leqq x \leqq h_1 \tag{31}$$

and

$$C_2(\mathrm{x}) = C^{(o)} + \frac{D_1(C^{(i)} - C^{(o)})}{h_1D_2 + h_2D_1}\,(h_1 + h_2 - x), \text{ for } h_1 \leqq x \leqq h_1 + h_2. \tag{32}$$

We now introduce an *equivalent permeability* $<P>$ which is defined by

$$J = <P>[C^{(i)} - C^{(o)}] \tag{33}$$

as a generalization of Eq. (29). To find $<P>$ we can apply Fick's law to either Eq. (31) or Eq. (32). We can also proceed more directly without using the expression for the concentration profiles: The declines in concentration through the plate can be written as $\Delta C_1 = J/P_1$ and $\Delta C_2 = J/P_2$, where ΔC_1 and ΔC_2 are the fall in concentration through medium (1) and medium (2), having permeabilities P_1 and P_2 respectively. Furthermore, we can put $\Delta C_1 + \Delta C_2 = \Delta C = J/<P>$. From this it follows that

$$\frac{J}{P_1} + \frac{J}{P_2} = \frac{J}{<P>}$$

or

$$\frac{1}{<P>} = \frac{1}{P_1} + \frac{1}{P_2}, \tag{34}$$

since the flux is continuous through the plate. This result can of course be generalized to represent a stratified plate consisting of n different media:

$$\frac{1}{<P>} = \frac{1}{P_1} + \frac{1}{P_2} + \dots \frac{1}{P_n}. \tag{35}$$

Thus the permeabilities for membranes and the conductances for electrical circuits both arranged in series are summed according to the same law.

It appears from Eq. (34) and (35) that transport through a compound membrane is limited by that region of the membrane with the lowest permeability. Equation (34) is also a useful expression, if for example we wish to evaluate the effect of incomplete mixing of the medium surrounding a membrane on the transport.

III. Time-Dependent Processes

It is impossible to establish a stationary state instantaneously. A certain time will always elapse before the system has become adjusted to a stationary state or to equilibrium. The time this takes depends entirely on the physical situation in question. In the following we will examine the time course of some characteristic situations.

In the examples given in section C. II, the diffusion equation degenerated to an ordinary differential equation of the second order. The solution of this involved two arbitrary constants A and B, which could, however, be determined because certain *boundary conditions* that the equation had to satisfy were specified at the same time. In solving the diffusion equation in its time-dependent form, however,

$$\frac{\partial C}{\partial t} = D \frac{\partial^2 C}{\partial x^2} \tag{19}$$

it is not enough to specify the boundary conditions of the problem in question. We must also have a given *initial* description for the *distribution of the substance* in space, *e. g.* of the form

$$C(x, 0) = f(x); \text{ for } t = 0.$$

This condition is called the *initial condition* for the problem. The time-dependent diffusion problem is thus solved when the functional relation is found between the independent variables, x and t, that satisfies not only the boundary conditions for the problem but also its initial condition. As an introduction to the problem, two simple, but for practical problems important, kinetic situations will be examined, in which the diffusion equation, Eq. (19), need not be solved directly.

1. Kinetics of Exchange between Two Phases Separated by a Membrane *(Quasistationary Processes)*

We consider two phases separated by a permeable membrane of thickness h. Both phases are so well mixed that a uniform concentration can be assumed in each phase. The concentrations of substance in phases (i) and (o) are respectively $C^{(i)}$ and $C^{(o)}$. If $C^{(i)} \neq C^{(o)}$, there will be an exchange of substance between the phases, during which the concentration gradient will gradually collapse. The assumption is now made that the exchange of substance between the phases takes place so slowly that the concentration profile in the membrane deviates from the stationary profile by only an infinitesimal degree (cf. Fig. 1). The flux in the direction (i) → (o) can therefore still be described by Eq. (29) as

$$J(t) = P[C^{(i)} - C^{(o)}], \tag{36}$$

but where $J(t)$ is now a function of time, because $C^{(i)}$ and $C^{(o)}$ are no longer maintained at constant values but move towards the equilibrium state $C^{(i)} = C^{(o)}$ for $t \to \infty$. We will start by examining the simplest situation.

a) One of the Phases is Infinitely Large

Let the volume $v^{(o)}$ of the outer phase be many times greater than that of the inner phase $v^{(i)}$. As a result, the outer concentration $C^{(o)}$ can, for all practical purposes, be regarded as being constant during the exchange process.

(i) *Outer concentration zero.* Let the concentration in phase (i) be $C_0^{(i)}$ at time $t = 0$ and $C^{(i)}$ at time t. We then have from Eq. (36)

$$J = PC^{(i)}.$$

The amount of substance, $dm^{(i)}$, removed in time dt from phase (i) by diffusion through the membrane is $dm^{(i)} = JAdt$, where A is the area of the membrane. The change in concentration in phase (i) in the time dt is for small values of dt

$$C^{(i)}(t + dt) - C^{(i)}(t) = \left(\frac{dC^{(i)}}{dt}\right) dt,$$

which in turn is equal to

$$- dm^{(i)}/v^{(i)} = - AJdt/v^{(i)} = - APC^{(i)}\, dt/v^{(i)},$$

from which it follows that

$$\frac{dC^{(i)}}{dt} = - \frac{AP}{v^{(i)}} C^{(i)} \tag{37}$$

or

$$\frac{dC^{(i)}}{dt} = - k C^{(i)}, \tag{38}$$

where the magnitude

$$k = \frac{AP}{v^{(i)}}, \tag{39}$$

is called the *rate constant* of the exchange process. The dimensions of k are $m^2 \cdot m \cdot s^{-1} \cdot m^{-3} = s^{-1}$. Equation (38) is integrated from $t = 0$ when the concentration is $C_o^{(i)}$ to time t when the concentration is $C^{(i)}$. This gives

$$\ln[C^{(i)}/C_o^{(i)}] = -kt,$$

or

$$C^{(i)} = C_o^{(i)} e^{-kt}. \tag{40}$$

The concentration in phase (i) thus decreases exponentially with time t. The rate at which this takes place depends on the magnitude of the rate constant k. The greater the rate constant, the more rapidly the substance will disappear from phase (i). The factors determining the magnitude of the rate constant are, according to Eq. (39), the ratio between the membrane area A and the volume $v^{(i)}$ of the phase (i), together with the permeability of the membrane. The rate constant is most easily determined by plotting the corresponding values of $\ln C^{(i)}$ and t, as this dependence is linear according to Eq. (40). Extrapolation to time $t = 0$ gives the initial concentration $C_o^{(i)}$, and the slope of the line is equal to $-k$. Furthermore, if the area A of the membrane and the volume $v^{(i)}$ of phase (i) are known, the membrane permeability for the substance in question can be determined.

(ii) *Outer Concentration Finite, Inner Concentration Initially Zero.* In this case the transport takes place from the outer phase (o) with constant concentration $C^{(o)}$ to the inner phase (i) with a concentration $C^{(i)}$ which is assumed to be initially zero. The flux into phase (i) at time t is then $J = P(C^{(o)} - C^{(i)})$. In time dt an amount $dm^{(i)} = JAdt$ is transported into phase (i), the concentration of which is thereby increased by the contribution $dC^{(i)} = dm^{(i)}/v^{(i)}$. We have therefore

$$dm^{(i)} = AP[C^{(o)} - C^{(i)}]\,dt = v^{(i)}\,dC^{(i)}.$$

The differential equation for the accumulation of substance in phase (i) from phase (o) is therefore

$$\frac{dC^{(i)}}{dt} = k(C^{(o)} - C^{(i)}), \tag{41}$$

where k is the rate constant given by Eq. (39). If Eq. (41) is integrated from $t = 0$ with concentration $C^{(i)} = 0$ to time t, when the concentration is $C^{(i)}$, we obtain

$$\ln[(C^{(o)}-C^{(i)})/C^{(o)}] = \ln[1-C^{(i)}/C^{(o)}] = -kt \tag{42}$$

or

$$C^{(i)} = C^{(o)}(1-e^{-kt}). \tag{43}$$

The concentration in the inner phase thus increases exponentially and asymptotically towards $C^{(o)}$, with the same rate constant k as in case (a). As appears from Eq. (42), there is a linear dependence between time t and $\ln(1-C^{(i)}/C^{(o)})$, and again with the slope $-k$.

b) Both Phases Comparable in Size

We then consider the situation in which both phases have finite volumes. In phase (i) with volume $v^{(i)}$ there is an amount of substance m_o dissolved at time $t=0$, while phase (o) with volume $v^{(o)}$ initially contains no dissolved substance. We now consider the situation at time t, when the concentration in the two phases are $C^{(i)}$ and $C^{(o)}$. Let the flux through the membrane in the direction (i) → (o) be $J(t) = AP(C^{(i)}-C^{(o)})$. In the time between t and $t+dt$ the phase (o) has received an amount of substance $dm^{(o)}$,

$$dm^{(o)} = AP(C^{(i)}-C^{(o)})dt. \tag{44}$$

The total amount of substance in the system remains constant, which implies that

$$m_o = m^{(i)} + m^{(o)},$$

where $m^{(i)}$ and $m^{(o)}$ are the amounts of substance in phases (i) and (o) at time t. The concentration in phase (i) can therefore be written as $C^{(i)} = m^{(i)}/v^{(i)} = (m_o - m^{(o)})/v^{(i)}$. Further, $C^{(o)} = m^{(o)}/v^{(o)}$.

If the values for $C^{(i)}$ and $C^{(o)}$ are inserted into Eq. (44), we obtain

$$\frac{dm^{(o)}}{dt} = AP\left\{\frac{m_o - m^{(o)}}{v^{(i)}} - \frac{m^{(o)}}{v^{(o)}}\right\},$$

which can be written in the form

$$\frac{dm^{(o)}}{dt} = k(b-m^{(o)}), \tag{45}$$

where

$$k = AP\,\frac{v^{(i)}+v^{(o)}}{v^{(i)}\cdot v^{(o)}} \quad \text{and} \quad b = \frac{m_o v^{(o)}}{v^{(i)}+v^{(o)}}. \tag{46}$$

The initial condition for phase (o) is $m^{(o)} = 0$ for $t=0$. If Eq. (45) is integrated from $t=0$ to t when the amount of substance is $m^{(o)}$, we obtain

$$m^{(o)} = b\,(1-e^{-kt}).$$

Since $C^{(o)} = m^{(o)}/v^{(o)}$, the above equation can also be written

$$C^{(o)} = C_{\infty}\,(1-e^{-kt}), \tag{47}$$

as $b/v^{(o)} = m_0/(v^{(i)} + v^{(o)}) = C_{\infty}$, which is the equilibrium concentration corresponding to the state where the amount of substance m_0 originally found in phase (i) has become uniformly distributed over the available space $v^{(i)} + v^{(o)}$. The concentration in phase (o) asymptotically approaches this value with a rate constant, Eq. (46), which differs from the expression given by Eq. (39) by containing the geometrical mean volume $1/v = 1/v^{(i)} + 1/v^{(o)}$ of the two phases instead of the actual volume of the outside phase, $v^{(o)}$.

The time course for the concentration $C^{(i)}$ in phase (i) can be determined from the time course for $C^{(o)}$, since at any time we have $m_0 = m^{(i)} + m^{(o)}$, or

$$m_0 = C^{(i)}\,v^{(i)} + C^{(o)}v^{(o)} = C_{x}\,(v^{(i)} + v^{(o)}).$$

If this expression for $C^{(o)}$ is solved, we obtain after inserting it into Eq. (47)

$$C^{(i)} = C_{\infty} + \frac{v^{(o)}}{v^{(i)}}\,C_{\infty}e^{-kt}. \tag{48}$$

The time course for establishment of equilibrium is illustrated in Figure 3. The two concentrations $C^{(i)}$ and $C^{(o)}$ approach the equilibrium concentration C_{∞}

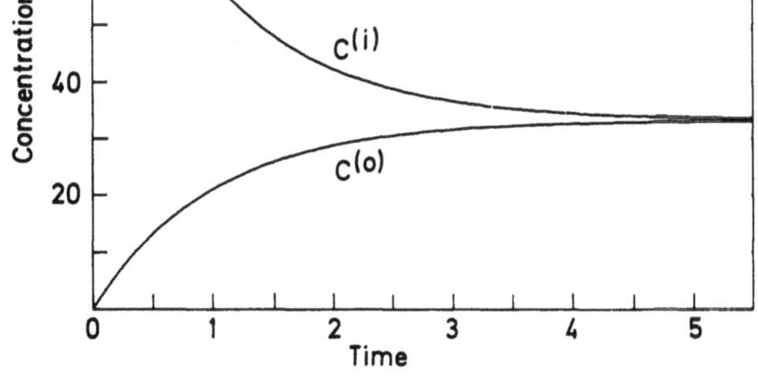

Fig. 3. Time course of the exchange between two compartments of finite size. $v^{(i)} : v^{(o)} = 2$. Initially $C^{(i)} = 100$. $C^{(o)} = 0$. Abscissa, time in units of the time constant $\tau = 1/k$. Ordinate, concentration

exponentially. The time courses are under the control of the *same* rate constant or *time constant* τ. This is defined by

$$\tau = \frac{1}{k} = \frac{1}{AP} \frac{v^{(i)} \cdot v^{(o)}}{v^{(i)} + v^{(o)}}. \tag{49}$$

The rate constant can be determined from a knowledge of the time course of $C^{(i)}$ or of $C^{(o)}$. For example, Eq. (47) can be written in the form

$$y = 1 - C^{(o)}/C_{\infty} = e^{-kt},$$

where y is a function of $C^{(o)}$. If ln y is then plotted as a function of time t, this gives a straight line with a slope equal to $-k$. If on the other hand we know $C^{(i)}$ instead, then rewriting Eq. (48) gives

$$y = C^{(i)}/C_{\infty} - 1 = \frac{v^{(o)}}{v^{(i)}} e^{-kt},$$

where y is now a function of $C^{(i)}$. Furthermore,

$$\ln y = \ln (v^{(o)}/v^{(i)}) - kt.$$

This also corresponds to a decreasing linear expression, with slope $-k$. The value of $\ln y$ extrapolated to $t=0$ gives the value of $\ln (v^{(o)}/v^{(i)})$.

It may happen that the experimental conditions are of such a nature that only a single set of associated values is available: $C_1^{(i)} = m_1^{(i)}/v^{(i)}$ and $C_1^{(o)} = m_1^{(o)}/v^{(o)}$ at time $t = t_1$, and $C_2^{(o)} = m_2^{(o)}/v^{(o)}$ at time $t + t_2$. Equation (45) is then integrated between these values, which gives

$$k(t_2 - t_1) = \ln \frac{b - m_1^{(o)}}{b - m_2^{(o)}} = \ln \frac{b - v^{(o)}C_1^{(o)}}{b - v^{(o)}C_2^{(o)}}.$$

If the values of k and b in Eq. (46) are inserted into this, the following expression for the determination of the permeability coefficient is obtained:

$$P = \frac{2.303\, v^{(i)}v^{(o)}}{(t_2 - t_1)A(v^{(i)} + v^{(o)})} \log \frac{m_o - C_1^{(o)}(v^{(i)} + v^{(o)})}{m_o - C_2^{(o)}(v^{(i)} + v^{(o)})}. \tag{50}$$

This expression, first derived by NORTHROP and ANSON (1929), corresponds to the condition that the total mass m_o is initially in phase (i). However, as the total amount of substance is constant, we have: $m_o = C_1^{(i)}v^{(i)} + C_1^{(o)}v^{(o)}$, which when inserted into Eq. (50) gives

$$P = \frac{2.303\, v^{(i)} \cdot v^{(o)}}{(t_2 - t_1)A(v^{(i)} + v^{(o)})} \log \frac{[C_1^{(i)} - C_1^{(o)}]\, v^{(i)}}{C_1^{(i)}v^{(i)} + C_1^{(o)}v^{(o)} - C_2^{(o)}(v^{(i)} + v^{(o)})}, \tag{51}$$

which was used by ROBBINS and MAURO (1960) to determine the water permeability in an artificial membrane. DAINTY and HOUSE (1966) start off from an arbitrary initial distribution

$$C^{(i)} = C_o^{(i)} \text{ and } C^{(o)} = C_o^{(o)} \text{ for } t = 0.$$

Inserting $m_o = C_o^{(i)} v^{(i)} + C_o^{(o)} v^{(o)}$ into Equation (50), we obtain

$$P = \frac{2.303\, v^{(i)} v^{(o)}}{(t_2 - t_1) A (v^{(i)} + v^{(o)})} \log \frac{C_o^{(i)} v^{(i)} + C_o^{(o)} v^{(o)} - C_1^{(o)} (v^{(i)} + v^{(o)})}{C_o^{(i)} v^{(i)} + C_o^{(o)} v^{(o)} - C_2^{(o)} (v^{(i)} + v^{(o)})}, \tag{52}$$

which is identical with the expression obtained by DAINTY and HOUSE.

c) Unidirectional Fluxes (USSING, 1949)

In the experimental situation in section C.III.1.a. (i), the *concentration in phase* (o) *is kept all the time at practically zero.* The result is that no molecule that has once wandered through the membrane from phase (i) to phase (o) will have the chance of wandering back as a result of a "random walk" (see Section D.III). To characterize the flux for a transport situation with such quite special boundary conditions a special designation for the flux has been introduced, namely the *unidirectional flux.* This is the net flux, corresponding to the condition that the concentration in the one phase is always kept at zero. Some also find it convenient to use this concept in situations where there is a final concentration of substance in both phases, for example as described in the previous Section III. When Eq. (36) is rewritten to give

$$J = PC^{(i)} - PC^{(o)} = J^{(io)} - J^{(oi)},$$

the flux in the direction (i) → (o) is considered to have arisen as the result of two unidirectional fluxes in opposite directions. The first of these

$$J^{(io)} = PC^{(i)} \tag{53}$$

in the direction (i) → (o) is called the "efflux" and the second

$$J^{(oi)} = PC^{(o)} \tag{54}$$

in the direction (o) → (i) is called the "influx". This purely formal manipulation of Fick's law, or of its possible generalizations, naturally gives us no extra insight into the transport process through the membrane. In certain formal calculations, however, it may be convenient to operate with "unidirectional fluxes", also even though the concentrations are finite in both phases. It should be borne in mind, however, that in this situation "unidirectional fluxes" have no physical significance. On the other hand, the unidirectional flux as described above is an operationally well-defined magnitude. As will be shown later, particularly simple relationships can be derived between the unidirectional fluxes, and these relationships are of value in evaluating the transport mechanism (passive versus active transport) through cell membranes.

2. Instantaneous Point Source (Green's Function)

There is one solution of the diffusion equation that is fundamental both in theory and in practice. Formulated one-dimensionally the problem can be expressed as follows. Let the space be free from substance at time $t < 0$. At time $t = 0$, in the plane corresponding to x_o an amount of substance equal to N mol per m^2 is produced instantaneously. How is this substance then distributed in space and time, if its diffusion coefficient in the medium is D? The mathematical formulation of this problem is as follows:

A solution is sought for the partial differential equation

$$\frac{\partial C}{\partial t} = D\frac{\partial^2 C}{\partial x^2} \tag{19}$$

in the unbounded space $-\infty < x < \infty$. $C(x, t)$, which is zero for $t < 0$ and all values of x, must satisfy the boundary conditions for $t \geqq 0$

$$C = \frac{\partial C}{\partial x} = 0, \text{ for } \begin{cases} x \rightarrow +\infty \\ x \rightarrow -\infty . \end{cases} \tag{I}$$

The point source with concentration N mol $\cdot$ m^{-2} is considered to be placed at $x_o = 0$. The initial condition can therefore be expressed.

$$C(x, 0) = N\delta(x), \text{ for } t = 0 \text{ and all } x, \tag{II}$$

where $\delta(x)$ is Dirac's delta function, with the characteristics

$$\delta(x - x_o) = 0, \text{ for } x \neq x_o$$

and

$$\int_{x_o-\varepsilon}^{x_o+\varepsilon} \delta(x - x_o)dx = 1$$

for any positive value of ε, no matter how small. From these two characteristics we then have that

$$\int_{-\infty}^{\infty} f(x)\,\delta(x - x_o)\,dx = f(x_o)\,.$$

The most direct way of solving the problem is to use the two-sided complex Fourier transform (see *e. g.* SNEDDON, 1972), defined by

$$\bar{C}_F = \int_{-\infty}^{\infty} C(x, t)\, e^{i\zeta x} dx.$$

Equation (19) is therefore multiplied by $e^{i\zeta x}$ and integrated from $-\infty$ to $+\infty$. This gives

$$\frac{d}{dt} \int_{-\infty}^{\infty} C(x, t)\, e^{i\zeta x}\, dx = D \int_{-\infty}^{\infty} \left(\frac{\partial^2 C}{\partial x^2}\right) e^{i\zeta x}\, dx$$

after exchanging the order of integration and differentiation on the left-hand side. The right-hand side is integrated by parts twice. This gives $-D\zeta^2 \bar{C}_F(\zeta,t)$ because of the boundary condition (I). The Fourier transformation of Eq. (19) therefore becomes

$$\frac{d\bar{C}_F(\zeta, t)}{dt} = -D\zeta^2 \bar{C}_F(\zeta, t)$$

the solution of which is

$$\bar{C}_F(\zeta, t) = A e^{-D\zeta^2 t}$$

where A is a constant whose magnitude is determined by the initial condition (II) corresponding to $t = 0$. The Fourier transformation of (II) is

$$\bar{C}_F(\zeta, 0) = N \int_{-\infty}^{\infty} e^{i\zeta x}\, \delta(x)\, dx = N.$$

The Fourier transformation of Eq. (19), which satisfies both the boundary condition (I) and the initial condition (II) is therefore

$$\bar{C}_F(\zeta, t) = N e^{-D\zeta^2 t}.$$

The inversion formula for $\bar{C}_F$ is

$$C(x, t) = \frac{1}{2\pi} \int_{-\infty}^{\infty} \bar{C}_F(\zeta, t)\, e^{-ix\zeta}\, dx$$

$$= \frac{N}{2\pi} \int_{-\infty}^{\infty} e^{-Dt\zeta^2 - ix\zeta}\, d\zeta.$$

The right-hand side can be solved as a complex contour integral, but as $C(x, t)$ is real, we also have

$$C(x,t) = \frac{N}{2\pi} \int_{-\infty}^{\infty} e^{-Dt\zeta^2} \cos(x\zeta)\, d\zeta.$$

It can be shown that

$$\int_{-\infty}^{\infty} e^{-\alpha^2 x^2} \cos(\varrho x)\, dx = \frac{\sqrt{\pi}}{\alpha} e^{-\varrho^2/4\alpha^2};$$

so that the final expression for the solution of the problem becomes

$$C(x,t) = \frac{N}{2\sqrt{\pi Dt}} e^{-x^2/4Dt}. \tag{55}$$

The spread of substance after release of the instantaneous plane source thus follows a Gaussian curve. In Figure 4 concentration profiles are shown corresponding to Eq. (55) for various times. Equation (55), which gives the response after the establishment of an instantaneous plane source of strength N, is also called *Green's function* for the one-dimensional diffusion process.

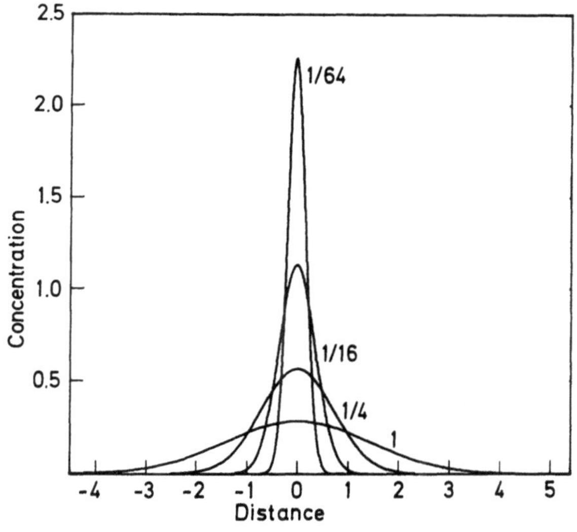

Fig. 4. Spread of a substance by diffusion after release of an instantaneous plane source at $x = 0$. Ordinate, concentration. Abscissa, distance in units of $(Dt)^{1/2}$. Number on each curve is corresponding value of Dt

a) Solutions of Diffusion Problems by Means of Green's Function

It is possible to solve a number of more complicated diffusion problems by means of Green's function. Let the initial distribution of concentration in unbounded space be given by

$$C(x,0) = f(x) \tag{I}$$

for $-\infty < x < \infty$. What is the concentration in the plane corresponding to x at a later time t? We consider a layer of thickness $d\xi$, whose distance from the initial point $x = 0$ is equal to ξ. The amount enclosed in this layer can be regarded as an instantaneous source of strength $f(\xi)\,d\xi$. The distance from this source to the point x is equal to $x - \xi$. At time t this source will contribute to the concentration in the plane at x by the concentration

$$\frac{1}{2\sqrt{\pi Dt}}\,e^{-(x-\xi)^2/4Dt} f(\xi)\,d\xi.$$

The contribution from the entire initial concentration profile $C(x,0) = f(x)$ to the concentration $C(x,t)$ in the plane at x at time t will therefore be

$$C(x,t) = \frac{1}{2\sqrt{\pi Dt}} \int_{-\infty}^{\infty} f(\xi)\,e^{-(x-\xi)^2/4Dt}\,d\xi. \tag{56}$$

This integral can be evaluated in closed form only in special cases. If this cannot be done, numerical approximations must be used, *e. g.* Simpson's method.

(i) *Initial Uniform Distribution in the One Half-Space.* As an example of the use of Eq. (56) we will consider the situation where the substance is found initially in the one half-space $-\infty < x \leqq 0$ with the constant concentration C_o. The initial condition is therefore

$$C(x,0) = \begin{cases} C_o \text{ for } x < 0 \\ 0 \text{ for } x > 0. \end{cases}$$

If this is inserted in Eq. (56) we obtain

$$C(x,t) = \frac{C_o}{2\sqrt{\pi Dt}} \int_{-\infty}^{0} e^{-(x-\xi)^2/4Dt}\,d\xi.$$

Introducing the substitution $u = (x-\xi)/2\sqrt{Dt}$, we obtain

$$C(x,t) = \frac{1}{2}\,C_o\,\frac{2}{\sqrt{\pi}} \int_{x/2\sqrt{Dt}}^{\infty} e^{-u^2}\,du = \frac{1}{2}\,C_o \operatorname{Erfc}\left\{\frac{x}{2\sqrt{Dt}}\right\} \tag{57}$$

where $\operatorname{Erfc}(y) = 1 - \operatorname{Erf}(y)$ is the *complementary error function,* and

$$\operatorname{Erf}(y) = \frac{2}{\sqrt{\pi}} \int_{0}^{y} e^{-u^2}\,du$$

is the *error function,* which is tabulated.

(ii) *The One Half-Space is Separated by an Impermeable Wall.* The wall is considered placed in the position $x = 0$ and the source in $x = \xi > 0$. As the flux is zero for $x = 0$, we here have $\partial C/\partial x = 0$. This situation can be produced by removing the wall and replacing it by another instantaneous source of the same strength in position $x = -\xi$. In this manner a concentration profile is produced, which is

$$C(x,t) = \frac{N}{2\sqrt{\pi D t}} \, e^{-(x-\xi)^2/4Dt} + \frac{N}{2\sqrt{\pi D t}} \, e^{-(x+\xi)^2/4Dt}, \tag{58}$$

which gives the solution sought for $x > 0$. If the two sources are brought together to position $x = 0$, we obtain, since $\xi \to 0$,

$$C(x,t) = \frac{N}{\sqrt{\pi D t}} \, e^{-x^2/4Dt}, \tag{59}$$

corresponding to the source now being placed on the wall, so that the material can now only diffuse into the positive half space.

(iii) *The Presence of an Absorbing Barrier.* Let there be an absorbing barrier (sink) in position $x = 0$, so that $C(0,t) = 0$ for all t. The instantaneous source is still considered placed at $x = \xi$. It can be seen immediately from equation (58) that the solution in the region $0 \leq x < \infty$ that corresponds to this situation is

$$C(x,t) = \frac{N}{2\sqrt{\pi D t}} \, e^{-(x-\xi)^2/4Dt} - \frac{N}{2\sqrt{\pi D t}} \, e^{-(x+\xi)^2/4Dt}. \tag{60}$$

(iv) *Variable Flux into the Half-Space.* Let a flux $J(t)$ take place through the plane corresponding to $x = 0$ into the half-space $0 \leq x < \infty$, that can be initially regarded as free from substance. The flux $J(t)$ is considered to vary with time for $t > 0$. In the time interval between τ and $\tau + d\tau$ an amount of material equal to $J(\tau)\,d\tau$ per m^2 will appear in the plane $x = 0$. At a later instant t this amount will contribute in the plane x with the concentration

$$\frac{J(\tau)\,d\tau}{\sqrt{\pi D(t-\tau)}} \, e^{-x^2/4D(t-\tau)},$$

in accordance with Eq. (59). The total concentration $C(x,t)$ is obtained by summing all these contributions from $J(\tau)$ from $\tau = 0$ to $\tau = t$. This gives

$$C(x,t) = \frac{1}{\sqrt{\pi D}} \int_0^t \frac{J(\tau)}{\sqrt{t-\tau}} \, e^{-x^2/4D(t-\tau)} \, d\tau \tag{61}$$

or

$$C(x,t) = \frac{1}{\sqrt{\pi D}} \int_0^t \frac{J(t-\tau)}{\sqrt{\tau}} \, e^{-x^2/4D\tau} d\tau. \tag{62}$$

These integrals, which only rarely can be evaluated in closed form, can easily be computed using numerical methods.

3. Diffusion Out of a Plate

Washout experiments from a tissue section are often performed, for example with radioactive ^{24}Na. On the basis of these data an attempt is made to evaluate which factors determine the rate of washout. What is sought in particular is the permeability of the cell membranes. As a first approximation all the cells are pooled together and the system is regarded as a two-chamber system as described in Section III.1.a. In many cases a linear relation will not be found between $\ln C$ and t, so that the model must be modified, possibly by the introduction of several compartments to characterize both the cellular and the intracellular space. In considerations of this kind it is useful to know as a starting point how a substance is washed out with time from a *homogeneous* plate in which the substance has a diffusion coefficient D.

a) Concentration Profiles

We consider a homogeneous plate of thickness h. The coordinate system is arranged with one surface of the plate at $x = 0$ and the other at $x = h$. Initially the plate is considered to contain a substance with uniform concentration C. At time $t = 0$ the concentration in the medium surrounding the plate is suddenly kept adjusted to zero and maintained there for $t > 0$. The problem is then to find how the concentration profile within the plate in the region $0 \leqq x \leqq h$ alters with time.

We must therefore find a solution of the diffusion equation

$$\frac{\partial C}{\partial t} = D \frac{\partial^2 C}{\partial x^2}, \tag{19}$$

with the initial condition

$$C = C_o \text{ for } 0 \leqq x \leqq h \text{ and } t = 0 \tag{I}$$

and the boundary condition

$$C = 0, \text{ for } = \begin{cases} x = 0 \text{ and } t > 0 \\ x = \text{h and } t > 0. \end{cases} \tag{II}$$

The simplest way to solve this problem is to use a finite Fourier transformation. Since the values of $C(x,t)$ are specified for $x = 0$ and $x = h$, the *sine transformation* must be used. This is defined for the interval $0 \leqq x \leqq h$ by

$$\bar{C}_s(k,t) = \int_0^h C(x,t) \sin\left(\frac{k\pi x}{h}\right) dx \tag{63}$$

and has the inversion formula

$$C(x,t) = \frac{2}{h}\sum_{k=1}^{\infty} \bar{C}_s(k,t)\sin\left(\frac{k\pi x}{h}\right) \tag{64}$$

(see *e.g.* SNEDDON [1972]). Equation (19) is multiplied by $\sin(k\pi x/h)$ and integrated from 0 to h with respect to x. This gives

$$\int_0^h \left(\frac{\partial C}{\partial t}\right)\sin\left(\frac{k\pi x}{h}\right)dx = D\int_0^h \left(\frac{\partial^2 C}{\partial x^2}\right)\sin\left(\frac{k\pi x}{h}\right)dx. \tag{19a}$$

On the left-hand side the order of integration and differentiation is interchanged. The right-hand side is integrated twice by integration by parts. If then the definition of $C_s(t,k)$ in Eq. (63) is employed, Eq. (19a) is converted to the following expression

$$\frac{d\bar{C}_s(k,t)}{dt} = -D\left\{\left(\frac{k\pi}{h}\right)^2 \bar{C}_s(k,t) + \frac{k\pi}{h}(C(h)\cos(k\pi) - C(\mathrm{o}))\right\}. \tag{65}$$

As the boundary values are $C(h) = C(\mathrm{o}) = 0$, the sine transformation of the diffusion equation takes on the form

$$\frac{d\bar{C}_s}{dt} = -D\left(\frac{k\pi}{h}\right)^2 \bar{C}_s,$$

the solution of which is

$$\bar{C}_s(k,t) = A\exp\left\{-D\left(\frac{k\pi}{h}\right)^2 t\right\},$$

where A is a constant, determined by the initial condition (I) of the problem. The sine transformation of this is

$$\bar{C}_s(k,0) = \int_0^h C_0\sin\left(\frac{k\pi}{h}x\right)dx$$

$$= \begin{cases} 0, \text{ for } k = 0, 2, 4, \ldots 2n \\ 2hC_0/k\pi, \text{ for } k = 1, 3, 5, \ldots 2n+1, \end{cases}$$

which in turn is equal to the constant A in the expression for $\bar{C}_s(k,t)$. The final expression then for the sine transformation of the diffusion equation that satisfies both the initial condition (I) and the boundary condition (II) is

$$\bar{C}_s(k,t) = \frac{2C_o h}{\pi(2n+1)} \exp\left\{-D\left(\frac{(2n+1)\pi}{h}\right)^2 t\right\}, n = 0, 1, 2, \ldots$$

If therefore the inversion formula of Eq. (64) is employed for the finite sine transformation, the solution to the problem becomes

$$C(x, t) = \frac{4C_o}{\pi} \sum_{n=0}^{\infty} \frac{\sin[(2n+1)\pi x/h]}{2n+1} \exp\left\{-D\left(\frac{(2n+1)\pi}{h}\right)^2 t\right\}. \tag{66}$$

When the dimensionless variables $X = x/h$ and $T = t/\tau_o$, where

$$\tau_o = h^2/\pi^2 D \tag{67}$$

are introduced, Eq. (66) takes on the form

$$\frac{C(X,T)}{C_o} = \frac{4}{\pi} \sum_{n=0}^{\infty} \frac{\sin[(2n+1)\pi X]}{2n+1} \exp\{-(2n+1)^2 T\}. \tag{68}$$

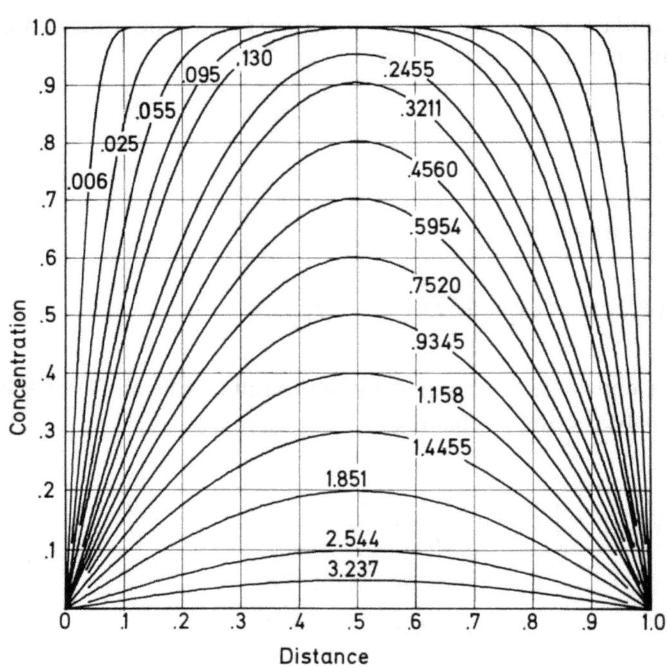

Fig. 5. Diffusion out of a plate. Ordinate, concentration in units of the initial concentration C_o. Abscissa, distance in units of plate thickness h. Numbers on the curves are the time in units of $h^2/\pi^2 D$

Figure 5 shows how the concentration profiles in the plate decrease with time. The curves are calculated from Eq. (68) with the times given on the individual curves expressed in units of the time constant τ_o in the exponential term corresponding to $n = 0$. The concentration profiles are seen to be symmetrical around the mid-plane of the plate corresponding to $X = 0.5$. Here $\partial C/\partial x = 0$. Figure 5 therefore also indicates the concentration profiles in a plate of thickness $h/2$, where one surface is impermeable to the substance.

b) The Time Constant for the Exchange of the Mean Concentration

In practice it is difficult to determine the concentration profiles in the plate. The easiest way to do this is by diffusion in crystals. What can be determined directly, however, is the mean concentration $<C>$ in the plate at time t. The analytic expression for this is

$$<C> = \frac{1}{h}\int_{o}^{h} C(x,t)\,dx = \int_{o}^{1} C(X,T)\,dX.$$

If Eq. (68) is inserted in this, we obtain

$$<C> = \frac{8C_o}{\pi^2}\sum_{n=o}^{\infty}\frac{\exp\{-(2n+1)^2T\}}{(2n+1)^2}, \tag{69}$$

an expression which was first derived by DÜNWALD and WAGNER (1934).

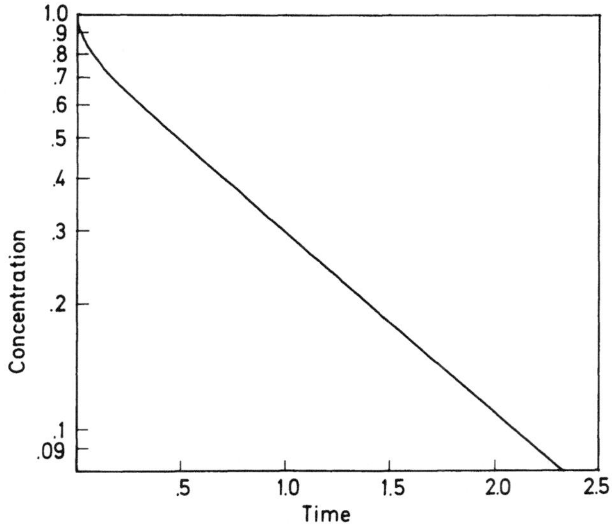

Fig. 6. Time course of the disappearance of the average concentration during washout from a plate of thickness h. Ordinate, concentration in units of the initial concentration C_o (log scale). Abscissa, time in units of $h^2/\pi^2 D$

Figure 6 shows the relation between ln$<C>$ and T. In contrast to the situation described in Section C. III.1., the logarithm of the mean concentration does not vary linearly with time over the whole range $0 \leqq T < \infty$. The reason for this becomes apparent from Eq. (69), as $<C>$ is determined by a sum of exponential terms, each with its time constant

$$\tau_n = h^2/(2n+1)^2 \pi^2 D,$$

i. e.

$$\tau_o, \tau_o/9, \tau_o/25, \ldots \tau_o/(2n+1)^2.$$

The first three terms in the series for $<C>$ expressed by the time constant τ_o are

$$<C>/C_o = 0.801 \, e^{-t/\tau_o} + 0.0901 \, e^{-9t/\tau_o} + 0.0324 \, e^{-25t/\tau_o}.$$

With increasing values of t, the first term will become increasingly dominant. For $t > 0.3\ \tau_o$, the time course of the mean concentration $<C>$ is thus for all practical purposes one which decreases monoexponentially with time and is given by

$$<C>/C_o = 0.8106 \exp\{-t/\tau_o\}. \tag{70}$$

The linear part of the curve for ln$<C>$ as a function of time thus has a slope $-1/\tau_o$. From this it is possible to determine the diffusion coefficient D for the substance in the plate by means of Eq. (67), if the thickness h of the plate is known. Note that this straight line does not give C_o by extrapolation to $t = 0$, but 0.81 C_o (cf. Section C. III.1.). The importance of the result of Eq. (79), however, goes further than indicating a not particularly convenient method of determining the diffusion coefficient. This result should also be kept in mind, if experiments are made with tissue sections with the aim of determining the permeability of cells to a given substance. In such a case it is tempting to consider all the cells combined into one given compartment, and in the first instance to regard the situation as being analogous to that discussed in Section C. III.1. At times, fortune favours the bold, but as a general rule boldness should be combined with caution. It should therefore be borne in mind that apart from quite brief periods the washout curve for a homogeneous plate decreases monoexponentially with time, with a time constant

$$\tau_o = h^2/\pi^2 D,$$

which increases with the square of the thickness of the tissue section. For a homogeneous plate of thickness 1 mm and a diffusion coefficient for the substance in question of 10^{-6} cm^2 s^{-1}, we have: $\tau_o = 10^{-2}/\pi^2 \times 10^{-6} = 16.9$ min, which is of the same order of magnitude as the time constant for the washout from many cell types. In this case, therefore, the homogeneous diffusion regime itself could be mistakenly regarded as one cellular compartment, which in reality does not exist. For this reason, therefore, the thickness of tissue sections used for washout experiments aimed at providing information on membrane permeabilities should be so small that the time constant τ_o for the homogeneous diffusion regime plays only an insignificant role in the rate of the exchange process.

4. Establishing the Stationary Concentration Profile

This section will deal with the problem which arose in Section C. III.1, namely an examination of those factors that determine the time taken to establish a stationary concentration profile. A membrane of thickness h is considered to be free from substance initially. At time $t = 0$ one side of the membrane is brought into contact with the substance, the concentration of which is maintained at C_o. The concentration on the other side of the membrane is maintained at zero. We know already that for large values of t the concentration profile is given by

$$C(x) = C_o(1 - x/h)$$

(cf. Eq. [24]). We shall now examine how the concentration $C(x, t)$ in the plate increases with time towards this stationary profile. We thus seek a solution of the diffusion equation

$$\frac{\partial C}{\partial t} = D \frac{\partial^2 C}{\partial x^2} \tag{19}$$

in the region $0 \leq x \leq h$, with the initial condition

$$C(x, 0) = 0, \text{ for } 0 \leq x \leq h \text{ and } t = 0 \tag{I}$$

and the boundary conditions

$$C = C_o \text{ for } x = 0 \text{ and } t > 0;$$

$$C = 0 \quad \text{for } x = h \text{ and } t > 0. \tag{II}$$

To solve this problem we again use a finite sine transform. Inserting the above boundary conditions (II) in Eq. (65), we obtain

$$\frac{d\bar{C}_s}{dt} = -D\left(\frac{k\pi}{h}\right)^2 \bar{C}_s + \left(\frac{k\pi}{h}\right) DC_o,$$

the solution of which is

$$\bar{C}_s(k, t) = A \exp\left\{-D\left(\frac{k\pi}{h}\right)^2 t\right\} + C_o h/k\pi,$$

where A is a constant determined by the initial condition (I), which when inserted in the above gives $A = -C_o h/k\pi$. The sine transform of the diffusion problem then becomes

$$\bar{C}_s(k,t) = C_o h/\pi k - \frac{C_o h}{\pi k} \exp\left\{-D\left(\frac{k\pi}{h}\right)^2 t\right\}.$$

Then, employing the inversion formula of Eq. (64) for the finite sine transform, we obtain

$$C(x,t) = \frac{2}{h}\sum_{k=1}^{\infty} \bar{C}_s(k,t)\sin(k\pi x/h)$$

$$= \frac{2C_o}{\pi}\sum_{k=1}^{\infty}\frac{\sin(k\pi x/h)}{k} - \frac{2C_o}{\pi}\sum_{k=1}^{\infty}\frac{\sin(k\pi x/h)}{k}\exp\left\{-D\left(\frac{k\pi}{h}\right)^2 t\right\}.$$

The first term is the Fourier series for the function $C_o(1-x/h)$ in the region $0 \leq x \leq h$. The above solution can therefore be written

$$C(x,t) = C_o\left(1-\frac{x}{h}\right) - \frac{2C_o}{\pi}\sum_{k=1}^{\infty}\frac{\sin(k\pi x/h)}{k}\exp\left\{-D\left(\frac{k\pi}{h}\right)^2 t\right\}, \tag{71}$$

which is an expression for the development of the concentration profile in the membrane. For $t \to \infty$, the last term disappears, and what remains is the stationary profile, already determined by the simpler methods in Section II.1. If Eq. (71) is expressed in the non-dimensional units

$$X = x/h \text{ and } T = t/\tau_o,$$

where

$$\tau_o = h^2/\pi^2 D \tag{67}$$

is once again the time constant in the exponential term for $k=1$, we obtain

$$C(X,T)/C_o = 1-X - \frac{2}{\pi}\sum_{k=1}^{\infty}\frac{\sin(k\pi X)}{k}\exp\left\{-k^2 T\right\}. \tag{72}$$

Figure 7 shows how the concentration increases with time to the stationary concentration profile. The curves are calculated from Eq. (72). A substance diffusing through a membrane can thus be ascribed a characteristic time, the *redistribution time*, t_{red}. This is the time taken for the substance to reach a new stationary concentration profile following a perturbation in the concentration in the surrounding medium. For $T = 3$, the concentration gradient $\partial C/\partial x$ for $C(x, t)$ deviates only insignificantly from the stationary gradient $-C_o/h$: it is therefore most convenient to put $T = \pi$, whereby the redistribution time becomes

$$t_{\mathrm{red}} = h^2/\pi D. \tag{73}$$

In the treatment of the exchange kinetics between two chambers we assumed that there were quasistationary conditions in the membrane separating the two chambers. This assumption is reasonable provided the time constant for the exchange between the chambers, $\tau = 1/k$, where k is the rate constant, is greater than the redistribution time for the substance in the membrane.

The reader will find further solutions of the diffusion equation in, for example, the texts by JACOBS (1967) and CRANK (1956) and the comprehensive treatise by CARSLAW and JAEGER (1959).

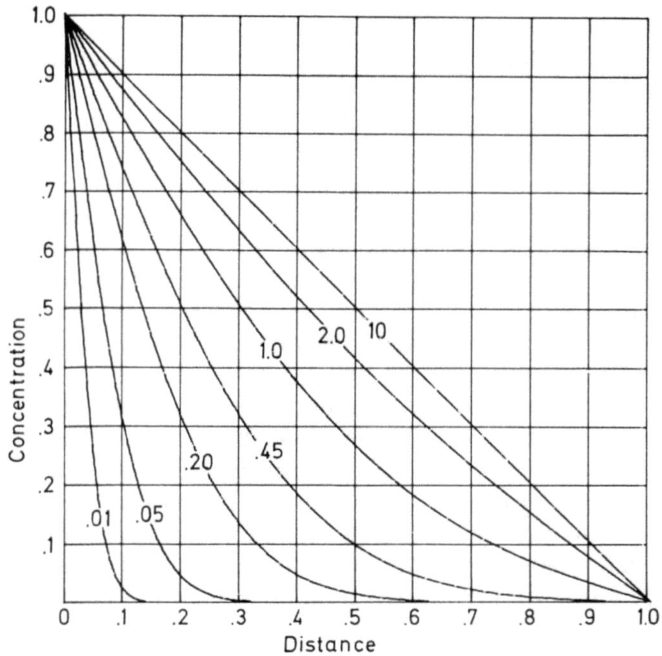

Fig. 7. Time course for establishment of the stationary concentration profile in a plate of thickness h. Ordinate, concentration in units of $C^{(i)}$. Abscissa, distance in units of plate thickness h. The numbers on the curves are times in units of $h^2/\pi^2 D$

D. Diffusion Processes: Microscopic Aspects

I. Brownian Movements

In Section B.III it was shown that the negative gradient of the chemical potential, $-d\mu/dx$, of the diffusing substance could be regarded as the "diffusional force" behind the diffusion process. But it was also pointed out that we were dealing with a pseudoforce, which might be useful to introduce in certain formal calculations. In this section the author will attempt to describe the basic mechanism which on the molecular level leads to a diffusion process with a flux as given empirically by Fick's law. The key to this understanding is that the kinetic energy of each molecule in the solution fluctuates unceasingly around a mean value $< \frac{1}{2} m v^2 > = (3/2)kT$, where k is Boltzmann's constant and T the temperature of the system. The fluctuation in the velocities of the liquid molecules will perpetually create local anisotropies of the impulses imparted to a given molecule. This impact will – if large enough – cause the molecule to undergo a displacement δ_1 in a certain direction, during which the additionally acquired kinetic energy is gradually distributed among the neighbouring molecules. A short moment later the "red molecule" in question will receive another impact and jump another distance δ_2, in a direction which is independent of the direction of the previous jump. This process will occur *ad infinitum*: all the molecules will incessantly bombard each other and from one time to another a certain fraction of the particles will undergo larger displacements (Platz-Wechslung) in a purely random manner. A reflection of these displacements can be observed experimentally under the microscope as the chaotic motion exhibited by a suspension of very small particles. This zig-zag motion of particles in suspension was first observed by the botanist Robert BROWN (1828) and has since been named *"Brownian movements"*. EINSTEIN (1905) and von SMOLUCHOWSKI (1906), working independently of each other, developed the theory for the Brownian movements. The basic idea behind the treatment of EINSTEIN and SMOLUCHOWSKI is that the Brownian movement displayed by a single suspended particle and the spreading of a swarm of particles in the available space by diffusion is a reflection of one and same stochastic process, which can be attributed to fluctuations in the thermal movements of the molecules. In their analysis EINSTEIN and SMOLUCHOWSKI introduced certain simplifying assumptions about the pattern of the molecular movements, and their theory is based on arguments of probability rather than on any special molecular model or the principles of particle dynamics. The papers by EINSTEIN and by SMOLUCHOWSKI on the theory of the Brownian movements have both been collected and edited (EINSTEIN, 1926; SMOLUCHOWSKI, 1923). Another mode of attack was later introduced by LANGEVIN (1908). His train of thought was taken up by UHLENBECK and ORNSTEIN (1930) and ORNSTEIN and VAN WIJK (1933), who developed a more complete dynamic theory. An excellent review of the whole theory has been given by CHANDRASEKHAR (1943). In this section a sketch will be given of the treatment by EINSTEIN and SMOLUCHOWSKI, because it gives a more vivid picture of the diffusion process brought about by the molecular chaos.

II. Smoluchowski's Treatment

1. Statistical Interpretation of the Diffusion Equation

The distribution of the particle concentration following the establishment of an instantaneous plane source of strength N molecules per m^2 at time $t = 0$ in the position $x = 0$ is given by

$$C(x,t) = \frac{N}{2\sqrt{\pi D t}} \exp\{-x^2/4Dt\}, \tag{55}$$

where $C(x,t)$ is the number of particles per m^3. The number of particles within the layers of unit area at x and $x + dx$ are: $dn = C(x,t)dx \cdot 1 \cdot 1$. This number constitutes the fraction

$$dP = dn/N = C(x,t)dx/N \tag{74}$$

of the total number of particles. This expression can also be regarded as the probability that particles which were initially ($t = 0$) at the position $x = 0$ will be found at time t between x and $x + dx$. We then have, from Eq. (55)

$$dP = \varphi(x,t)dx,$$

where

$$\varphi(x,t) = \frac{1}{2\sqrt{\pi D t}} \exp\{-x^2/4Dt\} \tag{75}$$

is the *probability density* for the displacement x in time t. Note the similarity between this expression and the probability density for the normal distribution (Gaussian distribution)

$$y = \frac{1}{\sqrt{2\pi}\sigma} \exp\{-x^2/2\sigma^2\},$$

where σ is the dispersion from the mean value $\langle x \rangle = 0$. Thus $\varphi(x,t)$ represents a Gaussian distribution for which

$$\sigma^2 = \langle x^2 \rangle = 2Dt. \tag{76}$$

This leads naturally to the concept that the displacement of the individual particles, $x(t)$, during a diffusion is characterized by a spectrum of displacements which are given by Eq. (75). That is to say, the displacements from the initial position are distributed normally with a mean value zero and a dispersion $\pm\sqrt{2Dt}$. Equation (76), which can also be derived from statistical considerations about the motion of the molecules, is often called the *Einstein-Smoluchowski equation.*

2. Random Walk in One Dimension

In his treatment of the diffusion process, SMOLUCHOWSKI (1906, 1916) gave an idealized description of the chaotic jumps of the dissolved molecules by solving the problem in the calculus of probability, which is now called "random walk" or "drunkard's walk" and which was first formulated explicitly by Karl PEARSON (1905). The simplest version of the problem formulated for the walk in one dimension is as follows. A particle exhibits a sequence of steps all of the same length λ, the steps being directed either forward (a positive step) or backward (a negative step). Each step has the same probability and each step is independent of the direction of the previous step. Thus each step has a probability of $1/2$. After having performed a total of N steps the particle could be found at any one of the points on the line

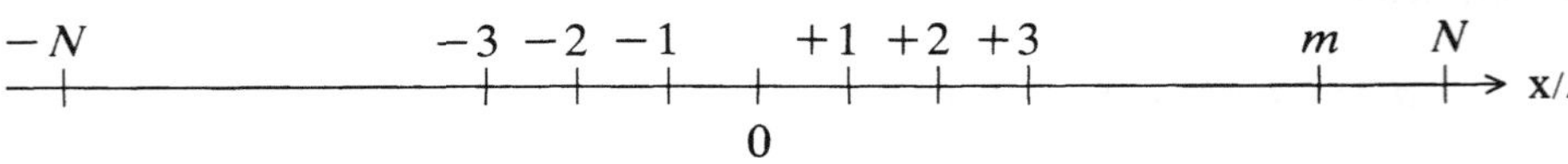

Where the points can be regarded as coordinates on the x-axis with the step length λ being the unit. Let m be an occupation number and x the corresponding position coordinate. In order to have arrived at the position $x = \lambda m$ after having performed N steps the particle must have made a certain number of steps, N_+, in the positive direction of the x-axis and another number of steps, N_-, in the opposite direction but in such a way that

$$N_+ - N_- = m \tag{77}$$

where the mutual sequence of each positive and negative step can be established in a certain number of ways which depend on how large m is relative to N. Furthermore,

$$N_+ + N_- = N, \tag{78}$$

from which it follows that

$$N_+ = (N+m)/2 \qquad \text{and} \qquad N_- = (N-m)/2\,. \tag{79}$$

It is seen that m is even for N even and odd for N odd. Further, since $m = 2N_+ - N$, m can only change in steps of two. One way of keeping track of the various probabilities of wandering is indicated in the scheme in Figure 8, below, which is taken from SMOLUCHOWSKI (1916).

The number of ways in which a given position m can be reached by a total of N steps (positive and negative) is equal to the sum of the ways in which the surrounding positions $m-1$ and $m+1$ can be reached in $N-1$ steps. In the scheme this number is given in the numerator of the fraction inside each circle. It is seen that this number varies in the same way as the coefficients in the binomial distribution. Thus, the number of ways in which a given position m

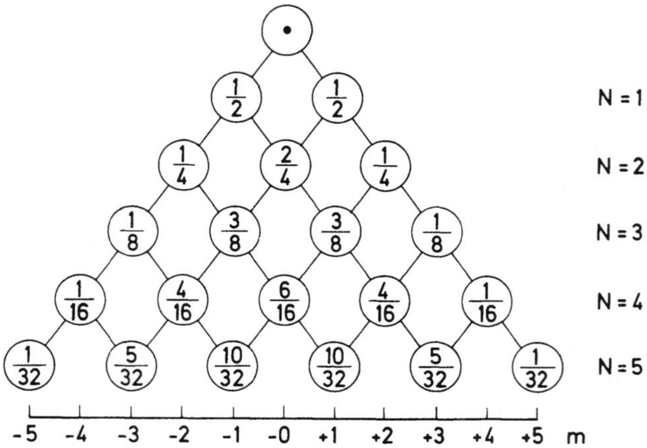

Fig. 8. Scheme for illustrating the way of calculating the probability for a random walk

which requires $N_+ = (N+m)/2$ positive steps out of a total of N steps can be reached, is

$$\binom{N}{N_+} = \frac{N!}{N_+!(N-N_+)!} = \frac{N!}{N_+!\cdot N_-!}\,.$$

The denominator gives the total number of ways, $n(N)$, in which all the available places can be occupied in N steps. This number is

$$n(N) = 2^N.$$

Hence the probability that the particle is in the position m after having performed a total of N steps is

$$W(m,N) = \frac{N!}{N_+!\cdot N_-!}\left(\frac{1}{2}\right)^N,$$

or using Eq. (79):

$$W(m,N) = \frac{N!}{\left(\frac{N+m}{2}\right)!\cdot\left(\frac{N-m}{2}\right)!}\left(\frac{1}{2}\right)^N. \tag{80}$$

Applying Stirling's approximation,

$$n! = \sqrt{2\pi n}\,(n/\mathrm{e})^n,$$

which is valid for large values of n, the above expression can be transformed into the following continuous expression:

$$W(m,N) = \sqrt{\frac{2}{\pi N}} \exp\left\{-\frac{m^2}{2N}\right\}, \tag{81}$$

which is valid with great precision when $N \gg 1$ and $m/N \ll 1$. To obtain the probability density for the random walk from Eq. (81) we introduce instead of the occupation number, m, the displacement $x = m\lambda$, where λ is the length of each single step. We now ask what the probability is that the particle will be found between the position $x = m\lambda$ and $x + dx = m\lambda + dm\lambda$. Let m be so small relative to N that $W(m,N)$ can be regarded as almost constant in the range between m and $m + dm$. The probability density is then equal to the probability of a walk ending up in the position $m = x/\lambda$, which is given by Eq. (81), multiplied by the number of available positions for the particle in the interval x to $x + dx$. Along the distance dx the number of positions is $dm = dx/\lambda$. The number of available positions along dx, however, is only $dx/2\lambda$, since m for a given value of N can only vary in steps of $\Delta m = 2$. Hence $W(m,k)dx/2\lambda$ is the probability density for finding the particle between x and $x + dx$, or

$$dP = \sqrt{\frac{2}{\pi N}} \exp\left\{-\frac{m^2}{2N}\right\} \frac{dx}{2\lambda}.$$

Let t be the time required to execute all the N steps. One single step will then on average last $\tau = t/N$. Further, the displacement, x, corresponding to the occupation number, m, is $x = m\lambda$. Elimination of N and m from the expression above yields

$$dP = \frac{1}{2\sqrt{\pi(\lambda^2/2\tau)t}} \exp\left\{-\frac{x^2}{4(\lambda^2/2\tau)t}\right\} dx.$$

Now, putting

$$D = \lambda^2/2\tau, \tag{82}$$

the probability density can be written as

$$dP = \varphi(x,t)dx = \frac{1}{2\sqrt{\pi Dt}} \exp\left\{-\frac{x^2}{4Dt}\right\} dx. \tag{83}$$

Imagine now that a large number – say N particles per m^2 – begin their random walk in the plane corresponding to the plane at $x = 0$ and at the time $t = 0$. At time t a certain fraction of the particles will be located between x and $x + dx$, the most likely number being $dn = N \cdot \varphi(x,t)\, dx \cdot 1 \cdot 1$. The particle concentration in this layer is $C(x,t) = dn/dx \cdot 1 \cdot 1$, or

$$C(x,t) = \frac{N}{2\sqrt{\pi Dt}} \exp\left\{-\frac{x^2}{4\,Dt}\right\}. \tag{84}$$

Thus, provided the ratio between the square of the unit step length, λ, and twice the execution time, τ, is equal to the diffusion coefficient D, random walk considerations and solution of the diffusion equation give the same distribution of particles in space as a function time. Both situations are characterized by mass conservation. In the first situation the phenomenological mechanism behind the mass transport was the presence of the diffusional flux as described by Fick's law:

$$J = -D(\partial C/\partial x).$$

In the second situation the same process of distribution was explained solely by the assumption that the particles incessantly perform chaotic movements where each step has the same probability and the probability of a step in one direction is independent of the direction of the previous step. This again leads to the concept that the diffusional flux is the result of the Brownian movements performed by all members of the particle swarm. At the same time it is no longer necessary to consider that the concentration gradient establishes a diffusional force that is equal to the negative gradient of the chemical potential.

3. The Einstein-Smoluchowski Equation

The distribution function $\varphi(x,t)$ is symmetrical around $x=0$. Thus the mean displacement $<x>$ of the particle is zero. But since there is a finite, non-zero, probability of finding the particle somewhere in the region $x \neq 0$ at the time t, it is useful to consider two different averages of the displacement in time.

Half the particles that were initially present at $x=0$ will at time t be located in the region $x \geqslant 0$, and the other half in the region $x \leqslant 0$. The mean displacement of each group of particles, *i. e.* the mean displacement in either the positive or negative direction, is

$$\bar{x} = \frac{1}{\sqrt{\pi D t}} \int_0^\infty x \mathrm{e}^{-x^2/4Dt} dx = 2\sqrt{Dt/\pi}. \tag{85}$$

The measure most used to describe the displacement of a particle or a collection of particles in the time t is the mean value of the square of the displacements $<x^2>$ which is

$$<x^2> = \int_{-\infty}^{\infty} x^2 \varphi(x,t) dx = \frac{1}{\sqrt{\pi D t}} \int_0^\infty x^2 \mathrm{e}^{-x^2/4Dt} dx = 2Dt.$$

This result is generally written as

$$D = \frac{<x^2>}{2t}, \tag{86}$$

and it is usually called the *Einstein-Smoluchowski equation,* which connects the mean of the square of the displacement to the time t — which is a reflection of the random walk executed by the particles — with the diffusion coefficient D of the particles, which is the fundamental parameter in the macroscopic description of the diffusion process. Equation (86) allows an estimation of the most likely motion of a collection of particles executing Brownian movements. If all the particles are initially present at $x = 0$ the fraction Z of the particles which at the time t are still confined to the region $-\sqrt{2Dt} < x < \sqrt{2Dt}$ is

$$Z = \frac{1}{2\sqrt{\pi Dt}} \int_{-\sqrt{2Dt}}^{\sqrt{2Dt}} e^{-x^2/4Dt}\, dx = \mathrm{Erf}\{1/\sqrt{2}\} = 0.683\,,$$

i. e. at the time t about two-thirds of the particles are still inside the space between the planes at $x = -\sqrt{2Dt}$ and $x = \sqrt{2Dt}$, while the rest will have moved farther away from the initial position. A comparison between Eqs. (85) and (86) gives $\bar{x} = 0.8\sqrt{\langle x^2 \rangle}$, *i. e.* the two measures for the displacements at the time t are nearly equal. The important point, however, is that the displacements are proportional to *the square root of the time t* and not to t itself. The significance of this can be illustrated in the following example: with $D = 10^{-6}\ \mathrm{cm^2/s^{-1}}$ and $\sqrt{\langle x^2 \rangle} = 1$ cm, Equation (86) gives $t = 5 \times 10^5$ s. If $\sqrt{\langle x^2 \rangle} = 1\ \mu$m, t is 5 ms and 0.5 μs with $\sqrt{\langle x^2 \rangle} = 100$ Å. A mass transport based upon Brownian movement will only occur with considerable speed if the distances in question are very small.

III. Random Walk and Fick's Law (Einstein)

In 1908, Einstein gave a simplified account of his theory of Brownian movement. His beautiful reasoning, which leads to Fick's law and the *Einstein-Smoluchowski equation,* is reproduced here. Einstein simplified the motion of a col-

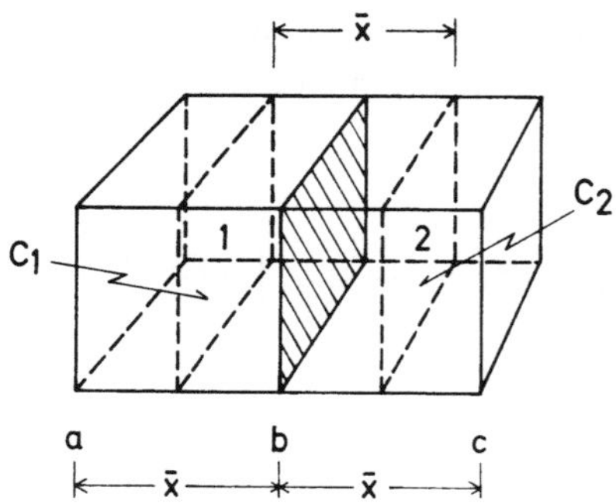

Fig. 9. To illustrate Einstein's simplified treatment of the random walk

lection of particles by replacing the whole spectrum of displacements in time t with a single displacement $\bar{x}$, *i. e. the mean displacement in one direction in time t.* He regarded the particles which are contained inside the particular volume element shown in Figure 9. The area A, placed perpendicularly to the direction of the concentration gradient at the position b, is the common end-surface for the two volume elements v_1 and v_2 with the side $\bar{x}$, *i. e.* the mean displacement in one direction in the time t. This special choice of dimensions guarantees that all the particles that are initially in one of the volume elements have a chance to cross the area A within the time t and end up somewhere inside the other volume element. Let C_1 and C_2 be the mean concentrations in the volumes v_1 and v_2. The number of particles initially present inside v_1 is $C_1A\bar{x}$. At time t, half of these particles will have moved out of v_1 in the leftward direction, while the other half will have moved towards the right and thus will have crossed the area A at b. The transport $M^{(12)}$ in the direction (1) → (2) through this area in the time t is $^1/_2\, C_1\, A\bar{x}$ and the transport $M^{(21)}$ through A in the direction (2) → (1) within the same time is $^1/_2\, C_2\, A\bar{x}$. The net transport through the plane A in the direction (1) → (2) is therefore

$$M = M^{(12)} - M^{(21)} = {}^1/_2\, A\bar{x}(C_1 - C_2).$$

The flux in this direction is $J = M/At = (C_1 - C_2)\bar{x}/2t$. For low values of $\bar{x}$, or with a linear concentration profile over the distance $\bar{x}$, we have: $C_2 = C_1 + (dC/dx)\bar{x}$. Thus the expression for the flux can be written

$$J = -\frac{(\bar{x})^2}{2t}\frac{dC}{dx}. \tag{87}$$

According to Eq. (85) the factor $(\bar{x})^2/2t$ is a constant. Thus, putting $D = (\bar{x})^2/2t$, Eq. (87) becomes identical with Fick's law. This value for D differs by 20 percent from that given by the *Einstein-Smoluchowski equation*, but considering the simplifications used in the present averaging process, complete agreement could never have been expected. However, the essential point is that Einstein succeeded in deducing Fick's law from the Brownian movement, and at the same time accounted for the mechanism which gives rise to a diffusional flux, without introducing any special unidirectional driving force behind the diffusion process. All the diffusing particles execute Brownian movement with equal probability for a forward and a backward step. These chaotic zig-zag movements of all the particles will give rise to a net transport in one direction if the density of the particles decreases in that direction. A layer having a greater density of particles than the neighbouring layer will deliver more particles per unit time to the neighbouring layer than it will receive in the same time from this layer. This will occur just because the probability density for the displacement is the same for each particle.

IV. The Smoluchowski Equation

In this section the equation of motion for a collection of particles will be derived on the basis of the Brownian movement executed by the dissolved particles. We consider a solution of particles, *e.g.* contained in a tube of infinite extension and cross-sectional area 1 m^2.

We want to calculate the accumulation of particles which occurs in the time τ from t to $t+\tau$ in the volume element ΔV_o, which is confined between the planes at x_o and $x_o + h$. This increase in numbers of particles is

$$\Delta n_\tau = h(\partial C(x_o, t)/\partial t)\tau$$

for $\tau \ll 1$. This increase can also be written as

$$\Delta n_\tau = \Delta n_\tau^{(in)} - \Delta n_\tau^{(out)},$$

where $\Delta n_\tau^{(in)}$ represents the number of particles received in time τ from the neighbouring regions at $x < x_o$ and $x > x_o + h$, and $\Delta n_\tau^{(out)}$ is the number of particles lost from ΔV_o in the same time. To calculate Δn_τ we introduce the probability density $\varphi_\tau(X)$ for the displacement X in time τ. This function is assumed (i) to be an even function, *i. e.* $\varphi_\tau(X) = \varphi_\tau(-X)$, (ii) to behave in principle as the function of Eq. (83) for $(X) \to \infty$, and (iii) since a particle must be somewhere in $-\infty < x < \infty$ at time τ, the integral of $\varphi_\tau(X)$ from $-\infty$ to $+\infty$ is equal to unity. A particle which at time t is at x will at time $t+\tau$ have the probability $\varphi_\tau(x_o - x) \cdot h$ of having moved into the planes between x_o and $x_o + h$. The number of particles which at time t are between x and $x + dx$ is $C(x, t)dx$. Among these a number $h \cdot C(x,t)\,\varphi_\tau(x_o - x)dx$ will have moved into ΔV_o at time $t+\tau$. Hence the number of particles $\Delta n_\tau^{(in)}$ which have moved into ΔV_o from the surrounding regions $x < x_o$ and $x > x_o + h$ is

$$\Delta n_\tau^{(in)} = h \int_{-\infty}^{\infty} C(x,t)\varphi_\tau(x_o - x)dx,$$

provided $h \ll 1$. Furthermore,

$$\Delta n_\tau^{(out)} = h\, C(x_o, t)$$

if τ is so large that standard deviations of the displacements $\sqrt{<\xi^2>}$ are much larger than $h/2$. Thus

$$\left(\frac{\partial C}{\partial t}\right)\tau = \int_{-\infty}^{\infty} C(x,t)\varphi_\tau(x_o - x)dx - C(x_o, t).$$

We now assume that the particles are in a constant external field of force. This field acts on each particle with the external force X, the result of which is that each particle will now have a higher probability of jumping in the direction of the force X than in the opposite direction, or alternatively the external force will superimpose a constant velocity component in the direction of the force

$$v = BX$$

on the Brownian movement. If v is much smaller than the velocities due to thermal motion, the shape of the probability density φ_τ^* with field and φ_τ without field will not differ significantly from each other, the only difference being a displacement of φ_τ^* with the distance $v\tau$ relative to the position of φ_τ. A particle which at time t is at the position x will therefore have a probability at time $t+\tau$ of being found inside the layers x_0 and $x_0 + h$, which is

$$dP = \varphi_\tau^*(x_0 - x)h = \varphi_\tau(x_0 - (x + v\tau))h,$$

since as a result of the external field the plane of symmetry now is at the position $x + v\tau$. The accumulation within the layers at x_0 and $x_0 + h$ with the field present is therefore

$$h\left(\frac{\partial C}{\partial t}\right)_{x_0} \tau = \left(\int_{-\infty}^{\infty} C(x, t)\ \varphi_\tau(x_0 - (x + v\tau))dx - C(x_0, t)\right) h.$$

Introducing the displacement $\xi = x + v\tau - x_0$, the above expression can be written

$$\left(\frac{\partial C}{\partial t}\right)_{x_0} \tau = \int_{-\infty}^{\infty} C(x_0 + \xi) - v\tau, t)\varphi_\tau(\xi)\, d\xi - C(x_0, t).$$

The integrand is expanded in a Taylor's series in powers of $(\xi - v\tau)$ about the value x_0. The result of inserting this expansion in the expression above becomes

$$\left(\frac{\partial C}{\partial t}\right)_{x_0} \tau = \sum_{n=1}^{\infty} \frac{1}{n!} \left(\frac{\partial^n C}{\partial x^n}\right)_{x_0} \int_{-\infty}^{\infty} (\xi - v\tau)^n \varphi_\tau(\xi) d\xi.$$

In this expansion all the integrals containing odd powers of n in the integrand $\xi^n\varphi_\tau(\xi)$ will vanish, since $\varphi_\tau(\xi) = \varphi_\tau(-\xi)$. When τ is macroscopically small it can be shown that all the higher terms of the integrals containing $\xi^n\varphi_\tau(\xi)$ for n even can be disregarded in comparison to the contribution from $\xi^2\varphi_\tau(\xi)$. With this approximation the above expression becomes, after ommission of the subscripts,

$$\frac{\partial C}{\partial t} = \frac{1}{2\tau}\int_{-\infty}^{\infty} \xi^2\, \varphi_\tau(\xi) d\xi \cdot \left(\frac{\partial^2 C}{\partial x^2}\right) - v\frac{\partial C}{\partial x} + \frac{(v\tau)^2}{2\tau}\,\frac{\partial^2 C}{\partial x^2}.$$

Now, the integral is equal to the mean value of the square of the displacements $<\xi^2>$ in the time τ. The above expression can be simplified further, if the drift $v\tau$ along the x-axis in time τ is much smaller than the mean of the Brownian displacements in the same time, *i. e.* $<\xi^2> \gg (v\tau)^2$. Hence the expression for the rate of accumulation can be written

$$\frac{\partial C}{\partial t} = \frac{<\xi^2>}{2\tau}\frac{\partial^2 C}{\partial x^2} - v\frac{\partial C}{\partial x}.$$

Putting $v = BX$ and

$$D = \frac{<\xi^2>}{2\tau}, \tag{88}$$

which is the *Einstein-Smoluchowski equation* we finally obtain

$$\frac{\partial C}{\partial t} = D\frac{\partial^2 C}{\partial x^2} - BX\frac{\partial C}{\partial x}, \tag{89}$$

which usually is called the *Smoluchowski equation,* as it was derived and used by SMOLUCHOWSKI (1913; 1915a and b; 1916) in his studies of Brownian motion in the presence of an external field of force. This equation, which is the basis for the description of the movements of a collection of particles which execute Brownian movement under the influence of an external force X, can also be written

$$\frac{\partial C}{\partial t} = -\frac{\partial}{\partial x}\left\{-D\frac{\partial C}{\partial x} + BXC\right\},$$

provided D and/or B do not change with the position. Mass conservation requires

$$\frac{\partial C}{\partial t} = -\frac{\partial J}{\partial x} \tag{18}$$

where J is the flux of the particle collection. Hence

$$J = -D\frac{\partial C}{\partial x} + BXC, \tag{90}$$

which is identical with Eq. (15), which was obtained by the less satisfying procedure of applying the principle of superposition on macroscopic considerations. The argument reproduced here to derive the *Smoluchowski equation* follows essentially that given by PLANCK (1917), which again was based upon an argument used by EINSTEIN (1905) to derive the diffusion equation.

V. Kramers' Equation

If the external force X is conservative, then the particles can be assigned a potential energy $U(x)$ everywhere in space, such that

$$X = -\frac{dU}{dx} \tag{91}$$

and Eq. (90) instead can be written

$$-\frac{J}{D} = \frac{dC}{dx} + \left(\frac{B}{D}\right) C \frac{dU}{dx}.$$

Since

$$\frac{d}{dx}\left\{C e^{U(x)B/D}\right\} = \frac{dC}{dx} e^{U(x)B/D} + \left(\frac{B}{D}\right) C e^{U(x)B/D} \frac{dU}{dx},$$

the above expression can be written as

$$J \cdot e^{U(x)B/D} = -D \frac{d}{dx}\left\{C e^{U(x)B/D}\right\},$$

which is an alternative expression for the flux corresponding to the Smoluchowski equation. If the process is stationary we have $\partial J/\partial x = 0$. The above equation can then be easily integrated between two points in space, *e.g.* x_o and x_1. The result is the following important equation, which was first derived by KRAMERS (1940):

$$J = -D \frac{C(x_1) e^{U(x_1)B/D} - C(x_o) e^{U(x_o)B/D}}{\int\limits_{x_o}^{x_1} e^{U(x)B/D} dx}. \tag{92}$$

This describes the flux for a collection of particles which perform their Brownian movement in a potential field $U(x)$.

VI. Diffusion Coefficient and Mobility

The diffusion coefficient D is related to the mean square displacement $<\xi^2>$ in time τ by the relation $D = <\xi^2>/2\tau$. The magnitude of $<\xi^2>$ must depend upon the length of the individual jumping distances caused by the asymmetrical bombardment from the neighbouring molecules. One parameter which, among others, determines this distance is the coefficient of friction f or the mobility

$1/f = B$. It is to be expected, therefore, that the two quantities D and B in the Smoluchowski equation must be related in some way. This extremely important relation was first derived by EINSTEIN (1905).

1. Einstein's Relation

We consider a collection of particles which perform their Brownian movement in a potential field $U(x)$, so that the particles are driven towards a reflecting barrier placed at x_o. Thus the force $X = -dU/dx$ tends to move all the particles to the position $x = x_o$, which corresponds to the position of lowest potential energy $U(x_o)$. The thermal movements, on the other hand, tend to distribute the particles uniformly throughout the available space. At a certain time a state of *equilibrium* will be established in which the concentration profile $C(x)$ has adjusted itself in such a way that the tendency of the particles to move in one direction due to the force X is exactly counteracted by the tendency of the particles to move in the opposite direction due to Brownian movement and the presence of the concentration gradient dC/dx. The equilibrium condition is $J = 0$ everywhere in space. We consider the two positions x_o and x with concentrations $C(x_o)$ and $C(x)$. To simplify matters, we put $U(x_o) = 0$. The equilibrium condition $J = 0$ implies that the numerator in Kramers' equation must be zero. This results in

$$C(x) = C(x_o)\,e^{-U(x)B/D}.$$

The statistical mechanical correlate to this equilibrium state is

$$C(x) = C(x_o)\,e^{-U(x)/kT}$$

where k is Boltzmann's constant (1.38×10^{-23} joule · K^{-1} per molecule). The two expressions above must be identical. This requires

$$D = kTB \qquad \text{(EINSTEIN, 1905)}, \tag{93}$$

which is the celebrated relation of Einstein connecting the coefficient of diffusion D for the particle with its mechanical mobility B. This relation implies that the Smoluchowski equation contains only one parameter, which may either be D or B according to convenience. In addition Kramers' equation can be written

$$J = -D\,\frac{C(x_1)\,e^{-U(x_1)/kT} - C(x_o)\,e^{-U(x_o)/kT}}{\int_{x_o}^{x_1} e^{U(x)/kT}\,dx} \tag{94}$$

2. Einstein-Stokes' Relation

Consider a spherical particle with radius r, which is suspended in a medium with the viscosity η. The particle moves with the stationary velocity v under the influence of an external force X and the opposing frictional force $-v/B$ of equal magnitude. These quantities are connected by Stokes' law:

$$X = 6\pi r \eta v \qquad \text{(Stokes, 1856).}$$

The mechanical mobility of the particle is $B = v/X = 1/6\pi r\eta$. Application of the Einstein relation gives $D = kTB = kT/6\pi r\eta$. Multiplication of the numerator and denominator by Avogadro's number $N_A = 6.023 \times 10^{23}$ gives

$$D = \frac{RT}{6\pi r \eta N_A}, \tag{95}$$

since the gas constant $R = N_A k = 8{,}314\ \text{J}\cdot\text{mol}^{-1}\cdot\text{K}^{-1}$. Equation (95) is called the *Einstein-Stokes' relation.* This relation was used by *Perrin* to determine Avogadro's number N_A from measurements of the displacements ξ which a Brownian particle of known dimension executed in time t. The diffusion coefficient was calculated from the relation $D = \langle\xi^2\rangle/2t$, and since the value of R was known the magnitude of N_A could be determined. Nowadays this relation is used to estimate the dimensions of molecules from measurements of their diffusion coefficients.

Equation (95) also explains why the diffusion coefficient for a particle varies so slowly with the molecular weight of the particle. The mass of a spherical particle with mass density ϱ and radius r is $m = 4\pi r^3\varrho/3$. Hence the molecular weight is $M = 4\pi r^3\varrho N_A/3$, from which it follows that $r \propto \sqrt[3]{M}$. Equation (95) tells that the product $D\cdot r$ is a constant. Accordingly, it should also be expected that

$$D\sqrt[3]{M} = \text{const.}, \tag{96}$$

provided the molecule is spherical and the molecular dimensions are large enough to ensure the validity of Stokes' law.

E. Diffusion and Superimposed Convection

As mentioned in Section B.II, transport of a substance can also take place in a system with uniform distribution, provided the entire system – *i. e.* the solvent and the dissolved particles – moves as a whole in a given direction. This type of transport is called *convection.* If the concentration of substance is C and the velocity at which the system is moving at right angles to a fixed plane is v, then according to section B.II.2, the convection flux of the substance is

$$J = vC.$$

If at the same time a concentration gradient exists in the system, a transport of substance by diffusion will also take place.

I. The Equation of Motion

Transport of matter through a fixed plane is the same, whether the system moves as a whole with a velocity v or whether the system is stationary, but where each separate particle is under the influence of a force X, which in combination with frictional forces imposes on the particle a steady-state velocity v. If, therefore, we put $BX = v$ in Eq. (90), the flux for the combined diffusion and convection process becomes

$$J = -D\frac{\partial C}{\partial x} + vC. \tag{98}$$

The equation of motion likewise becomes

$$\frac{\partial C}{\partial t} = D\frac{\partial^2 C}{\partial x^2} - v\frac{\partial C}{\partial x}. \tag{99}$$

As pointed out by SMOLUCHOWSKI (1915), this equation can be simplified by introducing a new variable, Y, defined by

$$C = Y\exp\left\{\frac{v}{2D}(x - x_o) - \frac{v^2 t}{4D}\right\}, \tag{100}$$

whereby Eq. (99) is transformed to

$$\frac{\partial Y}{\partial t} = D\frac{\partial^2 Y}{\partial x^2}.$$

This equation, which is of the same form as the one-dimensional diffusion equation is considerably easier to solve than Eq. (99). The solution of the diffusion equation corresponding to an instantaneous point source produced at time $t = 0$ in position $x_o = 0$ is according to Section C.III.2,

$$Y = \frac{N}{2\sqrt{\pi D t}}\,e^{-x^2/4Dt},$$

which when inserted in eq. (100)gives

$$C(x,t) = \frac{N}{2\sqrt{\pi D t}}\,e^{-(x-vt)^2/4Dt}. \tag{101}$$

The entire concentration profile is thus displaced with the constant speed v, while maintaining its symmetrical distribution around the peak of the profile, which at time t is localized in the position $x_0 = vt$. The time-dependent process corresponding to the transport through a slab can be solved with the aid of the solution given in section C.III.3. Here, however, we shall only consider the stationary state.

II. Steady-State Concentration Profile

In those cases where there are steady-state conditions, the concentration profile is determined by

$$\frac{d^2C}{dx^2} - \frac{v}{D}\frac{dC}{dx} = 0,$$

the general solution of which is

$$C = A e^{vx/D} + B, \tag{102}$$

where A and B are two constants whose magnitude depends on the boundary values of the physical system in question. The flux $J = -D(dC/dx)$ can then be determined from the above solution. However, it is simpler to use a method analogous to that contrived by Kramers (see Section D.V).

1. Stationary Transport through a Membrane

As an example we will determine the concentration profile and flux through a membrane of thickness h, corresponding to the case where the concentration is maintained constant on both sides while at the same time there is a convection flow with the velocity v through the membrane in the x-direction.

a) Determination of the Flux

We start from Eq. (98), written in the form

$$\frac{J}{D} = -\frac{dC}{dx} + \frac{v}{D}C.$$

A comparison with Section D.V reveals that the above expression can also be written as

$$J e^{-xv/D} = -D\frac{d}{dx}\left\{C e^{-vx/D}\right\}, \tag{103}$$

and as the process is stationary, $\partial C/\partial t = -\partial J/\partial x = 0$. Equation (103) can be easily integrated through the membrane from position $x = 0$ with concentration $C(\mathrm{o})$ to $x = h$ with concentration $C(h)$. We thus obtain the following expression for the flux

$$J = v\,\frac{C(\mathrm{o}) - C(h)\,\mathrm{e}^{-hv/D}}{1-\mathrm{e}^{-hv/D}}, \tag{104}$$

which was first derived by Gustav HERTZ (1923). If $hv \gg D$, the exponential term approaches zero and we obtain $J = v\,C(\mathrm{o})$, corresponding to the condition that convection dominates the transport. The same is the case for $|hv| \gg D$ but $v < 0$. This gives $J = vC(h)$. For $v = 0$, Equation (104) becomes, as expected, $J = D[C(\mathrm{o}) - C(h)]/h = P[C(\mathrm{o}) - C(h)]$, corresponding to the condition that the process is now completely controlled by diffusion. *The degree of rectification* can be defined as the ratio between fluxes, when v moves towards $+\infty$ and $-\infty$ respectively. This is therefore

$$J_{v=+\infty}/J_{v=-\infty} = C(\mathrm{o})/C(h). \tag{105}$$

b) Unidirectional Fluxes and Flux Ratio

In Section C.III.1. c, the unidirectional flux through a membrane was defined as the flux corresponding to the state where the concentration on the one side was maintained all the time at zero. For the case of a pure diffusion flux the unidirectional fluxes are given by Equations (53) and (54). If in the Hertz equation (104) we put $C(h) = 0$, the flux is now the unidirectional flux in the direction $(0) \rightarrow (h)$. The magnitude of this flux is

$$J^{(\mathrm{o}h)} = v\,\frac{C(\mathrm{o})\,\mathrm{e}^{hv/D}}{\mathrm{e}^{hv/D} - 1} \tag{106}$$

If instead we put $C(\mathrm{o}) = 0$, we obtain

$$J = -v\,\frac{C(h)}{\mathrm{e}^{hv/D} - 1},$$

where the minus sign appears, since the positive direction of the flux was chosen to be that of the positive direction of the x-axis. The unidirectional flux is always defined as the flux in the direction *towards* the phase with concentration zero. We have therefore $J^{(h\mathrm{o})} = -J$ in the above situation, or

$$J^{(h\mathrm{o})} = v\,\frac{C(h)}{\mathrm{e}^{vh/D} - 1}, \tag{107}$$

corresponding to the convention that unidirectional fluxes are always calculated positive, no matter which direction they have in relation to the positive direction of the reference system chosen. If Eq. (106) is now divided by Eq. (107), the following relationship is obtained between the unidirectional fluxes:

$$J^{(oh)}/J^{(ho)} = \frac{C(o)}{C(h)} e^{hv/D}, \tag{108}$$

which is a special case of a more general relation for the flux ratio, derived by USSING (1952). The above relation can be used to evaluate whether the transport of a substance through a membrane may be ascribed exclusively to a diffusion process with superimposed convection.

c) The Concentration Profile

As mentioned, the concentration profile can be determined by solving Eq. (102), corresponding to the boundary conditions $C = C(o)$ for $x = 0$ and $C = C(h)$ for $x = h$. However, the following procedure gives the required result more quickly: Equation (104) gives the flux through the membrane as a function of the concentrations $C(o)$ and $C(h)$. As the flux through the membrane is constant, the same flux is obtained by integrating Eq. (103) from $x = 0$ with concentration $C = C(o)$ to the position x with concentration $C(x)$. This gives

$$J = v \frac{C(o) - C(x) e^{-vx/D}}{1 - e^{-vx/D}}.$$

If this is equated with the expression for the flux, Eq. (104), we obtain

$$\frac{C(o) e^{hv/D} - C(h)}{e^{hv/D} - 1} = \frac{C(o) e^{vx/D} - C(x)}{e^{vx/D} - 1},$$

which when solved with respect to $C(x)$ gives

$$C(x) = \frac{C(o) e^{hv/D} - C(h) - (C(o) - C(h)) e^{vx/D}}{e^{hv/D} - 1} \tag{109}$$

It is seen from the above that the concentration profile is now no longer linear, except in the case where $v = 0$, which gives

$$C(x) = C(o) - \left[C(o) - C(h)\right] \cdot \frac{x}{h}$$

in agreement with Eq. (24), corresponding to the transport process, which is purely by diffusion. An appreciable deviation from linearity will occur if the number $\alpha = |vh/D| > 1$. Figure 10 shows a family of concentration profiles corresponding to different values of α. The profile is seen to resemble a sail stretched out between $C(o)$ and $C(h)$ under the influence of the wind blowing in the direction v. This configuration is a result of the condition that the flux $J = -D(dC/dx) + vC$ must be constant through the membrane. Thus for a high value of concentration C the contribution vC to the flux will be large, and accordingly the contribution $- DdC/dx$ must be correspondingly small. Conversely, the smallest value of vC is for positive values of v obtained where C is

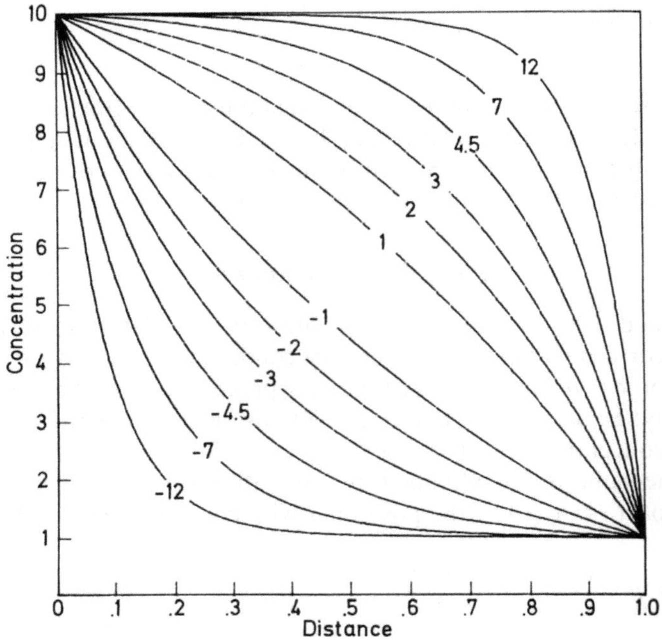

Fig. 10. Concentration profiles through a membrane when convection is superimposed upon the diffusion process. Ordinate, Concentration in units of $C(1)$. Abscissa, distance in units of membrane thickness h. Numbers on curves are values of parameter $\alpha = hv/D$

lowest, *i. e.* corresponding to $x = h$ in Figure 10, so that dC/dx here is maximal. For negative values of v we obtain the least value of vC at $x = 0$, where dC/dx is now maximal here.

F. Electrodiffusion

In comparison with that of uncharged molecules, the description of the passive transport of ions is more complicated. Most frequently there is an electrical potential difference through the solution or across the membrane. The ions will move therefore under the influence of both a concentration gradient and a potential gradient. The starting-off point for the discussion, therefore, is the more complicated *Smoluchowski equation* (Eq. 89) and not the diffusion equation (Eq. 19). Furthermore, the movements of anions and cations must be subject to extra constraints, if the electrical state of the system also has to be constant.

I. Conductance

When ions wander through a membrane, in addition to representing the transport of matter, they constitute at the same time an electrical current. In order to characterize the ionic transport through a membrane, therefore, it is often more convenient to use electrical magnitudes, *e. g. electrical current density* instead of flux, and *membrane resistance* or alternatively *membrane conductance* instead of membrane permeability. To describe these magnitudes, we consider an electrolyte solution of uniform concentration, in which an electrical field E (V per m) exists in the direction of the x-axis. Let N_k be the concentration (ions per m^3) of the ion type k, with the charge valency z_k. As a result of the electric field the ion is under the influence of a force

$$X_k = z_k q_e E, \tag{110}$$

where $q_e = 1.602 \times 10^{-19}$ coulomb is the *positive elementary charge.* This gives the ion a migration velocity $v_k = B_k X_k$, where B_k is the mechanical mobility of the ion (m · s^{-1} · N^{-1}). We therefore obtain

$$v_k = z_k q_e B_k E = u_k E, \tag{111}$$

where

$$u_k = z_k q_e B_k \tag{112}$$

is the *electrical mobility* of the ion, *i. e.* the steady-state velocity which the ion experiences under the influence of a field of 1 Vm^{-1}. In aqueous solution, the velocities of ions for fields of 10 Vm^{-1} are of the order of 9×10^{-7} ms^{-1}, so that the electrical mobility of ions is of the order of magnitude of 5×10^{-8} m^{2}s^{-1}V^{-1}.

The migration flux J_k is $v_k N_k$ or

$$J_k = u_k N_k E. \tag{113}$$

The *current density* I_k of the ion is

$$I_k = z_k q_e J_k \tag{114}$$

or

$$I_k = z_k q_e u_k N_k E.$$

This expression can also be written $I_k = z_k(q_e N_A) u_k (N_k/N_A) E$, where N_A is Avogadro's number. As N_k/N_A = the concentration C_k in mol m^{-3} and $N_A q_e = F =$ *Faraday's number* = 96492 coulombs per equivalent, the above equation can therefore also be written

$$I_k = z_k F u_k C_k E. \qquad (115)$$

If this is compared with the ordinary expression for Ohm's law $I = \varkappa E$, where $\varkappa$ is the *conductivity*, it is seen that the *partial conductivity* $\varkappa_k$, resulting from the ion type k, can be expressed as

$$\varkappa_k = z_k F u_k C_k. \qquad (116)$$

The *molar conductivity* is

$$\lambda_k = \varkappa_k / C_k = z_k F u_k. \qquad (117)$$

Tables most often give the molar conductivity λ_k^o corresponding to infinite dilution. The mobility corresponding to this is then

$$u_k^o = \lambda_k^o / F z_k. \qquad (118)$$

The total current I through the solution is

$$I = \sum I_k = \left(\sum z_k F u_k C_k\right) E \qquad (119)$$

so that the *total conductivity* of the solution is

$$\varkappa = \sum z_k F u_k C_k. \qquad (120)$$

Corresponding to this we have the *resistivity* $\varrho = 1/\varkappa$, *i. e.* the resistance of a cube of the electrolyte whose side has a length of 1 m. If the length of the cube in the direction of the current is not 1 m but h, this corresponds to a *resistance*

$$R = \varrho h = h/\varkappa \qquad (121)$$

and a *conductance*

$$G = \varkappa / h \qquad (122)$$

Since the units for ϱ are $\Omega \cdot m$, it follows that the units for R are $\Omega \cdot m^2$ and for G are then $\Omega^{-1} \cdot m^{-2} \equiv$ siemens $\cdot\, m^{-2}$.

If the electrolyte concentrations vary throughout the region under consideration, the conductance can be calculated if the concentration profile $C_k(x)$ is known for the individual ions. A layer of infinitesimal thickness dx has a resistance $dR = dx/\varkappa$. The total resistance through the entire layer of thickness h is then, according to Eq. (120),

$$R = \int_o^h \frac{dx}{\varkappa} = \int_o^h \frac{dx}{\Sigma z_k F u_k C_k}. \qquad (123)$$

The resistance R calculated from Eq. (123) is sometimes called the integral resistance, and the conductance $G = 1/R$ is called the integral conductance. In a corresponding manner, the resistance R_k and the conductance G_k for individual ions are calculated as

$$R_k = \int_o^h \frac{dx}{z_k F u_k C_k}. \tag{124}$$

It should be noted that the total conductance calculated as $G = \Sigma G_k$ is in general different from the total conductance calculated from Eq. (123). In the case of a homogeneous regime, the total conductance is given by Eq. (123) and the individual conductances calculated from Eq. (124) have only formal significance. On the other hand, if it is a case of a mosaic regime, where each separate ion migrates through its separate channel, the individual ion conductances R_k calculated from Eq. (124) have a real physical significance and the total conductance is $G = \Sigma G_k$.

II. The Nernst-Planck Equations

We shall now give the explicit expression for the magnitude of the ion flux J_k, when an ion of type k migrates under the influence of both a concentration gradient and an electrical potential gradient. The starting point is the *Smoluchowski equation* [Eq. (89) or Eq. (90)], but as the diffusion coefficient D_k and the mobility B_k are connected by the Einstein relation [Eq. (94)], a number of equivalent expressions for the ion flux can be derived, all designated as the *Nernst-Planck* equations.

1. Various Equivalent Forms

It is assumed that for each ion species the ionic flux J_k is described by the Smoluchowski equation

$$J_k = -D_k \frac{dC_k}{dx} + B_k C_k X_k, \tag{125}$$

where X_k is the driving force on the ion of species k, resulting from the presence of the electrical field E, which is connected to the *electrical potential profile* $\psi(x)$ by the relation

$$E = -\frac{d\psi(x)}{dx}. \tag{126}$$

We then have, according to Eq. (110): $X_k = z_k q_e E = -z_k q_e (d\psi/dx)$, which when inserted into Eq. (125) gives

$$J_k = -D_k \frac{dC_k}{dx} - z_k q_e B_k C_k \frac{d\psi}{dx} \tag{127}$$

as the flux equation for the ion k with a charge number z_k. Corresponding expressions hold for all the other types of ion present in the phase. The Einstein relation

$$D_k = kTB_k$$

holds for each ion species. One can then choose whether to eliminate either D_k or B_k from Eq. (127). If D_k is eliminated, then by introducing at the same time the electrical mobility of the ion, $u_k = z_k q_e B_k$ [cf. Eq. (112)], we obtain

$$J_k = -u_k \frac{kT}{z_k q_e} \frac{dC_k}{dx} - u_k C_k \frac{d\psi}{dx},$$

or multiplying the first right-hand term by N_A/N_A

$$J_k = -u_k \frac{RT}{z_k F} \frac{dC_k}{dx} - u_k C_k \frac{d\psi}{dx}, \tag{128}$$

since $R = N_A k$ and $F = N_A q_e$. This expression can also be written as

$$J_k = -u_k C_k \left\{ \frac{RT}{z_k F} \frac{d\ln C_k}{dx} + \frac{d\psi}{dx} \right\}, \tag{129}$$

where the term in brackets contains the "equivalent electrical potential gradient". If $1/z_k F$ is taken outside the brackets, we obtain

$$J_k = -\frac{u_k}{z_k F} C_k \left\{ RT \frac{d\ln C_k}{dx} + \frac{d(z_k F\psi)}{dx} \right\}.$$

But we have $u_k/z_k F = z_k q_e B_k / z_k q_e N_A = B_k/N_A = b_k$ = the *molar mechanical mobility* of the ion. In addition,

$$\bar{\mu}_k = \mu_k^o + RT\ln C_k + z_k F\psi \tag{130}$$

is *the electrochemical potential* of the ion in the position x. The above equation can therefore be written as

$$J_k = -b_k C_k \frac{d\bar{\mu}_k}{dx}, \tag{131}$$

which is analogous to Eq. (13), with the driving force per mol represented by the negative gradient of the electrochemical potential of the ion.

Finally, B_k can be eliminated instead of D_k in Eq. (127). This gives

$$-\frac{J_k}{D_k} = \frac{dC_k}{dx} + z_k \frac{q_e}{kT} C_k \frac{d\psi}{dx}.$$

It is often convenient to operate with a dimensionless potential

$$\varphi = \psi\left(\frac{kT}{q_e}\right)^{-1} \tag{132}$$

which is normalized in relation to kT/q_e. (At 25° C we have $kT/q_e = RT/F =$ 25,7 mV.) The above equation can as a result be written

$$-\frac{J_k}{D_k} = \frac{dC_k}{dx} + z_k C_k \frac{d\varphi}{dx}, \tag{133}$$

or with the help of Kramers' transformation in section D.V, as

$$J_k \exp\{z_k\varphi\} = -D_k \frac{d}{dx}\left\{C_k \exp\{z_k\varphi\}\right\}. \tag{134}$$

This expression, in which the right-hand side can be integrated directly, is very convenient in calculating ion-fluxes through a membrane.

2. The Poisson Equation

A problem in electrodiffusion is solved in principle when from a given set of boundary conditions the concentration profiles $C_k(x)$ for each ion and the potential profile $\psi(x)$ through the region under consideration have been determined. In order to calculate the fluxes it is necessary to set up one flux equation

$$J_k = -u_k \frac{RT}{z_k F} \frac{dC_k}{dx} - u_k C_k \frac{d\psi}{dx}; \qquad k = 1, 2, 3, \ldots m \tag{135}$$

for each of the m ion species present. This set of equations is insufficient to solve even the stationary problem, because we have only m equations and there are $m+1$ unknowns, namely the m concentrations $C_k(x)$ and the potential $\psi(x)$. But the set of Eq. (135) is incomplete, because the various concentrations $C_k(x)$ and the potential $\psi(x)$ are not independent of each other. The electrical field $\boldsymbol{E}$ at each point in space must satisfy *Poisson's equation*

$$\operatorname{div} \boldsymbol{E} = \varrho/\varepsilon_o K, \tag{136}$$

where ϱ is the *charge density* (coulomb per m^3), $\varepsilon_o = 8.85 \times 10^{-12}$ farad per m is the *permittivity in vacuum* and K is the *relative dielectric constant.* The charge density is

$$\varrho = F \sum (z_{+k} C_{+k} - |z_{-k}| C_{-k}) \tag{137}$$

where C_{+k} and C_{-k} represent respectively cations and anions. As $\boldsymbol{E} = -\,\mathrm{grad}\,\psi$ and $\mathrm{divgrad}\,\psi = \nabla^2\psi$, Poisson's equation can be written

$$\nabla^2\varphi = -F\sum(z_{+k}C_{+k} - |z_{-k}|C_{-k})/\varepsilon_0 K. \qquad (138)$$

This equation combined with the m equations in Eq. (135) now constitute a complete set of $m+1$ equations, in which the $m+1$ unknowns, namely the concentrations of the m different ions and the potential $\psi(x)$, can now be determined in principle. The mathematical complications encountered in this connection are usually enormous. Fortunately, in many situations it is permissible to use one of the following approximations, as a result of which the calculations are considerably facilitated.

a) Electroneutrality

We consider a single salt, *e.g.* HCl. Poisson's equation in its one-dimensional form is

$$\frac{d^2\psi}{dx^2} = -\frac{dE}{dx} = -\frac{F}{\varepsilon_0 K}(C_H - C_{Cl}).$$

In aqueous solution $K = 80$. Inserting the numerical values for F, ε_0 and K, we obtain

$$d^2\psi/dx^2 = -1.37 \times 10^{14}(C_H - C_{Cl}),$$

which means that with a deviation from electroneutrality in the system equal to $\triangle C = C_H - C_{Cl} = 1\ \mathrm{mol}\cdot\mathrm{m}^{-3} = 1\ \mathrm{mmol}\cdot\mathrm{l}^{-1}$, the field will change by an amount equal to $10^{14}\ \mathrm{V}\cdot\mathrm{m}^{-1}$ per m. Under laboratory conditions, potential differences between phases of the order of 200–300 mV are encountered. In many cases this potential difference is established over such a large distance that the field E and its change in space dE/dx is many orders of magnitude smaller than 10^{14}. NERNST (1888) and later PLANCK (1890) therefore assumed with confidence that the difference $C_H - C_{Cl}$ is immeasureable for all practical purposes. Thus one can put

$$C_H(x) = C_{Cl}(x) \qquad (139)$$

in the flux equations, without this involving any significant error in the subsequent calculations. Equation (139) and its generalization

$$\sum z_{+k}C_{+k} = \sum|z_{-k}|C_{-k} \qquad (140)$$

are called the *Nernst-Planck electroneutrality condition.* This condition will be employed later in the description of PLANCK's diffusion regime. In section F.III. 2b is treated a situation where the electroneutrality condition breaks down and, accordingly, the Poisson equation [Eq. (138)] must be invoked to calculate the concentration and potential profiles.

b) The Constant Field

Under certain experimental conditions, the electric field is always constant throughout the membrane. (Equal total concentrations on both sides of the membrane: see Section IV.3.9.) In other cases the fluxes can be calculated with reasonable accuracy by regarding the electric field in the membrane as constant (MacGILLIVRAY and HARE, 1969). The consequence in both cases is that Poisson's law can be automatically included in the set of Equation (135) by putting $d\psi/dx = -E =$ constant, so that the flux equations can now be integrated separately. The constant field assumption was first introduced by GOLDMAN (1943).

III. Membrane Equilibrium

We shall here discuss two types of equilibrium across an ion-permeable membrane, which are of significance also for the description of nonequilibrium situations.

1. Nonosmotic Equilibrium

Let a membrane of thickness h separate two electrolyte solutions (i) and (o). It is assumed that the membrane is impermeable to the solvent, so that the effect of an osmotic pressure difference between the phases can be ignored. The concentration of positive ions is $C_{+k}^{(i)}$, $C_{+k}^{(o)}$ and of negative ions $C_{-k}^{(i)}$, $C_{-k}^{(o)}$. When a state of equilibrium has been established, all ion fluxes must have disappeared, *i.e.*

$$J_{+k} = J_{-k} = 0, \qquad \text{for all ionic species.}$$

If the membrane is permeable to all ions, the equilibrium state will be established when all concentration differences between the phases have been equalized, and when there is no potential difference across the membrane.

In order to achieve a potential difference $\Delta\psi \neq 0$ across the membrane in equilibrium, the membrane must be impermeable to one or more of the ions on each side.

a) The Nernst Equation

The simplest case was discussed by NERNST (1888), in which he assumed that only one ion type could exchange between the two phases. This would correspond, for example, to an ion-selective permeable membrane surrounded by a simple salt solution, *e.g.* KCl, in different concentrations in the two phases. Initially, the permeable ion will migrate through the membrane as a result of concentration differences ΔC between the phases. Since the ion carries a charge, and its counter-ion cannot accompany it, a charge deficit will be established between the two phases (i) and (o), and thereby an electrical potential difference across the membrane, the polarity being such that the field E counteracts the transport of the ion by diffusion. At a certain time a potential difference and field will be established of such a magnitude that the flux comes to a stop and equilibrium is hereby achieved. The consequence of the equilibrium condition $J_k = 0$ is seen most easily from the *Nernst-Planck equation* in the form of Eq. (129), where only the term in brackets – the phenomenological driving force – can assume the value zero. When equilibrium is reached, therefore, we have

$$\frac{RT}{z_k F} \frac{d \ln C_k}{dx} + \frac{d\psi}{dx} = 0 .$$

This expression is multiplied by dx and integrated through the membrane from phase (i) with concentration $C_k^{(i)}$ and potential $\psi^{(i)}$ to phase (o) with concentration $C_k^{(o)}$ and potential $\psi^{(o)}$. This gives

$$\frac{RT}{z_k F} \int_{C_k^{(i)}}^{C_k^{(o)}} d \ln C_k = - \int_{\psi^{(i)}}^{\psi^{(o)}} d\psi$$

or

$$\psi^{(i)} - \psi^{(o)} = - \frac{RT}{z_k F} \ln \left(\frac{C_k^{(i)}}{C_k^{(o)}} \right), \tag{141}$$

which is the famous *Nernst equation*. This equation could of course also have been derived on thermodynamic principles by putting $\bar{\mu}_k^{(i)} = \bar{\mu}_k^{(o)}$. In this way the right-hand side would have appeared as the ratio between the individual ion activities $a_k^{(i)}/a_k^{(o)}$. However, with dilute solutions and with more or less the same ionic strength on both sides, no essential error will be made by putting $a_k^{(i)}/a_k^{(o)} = C_k^{(i)}/C_k^{(o)}$. If the membrane is permeable to several ion types of the same valency – either cations or anions – an equilibrium state like that described by Eq. (141) can only be established provided the various ions are initially present in the same ratio $C_k^{(i)}/C_k^{(o)} = C_j^{(i)}/C_j^{(o)}$, etc. On the other hand, the potential difference given by Eq. (141) can always be regarded as *the potential difference that has to be imposed across a membrane, if the ion flux is to be zero for the ion type* k, the concentration of which in the surrounding media is $C_k^{(i)}$ and $C_k^{(o)}$. For this reason, therefore, the potential difference

$$\psi^{(i)} - \psi^{(o)} = V_k^{(eq)} = - \frac{RT}{z_k F} \ln \frac{C_k^{(i)}}{C_k^{(o)}} \tag{142}$$

is designated the *equilibrium potential* for the ion in question, or the *Nernst potential* of the ion.

b) Equivalent Electrical Circuit for the Ion-Selective Membrane

As mentioned in the introduction to this section, it is often convenient to describe electrodiffusion in purely electrical terms. Let the membrane be permeable only to an ion of species k. The two phases are subjected to an electrical potential difference $V = \psi^{(i)} - \psi^{(o)}$, and we shall indicate the relationship between V and the ion current I_k, when the concentrations in the phases are $C_k^{(i)}$ and $C_k^{(o)}$. The ion current is

$$I_k = z_k F J_k . \tag{143}$$

As starting point for the flux J_k, we take the N-P equation in the form of Eq. (129). This gives

$$I_k = - z_k F u_k C_k \left\{ \frac{RT}{z_k F} \frac{d \ln C_k}{dx} + \frac{d\psi}{dx} \right\}$$

or

$$\frac{I_k}{z_k F u_k C_k} = - \frac{RT}{z_k F} \frac{d \ln C_k}{dx} - \frac{d\psi}{dx} .$$

This expression is multiplied by dx and integrated through the membrane from phase (i) at $x = 0$ with concentration $C_k^{(i)}$ and potential $\psi^{(i)}$ to phase (o) at $x = h$ with concentration $C^{(o)}$ and potential $\psi^{(o)}$. As a steady state is assumed to hold, we have $(dI_k/dx) = 0$, and consequently I_k can be put outside the integration sign. This gives

$$I_k \int_0^h \frac{dx}{z_k F u_k C_k} = - \frac{RT}{z_k F} \int_{C_k^{(i)}}^{C_k^{(o)}} d \ln C_k - \int_{\psi^{(i)}}^{\psi^{(o)}} d\psi$$

or

$$I_k \int_0^h \frac{dx}{z_k F u_k C_k} = \frac{RT}{z_k F} \ln \left(\frac{C_k^{(i)}}{C_k^{(o)}} \right) + \psi^{(i)} - \psi^{(o)} .$$

The first term on the right-hand side, according to Eq. (142), is equal to $- V_k^{(eq)}$, where $V_k^{(eq)}$ is the equilibrium potential for the ion, while the second term is equal to the impressed potential difference $V = \psi^{(i)} - \psi^{(o)}$ across the membrane. According to Eq. (123) the integral on the left-hand side is equal to the resistance R_k of the membrane for ion type k. The above may therefore be written as

$$I_k = (V - V_k^{(eq)}) / R_k$$

or, as the membrane conductance for the ion k is equal to $G_k = 1/R_k$,

$$I_k = G_k (V - V_k^{(eq)}) . \tag{144}$$

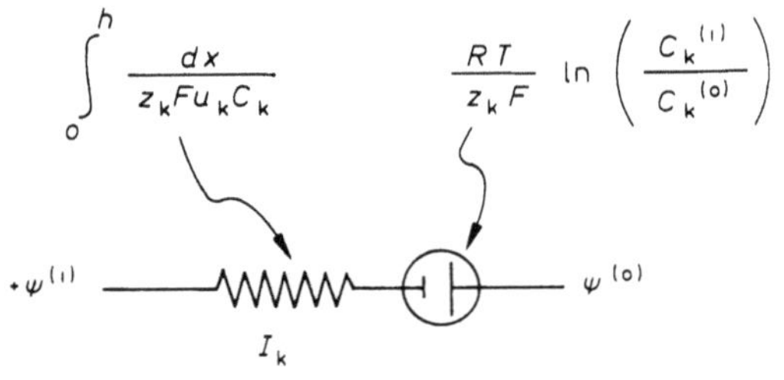

Fig. 11. Equivalent circuit for the ion-selective membrane

This relationship between the ion current I_k and the potential difference V by which it is generated can be represented by the equivalent electrical circuit illustrated in Figure 11. This circuit consists of a resistance R_k in series with an electromotive force, the magnitude of which is equal to the equilibrium potential of the ion. A positive value of I_k indicates a current in the direction (i) → (o). If the mobility of the ion in the membrane is known, and its concentration profile $C_k(x)$ in the region $0 \leqq x \leqq h$, the conductance G_k can be calculated from Eq. (123). Conversely, by measuring the related values of I_k and V, one can calculate G_k from Eq. (144).

2. Donnan Equilibrium

In this situation, the membrane is conceived of as being permeable to the solvent and to some – but not all – ions. The impermeable ions can be small ions or macromolecular polyelectrolytes. The equilibrium state in this situation was first studied by F. G. Donnan (1911), and is called the *Donnan equilibrium.* The membrane as such is not essential for this situation. A gel or any cross-linked macromolecular structure possessing fixed charges would also give rise to a distribution of the mobile ions corresponding to the Donnan type of distribution.

It is characteristic of the Donnan distribution that at equilibrium there is

(1) an uneven (asymmetrical) distribution of the diffusible ions between the two phases.

(2) An electrical potential difference between the phases. (Donnan potential).

(3) An osmotic pressure difference between the phases.

In what follows, an account will be given of both the macroscopic equilibrium state and the potential and concentration profiles connecting the equilibrium values in the two phases.

a) Thermodynamic Treatment

It will be instructive to consider the establishment of equilibrium by starting off from the following simple situation: In phase (i) there is initially only a polyelectrolyte HPo, which is completely dissociated: HPo → H^+ + Po^-. Phase (o) initially contains HCl. The initial concentrations of HPo and HCl are also supposed equal. To begin with there is no pressure or potential difference between the phases. Since the macroion Po^- cannot pass through the membrane, the initial state is characterized by the situation that H^+ and H_2O are in equilibrium because their mole fractions are identical in both phases; but Cl^- is not in equilibrium because its concentration in phase (i) is zero. Cl^- will therefore diffuse into phase (i), which thus obtains an excess negative charge density, while a corresponding positive charge density develops in phase (o). This separation of charge gives rise to an electric field in the direction (o) → (i), which will tend to check further migration of Cl^-. As a consequence of the electric field the H^+-ion is now no longer in equilibrium, and will be forced by the field to migrate into phase (i), thereby tending to neutralize the charges on the two sides of the membrane. As a result, more Cl^- ions and further H^+ ions will move from phase (o) to phase (i). The concentration of Cl^- in phase (i) will thus rise, while at the same time the concentration of H^+ in this phase will exceed the concentration in phase (o). The process will continue until just that potential profile and that concentration profile for H^+ and Cl^- is established, which corresponds to *zero flux* for both these ions. In this way the system arrives at equilibrium, providing that at the same time a pressure difference $P^{(i)} - P^{(o)} = \pi$ is established between the phases, where π is the osmotic pressure difference between the phases, which hinders the movement of water between the two phases.

The thermodynamic criterion for an equilibrium state is that the chemical potential for water and the electrochemical potential for each permeating ion have the same value in the two phases. For the present purpose the electrochemical potential is most conveniently written in the form

$$\bar{\mu}_k = \mu_k^o + P\bar{v}_k + RT\ln a_k + z_k F\psi,$$

where P and ψ are the pressure and electric potential in the phase and $\bar{v}_k$ is the partial molar volume of the k ion with charge number z_k. The equilibrium condition

$$\bar{\mu}_{H^+}^{(i)} = \bar{\mu}_{H^+}^{(o)}; \quad \bar{\mu}_{Cl^-}^{(i)} = \bar{\mu}_{Cl^-}^{(o)}; \quad \mu_w^{(i)} = \mu_w^{(o)}$$

leads to the following expressions:

$$\pi\bar{v}_{H^+} + FV_D = RT\ln\left(a_{H^+}^{(o)}/a_{H^+}^{(i)}\right) \tag{145}$$

$$\pi\bar{v}_{Cl^-} - FV_D = RT\ln\left(a_{Cl^-}^{(o)}/a_{Cl^-}^{(i)}\right) \tag{146}$$

$$\pi\bar{v}_w = RT\ln\left(a_w^{(o)}/a_w^{(i)}\right) \tag{147}$$

where $\pi = P^{(i)} - P^{(o)}$ is the osmotic pressure difference between the phases and $V_D = \psi^{(i)} - \psi^{(o)}$ is the potential difference between them, *the Donnan potential.* If Eq. (145) and Eq. (146) are added, we obtain

$$\frac{a_{H^+}^{(o)} \cdot a_{Cl^-}^{(o)}}{a_{H^+}^{(i)} \cdot a_{Cl^-}^{(i)}} = \exp\left\{\pi(\bar{v}_{H^+} + \bar{v}_{Cl^-})/RT\right\}.$$

Under almost all conditions encountered in biology, the osmotic pressure difference is so small that $\pi(\bar{v}_k + \bar{v}_{Cl}) \ll RT$. The above expression then reduced to

$$\frac{a_{H^+}^{(o)}}{a_{H^+}^{(i)}} = \frac{a_{Cl^-}^{(i)}}{a_{Cl^-}^{(o)}}, \tag{148}$$

which is called the *Gibbs-Donnan distribution condition.* With this approximation, in which the pressure contribution in Eq. (145) and Eq. (146) is ignored, the Donnan potential becomes identical with the Nernst potential for the permeating ions. We then have from Eq. (142)

$$V_D = -\frac{RT}{z_k F}\ln\left(\frac{C_k^{(i)}}{C_k^{(o)}}\right) = -\frac{RT}{z_j F}\ln\left(\frac{C_j^{(i)}}{C_j^{(o)}}\right) \tag{149}$$

or

$$\left(\frac{C_k^{(i)}}{C_k^{(o)}}\right)^{1/z_k} = \left(\frac{C_j^{(i)}}{C_j^{(o)}}\right)^{1/z_j}, \tag{148a}$$

which corresponds to Eq. (148), since in that case we have: $z_k = -z_j = 1$. It also follows that if there is a distribution ratio of 10:1 between monovalent cations, there will be a distribution ratio of 100:1 for the divalent cations.

The concentration of the permeating ions in phase (i), which contains the macroion, must be determined by the concentrations in the outer phase (o) and by the concentration C_{Po} of the macroion. This relationship will now be derived, assuming for the sake of simplicity that the permeating ions are monovalent. The macroion, on the other hand, is assumed to have a charge number z. A state of electroneutrality exists in the macroscopic part of the outer phase, so that

$$C_+^{(o)} = C_-^{(o)} = C_o. \tag{150}$$

The same is the case in the inner phase. Here, therefore,

$$C_+^{(i)} - C_-^{(i)} + zC_{Po} = 0 \tag{151}$$

From Eq. (149) we obtain the following relationship between the concentrations in the inner and the outer phase

$$C_+^{(i)} = C_o e^{-FV/RT} = C_o e^{-v} \tag{152}$$

and

$$C_-^{(i)} = C_o e^{FV/RT} = C_o e^{v}, \tag{153}$$

where $V = V_D$ is the Donnan potential and

$$v = \frac{V}{RT/F}$$

is the Donnan potential expressed in units of RT/F. If Eq. (152) and Eq. (153) are inserted into Eq. (151), we obtain

$$C_o(e^{v} - e^{-v}) = zC_{Po}.$$

The term in parenthesis is equal to 2sinh$v$. We have therefore

$$\sinh v = \sinh\left(\frac{FV}{RT}\right) = zC_{Po}/2C_o \tag{154}$$

or

$$V = \frac{RT}{F}\,\text{ar sinh}\left(\frac{zC_{Po}}{2C_o}\right), \tag{155}$$

which are convenient relations to determine one of the three magnitudes V, zC_{Po} and C_o, if the other two are known. It is also seen that the polarity of V is determined by the sign for z. If the macroion is positive, its phase will be positive in relation to the outer phase, and conversely if z is negative. We shall also consider two extreme situations:

(i) $zC_{Po} \ll C_o$. In this case sinh$v$ in Eq. (154) will be much less than 1. For $y \ll 1$, $\sinh y \simeq y$. We have therefore $v = zC_{Po}/2C_o$ or

$$V = \frac{RT}{2F}\,\frac{zC_{Po}}{C_o} \tag{156}$$

Furthermore, under the same conditions we have $e^{-y} \simeq 1-y$ and $e^{y} \simeq 1+y$. Using these relations in Eq. (152) and Eq. (153), we obtain

$$C_+^{(i)} = C_o - \frac{1}{2}zC_{Po} \tag{157}$$

$$C_-^{(i)} = C_o + \frac{1}{2}zC_{Po}, \tag{158}$$

which means that half of the charges of the nondiffusible polyelectrolyte are compensated for by an excess of counter-ions, *i. e.* ions with the opposite sign, and the other half by a deficit of ions of the same charge type (co-ions). The osmotic pressure difference is equal to

$$\pi = RT[C_+^{(i)} + C_-^{(i)} + C_{Po} - 2C_o] = RTC_{Po}, \tag{159}$$

i. e. in this case the osmotic pressure difference is determined exclusively by the concentration of the indiffusible polyelectrolyte.

(ii) $zC_{Po} \gg C_o$. For high values of $y = zC_{Po}/2C_o$ we have: $\operatorname{arsinh} y = \ln 2y$ for $y \gg 0$ and $\operatorname{arsinh} y = -\ln|2y|$ for $y \ll 0$. Equation (155) can therefore be written in the form

$$V = \pm \frac{RT}{F} \ln\left(\frac{|z|C_{Po}}{C_o}\right), \tag{160}$$

where the sign for V corresponds to the sign for z. If $V \gg RT/F$, Eq. (152) and Eq. (153) together with the electroneutrality condition Eq. (151) give

$$\left.\begin{aligned} C_+^{(i)} &\to 0 \\ C_-^{(i)} &\to zC_{Po} \end{aligned}\right\} \text{for } V \gg RT/F \text{ and } z \text{ positive.} \tag{161}$$

and

$$\left.\begin{aligned} C_+^{(i)} &\to |z|C_{Po} \\ C_+^{(i)} &\to 0 \end{aligned}\right\} \text{for } |V| \gg RT/F \text{ and } z \text{ negative.} \tag{162}$$

In other words, the charge on the polyelectrolyte is now almost completely compensated for by the counter-ions, and the system appears as if it were almost impermeable to co-ions. The osmotic pressure difference now becomes

$$\pi = RT(z+1)C_{Po} \tag{163}$$

since as a consequence of the condition $zC_{Po} \gg C_o$ the co-ions contribute only insignificantly to the osmotic pressure.

In general, the osmotic pressure is given by

$$\pi = RT[C_+^{(i)} + C_-^{(i)} + C_{Po} - 2C_o].$$

If we insert Eq. (152) and Eq. (153) in the above, the expression becomes

$$\pi = RT\left[4C_o \sinh^2\left(\frac{v}{2}\right) + C_{Po}\right], \tag{164}$$

which also holds for those situations lying between the two extreme cases treated. It is clear that Eq. (164) becomes Eq. (159) for $v \to 0$, and becomes Eq. (163) for $|v| \gg 1$. The article by Overbeek (1956) is strongly recommended to readers who wish to go beyond the present standard treatment of the Donnan equilibrium. As classical examples of applications to biological systems the reader is referred to the paper by Warburg (1922) on ionic distribution in erythrocytes, and to the papers by Boyle and Conway (1941) and Conway (1957) and by Hodgkin and Horowicz (1959) on distribution of ions in frog muscle fibres.

b) *Concentration and Potential Profiles between the Phases (The Poisson-Boltzmann Equation)*

In the thermodynamic treatment of the Donnan system, a determination is made of the macroscopic equilibrium concentrations and the potential difference between regions in the two phases where a condition of electroneutrality exists. However, at equilibrium there is an asymmetrical distribution between the diffusible anions and cations. Thus a separation must take place between the concentrations of anions and cations, so that the state of electroneutrality is broken. As a result free charges arise, which result in the development of a potential difference between the two phases. In this section the concentration and potential profiles in this transitional zone will be determined, and also the factors that determine the extent of this zone.

A complete solution for the distribution of ions and for the potential was first obtained by BARTLETT and KROMHOUT (1952). Their treatment, however, is not easy reading and their solutions are given in terms of elliptic functions, which may be rather unfamiliar to most biologists. In the present treatment the solution of the potential distribution will be obtained in terms of elementary functions. Neither in this treatment nor in that of BARTLETT and KROMHOUT (1952) is the influence of the changing pressure upon the distribution of ions and potential considered and neither is the pressure profile calculated.

The coordinate system is so placed that the plane $x = 0$ separates the Donnan system (i), present in the region $x < 0$, from the pure electrolyte phase (o) in the region $x > 0$. The region, viz. the membrane, that bounds the mobility of the macroions, is assumed to be infinitely thin. Let

x = The distance from the membrane, with positive direction from Donnan phase (i) to outer phase (o).

$\psi(x)$ = The potential profile. $\psi(x)$ assumes the constant value $\psi^{(o)}$ for $x \to +\infty$ and $\psi^{(i)}$ for $x \to -\infty$. However, in practice the potential variation will only occur within a range $-x_o \leq x \leq x_o$, so that we can put $\psi = \psi^{(o)}$ for $x > x_o$ and $\psi = \psi^{(i)}$ for $x < -x_o$. Arbitrarily, we put $\psi^{(o)} = 0$. The potential difference $\psi^{(i)} - \psi^{(o)} = \psi^{(i)}$ between the two macroscopic phases is the Donnan potential.

$\varphi(x)$ = $\psi/(kT/q_e) = \psi/(RT/F)$ is the normalized potential in units of RT/F = 25.69 mV at 25° C. The normalized Donnan potential is designated by v.

$C_+(x) = C_+(\varphi)$ = The concentration of the permeable monovalent cation in position x (mol · m^{-3}). For $x > x_o$, $C_+(x)$ takes the constant value $C_+^{(o)} = C_o$.

$C_-(x) = C_-(\varphi)$ = The concentration of the anion. For $x > x_o$, $C_-(x) = C_-^{(o)} = C_o$.

$M(x) = M(\varphi)$ = The concentration of the impermeable macroion with charge number z. For $x < -x_o$, $M(x)$ is constant and equal to M_v.

When the Donnan equilibrium has been established, the fluxes J_+, J_- and J_M are all zero. But in the transitional region $-x_o < x < x_o$ there will be just that distribution of space charges $\varrho(x)$ which establishes the equilibrium profile for the potential $\psi(x)$. These two magnitudes are linked through Poisson's equation

$$\frac{d^2\psi}{dx^2} = -\varrho(x)/\varepsilon_o K. \tag{165}$$

In the region $x \geqq 0$ the charge density is

$$\varrho = F[C_+(x) - C_-(x)],$$

while for $x \leqq 0$ it is

$$\varrho = F[C_+(x) - C_-(x) + zM(x)].$$

For example, the concentration profile for $C_+(x)$ in the region $-\infty < x < \infty$ can be determined from Eq. (134). As $J_+ = 0$, we immediately obtain:

$$C_+(x)\,e^{\varphi(x)} = A,$$

where A is a constant. For $\varphi(x) = 0$, $C_+(x) = C_+^{(o)} = C_o$. We then have

$$C_+(x) = C_o\,e^{-\varphi(x)}. \tag{166}$$

Correspondingly, we obtain

$$C_-(x) = C_o\,e^{\varphi(x)}. \tag{167}$$

Equations (166) and (167) could also have been derived either by saying that the electrochemical potential $\bar{\mu}_+(x)$ is continuous and constant throughout the entire region, or by expressing the equilibrium distribution by means of Boltzmann's distribution law. For the macroion we obtain in a corresponding manner in the region $x < 0$

$$M(x)\,e^{z\varphi(x)} = A.$$

For $x < -x_o$ we have $\varphi(x) = v$, and $M(x) = M_v$. This gives

$$M(x) = M_v e^{z(v-\varphi(x))}, \quad x < 0. \tag{168}$$

If these values for C_+, C_- and M are inserted in the expressions for the space charge, we obtain for the region $x > 0$

$$\varrho(x) = -FC_o[e^{\varphi(x)} - e^{-\varphi(x)}] = -2FC_o \sinh\varphi(x), \tag{169}$$

since $\sinh\varphi = (e^{\varphi} - e^{-\varphi})/2$, and for the region $x < 0$

$$\varrho(x) = -2FC_o\left[\sinh\varphi(x) - \frac{zM_v}{2c_o}\,e^{z(v-\varphi(x))}\right]. \tag{170}$$

Poisson's equation expressed by the normalized potential $\varphi = \psi/(RT/F)$ is

$$\frac{RT\varepsilon_0 K}{F} \cdot \frac{d^2\varphi}{dx^2} = -\varrho(x).$$

If Eqs. (169) and (170) are inserted, we obtain the following equations for the determination of the potential profile $\varphi(x)$. For the region $x > 0$

$$\lambda^2 \frac{d^2\varphi}{dx^2} = \sinh\varphi, \tag{171}$$

and for the region $x < 0$

$$\lambda^2 \frac{d^2\varphi}{dx^2} = \sinh\varphi - \frac{zM_v}{2C_0} e^{z(v-\varphi)}. \tag{172}$$

Equations of this type are called *Poisson-Boltzmann equations,* because the electrical state — which is determined from Poisson's equation — is given at the same time by the equilibrium state, which is described by Boltzmann's distribution law. The constant λ in these two equations

$$\lambda = \left(\frac{RT\varepsilon_0 K}{2F^2 C_0}\right)^{1/2}, \tag{173}$$

which has the dimension of a length, is called the *Debye length* for the region in which space charges are present. It is the magnitude of this length that is a measure of the extent of the space charge region. If the numerical values are inserted in this expression, we obtain

$$\lambda = 10.85 \times 10^{-10} \sqrt{K/C}\,(\mathrm{m}) = 10.85 \sqrt{K/C}\,(\text{Å}),$$

where C is in mol/m^{-3} or mmol/l^{-1}. In aqueous solution, $K = 80$. This gives $\lambda = 97.1/\sqrt{C}\,(\text{Å})$.

We shall now determine the potential profile $\varphi(x)$ by solving Eqs. (171) and (172). When $\varphi(x)$ is known, the concentration profiles in the transitional region can be found with the help of Eqs. (166), (167), and (168). We begin with the region $x > 0$.

Equation (171) is multiplied by $d\varphi/dx$. Since $(d^2\varphi/dx^2) \cdot (d\varphi/dx) = {}^1\!/_2 d(d\varphi/dx)^2/dx$, this gives

$$^1\!/_2 \lambda^2 \frac{\mathrm{d}}{dx}\left(\frac{d\varphi}{dx}\right)^2 = \sinh\varphi \cdot \left(\frac{d\varphi}{dx}\right)$$

or

$$^1\!/_2 \lambda^2 d\left(\frac{d\varphi}{dx}\right)^2 = \sinh\varphi d\varphi,$$

from which we obtain

$$\tfrac{1}{2}\lambda^2\left(\frac{d\varphi}{dx}\right)^2 = \cosh\varphi + A_1 .$$

The constant A_1 is determined from the boundary condition: $d\varphi/dx \to 0$ for $x \to \infty$. But $\varphi = 0$ for $x \to \infty$ (in practice for $x > x_o$). The boundary condition is therefore $d\varphi/dx \to 0$ for $\varphi \to 0$. When this is inserted in the above expressions we obtain $A_1 = -1$. We then have to solve

$$\tfrac{1}{2}\lambda^2\left(\frac{d\varphi}{dx}\right)^2 = \cosh\varphi - 1 = 2\sinh^2\left(\frac{\varphi}{2}\right) \tag{174}$$

or

$$\lambda\,\frac{d\varphi}{dx} = \pm 2\sinh(\varphi/2). \tag{174a}$$

The sign is determined by the polarity of $V = \psi^{(i)} - \psi^{(o)}$. If V is positive, *i.e.* corresponds to positive values of the charge number z for the macroion, the potential increases from the outer phase (o) towards the Donnan phase (i). A positive value of z thus results in $d\varphi/dx \leqq 0$ for the whole range $-\infty < x < \infty$. In what follows it will be assumed that z is positive. In Eq. (174) we introduce a new variable $\varphi/2$ and we integrate from $\varphi = \varphi(0_+)$ in the position $x = 0_+$ to φ in the position x. This gives

$$\int_{\varphi(0+)}^{\varphi} \frac{d(\varphi/2)}{\sinh(\varphi/2)} = -\frac{1}{\lambda}\int_{o}^{x} dx = -x/\lambda .$$

The indefinite integral of $1/\sinh u$ is $\ln\tanh(u/2)$. We then have

$$\ln\frac{\tanh(\varphi/4)}{\tanh(\varphi(0_+)/4)} = -x/\lambda$$

or

$$\varphi(x) = 4\,\mathrm{ar\,tanh}\left\{\tanh\left(\frac{\varphi(0_+)}{4}\right)\cdot e^{-x/\lambda}\right\}, \tag{175}$$

which is the expression for the potential profile in the region $x > o$. This expression contains an integration constant that is still undetermined, which is the potential value $\varphi(0_+)$ for $x = 0_+$. It can be determined by adjusting the solution of Eq. (175) with the solution of the differential equation for φ which is valid in the range $x < 0$. This solution will now be obtained.

In the range $x < 0$ the potential profile is determined by Eq. (172). We have already restricted z to be positive. Solutions to Eq. (172) can be found in terms of elementary functions only if $z = \pm 1$, so that the following solution will hold for $z = +1$. From Eq. (154) we have $M_\nu/2C_o = \sinh\nu$. This is inserted into Eq. (172), which when multiplied by $d\varphi/dx$ can then be integrated. This gives

$$\tfrac{1}{2}\lambda^2 \left(\frac{d\varphi}{dx}\right)^2 = \cosh\varphi + \sinh\nu \cdot e^{\nu-\varphi} + A_2$$

The integration constant A_2 can be determined, since $d\varphi/dx$ must have the value zero for $x \to -\infty$, where φ is equal to the Donnan potential ν (in practice for $x < -x_o$). If this condition is inserted in the above expression, we obtain $A_2 = -(\sinh\nu + \cosh\nu) = -e^\nu$. The Poisson-Boltzmann equation in the region $x < 0$ is therefore

$$\tfrac{1}{2}\lambda^2 \left(\frac{d\varphi}{dx}\right)^2 = \cosh\varphi + e^{\nu-\varphi} \cdot \sinh\nu - e^\nu \tag{176}$$

For the potential profile it must hold that both $\varphi(x)$ and $d\varphi/dx$ are continuous for $x = 0$, since it is assumed that the dielectric constant K is the same througout the region $-\infty < x < \infty$. The boundary conditions for $x = 0$ are therefore

$$\varphi(0_-) = \varphi(0_+) = \varphi(0)$$

and

$$(d\varphi/dx)_{x=0-} = (d\varphi/dx)_{x=0+}\,.$$

If these boundary conditions are applied to Eqs. (174) and Eqs. (176) we obtain

$$\cosh\varphi(0) - 1 = \cosh\varphi(0) + \sinh\nu \cdot e^{\nu-\varphi(0)} - e^\nu,$$

from which the potential $\varphi(0)$ for $x=0$ can be determined by the relation

$$\varphi(0) = \ln \frac{e^\nu + 1}{2}\,. \tag{177}$$

in which ν is the Donnan potential between the phases. With the value of $\varphi(0)$ known, the potential profile in the outer phase can be calculated by means of Eq. (175). For $y \ll 1$ we have $\tanh y \simeq y$ and therefore also ar $\tanh y \sim y$. For values of $\varphi(0)$ smaller than 0.6, Eq. (175) can be rewritten in the simpler form

$$\varphi(x) = \varphi(0)\, e^{-x/\lambda}, \tag{178}$$

which corresponds to the solution of the linearized form of the Poisson-Boltzmann equation. The physical significance of the Debye length expressed in this equation is that distance from the Donnan phase where the potential $\varphi(x)$ is $\varphi(0) \cdot e^{-1} = 0.37\, \varphi(0)$.

We can now find the solution of Eq. (176) corresponding to the boundary condition $\varphi(x) = \varphi(0)$ for $x=0$. Equation (176) can be rewritten

$$\lambda^2 \left(\frac{d\varphi}{dx}\right)^2 = e^{\varphi} + e^{2\upsilon} \cdot e^{-\varphi} - 2e^{\upsilon}.$$

If a new variable u is introduced, defined by $e^{\varphi} = u^2$, noting that $d\varphi/dx = 2(du/dx)/u$, the above expression takes the form

$$2\lambda\left(\frac{du}{dx}\right) = \pm(u^2 - u_\upsilon^2),$$

where $u_\upsilon^2 = e^{\upsilon}$. In the range $-\infty < x < 0$, $du/dx \leq 0$. Furthermore, $u \leq u_\upsilon$. It therefore follows that the positive sign should be used in the above differential equation which accordingly has the solution

$$\frac{1}{u_\upsilon} \operatorname{ar\,tanh}\left(\frac{u}{u_\upsilon}\right) = -\frac{x}{2\lambda} + B,$$

where the integration constant B is determined by $u = u(0) = \exp\{\varphi(0)/2\}$ for $x=0$. From this we obtain

$$\operatorname{ar\,tanh}\left(\frac{u}{u_\upsilon}\right) = \operatorname{ar\,tanh}\left(\frac{u(0)}{u_\upsilon}\right) - \frac{u_\upsilon}{2\lambda}x$$

or

$$u/u_\upsilon = \tanh\left\{\operatorname{ar\,tanh}\left(\frac{u(0)}{u_\upsilon}\right) - \frac{u_\upsilon}{2\lambda}x\right\}.$$

Since u is defined as $u^2 = e^{\varphi}$, it follows that $u_\upsilon = e^{\upsilon/2}$, $u/u_\upsilon = e^{(\varphi-\upsilon)/2}$ and $u(0)/u_\upsilon = e^{(\varphi(o)-\upsilon)/2}$. If these expressions are inserted into the above equation, we obtain

$$e^{(\varphi-\upsilon)/2} = \tanh\left\{\operatorname{ar\,tanh}(e^{(\varphi(o)-\upsilon)/2}) - \frac{e^{\upsilon/2}}{2\lambda}x\right\}$$

or

$$\varphi(x) = \upsilon = 2\ln\left[\tanh\left\{\operatorname{ar\,tanh}(e^{(\varphi(o)-\upsilon)/2}) - \frac{e^{\upsilon/2}}{2\lambda}x\right\}\right], \tag{179}$$

which is the expression for the potential profile in the range $x < 0$.

Fig. 12 shows the potential profile through the Donnan phase for $C_o = 100$ mM $\cdot$ l^{-1} and M = 235 mM $\cdot$ l^{-1}, with a corresponding value of $V_D = 25.69$ mV. The associated concentration profiles $C_+(x)$, $C_-(x)$ and $M(x)$ are calculated from the values of $\varphi(x)$ with the aid of Eq. (166), Eq. (167), and Eq. (168). Finally the charge density $\varrho(x)$, which brings about the potential profile, is calculated as $\varrho = F(C_+(x) - C_-(x))$ for $x > 0$ and $\varrho = F(C_+(x) + M(x) - C_-(x))$ for $x < 0$. It appears from the figure that it is only beyond a region 40 Å thick around $x = 0$ that the concentrations and the potential have

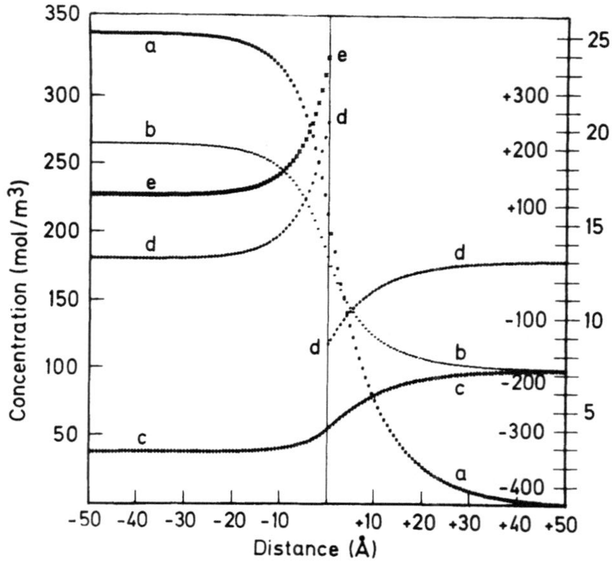

Fig. 12. Profiles through transition region of a Donnan system: (a) Potential $\psi(x)$. Curve (b) concentration $C_{-}(x)$ of negative ion. Curve (c) concentration of positive ions $C_{+}(x)$. Curve (d) space charge density. Curve (e) concentration of positive macroion M (x). Left ordinate: Concentration, mol m^{-3}. Outer right ordinate: Membrane potential, mV. Inner right ordinate, space charge density in units of Faraday's number per m^3. Abscissa, distance in Ångström

assumed those values required by the thermodynamic treatment, which in turn correspond to a state of electroneutrality in each of the phases. Within this region the deviation from electroneutrality is considerable, and accordingly the charge density is large enough to create the local electrical field that can balance the tendency of the ions to move down their concentration profiles by diffusion. For $x = 0$, the potential gradient is $d\psi/dx = -1.667 \times 10^7 \text{ V} \cdot \text{m}^{-1}$. Hence the electrical force on the positive ion $X_{el} = FE = 1\ 6 \times 10^{12} \text{ N} \cdot \text{mol}^{-1}$, which is the force required at this position to maintain the concentration gradients at the values $dC_{+}/dx = 3.49 \times 10^{10}$ and $dC_{-}/dx = 1.21 \times 10^{11} \text{ mol} \cdot \text{m}^{-3}$ per m. Corresponding to these gradients there is an equivalent "diffusion force", $-RTd\ln C/dx$ – which balances the oppositely directed electrical force X_{el}.

Finally, the total charge in the transitional zone will be calculated. We start from Poisson's equation:

$$\frac{d^2\psi}{dx^2} = -\varrho/\varepsilon_o K.$$

The charge dq, present in the layer between x and $x + dx$ for $x > 0$ and surface 1 m^2, is $dq = \varrho(x)dx$. We have

$$dq = -\varepsilon_o K \frac{d^2\psi}{dx^2} dx$$

and

$$q = -\varepsilon_0 K \int_0^\infty \left(\frac{d^2\psi}{dx^2}\right) dx = -\varepsilon_0 K \left[\frac{d\psi}{dx}\right]_0^\infty$$

or

$$q = \varepsilon_0 K \left(\frac{d\psi}{dx}\right)_{x=0}, \tag{180}$$

since $d\psi/dx \rightarrow 0$ for $x \rightarrow \infty$. According to Eq. (174a),

$$\left(\frac{d\psi}{dx}\right)_{x=0} = \frac{RT}{F}\left(\frac{d\varphi}{dx}\right)_{x=0} = -\frac{2RT}{F\lambda}\sinh(\varphi(0)/2).$$

When this is inserted in Eq. (180) together with the value for λ in Eq. (173), we obtain

$$q = -2\sqrt{2RT\varepsilon_0 KC_0}\sinh(\varphi(0)/2), \tag{181}$$

which is the total excess charge in the space $x > 0$. A corresponding quantity of opposite charge is present in the space $x < 0$. It should be mentioned in this connection that the potential profile described by Eq. (175) for $x > 0$ corresponds completely to that which would be obtained if, instead of the Donnan phase there was a membrane covered with fixed charges, with a surface charge density equivalent to that found in the space $x < 0$.

IV. Diffusion Potentials

In the Donnan system, the equilibirum state illustrated in Fig. 12 is maintained on the basis of the presence of the impermeable macroion in inner phase (i). If the membrane suddenly becomes permeable to the macroion, this will migrate into the outer phase (o). The space charges and thus the field in the transitional zone will thereby be reduced, and likewise it will no longer be possible to maintain the equilibrium concentration profiles for the small ions. The system will now move towards a new equilibrium state with a uniform distribution of all ions in both phases and no potential difference between the phases. Before the development of this state, it is possible – depending on the mutual magnitudes of the ionic mobilities – that a transient potential difference exists between the phases, decreasing in step with the equalization of the concentration. This potential difference, V_{dif}, which holds in a system not in equilibrium and which is due to differences between the mobilities of the individual types of ions, is called a *diffusion potential.* The mechanism of such a regime of electrodiffusion is perhaps vizualized most easily by considering the following situation: the inner phase (i) of a diffusion chamber contains a monovalent salt – *e.g.* HCl – with

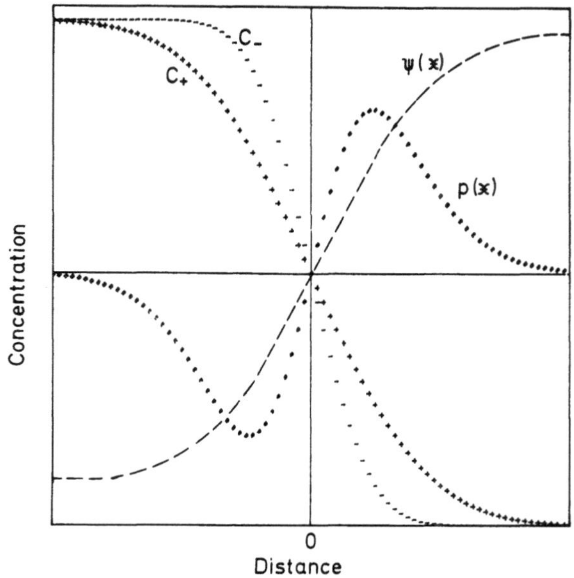

Fig. 13. To illustrate the establishment of the space charge regions responsible for the diffusion potential. Ordinate, concentration of positive (+) and negative (−) ions. Space charge density. Abscissa, distance in μm

concentration $C^{(i)}$, and the outer phase contains the same salt in concentration $C^{(o)} < C^{(i)}$. If the two phases are brought into contact, H^+ and Cl^- will diffuse from phase (i) into phase (o). However, since the H^+ ion is smaller than the Cl^- ion, the H^+ ion — in agreement with the Einstein-Stokes relation [Eq. (96)] — will diffuse more rapidly than the Cl^- ion. The two concentration profiles will therefore not overlap completely. From the transitional zone $x = 0$ into phase (o), we have $C_{H^+} > C_{Cl^-}$. As a result, a positive space charge $\varrho(x)$ will be built up in the region $x > 0$. Correspondingly, since $C_{H^+} < C_{Cl^-}$ in the region from the transitional zone into phase (i) a negative space charge will be built up here. As illustrated in Fig. 13, these space charges will create an electrical potential profile through the transitional zone. Phase (o) will become charged positively in relation to phase (i), because in the case under consideration it is the positive ion that moves more quickly by diffusion. Following this initial charging-up process, the result of the space charges and the associated electrical field E — in the direction (o) → (i) — will be that the rate of migration of H^+ becomes somewhat reduced and the rate of migration of the Cl^- ion will be correspondingly increased. The final result, therefore, is that the H^+ and Cl^- ions migrate with the same speed, and the basic diffusion process that then follows for H^+ and Cl^- takes place without any further separation of the concentration profiles. An electrodiffusion regime is thus characterized by two processes: (1) an initial charging process, in which a local electric field E is created that can be characterized by a *charging time* τ_q, followed by (2) a customary diffusion process, which can be characterized by a *redistribution time* τ_{red}, as described in section

C.III.4. In a nonstationary electrodiffusion regime, the ratio between these two times will determine what current passes through the system. It would therefore also be useful to have an estimate of the magnitude of the charging time τ_q. The following argument originates from PLANCK (1890a), but for the sake of clarity it has been written in vector form.

1. Charging Time and Redistribution Time

We consider a homogeneous membrane of thickness h, surrounded by a solution of an electrolyte consisting of an arbitrary number of monovalent ions. The concentrations have been maintained at the same value sufficiently long for equilibrium to exist throughout. At time $t=0$ the concentration is reduced in one phase and maintained at the new value. In the case of a positive ion of type k, we have in the layer at time t

$$\boldsymbol{J}_\mathrm{k}^+ = -\frac{RT}{F}\operatorname{grad}(u_\mathrm{k}^+ C_\mathrm{k}^+) + (u_\mathrm{k}^+ C_\mathrm{k}^+)\boldsymbol{E}, \tag{182}$$

where $\boldsymbol{E}$ is the field at time t. By summation over all positive ions, and setting, according to PLANCK,

$$U = \sum u_\mathrm{k}^+ C_\mathrm{k}^+ \tag{183}$$

we obtain

$$\sum \boldsymbol{J}_\mathrm{k}^+ = -\frac{RT}{F}\operatorname{grad} U + U\boldsymbol{E}. \tag{184}$$

Conservation of mass requires that $\partial C_\mathrm{k}^+/\partial t = -\operatorname{div} \boldsymbol{J}_\mathrm{k}$. By summation over all positive ions we obtain

$$\sum \frac{\partial C_\mathrm{k}^+}{\partial t} = \frac{\partial}{\partial t}\sum C_\mathrm{k}^+ = -\sum \operatorname{div} \boldsymbol{J}_\mathrm{k}^+ = -\operatorname{div}\sum \boldsymbol{J}_\mathrm{k}^+, \tag{185}$$

which when combined with Eq. (184) gives

$$\frac{\partial}{\partial t}\left(\sum C_\mathrm{k}^+\right) = \frac{RT}{F}\operatorname{div}\operatorname{grad} U - \operatorname{div}(U\boldsymbol{E}). \tag{186}$$

The corresponding expression for the negative ions becomes

$$\frac{\partial}{\partial t}\left(\sum C_\mathrm{k}^-\right) = \frac{RT}{F}\operatorname{div}\operatorname{grad} W + \operatorname{div}(W\boldsymbol{E}), \tag{187}$$

where

$$W = \sum u_\mathrm{k}^- C_\mathrm{k}^-. \tag{188}$$

The field is associated with the space charge density ϱ through Poisson's equation: div $\boldsymbol{E} = \varrho/\varepsilon_o K = \varrho/\varepsilon$, where $\varepsilon = K\varepsilon_o$ is is the permittivity in the solution. We will write Poisson's equation as

$$\varrho = \varepsilon \operatorname{div} \boldsymbol{E} = \operatorname{div}(\varepsilon \boldsymbol{E}),$$

since ε is considered constant. The space charge is

$$\varrho = F \sum (C_k^+ - C_k^+),$$

from which

$$\sum (C_k^+ - C_k^-) = \operatorname{div}(\varepsilon \boldsymbol{E})/F;$$

when differentiated with respect to time t this gives

$$\frac{\partial}{\partial t}\left\{\sum C_k^+ - \sum C_k^-\right\} = \frac{\partial}{\partial t}\left\{\operatorname{div}(\varepsilon \boldsymbol{E})\right\}\Big/F$$

$$= \operatorname{div}\left\{\frac{\partial(\varepsilon \boldsymbol{E})}{\partial t}\right\}\Big/F, \tag{189}$$

where the order of differentiation has been interchanged with respect to time and space. We now insert Eq. (186) and Eq. (187) in the left-hand side of Eq. (189). This gives

$$\operatorname{div}\left\{\frac{\partial(\varepsilon \boldsymbol{E})}{\partial t}\right\} = RT \operatorname{div}\operatorname{grad}(U - W) - \operatorname{div}\{(U + W)F\boldsymbol{E}\}$$

or

$$\operatorname{div}\left\{\frac{\partial(\varepsilon \boldsymbol{E})}{\partial t} + F(U + W)\boldsymbol{E} - RT \operatorname{grad}(U - W)\right\} = 0,$$

so that

$$\frac{\partial(\varepsilon \boldsymbol{E})}{\partial t} + F(U + W)\boldsymbol{E} - RT \operatorname{grad}(U - W) = A(t), \tag{190}$$

where A is an integration constant depending only on t. In his development, PLANCK now put $A(t) = 0$, since initially there is no local field. Furthermore, he introduced the fundamental assumption – the correctness of which must be evaluated when the argument is concluded – that the concentrations are not altered to any significant degree before the field has become established, *i. e.* $d\operatorname{grad}(U - W)/dt = 0$ for $t < \tau_q$. Hereby Eq. (190) becomes an ordinary differential equation

$$\frac{d\boldsymbol{E}}{dt} + \frac{F(U + W)}{\varepsilon}\boldsymbol{E} = \frac{RT}{\varepsilon} \operatorname{grad}(U - W). \tag{191}$$

The solution of this differential equation for $\boldsymbol{E}(t)$ corresponding to the initial condition $\boldsymbol{E}=0$ for $t=0$ is

$$\boldsymbol{E}(t) = \frac{RT}{F} \frac{\operatorname{grad}(U-W)}{U+W} [1 - \exp\{-F(U+W)t/\varepsilon\}]. \tag{192}$$

The local field will thus grow towards the value

$$\boldsymbol{E} = \frac{RT}{F} \frac{\operatorname{grad}(U-W)}{U+W} \tag{193}$$

with the time constant τ_q for the charging process – *i. e.* the charging time – equal to

$$\tau_q = \varepsilon / F(U+W). \tag{194}$$

For small ions at a concentration of 100 mmol·l^{-1}, a charging time of 10^{-10} s is obtained. This time should be compared with the redistribution time

$$\tau_{\text{red}} = \frac{h^2}{\pi D_k} = \frac{F}{RT} \frac{h^2}{\pi u_k}, \tag{73}$$

which for a layer of 1 cm thickness is of the order of magnitude of 10^5 s. However, the redistribution time is proportional to the square of the membrane thickness. The ratio between the two time constants is

$$\tau_{\text{red}}/\tau_q = 3.6 \times 10^{15} \times C \times h^2.$$

For $C = 100$ mol·m^{-3} and $h = 100$ Å $= 10^{-8}$ m, the ratio is approximately 40, so that it is still reasonable to regard the two processes as separate in time. In the case of thinner membranes this will no longer be the case, and the charging current could constitute a considerable fraction of the total current through the system.

2. The Henderson Regime

PLANCK's analysis of the charging time shows that for times greater than τ_q, a local field will be established whose magnitude is

$$E = -\frac{\partial \psi}{\partial x} = \frac{RT}{F} \frac{1}{U+W} \frac{\partial (U-W)}{\partial x} \tag{194}$$

However, this expression cannot be used in calculating the diffusion potential $V_{\text{dif}} = \psi^{(i)} - \psi^{(o)}$ between two adjacent phases, unless the concentration profiles are known for the individual ions through the transitional zone, which in reality is equivalent to knowing the potential profile. To overcome this difficulty and at the same time obtain an expression for the diffusion potential which was correct to a first order of approximation, HENDERSON (1907) assumed that the transitional zone consisted of a continuous mixture of the electrolytes in question, so that at any point in the transitional zone the concentrations are a linear mixture of the concentrations $C_k^{(i)}$ and $C_k^{(o)}$ in the two phases themselves. The concentration profiles therefore become

$$C_k(x) = C_k^{(i)} + (C_k^{(o)} - C_k^{(i)})\,x/h$$

By using a semithermodynamic argument similar to that used by W. THOMSON (Lord Kelvin) in his treatment of the thermo-electric effect, HENDERSON (1907) arrived at the following expression for monovalent ions:

$$\psi^{(i)} - \psi^{(o)} = \frac{RT}{F}\,\frac{(U^{(o)} - W^{(o)}) - (U^{(i)} - W^{(i)})}{(U^{(o)} + W^{(o)}) - (U^{(i)} + W^{(i)})}\ln\frac{U^{(o)} + W^{(o)}}{U^{(i)} + W^{(i)}}, \quad (195)$$

where U and W are defined by Eq. (183) and Eq. (188). Henderson's equation can also be derived directly from Eq. (194) by determining $U(x)$ and $W(x)$ from the concentration profiles and inserting them into Eq. (194), which can then be integrated. Henderson's expression is very useful in allowing an estimate of the magnitude of the diffusion potentials between two phases, in the case where it may be assumed that the transitional zone is well mixed.

3. The Planck Regime

In his analysis of the diffusion potential, which has ever since been the starting point for all subsequent theoretical studies of electrodiffusion, PLANCK (1890a, 1890b) started from the following situation: A homogeneous membrane of thickness h separates two electrolyte phases with concentrations $C_k^{(i)}$ and $C_k^{(o)}$. Both phases are well mixed, so that the concentrations are uniform in each of the two phases. All ions are assumed to be monovalent. Since the concentrations in the inner and outer phase are maintained at constant values, a stationary state will develop for the ion regime through the membrane, in which the flux for each ion type is given by the Nernst-Planck equation

$$J_k = -\frac{RT}{F}u_k\frac{dC_k}{dx} - u_k C_k\frac{d\psi}{dx} \quad (128)$$

but where neither the potential profile nor the concentration profiles are necessarily linear through the membrane. If the system is left to itself, *i.e.* if no current is forced through the system by means of external electrodes, the stationary state will be characterized by a *zero total ion current* through the membrane. PLANCK was therefore able to deduce the potential difference, *Planck's diffusion potential,* across the membrane corresponding to the above condition, by integrating the flux equations, with the assumption that the system in the membrane could be characterized approximately as satisfying the condition for *electroneutrality.*

a) Planck's General Relations

PLANCK (1890b) derived two fundamental relations which characterize his stationary diffusion regime. Eq. (128) is divided by RTu_k/F. This gives

$$-\frac{F}{RTu_k}J_k = \frac{dC_k}{dx} + C_k\frac{d\varphi}{dx},$$

where $\varphi = \psi/(RT/F)$ is the normalized potential. In the steady state the left-hand side is constant. Summation over all positive ion fluxes gives

$$\sum A_k = A = \frac{d}{dx}\left(\sum C_k^+\right) + \frac{d\varphi}{dx}\sum C_k^+, \tag{196}$$

where $A_k = -FJ_k^+/RTu_k^+$. In a corresponding manner, summation over all negative ion fluxes gives

$$\sum B_k = B = \frac{d}{dx}\left(\sum C_k^-\right) - \frac{d\varphi}{dx}\sum C_k^-, \tag{197}$$

where $B_k = -FJ_k^-/RTu_k^-$.

If Eq. (196) and Eq. (197) are added, we obtain

$$A + B = \frac{d}{dx}\left\{\sum C_k^+ + \sum C_k^-\right\} + \frac{d\varphi}{dx}\left\{\sum C_k^+ - \sum C_k^-\right\}$$

Since

$$\sum C_k^+ + \sum C_k^- = 2C(x), \tag{198}$$

where $C(x)$ is the total concentration of the ions in position x, and since

$$\sum C_k^+ - \sum C_k^- = 0$$

because of the electroneutrality condition, we then obtain *Planck's first relation*

$$\frac{dC(x)}{dx} = (A + B)/2 = \text{const}, \tag{199}$$

i. e. the total concentration C(x) varies linearly through the membrane. The integral of Eq. (199) corresponding to the boundary conditions $C(x) = C^{(i)}$ for $x=0$ and $C(x) = C^{(o)}$ for $x=h$ is

$$C(x) = [C^{(o)} - C^{(i)}]\frac{x}{h} + C^{(i)}. \tag{200}$$

However, this does not imply that the individual concentrations $C_k(x)$ also vary linearly through the range $0 \leqq x \leqq h$.

If Eq. (197) is subtracted from Eq. (196), we obtain

$$A-B=\frac{d}{dx}\left\{\sum C_{\mathrm{k}}^{+}-\sum C_{\mathrm{k}}^{-}\right\}+\frac{d\varphi}{dx}\left\{\sum C_{\mathrm{k}}^{+}+\sum C_{\mathrm{k}}^{-}\right\}.$$

If Eq. (198) is inserted into this, because of the electroneutrality condition, we obtain *Planck's second relation*

$$\frac{d\varphi}{dx}=\frac{A-B}{2C(x)}, \tag{201}$$

i. e. the *field* $E = -(d\varphi/dx)\cdot(RT/F)$ *is inversely proportional to the total concentration* $C(x)$ at the point in question. Inserting Eq. (200) into the above, we obtain

$$\frac{d\varphi}{dx}=\frac{1}{2}\,\frac{A-B}{[C^{(o)}-C^{(i)}](x/h)+C^{(i)}}. \tag{202}$$

If the total concentration of ions is the same on both sides of the membrane, $C^{(o)} = C^{(i)} = C_o$. This means that

$$\frac{d\varphi}{dx}=\frac{A-B}{2C_o}, \tag{203}$$

i. e. under these conditions the field throughout the membrane is always constant. This situation will be discussed in greater detail in Section V.

b) The Electrical Equivalent Circuit for the Planck Regime

We will consider a more general situation, where the total current I through the membrane is not necessarily zero. The flux for the positive ion type k is written in the form of Eq. (128)

$$J_{\mathrm{k}}^{+}=-\frac{RT}{F}u_{\mathrm{k}}^{+}\frac{dC_{\mathrm{k}}^{+}}{dx}-u_{\mathrm{k}}^{+}C_{\mathrm{k}}\frac{d\psi}{dx}.$$

The current carried by this ion is $I_{\mathrm{k}}^{+} = FJ_{\mathrm{k}}^{+}$, or

$$I_{\mathrm{k}}^{+}=-RT\frac{d(u_{\mathrm{k}}^{+}C_{\mathrm{k}}^{+})}{dx}-Fu_{\mathrm{k}}^{+}C_{\mathrm{k}}^{+}\frac{d\psi}{dx},$$

since u_{k}^{+} is assumed to be constant through the membrane. Summation of all currents carried by positive ions gives

$$\Sigma I_{\mathrm{k}}^{+}=I^{+}=-RT\frac{d}{dx}\sum u_{\mathrm{k}}^{+}C_{\mathrm{k}}^{+}-F\frac{d\psi}{dx}\sum u_{\mathrm{k}}^{+}C_{\mathrm{k}}^{+}.$$

If we put $\Sigma u_k^+ C_k^+ = U$, corresponding to Eq. (183), we get

$$I^+ = -RT\frac{dU}{dx} - FU\frac{d\psi}{dx}. \tag{204}$$

The flux for a negative ion is

$$J_k^- = -\frac{RT}{F}u_k^-\frac{dC_k^-}{dx} + u_k^- C_k^-\frac{d\psi}{dx}.$$

The current carried by this ion is $I_k^- = -FJ_k^-$. The expression for the total current carried by negative ions can therefore be written as

$$I^- = RT\frac{dW}{dx} - FW\frac{d\psi}{dx}, \tag{205}$$

where $W = \Sigma u_k^- C_k^-$. The total current carried by all ions is

$$I = I^+ + I^-.$$

If we insert Eq. (204) and Eq. (205) in this we get

$$I = -RT\frac{d(U-W)}{dx} - F(U+W)\frac{d\psi}{dx}, \tag{206}$$

which is the total current density through the membrane.

For $I=0$, the field has assumed the particular value which forces the ions to migrate through the membrane with just those mutual rates of migration so that there is no net charge transferred through the membrane. Putting $I=0$ in Eq. (206), we obtain

$$\left(\frac{d\psi}{dx}\right)_{I=0} = -\frac{RT}{F}\frac{1}{U+W}\frac{d(U-W)}{dx}, \tag{207}$$

which is once again Planck's expression for the local field in the membrane, based on those space charges resulting from the diffusion of the ions through the membrane. In the present treatment (Planck, 1890a), however, the expression is obtained by considering times far longer than the redistribution time τ_q.

Equation (206) allows an electrical equivalent for Planck's regime to be set up. The equation is rearranged as

$$\frac{1}{F(U+W)}I = -\frac{RT}{F}\frac{1}{U+W}\frac{d(U-W)}{dx} - \frac{d\psi}{dx}.$$

This equation is multiplied by dx and integrated through the membrane from $x=0$ to $x=h$. If a stationary state is present, $dI/dx = 0$. We then have

$$I\int_0^h \frac{dx}{F(U+W)} = -\frac{RT}{F}\int_0^h \frac{d(U-W)}{U+W} + \psi(0) - \psi(h), \tag{208}$$

where

$$\psi(0) - \psi(h) = V$$

is the potential difference across the membrane with a total current I. The diffusion potential V_{dif} is the value $V_{I=0}$, for which the system has zero current.

Putting $I=0$ in Eq. (208) we obtain

$$V_{\text{dif}} = \frac{RT}{F}\int_0^h \frac{d(U-W)}{U+W}, \tag{209}$$

which is Planck's formulation for calculating the diffusion potential if the concentration profile through the membrane is known for the separate ions. Equation (208) can therefore also be written as

$$IR = V - V_{\text{dif}}, \tag{210}$$

where, according to Eq. (123),

$$R = \int_0^h \frac{dx}{F(U+W)} \tag{211}$$

is the total membrane resistance produced by the different ion profiles through the membrane. The equivalent electrical circuit shown in Figure 14 thus corresponds to Planck's regime. Current I, which passes through the membrane, produces a fall in potential across the resistance R, which is in series with an electromotive force, the diffusion potential for the Planck regime. Thus this

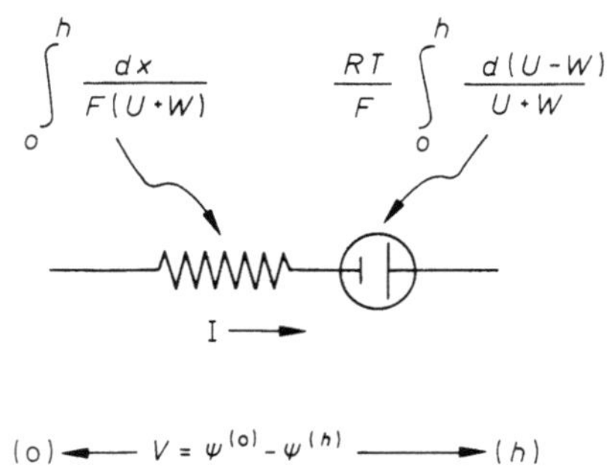

Fig. 14. The equivalent circuit of the Planck diffusion regime

electromotive force and the fall in potential IR across the total membrane resistance constitute the total fall in potential across the membrane. This circuit correctly reproduces the relation between the total current I and the potential fall V across the membrane, but does not give any information as to the individual ion fluxes.

c) Planck's Expression for the Diffusional Potential

PLANCK's (1890b) integration of the flux equations corresponding to the stationary diffusion regime gives the following final result for the diffusion potential $\psi^{(i)} - \psi^{(o)}$, which is defined by the parameter ξ through the relation

$$\psi^{(i)} = \psi^{(o)} = \frac{RT}{F} \ln \xi . \tag{212}$$

This parameter is determined by means of the following transcendental equation:

$$\frac{\xi U^{(i)} - U^{(o)}}{W^{(i)} - \xi W^{(o)}} = \frac{\ln(C^{(i)}/C^{(o)}) - \ln \xi}{\ln(C^{(i)}/C^{(o)}) + \ln \xi} \cdot \frac{\xi C^{(i)} - C^{(o)}}{C^{(i)} - \xi C^{(o)}}, \tag{213}$$

where $C^{(i)}$ and $C^{(o)}$ are the total concentrations on the two sides of the membrane and U and W have their usual significance. The details of the derivation will not be reproduced here, since they are not essential for an understanding of the later sections. A good account of Planck's derivation has been given by MCINNES (1961). The field in the region $0 \leq x \leq h$ is given by

$$-\frac{d\psi}{dx} = \frac{1}{h} \frac{\psi^{(i)} - \psi^{(o)}}{\ln(C^{(o)}/C^{(i)})} \cdot \frac{C^{(o)} - C^{(i)}}{(C^{(o)} - C^{(i)})(x/h) + C^{(i)}}, \tag{214}$$

and the potential profile $\psi(x)$ is therefore

$$\psi^{(i)} - \psi(x) = \frac{\psi^{(i)} - \psi^{(o)}}{\ln(C^{(o)}/C^{(i)})} \cdot \left[\ln\left(\frac{C^{(o)}}{C^{(i)}} - 1\right) \cdot \frac{x}{h} + 1\right]. \tag{215}$$

It appears from this that, in general, the potential profile varies logarithmically with the position in the membrane, viz. when $C^{(i)} \neq C^{(o)}$. For $C^{(o)} \rightarrow C^{(i)}$, in agreement with Eq. (203) the above expression becomes

$$\psi^{(i)} - \psi(x) = (\psi^{(i)} - \psi^{(o)})(x/h), \tag{216}$$

corresponding to a linear potential profile and a constant field.

The complete solution of the diffusion regime consists in first determining the parameter ξ from Eq. (213). This is most easily done graphically or by means of Newton's method. The diffusion potential $\psi^{(i)} - \psi^{(o)}$ is then determined from Eq. (212). As the potential – or the field – profile is now known, the concentration profile $C(x)$ can in principle be determined from the flux equation by the same means as shown in Section E.II.c. The stationary concentration profiles differ from the profiles of the Henderson regime by generally not being linear. Even in the simple case where the Planck regime contains only three types of ion, *e.g.* H^+, K^+ and Cl^-, one of the positive ion profiles can show a maximum at a point in the range $0 < x < h$ (PLANCK, 1930; PLETTIG, 1930; TEORELL, 1953).

PLANCK's treatment was extended by PLEIJEL (1910) to include multivalent ions. TEORELL (1953) treated the transport of monovalent ions across a membrane containing fixed charges. This analysis was extended by SCHLÖGL (1954) to include multivalent ions.

V. Electrodiffusion through Membranes

In Section F.III.1, an account was given of the equilibrium state across a membrane permeable only to either a cation or an anion. The present section will describe some simple stationary states which can be solved by means of Planck's theory, but without the need to use the completely general solution as described in the last section.

1. Single Salt

We will consider a single monovalent salt, *e.g.* HCl, surrounding a membrane of thickness h. The concentrations are maintained at the constant values $C = C^{(i)}$ for $x < 0$ and $C = C^{(o)}$ for $x > h$. The membrane is permeable to both ions, and a stationary state exists.

a) Diffusion Potential (NERNST, 1888; PLANCK, 1890a)

In the position $0 < x < h$, $C^+(x) = C^-(x) = C(x)$ as a result of the electroneutrality condition. The diffusion potential is calculated from Eq. (209)

$$-\frac{d\psi}{dx} = \frac{RT}{F}\,\frac{1}{U+W}\,\frac{d(U-W)}{dx}.$$

We have $U(x) = u^+ C(x)$ and $W(x) = u^- C(x)$; when these values are inserted in the above equation we obtain

$$-\frac{d\psi}{dx} = \frac{RT}{F}\,\frac{u^+ - u^-}{u^+ - u^-}\,\frac{1}{C}\,\frac{dC}{dx}. \qquad (217)$$

The equation is multiplied by dx and integrated through the membrane from $x = 0$, where $\psi(x) = \psi(0) = \psi^{(i)}$, and $C(x) = C^{(i)}$, to $x = h$, where $\psi(x) = \psi^{(o)}$ and $C(x) = C^{(o)}$. This gives

$$\psi^{(i)} - \psi^{(o)} = V_{\text{dif}} = \frac{RT}{F}\,\frac{u^+ - u^-}{u^+ + u^-}\ln\left(\frac{C^{(o)}}{C^{(i)}}\right). \qquad (218)$$

If $C^{(i)} > C^{(o)}$, ions will move into the outer phase (o). The equation shows that the electrical polarity of phase (o) with respect to phase (i) will be determined by the one of the two ions with greater mobility. If $u^+ > u^-$, $\psi^{(o)} > \psi^{(i)}$, and for $u^+ < u^-$, $\psi^{(i)} > \psi^{(o)}$. In the case of an HCl solution with $C^{(i)} = 100 \text{ mol} \cdot \text{m}^{-3}$ and $C^{(o)} = 10 \text{ mol} \cdot \text{m}^{-3}$ and $u^+ = 36 \times 10^{-8} \text{ Vm}^2 \cdot \text{s}^{-1}$ and $u^- = 7.9 \times 10^{-8}$ Vm^2s^{-1}, we obtain $\psi^{(i)} - \psi^{(o)} = -36.8$ mV. Equation (218) can also be obtained directly from Planck's transcendental equation [Eq. (213)] by inserting the values $U^{(i)} = u^+ C^{(i)}$, $U^{(o)} = u^+ C^{(o)}$, $W^{(i)} = u^- C^{(i)}$, and $W^{(o)} = u^- C^{(o)}$ and solving for ξ.

If the expression for the field, Eq. (217), is inserted in the flux equation, Eq. (128), for the positive ion, then allowing for the electroneutrality condition $C^+(x) = C^-(x) = C(x)$, we obtain

$$J^+ = -\frac{RT}{F} 2 \frac{u^+ \cdot u^-}{u^+ + u^-} \frac{dC}{dx} \tag{219}$$

and correspondingly for the negative ion,

$$J^- = -\frac{RT}{F} 2 \frac{u^+ \cdot u^-}{u^+ + u^-} \frac{dC}{dx} \tag{220}$$

i. e. that field given by Eq. (217) retards the movement of the ion with the greater mobility and increases the rate of migration of the ion with the lesser mobility by just such amounts that both ions, in the steady-state regime with no net current, move through the membrane with the same flux. Both ions move under the influence of the concentration gradient dC/dx with a common diffusion coefficient

$$D_\pm = 2 \frac{RT}{F} \frac{u^+ \cdot u^-}{u^+ + u^-} = 2 \frac{D^+ \cdot D^-}{D^+ + D^-} \tag{221}$$

The flux of the salt, $J_\pm$, through the membrane is therefore described by a single equation as a simple process of diffusion:

$$J_\pm = -D_\pm \frac{dC}{dx}, \tag{222}$$

corresponding to Fick's law. Equations (218), (221), and (222) were first derived by NERNST (1888).

b) Membrane Resistance

The membrane resistance can be calculated from Eq. (123)

$$R = \frac{1}{F} \int_0^h \frac{dx}{U + W}$$

if the concentration profiles of both ions through the membrane are known. The dependence of the total concentration on distance is, from Eq. (200),

$$C(x) = (C^{(o)} - C^{(i)})(x/h) + C^{(i)},$$

and since the condition of electroneutrality holds, $C^{+}(x) = C^{-}(x) = C(x)$. We then have: $U(x) = u^{+}C(x)$ and $W(x) = u^{-}C(x)$, and

$$U + W = (u^{+} + u^{-})\cdot C(x) = (u^{+} + u^{-})[(C^{(o)} - C^{(i)})(x/h) + C^{(i)}].$$

If this is inserted in the above expression for the membrane resistance we obtain

$$R = \frac{1}{F(u^{+} + u^{-})} \int_{0}^{h} \frac{dx}{(C^{(o)} - C^{(i)})(x/h) + C^{(i)}}.$$

The integrand is of the form $1/(a + bx)$, the indefinite integral of which is $\ln|a + bx|/b$. We have therefore

$$R = \frac{h}{F} \cdot \frac{\ln(C^{(o)}/C^{(i)})}{(u^{+} + u^{-})(C^{(o)} - C^{(i)})}. \tag{223}$$

For an HCl solution with $C^{(i)} = 100\ \mathrm{mol/m^{-3}}$ and $C^{(o)} = 10\ \mathrm{mol\ m^{-3}}$, $R = 0.6 \times h\ (\Omega \cdot \mathrm{m^2})$. A "pore" of radius 5 Å and length $h = 100$ Å therefore has a resistance

$$R_{\text{pore}} = 0.6\, \frac{10^{-8}}{\pi\,(5 \times 10^{-10})^2} = 7.6 \times 10^{9}\,\Omega.$$

The potential profile through the membrane is obtained by integration of Eq. (208) from $x = 0$ to the position x. This gives

$$\psi^{(i)} - \psi(x) = \frac{RT}{F}\, \frac{u^{+} - u^{-}}{u^{+} + u^{-}} \ln\left(\frac{C(x)}{C^{(i)}}\right).$$

Since $C(x) = (C^{(o)} - C^{(i)})\ (x/h) + C^{(i)}$ varies linearly with x through the membrane, it follows that the potential $\psi(x)$ varies logarithmically with x through the membrane. When $C(x)$ is inserted and the factor $RT \cdot (u^{+} - u^{-})/F \cdot (u^{+} + u^{-})$ is eliminated by means of Eq. (218), the above expression gives

$$\psi^{(i)} - \psi(x) = \frac{\psi^{(i)} - \psi^{(o)}}{\ln(C^{(o)}/C^{(i)})} \ln\left[\left(\frac{C^{(o)}}{C^{(i)}} - 1\right) \frac{x}{h} + 1\right], \tag{224}$$

in agreement with PLANCK's general expression [Eq. (215)]. The profiles of concentration, electric field and potential across a homogeneous membrane brought about by the diffusion of a single salt are shown in Figure 15.

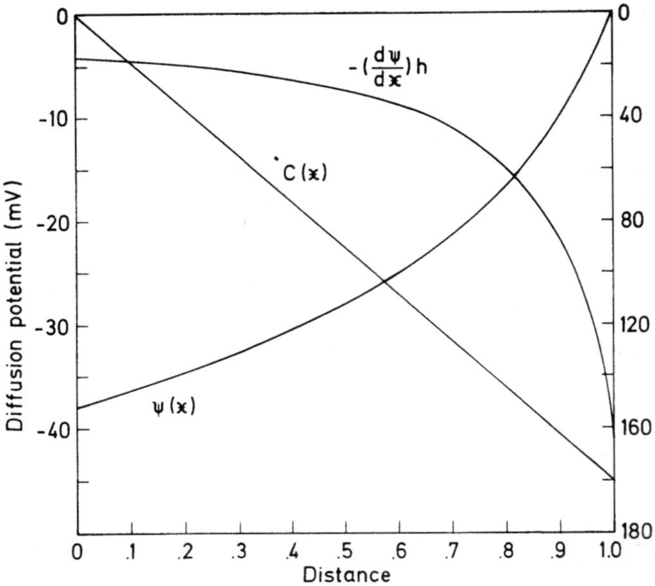

Fig. 15. The profiles of concentration, electric field and potential across a homogeneous membrane of thickness h brought about by the diffusion of HCl. $C(0) = 100\ \text{mol/m}^{-3}$; $C(h) = 10\ \text{mol/m}^{-3}$. Left ordinate, diffusion potential (mV). Right ordinate, field times membrane thickness (mV). Abscissa, distance in units of membrane thickness h

c) Equivalent Electrical Circuit

For the single salt, the relationship (Eq. 208) between the membrane current I and the membrane potential $V = \psi^{(i)} - \psi^{(o)}$ becomes

$$I \cdot \frac{h}{F} \frac{\ln(C^{(o)}/C^{(i)})}{(u^+ + u^-)(C^{(o)} - C^{(i)})} = V - \frac{RT}{F} \frac{u^+ - u^-}{u^+ + u^-} \ln\left(\frac{C^{(o)}}{C^{(i)}}\right). \tag{225}$$

The components of the equivalent circuit and the I-V characteristics are shown in Figure 16.

The particular feature which should be noted in the single salt regime is that the concentration profile $C(x)$ does not vary with the potential difference across the membrane and thereby with the current sent through the membrane. This diffusion regime therefore behaves electrically as a pure Ohmic regime, with a linear relationship between current and voltage. The short-circuit current I_{sc} is the current through the membrane which produces that potential fall across the membrane that is just compensated by the local electromotive force in the membrane, the diffusion potential. If we put $V=0$ in Eq. (225), we get the following expression for the short-circuit current:

$$I_{sc} = -RT(u^+ - u^-)(C^{(o)} - C^{(i)})/h. \tag{226}$$

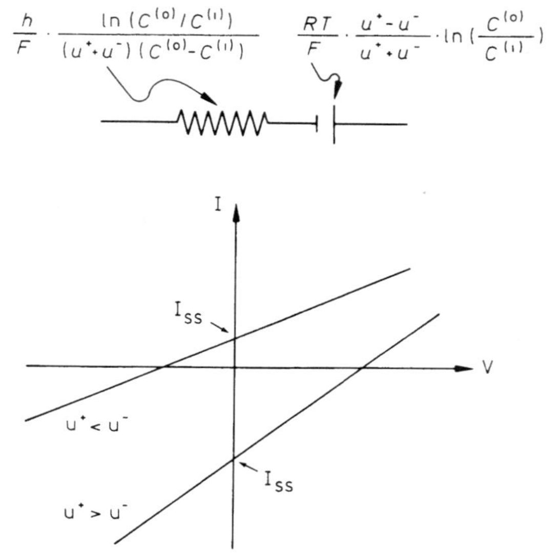

Fig. 16. The components of the equivalent electric circuit of the single-salt diffusion regime and the I-V characteristics corresponding to the situation $u^+ > u^-$ and $u^+ < u^-$

d) Electroneutrality

As is seen from Eq. (224), the potential through the membrane varies non-nearly. It will be instructive to make a quantitative evaluation of how good an approximation the electroneutrality condition is in this diffusion regime. From Poisson's equation (165), we have

$$\varrho(x) = F(C^+(x) - C^-(x)) = -K\varepsilon_o \frac{d^2\psi}{dx^2}$$

From Eq. (217) combined with $C(x) = (C^{(o)} - C^{(i)})\,\dfrac{x}{h} + C^{(i)}$ we have

$$\begin{aligned} \frac{d^2\psi}{dx^2} &= \frac{RT}{F}\,\frac{u^+ - u^-}{u^+ + u^-}\left[\frac{dC/dx}{C}\right]^2 \\ &= \frac{RT}{h^2F}\,\frac{u^+ - u^-}{u^+ + u^-}\left[\frac{C^{(o)} - C^{(i)}}{(C^{(o)} - C^{(i)})(x/h) - C^{(i)}}\right]^2. \end{aligned}$$

If this is inserted in Poisson's equation, we obtain

$$C^+(x) - C^-(x) = \frac{RTK\varepsilon_o}{(hF)^2}\,\frac{u^+ - u^-}{u^+ + u^-}\left[\frac{C^{(o)} - C^{(i)}}{(C^{(o)} - C^{(i)})(x/h) - C^{(i)}}\right]^2 \tag{227}$$

For a HCl solution with $C^{(o)} = 100\ \text{mol}\cdot\text{m}^{-3}$ and $C^{(i)} = 10\ \text{mol}\cdot\text{m}^{-3}$ we get, for $h = 10^{-2}\text{m}$, a maximum value $C^+(h) - C^-(h) = 9.5 \times 10^{-15}$

mol · m^{-3}, which is beyond the limit of any macroscopic measurement. In this case the electroneutrality condition is a good approximation. For $h = 10^{-8}$ m, Eq. (227) gives a deviation from electroneutrality of the same order of magnitude as $C^{(o)}$. In this situation, the diffusion regime must be solved by invoking Poisson's equation (BASS, 1964). The result of this analysis shows that the field is no longer given by Eq. (217), but is a solution of a second-order differential equation of Airy's type. The diffusion potential Eq. (218) appears as the asymptotic solution, which holds for $(h/\lambda)^2 \gg 1$, where λ is the greatest Debye length [Eq. (173)] in the layer. For $(h/\lambda) \to 1$, the diffusion potential is overestimated in PLANCK's formula. But for $C^{(o)} = 100$ mol · m^{-3}, $\lambda = 10$ Å, and for $h = 100$ Å PLANCK's expression is still a reasonably good approximation.

2. Ion-Selective Membrane

Once again we consider a membrane whose permeability to cations and permeability to anions deviate so much from each other that for practical purposes the membrane may be considered *only* permeable to ions of the one charge-type (positive or negative). The equilibrium situation for a single ion was treated in Section F. III.1. If several different permeable ions are present in the surrounding solutions, equilibrium will only exist provided the concentration ratio $C_k^{(i)}/C_k^{(o)}$ for each separate ion corresponds to the same equilibrium potential. If this is not the case, we can no longer talk about an equilibrium system, but only about a stationary system which moves towards an equilibrium state. We will now derive the magnitude of the diffusion potential under these conditions. Consider a membrane of thickness h surrounded by ions that are all monovalent. According to Eq. (204) and Eq. (205), the ionic current is

$$I^+ = -RT\frac{du}{dx} - FU\frac{d\psi}{dx} \tag{204}$$

for the positive ions, and

$$I^- = RT\frac{dW}{dx} - FW\frac{d\psi}{dx} \tag{205}$$

for the negative ions. If the membrane is only permeable to the positive ions, $I^- = 0$ for all values of the potential difference $\psi^{(i)} - \psi^{(o)}$ across the membrane. If the system is left to itself, the component currents carried by the positive ions will adjust themselves so that $I^+ = 0$. This stationary state is therefore characterized by

$$-\frac{d\psi}{dx} = \frac{RT}{F}\frac{1}{U}\frac{dU}{dx} = \frac{RT}{F}\frac{d\ln U}{dx}. \tag{228}$$

This expression is multiplied by dx and integrated from $x = 0$, where $\psi = \psi^{(i)}$ and $U = U^{(i)}$, to $x = h$ where $\psi = \psi^{(o)}$ and $U = U^{(o)}$. We thus obtain

$$\psi^{(i)} - \psi^{(o)} = V_{\text{dif}} = \frac{RT}{F}\ln\left(\frac{U^{(o)}}{U^{(i)}}\right)$$

or, since $U = \Sigma u_k^+ C_k^+ = \Sigma u_k C_k$, as the summation is only over the positive ions,

$$\psi^{(i)} - \psi^{(o)} = \frac{RT}{F}\ln\frac{u_{+1}C_{+1}^{(o)} + u_{+2}C_{+2}^{(o)} + \ldots u_{+n}C_{+n}^{(o)}}{u_{+1}C_{+1}^{(i)} + u_{+2}C_{+2}^{(i)} + \ldots u_{+n}C_{+n}^{(i)}}. \tag{229}$$

Since the ionic mobility u_k is proportional to the diffusion coefficient D_k and to the membrane permeabilities $P_k = D_k/h$, Eq. (229) can also be written as

$$\psi^{(i)} - \psi^{(o)} = \frac{RT}{F}\ln\frac{P_{+1}C_{+1}^{(o)} + P_{+2}C_{+2}^{(o)} + \ldots P_{+n}C_{+n}^{(o)}}{P_{+1}C_{+1}^{(i)} + P_{+2}C_{+2}^{(i)} + \ldots P_{+n}C_{+n}^{(i)}} \tag{230}$$

On the other hand, if the membrane is only permeable to the *negative* ions, the field is determined from Eq. (205) by

$$\frac{d\psi}{dx} = \frac{RT}{F}\frac{1}{W}\frac{dW}{dx}$$

and the expression for the diffusion potential corresponding to Eq. (230) becomes

$$\psi^{(i)} - \psi^{(o)} = \frac{RT}{F}\ln\frac{P_{-1}C_{-1}^{(i)} + P_{-2}C_{-2}^{(i)} + \ldots P_{-n}C_{-n}^{(i)}}{P_{-1}C_{-1}^{(o)} + P_{-2}C_{-2}^{(o)} + \ldots P_{-n}C_{-n}^{(o)}}. \tag{231}$$

For example, let the surrounding solutions be a mixture of NaCl and KCl, and in addition let the membrane be permeable to cations. The membrane potential is then

$$\psi^{(i)} - \psi^{(o)} = \frac{RT}{F}\ln\frac{P_K C_K^{(o)} + P_{Na}C_{Na}^{(o)}}{P_K C_K^{(i)} + P_{Na}C_{Na}^{(i)}}. \tag{232}$$

The same expression will hold if the membrane is also permeable to Cl^-, but only if the concentrations $C_{Cl}^{(o)}$ and $C_{Cl}^{(i)}$ are such that the membrane potential $\psi^{(i)} - \psi^{(o)}$ is equal to the equilibrium potential $V_{Cl}^{(eq)}$ for the Cl ion. Under these conditions $I_{Cl} = 0$, and the potential profile is again determined by Eq. (228).

Equation (232) also presents a convenient way of determining the ratio between the permeabilities of the permeating ions. For the special boundary conditions $C_K^{(o)} = C_{Na}^{(i)} = 0$ and $C_K^{(i)} = C_{Na}^{(o)} = C$, Eq. (232) becomes

$$\psi^{(i)} - \psi^{(o)} = \frac{RT}{F}\ln\left(\frac{P_{Na}}{P_K}\right), \tag{233}$$

i. e., a measurement of the membrane potential under these conditions gives the magnitude of P_{Na}/P_K, which is also designated as the *selectivity ratio* of the membrane for the ions in question.

3. Membrane Separating Electrolytes Having a Common Ion

We shall now consider the slightly more complicated situation in which the membrane is permeable to both anions and cations. The surrounding medium is considered to contain three ions, *e. g.* Na^+, K^+, and Cl^-, the concentrations of which in the inner phase are $C_{Na}^{(i)}$, $C_K^{(i)}$, and $C_{Cl}^{(i)}$ and in the outer phase $C_{Na}^{(o)}$, $C_K^{(o)}$ and $C_{Cl}^{(o)}$. Once the system has reached its steady state, a potential difference $\psi^{(i)} - \psi^{(o)} = V$ will have been established, which is determined by the surrounding concentrations, and the three ions will migrate through the membrane, with zero total membrane current, if no external potential difference different from V is applied. This diffusion regime is solved once we have determined the membrane potential, the potential profile $\psi(x)$ and the concentration profiles $C_k(x)$. From these, the individual ion fluxes J_x can then be obtained.

a) Flux Ratio (Ussing, 1949)

The difficulties presented by even this simple system can be illustrated most easily by considering the individual ion flux expressed as by Kramers (1940):

$$J_k \exp\{z_k \varphi(x)\} = -D_k \frac{d}{dx}\left\{C_k \exp\{z_k \varphi(x)\}\right\}, \tag{134}$$

where $\varphi(x)$ is the normalized potential and C_k the concentration at position x. Let the thickness of the membrane be h. The above equation is integrated from $x = 0$, where $\varphi(x) = \varphi^{(i)}$ and $C_k(x) = C_k(0)$, to $x = h$, where $\varphi(x) = \varphi^{(o)}$ and $C_k(x) = C_k(h)$. This gives

$$J_k = D_k \cdot \frac{C_k(0)\exp\{z_k\varphi^{(i)}\} - C_k(h)\exp\{z_k\varphi^{(o)}\}}{\int_0^h \exp\{z_k\varphi(x)\,dx\}},$$

since the process is stationary ($\partial J_k/\partial x = 0$), or

$$J_k = D_k \frac{C_k(0)\exp\{z_k v\} - C_k(h)}{\int_0^h \exp\{z_k\varphi(x)\}\,dx}, \tag{234}$$

where $\varphi^{(i)} - \varphi^{(o)} = v$ and $\varphi^{(o)}$ is arbitrarily put equal to zero. Thus v is the normalized membrane potential $v = V/(RT/F)$. An expression essentially similar to that of Eq. (234) was already derived in 1897 by Behn (1897). It appears from the above equation that the flux J_k can be calculated only if the dependence of the potential profile on the position x in the membrane is known. In the general case, it is necessary to go back to PLANCK's general scheme to determine the potential profile $\varphi(x)$. Under certain conditions, the potential profile is linear, so that the denominator in Eq. (234) can easily be calculated. On the other hand, there is always a simple expression for the ratio between the unidirectional fluxes, which holds independent of the form of the potential profile. An isotope of the ion k is added to the inner phase (i) in such a small amount that it does not interfere with the diffusion regime. The concentration of the isotope is kept at zero in the outer phase. Then according to Eq. (234) the unidirectional flux for the isotope in the direction (i) → (o) is

$$*J^{(io)} = D_k^* C_k(0) \exp\{z_k v\}/A_k ,$$

where $*C_k^{(o)}$ is the concentration of the isotope at $x = 0$ and A_k is the integral in the denominator of Eq. (234). If the same isotope is then added to the outer phase, and the concentration of the isotope kept at zero in the inner phase, the unidirectional flux of the isotope in the direction (oi) is

$$*J^{(oi)} = D_k^* C_k(h)/A_k ,$$

bearing in mind that unidirectional fluxes are reckoned as positive regardless of their direction. Since the amounts of the isotope added are so small that the potential profile is not affected, the denominator A_k is the same in both cases. The ratio $*J^{(oi)}/*J^{(oi)}$ is now

$$\frac{*J_k^{(io)}}{*J_k^{(oi)}} = \frac{*C_k(0)}{*C_k(h)} \exp\{z_k v\} = \frac{*C_k^{(o)}}{*C_k^{(i)}} \exp\{z_k v\} = \frac{*C_k^{(i)}}{*C_k^{(o)}} \exp\{z_k V/RT\}, \tag{235}$$

where $V = \psi^{(i)} - \psi^{(o)}$. Assuming $*C_k(0) = \alpha_k^* C_k^{(i)}$ and $*C_k(h) = \alpha_k^* C_k^{(o)}$, where α_k is the distribution coefficient of the ion, Eq. (235) is then the celebrated flux ratio relation which was first derived by USSING (1949), and which has since been such a useful tool in deciding whether an ion migrates through biological membranes by simple electrodiffusion. A more general derivation of the flux ratio relation which is valid for non-stationary, time dependent fluxes through a stratified medium has recently been worked out by STEN-KNUDSEN and USSING (in preparation).

b) The Goldman Regime

Equation (234) can readily be integrated if the potential profile $\psi(x)$ is linear through the membrane. This requires that the field $E = - d\psi/dx$ is constant through the membrane. As Eq. (203) shows, this is always the case if the total concentrations $\Sigma C_k^{(i)}$ and $\Sigma C_k^{(o)}$ have the same value on both sides of the membrane. GOLDMAN (1943) assumed *a priori* that the field through the membrane was constant, and then integrated the flux equations. Whether this assumption is justified depends upon the ratio of a Debye length of the same form as that of Eq. (173) and the membrane thickness (MacGILLIVRAY and HARE, 1969) The results that can be derived from this approximation have since been applied very often in the description of the passive ion transport across biological membranes.

(i) *The Separate Ionic Currents and the Diffusion Potential.* GOLDMAN (1943) started with the following situation, which is shown in Figure 17:

(α) The thickness of the membrane is h. (β) The concentrations in the membrane at the boundaries $x = 0$ and $x = h$ are $C_k(0)$ and $C_k(h)$. These are related to the outer concentrations $C_k^{(i)}$ and $C_k^{(o)}$ by

$$C_k(0) = \alpha_k C_k^{(i)},\ C(h) = \alpha_k C_k^{(o)}, \tag{236}$$

where α_k is the distribution coefficient for the ion. (γ) The field in the membrane is assumed to be constant, *i. e.*

$$\frac{d\varphi}{dx} = \frac{\varphi(h) - \varphi(0)}{h} = -v/h \tag{237}$$

where $v = \varphi(0) - \varphi(h) = \varphi^{(i)} - \varphi^{(o)}$, and furthermore, $\varphi^{(o)} = 0$, by convention. (δ) In what follows it is assumed that the membrane current is carried by the ions K^+, Na^+ and Cl^- only.

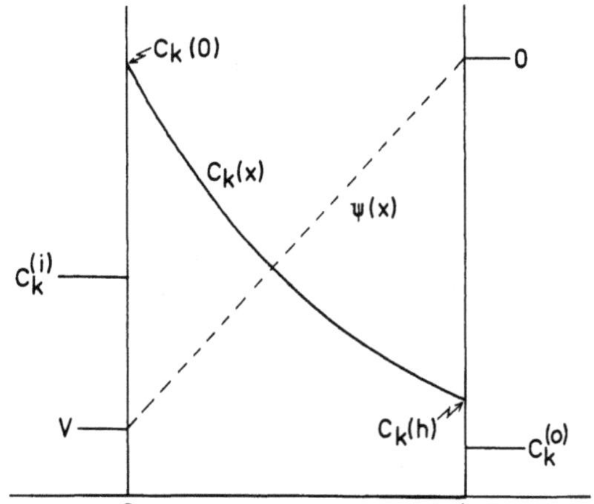

Fig. 17. To illustrate the Goldman regime

As a result of the condition (γ) and $\varphi = v$ for $x = 0$ and $\varphi = 0$ for $x = h$, the integral in the denominator of *Kramers' equation* [Eq. (234)] becomes

$$\int_0^h \exp\{z_k\varphi(x)\}\,dx = \int_v^0 \exp\{z_k\varphi(x)\}\left(\frac{dx}{d\varphi}\right)d\varphi$$

$$= -\frac{h}{v}\int_v^0 \exp\{z_k\varphi(x)\}\,d\varphi = -\frac{h}{z_k v}(1-\exp\{z_k v)\}$$

Inserting this into Eq. (234) the expression for the flux of ion k becomes

$$J_k = z_k v \left(\frac{D_k}{h}\right) \frac{C_k(h) - C_k(0)\exp\{z_k v\}}{1 - \exp\{z_k v\}}.$$

If we then introduce the outer concentrations $C_k^{(i)}$ and $C_k^{(o)}$ from Eq. (236), we obtain

$$J_k = z_k P_k \frac{v}{1 - \exp\{z_k v\}} \left\{C_k^{(o)} - C_k^{(i)} \exp\{z_k v\}\right\}, \tag{238}$$

where the permeability P_k is defined as

$$P_k = \alpha_k D_k / \mathrm{h} \tag{239}$$

in agreement with the original definition of Eq. (28).

The membrane potential $v_k^{(eq)}$ corresponding to $J_k = 0$ is determined by

$$C_k^{(o)} - C_k^{(i)} \exp\{z_k v_k^{(eq)}\} = 0$$

or

$$v_k^{(eq)} = \frac{1}{z_k} \ln\left(\frac{C_k^{(o)}}{C_k^{(i)}}\right), \tag{240}$$

which is Nernst's equation, [Eq. (142)], for the equilibrium potential. If the membrane is short-circuited ($v = 0$), we obtain, since $z_k v/(1 - \exp\{z_k v\}) \to -1$ for $v \to 0$,

$$J_k = P_k (C_k^{(i)} - C_k^{(o)}), \tag{241}$$

corresponding to the stationary flux Eq. (29) for an uncharged particle.

At the membrane potential v, the fluxes for the three ions K^+, Na^+ and Cl^- are

$$\left.\begin{aligned} J_K &= P_K v \frac{C_K^{(o)} - C_K^{(i)} e^v}{1 - e^v} \\ J_{Na} &= P_{Na} v \frac{C_{Na}^{(o)} - C_{Na}^{(i)} e^v}{1 - e^v} \\ J_{Cl} &= -P_{Cl} v \frac{C_{Cl}^{(i)} - C_{Cl}^{(o)} e^v}{1 - e^v} \end{aligned}\right\}. \tag{242}$$

To these correspond the ionic currents

$$\left.\begin{aligned} I_K &= F J_K \\ I_{Na} &= F J_{Na} \\ I_{Cl} &= -F J_{Cl} \end{aligned}\right\}. \tag{243}$$

The total current is $I = I_K + I_{Na} + I_{Cl}$ · Combination of Eq. (243) with Eq. (242) gives

$$I = \left\{\frac{P_K C_K^{(o)} + P_{Na} C_{Na}^{(o)} + P_{Cl} C_{Cl}^{(i)} - [P_K C_K^{(i)} + P_{Na} C_{Na}^{(i)} + P_{Cl} C_{Cl}^{(o)}] e^{v}}{1 - e^{v}}\right\} v. \tag{244}$$

The total current thus varies in a complicated manner with the membrane potential across the membrane. For large positive values of v we obtain

$$I_{v \gg o} = F(P_K C_K^{(i)} + P_{Na} C_{Na}^{(i)} + P_{Cl} C_{Cl}^{(o)})\, v,$$

which corresponds to a linear dependence between current in the direction (i) → (o) and v. For $v \ll 0$, a corresponding linear dependence is obtained:

$$I_{v \ll o} = F(P_K C_K^{(o)} + P_{Na} C_{Na}^{(o)} + P_{Cl} C_{Cl}^{(i)})\, v,$$

where the current is now in the direction (o) → (i). The membrane thus shows a rectifying effect, and the rectification ratio is

$$J_{v \gg o}/J_{v \ll o} = \frac{P_K C_K^{(i)} + P_{Na} C_{Na}^{(i)} + P_{Cl} C_{Cl}^{(o)}}{P_K C_K^{(o)} + P_{Na} C_{Na}^{(i)} + P_{Cl} C_{Cl}^{(i)}} \tag{245}$$

The diffusion potential in the Goldman regime is the value v_m at which the system is current-free, *i. e.* corresponding to $I = 0$ in Eq. (244). This does not correspond to $v = 0$, since $v/(1 - e^v) \to -1$ for $v \to 0$. The numerator in the bracketed term of Eq. (244) must therefore vanish for $v = v_m$. As a result,

$$e^{v_m} = \frac{P_K C_K^{(o)} + P_{Na} C_{Na}^{(o)} + P_{Cl} C_{Cl}^{(i)}}{P_K C_K^{(i)} + P_{Na} C_{Na}^{(i)} + P_{Cl} C_{Cl}^{(o)}}$$

or since $\psi^{(i)} - \psi^{(o)} = V_m = v_m RT/F$,

$$V_m = \frac{RT}{F} \ln \frac{P_K C_K^{(o)} + P_{Na} C_{Na}^{(o)} + P_{Cl} C_{Cl}^{(i)}}{P_K C_K^{(i)} + P_{Na} C_{Na}^{(i)} + P_{Cl} C_{Cl}^{(o)}} \tag{246}$$

which is the celebrated expression derived by HODGKIN and KATZ (1949) on the assumption of a constant field in the membrane. Their permeabilities were defined by $P_k = \alpha_k RT u_k / Fh$, but as $u_k = q_e B_k = q_e D_k / kT = F D_k / RT$, we obtain $P_k = \alpha_k D_k / h$, which is identical with the ionic permeability P_k defined here.

It might be mentioned that PLANCK (1890b) had already derived the expression for the diffusion potential under a constant field in the membrane. Planck's condition for a constant field is that the total concentrations in both phases are identical. He inserted this condition in his general expression Eq. (213). A few elementary calculations will thus lead to a result identical with Eq. (246).

(ii) *Total Membrane Current and Membrane Potential.* In Eq. (244) the relation between the total current and the potential difference across the membrane is expressed with the help of the separate ionic permeabilities P_k. In this way the expression for the stationary current-free diffusion potential, Eq. (246), was obtained in the form usually employed. The I–V relation will now be derived with the help of the separate ionic mobilities u_k. We have $P_k = \alpha_k D_k/h$. Using Einstein's relation, we obtain

$$P_k = \alpha_k kTB_k/h = \alpha_k kTq_e B_k/q_e h = \alpha_k RTu_k/Fh,$$

since $q_e B_k = u_k$. The normalized membrane potential is $v = VF/RT$. We therefore have $v \cdot P_k = \alpha_k \; Vu_k/h$, which when inserted into Eq. (244) gives

$$I = F\left(\frac{V}{h}\right)\left\{\frac{u_K\alpha_K C_K^{(o)} + u_{Na}\alpha_{Na}C_{Na}^{(o)} + u_{Cl}\alpha_{Cl}C_{Cl}^{(i)}}{1 - e^{VF/RT}}\right.$$

$$\left. - \frac{[u_K\alpha_K C_K^{(i)} + u_{Na}\alpha_{Na}C_{Na}^{(i)} + u_{Cl}\alpha_{Cl}C_{Cl}^{(o)}]e^{VF/RT}}{1 - e^{VF/RT}}\right\}$$

In the case of Planck's regime corresponding to a constant field in the membrane, $C_{Cl}^{(i)} = C_{Cl}^{(o)}$. Now, according to Eq. (120),

$$\varkappa^{(o)} = F(u_K\alpha_K C_K^{(o)} + u_{Na}\alpha_{Na}C_{Na}^{(o)} + u_{Cl}\alpha_{Cl}C_{Cl}^{(o)}),$$

which is the total conductivity of the membrane if it were surrounded on both sides with the solution of phase (o). Correspondingly,

$$\varkappa^{(i)} = F(u_K\alpha_K C_K^{(i)} + u_{Na}\alpha_{Na}C_{Na}^{(i)} + u_{Cl}\alpha_{Cl}C_{Cl}^{(i)})$$

is the total membrane conductivity corresponding to the ions of the solution in phase (i).

We therefore have

$$I = \frac{\varkappa^{(o)} - \varkappa^{(i)}e^{VF/RT}}{1 - e^{VF/RT}}\left(\frac{V}{h}\right) \tag{247}$$

For large positive values of $\psi^{(i)} - \psi^{(o)} = V$,

$$I = \varkappa^{(i)}\left(\frac{V}{h}\right) = \varkappa^{(i)}E. \tag{248}$$

Since V/h equals the field E [see Eq. (237)] the conductivity in the membrane is thus completely determined by the ions in the inner phase. The field, which has the direction (i) → (o), will drive K^+ and Na^+ from the inner phase through the membrane. If the field is sufficiently strong, it will fill the membrane completely with ions originating from the inner phase, and the concentration profiles of these, $C_K(x) = C_K^{(i)}$ and $C_{Na}(x) = C_{Na}^{(i)}$, will not change through the membrane. If, on the other hand, V assumes large negative values, we obtain

$$I = \varkappa^{(o)}\left(\frac{V}{h}\right) = \varkappa^{(o)}E, \tag{249}$$

where the conductivity of the membrane is now determined by the concentrations of K^+ and Na^+ in the outer phase. The rectification ratio can therefore also be written as

$$I_{v\gg 0}/I_{v\ll 0} = \varkappa^{(i)}/\varkappa^{(o)}. \tag{250}$$

Two examples of I–V relations calculated from Eq. (247) are shown in Fig. 18.

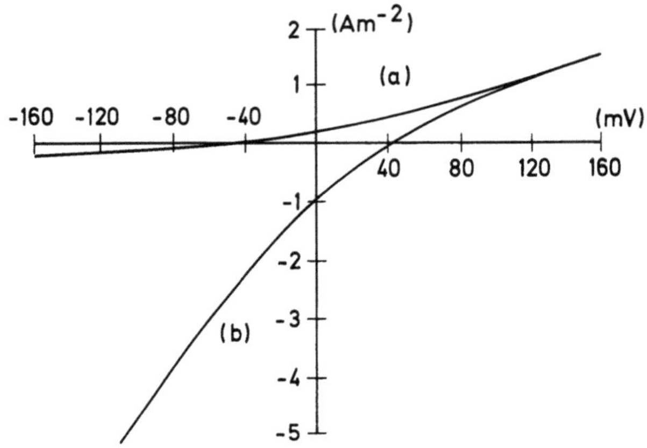

Fig. 18. Two examples of total current-membrane potential relations calculated from Eq. (247). $C_{Cl}{}^{(o)} = C_{Cl}{}^{(i)} = 165$ mol/m^{-3}, $C_{Na}{}^{(i)} = 15$, $C_{Na}{}^{(o)} = 160$. $C_K{}^{(o)} = 5$, $C_K{}^{(i)} = 150$. $P_K = 1.5 \times 10^{-8}$ m/s^{-1}. Curve (a): $P_{Na} = 0.01\ P_K$. $P_{Cl} = 0.1\ P_K$. Curve (b): $P_{Na} = 5\ P_K$. $P_{Cl} = 0.1\ P_K$

(iii) *Concentration Profiles and Membrane Potential.* These are determined in principle in the same way as in section E.II.1.c. We start from the Nernst-Planck equation in Kramers' form [Eq. (134)]

$$-\frac{J_k}{D_k}\exp\{z_k\varphi(x)\}\,dx = -\frac{J_k}{D_k}\exp\{z_k\varphi(x)\}\left(\frac{dx}{d\varphi}\right)d\varphi = d\,(C_k(x)\exp\{z_k\varphi(x)\}) \tag{134}$$

The form of the potential profile is $\varphi(x) = v(1-x/h)$, so that $dx/d\varphi = -h/v$. We then have

$$\frac{hJ_k}{D_k v}\exp\{z_k\varphi(x)\}\,d\varphi = d\,(C_k(x)\exp\{z_k\varphi(x)\})$$

which when integrated from $x=0$, where $\varphi = v$ and $C_k(x) = \alpha_k C_k^{(i)}$ to $x = h$ where $\varphi = 0$ and $C(x) = C_k(h) = \alpha_k C_k{}^{(o)}$, gives

$$\frac{hJ_k}{z_k D_k v}(1-\exp\{z_k v\}) = (C_k^{(o)} - C_k^{(i)}\exp\{z_k v\})\alpha_k \tag{251}$$

If instead the flux equation is integrated from position x with potential $\varphi(x)$ and concentration $C_k(x)$, we obtain

$$\frac{hJ_k}{z_k D_k v}(1 - \exp\{z_k\varphi(x)\}) = \alpha_k\, C^{(o)} - C_k(x)\exp\{\, z_k\varphi(x)\}) \tag{252}$$

Dividing Eq. (252) by Eq. (251) gives

$$\frac{1-\exp\{\, z_k\varphi(x)\}}{1-\exp\{\, z_k v\}} = \frac{\alpha_k C_k^{(o)} - C_k(x)\exp\{\, z_k\varphi(x)\}}{(C_k^{(o)} - C_k^{(i)}\exp\{\, z_k v\}\,)\alpha_k}$$

Replacing $\varphi(x)$ by $v(1-x/h) = FV(1-x/h)/RT$ and solving with respect to $C_k(x)$ results in

$$C_k(x) = \alpha_k \frac{C_k^{(i)}\exp\{\, z_k FV/RT\} - C_k^{(o)} - [C_k^{(i)} - C_k^{(o)}]\exp\{\, z_k FVx/RTh\}}{\exp\{\, z_k FV/RT\} - 1} \tag{253}$$

This result could also be obtained directly from Eq. (109) by substituting $B_k q_e V/h$ for the stationary convection velocity. Figure 19 shows the stationary concentration profiles through the membrane for two different values of the membrane potential V corresponding to the concentrations: $C_K^{(i)} = 150$, $C_K^{(o)} = 5$; $C_{Na}^{(i)} = 15$, $C_{Na}^{(o)} = 160$; $C_{Cl}^{(i)} = C_{Cl}^{(o)} = 165$. Note the similarity of the concentration profiles to those for the diffusion process with superimposed

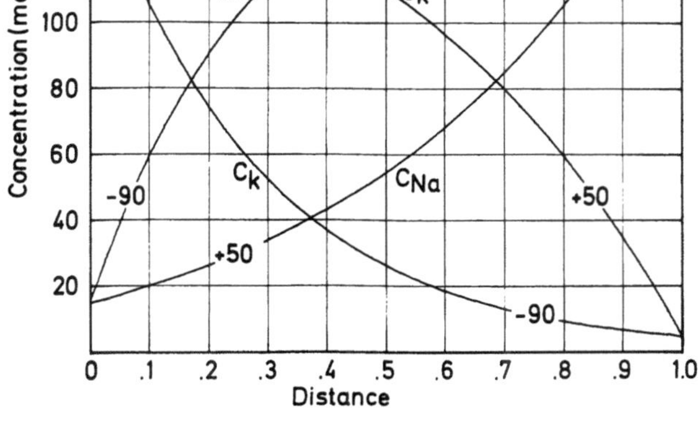

Fig. 19. Concentration profiles across the membrane of thickness h for two different values (-90 and $+50$ mV) of membrane potential

convective flow [cf. Eq. (109)]. The time course of the transition from one stationary concentration profile to the other following a sudden change in the membrane potential has been obtained by COHEN and COOLEY (1965) by solving the time-dependent Nernst-Planck equations, *i. e.* the continuity equations $\partial C_k/\partial t = -\partial J_k/\partial x$, together with the Poisson equation $\partial^2\psi/\partial x^2 = -\varrho/K\varepsilon_o$.

(iv) *Ionic Conductances and Membrane Potential.* Once the concentration profile $C_k(x)$ has been determined the separate ionic conductances can be calculated from Eq. (124):

$$1/G_k = R_k = \int_o^h \frac{dx}{z_k F u_k C_k(x)}, \tag{124}$$

but since the fluxes are already known, another way of calculating G_k is to express the ionic current in the Goldman regime as

$$I_k = G_k\,(V - V_k^{(eq)}) \tag{254}$$

[cf. Eq. (144)]. We shall begin by considering the situation when an ion of type k is very near or at equilibrium. The simplest equilibrium situation corresponds to no potential difference across the membrane and a horizontal concentration profile across the membrane, *i. e.* $C_k(x) = C_k(o) = C_k(h)$. We have from Eq. (124), since $C_k(x)$ is constant,

$$G_k = z_k F u_k C_k(x)/h.$$

Substituting $u_k = z_k q_e B_k = z_k q_e D_k/kT = z_k F D_k/RT$ and $C_k(x) = \alpha_k C_k^{(i)} = \alpha_k C_k^{(o)}$ gives

$$G_k = \frac{(z_k F)^2}{RT} P_k C_k^{(i)} = \frac{(z_k F)^2}{RT} P_k C_k^{(o)}, \tag{255}$$

when $P_k = \alpha_k D_k/h$. Using the definition for the unidirectional fluxes [Eqs. (53) and (54)] the above expression takes the alternative, although rather artificial, form

$$G_k = \frac{(zF)^2}{RT} J^{(io)} = \frac{(zF)^2}{RT} J^{(oi)}, \tag{256}$$

which was first derived by HODGKIN (1951). See also USSING and ZERAHN (1951).

If $C_k^{(o)} \neq C_k^{(i)}$ and $\psi^{(i)} \neq \psi^{(o)}$ the expression for the ionic current in the Goldman regime is

$$I_k = z_k^2 F P_k v \frac{C_k^{(o)} - C_k^{(i)} \exp\{z_k v\}}{1 - \exp\{z_k v\}}, \tag{257}$$

which bears little resemblance to the expression in Eq. (254). However, since it is assumed that the ion in question is near equilibrium, Eq. (257) can be transformed by linearization. The concentrations $C_k^{(o)}$ and $C_k^{(i)}$ are associated with the equilibrium potential $v_k^{(eq)}$ for the ion k by Eq. (241), or

$$C_k^{(i)} = C_k^{(o)} \exp\{z_k v_k^{(eq)}\},$$

which when inserted into Eq. (257) gives

$$I_k = z_k^2 F P_k v C_k^{(o)} \frac{1-\exp\{z_k(v-v_k^{(eq)})\}}{1-\exp\{z_k v\}}.$$

Since the system is near equilibrium, $z_k(v-v_k^{(eq)}) \ll 1$. We then have approximately

$$1-\exp\{z_k(v-v_k^{(eq)})\} = -z_k(v-v_k^{(eq)}).$$

Substituting in the above expression gives

$$I_k = -z_k^3 F P_k C_k^{(o)} \frac{v}{1-\exp\{z_k v\}}(v-v_k^{(eq)}).$$

Now $v = FV/RT$ and $v_k^{(eq)} = FV_k^{(eq)}/RT$, which when inserted in the above expression give

$$I_k = -\frac{z_k^3 F_k^3 P_k C_k^{(o)} V}{(RT)^2(1-\exp\{z_k FV/RT\})}(V-V_k^{(eq)}),$$

an equation that is identical with Eq. (254), for

$$G_k = -\frac{z_k^3 F^3 P_k C_k^{(o)} V}{(RT)^2(1-\exp\{z_k FV/RT\})}. \tag{258}$$

This is the expression for the conductance of the individual ion, provided the membrane potential V does not deviate more than a few mV from the equilibrium potential $V_k^{(eq)}$. If the ion is in fact in equilibrium, $V = V_k^{(eq)}$, *i.e.*

$$V = \frac{RT}{z_k F}\ln\left(\frac{C_k^{(o)}}{C_k^{(i)}}\right) \text{ and } \exp\{z_k VF/RT\} = C_k^{(o)}/C_k^{(i)},$$

which when inserted into Eq. (258) gives the alternative expression

$$G = -\frac{z_k^2 F^2 P_k C_k^{(i)} C_k^{(o)}}{RT(C_k^{(i)}-C_k^{(o)})}\ln\frac{C_k^{(o)}}{C_k^{(i)}}. \tag{259}$$

Since the ion is in equilibrium, the concentration profile is given by Eq. (166)

$$C_k(x) = C_k^{(o)} \exp\{-z_k \varphi(x)\}, \tag{166}$$

with $\varphi = 0$ for $x=h$ and $\varphi = v$ for $x=0$. Furthermore, the potential profile $\varphi(x) = v-(v/h)x$, in which case the expression for the concentration becomes

$$C_k(x) = C_k^{(o)} \exp\{-z_k v\} \cdot \exp\{z_k(v/h)x\} = C_k^{(i)} \exp\{z_k(v/h)x\}.$$

Inserting this expression into Eq. (124) leads once again, after integration, to the expression of Eq. (258) for the ionic conductance. The method used in this section to derive Eq. (258), however, has the advantage that it does provide some insight into statements such as "this equation is valid when the ion is near equilibrium" or "for a very small current we have the following relation". Note that Eq. (255) can be obtained from Eq. (258) by putting $V_k^{(eq)} = 0$, or from Eq. (259) by putting $C_k^{(i)} = C_k^{(o)}$.

We now consider the situation in which $V \neq V_k^{(eq)}$. We start with the expression for the ionic current, Eq. (257),

$$I_k = \frac{(z_k F)^2 P_k V(C_k^{(o)} - C_k^{(i)} \exp\{z_k FV/RT\})}{RT(1-\exp\{z_k FV/RT\})},$$

where the normalized potential $v = FV/RT$ is replaced by the membrane potential V. This equation can be put into the form $I_k = G_k(V-V_k^{(eq)})$ by multiplication by $(V-V_k^{(eq)})/V-V_k^{(eq)})$. This gives

$$I_k = \frac{(z_k F)^2 P_k V(C_k^{(o)} - C_k^{(i)} \exp\{z_k FV/RT\})}{(RT)(1-\exp\{z_k FV/RT\})(V-V_k^{(eq)})}(V-V_k^{(eq)})$$

from which it follows that

$$G_k = \frac{(z_k F)^2 P_k V(C_k^{(o)} - C_k^{(i)} \exp\{z_k FV/RT\})}{RT(1-\exp\{z_k FV/RT\})(V-V_k^{(eq)})}, \tag{260}$$

represents the expression for the individual ionic conductance *at the membrane potential V*. If $C_k^{(i)} = C_k^{(o)} \exp\{-z_k FV_k^{(eq)}/RT\}$ is used to calculate the equilibrium potential $V_k^{(eq)}$ for the ion of type k, alternative expressions for the conductance can be obtained, *e.g.*

$$G_k = \frac{(z_k F)^2 P_k C_k^{(o)} V(1-\exp\{z_k F(V-V^{(eq)})/RT\})}{RT(1-\exp\{z_k FV/RT\})(V-V_k^{(eq)})}, \tag{261}$$

or

$$G_k = \frac{(z_k F)^2 P_k C_k^{(o)} V(1-\exp\{z_k F(V-V_k^{(eq)})/RT\})}{(RT)^2(1-\exp\{z_k FV/RT\})(z_k FV/RT+\ln(C_k^{(i)}/C_k^{(o)})}. \tag{262}$$

If, in Eq. (261), $V \rightarrow V_k^{(eq)}$, the expression for the conductance again assumes the form [Eq. (258)] already derived for the ion conductance near equilibrium. Note also that Eq. (255) is the limiting case of Eq. (262) corresponding to $V \rightarrow 0$ and $C_k^{(i)} \rightarrow C_k^{(o)}$.

c) Equivalent Electrical Circuits

The total current through the membrane can now in principle be expressed in two ways:

(i) Equation (247) gives the direct relation between the total current and the membrane potential. This relation can also be expressed as

$$V = IR_{\mathrm{tot}} + V_{\mathrm{dif}}, \tag{263}$$

where R_{tot} and V_{dif} are calculated from Eq. (209) and Eq. (211) corresponding to the diffusion regime of Planck. The equivalent circuit corresponds to that shown in Figure 14. However, since the concentration profiles depend on V, this will in general also hold for the magnitude of V_{dif}.

(ii) The individual ionic currents are expressed in the form

$$I_k = G_k(V - V_k^{(\mathrm{eq})}), \tag{264}$$

where the factors determining the magnitude of the separate ion conductances are given by Eq. (262). Summating Eq. (264) over all currents gives

$$V = \frac{1}{\Sigma G_k} I + \frac{\Sigma G_k V_k^{(\mathrm{eq})}}{\Sigma G_k}, \tag{265}$$

whose equivalent circuit is shown in Figure 20, below.

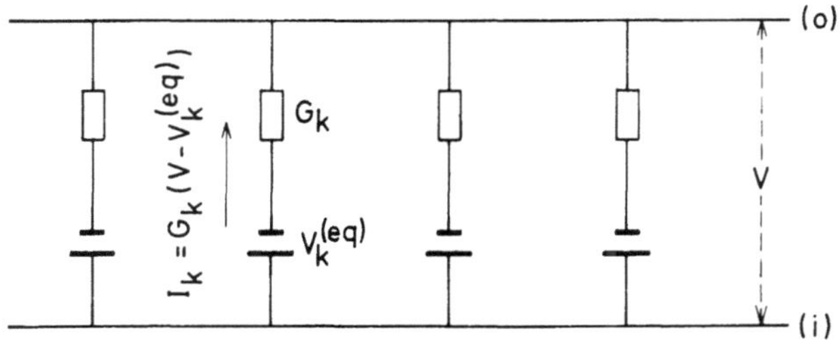

Fig. 20. The equivalent diagram for a mosaic membrane

Both representations indicate the correct relation between I and V. However, in the latter representation, the total conductance ΣG_k is not in general equal to the membrane conductance calculated on the basis of the homogeneous Planck regime. The question is, therefore, which of the two representations has physical significance, and which is a formal representation. The answer depends on which diffusion regime actually operates in the membrane (FINKELSTEIN and MAURO, 1963). If the membrane is homogeneous, and the whole system can be considered as a pure Planck regime, then it is the first representation that gives the correct membrane conductance, while the individual ion conductances are formal magnitudes, giving the correct values for the fluxes. If we are dealing with a mosaic membrane, in which the individual ion types pass through *separate* pores, it is the second representation that is the physically correct one, and in addition the separate ion conductances are real physical magnitudes, whose expression is given by Eqs. (255), (258), (259), and (260)–(262), provided the field in the membrane can be assumed to be constant.

Acknowledgements

The present chapter owes much of its existence to the friendship and stimulating thoughts which professor *Alexander Mauro* has given the author over many years.
I wish to acknowledge my indebtedness to Mrs. *Jette Jensen* and Mrs. *Julia Ann Halkier* for devoting their skill and patience to the preparation of a difficult manuscript. I am also very grateful to Dr. *Leon Pape* for critical examination of the text and helpful suggestions. The first version of this chapter was written in october 1975 at the cloister *San Cataldo* near Amalfi. The grant which I received to this end from the committee of The Institution San Cataldo is gratefully acknowledged. This work was also supported by The Danish Natural Science Research Council.

List of Symbols

The more important symbols used in the text are listed below. Where it is useful, the number of the equation (in *italics*) or section in which the symbol is introduced is given in brackets.

A	area
a_k	activity of chemical component k
B	mechanical mobility (*4*)
b	molar mechanical mobility (B.III.3.b)
C	concentration
$\bar{C}_F$	complex Fourier transform (C.III.2)
$\bar{C}_s$	finite sine Fourier transform (*63*)
D	diffusion coefficient (B.III.3.a)
E	electrical field (F.I)
F	*Faraday's number*
f	coefficient of friction (B.III.1)
G	conductance (*122*)
h	thickness of membrane
I	current density
J	flux (B.I)
$J^{(io)}$	undirectional flux in the direction from phase (i) to phase (o) (*53*)
K	relative dielectric constant (*136*)
k	Boltzmann's constant

k	rate constant of exchange process (*39*)
k	subscript indicating type of chemical component
k	transform parameter of finite Fourier transform (*63*)
M	molecular weight
m	mass
m	amount of substance
N	Number of particles per unit volume, sometimes also used to represent surface density, *e.g.* mol per unit area
N_A	Avogadro's number
n	integer 0,1,2,3...
P	permeability coefficient (*28*)
P	pressure
q	charge
q_e	elementary positive charge
R	gas constant
r	radius
T	absolute temperature
T	dimensionless time variable (C.III.3.a)
t	time
$U(x)$	potential energy at position x
U	generalized cation concentration introduced by Planck (*183*)
u_k	electrical mobility of the ion of type k (*112*)
V	electrical potential difference across the membrane
$V_k^{(eq)}$	equilibrium potential of the ion of type k (*142*)
V_D	Donnan potential (*149*)
v	dimensionless potential normalized in units of kT/q_e (F.III.2.a)
v	migration or convection velocity
v	volume of a phase
$\bar{v}$	partial molar volume
W	generalized anion concentration introduced by Planck (*188*)
X	force
X	dimensionless distance variable (C.III.3.a)
x	position coordinate along the x-axis
z	charge number of an ion (F.I)
α	distribution coefficient (C.II.1)
α	parameter $= vh/D$ (E.II.1.c)
Δ	change in a quantity
δ	Dirac's delta function
ε_o	permittivity in vacuum
ζ	transform parameter of complex Fourier transform (C.III.2)
η	viscosity
$\varkappa$	conductivity (*120*)
λ	molar conductivity (*117*)
λ	Debye length (*173*)
μ	molar chemical potential
$\bar{\mu}$	molar electrochemical potential
ξ	running position coordinate when position x is fixed

ξ	parameter introduced by Planck (*212*)
π	osmotic pressure difference
ϱ	mass density
ϱ	charge density
ϱ	resistivity $= 1/\varkappa$ (*120*)
τ	time constant (*49*)
τ	time coordinate, *e.g.* used as running variable when t is fixed
τ_{red}	redistribution time (*73*)
τ_q	charging time (*194*)
$\varphi(x, t)$	probability density (*75*)
φ	normalized electrical potential in units of kT/q_e (*132*)
ψ	electrical potential
∇^2	Laplacian operator

Subscripts

k	chemical component
+k	cation of type k
−k	anion of type k
o	initial state

Superscripts

(i)	inner phase
(o)	outer phase
(eq)	equilibrium state
+	refer to cation
−	refer to anion

References

Bartlett, J. H., Kromhout, R. A.: Bull. Math. Biophys. **14**, 385 (1952).
Bass, L.: Trans. Faraday Soc. **60**, 1914 (1964).
Behn, U.: Ann. Phys. Chem. **62**, 54 (1897).
Boyle, P., Conway, E. J.: J. Physiol. **100**, 1 (1941).
Brown, R.: Phil. Mag. **4**, 161 (1828).
Carslaw, H. S., Jaeger, J. C.: Conduction of Heat in Solids. London: Oxford University Press 1959.
Chandrasekhar, S.: Rev. mod. Phys. **15**, 1 (1943).
Cohen, H., Cooley, J. W.: Biophys. J. **5**, 145 (1965).
Conway, E. J.: Physiol. Rev. **37**, 84 (1957).
Crank, J.: The Mathematics of Diffusion. London: Oxford University Press 1956.
Dainty, J., House, C. R.: J. Physiol. **185**, 172 (1966).
Donnan, F. G.: Z. Elektrochem. **17**, 572 (1911).
Dünwald, H., Wagner, C.: Z. physik. Chem. **B24**, 53 (1934).
Einstein, A.: Ann. Physik. **17**, 549 (1905).
Einstein, A.: Z. Elektrochem. **14**, 235 (1908).
Einstein, A.: Investigations on the Theory of the Brownian Movements. London: Methuen 1926 (new ed., New York: Dover 1956).
Fick, A.: Ann. Physik. **94**, 59; Phil. Mag. **10**, 30 (1855).

FINKELSTEIN, A., MAURO, A.: Biophys. J. **3**, 215 (1963).
FOURIER, J.: Théorie Analytique de la Chaleur. Translated with notes by A. Freeman (1878). New ed., New York: Dover 1955.
GOLDMAN, D. E.: J. gen. Physiol. **27**, 37 (1943).
HENDERSON, P.: Z. physik. Chem. **59**, 118 (1907).
HERTZ, G.: Z. Phys. **13**, 35 (1923).
HODGKIN, A. L.: Biol. Rev. **26**, 339 (1951).
HODGKIN, A. L., HOROWICZ, P.: J. Physiol. **148**, 127 (1959).
HODGKIN, A. L., KATZ, B.: J. Physiol. **108**, 37 (1949).
JACOBS, M. H.: Ergeb. Biol. **12**, (1935) (new ed., Berlin–Heidelberg–New York: Springer 1967.
KRAMERS, H. A.: Physica **7**, 284 (1940).
LANGEVIN, P.: C. R. Mebd. Séanc. Acad. Sci., Paris **146**, 530 (1908).
MacGILLIVRAY, A. D., HARE, D.: J. theoret. Biol. **25**, 113 (1969).
McINNES, D. D.: The Principles of Electrochemistry. New York: Dover 1961.
NERNST, W.: Z. phys. Chem. **2**, 613 (1888).
NORTHROP, J. H., ANSON, M. V.: J. gen. Physiol. **12**, 543 (1929).
ORNSTEIN, L. S., WIJK, W. R. van: Physica **1**, 235 (1933).
OVERBEEK, T. T.: Progr. Biophys. **6**, 57 (1956).
PEARSON, K.: Nature (Lond.) **77**, 294 (1905).
PLANCK, M.: Ann. Physik u. Chem. N. F. **39**, 161 (1890a).
PLANCK, M.: Ann. Physik u. Chem. N. F. **35**, 561 (1890b).
PLANCK, M.: Sitzungsber. Preuss. Akad. d. Wiss. Phys.-math. Kl. 324 (1930).
PLEIJEL, H.: Z. phys. Chem. **72**, 1 (1910).
PLETTIG. V.: Ann. Phys. Ser. 5 **5**, 735 (1930).
ROBBINS, E., MAURO, A.: J. gen. Physiol. **43**, 523 (1960).
SCHLÖGL, R.: Z. physik. Chem. N. F. **1**, 305 (1954).
SMOLUCHOWSKI, M. v.: Ann. Physik **21**, 756 (1906).
SMOLUCHOWSKI, M. v.: (1913) Bull. Acad. Cracovie A 418 (1913). (See also SMOLUCHOWSKI [1923]).
SMOLUCHOWSKI, M. v.: Physik. Zschr. **16**, 318 (1915a).
SMOLUCHOWSKI, M. v.: Ann. Physik **48**, 1103 (1915b).
SMOLUCHOWSKI, M. v.: Physik. Zschr. **17**, 557, 585 (1916).
SMOLUCHOWSKI, M. v.: Abhandlungen über die Brownske Bewegung und verwandte Erscheinungen. In: Ostwalds Klassiker der exakten Wissenschaften. Leipzig: Akademische Verlagsgesellschaft 1923.
SNEDDON, I. N.: The Use of Integral Transform. New York: McGraw-Hill, 1972.
STOKES, G. G.: Trans. Camb. Phil. Soc. **9**, 8 (1856).
TEORELL, T.: Progr. Biophys. **3**, 305 (1953).
UHLENBECK, G. E., ORNSTEIN, L. S.: Phys. Rev. **36**, 823 (1930).
USSING, H. H.: Acta physiol. scand. **17**, 1 (1949).
USSING, H. H.: Advanc. Enzymol. **13**, 21 (1952).
USSING, H. H., ZERAHN, K.: Acta physiol. scand. **23**, 110 (1951).
WARBURG, E. J.: Biochem. J. **25**, 153 (1922).

Chapter 3

Interpretation of Tracer Fluxes

H. H. USSING

A. Introduction

When isotopes became available as tools in biological research, the tracer method was initially used mostly in experiments on whole organisms. This pioneering work, initiated by de Hevesy (see HEVESY, 1948), led to the concept of the dynamic state of body constituents (SCHOENHEIMER, 1946). In the membrane transport field the main result was the realization that a constant turnover of inorganic ions is going on between cells and their surroundings, even in cases where no transport can be detected by chemical analysis (cf. KROGH, 1946). It was also realized, however, that the interpretation of the results in terms of membrane permeabilities presented enormous difficulties. Often it was not even possible to decide whether the isotopic exchange process was limited by blood flow, diffusion resistances in extracellular tissue spaces, or processes in the cell membranes.

Thus interest shifted towards the use of isotopic tracers in the study of isolated cells, tissues, and organs, where the boundary conditions could be better controlled. With proper preparations the difficulties encountered in the determination of tracer permeabilities are no more and no less than those encountered in determining permeabilities to other substances. In this chapter we shall be mainly concerned with tracers for substances already present in the system.

B. Fundamental Concepts

A tracer can be considered from two fundamentally different points of view. One involves letting the tracer molecules represent the species for which they are "tracing". Alternatively, the tracer molecules can be considered as belonging to a species which just happens to have most physical and chemical properties in common with the "macro-species". Both points of view are fundamentally correct and equally "good", but it is absolutely necessary to keep to one point of view at a time. Much confusion has resulted from the use of concepts belonging to one of these viewpoints in connection with interpretations belonging to the other. We shall return to this problem later.

C. Tracer Permeability Coefficients

I. Measurement of Tracer Permeability Coefficients

No matter which tracer-concept we choose, the experimental approach and the type of data collected are the same. In the best cases, figures having the formal nature of permeability coefficients can be obtained. The procedures for measuring permeability coefficients have been discussed extensively in the foregoing chapter. Systems of higher degrees of complexity than are discussed in that chapter can be handled according to methods described for instance in treatises by CRANK (1956) and JACQUEZ (1972). Many useful special cases of tracer application to permeability problems have been compiled, for instance by USSING (1952) and KOTYK and JANACÉK (1975).

As already mentioned, the measurement of tracer permeability coefficients makes use of the same theories and the same mathematical apparatus as are used in the determination of ordinary permeability coefficients. Often it is possible to choose experimental conditions where the tracer kinetics are simpler than the kinetics for net transport. Let us consider the determination of the tracer permeability for, say Na^+, of the fibre membrane of an isolated muscle.

A standard procedure (LEVI and USSING, 1948) is to preload the muscle with ^{24}Na in Ringer solution for sufficient time to give isotope equilibrium, and then follow the kinetics of the washout of activity into inactive Ringer. During this process we can assume that, in contrast to chemical washout of sodium from the muscle, the physicochemical conditions along the transport path are strictly independent of time.

Since the process at all levels is simply an exchange of virtually identical ions which are exposed to identical forces, the process of mixing can be handled by means of equations derived for ideal diffusion of uncharged molecules. In the example chosen the process of washing out the radiosodium can be resolved into two first-order processes: a fast one with a half-time of about two minutes and a slower one with a half-time of about 40 minutes. The short half-time could be identified with the washout of the extracellular spaces, whereas the longer half-time was assumed to describe the exchange across the fibre membrane. It should be emphasized that such a relatively simple interpretation of the washout kinetics depends on the fact that the rate constant for the washout of the extracellular spaces is much higher than that for the exchange across the fibre membrane. This means that radiosodium which has escaped the fibre is not likely to enter the sodium pool of other fibres, or even to be recycled through more fibres. In the general case of washout kinetics such recycling must be considered. HARRIS and BURN (1949), studying the washout kinetics of ^{42}K from preloaded muscles, gave the general solution to the problem together with suitable approximative solutions for cases of particular importance. This paper should be studied in detail by anyone working with washout kinetics. In the case of potassium, the kinetic situation is quite different from that governing sodium exchange. The exchange across the fibre membrane is faster than the sodium exchange, but much worse: the washout from the interspaces is relatively slow because the potassium concentration in the extracellular space is only 2 mM as opposed to the sodium concentration of 115 mM. Thus the chance of recycling *en route* is quite high, especially for ^{42}K leaving the central muscle fibres. Consequently, the slow component of the washout curve may grossly underestimate the true exchange across the fibre membranes.

The above treatment can also be used, *mutatis mutandis,* for tissue slices and other flat slabs of tissue. The recycling of potassium increases with increasing thickness of the tissue and with decreasing width of the intercellular spaces. The point can easily be reached where the exchange between cells and tissue space is practically complete at all times, so that the deeper layers cannot be "seen" kinetically. In such cases the true cellular exchange rate cannot be measured.

Loading – and washout – kinetic studies of the type discussed above thus, with luck, yield tracer permeability coefficients for cell membranes. Even if the mathematical apparatus necessary looks impressive enough, it is simpler than the one which would have been necessary in a study of net transport where time-dependent changes in potential, concentration and resistance along the route would have had to be considered. It is equally clear, however, that the gain in simplicity is obtained at the expense of information about the parameters which are disregarded: as a general rule, the tracer permeability coefficients obtained do not convey any information about the forces bringing about the exchange. Likewise, tracer experiments performed with different concentrations of tracer, but with maintained chemical composition of the system, give absolutely no additional information.

II. Multicompartment Systems

Exchange kinetics experiments of the types mentioned above may yield a very valuable "by-product" in the form of "spaces" or "pools" of the substance in question. Frequently the analysis in terms of pools and exchanges between them, presupposes a kinetic model where exchange between well-stirred compartments dominates the picture, whereas "long-distance diffusion" and convection can be disregarded. It is obvious, however, that if the number of compartments exceeds three or four, the number of sets of solutions increases rapidly, especially considering the inevitable scatter of measurements of radioactivity. Thus agreement with such elaborate models can only be considered circumstantial evidence in favour of any model considered. This is even more true when the condition of constant chemical composition of the experimental object during the tracer experiment is not fulfilled. Analytical solutions for certain simple cases have been derived, and computer simulation is possible for somewhat more involved cases, but it is rarely worthwhile to use tracer methods on very involved kinetic situations. Neither can tracer methods compete with, say, electrical or optical methods for analysing very fast processes.

D. The Concept of Unidirectional Flux

I. Unidirectional Fluxes

The tracer permeability coefficients thus obtained differ from real permeability coefficients in that they mostly attain different values for the "inward" and "outward" directions. To illustrate the situation, let us consider an isolated cell in quasi-equilibrium with Ringer solution. The potassium concentration in the cytoplasm is, say, 100 mM per liter, whereas the K^+ concentration of the Ringer solution is 2 mM per liter. Since the cellular potassium concentration does not change, the ion must enter as fast as it leaves. This means, however, that $*P_K^{in} \times C_{K(o)} = *P_K^{out} \times C_{K(i)}$. In other words $*P_K^{in}$ must be 50 times $*P_K^{out}$. Neither $*P_K^{in}$ nor $*P_K^{out}$, considered alone, gives any useful information about the future development of the system. Even if both $*P$ values are known, we need information about the K^+ concentrations in the inside and outside phases in order to ascertain that the cell is neither gaining nor losing potassium.

This information can be expressed by stating that the *influx* of K^+, $J_K^{(oi)}$, or the *unidirectional flux* of K^+ in the direction (oi) is equal ot the outflux or outward unidirectional flux $J_K^{(io)}$. In the example chosen, $J_K^{(oi)} = P_K^{out} \times C_K^{(i)}$. The *unidirectional flux* (USSING, 1948) can be visualized as the amount of ions of the species in question passing in the direction indicated through unit area of membrane, provided that the concentration of tracer ions is 100 percent at the side of origin of the flux, and that the concentration of tracer is maintained equal to zero on the opposite side of the membrane. It will be seen that this definition of unidirectional flux is based on the assumption that the tracer molecules or ions *represent* all members of the species.

II. Isotope Effects

This is more or less tacitly assumed in most tracer applications. In other words, it is assumed that such properties as diffusion rate, solubility, adsorption and chemical binding are the same for the normally occurring species and the one used as tracer. In general the assumption can be considered correct within the accuracy of measurement. As a rule of thumb one can say that the percentage difference in molecular weight between two species of an element gives a measure of the likelihood of encountering differences in behaviour ("isotope effects"). In case it is desired to ascertain whether or not isotope effects can be expected in a particular case, a simple test can often be applied (cf. HOSHIKO and LINDLEY, 1970). To find whether or not ^{24}Na is an ideal tracer for sodium in the study of active transport through, say, frog skin, one can add ^{24}Na and ^{22}Na simultaneously to the outside bathing medium and measure their rates of appearance in the inside bathing solution.

If the two tracers come out even in this race, it is clear that they must both trace ideally for the naturally occurring species ^{23}Na. If a difference is encountered, on the other hand, this will reveal information about the transport process! In general, isotope separation becomes more pronounced as the processes involved progress towards irreversibility. Thus, careful studies of tracer separation might become a useful tool in the study of transport processes. So far, however, this possibility has not been exploited seriously.

III. Associated Unidirectional Fluxes

The two associated unidirectional fluxes describing the permeation of a given membrane by a given substance under a given set of stationary conditions were originally named influx and outflux (USSING, 1948). Later MAIZELS (1954) pointed out that outflux is a cross between English and Latin and proposed the term efflux instead. Both terms, outflux and efflux, are still in use.

In this paper unidirectional fluxes are written $J_j^{(io)}$ or $J_j^{(12)}$ etc., the bracketed symbols indicating origin and terminal for the flux. Net fluxes are written J_j.

The introduction of the concept of unidirectional flux was dictated in part by the need for a term which did not imply any assumptions as to the physicochemical reasons for the transfer. If we consider again a cell in dynamic equilibrium with Ringer solution, the outward and inward fluxes of potassium may be equal, because potassium happens to be in Donnan equilibrium in that system. The inward and outward fluxes of sodium may also be the same, but here the outward flux may be due almost exclusively to active transport, whereas the inward flux may be due to a leak current driven by the combined effects of potential and concentration difference. A description by way of inward and outward permeability constants would have been tainted with the traditional use of the term permeability in connection with diffusion processes. If two suitable tracers are available for the substance studied, the associated unidirectional fluxes can be determined simultaneously by double-labelling (LEVI and USSING, 1949). Alternatively one unidirectional flux plus chemical determination of the net flux will suffice since $J_j = J_j^{(12)} - J_j^{(21)}$. Often satisfactory results can be obtained by determining $J_j^{(12)}$ and $J_j^{(21)}$ in parallel experiments.

IV. The Relation of Tracer Fluxes to Active and Passive Transport

At this juncture one might well ask what the flux values measured with isotopes can be used for, if they cannot be used to determine membrane permeabilities.

To answer this question it may be appropriate to start with the considerations advanced by August KROGH (1946) in his famous Croonian Lecture. He pointed out that, as far as inorganic ions go, the constant exchange between cells and surroundings must mean a constant active transport of certain (or maybe all) species in one direction and leakage in the opposite direction. If for a given ionic species, the concentration drop across the cell membrane was sufficiently large, the direction of active and passive transport could be inferred. In such cases the ionic exchange, as measured with isotopes, would be a direct measure not only of the permeability of the leak path but also of the active transport by way of a special, energy-consuming transport path.

The real importance of KROGH's paper was that it focussed interest on the possibility of using isotopes for distinguishing passive and active transport pathways.

V. Effects of Membrane Potentials on Ionic Fluxes

In the years that followed KROGH's proposal, it turned out that the problem was not quite as simple as at first assumed. In the first place, active transport need not be invoked in cases where a high potassium concentration in the cell versus a low one in the medium is balanced by an appropriated membrane potential, as the case may be for Donnan distribution of potassium. Thus a knowledge of the trans-membrane potentials became vitally important for deciding whether or not the transport in question could be passive (cf. USSING, 1947; ROSENBERG, 1948; USSING, 1949a).

VI. Exchange Diffusion

However, even if it could be demonstrated that a flux of some ion went against an electrochemical potential gradient, one could not be sure that the whole flux was due to active transport. This point was brought to the fore in the study of sodium exchange in isolated sartorius muscle mentioned above (USSING, 1947; LEVI and USSING, 1948).

Following the washout kinetics of radioactivity for a ^{24}Na-preloaded muscle made it possible to calculate the outward flux across the fibre membrane. This flux manifestly took place against a very steep electrochemical potential gradient. When electro-diffusion of free ions in the membrane phase was assumed (cf. GOLDMAN, 1943) it could be demonstrated that only an insignificant part of the flux could be due to passive electro-diffusion of sodium. The flux was indeed so large that it would require an uncannily large part of total muscle metabolic energy to bring about an active sodium efflux of the observed magnitude. A scrutiny of alternative possibilities led to the realization that a one-to-one exchange of an ion across a membrane without the consumption of metabolic energy is physically possible (USSING, 1947). Two possible mechanisms for such a hypothetical exchange were proposed in the paper mentioned. If the ion-impermeable lipid phase of the cell membrane contains sodium-binding carrier molecules which can cross the lipid from one boundary to the other only when loaded with sodium, no net transport can take place, and yet a one-to-one exchange is possible. Alternatively, the same results could be obtained if sodium-binding groups in the membrane could flip-flop between two positions, one of which gave contact with the cytoplasm and the other with the medium. The experimental data did not *prove* that such an exchange diffusion was actually responsible for the large sodium exchange. Much less did it *prove* the existence of membrane carriers. But it did prove that such a cost-free exchange did not violate the second law of thermodynamics. It *is* permissible to lift an ion from a lower to a higher energy level, if this movement is compulsorily coupled to the movement of an identical ion from the higher to the lower energy level. It should be stressed that the exchange-diffusion phenomenon would be seen even if both loaded and unloaded carrier could pass the membrane, provided that the substance transported saturated the carrier system on both sides of the membrane. Such special membrane structures would mostly escape discovery without the tool provided by isotopic tracers.

VII. Limitations for Integration of Flux Equations

By this stage it was intuitively clear that at least two classes of phenomena, active transport and exchange diffusion, might give rise to "anomalous" isotope fluxes. Thus it became of the utmost importance to define "normal" behaviour of the flux. The choice is easy as far as uncharged substances are concerned; in their case behaviour according to Fick's law of diffusion would be considered "normal".

For ions the situation is different. If an ion is moving under the combined effects of a chemical and electrical gradient, the flux equation (see Chapter 2) can only be solved under the assumption of certain properties of the transport path. The assumption most commonly used is that of a constant electrical field across the membrane phase. Thus the well known Goldman equation (GOLDMAN, 1943) is based on this assumption (see Chapter 2).

If the membrane under study has a sandwich structure or if several cell membranes and other barriers are placed in series in the object under study, the constant-field assumption is invalid and a calculation of the flux equation becomes virtually impossible. This problem was a stumbling block when it became desirable to use the isotope method for studying transepithelial ion transport. The interest of our group in epithelial transport was understandable. In the mid-forties many scientists, including many physico-chemists, doubted the existence of active transport in general and active transport of alkali metal ions in particular. The realization that apparent uphill transport of ions under steady-state conditions might be due to exchange diffusion (see above) weakened the argument for using such transport as evidence for active transport. At the same time there was a colloid-chemical school who maintained that uneven ionic distribution between cells and their surroundings depended on specific binding in the cytoplasm rather than on membrane processes. Therefore the existence of massive net transport of ions across many epithelia might provide a better chance for studying something that was unquestionably active transport.

Here, however, the impossibility of integrating the flux equation for the total thickness of the experimental subject seemed to make the use of the isotope method suspect. The way out was to use the flux ratio rather than the flux itself.

E. Flux Ratio Analysis

I. The Flux Ratio Equation

The rationale for studying the properties of the flux ratio was initially the analogy between a transport process and a chemical reaction. Considered from such a point of view, the flux ratio, or rather its logarithm, would be a measure of the affinity of the process, whereas the difference between the fluxes would be a measure of the net reaction rate.

It was shown (USSING, 1949b) that, for an ionic species which does not interact with other moving molecules or groups in the membrane phase, the flux ratio $J_j^{(12)}/J_j^{(21)}$ is independent of the membrane structure. The flux ratio for an ionic species in such cases is determined solely by the difference between its electrochemical potentials in the two bathing solutions.

Thus we have for a passively diffusing ion, (j), which does not interact with other moving particles during its passage through the membrane:

$$RT \ln \left(J_j^{(12)}/J_j^{(21)}\right) = \bar{\mu}_j^{(1)} - \bar{\mu}_j^{(2)} \tag{1}$$

where $J_j^{(12)}$ and $J_j^{(21)}$ are the unidirectional fluxes from solution 1 to solution 2 and from solution 2 to solution 1, respectively, and $\bar{\mu}_j^{(1)}$ and $\bar{\mu}_j^{(2)}$ are the electrochemical potentials of the ion in the two bathing solutions.
The equation can also be written:

$$J_j^{(12)}/J_j^{(21)} = \left(a_j^{(1)}/a_j^{(2)}\right) \exp\left(zF(\psi^{(1)} - \psi^{(2)})/RT\right) \tag{1a}$$

where $a_j^{(1)}$ and $a_j^{(2)}$ are the chemical activities of the ion in the bathing solutions and $(\psi^{(1)} - \psi^{(2)})$ is the electrical potential difference across the membrane. *z, F, R, and T* have their usual meanings.

The equation can be derived in different ways. Thus it can be seen as a consequence of Eyring's theory of rate processes (see JOHNSON et al., 1954). An equation of the same form was obtained by TEORELL (1949), but his derivation was restricted to a homogeneous membrane, like the derivation shown in this volume, Chapter 2. The general application of the equation depends, however, on the fact that it is valid for any number of layers and for any shape of the concentration, potential and resistance profiles across the membrane. The equation is actually valid for fluxes passing barriers of arbitrary shape and with properties varying in three dimensions (SCHWARTZ, 1971). Indeed it can be shown that the equation is valid for substances formed or consumed *en route*

(USSING, 1952). However, the equation in the form given above assumes that the ions in question are only under the influence of the "diffusion force" and the electrical field. The complications arising from interactions with the flow of solvent ("solvent drag") are discussed below. As discussed elsewhere (see volume II, Chapter 3 and 5), there is mounting evidence that most transport processes, passive as well as active, take place by way of highly specific channels. Therefore interaction between water flow and ionic flow may well be negligible in most cases of transport through plasma membranes and tight epithelia.

II. Derivation of the Flux Ratio Equation

Originally (USSING, 1949b) the flux ratio equation was derived for steady-state fluxes only. The derivation below (cf. USSING, 1972) demonstrates that it is also valid under certain frequently encountered non-steady state conditions.

In the rest of this Chapter the word membrane is used in a wider sense, meaning an unstirred sheet consisting of an arbitrary number of layers, whose permeability properties and ionic compositions may differ in any conceivable way.

The rate of passage of the substance considered through a plane within the membrane and parallel to its surface is given by

$$\frac{dn}{dt} = -A \cdot u \cdot C \cdot \frac{d\bar{\mu}}{dx} \tag{2}$$

where dn is the amount passing the plane during the time dt, A is the area, u is the (variable) mobility of the substance in question, C the concentration of the substance, $\bar{\mu}$ its electrochemical potential, and x the distance of the plane from the surface bathed with the well-stirred solution 1. If there is no interaction with other moving particles, no pressure gradient, and no active transport, we have

$$\frac{d\bar{\mu}}{dx} = RT\frac{d\ln C}{dx} + RT\frac{d\ln\gamma}{dx} + zF\frac{d\psi}{dx} \tag{3}$$

where R, T, z, and F have their usual meanings, γ is the activity coefficient of the ion and ψ is the electrical potential at position x.

A function $\bar{a}$ (the electrochemical activity) can be defined by the following equation:

$$RT\ln\bar{a} = RT\ln C + RT\ln\gamma + zF\psi = \bar{\mu} - \bar{\mu}_{\text{standard}}. \tag{4}$$

Solving for C in (4), we obtain

$$C = \frac{\bar{a}}{\gamma}\exp(-zF\psi/RT). \tag{5}$$

Combination of (2) and (5) yields

$$\frac{dn}{dt} = -uART \cdot \frac{\bar{a}}{\gamma \exp\left(\frac{zF\psi}{RT}\right)} \cdot \frac{d\ln\bar{a}}{dx}$$

or

$$\frac{dn}{dt} = \left[-\frac{uART}{\gamma \exp\left(\frac{zF\psi}{RT}\right)}\right] \cdot \frac{d\bar{a}}{dx}\,; \tag{6}$$

for convenience we can put the bracketed expression in (6) equal to $B(x,t)$. Thus we can write (6) as

$$\frac{dn}{dt} = B(x,t) \cdot \frac{d\bar{a}}{dx}\,. \tag{6a}$$

Let us consider the movement through the membrane of two different ideal tracers for the substance in question. Subscripts I and II will be used as symbols for tracer I and II respectively. Initially, tracer I is added to solution 1 and tracer II to solution 2. We now want an expression for the ratio of the fluxes of the two isotopes across the plane defined by the abscissa x. An equation of the form given as (6a) can be written for each isotope. The ratio between the left-hand sides of these equations is set equal to the ratio between the right hand sides. We now notice that the two B-functions cancel, because they contain only variables that describe the (unknown) physical conditions at x, which are common to both isotopic species.

Thus we obtain

$$\frac{\frac{dn_{\mathrm{I}}}{dt}}{\frac{dn_{\mathrm{II}}}{dt}} = \frac{\frac{d\bar{a}_{\mathrm{I}}}{dx}}{\frac{d\bar{a}_{\mathrm{II}}}{dx}}\,. \tag{7}$$

Clearly, certain conditions have to be fulfilled in order to make elimination of the B-functions permissible. As already mentioned, u, ψ, A and γ may be functions not only of x but also of t. This can give rise to problems. Statistically, the plane we are considering at x will be reached at different times by ions starting simultaneously from solution 1 and solution 2. Thus, if the potential and resistance profiles through the membrane are time-dependent, $B_{\mathrm{I}}(x,t)$ is different from $B_{\mathrm{II}}(x,t)$, and thus the B functions do not cancel.

For the following treatment we shall therefore consider a period during which the B-function varies with x only.

By the above procedure we have managed to get rid of all the unknown functions describing the properties of the interior of the membrane and we have obtained a separation of the variables so that the equation can be integrated. For this purpose we can rewrite (7) as

$$\frac{dn_{\mathrm{I}}}{dt} \cdot \frac{d\bar{a}_{\mathrm{II}}}{dx} = \frac{dn_{\mathrm{II}}}{dt} \cdot \frac{d\bar{a}_{\mathrm{I}}}{dx}. \tag{7a}$$

This differential equation can be solved by performing double integration on both sides of the equality sign:

$$\int_{\tau_1}^{\tau_2} dt \int_0^h \frac{dn_{\mathrm{I}}}{dt} \cdot \frac{d\bar{a}_{\mathrm{II}}}{dx}\, dx = \int_{\tau_1}^{\tau_2} dt \int_0^h \frac{dn_{\mathrm{II}}}{dt} \cdot \frac{d\bar{a}_{\mathrm{I}}}{dx}\, dx. \tag{8}$$

The n-functions are integrated between the times τ_1 and τ_2. For the $\bar{a}$-functions we shall use the following boundary conditions (h being the total thickness of the membrane):

for $x = 0$: $\bar{a}_{\mathrm{I}} = \bar{a}_{\mathrm{I}}^{(1)}$ and $\bar{a}_{\mathrm{II}} = 0$
for $x = h$: $\bar{a}_{\mathrm{I}} = 0$ and $\bar{a}_{\mathrm{II}} = a_{\mathrm{II}}^{(2)}$

where $\bar{a}_{\mathrm{I}}^{(1)}$ and $\bar{a}_{\mathrm{II}}^{(2)}$ are the (constant) values of the electrochemical activities of the ion in question in solutions 1 and 2, respectively.

These boundary conditions imply that the bathing solutions are of such volume and so well stirred that no back diffusion of either isotope is possible.

Initially after the addition of the two isotopes to the bathing solutions, n_{I} and n_{II} may be functions of both x and t. However, within a very short time (the redistribution time, see Chapter 2 Section III) the situation becomes quasistationary. Thus $\frac{dn_{\mathrm{I}}}{dt}$ and $\frac{dn_{\mathrm{II}}}{dt}$ will become functions of t only, *i.e.* $\frac{\partial}{\partial x}\left(\frac{dn}{dt}\right) = 0$, valid for both isotopes, and we can rewrite (8) as follows:

$$\int_{\tau_1}^{\tau_2} \frac{dn_{\mathrm{I}}}{dt} \cdot dt \int_{\bar{a}_{\mathrm{II}}^{(2)}}^{0} d\bar{a}_{\mathrm{II}} = \int_{\tau_1}^{\tau_2} \frac{dn_{\mathrm{II}}}{dt} \cdot dt \int_0^{\bar{a}_{\mathrm{I}}^{(1)}} d\bar{a}_{\mathrm{I}}. \tag{8a}$$

Since $\bar{a}_{\mathrm{I}}^{(1)}$ and $\bar{a}_{\mathrm{II}}^{(2)}$ are constants, we finally obtain

$$(-)\,[n_{\mathrm{I}}(\tau_2) - n_{\mathrm{I}}(\tau_1)] \cdot \bar{a}_{\mathrm{II}}^{(2)} = [n_{\mathrm{II}}(\tau_2) - n_{\mathrm{II}}(\tau_1)] \cdot \bar{a}_{\mathrm{I}}^{(1)} \tag{8b}$$

or by rearrangement

$$(-)\,\frac{n_{\mathrm{I}}(\tau_2) - n_{\mathrm{I}}(\tau_1)}{n_{\mathrm{II}}(\tau_2) - n_{\mathrm{II}}(\tau_1)} = \frac{C_{\mathrm{I}}^{(1)}\gamma_{\mathrm{I}}^{(1)}}{C_{\mathrm{II}}^{(2)}\gamma_{\mathrm{II}}^{(2)}} \cdot \exp\frac{zF(\psi^{(1)} - \psi^{(2)})}{RT}. \tag{8c}$$

The minus sign in parentheses in front of the equation refers to the fact that, if the direction from solution 1 to solution 2 is considered positive, movement in the opposite direction comes out negative in the procedure of integration. Conventionally, however, unidirectional fluxes are always considered positive. The equation obtained is formally identical to the flux ratio equation in its original form (USSING, 1949b). Alternatively, however we can interpret the symbols slightly differently. Let $C_{I}^{(1)}$ and $C_{II}^{(2)}$ mean the concentrations of the two tracers in the bathing solutions in arbitrary units. The equation is then taken to mean that the ratio between the cumulative counts of the two tracers passing the membrane in the two directions between times τ_1 and τ_2 is independent of the length of this period, and is constant from time of first appearance of the two isotopes on the opposite sides of the membranes. For practical purposes this time coincides with the redistribution times. Thus one does not have to wait for the establishment of steady-state fluxes in order to obtain the isotope flux ratio. This result is of course completely uninteresting for work with single cell plasma membranes, because steady-state fluxes are established within milliseconds. For work with epithelia and other composite structures, however, steady state isotope fluxes may require minutes or even hours. Thus in toad bladder and frog skin the potassium and rubidium permeabilities are so small and the pools so large that a true steady state may be difficult to obtain. In such cases the non-steady state isotope fluxes may give a good estimate of the steady state isotope flux ratio. If the non-steady state isotope flux ratio changes with time, it may mean either that the substance in question can pass the system by way of two or more pathways with different passage times, or that the preparation is changing its properties with time. The latter case is usually associated with a change in the B-function. If, however, the sampling time for the isotopes is made short compared to the rate of change in the B-function, the flux-ratio equation is still valid.

The isotope flux ratio evidently is equal to $^*P^{(12)}/^*P^{(21)}$ where the asterisks refer to the fact that we are considering the tracers only.

Assuming that we are dealing with ideal tracers which are both representative of the "macrospecies", we have

$$J^{(12)}/J^{(21)} = {}^*P^{(12)} \times C^{(1)}/{}^*P^{(21)}\,C^{(2)}. \tag{9}$$

III. Estimation of Electrochemical Potential Differences

As already mentioned, Eq. (8b) states that the flux ratio for an independently moving ion species is independent of the transport path and is determined solely by the difference between its electrochemical potentials in the two bathing solutions.

If specific ion-selective electrodes were available for the ions studied, the electrochemical potential difference between the solutions could be measured directly. Mostly, however, the electrochemical activity ratio must be calculated from measurements of concentrations and potential difference, combined with an estimate of the ratio between the activity coefficients. The necessity of making this estimate has given rise to many misgivings. According to classical electrolyte theory, activity coefficients belong to the salt and cannot be distributed between the individual ions. In dilute solutions of mono-monovalent salts, no great error can arise from assuming the activity coefficient of anion and cation to be equal.

In recent years methods have been developed which yield reasonably safe estimates for single ion activities in simple dilute solutions (cf. LEV and ARMSTRONG, 1975).

The uncertainty with respect to single ion activities has its corrolary as regards the measurement of the electric potential differences between solutions of different composition. However, the use of proper salt bridges usually suffices to reduce the uncertainties with respect to diffusion potentials to an acceptable minimum (see Chapters 2 and 4). In any case, the uncertainties with respect to the measurement of activity coefficients and diffusion potentials can be circumvented if it is possible to use identical solutions on both sides of the membrane.

IV. The Short-Circuiting Method

A particularly simple case arises if both bathing solutions are identical and at the same time the electric potential difference across the preparation is reduced to zero by passing a suitable electric current *via* a separate set of electrodes (USSING and ZERAHN, 1951). Due to the potential drop in the bathing solutions, the transmembrane potential has to be measured with electrodes in the immediate vicinity of the membrane or a correction must be made for the potential drop between the membrane and the electrode tips.

Under these conditions, as seen from the flux-ratio equation, the flux ratio for passively moving ions should be one. If any ionic species should undergo exchange diffusion, its flux ratio would also be one. Only actively transported ions would give flux ratios different from one. Moreover, the sum of the net transports of actively transported ions, measured in electrical units, should be equal to the electric current passing through the preparation.

V. Flux Ratio with Solvent Drag

If there is net solvent flow through the membrane, one must consider the possibility that interaction between the solvent and the solute studied may influence the flux ratio for the solute. In particular, such an interaction is to be expected if penetration takes place through pores or water-filled channels. Clearly, particles moving downstream will be speeded up and those moving upstream will be slowed down.

If the pores were so large that interference with the walls could be disregarded, it would be natural to ascribe a certain bulk velocity to the solution. This is the device introduced by ONSAGER (1945), in his treatment of liquid diffusion.

Compare also KEDEM and KATCHALSKY (1958 and 1963 a–c). In biological membranes, however, the pores, if present at all, must be very small, and solvent molecules must be impeded in their movements by interactions with the pore walls. At least it is a fact that in living membranes the permeability coefficient for water is mostly several orders of magnitude higher than those for hydrophilic solutes.

For this reason it may be more realistic to see the effect of the solvent flow as a force acting on solute molecules which are almost stationary relative to the membrane phase. This consideration is the basis for the derivation of the flux ratio with solvent drag given below; it follows the treatment given by USSING (1952); see also KOEFOED-JOHNSEN and USSING (1953) and MEARES and USSING (1959a, b). Not unexpectedly, however, treatments based on the principles of irreversible thermodynamics (HOSHIKO and LINDLEY, 1964; KEDEM and ESSIG, 1965) lead to equations of exactly the same form as those derived by USSING et al. (see above).

We shall assume that the force, acting upon 1 mol of water in the direction of flow is $-V_w\,dP/dx$. If f_w' is the friction exerted on 1 mol of water, moving in the membrane at unit velocity, the linear rate of flow must be $-(V_w dP/dx)(1/f_w')$. The force which acts upon one mole of solute is therefore $-(V_w\,dP/dx)(f_j/f_w')$, where f_j is the friction between 1 mol of solute and water at unit velocity difference. f_j is assumed to be the same in the pore and in free solution, since it describes the interaction between solute and solvent. Thus f_j can be obtained from the free diffusion coefficient, D_j, *viz*: $f_j = RT/D_j$.

The flux, J, across unit area normal to the direction of flow is assumed to be proportional to the force per mol, to the concentration of the diffusing species, and to the fraction, A, of unit area which is accessible for the species in question. Furthermore, the flux is assumed to be proportional to the variable mobility u_j, of the species while inside the membrane phase. Theoretically, one more term should be included in the expression for the flux, *viz* that exerted on the activity of the ion by the hydrostatic pressure. The term has the form $V_j\,dP/dx$, where V_j is the molar volume of the species. This term is usually quite small and vanishes completely in the final expression if the hydrostatic pressure is the same on both sides of the membrane. Thus it will be noticed that the hydrostatic pressure gradient has two quite distinct effects. One is to bring about filtration of the solution through pores and channels. The other one has to do with the "escaping tendency" of all molecules in the solution, which increases with increasing pressure.

For the pressure gradients tolerated by animal tissues the effect is virtually nil, whereas it may be of importance in connection with the extremely high turgor pressures encountered in some plant tissues. The equation for the flux is

$$J = \frac{dn}{dt} = -A u_j C\Big(RT\frac{d\ln C}{dx} + RT d\ln\gamma/dx + zF\frac{d\psi}{dx} + V_w\frac{dP}{dx}\frac{f_j}{f_w'} + V_j\frac{dP}{dx}\Big)\,. \tag{10}$$

A, u_j, c, ψ, P and f_w' are all unknown functions of x. In the expression for the flux ratio, however, most of these unknowns cancel (cf. Section D), and we obtain the following expression (assuming no pressure difference across the membrane):

$$\ln(J^{(12)}/J^{(21)}) = \ln(a^{(1)}/a^{(2)}) + (zF/RT)(\psi^1 - \psi^2) + \frac{f_j}{RT}\int_0^h \frac{V_w}{f_w'}\left(\frac{dP}{dx}\right)dx \tag{11}$$

or

$$\ln(J^{(12)}/J^{(21)}) = \ln(a^{(1)}/a^{(2)}) + (zF/RT)(\psi^1 - \psi^2) + \frac{1}{D}\int_0^h \frac{V_w}{f_w'}\cdot\left(\frac{dP}{dx}\right)dx. \tag{12}$$

The linear rate of flow, $(dP/dx)\,(V_w/f_w')$ may vary in an unknown way with x. But the volume rate must be the same for all values of x, since water is non-compressible.

If the volume rate of flow is called J_w, we obtain:

$$J_w = A\cdot\frac{V_w}{f_w'}\left(\frac{dP}{dx}\right) \tag{13}$$

or

$$\frac{J_w}{A} = \frac{V_w}{f_w'}\left(\frac{dP}{dx}\right).$$

Combining Eqs. (12) and (13) we finally obtain

$$\ln(J^{(12)}/J^{(21)}) = \ln(a^{(1)}/a^{(2)}) + \frac{zF}{RT}(\psi^1 - \psi^2) + \frac{J_w}{D_j}\int_0^h \frac{1}{A}\,dx \tag{14}$$

or

$$J^{(12)}/J^{(21)} = (a^{(1)}/a^{(2)})\exp\left[\frac{zF}{RT}(\psi^1 - \psi^2) + \frac{J_w}{D_j}\int_0^h \frac{1}{A}\,dx\right]. \tag{15}$$

It should be noticed that $J^{(12)}$ and $J^{(21)}$ are unidirectional fluxes (always positive), and J_w is the net water flux which is considered positive in the x-direction.

The integral occurring in the last term of Eqs. (14) and (15) cannot be evaluated directly, but it will be noticed that it is independent of the transported species, being only a function of the generalized shape of the transport path. Therefore it is the same for all species using that pathway. Thus, if it is evaluated for one or more substances, assumed not to interact with other moving species, except water, the values for the integral obtained can now be used for other substances of similar size to evaluate their theoretical flux ratios. For homogeneous membranes the integration can be carried through and yields $J_w h / D$.

VI. Solvent Drag on Non-Electrolytes and Water

For non-electrolytes the potential term vanishes and the expression reduces to

$$\ln (J^{(12)} / J^{(21)}) = \ln \frac{a^{(1)}}{a^{(2)}} + \frac{J_w}{D_j} \int_0^h \frac{1}{A} \, dx. \tag{16}$$

This equation is also applicable if the fluxes are determined for water itself, using isotope-labelled water.

For reasonably high rates of osmotic flow, the ratio between water activities $a_w^{(1)} / a_w^{(2)}$ becomes insignificant, relative to the term containing the integral, so it will be the latter which determines the flux ratio. It should be remembered, however, that the object for which the equation is valid is the whole thickness of the slab separating two well stirred solutions. Thus, the "membrane" for which the equation is valid includes the unstirred layers on either side of the membrane proper. For most solutes this condition is unimportant, since the unstirred layers contribute only a small fraction to the diffusion resistance to solutes. For water diffusion and the diffusion of gases the situation may be different, because cell membranes are relatively permeable to these substances.

In the above treatment the forces bringing about the water flow are not considered. It is assumed that the solvent drag will be the same whether the water flow is due to osmosis, electro-osmosis or (hypothetical) active transport.

The above treatment is simplified in the sense that interactions other than those between solute and solvent are disregarded. The justification for this simplification is the fact that the solvent is present in a very much higher concentration than the other solutes. For this reason the term solvent drag rather than solution drag was introduced. Formally one can introduce terms describing all conceivable solute-solute interactions. This was done by Meares and Ussing (1959 a and b), Hoshiko and Lindley (1964, 1967) and Kedem and Essig (1965), the two last-mentioned groups using the formalism of irreversible thermodynamics. The determination of all the necessary cross-coefficients between solutes is quite a job, however, even for simple artificial membranes, and for composite biological membranes the situation is even more complex. The real problem is that biological membranes, even the simplest ones, are probably "multichannel systems".

For one thing, in contrast to most hydrophilic substances in general and ions in particular, water can penetrate the lipid phase of most membranes, besides using other pathways. Thus the total water flux is rarely passing any specific pathway. Therefore a calculation based on Eq. (15) may over-emphasize the solvent drag to be expected. A proper treatment of the system would require that the fluxes were resolved in sets of influx and efflux belonging to each individual pathway. A meaningful resolution is hardly possible if the number of pathways exceeds two or three.

VII. Solute-Solute Interactions

In the above treatment it was assumed that only solvent drag interfered with the behaviour according to the simple flux-ratio equation. It is conceivable, however, that other specific fluxes than that of water might modify the fluxes of a given substance. Thus van BRUGGEN and collaborators (FRANZ and van BRUGGEN, 1967) suggested that one solute diffusing through a membrane might exert a "solute drag" on another solute. Such cases and cases of more specific coupling can be treated according to a formalism developed by KEDEM and ESSIG (1965). For one interacting flux, the equation takes the following shape:

$$\frac{J_i^{(12)}}{J_i^{(21)}} = \frac{\text{influx}}{\text{efflux}} = \frac{C_i^{(1)}}{C_i^{(2)}} \cdot \exp\left[\frac{zF}{RT} \cdot (\psi^{(1)} - \psi^{(2)}) + \frac{1}{RT} \int_0^h r_j \cdot J_j \cdot dx\right] \tag{17}$$

where J_j is a flux interacting with the fluxes of the substance under study and r_j is the coefficient of coupling between the fluxes of the species under study and species j. The formal similarity between Eq. (17) and Eq. (15) is apparent.

VIII. Meaning of the Term "Interaction"

It is timely to recall (see p. 115) that a tracer experiment can be interpreted as describing the behaviour of representative members of the species, in which case one obtains the unidirectional flux of all molecules of the species studied. Alternatively the tracer can be considered as a different "microspecies" whose properties happen to coincide with those of the "macrospecies". As already mentioned, both points of view are equally valid. In the foregoing we have stuck mainly to the first interpretation. In the KEDEM-ESSIG treatment the second interpretation is used. Thus the fluxes $J_i^{(12)}$ and $J_i^{(21)}$ are tracer fluxes of the microspecies whereas J_j is a net flux of a macrospecies. In the KEDEM-ESSIG vocabulary it is possible (and correct) to speak about interaction between microspecies and macrospecies of the same element. Such an interaction can be described by replacing J_j with J_i in the integral term in Eq. (17). If we are considering ideal tracers where members of the micro- and macrospecies have identical properties, the meaning of the "interaction" is hard to express in physical terms. "Interaction" in this context is a purely phenomenological description: It "looks" as if the macro-flux interferes with the fluxes of the microspecies. On the molecular level this means that peculiar properties of the membrane force all members of the species considered to behave in a way that cannot be predicted from simple diffusion and electro-diffusion theory.

If we apply the representative tracer concept, the term interaction between micro- and macrospecies obviously is nonsensical. If such an interaction existed, the tracer would not be representative. However, the restraints on the behaviour of the species called interaction in the KEDEM-ESSIG treatment manifests itself in typical deviations from ideality of the flux ratio.

IX. Interpretation of Deviations from the Flux Ratio Equation

In a purely formal way, deviations from "simple passive behaviour" can be divided in two classes, characterized by the relationships

$$\text{a)} \qquad |RT\ln(J^{(12)}/J^{(21)})| < |\bar{\mu}^{(1)} - \bar{\mu}^{(2)}| \tag{18}$$

and

$$\text{b)} \qquad |RT\ln(J^{(12)}/J^{(21)})| > |\bar{\mu}^{(1)} - \bar{\mu}^{(2)}|. \tag{19}$$

We shall first consider cases where solvent flow is absent or so small that solvent drag can be considered insignificant.

Both types of deviation from idealities are encountered frequently in biological systems. Examples of the one-to-one exchange diffusion process discussed on p. 133 in the limiting case leads to a flux ratio of one, quite independent of the electrochemical-potential gradient across the membrane. This situation obtains if the hypothetical carrier can cross the membrane only in the loaded state, and if the transported species cannot cross at all, unless attached to the carrier. If unloaded carrier molecules can cross the membrane to some extent and/or if the transported species can diffuse unaided by the carrier, the deviations from the flux-ratio equation will be less pronounced, but the findings will still belong to category (a).

The different types of carrier transport kinetics that can be anticipated for assumed sets of diffusion coefficients, binding constants, and rates of formation etc. have been the subject of many studies over the last few decades. For references, see ROSENBERG and WILBRANDT (1955), CHRISTENSEN (1960), LIEB and STEIN (1974a and b), WYSSBROD et al. (1971), and KOTYK and JANAĆEK (1975). LeFEVRE (1975) has given a penetrating analysis of the carrier concept as well as a valuable bibliography on the subject. Practically all solutions treat only uncharged carriers and "passengers". Thus most solutions can only be used with the greatest caution in ion transport studies.

For non-electrolytes the kinetic models so far described often give a very satisfactory description of both unidirectional fluxes and net fluxes in systems performing passive, but selective transport. Thus exchange diffusion and counter transport (ROSENBERG and WILBRANDT, 1957) appear as special cases of the general solutions.

Clearly, this formal agreement with theory does not prove that diffusing membrane carriers exist. The entity which moves may be a cavity in a molecule or a binding site that is translocated due to flip-flopping of a molecule or due to cooperative molecular events. All one can say is that deviations from the flux-ratio equation of category (a) indicate highly specific interactions with membrane elements, of a nature which is not encountered in homogeneous systems, where the laws of electro-diffusion are valid. Deviations from the flux-ratio equation of category (a) are not common in artificial membranes, but a few examples have been reported by DE SOUSA et al. (1971) and GOTTLIEB and SOLLNER (1968).

The molecular basis for the abnormal behaviour of the membranes studied is unknown. It should also be pointed out that the exchange diffusion components of the fluxes through those membranes were modest and do not compare with the exchange diffusion found in many biological systems.

It is characteristic of the one-to-one exchange diffusion that it does not require metabolic energy, and thus need not have anything in common with the active transport mechanisms. Nevertheless it is conceivable that elements of the active pumps may provide the basis for exchange diffusion. Indeed it can be argued that a pump, operating in a region where its driving force is just balancing the electrochemical potential gradient of the transported species, may serve as a pathway for a one-to-one exchange, with no net consumption of energy (compare ESSIG, 1968).

Deviation from the flux ratio equation of type (b), usually suggests the involvement of active transport. This conclusion is rather trivial in cases where the net transport takes place from a lower to a higher electrochemical potential of the transported species. Likewise, the conclusion is safe when transport takes place between identical and equipotential solution as is the case for short-circuited systems.

F. Examples

I. The Short-Circuited Frog Skin

As an example of the short-circuiting technique, we shall briefly discuss its use with the isolated frog skin (USSING and ZERAHN, 1951).

The method was developed in order to demonstrate that active sodium transport was the sole process responsible for the electric asymmetry between the outside and inside of the preparation.

At that time it had already been demonstrated (cf. Chapter 6, vol. III) that the sodium transport inward must be due to active transport, since it went against the electrochemical potential difference for that ion, and since the flux ratio for sodium was many times higher than that predicted form the electrochemical potential difference for the ion, as measured between the outside and inside bathing solutions. Since, however, many theories for the origin of the skin potential had involved hydrogen and bicarbonate ions, whose flux ratios cannot be measured with isotopes, it was important to ascertain whether or not other processes than the active sodium transport contributed measurably to the electric asymmetry. As mentioned above (p. 127) it became clear to us that the short-circuited state would offer unique possibilities for revealing active transport. Determinations of the current that can be drawn from a frog skin partly short-circuited through reversible electrodes had been made previously by FRANCIS (1933) and by STAPP (1941) and LUND and STAPP (1947). However, none of these investigators had related the current to active ion transport, still less performed determinations of the Na^+-transport.

From the data obtained by LUND and STAPP (l. c.) it could be calculated, however, that a bull-frog skin partially short-circuited *via* a pair of lead chloride electrodes could produce a maximum of 0.06 coul. cm^{-2} $hour^{-1}$. With open-circuit Rana temporaria skins we had obtained figures for sodium net influx up to 0.1 coul. cm^{-2} $hour^{-1}$ (USSING, 1949a). This indicated that a complete short-circuit might well give identical values for short-circuit current and net sodium transport. Of course this suggestion could not be stressed unduly, since the values were obtained on different species and under different experimental conditions. As it turned out, however, the working hypothesis was correct.

Due to the external resistances a complete short-circuit cannot be brought about merely by using reversible metal, metal chloride electrodes. In order to overcome the resistance of electrodes and bathing solutions, an additional E. M. F. has to be applied in series with the skin in the current circuit. From an electrical point of view this principle is identical to the voltage clamp procedure developed simultaneously and independently by HODGKIN et al. (1952) for measuring membrane currents in nerve. The purpose of their method, however,

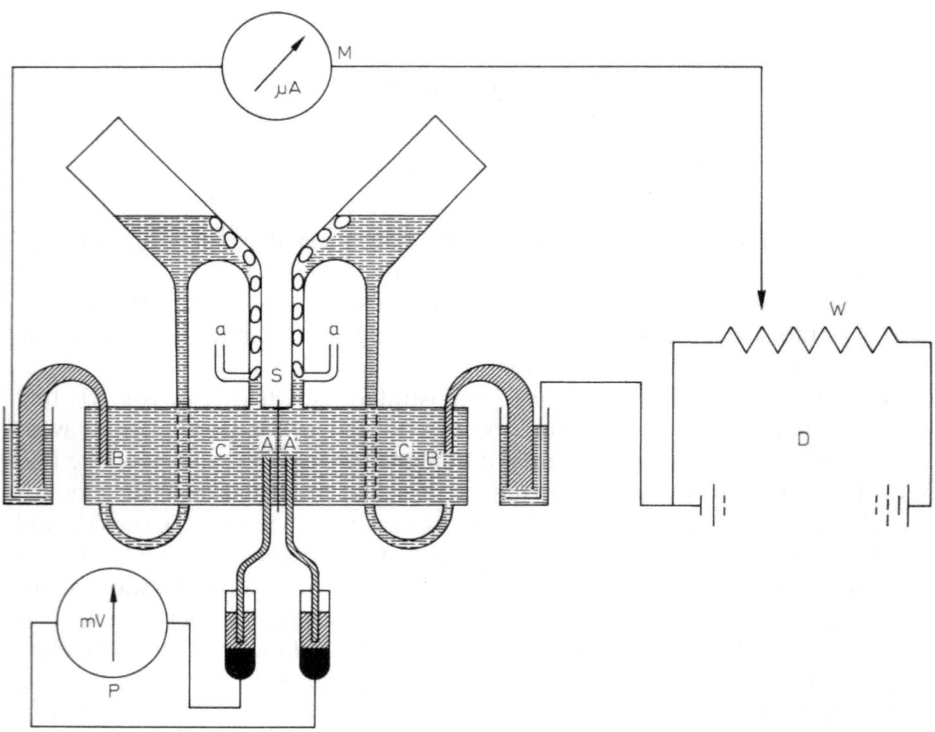

Fig. 1. Diagram of setup for determining ionic fluxes and short-circuit current (USSING and ZERAHN, 1951). S: Frog skin. C: Two half-chambers between which the skin is clamped. a: Inlets for air. A and A′: Agar-Ringer bridges, connecting the outside and inside bathing solutions, respectively, with the calomel electrodes. B and B′: Agar-Ringer bridges used for applying outside E. M. F. D: Battery; W: Potential divider; M: Microammeter; P: High-impedance millivoltmeter

was to measure passive ion flows created by ionic concentration differences across the membrane, whereas the short-circuit method was designed to measure active transport currents between identical solutions.

A diagram of the apparatus used is shown in Figure 1. The skin S, is placed as a diaphragm, separating the two lucite half chambers marked C. Two narrow agar-Ringer bridges A and A'open on either side, a few millimeters from the skin. The outer ends of A and A' make contact with saturated KCl-calomel electrodes. The potential difference between the latter is measured by a high impedance millivoltmeter.

Another pair of agar-Ringer bridges open in each half chamber, as far as possible from the skin, to insure a uniform electrical field in the solutions. The outer ends of these bridges dip in beakers with saturated KCl. Spirals of stout silver wire, immersed in these beakers, are used as electrodes through which the outer E. M. F. can be applied. The short-circuit is completed by a microammeter, M, and the voltage supply, consisting of a battery, D, and a potential-divider, W, connected in series with the silver electrodes. During operation, the applied E. M. F. is adjusted by the potential-divider so that the potential drop across the skin is maintained at zero as read on the millivoltmeter. (In principle, a correction has to be made for the potential drop in the solutions separating the skin from the electrodes (cf. USSING and ZERAHN, 1951), but with the frog skins this correction is usually insignificant. For low-impedance structures such as small intestine etc., however, the correction may assume decisive importance). When the potential drop is adjusted to zero, the skin is, by definition, short-circuited: the current generated by the skin passes through an outer circuit of zero effective resistance and can be read on the microammeter. The bathing solutions are aerated and stirred during the experiment by bubble lifts with air inlets marked "a". Thus when the sodium isotopes ^{22}Na and ^{24}Na are added to the outside and inside bathing solutions, respectively, the sodium fluxes in both directions can be measured simultaneously during the experiment. Alternatively, influx and efflux can be measured on skin halves from the same animal. It then turns out that within the accuracy of the experiments the net inward transport of sodium (in electrical units) is equal to the short-circuit current (compare Table 1). The flux ratio which, for a passive ion ought to be one, is always very high, substantiating the active nature of the sodium transport. If the chloride fluxes are determined with the isotopes ^{36}Cl and ^{38}Cl, the flux ratio is very close to one (see Table 2), indicating that this ion is mainly or exclusively passive under the conditions of the experiment and that, consequently, it does not contribute to the short-circuit current. The uniqueness of sodium transport in the preparation is emphasized by the fact that chloride can be replaced by any number of anions, even non-penetrating ones like gluconate, without stopping the net sodium transport and the short-circuit current. Among cations, on the other hand, only lithium (ZERAHN, 1955) can partly replace sodium in the process of producing a short-circuit current. The ion transport processes in frog skin and other epithelia are discussed in greater detail in Vol. III.

It is important to remember that the short-circuit and the flux ratio method are "black-box" procedures where the whole sheet between the well-stirred bathing solutions is considered as the "membrane". Only under these conditions

Table 1. Data showing that active sodium transport accounts for the total short-circuit current of isolated frog skin. Sodium fluxes (double labelling with ^{22}Na and ^{24}Na) and short-circuit current for experiments with 9 isolated frog skins (R. temp.). Ringer solution on both sides. One-hour experimental periods. Fluxes and electric current calculated as μAmps/cm^{-2}. (Data from KOEFOED-JOHNSEN et al., 1952)

$J_{Na}^{(oi)}$	$J_{Na}^{(io)}$	J_{Na}	Short-circuit current
19.3	0.9	18.4	20.2
11.8	3.3	8.5	9.5
21.1	1.4	19.7	23.2
29.7	0.9	28.8	29.3
23.9	1.4	22.5	21.2
39.0	0.9	38.6	33.3
29.8	2.5	27.3	24.7
43.6	1.3	42.3	38.8
26.4	2.1	24.3	25.6

Table 2. Data showing that chloride ions do not contribute to short-circuit current of isolated frog skin. Chloride fluxes (double labelling with ^{36}Cl and ^{38}Cl) and short-circuit current for a number of experiments with isolated frog skins (R. temp.). Ringer solution on both sides. One-hour experimental periods. Fluxes and electric current given in μAmps/cm^{-2}. (Data from KOEFOED-JOHNSEN et al., 1952

Exp. No.	$J_{Cl}^{(oi)}$	$J_{Cl}^{(io)}$	J_{Cl}	Short-circuit current
I	16.9	14.9	−2.0	45
	12.0	11.5	−0.5	41
II	6.7	6.1	−0.6	57
	4.4	5.3	0.9	59
III	31.3	32.3	1.0	43

can the flux ratios for the different ions be compared with the short-circuit current. These conditions also set a limit for the amount of information which can be extracted from the experiments. The analysis only tells whether or not there is active transport of a particular species somewhere along the transport path. Also it should be emphasized that only substances which, by active or passive means, can penetrate all boundaries in the "membrane" studied will reveal their participation in active transport processes. Thus, since the outward-facing membrane of the isolated frog skin is virtually tight to potassium, the short-circuit method with the flux determinations demonstrate the presence of the inwardly directed active transport of sodium, whereas the outward directed active potassium transport (see Chapter 6, Vol. III), is only revealed when the outward-facing membrane is made leaky to potassium with polyene antibiotics (NIELSEN, 1971).

REHM (1975), in a recent discussion of the virtues and shortcommings of the short-circuiting technique, admits that if the important ions of the system all pass through conductive channels, the analysis is correct without limitations. He claims, however, that if the true active process in, say, frog skin, were the coupled transport of sodium and chloride, the chloride transport might pass unnoticed with the short-circuiting technique. This would be true if the investigators did not combine the current measurements with flux-ratio measurements for the ions concerned. One example he mentions (ANDERSEN and USSING, 1957) is very ill-chosen, however, since the preparations in question had been studied extensively in our institute with flux ratio measurements for both Na^+ and Cl^-, so that we knew that under the conditions of the experiments chloride did not contribute to the short-circuit current, while sodium accounted for all the short-circuit current within the accuracy of measurement.

Quite apart from this, however, a sodium-independent active chloride transport can be demonstrated under proper circumstances in both frog skin and toad skin (see Chapter 6, vol. III).

II. Single-File Diffusion

Even in cases where the transport is "downhill", with respect to the electrochemical potential gradient an unexpectedly high flux ratio may indicate active transport.

For downhill transport, however, abnormally high flux ratios can be due to a special type of passive transport, called single-file diffusion by HODGKIN and KEYNES (1955). These workers studied the leakage of potassium from dinitrophenol-poisoned decapod giant axons. Under these conditions the sodium potassium exchange pump is non-operative, and it turned out that the inward and outward fluxes of potassium, as determined with ^{42}K, could be described by a function of the form

$$RT \ln (J^{(io)} / J^{(oi)}) = n(\bar{\mu}^{(o)} - \bar{\mu}^{(i)}) \tag{20}$$

where n has a value between 2 and 3. Thus, if the flux ratio equation were to be valid, during its passage through the membrane potassium must behave as if it had two to three positive charges. The authors pointed out that this peculiar behaviour could be explained, on the assumption that the transport path for potassium had two to three sites in succession, all of which must be occupied by potassium ions. Furthermore, it was assumed that all ions in the row must move as one unit if they were to move at all. Statistical considerations led HODGKIN and KEYNES to the conclusion that such a single-file movement would be governed by an equation of the type given above. The single-file case has been treated in more detail by HECKMANN (1965) and by ESSIG et al. (1966). The theoretical possibility of such a type of penetration can hardly be disputed. Apart from the experimental condition for which it was originally suggested, few cases of manifest single-file behaviour have been described.

III. Solvent Drag Effects

Well-documented examples of solvent drag effects on transport of substances through cell membranes are few. This fact may be due to technical difficulties. Inspection of equation 16 shows that the effect is strongly dependent on the length of the path over which the effect is exerted. For plasma membranes, the effect would show only at very high flow rates. More likely, however, the lack of good examples is due to the fact, already mentioned above, that transport of most substances through plasma membranes takes place through highly specific pathways, possibly associated with carrier systems which need not allow water flow at all.

Flux ratio studies of epithelial transport, on the other hand, have yielded abundant evidence for anomalies which can best be explained in terms of solvent drag. To mention a few examples, osmotic inward flow of water through isolated toad skin gives rise to anomalous increased flux ratios for water (KOEFOED-JOHNSEN and USSING, 1953) thiourea, acetamide (ANDERSEN and USSING, 1957), glycerol, and sucrose (USSING, 1973).

The osmotic water flow can be replaced by hydraulic flow (USSING, 1969). The very nature of the method makes it impossible to localize the site or sites of interaction between the flow of water and the test substance in question. The site may be a pore in the plasma membrane or membranes. However, it may equally well be regions of rapid solvent flow in lateral spaces between the cells or tubular systems inside the cells. In this context it is pertinent to mention a semantic problem.

A solvent drag exerted upon some substance in the lateral species or in the intracellular tubular systems could also be classified as unstirred-layer effects (compare DAINTY and HOUSE, 1966; HAYS, 1972). The obvious point is that the flux ratio tells that there is interaction between solvent and solute, but additional information is necessary to decide whether or not this interaction is located in a cell membrane.

In heteroporous systems the flow pattern for the solvent may become quite intricate, including actual circulation of the solvent. In such cases the direction of the solvent drag may be opposite to that expected from the direction of net osmotic flow. The net inward transport of a whole series of solutes through frog skins exposed to hypertonic solutions on the outside (USSING, 1966; FRANZ and VAN BRUGGEN, 1967), may well be due to such an anomalous solvent drag (USSING, 1969) although FRANZ and VAN BRUGGEN assumed it to be due to direct solute-solute coupling ("solute drag").

GALEY and VAN BRUGGEN (1970) have performed model experiments on artificial membranes which seemed to support their solute drag hypothesis. However, PATLAK and RAPOPORT (1971) were able to show that most of the cases of apparent solute drag could be explained in terms of solvent circulation in a heteroporous system.

G. Concluding Remarks

The tracer method has become an indispensable tool in the study of biological transport. It should always be remembered, however, that the determination of the tracer permeability, or the tracer flux, in one direction only, yields virtually no information as to the forces bringing about the transport. Such data are certainly less informative than chemical determinations of net fluxes.

However, if fluxes in both directions are measured, or if the isotope fluxes are combined with net flux determinations, one obtains a type of information about the membrane processes which could hardly be obtained in any other way.

References

ANDERSEN, B., USSING, H. H.: Acta physiol. scand. **39**, 228 (1957).
CHRISTENSEN, H. N.: Advanc. Protein Chem. **15**, 239 (1960).
CRANK, J.: In: The Mathematics of Diffusion. Oxford: Clarendon 1956.
DAINTY, J., HOUSE, C. R.: J. Physiol. **185**, 172 (1966).
DE SOUSA, R. C., LI, J. H., ESSIG, A.: Nature **231**, 44 (1971).
ESSIG, A.: Biophys. J. **8**, 53 (1968).
ESSIG, A., KEDEM, O., HILL, T. L.: J. theoret. Biol. **13**, 72 (1966).
FRANCIS, W. L.: Nature (Lond.) **131**, 805 (1933).
FRANZ, T. J., VAN BRUGGEN, J. T.: J. gen. Physiol. **50**, 933 (1967).
GALEY, W. R., VAN BRUGGEN, J. T.: J. gen. Physiol. **55**, 220 (1970).
GOLDMAN, D. E.: J. gen. Physiol., **27**, 37 (1943).
GOTTLIEB, M. H., SOLLNER, K.: Biophys. J. **8**, 515 (1968).
HARRIS, E. J., BURN, G. P.: Trans. Faraday Soc. **45**, 508 (1949).
HAYS, R. M.: In: Current Topics in Membranes and Transport (F. Bronner and A. Kleinzeller, Eds), Vol. 3. New York: Academic Press 1972, pp. 339–366
HECKMANN, K.: Zschr. phys. Chemie, N. F. **44**, 184 (1965).
HEVESY, G. v.: In: Radioactive Indicators. New York: Interscience 1948.
HODGKIN, A. L., KEYNES, R. D.: J. Physiol. **128**, 61 (1955).
HODGKIN, A. L., HUXLEY, A. F., KATZ, B.: J. Physiol. **116**, 424 (1952).
HOSHIKO, T., LINDLEY, B. D.: Biochim. biophys. Acta **79**, 301 (1964).
HOSHIKO, T., LINDLEY, B. D.: J. gen. Physiol. **50**, 729 (1967).
HOSHIKO, T., LINDLEY, B. D.: J. theoret. Biol. **26**, 315 (1970).
JACQUEZ, J. A.: In: Compartmental Analysis in Biology and Medicine. New York: Elsevier, Amsterdam, London, New York 1972.
JOHNSON, F. H., EYRING, H., POLISSAR, M. J.: In: The Kinetic Basis of Molecular Biology. New York: Wiley 1954.
KEDEM, O., ESSIG, A.: J. gen. Physiol., **48**, 1047 (1965).
KEDEM, O., KATCHALSKY, A.: Biochim. biophys. Acta **27**, 229 (1958).
KEDEM, O., KATCHALSKY, A.: Trans. Faraday Soc. **59**, 1918 (1963a).
KEDEM, O., KATCHALSKY, A.: Trans. Faraday Soc. **59**, 1931 (1963b).
KEDEM, O., KATCHALSKY, A.: Trans. Faraday Soc. **59**, 1941 (1963c).
KOEFOED-JOHNSEN, V., USSING, H. H.: Acta physiol. scand. **28**, 60 (1953).
KOEFOED-JOHNSEN, V., USSING, H. H., ZERAHN, K.: Acta physiol. scand. **27**, 38 (1952).
KOTYK, A., JANÁCEK, K.: Cell Membrane Transport. Principles and Techniques, 2nd ed. New York: Plenum 1975.
KROGH, A.: Proc. roy. Soc. Lond. B. **133**, 140 (1946).

LE FEVRE, P. G.: In: Current Topics in Membranes and Transport, Vol. 7 (F. Bronner and A. Kleinzeller, Eds). New York: Academic Press 1975, p. 109.
LEV, A. A., ARMSTRONG, W. McD.: In: Current Topics in Membranes and Transport, Vol. 6 (F. Bronner and A. Kleinzeller, Eds). New York: Academic Press 1975, p. 59.
LEVI, H., USSING, H. H.: Acta physiol. scand. **16**, 232 (1948).
LEVI, H., USSING, H. H.: Nature (Lond.) **164**, 928 (1949).
LIEB, W. R., STEIN, W. D.: Biochim. biophys. Acta **373**, 165 (1974a).
LIEB, W. R., STEIN, W. D.: Biochim. biophys. Acta **373**, 178 (1974b).
LUND, E. J., STAPP, P.: In: Bioelectric Fields and Growth. By E. J. Lund: Austin University, Texas Press 1947, p. 235.
MAIZELS, M.: Symp. Soc. exper. Biol., **8**, 202 (1954).
MEARES, P., USSING, H. H.: Trans. Faraday Soc. **55**, 142 (1959a).
MEARES, P., USSING, H. H.: Trans. Faraday Soc. **55**, 244 (1959b).
NIELSEN, R.: Acta physiol. scand. **83**, 106 (1971).
ONSAGER, L.: Ann. N. Y. Acad. Sci. **46**, 241 (1945).
PATLAK, C. S., RAPOPORT, S. I.: J. gen. Physiol. **57**, 113 (1971).
REHM, W. S.: In: Current Topics in Membranes and Transport. Vol. 7 (F. Bronner and A. Kleinzeller, Eds). New York: Academic Press 1975, p. 217.
ROSENBERG, T.: Acta chem. scand. **2**, 14 (1948).
ROSENBERG, T., WILBRANDT, W.: Exp. Cell Res. **9**, 49 (1955).
ROSENBERG, T., WILBRANDT, W.: J. gen. Physiol. **41**, 289 (1957).
SCHOENHEIMER, R.: The Dynamic State of Body Constituents. Cambridge: Harvard University Press 1946.
SCHWARTZ, T. L.: Biophys. J. **11**, 596 (1971).
STAPP, P.: Proc. Soc. exp. Biol. (N. Y.) **46**, 382 (1941).
TEORELL, T.: Arch. Sci. Physiol. **3**, 205 (1949).
USSING, H. H.: Nature **160**, 262 (1947).
USSING, H. H.: Cold Spring Harb. Symp. quant. Biol. **13**, 193 (1948).
USSING, H. H.: Acta physiol. scand. **17**, 1 (1949a).
USSING, H. H.: Acta physiol. scand. **19**, 43 (1949b).
USSING, H. H.: Advanc. Enzymol. **13**, 21 (1952).
USSING, H. H.: Ann. N. Y. Acad. Sci. **137**, 543 (1966).
USSING, H. H.: Quart. Rev. Biophys. **1**, 365 (1969).
USSING, H. H.: In: Perspectives in Membrane Biophysics. A Tribute to Kenneth S. Cole. (D. P. Agin, Ed.). New York, London, Paris: Gordon and Breach Science Publishers 1972.
USSING, H. H., THORN, N. A. (Eds): Transport Mechanisms in Epithelia. Alfred Benzon Symposium V. Copenhagen: Munksgaard, New York: Academic Press 1973.
USSING, H. H., ZERAHN, K.: Acta physiol. scand. **23**, 110 (1951).
WYSSBROD, H. R., SCOTT, W. N., BRODSKY, W. A., SCHWARTZ, T. L.: In: Handbook of Neurochemistry. Vol. **5**, Part B (A. Lajtha, Ed.). New York: Plenum 1971, p. 683.
ZERAHN, K.: Acta physiol. scand. **33**, 347 (1955).

Chapter 4

Nonequilibrium Thermodynamics of Isotope Flow through Membranes

F. A. SAUER

A. Introduction

This article aims to present a theoretical description of isotope transport across membranes in phenomenological terms. Thermodynamic considerations play a crucial role in such a description. Following the pioneering work of USSING (1949) on the flux ratio, further applications of the methods of nonequilibrium thermodynamics were worked out by NIMS (1962) and KEDEM and ESSIG (1965). All thermodynamic descriptions of isotope transport suffer from an inherent difficulty, when one considers unidirectional flows, caused by the singularity ($\pm\infty$) in the conjugated driving force for an isotope. This article will re-examine this question and try to clarify the extra nonthermodynamic assumptions one has to invoke to obtain a consistent description of unidirectional flows. Besides, the general approach to be outlined will also apply to membrane systems that show large deviations from thermodynamic equilibrium. The linear laws usually considered, which sometimes fail to describe biological membranes, turn out to be special limiting cases of the more general laws.

B. The System: Definitions and Mathematical Techniques

In this section a system (Fig. 1) is considered in which a membrane in a steady state separates two homogeneous phases ′ and ″, which are in internal thermodynamic equilibrium. The membrane itself is characterized by its two sides a and b. At least one chemical reaction, while inhibited in the phases ′ and ″, is postulated to take place inside the membrane. This reaction can be described by

$$\sum_{k=1}^{n} \nu_k \{B_k\} = 0\,, \tag{B 1}$$

where ν_k represents the stoichiometric numbers and $\{B_k\}$ the chemical species participating in the chemical reaction. Components that do not participate in the reaction are characterized by zero stoichiometric numbers ($\nu_k \equiv 0$).

We assume that both outer phases ′ and ″ contain n permeable chemical components and possibly m impermeable components. Furthermore, we assume

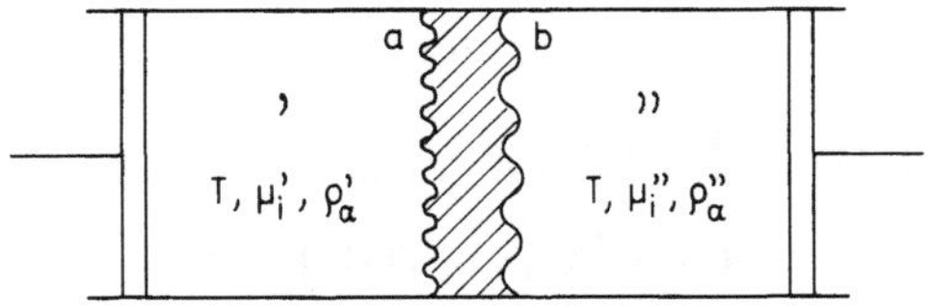

Fig. 1. The system

that there are two isotopes, α and β, of the nth component, which can be distinguished by experimental means and have the same physicochemical properties as component n. Component n is assumed not to participate in the chemical reaction ($\nu_n = 0$). In the discussion that follows it will be important to distinguish between two nonequilibrium situations, which call for two different modes of description. In one the component n alone is considered. This description is called the chemical picture. If we distinguish between the two isotopes, we speak of the isotope picture.

The assumption that the isotopes α and β have the same physicochemical properties implies that all possible isotope effects are neglected as small. This is a reasonable assumption at room temperature. Therefore, isotope separation falls outside the scope of the present formalism. When these assumptions hold, there is a very close connection between the chemical picture and the isotope picture. The concentrations c_n of the component n and those of the isotopes α and β are related by

$$c_n = c_\alpha + c_\beta \,. \tag{B 2}$$

The Gibbs-Duhem equation relates the chemical potentials μ_n, μ_α, and μ_β:

$$c_n \, d\mu_n = c_\alpha \, d\mu_\alpha + c_\beta \, d\mu_\beta \; (T = \text{const}) \,. \tag{B 3}$$

This equation implies that μ_α and μ_β have the form

$$\mu_\alpha = \mu_n + RT \ln \varrho_\alpha \,; \qquad \mu_\beta = \mu_n + RT \ln \varrho_\beta \,, \tag{B 4}$$

where μ_n is independent of the isotope composition.
The ϱ_α and ϱ_β are the specific activities of α and β defined by

$$\varrho_\alpha \equiv \frac{c_\alpha}{c_n} \,; \qquad \varrho_\beta \equiv \frac{c_\beta}{c_n} \,, \tag{B 5}$$

with the further requirement that

$$\varrho_\alpha + \varrho_\beta = 1 \,. \tag{B 6}$$

If the components n, α and β are charged, the μ's in (B.3) and (B.4) should be replaced by the electrochemical potentials of the components n, α and β.

An average specific activity $\tilde{\varrho}_\alpha$ of ϱ_α' and ϱ_α'' is now defined so that the relation

$$\Delta\mu_n = \tilde{\varrho}_\alpha \Delta\mu_\alpha + (1 - \tilde{\varrho}_\alpha)\Delta\mu_\beta \tag{B 7}$$

is identically satisfied for finite differences of chemical potentials $\Delta\mu_n \equiv \mu_n' - \mu_n''$, $\Delta\mu_\alpha \equiv \mu_\alpha' - \mu_\alpha''$ and $\Delta\mu_\beta \equiv \mu_\beta' - \mu_\beta''$. Using Eq. (B.4), one finds

$$\tilde{\varrho}_\alpha = \left(1 - \frac{\Delta \ln \varrho_\alpha}{\Delta \ln (1 - \varrho_\alpha)}\right)^{-1}, \tag{B 8}$$

where $\Delta \ln \varrho_\alpha = \ln \dfrac{\varrho_\alpha'}{\varrho_\alpha''}$ and $\Delta \ln (1 - \varrho_\alpha) = \ln \left(\dfrac{1 - \varrho_\alpha'}{1 - \varrho_\alpha''}\right)$.

Some properties of this function will now be discussed. For $\varrho_\alpha' = \varrho_\alpha''$ the logarithmic mean $\tilde{\varrho}_\alpha$ becomes

$$\tilde{\varrho}_\alpha = \varrho_\alpha' = \varrho_\alpha'', \tag{B 9}$$

while for

$$\tilde{\varrho}_\alpha' = 0 \text{ or } \varrho_\alpha'' = 0 \text{ we obtain}$$
$$\tilde{\varrho}_\alpha = 0. \tag{B 10}$$

For values of ϱ_α' and ϱ_α'' obeying the condition

$$\varrho_\alpha' + \varrho_\alpha'' = 1 \tag{B 11}$$

$\tilde{\varrho}_\alpha$ becomes

$$\tilde{\varrho}_\alpha = \frac{1}{2}$$

and coincides with the arithmetic mean

$$\bar{\varrho}_\alpha = \frac{\varrho_\alpha' + \varrho_\alpha''}{2}, \tag{B 12}$$

which also equals $1/2$ when the condition (B.11) is imposed. Introduction of the new variables $\bar{\varrho}_\alpha$ and $\Delta\varrho_\alpha$, given by

$$\bar{\varrho}_\alpha = \frac{\varrho_\alpha' + \varrho_\alpha''}{2} \text{ and } \Delta\varrho_\alpha = \varrho_\alpha' - \varrho_\alpha'', \tag{B 13}$$

which are related to ϱ_α' and ϱ_α'' by the transformations

$$\varrho_\alpha' = \bar{\varrho}_\alpha + \frac{\Delta\varrho_\alpha}{2}; \qquad \varrho_\alpha'' = \bar{\varrho}_\alpha - \frac{\Delta\varrho_\alpha}{2}, \tag{B 14}$$

into (B.8), yields the following Taylor expansion of $\tilde{\varrho}_\alpha$, which for small values of

$$\Delta\varrho_\alpha \left(\frac{|\Delta\varrho_\alpha|}{2\bar{\varrho}_\alpha} << 1 \text{ and } \frac{|\Delta\varrho_\alpha|}{2(1-\bar{\varrho}_\alpha)} << 1\right)$$

$$\tilde{\varrho}_\alpha = \bar{\varrho}_\alpha \left(1 - \frac{(1-2\bar{\varrho}_\alpha)}{3(1-\bar{\varrho}_\alpha)} \left(\frac{\Delta\varrho_\alpha}{2\bar{\varrho}_\alpha}\right)^2\right) + 0\left(\left(\frac{\Delta\varrho_\alpha}{2\bar{\varrho}_\alpha}\right)^4\right). \quad \text{(B 15)}$$

Thus, the logarithmic and the arithmetic means become nearly equal for small values of $\Delta\varrho_\alpha$ and the deviation is of the order of $\left(\frac{\Delta\varrho_\alpha}{2\bar{\varrho}_\alpha}\right)^2$, i.e. $0\left(\frac{\Delta\varrho_\alpha}{2\bar{\varrho}_\alpha}\right)^2$.

In a typical black-box description of the membrane system, such as the one envisaged here, at steady state all observables are related to "events" in the outer phases ′ and ″. The nonequilibrium state of the membrane system is characterized by the flows of the components through the membrane and the rate of reaction inside the membrane. Here one must distinguish between the two pictures referred to earlier. In the chemical picture the n flows of the components relative to the membrane are defined by

$$J_i^{\mathrm{a}} \equiv \frac{dn_i'}{dt} \quad \text{and} \quad J_i^{\mathrm{b}} \equiv \frac{dn_i''}{dt}\,; (i = 1 \ldots n), \quad \text{(B 16)}$$

where J_i^{a} is the flow of the ith component flowing out of the a side of the membrane and J_i^{b} is that from the b side. The n_i are the mole numbers of the permeable components. If we have ions with electrical current I through the membrane Eq. (B 16) must be modified for the key ion ($i = 1$, for example) to:

$$J_1^{\mathrm{b}} = \frac{dn_1''}{dt} - \frac{I}{e_1}, \quad \text{(B 16a)}$$

where e_1 is the charge per mole of the key ion.

When the membrane is in a steady state, J_i^{a} and J_i^{b} are connected with the rate of reaction J_{r} by the steady-state condition:

$$J_i^{\mathrm{a}} + J_i^{\mathrm{b}} = \nu_i J_{\mathrm{r}}; \qquad (i = 1, \ldots n). \quad \text{(B 17)}$$

J_{r} can be determined by means of this equation, if the flows are known. Finally, one has n independent flows (the J_i^{b}, for example) and J_{r} for the description of the nonequilibrium state. In a steady state these quantities depend on the thermodynamic state of the ′ and ″ phases. The thermodynamic state variables of the outer phases are given by

$$\begin{aligned} &T, \\ &{}' \equiv (\mu_1', \ldots \mu_n', \mu_{n+1}', \ldots \mu_{n+m}), \\ &{}'' \equiv (\mu_1'', \ldots \mu_n'', \mu_{n+1}'', \ldots \mu_{n+m}''), \end{aligned} \quad \text{(B 18)}$$

where T is the absolute temperature (assumed to be the same in $'$ and $''$) and the μ's are the chemical potentials of the components.

Thus, in the chemical picture one can regard the J_i^{b} and J_{r} as functions of these $2(n + m) + 1$ variables. The thermodynamic treatment shows that these functions become zero in thermodynamic equilibrium, which is characterized by the requirements that

$$X_i \equiv \Delta\mu_i \equiv \mu_i' - \mu_i'', \qquad (i = 1, \dots n) \tag{B 19}$$

and the affinity of the chemical reaction

$$A' \equiv -\sum_{i=1}^{n} \nu_i \mu_i' \tag{B 20}$$

are identically zero. Therefore, it is meaningful to introduce the $n + 1$ quantities (B. 19 and B. 20) as new independent variables. This can be done by eliminating some of the variables (B.18) by means of the following transformations

$$\begin{aligned} &X_i \equiv \mu_i' - \mu_i'' \qquad (i = 1, \dots n); \\ &A' \equiv -\sum_{i=1}^{n-1} \nu_i \mu_i'; \\ &s_l = s_l(\mu_1' \dots \mu_m', \mu_1'' \dots \mu_m'') \qquad (l = 2, \dots, n+2m); \\ &T = T. \end{aligned} \tag{B 21}$$

In the sequel the temperature will not appear explicitly. $s_l = s_l(', '')$ is an abbreviation for the "reference state." The choice of the reference state is arbitrary. There is an infinite number of analytical forms for the reference state. For convenience, if $'$ and $''$ were the same, we would choose the reference state to be

$$s_l(', ') = \mu_l' \qquad (l = 2, \dots, n). \tag{B 22}$$

For theoretical considerations it is more useful to introduce a symmetrical reference state. This is a set of functions s_l, which are symmetrical in the $'$ and $''$ variables. From the point of view of practical applications, the most convenient reference state would be the one that remains constant while the driving forces X_l are varied. In simple cases without impermeable components, where the average thermodynamic state of the $'$ and $''$ phases is kept constant during the experiment, the arithmetic mean

$$s_l = \frac{\mu_l' + \mu_l''}{2}, \qquad (l = 2, \dots, n) \tag{B 23}$$

is very often used as the reference state. Finally we have

$$J_i^b = J_i^b(s_l, X_k, A') \qquad (i, k = 1, \dots, n);$$

$$J_r = J_r(s_l, X_k, A') \qquad (l = 2, \dots n+2m). \tag{B 24}$$

Clearly the independent flows J_i^b and the reaction rate J_r depend on the reference state s_l and the "driving forces" X_k and A'. Special properties of these functions with regard to their dependence on X_k and A' are discussed in the nonequilibrium thermodynamics part of this article. Furthermore, in view of (B 24), a complete description of the nonequilibrium state of the system follows on assignment of the variables s_l, J_i^b and J_r.

In the isotope picture the flows

$$J_\alpha^b \equiv \frac{dn_\alpha''}{dt} \quad \text{and} \qquad J_\beta^b \equiv \frac{dn_\beta''}{dt} \tag{B 25}$$

are introduced instead of J_n^b. Here n_α'' and n_β'' are the mole numbers of α and β. Because one assumes that the chemical component n does not participate in the chemical reaction, the flows emerging from the "a" side of the membrane in the steady state are given by

$$J_\alpha^a = -J_\alpha^b \text{ and} \qquad J_\beta^a = -J_\beta^b. \tag{B 26}$$

The isotope flows J_α^b and J_β^b obviously are functions of the variables s_l, X_k, A'. But in addition they depend on the isotopic composition of the $'$ and $''$ phases. On choosing the independent variables ϱ_α' and ϱ_α'' to specify the isotopic composition, we have

$$J_\alpha^b = J_\alpha^b(s_l, X_k, A', \varrho_\alpha', \varrho_\alpha'')$$

$$J_\beta^b = J_\beta^b(s_l, X_k, A' \varrho_\alpha, \varrho_\alpha''). \tag{B 27}$$

The flows J_k^b of the other chemical components k ($k = 1, \dots, n-1$) and J_r do not depend on the isotope variable, because our treatment disregards isotope effects.

There is a further relation between the functions J_α^b and J_β^b, namely

$$J_\alpha^b(s_l, X_k, A', 1-\varrho_\alpha', 1-\varrho_\alpha'') = J_\beta^b(s_l, X_k, A', \varrho_\alpha', \varrho_\alpha''). \tag{B 28}$$

This means the flow of the isotope β corresponding to isotopic compositions ϱ_α' and ϱ_α'' is equal to the flow of the isotope α corresponding to isotopic compositions $1 - \varrho_\alpha'$ and $1 - \varrho_\alpha''$, when the other variables are held fixed.

The net flow J_n^b of the chemical component n equals the sum of the isotope flows. Therefore

$$J_n^b = J_\alpha^b + J_\beta^b. \tag{B 29}$$

(B 28) and (B 29) together yield

$$J_n^b = J_\alpha^b(s_l, X_k, A', \varrho_\alpha', \varrho_\alpha'') + J_\alpha^b(s_l, X_k, A', 1 - \varrho_\alpha', 1 - \varrho_\alpha''). \tag{B 30}$$

Since J_n^b on the left side of this equation is independent of the isotopic compositions, (B 30) is a condition concerning the dependence of J_α^b on the isotope compositions. How rigid this condition is will be discussed later.

Sometimes it will be more convenient to consider J_α^b as a function of the variables

$$s_l, (l = 2, \ldots n+2m);\ J_k, (k = 1, \ldots n);\ J_r;\ \varrho_\alpha';\ \varrho_\alpha''. \tag{B 31}$$

The rest of this article is devoted to a discussion of the properties of the isotope flows J_α^b as functions of the variables (B 31). The mathematical content of this discussion is quite elementary; nevertheless, a few explanatory remarks may be appropriate at this point. These remarks have to do with the fact that one deals with functions

$$f(y_l, X_k) \tag{B 32}$$

of two sets of variables, y_l and X_k, where the X_k's are sometimes small ($|X_k| << 1$). The zero- and first-order terms of the Taylor expansion of f around $X_k = 0$ then yield a fairly good approximation of f. This expansion reads

$$f(y_l, X_k) = f(y_l, X_k = 0) + \sum_{k=1}^{n} f_k(y_l)\, X_k + \text{remainder} \tag{B 33}$$

with

$$f_k(y_l) = \left(\frac{\partial f}{\partial X_k}\right)_{X_i=0}. \tag{B 34}$$

The remainder of this Taylor series depends on the y_l and X_k. For a given function f, it is possible to find an estimate of the remainder. The remainder can be neglected if it is small. On the other hand, if the experimental data can be adequately described by the zero- and first-order terms of Eq. (B 33) with a good degree of accuracy, this equation can be used to determine the expansion coefficients f_k. Since the expansion coefficients and the remainder of the Taylor series depend on the y_l's, the accuracy of the linear approximation is clearly influenced by the choice of the parameters y_l. This point must be borne in mind when the usefulness of the linear approximation in the thermodynamic approach is discussed.

C. Nonthermodynamic Considerations of Isotope Flow

Since measurements of isotope flows and the interpretation of the data are widely used in attempts to gain information about biological transport systems, it is worthwhile to continue the discussion of the properties of the isotope flows. The title of this article notwithstanding, it is necessary to pay attention to some nonthermodynamic considerations that are usually overlooked. With this in view, we return to Eq. (B 30) cast in the form

$$J_n^{\mathrm{b}}(s_l, J_k^{\mathrm{b}}, J_{\mathrm{r}}) = J_\alpha{}^{\mathrm{b}}(s_l, J_k^{\mathrm{b}}, J_{\mathrm{r}}, \varrho_\alpha', \varrho_\alpha'') + J_\alpha^{\mathrm{b}}(s_l, J_k^{\mathrm{b}}, J_{\mathrm{r}}, 1-\varrho_\alpha', 1-\varrho_\alpha''). \tag{C 1}$$

Here the variables J_k^{b}, J_{r} are used in the place of X_k, A'. As already mentioned, this equation is essentially a condition with regard to the dependence of the isotope flow on the isotope compositions, the left-hand side of this equation being independent of the isotope compositions. A linear dependence of J_α^{b} on ϱ_α' and ϱ_α'' is sufficient to fulfill Eq. (C 1). This gives

$$J_\alpha^{\mathrm{b}}(s_l, J_k^{\mathrm{b}}, J_{\mathrm{r}}, \varrho_\alpha', \varrho_\alpha'') = F_+(s_l, J_k^{\mathrm{b}}, J_{\mathrm{r}}) \cdot \varrho_\alpha' + F_-(s_l, J_k^{\mathrm{b}}, J_{\mathrm{r}}) \cdot \varrho_\alpha'', \tag{C 2}$$

where F_+ and F_- are independent of the isotopic compositions. In fact, there are an infinite number of nonlinear functions of ϱ_α' and ϱ_α'' that also fulfill (C 1) (in addition to the one proposed above). Any arbitrary function $G(x, y)$ with the properties

$$G(x,y) = -G(-x,-y)$$
$$G\left(-\frac{1}{2}, -\frac{1}{2}\right) = -\frac{J_n^{\mathrm{b}}}{2} \tag{C 3}$$

leads to

$$J_\alpha^{\mathrm{b}} = \frac{J_n^{\mathrm{b}}}{2} + G\left(s_l, J_k^{\mathrm{b}}, J_{\mathrm{r}}, x = \varrho_\alpha' - \frac{1}{2}, y = \varrho_\alpha'' - \frac{1}{2}\right), \tag{C 4}$$

which fulfills (C 1). Note that (C 2) is a special case of (C 4). Statistical mechanical and hydrodynamic models in general lead to the linear form. There is also overwhelming experimental evidence to support the proposed linear form (C 2). One observes that this is really not an approximation restricted to small values of the ϱ's; there are sound reasons to postulate the validity of (C 2) for all accessible values of ϱ_α' and ϱ_α'', namely

$$0 \leq \varrho_\alpha', \qquad \varrho_\alpha'' \leq 1. \tag{C 5}$$

One can thus say that the proposed linear form is not only sufficient from a mathematical point of view, but also necessary from a physical point of view.

The quantities F_+ and F_- that were introduced into (C 2) are actually the unidirectional flows, because one has

$$J_\alpha^{\mathrm{b}} = F_+ \cdot \varrho_\alpha' \qquad \text{if } \varrho_\alpha'' = 0$$

and (C 6)

$$J_\alpha^{\mathrm{b}} = F_- \cdot \varrho_\alpha'' \qquad \text{if } \varrho_\alpha' = 0\,.$$

It has already been mentioned that the unidirectional flows are independent of the isotopic compositions but that they depend in general on the s_l, J_k^{b} and J_{r}. The discussion of this dependence will be postponed to the section of this article that deals with thermodynamic aspects. Nevertheless, some essential properties of F_+ and F_- emerge without reference to this dependence. With the adopted convention for the signs of the flows it follows from (C 6) (positive flows point outward from the membrane) that F_+ must be positive or zero

$$F_+ \geqq 0 \tag{C 7}$$

and F_- negative or zero

$$F_- \leqq 0. \tag{C 8}$$

The indices + and − have accordingly been chosen. Usually the F_+ is called the influx and $-F_-$ the efflux.

Because of (B 28) we get

$$J_\beta^{\mathrm{b}} = F_+ \cdot (1 - \varrho_\alpha') + F_- \cdot (1 - \varrho_\alpha'') \tag{C 9}$$

with the same F_+ and F_- as in (C 2).

Introduction of (C 2) and (C 9) into (B 29) yields

$$J_n^{\mathrm{b}} = F_+ + F_- \tag{C 10}$$

i. e., the sum of the unidirectional flows is equal to the net flow J_n^{b} (Ussing's first law for unidirectional flows). Extensive use has been made of this relation for the determination of net flows from measurements of both unidirectional flows. It should be mentioned that an essential assumption in the derivation of (C 10) is the assumed linearity in ϱ_α', ϱ_α'' of (C 2).

Further, one can say something about the behavior of F_+ and F_- for large positive or large negative net flows J_n^{b}. From the inequalities of (C 7) and (C 8) and Eq. (C 10), one arrives at the following asymptotic behavior:

$$J_n^{\mathrm{b}} > 0 \qquad \lim_{J_n^{\mathrm{b}} = +\infty} F_+ = J_n^{\mathrm{b}}; \qquad \lim_{J_n^{\mathrm{b}} = +\infty} F_- = 0$$

$$J_n^{\mathrm{b}} < 0 \quad \lim F_+ = 0; \quad \lim F_- = J_n^{\mathrm{b}}$$
$$J_n^{\mathrm{b}} = -\infty \qquad J_n^{\mathrm{b}} = -\infty$$

Eq. (C 7), (C 8), (C 10), and (C 11) suggest the following analytical forms for the unidirectional flows

$$F_+ = \frac{J_n \exp(BJ_n)}{\exp(BJ_n) - 1}; \qquad F_- = -\frac{J_n}{\exp(BJ_n) - 1}, \tag{C 12}$$

B being an unspecified positive function of s_l, J_k, J_r

$$B = B(s_l, J_k, J_r) > 0. \tag{C 13}$$

Here the index b has been omitted.

Eq. (C 12) are generalizations of equations derived by KEDEM and ESSIG (1965) from a continuum model of a system close to thermodynamic equilibrium. While the Kedem-Essig equations are valid only for near-equilibrium situations, Equations (C 12) are valid both for large net flows J_k and large reaction rate J_r. They fulfill (C 7), (C 8), (C 10) and (C 11) automatically. Thanks to the arbitrary nature of B they are quite general. Knowledge of B determines both unidirectional flows.

If the net flow J_n becomes zero, one obtains from (C 12)

$$F_+(J_n = 0) = \frac{1}{B(J_n = 0)}; \qquad F_- = -\frac{1}{B(J_n = 0)}. \tag{C 14}$$

For $J_n = 0$ (C 10) yields

$$F_+ = -F_-. \tag{C 15}$$

Introduction of (C 12) into (C 2) gives

$$J_\alpha^{\mathrm{b}} = \frac{J_n}{\exp(BJ_n) - 1}(\varrho_\alpha' \exp(BJ_n) - \varrho_\alpha'') \tag{C 16}$$

or

$$J_\alpha^{\mathrm{b}} = \bar{\varrho}_\alpha J_n + \frac{\Delta\varrho_\alpha}{2} \frac{\exp(BJ_n) + 1}{\exp(BJ_n) - 1} J_n \tag{C 17}$$

with $\bar{\varrho}_\alpha \equiv \frac{\varrho_\alpha' + \varrho_\alpha''}{2}$ and $\Delta\varrho_\alpha \equiv \varrho_\alpha' - \varrho_\alpha''$.

For $\Delta\varrho_\alpha = 0$, (C 17) becomes

$$J_\alpha^{\mathrm{b}} = \bar{\varrho}_\alpha J_n^{\mathrm{b}}. \tag{C 18}$$

Thus, isotope flow measurements can be used to determine the net flow J_n^{b} if the difference of the specific activities is zero.

The discussion of data from isotope experiments uses the so-called flux ratio r, defined by

$$r \equiv -\frac{F_+}{F_-}\,. \tag{C 19}$$

Using (C 12), one finds

$$r = \exp(BJ_n)\,. \tag{C 20}$$

Again USSING (1949) has shown that for very simple systems the flux ratio has the form (ideal flux ratio r_{id})

$$r = r_{\mathrm{id}} \equiv \exp\frac{X_{\mathrm{n}}}{RT} \quad \text{(Ussing's "second law")} \tag{C 21}$$

where X_n is the difference of the electrochemical potential of the nth component, R the gas constant, and T the absolute temperature. Thermodynamic considerations published by KEDEM and ESSIG (1965) have revealed that this relation cannot be maintained for systems close to thermodynamic equilibrium in the presence of any kind of coupling between the flow of the nth component and other processes or isotope coupling. LI and ESSIG (1976) have further pointed out that any heterogeneity in the membrane might lead to deviations from (C 21). If the thermodynamic system itself is far from equilibrium and nonlinear effects occur, one again expects deviations of the flux ratio from r_{id}, because then B in (C 20) depends on the J_k and J_r.

To throw some light on the properties of composite membrane systems, two simple arrangements of two membranes v and w (see Fig. 2) with different transport properties are treated. If each of the membranes in the series arrangement is in the steady state, one could write

$$J_\alpha^{\mathrm{a}} = -F_+^{\mathrm{v}}\varrho_\alpha' - F_-^{\mathrm{v}}\varrho_\alpha^{\mathrm{o}}$$

$$J_\alpha^{\mathrm{b}} = F_+^{\mathrm{w}}\varrho_\alpha^{\mathrm{o}} + F_-^{\mathrm{w}}\varrho_\alpha''\,. \tag{C 22}$$

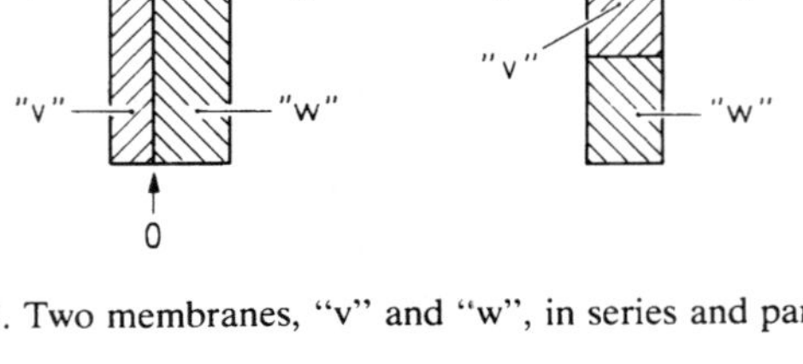

Fig. 2. Two membranes, "v" and "w", in series and parallel

Because of the assumed steady state, the relation

$$J_\alpha^a = -J_\alpha^b \tag{C 23}$$

holds. Elimination of ϱ_α^o yields

$$J_\alpha^b = \frac{F_+^w F_+^v}{F_+^w + F_+^w - J_n^b} \varrho_\alpha' + \frac{F_-^w F_-^v}{F_-^w + F_-^v - J_n^b} \varrho_\alpha'' . \tag{C 24}$$

Where use has been made of

$$J_n^b = F_+^v + F_-^v = F_+^w + F_-^w . \tag{C 25}$$

Therefore, one has for the composite membrane

$$F_+^{vw} = \frac{F_+^w F_+^v}{F_+^w + F_+^v - J_n^b} ; \qquad F_-^{vw} = \frac{F_-^w F_-^v}{F_-^w + F_-^v - J_n^b} . \tag{C 26}$$

The relations between $F_\pm^{vw}$ and $F_\pm^w$, $F_\pm^v$ are analogous in form to that between the conductivity of a composite membrane (made up of two membranes in series) and those of its constituent membranes, apart from the J_n^b in the denominator.

For the parallel arrangement one has

$$J_\alpha^{vb} = F_+^v \varrho_\alpha' + F_-^v \varrho_\alpha''$$

$$J_\alpha^{wb} = F_+^w \varrho_\alpha' + F_-^w \varrho_\alpha'' \tag{C 27}$$

and

$$J_\alpha^b = J_\alpha^{vb} + J_\alpha^{wb} \tag{C 28}$$

Putting these together, one finds

$$J_\alpha^b = (F_+^w + F_+^v) \varrho_\alpha' + (F_-^w + F_-^v) \varrho_\alpha'' . \tag{C 29}$$

Therefore,

$$F_+^{vw} = (F_+^w + F_+^v); \qquad F_-^{vw} = (F_-^w + F_-^v) . \tag{C 30}$$

The relations above between the unidirectional flows again bear a formal resemblance to that between the conductivity of a composite membrane (made up of two membranes in parallel) and the conductivities of the constituent membrane.

An essential feature in the description of membrane transport is the investigation of the behavior of the membrane system under reflection of the membrane (see Fig. 3). The transport through the membrane in the original configuration is then compared with that in the reflected configuration. In the original configu-

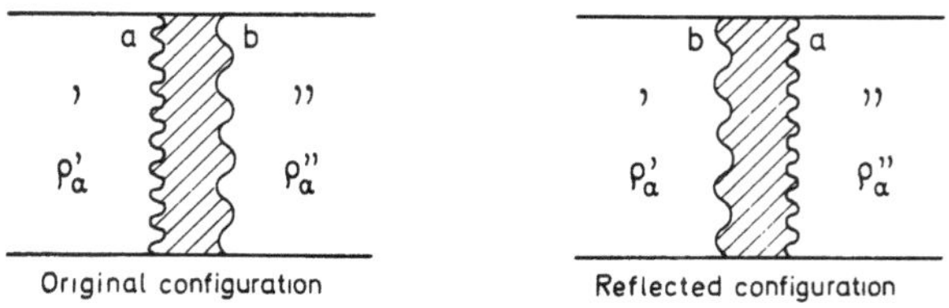

Fig. 3. Reflection of membrane

ration, the *a* side of the membrane faces the ′ phase and the *b* side the ″ phase, while in the reflected configuration, the *a* side faces the ″ phase and the *b* side the ′ phase. The flows through the membrane in the reflected configuration will be marked by a bar. For the membrane in the original configuration, the steady-state conditions read

$$J_\alpha^a + J_\alpha^b = 0; \qquad J_\beta^a + J_\beta^b = 0$$

$$J_k^a + J_k^b = \nu_k J_r \qquad (k = 1, \dots n-1); \qquad \text{(C 31)}$$

while those for the reflected configuration read

$$\bar{J}_\alpha^a + \bar{J}_\alpha^b = 0; \qquad \bar{J}_\beta^a + \bar{J}_\beta^b = 0$$

$$\bar{J}_k^a + \bar{J}_k^b = \nu_k \bar{J}_r \qquad (k = 1, \dots n-1). \qquad \text{(C32)}$$

The reference state becomes

$$\bar{s}_l = s_l('', ') \qquad (l = 2, \dots n+2m). \qquad \text{(C 33)}$$

A function symmetric in the ′ and ″ variables is subsequently chosen to represent the reference state. Thus

$$\bar{s}_l = s_l \qquad (l = 2, \dots, n+2m) \qquad \text{(C 34)}$$

The foregoing considerations enable one to derive the properties of an ideal membrane, one which possesses structural symmetry *e. g.* a structurally homogeneous membrane with plane surfaces; however, this is not the only example of a structurally symmetric membrane. Any membrane having a plane of symmetry in the middle is structurally symmetric. For such membranes one has the following relations between the flows in the two "situations," namely

$$\bar{J}_\alpha^a = J_\alpha^b; \qquad \bar{J}_\beta^a = J_\beta^b$$

$$\bar{J}_k^a = J_k^b; \qquad \bar{J}_k^b = J_k^a \qquad (k = 1, \dots n-1) \qquad \text{(C 35)}$$

This means that the flow emerging from the side of the membrane in the reflected configuration equals the flow from the *b* side of the membranes in the original configuration and *vice versa*. By putting together the steady-state conditions (C 31) and (C 32) one derives

$$\bar{J}_r = J_r \tag{C 36}$$

Therefore

$$\bar{J}_k^b = -J_k^b + \nu_k J_r \qquad (k = 1, \ldots, n-1) \tag{C 37}$$

Further, one obtains

$$\bar{J}_\alpha^b = -J_\alpha^b \tag{C 38}$$

In view of the postulated relations

$$J_\alpha^b = F_+(s_l, J_k^b, J_r)\,\varrho_\alpha' + F_-(s_l, J_k^b, J_r)\,\varrho_\alpha''$$

and (C 39)

$$\bar{J}_\alpha^b = F_+(s_l, \bar{J}_k^b, \bar{J}_r)\,\varrho_\alpha'' + F_-(s_l, \bar{J}_k^b, \bar{J}_r)\,\varrho_\alpha'$$

Equation (C 38) leads to

$$F_+(s_l, J_k^b, J_r) = -F_-(s_l, \bar{J}_k^b, \bar{J}_r)$$

and (C 40)

$$F_-(s_l, J_k^b, J_r) = -F_+(s_l, \bar{J}_k^b, \bar{J}_r)$$

The use of (C 10), (C 36), and (C 37) yields

$$F_+(s_l, J_k^b, J_r) - F_+(s_l, -J_k^b + \nu_k J_r, J_r) = J_n^b$$

and (C 41)

$$F_-(s_l, J_k^b, J_r) - F_-(s_l, -J_k^b + \nu_k J_r, J_r) = -J_n^b$$

These are functional equations for the unidirectional flows through membranes with structural symmetry. The equations are not restricted to small net flows and reaction rates. At thermodynamic equilibrium ($J_k^b \equiv 0$ and $J_r \equiv 0$), they are identically fulfilled. Conclusions can thus be drawn from (C 41) only for non-equilibrium "situations." In many applications, it is useful to discuss the properties of F_+ through a Taylor expansion of this function in the variables J_k^b and J_r for fixed values of s_l. This expansion in the J_k^b's and J_r up to the first order has the form

$$F_+(s_l, J_k^b, J_r) = F_{+o}(s_l) + \sum_{k=1}^{n} F_{+k}(s_l) \cdot J_k^b + F_{+r}(s_l) \cdot J_r \tag{C 42}$$
$$+ \text{ higher-order terms}$$

with

$$F_{+\mathrm{o}}(s_l) = F_+(s_l, 0, 0), \qquad F_{+k}(s_l) = \left(\frac{\partial F_+}{\partial J_k^{\mathrm{b}}}\right)_{J_i^{\mathrm{b}}, J_{\mathrm{r}} = 0} \quad \text{and}$$

$$F_{+\mathrm{r}}(s_l) = \left(\frac{\partial F_+}{\partial J_{\mathrm{r}}}\right)_{J_i^{\mathrm{b}}, J_{\mathrm{r}} = 0}$$

Two questions arise in connection with the expansion (C 42). One concerns the accuracy of the first-order expansion, or the question of how small the J_k^{b}'s and J_{r} must be, in order that the first order expansion may serve as a good approximation. In general, the validity of (C 42) must be tested by experiments. Experimental data must provide the justification for the omission of the higher-order terms. If, on the other hand, the function F_+ is already known, one can obtain an estimate of the higher-order terms. The other question concerns the form and structure of the expansion coefficients in (C 42). For known model systems the expansion coefficients can be calculated, but, in general, they must be determined by experiments. This, in fact, is the general "philosophy" of a phenomenological approach. Likewise, on making a Taylor expansion of the function B in (C 13), one obtains

$$B(s_l, J_k^{\mathrm{b}}, J_{\mathrm{r}}) = B_{\mathrm{o}}(s_l) + \sum_{k=1}^{n} B_k(s_l)\, J_k^{\mathrm{b}} + B_{\mathrm{r}}(s_l) \cdot J_{\mathrm{r}} + \text{higher-order terms.} \tag{C 43}$$

On introducing this expansion into (C 12), by discarding terms beyond the first order one obtains

$$F_+ = \frac{1}{B_{\mathrm{o}}} + \frac{J_n^{\mathrm{b}}}{2} - \sum_{k=1}^{n} \frac{B_k}{B_{\mathrm{o}}^2} J_k^{\mathrm{b}} - \frac{B_{\mathrm{r}}}{B_{\mathrm{o}}^2} J_{\mathrm{r}} + \text{higher-order terms} \tag{C 44}$$

Comparison of (C 44) with (C 42) leads to

$$F_{+\mathrm{o}} = \frac{1}{B_{\mathrm{o}}}; \qquad F_{+k} = -\frac{B_k}{B_{\mathrm{o}}^2} \qquad (k = 1, \ldots, n-1);$$

$$F_{+n} = \frac{1}{2} - \frac{B_n}{B_{\mathrm{o}}^2}; \qquad F_{+\mathrm{r}} = -\frac{B_{\mathrm{r}}}{B_{\mathrm{o}}^2}. \tag{C 45}$$

In the case of a structurally symmetric membrane, some of the first-order expansion coefficients can be determined from (C 41). If one differentiates (C 41) with respect to J_k^{b} and puts J_k^{b} and J_{r}'s equal to zero one obtains

$$F_{+k} = \frac{1}{2}\delta_{kn} \tag{C 46}$$

where $\delta_{kn} = 0$ if $k \neq n$ and $\delta_{kn} = 1$ if $k = n$. Therefore, the expansion of F_+, for a structurally symmetric membrane, up to the first-order terms reads

$$F_+ = F_{+\mathrm{o}}(s_l) + \frac{1}{2} J_n^{\mathrm{b}} + F_{+\mathrm{r}}(s_l) \cdot J_\mathrm{r} + \text{higher-order terms.} \qquad (\text{C } 47)$$

Moreover, one obtains from (C 10)

$$F_- = - F_{+\mathrm{o}}(s_l) + \frac{1}{2} J_n^{\mathrm{b}} - F_{+\mathrm{r}}(s_l) \cdot J_\mathrm{r} + \text{higher-order terms.} \qquad (\text{C } 48)$$

This means that in the first-order approximation the unidirectional flow in a structurally symmetric membrane depends on J_n^{b} and J_r alone. The dependence on J_n^{b} is known (factor $1/2$), while that on J_r in general is not. This fact must be kept in mind when (C 48) is used to determine the tracer permeability. Equation (C 41) can also be used to determine some of the higher-order expansion coefficients by repeated differentiations. To cite an example, it turns out that all coefficients of odd powers of the J_k^{b}'s are zero. The flux ratio r, defined by (C 19), can also be expanded. Retaining terms up to first order one finds

$$r = 1 + \frac{1}{F_{+\mathrm{o}}(s_l)} J_n^{\mathrm{b}} + \text{higher-order terms.} \qquad (\text{C } 49)$$

The differences between symmetric and asymmetric membranes show up only in the higher-order terms, irrespective of the symmetry of the membrane. In addition, it should be observed that no first-order terms in J_r and J_k^{b}'s with $k \neq n$ appear in (C 49).

To summarize the nonthermodynamic considerations, the mere assumption of the existence of a thermodynamic equilibrium enables one to derive many properties of unidirectional flows for membrane systems with small and large deviations from thermodynamic equilibrium. The objective in the following sections is to close the gap between this approach and the usual nonequilibrium thermodynamic treatment of isotope flow. In particular, expressions will be derived that relate the thermodynamic transport coefficients with the expansion coefficients of F_+ in (C 42).

D. The Nonequilibrium Thermodynamic Approach

In this approach the system (Fig. 1) is viewed as a thermodynamic system with the membrane in the steady state. Standard thermodynamic treatment (see for example KATCHALSKY and CURRAN, 1965; or SAUER, 1973) leads to an entropy balance equation from which an expression for the entropy production is derived. If one assumes that the temperature difference $\Delta T = T' - T''$ across the membrane is zero, in the chemical picture one obtains the following bilinear form in the flows and driving forces ($X_k \equiv \Delta\mu_k$ and A') for the entropy production $\frac{dS_{\text{int}}}{dt}$:

$$T\frac{dS_{\text{int}}}{dt} = \sum_{k=1}^{n-1} J_k^{\text{b}} \cdot \Delta\mu_k + J_n^{\text{b}} \cdot \Delta\mu_k + J_{\text{r}} \cdot A' \tag{D 1}$$

The statement of the second law of thermodynamics is that the entropy production must be positive for all nonequilibrium "situations" no matter how large the deviations are from equilibrium, and that it is identically zero in thermodynamic equilibrium ($X_k \equiv 0$ and $A' \equiv 0$). For systems close to equilibrium defined by the requirements that $\left|\frac{X_k}{RT}\right| << 1$ and $\left|\frac{A'}{RT}\right| << 1$,
ONSAGER (1931) proposed the existence of linear relations of the form

$$J_i^{\text{b}} = \sum_{k=1}^{n-1} L_{ik}^{\text{o}}(s_l) \cdot X_k + L_{in}^{\text{o}}(s_l) \cdot X_n + L_{i\text{r}}(s_l) \cdot A' \qquad (i,\, k = 1, \ldots, n\text{–}1)$$

$$J_n^{\text{b}} = \sum_{k=1}^{n-1} L_{nk}^{\text{o}}(s_l) \cdot X_k + L_{nn}^{\text{o}}(\text{s}_1) \cdot X_n + L_{n\text{r}}^{\text{o}}(s_l) \cdot A' \tag{D 2}$$

$$J_{\text{r}} = \sum_{k-1}^{n=1} L_{\text{r}k}^{\text{o}}(s_l) \cdot X_k + L_{\text{r}n}^{\text{o}}(s_l) \cdot X_n + L_{\text{rr}}^{\text{o}}(s_l) \cdot A'$$

where the L^{o} coefficients depend on the reference state variables alone. The validity of (D 2) must again be checked by experiments. Further, ONSAGER (1931) proved the following reciprocity relations between the transport coefficients:

$$L_{ik}^{\text{o}} = L_{ki}^{\text{o}},\, (i, k = 1, \ldots, n); \qquad L_{i\text{r}}^{\text{o}} = L_{\text{r}i}^{\text{o}},\, (i = 1, \ldots, n) \tag{D 3}$$

In the isotope picture one must introduce the flows J_α^{b} and J_β^{b} in the place of J_n^{b}. Using an approach for the discontinuous systems (Fig. 1) similar to the one used by KEDEM and ESSIG (1965) for continuous systems, one obtains for $\frac{dS^*_{\text{int}}}{dt}$, the entropy production, the following:

$$T\frac{dS^*_{\text{int}}}{dt} = \sum_{k=1}^{n-1} J_k^{\text{b}} \cdot X_k + J_\alpha^{\text{b}} \cdot X_\alpha + J_\beta^{\text{b}} \cdot X_\beta + J_{\text{r}} \cdot A' \tag{D 4}$$

with $X_\alpha \equiv \Delta\mu_\alpha$ and $X_\beta \equiv \Delta\mu_\beta$. From (B 4) and (B 7) one has

$$X_\alpha = X_n + RT\Delta \ln \varrho_\alpha$$
$$X_\beta = X_n + RT\Delta \ln (1 - \varrho_\alpha) \tag{D 5}$$

and

$$X_n = \tilde{\varrho}_\alpha X_\alpha + (1 - \tilde{\varrho}_\alpha) X_\beta$$

where $\tilde{\varrho}_\alpha$ is defined in (B 8). For systems close to equilibrium, the following inequalities should hold in the isotope picture:

$$\left|\frac{X_k}{RT}\right| \ll 1; \qquad \left|\frac{X_\alpha}{RT}\right| \ll 1; \qquad \left|\frac{X_\beta}{RT}\right| \ll 1; \qquad \left|\frac{A'}{RT}\right| \ll 1. \tag{D 6}$$

Consequently, one again expects the existence of linear laws. Taking into account the conditions (D 6) and the relations (D 5), this means that

$$\left|\frac{\Delta\varrho_\alpha}{\tilde{\varrho}_\alpha}\right| \ll 1 \quad \text{and} \quad \left|\frac{\Delta\varrho_\alpha}{1 - \tilde{\varrho}_\alpha}\right| \ll 1. \tag{D 7}$$

Clearly these conditions are not fulfilled in experiments where one measures unidirectional flows, because in this case ϱ_α', for example, is zero. This leads to a singular driving force X_α, actually X_α becomes $-\infty$. Therefore linear laws for the isotope flows cannot be applied directly to unidirectional flows. It will be shown later how one can bypass this restriction with the aid of additional assumptions introduced in section C. Imposing the conditions (D 6), one writes the following linear relations

$$J_\alpha^b = \sum_{k=1}^{n-1} L^o_{\alpha k} \cdot X_k + L^o_{\alpha\alpha} \cdot X_\alpha + L^o_{\alpha\beta} + X_\beta + L^o_{\alpha r} \cdot A'$$

$$J_\beta^b = \sum_{k=1}^{n-1} L^o_{\beta k} \cdot X_k + L^o_{\beta\alpha} \cdot X_\alpha + L^o_{\beta\beta} \cdot X_\beta + L^o_{\beta r} \cdot A'$$

$$J_i^b = \sum_{k=1}^{n-1} L^o_{ik} \cdot X_k + L^o_{i\alpha} \cdot X_\alpha + L^o_{i\beta} \cdot X_\beta + L^o_{ir} \cdot A' \qquad (i = 1, \ldots n-1) \tag{D 8}$$

$$J_r = \sum_{k=1}^{n-1} L^o_{rk} \cdot X_k + L^o_{r\alpha} \cdot X_\alpha + L^o_{r\beta} \cdot X_\beta + L^o_{rr} \cdot A'.$$

The L coefficients in (D 8) are functions of the reference state variables s_l and the average isotopic composition $\tilde{\varrho}_\alpha$. They obey ONSAGER reciprocity

$$L^o_{\alpha\beta} = L^o_{\beta\alpha}; \qquad L^o_{\alpha k} = L^o_{k\alpha}; \qquad L^o_{\beta k} = L^o_{k\beta} \qquad (k = 1, \ldots, n-1)$$

$$L^o_{\alpha r} = L^o_{r\alpha}; \qquad L^o_{\beta r} = L^o_{r\beta} \tag{D 9}$$

For the other coefficients (D 3) holds. In order to determine the $\tilde{\varrho}_\alpha$ dependence of the coefficients, one has to compare the linear laws in the chemical picture (D 2) with those in the isotope picture (D 8). Recalling that $J_n^b = J_\alpha^b + J_\beta^b$, and using (D 5), one obtains relations between the transport coefficients appearing in (D 2) and (D 8). These relations are

$$\begin{aligned} \tilde{\varrho}_\alpha L^\circ_{nn} &= L^\circ_{\alpha\alpha} + L^\circ_{\alpha\beta}; & \tilde{\varrho}_\beta L^\circ_{nn} &= L^\circ_{\beta\beta} + L^\circ_{\alpha\beta} \\ \tilde{\varrho}_\alpha L^\circ_{ni} &= L^\circ_{\alpha i}; & \tilde{\varrho}_\beta L^\circ_{ni} &= L^\circ_{\beta i} \\ \tilde{\varrho}_\alpha L^\circ_{nr} &= L^\circ_{\alpha r}; & \tilde{\varrho}_\beta L^\circ_{nr} &= L^\circ_{\beta r} \end{aligned} \tag{D 10}$$

with $\tilde{\varrho}_\beta = 1 - \tilde{\varrho}_\alpha$.

These equations enable one to express nearly all transport coefficients of the isotope picture in terms of transport coefficients of the chemical picture with a simple linear dependence on the $\tilde{\varrho}_\alpha$. This is in general not the case with $L^\circ_{\alpha\alpha}$, $L^\circ_{\beta\beta}$ and $L^\circ_{\alpha\beta}$. Here the first two relations in (D 10) suggest the following dependence on $\tilde{\varrho}_\alpha$

$$\begin{aligned} L^\circ_{\alpha\alpha} &= h_n(s_l)\,\tilde{\varrho}_\alpha + g_n(s_l)\,\tilde{\varrho}_\alpha^2 \\ L^\circ_{\beta\beta} &= h_n(s_l)\,(1-\tilde{\varrho}_\alpha) + g_n(s_l)\,(1-\tilde{\varrho}_\alpha)^2 \\ L^\circ_{\alpha\beta} &= g_n(s_l)\,\tilde{\varrho}_\alpha(1-\tilde{\varrho}_\alpha) \end{aligned} \tag{D 11}$$

with two functions h_n and g_n of the chemical reference state s_l. From (D 10) one obtains

$$h_n(s_l) + g_n(s_l) = L^\circ_{nn}(s_l) \tag{D 12}$$

and

$$\begin{aligned} l^\circ_{\alpha\alpha} &= L^\circ_{nn}\tilde{\varrho}_\alpha - g_n(s_l)\,\tilde{\varrho}_\alpha(1-\tilde{\varrho}_\alpha) \\ L^\circ_{\beta\beta} &= L^\circ_{nn}(1-\tilde{\varrho}_\alpha) - g_n(s_l)\,\tilde{\varrho}_\alpha(1-\tilde{\varrho}_\alpha) \end{aligned} \tag{D 13}$$

Generally, as long as $g_n \neq 0$, the $L^\circ_{\alpha\alpha}$ and $L^\circ_{\beta\beta}$ cannot be expressed in terms of the transport coefficients of the chemical picture alone. Since g_n is connected to $L^\circ_{\alpha\beta}$, the isotope − isotope cross-coefficient, it has a clear-cut meaning in the formal development of the phenomenological theory. This so-called tracer coupling had already been introduced by NIMS (1962) and KEDEM and ESSIG (1965). From a kinetic point of view appropriate to a fluid phase, a finite g_n or $L^\circ_{\alpha\beta}$ will result if there are a few β molecules with finite interaction in the immediate vicinity of an α molecule. This is particularly true of concentrated solutions of the chemical component n. From considerations of this kind, one can deduce the following limiting behavior of g_n namely

$$\lim_{\bar{c}_n = 0} \frac{g_n(s_l)}{\bar{c}_n} = 0 \tag{D 14}$$

which is valid for membrane systems as well. Because $\frac{L^{\circ}_{nn}}{\bar{c}_n}$ remains finite if $\bar{c}_n$ approaches zero one can neglect tracer coupling in dilute solutions. How small $\bar{c}_n$ must become has to be tested, again, by experiment. In asymmetric membranes in particular, interaction with the membrane component might mediate isotope interaction. A more detailed discussion falls outside the scope of this article.

When (D 2) and (D 8) are combined with (D 10) one obtains the relation

$$J^{\mathrm{b}}_{\alpha} = \tilde{\varrho}_{\alpha} J^{\mathrm{b}}_{n} + (\tilde{\varrho}_{\alpha}(1 - \tilde{\varrho}_{\alpha}) L^{\circ}_{nn} - L^{\circ}_{\alpha\beta})(X_{\alpha} - X_{\beta}) \tag{D 15}$$

In view of (D 5), $X_{\alpha} - X_{\beta}$ depends only on ϱ'_{α} and ϱ''_{α}. Thus

$$X_{\alpha} - X_{\beta} = RT\Delta \ln \frac{\varrho_{\alpha}}{1 - \varrho_{\alpha}} \tag{D 16}$$

for small $\Delta\varrho_{\alpha}$ $\left(|\frac{\Delta\varrho_{\alpha}}{\bar{\varrho}_{\alpha}}| << 1 \text{ and } |\frac{\Delta\varrho_{\alpha}}{1 - \bar{\varrho}_{\alpha}}| << 1\right)$ one finds on discarding higher-order terms $(0(\Delta\varrho_{\alpha})^2$ and higher)

$$X_{\alpha} - X_{\beta} = \frac{RT}{\bar{\varrho}_{\alpha}(1 - \bar{\varrho}_{\alpha})} \Delta\varrho_{\alpha} \tag{D 17}$$

Because $\Delta\varrho_{\alpha}$ is assumed to be small, one can make use of (B 15) to replace the logarithmic mean $\tilde{\varrho}_{\alpha}$ in (D 15) with the arithmetic mean $\bar{\varrho}_{\alpha}$. Thus

$$J^{\mathrm{b}}_{\alpha} = \bar{\varrho}_{\alpha} J^{\mathrm{b}}_{n} + RT(L^{\circ}_{nn} - g_n)\Delta\varrho_{\alpha} \tag{D 18}$$

with $g_n = \frac{L^{\circ}_{\alpha\beta}}{\bar{\varrho}_{\alpha}(1 - \bar{\varrho}_{\alpha})}$. This is an expression for J^{b}_{α} linear in J^{b}_{n} and $\Delta\varrho_{\alpha}$. To facilitate comparison of (D 18) with the relation (C 2) for $J_{\alpha}{}^{\mathrm{b}}$, the same approximation is employed in the latter. Equation (C 2) then reads

$$J^{\mathrm{b}}_{\alpha} = \bar{\varrho}_{\alpha} J^{\mathrm{b}}_{n} + F_{+\mathrm{o}}(s_l) \cdot \Delta\varrho_{\alpha} \tag{D 19}$$

Comparison of (D 18) and (D 19) yields

$$F_{+\mathrm{o}}(s_l) = RT(L^{\circ}_{nn}(s_l) - g_n(s_l)) \tag{D 20}$$

This equation relates the zero-order unidirectional flow $F_{+\mathrm{o}}$ to the diagonal coefficient L°_{nn}. The right-hand side of (D 20) is positive irrespective of the sign of g_n, which could be positive or negative. The L coefficient must obey certain inequalities since the entropy production must remain positive. In what follows only a few of them are needed, namely

$$L^{\circ}_{nn} > 0; \quad L^{\circ}_{\alpha\alpha} > 0; \quad L^{\circ}_{\beta\beta} > 0 \quad \text{and} \quad L^{\circ}_{\alpha\alpha} L^{\circ}_{\beta\beta} - L^{\circ 2}_{\alpha\beta} > 0 \tag{D 21}$$

Using (D 10), the last inequality can be rewritten in the form

$$(L^{\circ}_{nn}\tilde{\varrho}_{\alpha} - L^{\circ}_{\alpha\beta})(L^{\circ}_{nn}\tilde{\varrho}_{\beta} - L^{\circ}_{\alpha\beta}) - L^{\circ 2}_{\alpha\beta} > 0$$

or (D 22)

$$\tilde{\varrho}_{\alpha}\tilde{\varrho}_{\beta} L^{\circ}_{nn}\left(L^{\circ}_{nn} - \frac{L^{\circ}_{\alpha\beta}}{\tilde{\varrho}_{\alpha}\tilde{\varrho}_{\beta}}\right) > 0$$

Therefore, it follows that

$$L^{\circ}_{nn} - \frac{L^{\circ}_{\alpha\beta}}{\tilde{\varrho}_{\alpha}\tilde{\varrho}_{\beta}} > 0 \qquad \text{(D 23)}$$

because $\tilde{\varrho}_{\alpha}\tilde{\varrho}_{\beta} L^{\circ}_{nn}$ is positive.

The relation (D 20) and the inequality (D 23) are the principal results of the nonequilibrium thermodynamic treatment. Although it is difficult to apply this approach directly to unidirectional flows, one can, nevertheless, bypass this difficulty as shown above; it should be stressed that the results contain information about the unidirectional flows only in thermodynamic equilibrium. To gain information about the first-order term of the unidirectional flows it is necessary to consider the approximation used in (D 18). This will be done for a few model systems in the next section.

A re-examination of the flux-ratio equation (C 20) is appropriate at this stage. Discarding quadratic and higher-order terms in the J's in the argument of the exponential function, and approximating $B(s_l, J^{\mathrm{b}}_n, J_{\mathrm{r}})$ by $B_{\mathrm{o}}(s_l) = \dfrac{1}{F_{+\mathrm{o}}}$, one obtains

$$r = \exp \frac{J^{\mathrm{b}}_n}{F_{+\mathrm{o}}} . \qquad \text{(D 24)}$$

Substitution of (D 2) and (D 20) in (D 24) yields

$$r = \exp\left(\frac{L^{\circ}_{nn}}{L^{\circ}_{nn} - g_n}\frac{X_n}{RT} + \sum_{k=1}^{n-1}\frac{L^{\circ}_{nk}}{L^{\circ}_{nn} - g_n}\frac{X_k}{RT} + \frac{L^{\circ}_{n\mathrm{r}}}{L^{\circ}_{nn} - g_n}\frac{A'}{RT}\right) \qquad \text{(D 25)}$$

or

$$r = r_{\mathrm{id}}\exp\left(\frac{g_n}{L^{\circ}_{nn} - g_n}\frac{X_n}{RT} + \sum_{k=1}^{n-1}\frac{L^{\circ}_{nk}}{L^{\circ}_{nn} - g_n}\frac{X_k}{RT}\right.$$

$$\left. + \frac{L^{\circ}_{n\mathrm{r}}}{L^{\circ}_{nn} - g_n}\frac{A'}{RT}\right) \qquad \text{(D 26)}$$

This means that in the linear approximation any kind of coupling (tracer coupling, coupling of the net flows and the reaction rate) would lead to deviations from the ideal flux ratio. Again, had the nonlinear contributions been retained, they would have led to deviations from the ideal flux ratio as well.

E. Applications to Model Systems

The purpose of this section is to apply the formalism developed in the previous sections to some model systems to gain more insight into the concepts employed. First a simple "diffuse" membrane system is discussed, with the following assumptions. One considers a nonelectrolyte component n without a chemical reaction in the membrane. Coupling phenomena between the net flows are neglected. One employs the usual formula for the chemical potential $\mu_n = \mu_{\text{standard}} + RT \ln c_n$ appropriate to dilute solutions. The diagonal coefficient L°_{nn} is taken to be linear in the average concentration $\bar{c}_n = \dfrac{c'_n + c''_n}{2}$ and the membrane is assumed to be symmetric. With these assumptions one obtains

$$J^{\text{b}}_n = L^{\circ}_{nn} \cdot X_n \qquad X_n = \Delta\mu_n = \frac{RT}{c_n} \Delta c_n. \tag{E 1}$$

Writing $L^{\circ}_{nn} = \dfrac{P_n}{RT} \bar{c}_n$, one obtains the desired expression for J^{b}_n in the form

$$J^{\text{b}}_n = P_n \cdot \Delta c_n \tag{E 2}$$

where P_n is the chemical permeability of component n.

The zero-order approximation of the unidirectional flow becomes

$$F_{+\text{o}} = RT(L^{\circ}_{nn} - g_n) = P_n \bar{c}_n \left(1 - \frac{g_n RT}{P_n \bar{c}_n}\right) \tag{E 3}$$

The expression $P_n \left(1 - \dfrac{g_n RT}{P_n \bar{c}_n}\right)$ is called the tracer permeability P_α. It deviates from P_n if $g_n \neq 0$, the deviation being a measure of tracer coupling. When $\bar{c}_n$ approaches zero $\dfrac{g_n}{\bar{c}_n}$ becomes small and P_n coincides with P_α. For a symmetric membrane the first-order approximation of the unidirectional flow is

$$F_{+} = F_{+\text{o}} + \frac{1}{2} J_n \tag{E 4}$$

which for the diffuse membrane reads

$$F_{+} = P_n \bar{c}_n \left(1 - \frac{g_n RT}{P_n \bar{c}_n}\right) + \frac{1}{2} P_n \Delta c_n \tag{E 6}$$

After a slight rearrangement of terms one obtains

$$F_+ = P_n c_n' - g_n RT \tag{E 7}$$

In many kinetic models the first term alone in (E 7) is used, and the influence of tracer coupling is neglected. For ions the chemical potential difference in the driving force must be replaced by the electrochemical potential difference $\Delta\eta_n$

$$X_n = \Delta\eta_n = \Delta\mu_n + z\tilde{F}\Delta\varphi = \frac{RT}{\bar{c}_n}\Delta c_n + z\tilde{F}\Delta\varphi \tag{E 8}$$

where $\tilde{F}$ is the Faraday constant, z the charge number, and $\Delta\varphi$ the so-called membrane potential difference, which, in principle, is not a measurable quantity. Then (E 7) can be transformed to

$$F_+ = P_n c_n' - g_n RT + \frac{1}{2} P_n \bar{c}_n \frac{z\tilde{F}}{RT}\Delta\varphi \tag{E 9}$$

This is a good approximation for F_+, provided that $\left|\frac{z\tilde{F}}{RT} \cdot \Delta\varphi\right| \ll 1$. For larger values of this quantity it is necessary to return to (C 12) and approximate B by B_o. When tracer coupling is neglected and it is assumed that $\Delta c_n = 0$, it follows that

$$F_+ = P_n \bar{c}_n \frac{z\tilde{F}\Delta\varphi}{RT} \frac{\exp\frac{z\tilde{F}\Delta\varphi}{RT}}{\exp\frac{z\tilde{F}\Delta\varphi}{RT} - 1} \tag{E 10}$$

This relation is very frequently used in the interpretation of isotope-flow measurements. Since the influence of tracer coupling and the higher-order terms in the B-function have been neglected in its derivation, its range of validity is necessarily limited. From the assumed symmetry of the membrane, one expects B to be of the form (omitting the index b)

$$B = B_o + \frac{d}{2} J_n^2 + 0(J_n^4) \tag{E 11}$$

This leads to the following expansion for F_+, whose expression is given by (C 12):

$$F_+ = F_{+o} + \frac{1}{2} J_n - \frac{1}{6F_{+o}}(1 + 3dF_{+o}^3) J_n^2 \tag{E 12}$$

It turns out that the second-order terms in (E 12) already contain the nonlinear contribution of the B function. This should not be overlooked when one applies (E 10) to larger values of $\frac{z\tilde{F}}{RT} \cdot \Delta\varphi$.

The second model to be discussed is an asymmetric membrane composed of two different membranes, v and w, in series (see Fig. 2). It is assumed that the single membranes are properly described by the relations

$$F_+^{\mathrm{v}} = F_{+\mathrm{o}}^{\mathrm{v}}\left(\frac{\mu_n' + \mu_n^{\mathrm{o}}}{2}\right) + \frac{J_n^{\mathrm{b}}}{2}$$

$$F_+^{\mathrm{w}} = F_{+\mathrm{o}}^{\mathrm{w}}\left(\frac{\mu_n'' + \mu_n^{\mathrm{o}}}{2}\right) + \frac{J_n^{\mathrm{b}}}{2}$$

and that

$$J_n^{\mathrm{b}} = L_{nn}^{\mathrm{ov}}(\mu_n' - \mu_n{}^{\mathrm{o}}) = L_{nn}^{\mathrm{ow}}(\mu_n^{\mathrm{o}} - \mu_n'') \qquad \text{(E 14)}$$

From (C 26), for the unidirectional flow for the series arrangement of two membranes one has the expression

$$F_+^{\mathrm{vw}} = \frac{F_+^{\mathrm{v}} \cdot F_+^{3}}{F_+^{\mathrm{v}} + F_+^{3} - J_n^{\mathrm{b}}} \qquad \text{(E 15)}$$

In the sequence, the average chemical potentials $\bar{\mu}_n = \dfrac{\mu'_n + \mu''_n}{2}$ of the composite membrane and J_n^{b} serve as new variables.

Using (E 11) one obtains

$$\frac{(\mu_n' + \mu_n^{\mathrm{o}})}{2} = \bar{\mu}_n + \frac{1}{2L_{nn}^{\mathrm{ow}}} J_n^{\mathrm{b}}$$

and (E 16)

$$\frac{\mu_n'' + \mu_n^{\mathrm{o}}}{2} = \bar{\mu}_n - \frac{1}{2L_{nn}^{\mathrm{ov}}} J_n^{\mathrm{b}}$$

Hence (E 15) can be cast in the form

$$F_+^{\mathrm{vw}}(\bar{\mu}_n, J_n^{\mathrm{b}}) = \frac{\left(F_{+\mathrm{o}}^{\mathrm{v}}\left(\bar{\mu}_n + \frac{J_n^{\mathrm{b}}}{2L_{nn}^{\mathrm{ow}}}\right) + \frac{J_n^{\mathrm{b}}}{2}\right)\left(F_{+\mathrm{o}}^{\mathrm{w}}\left(\bar{\mu}_n - \frac{J_n^{\mathrm{b}}}{2L_{nn}^{\mathrm{ov}}}\right) + \frac{J_n^{\mathrm{b}}}{2}\right)}{F_{+\mathrm{o}}^{\mathrm{v}}\left(\bar{\mu}_n + \frac{J_n^{\mathrm{b}}}{2L_{nn}^{\mathrm{ow}}}\right) + F_{+\mathrm{o}}^{\mathrm{w}}\left(\bar{\mu}_n - \frac{J_n^{\mathrm{b}}}{2L_{nn}^{\mathrm{ov}}}\right)} \qquad \text{(E 17)}$$

An expansion of the right-hand side of (E 17) to first order, for fixed values of $\bar{\mu}_n$, gives

$$F_+^{\mathrm{vw}}(\bar{\mu}_n, J_n^{\mathrm{b}}) = \frac{F_{+\mathrm{o}}^{\mathrm{v}}(\bar{\mu}_n) \cdot F_{+\mathrm{o}}^{\mathrm{w}}(\bar{\mu}_n)}{F_{+\mathrm{o}}^{\mathrm{v}}(\bar{\mu}_n) + F_{+\mathrm{o}}^{\mathrm{w}}(\bar{\mu}_n)} + \frac{1}{2} J_n^{\mathrm{b}}$$

$$+ \frac{F_{+\mathrm{o}}^{\mathrm{v}}(\bar{\mu}_n) \cdot F_{+\mathrm{o}}^{\mathrm{w}}(\bar{\mu}_n)}{F_{+\mathrm{o}}^{\mathrm{v}}(\bar{\mu}_n) + F_{+\mathrm{o}}^{\mathrm{w}}(\bar{\mu}_n)} \frac{RT}{2} \frac{\partial}{\partial \bar{\mu}_n} \ln \frac{F_{+\mathrm{o}}^{\mathrm{v}}(\bar{\mu}_n)}{F_{+\mathrm{o}}^{\mathrm{w}}(\bar{\mu}_n)} \cdot J_n^{\mathrm{b}} \qquad \text{(E 18)}$$

The first two terms in (E 18) give the well-known result for symmetric membranes. The last term owes its existence to the fact that the composite membrane is asymmetric. Through (E 18), the expansion coefficient F_{+n} of (C 42) has in fact been determined. Again the influence of tracer coupling has been neglected. From (E 18) it is seen that for asymmetric membranes there occur terms that are related to the dependence of the transport coefficients on the reference state. Only when nonlinear contributions are taken into account do terms of this kind occur in the net flows. Thus, measurements of both unidirectional flows and net flows, even in the first-order approximation, provide a good deal of information about membrane properties.

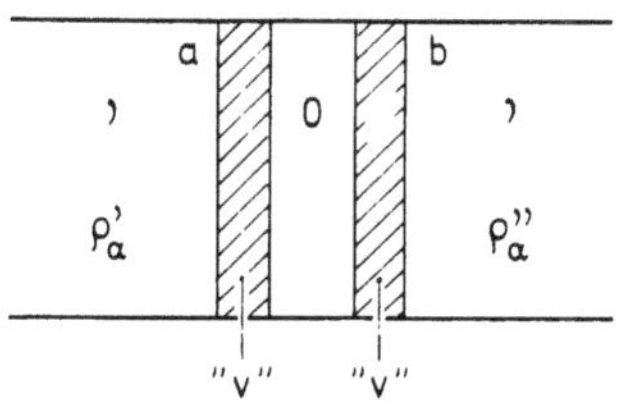

Fig. 4. Sandwich membrane with chemical reaction and two identical membranes "v"

The third model consists of a series arrangement of two identical membranes separated by a homogeneous phase 0, where a chemical reaction is postulated to occur (see Fig. 4). For simplicity it is assumed that the phases ′ and ″ are identical in chemical composition, but that the affinity A' of the chemical reaction is nonzero. Since the membrane arrangement is symmetric, one has for the net flows (′ = ″ is assumed)

$$J_k^{\mathrm{b}} = \frac{\nu_k}{2} J_{\mathrm{r}} \qquad (k = 1, \dots n-1) \qquad \text{(E 19)}$$

while J_n^{b} vanishes because ν_n is zero by assumption. One further assumes that the unidirectional flow of the single "v" membrane is independent of the flows J_k^{b} and that its dependence on the reference state is solely a dependence on the variable

$$\bar{\mu}_n^{\mathrm{v}} = \frac{\mu_n' + \mu_n^{\mathrm{o}}}{2}$$

Hence,

$$F_+^{\mathrm{v}} = F_{+\mathrm{o}}^{\mathrm{v}}(\bar{\mu}_n^{\mathrm{v}}) \qquad \text{(E 20)}$$

For the series arrangement,

$$F_+^{\mathrm{vv}} = \frac{1}{2} F_{+\mathrm{o}}^{\mathrm{v}}(\bar{\mu}_n^{\mathrm{v}}) \qquad \text{(E 21)}$$

The average chemical potential of the "v" membrane can be expressed in terms of the average chemical potential of the series membrane, which is just μ_n', and the driving force $X_n^{\mathrm{v}} = \mu_n' - \mu_n^{\mathrm{o}}$. The relation between these quantities reads

$$\bar{\mu}_n^{\mathrm{v}} = \mu_n' - \frac{X_n^{\mathrm{v}}}{2} \,. \tag{E 22}$$

Therefore,

$$F_+^{\mathrm{vv}}(\mu_n', J_\mathrm{r}) = \frac{1}{2} F_{+\mathrm{o}}^{\mathrm{v}}\left(\mu_n' - \frac{x_n^{\mathrm{v}}}{2}\right) . \tag{E 23}$$

Because A' is the sole driving force for the series membrane and relation (E 19) holds, X_n^{v} is proportional to J_r if the "v" membrane is described by linear laws:

$$J_i^{\mathrm{b}} = \sum_{k=1}^{n} L_{ik}^{\mathrm{o}} \cdot X_k^{\mathrm{v}} . \tag{E 24}$$

One obtains

$$X_n^{\mathrm{v}} = \sum_{k=1}^{n-1} R_{nk}^{\mathrm{ov}} \cdot \frac{\nu_k}{2} J_\mathrm{r} . \tag{E 25}$$

The R_{ik}^{ov} coefficients are the elements of the inverse of the matrix defined in (E 24).

An expansion of (E 23) to first order in J_r has the form

$$F_+^{\mathrm{vv}}(\mu_a', J_\mathrm{r}) = \frac{1}{2} F_{+\mathrm{o}}^{\mathrm{v}}(\mu_n') - \frac{1}{8} \frac{\partial F_{+\mathrm{o}}^{\mathrm{v}}(\mu_n')}{\partial \mu_n'} \sum_{k=1}^{n-1} R_{nk}^{\mathrm{o}}(\mu_n')\, \nu_k J_\mathrm{r} \tag{E 26}$$

The first term on the right-hand side is the unidirectional flow of the series membrane in thermodynamic equilibrium. The second term arises because of the chemical reaction occurring within the series membrane, which changes the forces across the single membranes. Equation (E 26) determines the expansion coefficient $F_{+\mathrm{r}}$, which was introduced in (C 42). This provides a simple example of the influence of chemical reactions in the membrane on the unidirectional flow. It should be stressed that this effect has nothing to do with active transport of the component n, because due to the assumed symmetry of the series membrane, the active transport is abolished.

F. Summary

The last section has dealt with a few model systems that are sufficiently close to thermodynamic equilibrium for linear approximations to apply. The linear laws provide a powerful tool for a cogent description of membrane transport phenomena, particularly those involving coupling. The dependence of the unidirectional flows on net flows and on chemical reaction rate and the role played by the membrane structure all form part of such a description. Deviations of the flux ratio from the ideal flux ratio r_{id} can only be attributed to coupling phenomena and the membrane structure. For systems far from thermodynamic equilibrium a detailed discussion is virtually impossible. Here thermodynamic considerations provide nothing more than a skeleton that could be basic to all model systems. For example, the condition that the entropy production $\frac{dS^*_{int}}{dt}$ in (D 4) remains positive is model-independent and universal. The isotope flows can be discussed in terms of the same driving forces $\Delta\mu_\alpha$ and $\Delta\mu_\beta$ also in the nonlinear case: Eqs. (B 8), (C 2) and (D 5) lead to

$$J_a^b = [F_+ (s_l, J_k, J_r) \, exp\,(\tilde{\varrho}_\beta \cdot U) + F_- (s_l, J_k, J_r)] \, \frac{exp\,(\tilde{\varrho}_\alpha \cdot U)}{exp\,(\tilde{\varrho}_\alpha \cdot U) - exp\,(\tilde{\varrho}_\beta \cdot U)} \tag{F 1}$$

with $U = \frac{1}{RT} (\Delta\mu_\alpha - \Delta\mu_\beta)$.

This is a highly nonlinear relation for $|U| >> 1$. A mathematical discussion of this relation, despite its formal advantages, is somewhat inconsistent with the scope of this article and will therefore not be attempted. It is interesting to note that a good deal of information can be obtained from nonthermodynamic considerations, as in Section B of this Chapter. For experimental facts and their relation to theoretical considerations of model systems, the reader might consult the detailed treatment by STEN-KNUDSEN (Chapter 2, this volume).

Acknowledgements

The author wishes to express his thanks to colleagues at the Max Planck Institute for discussions on the subject of this paper. Thanks are also due to Mrs. Olga Woinoff for her excellent typing of complicated formulas.

List of Symbols

A'	Affinity of chemical reaction
a	One side of the membrane
(B_k)	Symbol for chemical component in reaction
B	Symbol for special function
b	Other side of the membrane

c_i	Concentration of the ith component
e_1	Charge per mole of the first ion
F_+, F_-	Unidirectional flows
$\tilde{F}$	Faraday constant
I	Electrical current
J_i^a	Net flow of the ith component out of the a side
J_i^b	Net flow of the ith component out of the b side
J_r	Chemical reaction rate
L^o	Onsager matrix
L_{ik}^o	*ik* component of the Onsager matrix
n_i	Mol number of the ith component
R_{ik}^o	*ik* component of the inverse Onsager matrix
$\frac{dS_{int}^*}{dt}$	Entropy production of the chemical system
$\frac{dS_{int}}{dt}$	Entropy production of the isotope system
T	Absolute temperature
t	Time
X_i	i^{th} component of the "driving Force"
α	Index for an isotope
β	Index for an isotope
Δ	Difference of $' - ''$
ν_i	Stoichiometric number of the ith component in the reaction
μ_i', μ_i''	Chemical potential of the ith component in the $'$ and $''$ phase

References

Katchalsky, A., Curran, P. F.: Nonequilibrium Thermodynamics in Biophysics. Cambridge, Mass.: Harvard University Press 1965.

Kedem, O., Essig, A.: J. gen. Physiol. **48**, 1047 (1965).

Li, T. H., Essig, A.: J. Membrane Biol. **29**, 255 (1976).

Nims, L. F.: Science **137**, 130 (1962).

Onsager, L.: Phys. Rev. **37**, 405 (1931).

Sauer, F.: Appendix to: Handbook of Physiology, Section 8 (J. Orloff and R. W. Berliner, Eds). Baltimore, Md: Williams and Wilkins 1973.

Ussing, H. H.: Acta physiol. scand. **17**, 1 (1949).

Chapter 5

Use of Microelectrodes for Measurement of Membrane Potentials

U. V. Lassen and B. E. Rasmussen

A. Introduction

"– one must be critical of experimental conditions, requirements, and the results obtained with any particular arrangement –"

(Moore and Cole, 1963)

The primary aim of measuring biological transport of charged molecules and correlated electrical phenomena is to learn about properties of biological membranes under normal and pathologic conditions. In the present context a biological membrane is synonymous with the plasma membrane, even though the same words have been used with equal right to designate a layer of cells separating one compartment in the organism from another. Detailed knowledge of the plasma membrane is necessary in order to understand the function of cells in the integrated organism.

In the preceding chapters, the theory of transport of molecules, charged or uncharged, has been considered in some detail. To carry out the necessary calculations it is customary to put certain constraints (mostly simplifying) on the particular system under consideration. Such procedures have the great advantage of defining the conditions under which a certain expression is valid, a fact that is occasionally overlooked by biologists attempting to draw conclusions from experimental data. Furthermore, speculations concerning the mechanism of a biological transport system can only be fruitful if they are based on firm experimental grounds. This fact, obvious as it is, is frequently neglected in electrophysiological studies on small cells where serious experimental difficulties are encountered. For these reasons we have found it justified to discuss some elementary problems associated with measurements of membrane potential with microelectrodes. Although it appears that such measurements are simple in principle if the membrane potential is defined as the difference in potential between two nonpolarizable electrodes, one inside and one outside the cell, the potential difference between the two electrodes has to be carefully analyzed. For example, the media to which the electrodes are exposed are different. This may lead to differences in the junction potentials of the reference electrode outside the cell and the microelectrode tip when placed in the cytosol. Since electrical measurements report solely the difference in potential between the two metallic conductors from the electrodes, the recorded potential should as far as possible be resolved into its components. A microelectrode filled with concentrated salt solution will lose salt during the measurements. This diffusion from the microelectrode tip is calculated in order to estimate to what extent the electrolyte concentrations in the cell will change during the measurements. Other problems and properties of microelectrodes will be considered. Practical procedures for making microelectrodes and for penetration of the cell membrane will also be discussed. In close connection with the latter process, the effects of even a small leak around the electrode tip will also be considered. Whereas very large cells or interconnected cells are only slightly influenced by the micropuncture, penetration of the membrane of a small cell may in the worst case lead to potential measurements with little or no relation to the membrane potential of the unperturbed cell. In selecting examples of experiments where the conclusions in our opinion are questionable, our purpose is not to sit in judgement, but to illustrate the need for caution toward the validity of the data obtained and conclusions drawn from measurements of membrane potential in biological systems.

We have deliberately chosen a simple and pragmatic presentation, and hope that the chapter will thus be of value for the reader who wants a brief introduction to the use of microelectrodes. At the same time we want to emphasize that some of the problems, especially concerning the leak around the microelectrode tip have not been given appropriate attention, even in some recent publications.

B. Principles of Bioelectric Recording

All cells are surrounded by a plasma membrane that modifies and limits the movement of substances in and out of the cell. Only in few, exceptionally large cells such as squid axons, or in geometrically favorable structures such as myelinated nerves it is possible to obtain information about membrane potential and current without mechanical penetration of the membrane in an area facing the medium constituting the potential reference outside the cell. The common experimental approach is therefore to introduce an electrode through the membrane and measure the potential change. Even though this procedure is simple in principle, the actual performance and subsequent interpretation of the measurements is far from trivial.

The main requirements for recording the membrane potentials of living cells are obviously that the measurements impose minimal damage to the cells and that the potentials are faithfully recorded. However, especially in the case of small isolated cells and organelles, the experimenter faces serious or invalidating problems when using conventional techniques of *e.g.* micropuncture. We consider below some general problems of bioelectric recording, as a background for the more explicit discussion of properties and use of microelectrodes in the later sections of this chapter.

I. Electrode Chains and Junction Potentials

In order to estimate membrane potentials of cells in steady state, it is usually necessary to perform d.c. measurements. This means that thermodynamically reversible components for the solid/liquid junction should be chosen. At this point it may be appropriate to stress that measurements of "stable" membrane potentials which involve irreversible electrodes such as polarizable metal/liquid junctions must be considered with great caution before conclusions are drawn from the data. For a competent review of problems of electrode polarization, see BOCKRIS and REDDY (1973).

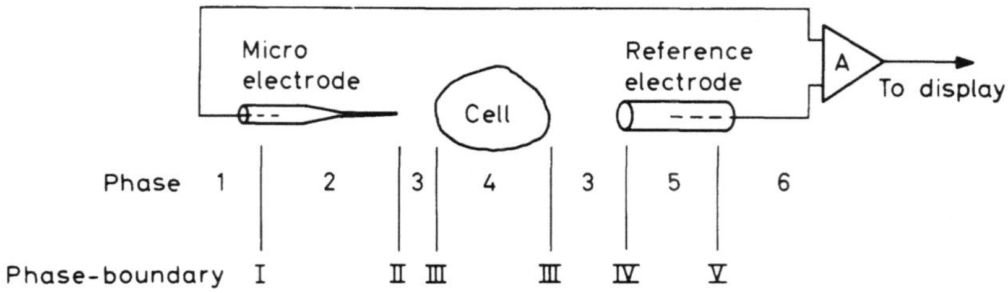

Fig. 1. Schematic diagram of the measuring chain employed in estimates of membrane potential with microelectrodes. For further details, see text

In bioelectric measurements it is customary to use systems that contain electrochemical cells with transference and two similar electrodes (see *e. g.* GUGGENHEIM, 1957, and MOORE, 1972, for definitions) for the solid/liquid boundaries. To discuss some of the basic problems involved it is convenient to depict the measuring situation as shown in Figure 1. The only solid/liquid junctions to be considered in this presentation are silver rods covered with a layer of AgCl, which in turn are in contact with a Cl ion-containing solution. The measured potential difference V across the schematic electrochemical cell can be thought of as a sum of terms

$$V = {}^1\Delta^2\varphi + {}^2\Delta^5\varphi + {}^5\Delta^6\varphi = {}^1\Delta^2\varphi + {}^2\Delta^5\varphi - {}^6\Delta^5\varphi,$$

where ${}^2\Delta^5\varphi = \varphi_2 - \varphi_5$ is the difference in potentials of bulk phases 2 and 5. The potential differences ${}^1\Delta^2\varphi$ and ${}^6\Delta^5\varphi$ are inaccessible from electrochemical measurements, since they are differences across a single interface. However, under the usual experimental conditions, where the compositions of bulk phases 2 and 5 are almost identical and the current through the system is small, there are strong reasons to believe that the difference ${}^1\Delta^2\varphi - {}^6\Delta^5\varphi$ will vanish. This follows from the symmetry of the experimental set-up, and will be discussed in more detail shortly. Under these conditions the measured potential difference V equals ${}^2\Delta^5\varphi$, which in turn can be thought of as a sum of terms (with the microelectrode inside the cell)

$$
{}^2\Delta^5\varphi = {}^2\Delta^4\varphi + {}^4\Delta^3\varphi + {}^3\Delta^5\varphi + \text{ ohmic terms } = {}^2\Delta^4\varphi + V^M - {}^5\Delta^3\varphi + \text{ ohmic terms.}
$$

The quantities ${}^2\Delta^4\varphi$ and ${}^5\Delta^3\varphi$ are the diffusion potentials across the liquid junctions II and IV respectively, $V^M = {}^4\Delta^3\varphi$ is the membrane potential, and the ohmic terms are the potential drops in the bulk solutions caused by the small but finite current through the system, which is necessary to drive the amplifier A (Fig. 1). Since modern amplifiers are available with input bias currents of ~ 10^{-12} A or smaller, the ohmic terms can usually be ignored.

With regard to the two diffusion potentials, their absolute value cannot be measured. If, however, a high concentration of KCl is present in the salt bridge and the electrode shaft, the diffusion potentials will probably be small, due to the similar mobilities of the dominating ions, potassium and chloride. A rough estimate can be calculated for example from the Henderson equation, as discussed by STEN-KNUDSEN in chapter 2 of this volume. With high concentrations of KCl on one side of the liquid junction, diffusion potentials calculated in this way seldom exceed a few millivolts. The presence of charged macromolecules in the solution may invalidate such a simple calculation, since it may lead to significant changes in ionic activity as well as mobility of the small inorganic ions ("the suspension effect"). At present there is no generally accepted way of treating this problem.

Strictly speaking, the electrochemical system as a whole is not in equilibrium. Irreversible processes go on at the liquid junctions (diffusion) and the small electric net current will be accompanied by heat production. Working with nonpolarizable electrodes means, however, that the potential differences ${}^1\Delta^2\varphi$ and ${}^6\Delta^5\varphi$ are independent of the small currents through the system and can be calculated from equilibrium conditions. Diffusion across the liquid junctions may alter the ionic activities close to the solid-liquid interface if the electrode is left in the cell for an extended period of time. However, the diffusional loss of salt is a slow process and the charge transfer reactions at the solid-liquid interfaces I and V can therefore in practice be considered as being at equilibrium.

$$
Ag^+ + e \rightleftarrows Ag
$$

and therefore

$$
\mu_{Ag}^{(1)} = \mu_e^{(1)} - F \cdot \varphi_1 + \mu_{Ag^+}^{(2)} + F \cdot \varphi_2 = \mu_e^{(1)} + \mu_{Ag^+}^{(2)} - F \cdot {}^1\Delta^2\varphi, \qquad \text{(B 1)}
$$

where $\mu_a^{(i)}$ is the chemical potential of substance a in the phase i.

Similarly, for the charge transfer reactions at the phase boundary V:

$$
\mu_{Ag}^{(6)} = \mu_e^{(6)} + \mu_{Ag^+}^{(5)} - F \cdot {}^6\Delta^5\varphi. \qquad \text{(B 2)}
$$

Subtracting Eq. (B 1) from (B 2) gives:

$$
F \cdot ({}^1\Delta^2\varphi - {}^6\Delta^5\varphi) = \mu_{Ag^+}^{(2)} - \mu_{Ag^+}^{(5)},
$$

since the chemical potential of solid silver and the electrons is the same in phases 1 and 6. The concentration of silver ions in phases 2 and 5 in turn is controlled by the chemical reaction:

$$AgCl \rightleftarrows Ag^+ + Cl^-,$$

for which

$$\mu^{(1)}_{AgCl} = \mu^{(2)}_{Ag^-} + \mu^{(2)}_{Cl^-}$$

$$\mu^{(6)}_{AgCl} = \mu^{(5)}_{Ag^-} + \mu^{(5)}_{Cl^-}$$

under equilibrium conditions. Therefore

$$\mu^{(2)}_{Ag^-} - \mu^{(5)}_{Ag^-} = \mu^{(5)}_{Cl^-} - \mu^{(2)}_{Cl^-},$$

which implies that

$$^1\Delta^2\varphi - {}^6\Delta^5\varphi = \frac{1}{F} \cdot (\mu^{(5)}_{Cl^-} - \mu^{(2)}_{Cl^-})$$

or in terms of single ion activities

$$^1\Delta^2\varphi - {}^6\Delta^5\varphi = \frac{RT}{F} \cdot \ln\left(\frac{a^{(5)}_{Cl^-}}{a^{(2)}_{Cl^-}}\right) \qquad \text{(B 3)}$$

Under symmetrical conditions the right-hand side of Eq. (B 3) must vanish, as stated previously, whereas under nonsymmetrical conditions the ratio of the single ion activities is usually approximated by that of the mean salt activities.

The coating of the silver rod with AgCl renders the electrode reversible (nonpolarizable), which is equivalent to almost abolishing the capacitive component of the electrode impedance. Thus in contrast to an Ag rod with typically high pass properties, the AgCl-coated Ag rod is almost independent of frequency from d.c. to well above 30 kHz (Geddes and Baker, 1967). Furthermore, the increased area of the corrugated or granular surface of the AgCl reduces the ohmic resistance. This last statement is true with qualifications. It should be borne in mind that AgCl has a very high specific resistance (in the order of 10^7 Ωcm, Janz and Ives, 1968). Thus excessive coating with AgCl will cause the junctional resistance to rise. This is undesirable, especially in the case of the reference electrode (see below).

The fluid in the shaft of the microelectrode or microelectrode holder (Fig. 1, phase 2) is usually an electrolyte solution similar to that in contact with the reference Ag/AgCl electrode. In almost all cases the fluid in the tip of the microelectrode (micropipette) is concentrated KCl (Nastuk and Hodgkin, 1950), which decreases electrode resistance and the diffusion potential in the tip. In cases where solutions other than concentrated KCl are introduced into the electrode shaft or holder, the ensuing phase boundary must be taken into account. Also, diffusion of KCl may slowly change the Cl activity around the

Ag/AgCl electrode, causing d.c. drift. There is no universally accepted solution to this problem but each investigator must design the system according to his or her needs.

The next section in this chapter will consider the microelectrode tip (Fig. 1, phase boundary II) in greater detail. Suffice it here to mention that the majority of problems encountered with the use of microelectrodes are associated with the microelectrode tip. Not only does the tip have a high resistance, but also by the very nature of the measurements, it will be exposed first to the extracellular and then to the intracellular phases.

The extracellular phase in a given experiment is determined by the experimental conditions. Especially in *in-vitro* experiments the composition of this salt solution (Fig. 1, phase III) is often varied to give information about properties of the cell membrane. Such shifts between electrolytes have pronounced effects at liquid junctions, with ionic strength similar to that of the extracellular phase in contact with the cells. This has particular relevance to the choice of fluid in the reference electrode system (Fig. 1, phase boundary IV, and phase 5). If one wants to minimize the changes in liquid junction potential at the reference electrode (Fig. 1, phase boundary IV), a concentrated KCl solution can be used in this electrode (Fig. 1, phase 5). Alternatively it is possible to introduce an agar bridge with 3 M KCl between the solution bathing the cell (Fig. 1, phase 3) and the reference electrode. But in both cases a considerable leak of KCl out of the electrode will occur, which has to be taken into account. Therefore the use of reference electrode or bridges with concentrated salt solutions should be limited to experimental conditions with large extracellular fluid volumes or where fluid is constantly flowing through the measuring chamber.

When the salt solution in the reference electrode is dilute, a change in the composition of the extracellular phase must result in a change of the liquid junction potential. The absolute magnitude of such changes can be estimated from the Henderson equation (see Sten-Knudsen, Eq. 195, Chapter 2, this Volume) which assumes constant concentration gradients throughout the junction. Such a correction is only necessary, however, when the shift in extracellular fluid occurs with the microelectrode tip present *in situ* in the cell. Otherwise the change in junction potential at the reference electrode will merely be noted as a shift in the reference potential before penetration of the cell membrane with the microelectrode tip.

In the above discussion we have deliberately used "single ion activities," despite the fact that only mean activities of salts in solution can be measured by thermodynamical methods. Because of this, the validity of any measurement of membrane potential has been questioned. Guggenheim (1957) states that "the electric potential difference between two points in different media can never be measured and has not yet been defined in terms of physical realities", and Tasaki and Singer (1968) stress that there will always be some ambiguity in the interpretation of bioelectric recordings. But the latter authors also point out that physiologically relevant information on changes in membrane properties can be obtained from such measurements, as long as the limitations of the system are carefully considered. The fundamental problem resides in the conflict of assuming electroneutrality at the same time as an electric field in a given domain indicates imbalance between positive and negative charges. Similarly the work function (free energy) of an ion species cannot be calculated by classic thermodynamical methods, since the movement of this ion species strongly influences the state of the system due to the ensuing charge separation. Consequently the change in free energy of a particular ion species cannot be isolated from the rest of the system. However, as discussed by STEN-KNUDSEN (this Volume) and several others, the deviations from equality of concentrations of positive and negative charge necessary to give potentials in the range encountered in biology are so small that electroneutrality is still a very good approximation except at distances very close to phase boundaries. In a recent account of ionic activity in cells LEV and ARMSTRONG (1975) have treated this problem in detail. They conclude that only moderate corrections are necessary when single ion activities are used instead of weighted mean activities. They also conclude that estimates of single ion activities in polyelectrolyte solutions are no better or worse than in the case of simple salt solutions. This is of importance for the discussion of "the suspension effect" (Pallmann effect) to be given below.

The half-cell potential of an Ag/AgCl electrode can only be measured with reference to another half cell. Therefore the only quantity which is measured is the total difference in EMF between the electrode pair. However, with nonthermodynamical methods it should be possible to determine the absolute magnitude of a half-cell potential (KIRKWOOD and OPPENHEIM, 1961).

From the above discussion it might seem that measurements of steady-state membrane potentials of living cells are of very limited value. Fortunately problems with the theoretical aspects of interpretation have not kept physiologists and biophysicists from performing electrical measurements. The state of the art can probably best be described by the words of COLE and MOORE (1960):

"It is most unlikely that attempts to measure membrane potentials and to assign a physical significance to them will not continue for a while longer. Nor is there any real prospect that a completely rigorous meaning for the results can be found in the near future. Consequently it is important that the imperfections of both rigor and expediency be isolated, identified, and to whatever extent possible understood".

II. Comments on Electronic Equipment

It is not the object of this presentation to go into a detailed discussion of the numerous electronic problems related to recording potentials between the leads from the two electrodes at the solid/liquid phase boundaries. However, in the biological literature it is common that the performance of the electronic equipment used is very sparingly described. This fact often limits the evaluation of the reported data. This section will therefore treat a few crucial points concerning electronics, which should be taken into consideration when measuring membrane potential of cells. Going back to the simplified presentation of the measuring chain as shown in Figure 1, the issue under consideration is the amplifier (denoted A) and its relation to the electrode chain (and the display).

One of the amplifier inputs (usually the "positive" terminal) is connected to the microelectrode. As discussed in the next section, a microelectrode represents a relatively large ohmic resistance, in the order of 10^7 Ω or more. The amplifier must have an input resistance that is at least 10, and preferably 100 or more, times larger than the electrode resistance. Otherwise the measured potential will be attenuated by voltage division between electrode and amplifier. Furthermore, this error will depend on the relative magnitude of the electrode resistance in comparison to the input resistance of the amplifier. As the effective resistance of the electrode chain usually increases when the microelectrode tip is introduced into the cell, the attenuation due to insufficient input resistance of the amplifier will become larger, leading to a further underestimation of the membrane potential. In the early days of electrophysiology, the construction of high input impedance amplifiers was a large problem, often left to be solved by the individual investigator. This situation has changed dramatically in the era of advanced semiconductor technology. Cheap, precise operational amplifiers with a low current consumption are commercially available. Such integrated amplifiers with field effect transistors in the input can solve the problem of a high-input resistance, as well as some other problems discussed below, quite simply. A typical high-quality operational amplifier (*e. g.* Analog Devices AD 515 L) will have a common mode input resistance of about 10^{15} Ω. This is some eight orders of magnitude higher than the resistance of the measuring chain and thus more than adequate for most bioelectric measurements.

To avoid perturbation of the membrane potential with the electrode inside the cell and to avoid changing potential drops in the electrode it is necessary to use amplifiers with minimal "leakage currents" from input to ground. In this respect, the modern semiconductor devices will often meet requirements. Input bias currents as low as 10^{-13} A are not uncommon. In the case of a microelectrode resistance of 15 MΩ, the IV drop from this current will be 1.5 μV, or far below the normal resting potentials in living cells. By careful selection of the amplifier and other components this figure might be brought even lower.

To follow bioelectric events, it is necessary to have electronic equipment with an appropriate bandwidth. This requirement is closely related to the various capacitances in the recording system, especially between the amplifier input leads (including the microelectrode) and ground. Such "stray" capacitances will tend to attenuate changes in the potential of the microelectrode interior. A simple case would be the voltage division between the cell membrane capacitance and the stray capacitance immediately after penetration of the membrane. The initial potential change is of importance for estimating the steady state membrane potential in single cells (see *e. g.* Lassen et al., 1971). For this and other reasons it is desirable to reduce the effective stray capacitance as much as possible. This can be done in a number of ways, of which we will deal briefly with two important configurations: the driven shield, and the "negative capacitance" amplifier.

Figure 2 A depicts an amplifier with a shielded input cable. The shield has the primary purpose of reducing the influence of external fields on the potential of the input lead. The cable shield, however, has a relatively large capacitance to the input lead itself. Therefore a simple connection of the shield to earth would

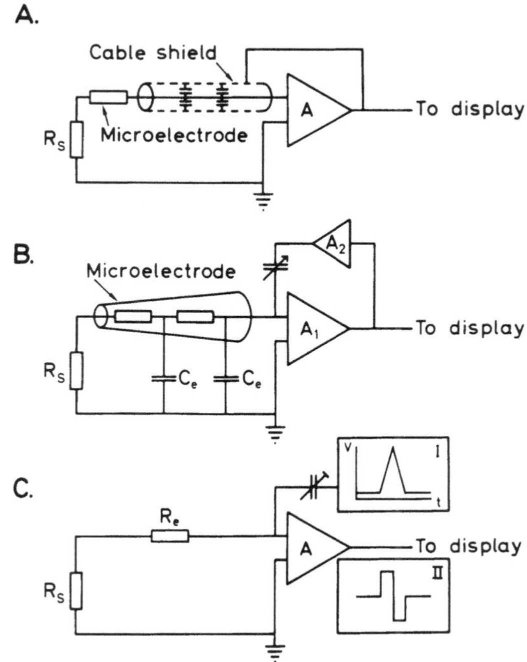

Fig. 2. Essential features of electronics commonly used for bioelectric measurements. Triangles (A, A_1, or A_2) symbolize amplifiers. Frame A shows an example of a "driven shield". Frame B shows features of a "negative capacitance" amplifier used to compensate for input capacitance of *e.g.* microelectrode. Frame C summarizes the method of LETTVIN et al. (1958) for generation of diphasic current pulses through electrode. For further details, see text.

aggravate the undesired effects of the stray capacitance. If, instead, the shield at any instant is held at a potential very close to that of the input lead, the capacitive current from input to shield is almost negligible (depending on the requirements in a given experimental situation). In Figure 2A the cable shield is connected to the in-phase output of the amplifier. Provided that the amplifier has a low impedance output and unity gain, the shield potential will closely follow the potential at the input of the amplifier. Innumerable variations on this theme have been presented in the literature from the earliest days of the application of amplifiers to bioelectric measurements. We have purposely not mentioned the input capacitance of the amplifier itself as this presents a minor problem with integrated solid-state amplifiers (C_{in} typically less than 0.2 pF).

In measurements with microelectrodes the driven shield is usually not sufficient to ensure a good overall bandwidth of the recording system. The microelectrode can be represented as a resistive conductor with distributed stray capacitance to ground (see Fig. 2B). The magnitude of this capacitance is about 1 pF per mm of immersion of the tip of the electrode (MOORE, 1971). For technical reasons it is not possible to enclose the microelectrode tip in a conductive shield, as discussed in relation to the input cable. An alternative and often supplementary method of reducing the effect of input capacitance is the negative capacity amplifier (AMATNIEK, 1958; GULD, 1962). Figure 2 B summarizes the principle of this configuration. The output signal from the main amplifier is fed back to the input lead itself (not the shield) through another amplifier and a capacitor (positive feedback). Upon changes in the potential of the input, the ac-coupled positive feedback loop will cause additional charge displacement, proportional to the original rate of change. Obviously this can cause a regenerative response (oscillation) in the loop containing the two amplifiers A_1 and A_2, but if properly adjusted, *e. g.* by adjusting the magnitude of C_f, the feedback will be, in the ideal case, just enough to compensate for the charge flow through the stray capacitances. A full compensation is never possible due to the relatively high resistance of the microelectrode interior, which is in series with the stray capacitance at the tip. However, it is possible to obtain rise times in the order of 25–50 μs. This allows measurements on relatively large single cells but is insufficient for estimates of the membrane potential of small cells and organelles, as discussed in more detail below. AMATNIEK (1958) gives an extensive review of the properties of the negative capacitance amplifier.

Frequently it is convenient to have repeated measurements of the electrode resistance before, during, and after impalement of a cell. This can be performed by the technique of LETTVIN et al. (1958), in which a triangular voltage pulse is differentiated by a small capacitor connected directly to the input lead of the amplifier (Fig. 2C). Provided that the change in potential of the input lead is small in comparison to the total magnitude of the triangle pulse, the small capacitor will be a good differentiator, causing the injection of diphasic constant current pulses at the amplifier input. The recorded diphasic voltage pulses during the current pulses will in essence be proportional to the magnitude of the microelectrode resistance. An increase in resistance will be observed when the electrode tip has penetrated the cell membrane. This "extra" resistance is in some cases a direct measure of the membrane resistance of the cell. In other cases, the leak in the cell membrane caused by the electrode tip has a much lower resistance than the cell membrane. In such instances, the rise in resistance will serve merely as a check on the intracellular location of the electrode tip.

The above remarks on electronics for measuring membrane potential are very fragmentary. The interested reader is referred to comprehensive accounts on this and related subjects by MOORE and COLE (1963) and MOORE (1971). CLAYTON (1975) has given an extensive discussion of the use and advantages of differential amplifiers in contrast to the "single-ended" amplifiers presented in Figure 1. Examples of circuits with monolithic amplifiers are excellently presented by JUNG (1974).

C. The Glass Capillary Microelectrode

The typical glass microelectrode for recording intracellular potential is made from Pyrex or similar glass borosilicate tubing. This glass is used because of its mechanical strength and resistance to the thermal shock that occurs during pulling of the capillary when making the electrodes. Furthermore, the thin wall of the electrode tip is less conductive and bears less surface charges than soda-lime glass micropipettes of comparable dimensions (LAVALLÉE and SZABO, 1969). The low surface charge and low solubility of the borosilicate glass seem to be particularly involved in minimizing the so-called tip potential to be discussed below.

The outer diameter of the glass tubing used for microelectrodes is usually in the range 0.7 to 2 mm. The microelectrode is formed by heating the mid-section of a glass tube by flame or by a heating coil in one of the numerous types of automatic or semiautomatic pulling machines, and then, when the tubing has reached the necessary plasticity, pulling the two ends apart. In the pulling machines this process usually occurs in two steps: a gentle initial pull and a final forceful pull. In most pulling machines the position of one end of the tube is fixed with regard to the heating coil. Consequently, the softened part of the glass tubing is drawn away from the coil during the final pull. Even though the current for the heating is usually switched off at the same time, the removal of the tip from the zone of heating seems to be important if occlusion of the tips is to be avoided. The breakage of the tube occurs at the point of stiffening of the glass, thus improving the chances for sharp edges on the electrode tip, which in turn will facilitate penetration of the cell membrane. The technical details of how the various parameters of the pulling procedure influence the final tip diameter, tapering angle of the tip, and overall shape of the electrode are presented by FRANK and BECKER (1964) and reviewed in several books dealing with fabrication and use of microelectrodes (*e. g.* GEDDES, 1972; FERRIS, 1974; LAVALLÉE et al., 1969). In the present context it is sufficient to note that commonly used microelectrodes have tip diameters in the range 0.1 to 0.5 μm, and tapering angles of the tips (see Fig. 3) of some 5° to 15°. Since tip diameters of this magnitude are below the resolution of the optical microscope, many researchers resort to estimating tip diameters from the resistances of the electrodes. This procedure may give the highly erroneous impression that high-resistance electrodes necessarily have fine tips. The only safe way of estimating the typical tip diameter of a given microelectrode batch is by electronmicroscopy. To avoid corona formation around the tip and consequent distortion of the electron microscope image, relatively heavy coating with gold (or other conductive material) is necessary.

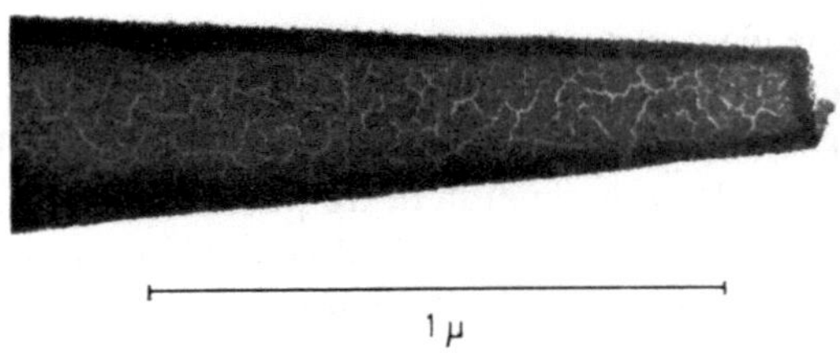

Fig. 3. Electron micrograph of typical microelectrode tip (before filling). Cracked appearance is due to heavy gold plating (400 Å) necessary to avoid corona formation when tip is placed in electron beam of microscope. Horizontal bar indicates 1 μm

Filling of the micropipette is a crucial step in the fabrication of usable microelectrodes. The original paper by LING and GERARD (1949) describes how air in the pipettes was replaced by KCl solution during vigorous boiling and gradual evaporation of water from the solution in which the electrodes were immersed. A final addition of cold KCl solution completed the process. Microelectrodes prepared in this way show variable resistances and considerable tip potentials. For the latter reason NASTUK (1953) filled electrodes by first immersing them in well-filtered 3 M KCl for two hours. The fine tip of the electrode filled by capillary action and any remaining air bubbles were removed by poking with a glass thread. The use of 3 M KCl, as introduced by NASTUK and HODGKIN (1950), is meant to reduce the resistance of the microelectrode and to reduce the changes in liquid junction potentials at the tip as the electrode is moved from the extracellular fluid into the cytoplasm. KCl solutions of 2.5 to 3 M are now in general use, and the subsequent considerations will only take microelectrodes with concentrated KCl solution into account.

Due to the nearly equal mobility of potassium and chloride ions in solution (and to the high concentration of KCl in the pipettes) the expected liquid junction potential of the tip is small (see *e.g.* LASSEN and STEN-KNUDSEN, 1968). However, filling procedures that involve boiling of the electrodes and or prolonged storage of the electrodes in salt solution result in the appearance of a tip potential. This phenomenon is operationally defined as the difference in potential recorded with an intact electrode and the potential after breaking the tip away. Usually such tip potentials are quite large when measurements are made with the tip immersed in dilute salt solutions, the magnitude being comparable to that of the membrane potentials to be recorded (ADRIAN, 1956; AGIN and HOLTZMAN, 1966; AGIN, 1969). Furthermore, the larger the resistance of the electrode the higher the tip potential. ADRIAN (1956) devised procedures for the selection of electrodes with minimal tip potentials and ways of correcting for the fact that in a given preparation, electrodes with large tip potentials gave low values for the membrane potential. He also pointed out that electrodes with large tip potentials behaved like ion-selective electrodes having a high K^+ sensitivity but a low sensitivity to Na^+ and especially to Cl^-. ADRIAN (1956) concludes that it seems as if the electrode tip becomes blocked with "some substance" having ion exchange properties. AGIN and HOLTZMANN (1966) and LAVALLÉE and SZABO (1969) have investigated the nature of the tip potential and report that thorium can reduce the magnitude or reverse the sign of the potential. From these and other reports (SNELL, 1969) it seems likely that conduction along the glass surfaces (or conduction through the glass wall of the tip) plays a role in the generation of tip potentials.

Though there is some theoretical interest in the processes by which tip potentials of microelectrodes are generated, tip potentials can distort results obtained with microelectrodes. Consequently it is highly desirable to use a filling procedure that yields electrodes with minimal tip potentials. Over a period of several years we have tested most of the published methods of filling microelectrodes, with limited success, and like everybody else working with microelectrodes have devised a number of modifications. In 1971 ZEUTHEN published a method for filling electrodes, in which water is distilled from the shaft of the electrode by heating with a coil of tungsten wire. If the electrode tip is located just outside the heating coil, the water vapor will condense in the tip. By adjusted movement of the electrode it is possible to get most of the tapering part of the electrode filled with water in less than one minute. The shaft of the electrode is then filled with concentrated KCl solution and the whole electrode finally submerged in membrane-filtered 2.5 to 3 M KCl. After an equilibration time of 2–3 hours, the resistance is constant and the electrode is ready for use. In our hands this method has a yield of more than 90 percent usable electrodes with tip potentials of less than 3 mV. Furthermore the resistances are uniform in a given batch of electrodes, usually being in the range 10–16 MΩ when immersed in 0.1 M KCl or NaCl. The corresponding outer diameters of the electrodes are in the range 0.18 to 0.22 μm. The low tip potential and relatively low resistance are both necessary for measurements in single cells without electrical connection to other cells. The success of ZEUTHEN's method (1971) is probably related to two simple facts: 1) the tip of the electrode is never in contact with hot or boiling saline, nor is it exposed to concentrated salt solutions for extended periods of time (one day or more); and 2) dust particles or other material inside the electrode tip are flushed away from the tip into the large-diameter shaft, rendering clogging a rare event. Since the internal tip diameter (some 0.05 to 0.1 μm) is orders of magnitude larger than the Debye length of the salt solutions involved it is unlikely that surface charge from the glass wall contributes to the overall properties of the electrode tip (for a quantitative approach to this problem, see SNELL, 1969).

In 1968 TASAKI et al. reported another method for filling microelectrodes without bringing them in contact with hot electrolyte solution. Thin glass fibers are inserted into the glass tubing prior to pulling the pipette. After the pulling, the electrodes can be filled simply by injection of concentrated saline into the back end of the tapered portion. Capillary action will then rapidly cause filling of the tip. A similar method where the fiber technique is combined with centrifugation of the electrodes has recently been described by PLAMONDON et al. (1976). With proper precautions during rinsing of the glass tubing prior to pulling, HIGGINS et al. (1977) have successfully employed a version of the fiber technique in the preparation of electrodes used in a study of electrical properties of epithelia. In such measurements there are strict requirements for low tip potentials and clean outer surfaces of the electrodes (to allow sealing between cell membrane and electrode wall).

I. The Suspension Effect

Even though liquid junction potentials at the electrode tip presumably are small with 2–3 M KCl filled electrodes, provided that tip potentials are minimized with proper precautions, the so-called "suspension effect" may still influence the interpretation of microelectrode measurements. This phenomenon was first described in 1930 by PALLMANN, who observed a difference in pH of up to two units between a saline solution and a suspension of highly charged clay particles in equilibrium with the saline. Some 20 years later the effect was the subject of a considerable amount of discussion and experimentation (for references, see PFISTER and PAULY, 1969). Even though the pioneering work was done in relation to earth sciences, the suspension effect has also to be considered when assessing potential differences across cell membranes. This is necessary because of the large concentration of charged macromolecules such as protein and nucleic acids in the intracellular compartments. The fixed negative charge of such molecules will, as demonstrated by OVERBEEK (1956) and MÖLLER et al. (1961a and b) give rise to a marked reduction in both activity coefficient and mobility of the counterions (cations). The effect on small coions (*e. g.* chloride) is to create an "exclusion zone" around each of the charged macromolecules, leaving the mobility and activity coefficient in the available solvent essentially unaltered. Qualitatively the suspension effect is similar to the effects of surface charges on biological and other membranes.

TASAKI and SINGER (1968) pointed out that the suspension effect, although well elucidated in physical chemistry, was not taken into account in the evaluation of measurements in biological systems. To demonstrate the effect in a straightforward way they determined the potential difference between a suspension of finely powdered cation exchange resin and saline in equilibrium with the resin. The potential, as measured with two calomel electrodes, was a function of the salt concentration of the fluid, being about 20 mV (resin suspension-negative) at 0.1 M NaCl. At lower salt concentrations (1 mM) the potential difference could exceed 100 mV. Similar results were obtained with polyglutamic acid having fixed negative charges and also with conventional microelectrodes instead of the calomel electrodes. On the basis of these observations TASAKI and SINGER (1968) stress that the presence of polyelectrolytes within most cells makes the potentials induced by the suspension effect very important. However, as also pointed out by these authors, their "model" system has no limiting membrane between the phase with fixed charges and the supernatant. The measured potential difference is thus the sum of the junction potentials at this phase boundary and at the liquid junction at the electrode tip. It is not clear whether the major potential change occurs at the electrode tip (as implicit in the argument of TASAKI and SINGER, 1968) or at the saline/resin suspension boundary. The suspension effect may explain the fact that different potentials can be measured at different points in the same cell (BINGLEY and THOMPSON, 1962; ZEUTHEN, 1977), provided that there is a variation in the density and nature of the fixed charges within the cell. However, the wealth of consistent data, especially from excitable tissues, makes it seem rather unlikely that fundamental errors have been made in estimates of membrane potential. This would suggest that even though the suspension effect must be taken into account it is at present not obvious that measurements of steady-state membrane potentials are invalidated by the effect, as occasionally claimed in the literature.

II. Diffusion Regime of the Microelectrode Tip

The analysis of d.c. potential measurements obtained with electrolyte-filled micropipettes in the electrode chain is often hampered by the following properties of the microelectrodes:

1. Tip potentials in the order 50 mV or more are sometimes recorded. When instabilities at the metal solution interface can be excluded, such high potentials are usually strongly dependent on pH, indicating the presence of surface charges on the glass (LAVALLÉE and SZABO, 1969). As discussed previously it is possible to manufacture microelectrodes with vanishing tip potentials over a wide range of pH. In this case the difference in potential between the electrode shaft and the surrounding bulk solution is believed to be a pure liquid junction potential resulting from the passage of ions with dissimilar ionic mobilities between phases of different ionic activities. Even in this case factors which affect the outside ionic mobility or activity will likewise affect the junction potentials (the suspension effect) as discussed above. Therefore the recorded potential change on penetration of a cell membrane may contain a term that has no direct relation to the membrane potential.
2. The measured resistance of the electrode, usually being of the order 10 MΩ, depends strongly on the resistivity of the surrounding solution when the salt concentration in the shaft is high and the concentration in the surrounding solution is in the physiological range. The explanation is that diffusion out of the electrode leads to a dilution inside the electrode near the tip thereby increasing the resistivity in the domain which is of dominant importance for the magnitude of the total resistance. Measurements of this effect have been discussed by LANTHIER and SCHANNE (1966) and LASSEN and STEN-KNUDSEN (1968).
3. The leak of salt by diffusion out of the electrode must often be taken into account. Recent measurements by EHRENFELD et al. (1978) show this loss to be of the order 5×10^{-14} moles/s^{-1} for an electrode filled with 3 M KCl depending of course on the geometry of the electrode. The corresponding increase in KCl concentration in a spherical cell of diameter ~ 20 μm will be ~ 10 mM/s^{-1}!

Attempts have been made to explain the magnitude of the electrode resistance and the salt loss from electrodiffusion theory (KRNJEVIĆ et al., 1963; LANTHIER and SCHANNE, 1966; GEISLER et al., 1972). Common to these approaches is an attempt to determine the full concentration distribution inside and outside the micropipette. The flux is then determined by differentiation and the resistance by integration. This procedure is difficult if one wants to include the possibility of concentration-dependent transport coefficients and the physical boundary condition that the ion flux densities must be parallel to the surface everywhere near the surface.

In this section we present an approach that allows an approximate evaluation of the current-voltage relationship without the need of a solution for the concentration distribution. We consider only the simplest case of a micropipette filled with a single strong 1:1 electrolyte of concentration C^e dipped into a

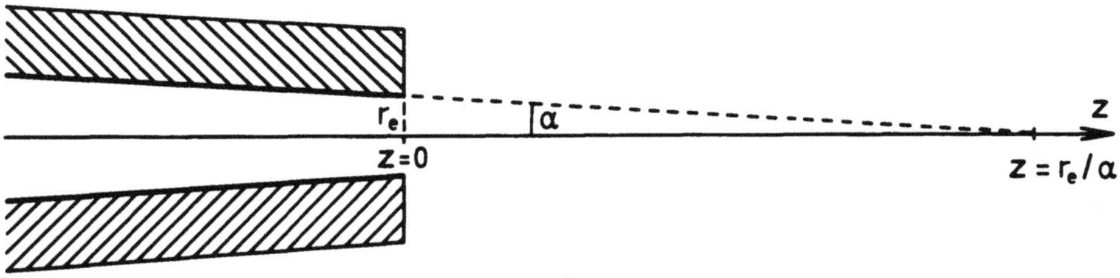

Fig. 4. Schematic representation of microelectrode tip. For definition of symbols, see text. Cross-hatched area indicates glass wall of electrode

solution of the same electrolyte of concentration C^s. General expressions will be derived for the total ionic fluxes j_+ and j_-, for the resistance of the electrode R and for the diffusion potential V_{junc} created across the liquid junction connecting the two solutions. Approximate expressions for the fluxes and the resistance are derived under the assumption that the electrode tip has the form of a truncated cone with a circular aperture of radius r_e and half top angle $\alpha << 1$, as shown in Figure 4. The derivation is based on an integration of the electrodiffusion equations

$$\bar{J}_+ = -b_+ C_+ \operatorname{grad}(\mu_+ + F\Psi)$$
$$\bar{J}_- = -b_- C_- \operatorname{grad}(\mu_- - F\Psi) \qquad \text{(C 1)}$$

where $\bar{J}$ is the flux density, b the molar mechanical mobility, μ the chemical potential subscripted to indicate positive or negative ionic species; Ψ is the electrical potential and the other symbols have their usual meaning. The integration of Eqs. (C 1) will be carried out in the Planck regime, *i. e.* rather than finding the potential distribution as a solution to the Poisson equation, it will be assumed that the electrical potential distributes in such a way that local electroneutrality is ensured everywhere (cf. the chapter by STEN-KNUDSEN, Section F.II.2.a). The results are limited, strictly speaking, to situations where volume flow can be ignored. Steady state in which a constant current i is passed through the pipette will be assumed throughout

$$i/F = j_+ - j_- \qquad \text{(C 2)}$$

The only geometrical condition imposed on the system is that it has rotational symmetry around some coordinate axis z, with its origin coinciding with the center of the tip (Fig. 4).

For a single 1:1 electrolyte, Eqs. (C 1) read, when projected on the z-axis:

$$J_+ = -b_+ C \frac{d}{dz}(\mu_+ + F\Psi) \qquad \text{(C 3)}$$

$$J_- = -b_- C \frac{d}{dz}(\mu_- - F\Psi) \qquad \text{(C 4)}$$

where the electroneutrality condition $C_+ = C_- = C$ has been used. At some point z on the symmetry axis $J_+(z)$ must be the same fraction of the total flux j_+ as $J_-(z)$ is of j_-. Thereby an area $A(z)$ is defined as a function of z:

$$A(z) = \frac{j_+}{J_+(z)} = \frac{j_-}{J_-(z)} \tag{C 5}$$

$A(z)$ is expected to be a purely geometrical quantity, *i. e.* independent of concentrations and mobilities.

Equations (C 2), (C 3) and (C 4) are conveniently rearranged

$$j_+ = \frac{t_+}{F} \cdot i - A(z) \cdot D \cdot \frac{dC}{dz} \tag{C 6}$$

$$j_- = -\frac{t_-}{F} \cdot i - A(z) \cdot D \cdot \frac{dC}{dz} \tag{C 7}$$

$$F \cdot \frac{d\psi}{dz} = \frac{-\frac{j_-}{b_-} \cdot \frac{d\mu_+}{dz} + \frac{j_+}{b_+} \cdot \frac{d\mu_-}{dz}}{\frac{j_+}{b_+} + \frac{j_-}{b_-}} \tag{C 8}$$

where we have introduced the transference numbers t_+, t_- and the conventional diffusion coefficient D as defined by Fick's law of diffusion.

$$t_+ = \frac{b_+}{b_+ + b_-}$$

$$t_- = \frac{b_-}{b_+ + b_-}$$

$$D = \frac{b_+ \cdot b_-}{b_+ + b_-} \cdot \frac{d(\mu_+ + \mu_-)}{d \ln C}$$

Since transference numbers for KCl are known to be independent of concentration within 0.5 percent from infinite dilution up to 1 M (ROBINSON and STOKES, 1959), integration of Eqs. (C 6) and (C 7) gives to a very good approximation:

$$j_+ = \frac{t_+}{F} \cdot i - \frac{\int_{\mathrm{e}}^{\mathrm{s}} D \cdot dC}{\int_{-\infty}^{\infty} \frac{dz}{A(z)}} \tag{C 9}$$

$$j_- = -\frac{t_-}{F} \cdot i - \frac{\int_{\mathrm{e}}^{\mathrm{s}} D \cdot dC}{\int_{-\infty}^{\infty} \frac{dz}{A(z)}} \tag{C 10}$$

When these expressions are inserted into Eq. (C 8) one finds

$$\frac{d\Psi}{dz} = -i \cdot \frac{\int_{-\infty}^{\infty} \frac{dz}{A(z)}}{\int_{e}^{s} D dC} \cdot \frac{D}{\lambda} \cdot \frac{d\ln C}{dz} + \frac{1}{F} \cdot \left(t_{+} \frac{d\mu_{+}}{dz} - t_{-} \frac{d\mu_{-}}{dz}\right) \tag{C 11}$$

where the equivalent conductivity λ has been introduced:

$$\lambda = F^2 (b_{+} + b_{-})$$

Integration of Eq. (C 11) gives an equation of the form

$$V = R \cdot i + V^{\text{junc}}$$

where V is the bulk to bulk potential difference

$$V = \Psi(-\infty) - \Psi(\infty)$$

and where the resistance R and the junction potential V^{junc} are given by

$$R = \frac{\int_{e}^{s} \frac{D}{\lambda} d\ln C}{\int_{e}^{s} D dC} \cdot \int_{-\infty}^{\infty} \frac{dz}{A(z)} \tag{C 12}$$

$$V^{\text{junc}} = \frac{1}{F} \cdot \int_{e}^{s} (t_{+} d\mu_{+} - t_{-} d\mu_{-}) \tag{C 13}$$

As will be discussed below, the integral $\int_{-\infty}^{\infty} \frac{dz}{A(z)}$ is easily estimated for the simple geometry of Figure 4. Equations (C 9), (C 10), and (C 12) therefore permit a numerical evaluation of the fluxes and the resistance, with tabulated values of D and λ as functions of concentration (ROBINSON and STOKES, 1959). To evaluate V^{junc} according to Eq. (C 13) one must know not only the concentration-dependence of the transference numbers, but also the concentration-dependence of the single ion activity coefficients. The latter are inaccessible from thermodynamic measurements, but can of course be calculated from theory. The common practice is however to assume

$$d\mu_{+} = d\mu_{-} = RT \cdot d\ln \gamma_{\pm} C$$

where $\gamma_\pm$ is the mean activity coefficient of the electrolyte. With this approximation, Eq. (C 13) gives for concentration-independent transference numbers:

$$V^{\text{junc}} = \frac{RT}{F} \cdot \frac{b_+ - b_-}{b_+ + b_-} \cdot \ln \frac{\gamma_\pm^s C^s}{\gamma_\pm^e C^e}$$

As discussed by OVERBEEK (1956), one may expect significant corrections to this formula if large polyelectrolytes are present in the external electrolyte.

If D and λ were concentration-independent, the expression for the resistance would simplify to

$$R = \frac{1}{\lambda} \cdot \frac{\ln C^s/C^e}{C^s - C^e} \cdot \int_{-\infty}^{\infty} \frac{dz}{A(z)} \tag{C 14}$$

However λ is known to depend rather strongly on concentration decreasing by about 50 percent from infinite dilution to 3 M for a KCl solution. To take this fact into account, we suggest that the λ to be used in Eq. (C 14) is half the sum of its value in the electrode shaft and in the bulk solution.

$$\lambda = \frac{\lambda^e + \lambda^s}{2}$$

A similar averaging procedure should be applied to the diffusion coefficient in Eqs. (C 9) and (C 10). In this case, however, the variation with concentration is much smaller (only ~ 10%) and we suggest that the limiting value at infinite dilution is the appropriate value to use.

The limiting value for the resistance when $C^s \to C^e$ is

$$R_{\text{lim}} = \varrho^e \cdot \int_{-\infty}^{\infty} \frac{dz}{A(z)} \tag{C 15}$$

where the specific resistance ϱ^e of the electrolyte in the electrode shaft has been introduced:

$$\varrho^e = \frac{1}{\lambda^e C^e} .$$

In the nonlimited case one finds from Eq. (C 14):

$$R = R_{\text{lim}} \cdot \frac{2\lambda^e}{\lambda_e + \lambda^s} \cdot \frac{\ln C^e/C^s}{1 - C^s/C^e} \tag{C 16}$$

Before we turn to the evaluation of $\int_{-\infty}^{\infty} \frac{dz}{A(z)}$ let us mention one important consequence of these general results. From Eqs. (C 12), (C 9), and (C 10) *it is to be expected that the product of the resistance and the total salt flux is independent of the geometry of the electrode under conditions of zero electric current.*

$$R \cdot j = -\int_{e}^{s} \frac{D}{\lambda}\, d\ln C \simeq \frac{2D}{\lambda^{e} + \lambda^{s}} \cdot \ln \frac{C^{e}}{C^{s}}\,. \qquad \text{(C 17)}$$

The prediction of Eq. (C 17) is tabulated in Table 1 for a KCl-filled micropipette.

Table 1. (See text for explanation)

KCl, 25°C, $C^{e} = 3$ M	
C^{s} (M)	$R \cdot j \cdot 10^{7}$ ($\Omega \cdot$ mol s^{-1})
0.1	5.94
0.2	4.83
0.5	3.30
1.0	2.07

Consider, finally, Figure 4. Within the electrode it is reasonable to assume spherical symmetry around the point $z = r_{e}/\alpha$. The flux density $\bar{J}_{+}$ then points everywhere towards this point and has the same magnitude everywhere on the surface formed by intersection of the cone and a sphere with center at $z = r_{e}/\alpha$. Thus the area $A(z)$, as defined by Eq. (C 5) is

$$A(z) \simeq \alpha^{2}\pi\,(z - r_{e}/\alpha)^{2}; \qquad z < o$$

and therefore

$$\int_{-\infty}^{o} \frac{dz}{A(z)} \simeq \frac{1}{\alpha\pi r_{e}}$$

Whatever the precise functional form of $A(z)$ just outside the tip one always finds the integral from zero to infinity to be of zero order in α and therefore small compared to the integral from minus infinity to zero. In conclusion we shall use the approximation

$$\int_{-\infty}^{\infty} \frac{dz}{A(z)} \simeq \frac{1}{\pi\alpha r_{e}} \qquad \text{(C 18)}$$

Since typical values of α never exceed 0.1 it is likely that the approximation of Eq. (C 18) is not wrong by more than 10 percent and is definitely better for smaller values of α. More accurate estimates of the integral could easily be devised. However, since the internal diameter of the tip cannot be determined with certainty such procedures will be of limited value.

The approximate expressions for the fluxes and the resistance can finally be written as

$$j_+ = \frac{t_+}{F} \cdot i + \alpha\pi r_e D \cdot (C^e - C^s) \tag{C 19}$$

$$j_- = -\frac{t_-}{F} \cdot i + \alpha\pi r_e D \cdot (C^e - C^s) \tag{C 20}$$

$$R = R_{lim} \cdot \frac{2\lambda^e}{\lambda^e + \lambda^s} \cdot \frac{\ln C^e/C^s}{1 - C^s/C^e} \tag{C 21}$$

$$R_{lim} = \frac{\varrho_e}{\alpha\pi r_e} \tag{C 22}$$

In Figure 5 the prediction of Eq. (C 21) is compared with the experimental values as measured by LASSEN and STEN-KNUDSEN (1968). The micropipettes used in their experiments were filled with 3 M KCl. The electrodes had a half-top angle $\alpha \simeq 0.062$ and an outer tip diameter of $\simeq 0.18$ µm, as judged from an electronmicroscopic picture. The value used for r_e in Eq. (C 22) to give the best fit was $r_e \simeq 0.032$ µm. We do not think that these figures are conflicting, since a glass wall thickness of ~ 600 Å does not seem unreasonable. This value finds some support in the investigations of NASTUK and HODGKIN (1950), who estimated a glass wall thickness of ~ 750 Å for a pipette with an outer tip diameter of ~ 0.4 µm.

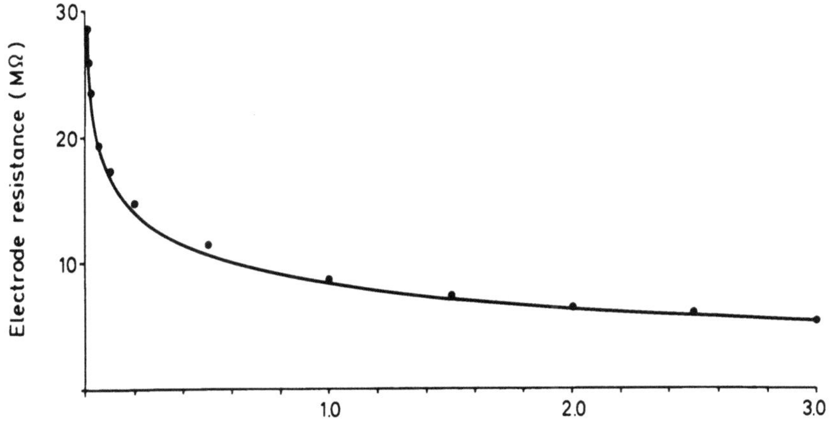

Fig. 5. Resistance of a microelectrode as a function of KCl concentration of medium surrounding electrode tip. Experimental values reported by LASSEN and STEN-KNUDSEN (1968) plotted as dots. Solid line is drawn according to Equations (C 21) and (C 22) with $\alpha = 0.062$ and $r_e = 0.032$ µm, tabulated values being used for equivalent conductivities of KCl. (ROBINSON and STOKES, 1959)

COOMBS et al. (1955) noticed that when a microelectrode filled with 3 M KCl was moved from a 3 M KCl solution and immersed in a 0.15 M KCl solution "the resistance rises within a second or so to a steady level almost three times the value measured in a 3 M KCl solution." The increase expected from Eq. (C 21) is a factor of 2.8.

Recent measurements by EHRENFELD et al. (1978) show the KCl loss to be $\sim 6 \cdot 10^{-14}$ mol s^{-1} from a micropipette filled with 3 M KCl and with a resistance of 4.3 MΩ. Unfortunately, however, the bathing solution in which the resistance was determined was not specified. From Eqs. (C 19) and (C 22) we estimate the *maximal* KCl loss by diffusion to be $\sim 6 \cdot 10^{-14}$ mol s^{-1} for a pipette filled with 3 M KCl and with a limiting resistance of $\sim$ 3.2 MΩ.

The introduction of the function $A(z)$, a purely geometrical quantity, allows the necessary integrations to be carried out without prior knowledge of the concentration distribution near the tip. As a result, the concentration-dependence of the fluxes and the resistance is believed to be well described. However, their magnitude could be subject to a small concentration-independent correction. This uncertainty reflects the fact that it is very difficult to calculate the concentration distribution accurately near the opening of the pipette, which is the region of paramount importance for the magnitude of the resistance.

D. Potential Recording with Microelectrodes

The following account will concentrate on practical procedures and limitations when microelectrodes are used for measurements of membrane potential. Some examples of experimental data will be discussed from the point of view of illustrating methodological problems.

I. Penetration of the Cell Membrane

A microelectrode measurement operationally consists of three phases: recording of extracellular potential difference between the electrodes (usually adjusted to zero), penetration of the cell membrane, and finally recording of the potential difference with the electrode located in the cytosol. To check that the electrode has not become clogged or otherwise damaged during this procedure it is customary to check the potential (and electrode resistance) after withdrawal of the electrode tip from the cell. Repeated puncture of cell membranes with the same electrode in some cases tends to increase the tip potentials with ensuing uncertainties in the interpretation of the values obtained (REDMANN and KALKOFF, 1968; TASAKI and SINGER, 1968). With optimal visual control of the penetration (*e. g.* LASSEN et al., 1971) it is often possible to perform more than 10 consecutive measurements before measurable alteration of electrode characteristics occurs.

Although the measurements are simple in principle, practical difficulties often accompany performance of the crucial step: penetration of the cell membrane. In the case of large cells such as striated muscle or giant neurons, dimpling of the cell membrane with the electrode tip followed by a slight vibration of the equipment (like jarring the table top) produces enough shear to cause the electrode to impale the cell. If the cell in question is buried in a tissue or is covered with a collagen sheath, breakage and/or clogging is frequently encountered. Motor-driven microelectrode holders seem to solve some of the problems (*e.g.* ANDERSEN and LAURSEN, 1959). Recently several authors have reported that bevelling the tips of glass microelectrodes leads to less breakage and clogging. Tips with diameters less than 1 µm can be manufactured after appropriate grinding (BARRET and WHITLOCK, 1973; BROWN and FLAMING, 1974; CHANG, 1975; CLEMENTS and GRAMPP; 1976; KRIPKE and OGDEN, 1974). Such electrodes are reported to be easier to introduce and to cause less tissue damage. From simple mechanical considerations bevelled electrodes seem to present an advantage over the common serrated-edge tips. However more experimental evidence is needed to justify this optimism.

Measurements of ionic (*e.g.* tracer) fluxes across cell membranes are often performed with a suspension of single cells, such as erythrocytes, leucocytes or Ehrlich ascites cells. The uniform exposure of the membranes of such cells to the suspending medium in many cases leads to a more clear-cut interpretation of the data than is possible with single- or multilayered sheets of cells. An obvious disadvantage of the suspended cell preparation is the difficulty of obtaining measurements of membrane potential. It is rarely possible, if ever, to penetrate the cell membrane in a controlled fashion by merely advancing the electrode tip towards the cell. In the past this problem has been overcome by embedding the cells in a high-viscosity medium like agar- or methylcellulose-containing Ringer's. Preparation of cells in agar takes some time, with consequent risk of cell deterioration, especially in the case of cells with a high rate of metabolism, *e.g.* ascites cells. The viscosity of methylcellulose in the concentrations commonly used (*e.g.* LASSEN et al., 1971) is insufficient to give total fixation of the cells, which will settle in the experimental chamber. When lying on the bottom of the chamber, the cells are accessible for micropuncture. However, this case also requires special procedures to cause the tip of the electrode to penetrate the membrane, as the cells will tend to bounce away upon advancement of the electrode. Coating the bottom of the chamber with multivalent cations like La^{3+} (SMITH and LEVINSON, 1975) seems to give sufficient fixation to allow impalement but may at the same time influence the membrane properties.

A different approach to the problem was chosen by LASSEN and STEN-KNUDSEN (1968), who used a piezoelectric electromechanical transducer to give a rapid and precise advancement of the electrode tip. If the tip was placed close to the cell prior to the controlled advance (*i.e.* voltage application to the piezoelectric elements), penetration of cells could be obtained. As shown in Figure 6, the piezoelectric "pistol" consisted of a metal capsule with flexible mounts in the form of a cross for the four piezoelectric bender elements. A step voltage of 50–250 V applied to the benders caused the microelectrode to move forward over a distance of 10–50 µm in about 2 ms. Even though this transducer made

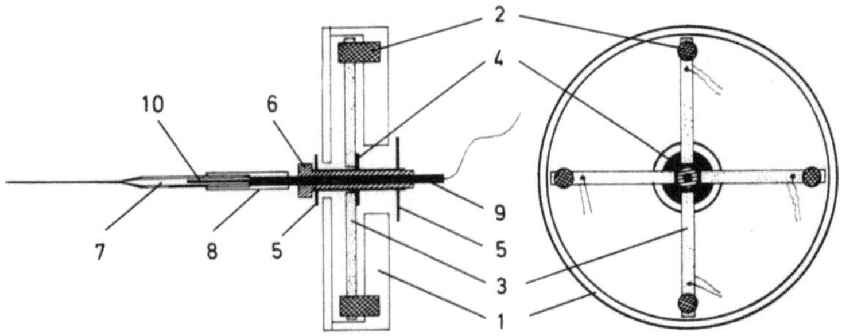

Fig. 6. Electromechanical transducer used by LASSEN and STEN-KNUDSEN (1968) for advancement of microelectrodes. Aluminium casing (1) with four holes for polyvinylchloride holders (2) for the four piezoelectric bender elements (3), glued at opposite ends to a flange (4) on brass tubing holding two shield plates (5). Teflon tube (6) for insulation of stainless steel electrode support (9). Microelectrode (7) was held in a polyethylene tube (8). Contact with fluid in electrode was mediated by silver rod (10) coated with silver chloride. Diameter of assembly about 50 mm

penetration of the plasma membrane of human red cells in suspension possible, the flaccid mounting of the benders and consequently of the electrode made precise "aiming" very difficult. The low resonant frequency of the assembly (about 200 Hz) left the investigator with the choice of either critical damping, and thus a slow advancement, or undamped, rapid advancement with ensuing large axial oscillations. These properties made it desirable to redesign the transducer for subsequent measurements of membrane potential of Ehrlich ascites cells (LASSEN et al., 1971). In the new version (Fig. 7) the piezoelectric bender elements were arranged in two parallel rows with 10 elements in each row, one row above and one below the movable perspex rod carrying the microelectrode holder. This design gave a doubling of the advancement velocity of the electrode, an almost total shielding of the electrode from the high-voltage pulse to the bender elements, and a better stiffness (resonant frequency about 900 Hz). With this construction the limiting factor for the precision of advancement seemed to reside mainly in the electrode tip. The movement of the electrode with a step of 400 V to the bender elements was 20 to 35 μm in less than 1 ms (with subcritical damping). Ehrlich ascites tumor cells are 10 to 20 μm in diameter (SELBY et al., 1956). The long stroke of the piezoelectric pistol in relation to the size of cells aided in penetrating the highly microvillous surface. But at the same time, the long range required some skill on the part of the experimenter. The advance had to be released with the electrode tip about 10 μm from the cell to avoid disruption of the cell upon penetration. Smaller total strokes (decrease in voltage to the piezoelectric elements) resulted in a decrease in number of successful penetrations. Despite these shortcomings, the piezoelectric driver shown in Figure 7 allowed direct measurements in Ehrlich cells, thereby solving a discrepancy between the magnitude of the equilibrium potential for the (presumably) passively distributed Cl^- and the measured potentials, the latter being one-half to one-third of E_{Cl^-}. This matter has been discussed in detail by LASSEN et al. (1971) and by SIMONSEN and NIELSEN (1971). As discussed in the next

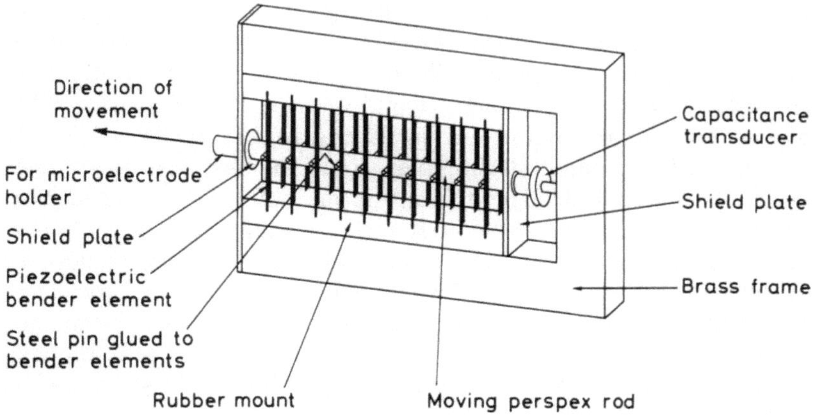

Fig. 7. Electromechanical transducer used by LASSEN et al. (1971) for advancement of microelectrodes in study of membrane potential of Ehrlich ascites cells. High voltage (up to 400 V) was applied simultaneously to all 20 bender elements, causing them to bend in same direction. Movable perspex rod suspended between bender elements by short stainless steel pins was moved forward as a result of bending of piezoelectric elements. Axial movements monitored by capacitance-type transducer at back end of the movable perspex rod. For further details see text. Total length of assembly 84 mm

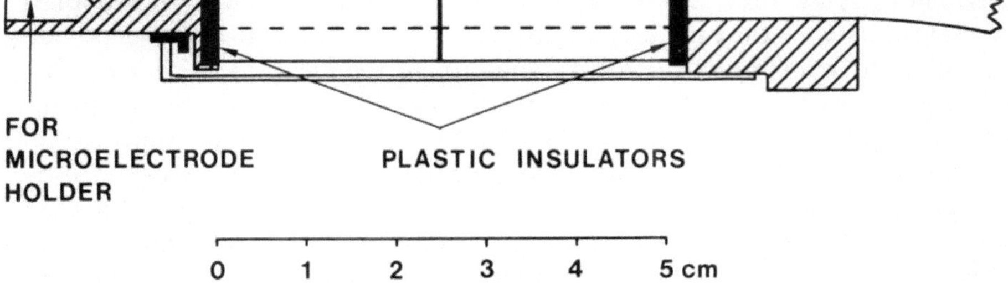

Fig. 8. Piezoelectric driver used for impalements of *Amphiuma* red cells with microelectrodes (as described further in text). Tubular piezoelectric elements give lengthwise expansion as high voltage (400 V) is applied between outside and inside of elements. This device has greater mechanical stability and shorter range of movement than transducers shown in Fig. 6 and Fig. 7

section, a leak around the microelectrode after penetration has to be taken into account. In order to circumvent this problem giant red cells of the salamander, *Amphiuma means,* were used in a preliminary study of the electrical properties of the erythrocyte membrane (HOFFMAN and LASSEN, 1971; LASSEN, 1972).

Amphiuma red cells are discoid and have a long axis of 50–80 μm. When lying on the bottom of a chamber where they could be observed in an inverted microscope, the microelectrode tip could be made to dimple the membrane before penetration. Penetration without further fixation was possible upon a short, rapid advancement. This was produced by an electromechanical transducer of still another design (Fig. 8). Two tubular ceramic piezoelectric elements were glued end to end, one end of the assembly being fixed to a micromanipulator, the other carrying the microelectrode holder. Application of a step voltage of 400 V causes an advancement of 1 to 2 μm, depending on the specifications of the piezoelectric material. The resonance of the transducer is complex but has a major component around 10 kHz. When critically damped, the movement seems too slow to cause entry of the microelectrode tip into the cell. By adjusting the damping to fit the individual microelectrode it is possible to obtain successful penetration in more than 50 percent of attempts. This device has been applied in several studies on the electrical properties of the *Amphiuma* red cell membrane (which seem to be common to red cells in general, see review by LASSEN, 1977).

II. Microelectrodes and Leaks in the Membrane

As a microelectrode penetrates the cell membrane, the hydrated glass surface of the tip will provide an electrical leak pathway between the extracellular and intracellular phases. Furthermore, the funnel-shaped dimpling of the membrane prior to penetration must tend to pull material from the outer face of the membrane along with the electrode, thus creating an additional leak. In most of the numerous studies of membrane properties by means of microelectrodes, the presence of a leak pathway and its influence on the experimental results have not been taken into account. It is often noted that potential recordings in a given experiment started after achievement of a stable resting potential. This implies that an instability of the potential immediately after the puncture was succeeded by a satisfactory sealing of the membrane around the electrode. When dealing with large or interconnected cells where the input resistance of the cell membranes is small, it is less likely that the measurements are seriously affected by the maintained presence of a small leak pathway. This notion is supported by the wealth of reproducible data obtained in experiments with striated muscle fibers, giant salivary gland cells from insects (which are electrically interconnected; see *e.g.* LOEWENSTEIN, 1966), or giant nerve cells in molluscs. In such cases the microelectrode measurements obviously are subjected to the "usual" limitations as discussed above in relation to KCl leak from the electrode and to the changed environment of the electrode tip when moved from extra- to intracellular fluid.

In the case of single cells without electrical connection, relatively simple and clear-cut examples of the problems encountered come from studies of membrane potentials in Ehrlich cells and red cells.

The consistent discrepancy between the equilibrium potential for Cl^- and measured membrane potentials in Ehrlich cells (as mentioned above) led LASSEN et al. (1971) to investigate the change in potential in immediately following the penetration of the membrane by the microelectrode tip. An initial large potential drop followed by decay to stable and less negative potential values was detected, the potential peak being obtained within 200 μs after the penetration. Figure 9 shows this type of recording. An oscilloscope with two time-base generators and two dual-trace differential *Y*-axes was used. In this way, both the fast response and the stable potential difference across the same cell could be recorded. In the figure the above mentioned potential peak is clearly seen in trace 1, whereas trace 2 depicts the stable potential level finally achieved. Upon retraction of the electrode (positive deflection in trace 2) the potential returns to zero (extracellular level). LASSEN et al. (1971) interpreted this pattern as a consequence of a leak in the cell membrane caused by the impalement. The stable level was then the "simple" diffusion potential between the cell contents and the extracellular fluid. The magnitude of the initial potential peak was consistent with the magnitude of the Nernst potentials for Cl^- under a number of experimental conditions. With this type of circular argument, it was concluded that the peak potential was a measure of the membrane potential of the unperturbed cell and that the decay from this value was a discharge of the membrane potential due to the leak around the electrode. On the basis of this pattern a simple model was formulated in which the initial potential drop was followed by the decay of the original membrane potential to a less negative value.

The recorded peak potential after penetration must be numerically smaller than the actual membrane potential of the unperturbed cell. In order to estimate the influence of various time constants of the electrode-amplifier system at a

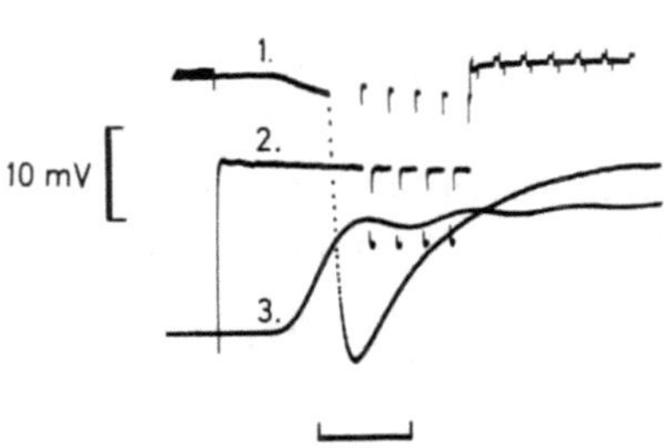

Fig. 9. Typical decaying potential obtained upon impalement of Ehrlich ascites cell in Na-Ringer. Traces 1 and 3 show immediate events upon impalement. Horizontal time mark equals 1 ms for these two traces. Trace 1 is potential "seen" by microelectrode. Trace 3 shows axial movements of electrode (total advancement about 25 μm mediated by transducer in Fig. 7). Trace 2 shows potential of microelectrode but with a timebase 40 times slower than traces 1 and 3 (horizontal time mark equals 40 ms). Vertical calibration mark (10 mV) is common for traces 1 and 2 (trace 3 not calibrated). Train of diphasic pulses for resistance measurement started about 50 ms after advancement of microelectrode. (From LASSEN et al., 1971)

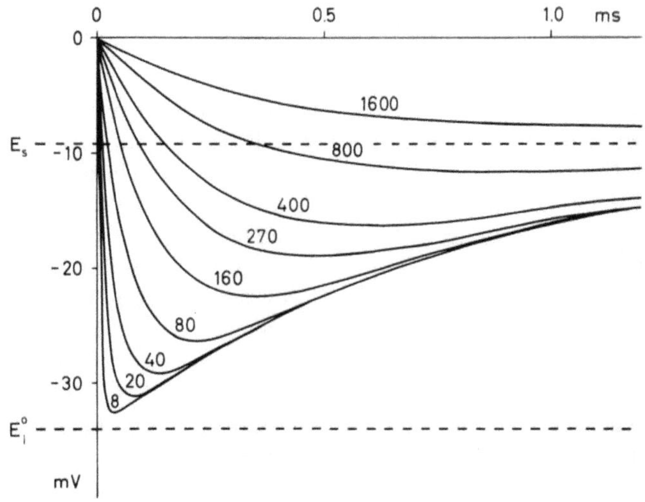

Fig. 10. Family of curves calculated from Equation (D 3) for different time constants of microelectrode-amplifier system (τ_1). These time constants (in µs) are indicated for each curve. Ordinate is potential in mV. Abscissa is time in ms. E_s is stable potential (= − 9.2 mV) and E_o^o (= − 34 mV) is calculated membrane potential of unperturbed cell. See text for further details. (From LASSEN et al., 1971)

given time constant for the decay through the leak, the following simplifying equations were used:

$$P = E_i\,[1 - \exp(-t/\tau_1)] \qquad \text{(D 1)}$$

and

$$E_i = (E_i^o - E_s)\exp(-t/\tau_2) + E_s \qquad \text{(D 2)}$$

where P is the recorded potential at any given time after the micropuncture, E_i the intracellular potential at time t, E_i^o is the membrane potential of the nonperturbed cell and E_s the stable potential (see Fig. 9, trace 2). τ_1 and τ_2 are the time constants for the recording system and the discharge of the potential respectively. Combination of the above two equations gives:

$$P = [(E_i^o - E_s)\exp(-t/\tau_2) + E_s]\,[1 - \exp(-t/\tau_1)]\,. \qquad \text{(D 3)}$$

This relationship is expressed graphically in Figure 10 with insertion of τ_2 = 800 µs (mean from a number of experiments on Ehrlich cells). It is evident from the figure that even with fast recording systems having time constants of 20–40 µs the recorded peak potentials are numerically some 5–10 percent smaller than the original membrane potential (E_i^o). However, Figure 10 does not necessarily give the right magnitude of the peak potential with regard to the true intracellular potential, since the electrode was still in forward motion during the sharp potential drop to the peak value. This would tend to reduce the leak as the wedge-shaped electrode was forced into the hole in the cell membrane. This impression was strongly supported by measurements of membrane potential of *Amphiuma* red cells, where the most negative potential was frequently maintained some hundreds of microseconds before a decay similar to that observed in ascites cells occurred.

If a similar model of discharge of the membrane potential is applied to smaller cells, such as human red cells, or to the even smaller mitochondria, the inevitable conclusion is that until the appearance of as yet unknown procedures for obtaining sealing between the electrode and the cell wall, the results are bound to be misleading. In these cases the recorded potential is solely the stable junction potential irrespective of the original membrane potential (LASSEN et al., 1971).

In the Ehrlich cell a major part of the membrane conductance is presumably due to Cl^- passage. If this is the case, measurements of membrane resistance by current passage support the notion of a leak in the membrane due to the micropuncture. The measured, corrected, specific membrane resistance in Ehrlich cells is some 70 Ωcm^2, as compared with a "chloride resistance" of 4000 Ωcm^2 calculated from tracer fluxes. Thus the measured resistance is some two orders of magnitude smaller than expected under the assumption that Cl^- moves solely by diffusion and not by an electrically silent exchange mechanism. There is, however, good evidence that a major fraction of the Cl transport proceeds *via* an exchange mechanism (KROMPHARDT, 1968; HOFFMANN et al., 1978). This fact stresses even more the notion that the apparent input resistance of the cells may well be totally dominated by the leak pathway.

Even with the above-mentioned technique for advancing the microelectrode, Ehrlich cells are difficult to handle experimentally, mainly because their spherical shape allows them to roll away from the electrode. This is not a problem in the case of *Amphiuma* red cells. This may be one of the reasons for the greater reproducibility of measurements in these cells (see LASSEN, 1977). The following indirect confirmation of the assumption of a leak can be noted (LASSEN, unpubl.). Figure 11 shows the potential trace upon penetration of an *Amphiuma* red cell. The familiar pattern of an initial potential drop followed by a potential decay is seen. After attainment of a stable potential, current is injected through the electrode (LETTVIN et al., 1958) and a charging of the membrane to a new potential takes place. If the leak stayed essentially constant throughout the recording, the time constant of the discharge (τ_1) should be similar to that of the current-induced recharging (τ_2). As seen in Figure 12, this is in fact the case. The slope of the regression line is not significantly different from one. If anything, τ_2 is slightly smaller than τ_1, indicating that the leak may increase slightly with time after the micropuncture.

As indicated in a number of studies of membrane potential in *Amphiuma* red cells (see LASSEN, 1977) reasonable estimates of membrane potentials are possible despite the induced leak. Even with the leak present, measurements of membrane resistance gave maximal values in excess of 2000 Ωcm^2. As the corresponding resistance calculated from equilibrium exchange of $^{36}Cl^-$ is 0.7 Ωcm^2 (LASSEN et al., 1973), the data gave the first direct proof of the exchange of Cl^- proceeding in an electrically silent fashion. This confirmed the results of

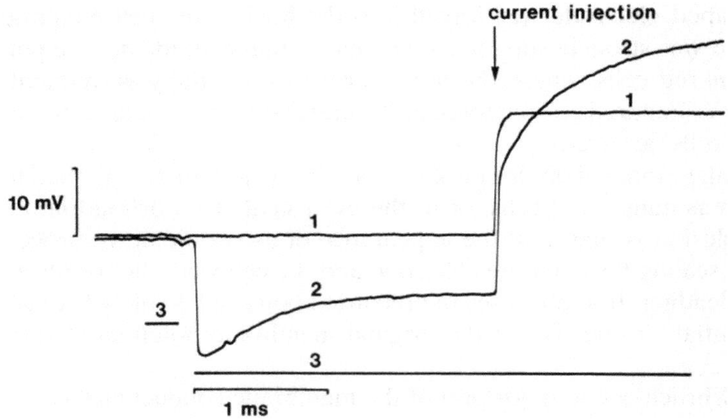

Fig. 11. Potential traces in relation to advancement of microelectrode by electromechanical transducer shown in Fig. 8. Trace 1 shows potential of microelectrode when high voltage is applied to transducer (break in trace 3). Apart from a small artifact due to direct electrical coupling, advancement does not cause any change in potential when electrode is in free fluid (trace 1). At time indicated by arrow, constant current is injected through electrode by method of LETTVIN et al. (1958). This results in positive deflection of potential of electrode due to the IV drop across the electrode resistance. When electrode is held against an *Amphiuma* red cell (trace 2) there is a rapid drop in potential following forward movement of the microelectrode, followed by slower decay of potential to stable value (time constant for the decay τ_1). Subsequent current injection causes charging of cell membrane capacitance (time constant for recharging τ_2); for rationale, see text. Vertical bar indicates 10 mV for traces 1 and 2 (trace 3 uncalibrated). Horizontal bar indicates 1 ms

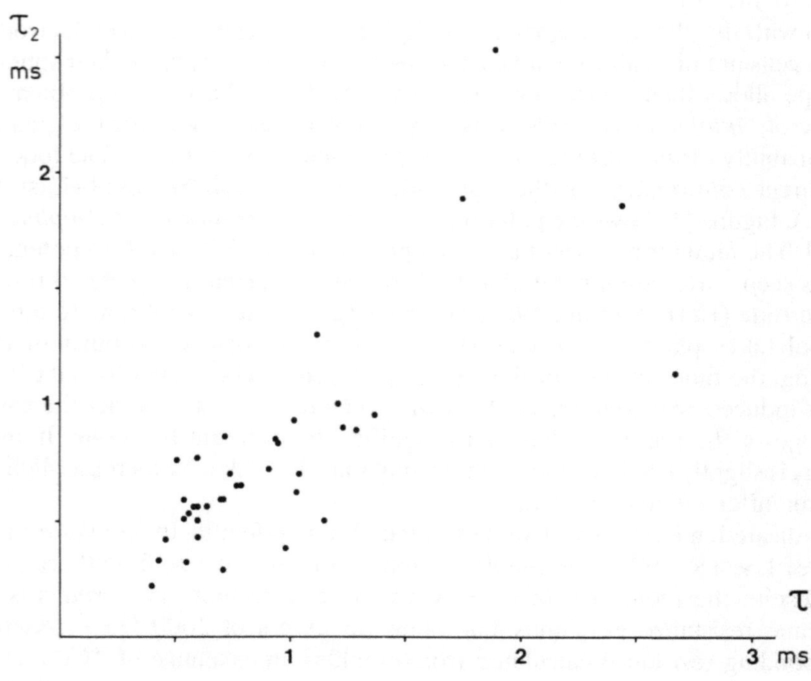

Fig. 12. Relation between time constants, τ_1 and τ_2 obtained from experiments similar to those presented in Fig. 11. Both axes in ms. See text for further explanation

HUNTER (1971) in human red cells; in this study valinomycin was used to change the K permeability of the membrane. In the case of *Amphiuma* red cells, LASSEN et al. (1975, 1978) and STONER and KREGENOW (1976) have given evidence for the same conclusion. It is interesting to note that an electrically silent Cl^- exchange has been observed in protein-free black lipid membranes (TOYOSHIMA and THOMPSON, 1975) and in black lipid membranes modified with organo-tin compounds (TIEFFENBERG and WIETH, pers. comm.).

The specific membrane resistance of *Amphiuma* red cells has been estimated to be in the order of 10^6 Ωcm² (LASSEN et al., 1978). With a typical cell surface of 5000 μm², the total input resistance of a single cell is 2×10^{10} Ω. Such values have never been observed with *Amphiuma* red cells or with any other cell when using microelectrodes. The potential recording if such a value was actually achieved would be extremely unstable due to spontaneous fluctuations in microelectrode properties (DEFELICE and FIRTH, 1971). Returning to the case of potential measurements in *Amphiuma* erythrocytes, the above considerations of time-course of the recorded potential and magnitude of the observed resistance increase make it almost certain that the leak pathway around the electrode is several orders of magnitude more conductive than the unperturbed membrane.

What would be the approximate dimension of a cylindrical leak (thickness Δr_e) around the electrode that would result in a typical leak between the cell and its surroundings? If we are dealing with a Ringer's-filled channel (specific resistance = ϱ) and a radius of the electrode at the point of contact with the membrane of r_e, the resistance of a channel (R_L) of length l can be calculated:

$$R_L = l\varrho/(\pi(r_e + \Delta r_e)^2 - r_e^2). \tag{D 4}$$

If we insert $l = 100$ Å $= 10^{-6}$ cm, $\varrho = 100$ Ωcm, $r_e = 0.25$ μm $= 2.5 \times 10^{-5}$ cm and $R_L = 10 \times 10^6$ Ω (a value of this order of magnitude is frequently measured, see *e.g.* Fig. 11), the resulting thickness of the conducting pathway around the electrode tip (Δr_e) is 6 Å. This small dimension of a leak pathway makes its presence very likely. However, the value of l, which is the distance of contact between the electrode and the cell membrane, is presumably grossly underestimated. Frequently there is a visible dimbling of the cell membrane at the point of impalement. Considering the resolution of the optical microscope this could easily correspond to $l = 0.5$ μm. In this case, the above equation will give a value for the thickness of a cylindrical leak pathway (Δr_e) of 0.3 μm, which is probably a more realistic figure. Subsequent sealing of the cell membrane to the wall of the electrode may take place without a change in l, with a resulting increase in the measured resistance. If these considerations are correct, they also explain why STONER and KREGENOW (1976), using a method with inherent mechanical stability of the electrode with respect to the cell, were able to maintain unchanged potentials (without decay) for seconds to minutes as compared to the millisecond range in the studies by the present group. Unfortunately the method of STONER and KREGENOW (1976) does not seem applicable when large net fluxes occur, as these will inevitably change the ionic concentrations in the narrow space between the cells and the glass capillary in which they are held.

Microelectrode measurements in epithelia, may also be influenced by cell damage following the impalement. The following example can be considered as typical for the problem. REUSS and FINN (1975) reported that the serosal membrane potential of toad bladder epithelium responds almost instantaneously to changes of the mucosal membrane potential. In their experiments a microelectrode was advanced through the mucosal cellular border. A change in mucosal membrane potential induced by a shift in composition of the mucosal bathing fluid caused an apparent simultaneous change in serosal membrane potential. LINDEMANN (1975) has analyzed the experimental situation carefully and concludes that a leak resistance of $7.8 \times 10^8\ \Omega$ in an apical cell membrane of $7 \times 7\ \mu m^2$ area may provide a simple explanation for the apparent change in serosal membrane potential. Thus the conclusion that there is a coupling of the potentials at the mucosal and serosal potentials (REUSS and FINN, 1975) does not appear to be justified. In his argument, LINDEMANN (1975) restricted his attention to stationary diffusion potentials and resistive networks. Active transport, degradation of ionic gradients, coupling between cells, and capacitive phenomena were not taken into account. Despite these limitations the leakage pathway can adequately explain the observed phenomena. Consequently, in the case of interconnected cells such as an epithelium, even a small leak has also to be considered an important factor when interpreting experimental data. A further complication stems from the leak of KCl from the micropipette, as discussed in a previous section of this chapter (see also EHRENFELD et al., 1978).

HIGGINS et al. (1977) and FRÖMTER and GEBLER (1977) have studied the cell potential profile and cell membrane resistance in amphibian urinary bladder. These authors point out that the microelectrode measurements may have been affected by leak artifacts despite careful experimental techniques. The luminal plasma membrane was estimated to have a specific resistance of about 20 $k\Omega cm^2$ of epithelium. This corresponds to a resistance of $5 \times 10^9\ \Omega$ for the luminal membrane of a single cell (*Necturus* urinary bladder). This value increases to $5 \times 10^{10}\ \Omega$ in the presence of amiloride (FRÖMTER and GEBLER, 1977). If a leak resistance to membrane resistance ratio of 10:1 is tolerable, the leak resistance must be at least $5 \times 10^{11}\ \Omega$. HIGGINS et al. (1977) point out that this is in the same range as the resistance of a single gramicidin channel in an artificial lipid membrane (HLADKY and HAYDON, 1972). The requirements for sealing between microelectrode and cell membrane are less strict if the measurements are performed on a membrane with lower specific resistance. In an elegant series of experiments HIGGINS et al. (1977) were able to show that micropuncture of *Necturus* urinary bladder cells from the serosal surface yielded more reproducible data than micropuncture of cells from the luminal surface. The improvement was related to the fact that the specific resistance of the serosal plasma membrane was some 13 times lower than that of the luminal membrane. HIGGINS et al. (1977) stress that many of the published observations on membrane potentials in epithelia may have been seriously complicated by leak artifacts (in line with the comments by LINDEMANN, 1975, as presented above).

E. Epilogue

Any new technique that solves important problems is gratefully received by experimenters in need of solutions to problems. Even though glass micropipettes have been available since 1925 (Ettish and Peterfi, see GEDDES, 1972, p. 156), it was the paper by LING and GERARD (1949), together with the rapid improvements in electronics, that brought about the common use of microelectrodes for measurement of membrane potential of living cells. The classic description by Hodgkin and Huxley of the electrical properties of the squid axon intensified the desire to quantitate membrane potential and membrane conductance. However, even after more than two decades many essential problems have remained unsolved. As discussed above, the limitations for measurements with microelectrodes seem to reside in the properties of the electrodes themselves and not in the electronic equipment used for recording or for current injection through the electrodes. Regardless of the tissue under study, sealing of the electrode wall to the cell membrane is a major problem to which no good solution has yet been found. And even if the leaks around the electrode were properly reduced, we would still face the fact that diffusion of salt from the electrode tip might influence the concentration of (*e. g.*) KCl in the cell. If the electrode is filled with isotonic solution, problems with changes in junction potentials must be expected.

Considering this state of affairs, it is not surprising that there is a constant search for alternative methods of estimating membrane potential. An obvious solution would be to measure the equilibrium distribution of an ion that is passively distributed and has a sufficiently high permeability to ensure that changes in membrane potential are immediately followed by redistribution. A good deal of optimism followed the reports by DAVILA et al. (1973), HOFFMAN and LARIS (1974) and SIMS et al. (1974) concerning the use of carbocyanine dyes to monitor membrane potential. However, recent work by HLADKY and RINK (1976) makes it very clear that these dyes do not give a simple solution to the problem of membrane potential measurement. The fluorescence of some carbocyanine dyes is a function of the potential difference across the membrane (with a relatively poor signal-to-noise ratio) as well as a function of binding to cell constituents and of formation of di- or polymers with low or absent fluorescence. At present there is little doubt that the application of these dyes is here to stay, but it is equally clear that the quantitative evaluation of fluorescence or absorbance chances has to be carried out with great care. Furthermore, in contrast to dye molecules, the tip of a microelectrode is much larger than the usual distance between protein molecules in cells, and the electrode will therefore average out local changes in the field close to the individual protein molecules. Thus it is not immediately obvious that a close agreement can be expected between microelectrode measurements and estimates of membrane potential from partition of permeable ions (*e. g.* carbocyanine dyes).

Despite the drawbacks in the use of microelectrodes, which we have discussed at length in this chapter, these electrodes still are the most direct approach to measurement of the electrical parameters of the membranes of single cells. Provided that experimenters are aware of, and pay attention to the limitations of microelectrode measurements, much valuable information can still be obtained.

Acknowledgements

Elisabet Krenchel, Hanne Olesen, and Villy Rasmussen have given valued assistance in the preparation of the manuscript. Dr. Leon Pape is gratefully acknowledged for stimulating discussions and for critical examination of the text.

References

Adrian, R. H.: J. Physiol. **133**, 631 (1956).
Agin, D. P.: In: Glass Microelectrodes (M. Lavallée, O. F. Schanne, and N. C. Hébert, Eds). New York: Wiley 1969, p. 62.
Agin, D. P., Holtzman, D.: Nature **211**, 1194 (1966).
Amatniek, E.: I. R. E. Trans. med. electron. **PGME-10**, 3 (1958).
Andersen, V. O., Laursen, A. M.: Electroenceph. clin. Neurophysiol. J. **11**, 172 (1959).
Barret, J. N., Whitlock, D. G.: In: Intracellular Staining in Neurobiology. (S. B. Kater and C. Nicholson, Eds). New York: Springer 1973, p. 297.
Bingley, M. S., Thompson, C. M.: J. theoret. Biol. **2**, 16 (1962).
Bockris, J. O'M., Reddy, A. K. N.: Modern Electrochemistry, **Vol. 2**. New York: Plenum 1973.
Brown, K. T., Flaming, D. G.: Science **185**, 693 (1974).
Chang, J. J.: Comp. Biochem. Physiol. **52 A**, 567 (1975).
Clayton, G. B.: In: Linear Integrated Circuit Applications. London: Macmillan 1975, p. 17.
Clements, B., Grampp, W.: Acta physiol. scand. **96**, 286 (1976).
Cole, K. S., Moore, J. W.: J. gen. Physiol. **43**, 971 (1960).
Coombs, J. S., Eccles, J. C., Fatt, P.: J. Physiol. **130**, 326 (1955).
Davila, H. V., Salzberg, B. M., Cohen, L. B., Waggoner, A. S.: Nature New Biol. **241**, 159 (1973).
DeFelice, L. J., Firth, D. R.: IEEE Trans. Bio-Med. Eng. **BME-18**, 339 (1971).
Ehrenfeld, J., Nelson, D. J., Lindemann, B.: (in press) (1978)
Ferris, C. D.: Introduction to Bioelectrodes. New York, London: Plenum 1974.
Frank, K., Becker, M. C.: In: Physical Techniques in Biological Research (W. L. Nastuk, Ed.), Chap. 2. New York: Academic Press 1964.
Frömter, E., Gebler, B.: Pflügers Arch. **371**, 99 (1977).
Geddes, L. A.: Electrodes and the Measurement of Bioelectric Events. New York: Wiley-Interscience 1972.
Geddes, L. A., Baker, L. E.: Med. Res. Eng. **6**, 33 (1967).
Geisler, C. D., Lightfoot, E. N., Schmidt, F. P., Sy, F.: IEEE Transact. Biomed. Eng. **BME-19**, 372 (1972).
Guggenheim, E. A.: Thermodynamics. Amsterdam: North-Holland 1957.
Guld, C.: Proc. IRE **50**, 1912 (1962).
Higgins, J. T., Gebler, B., Frömter, E.: Pflügers Arch. **371**, 87 (1977).
Hladky, S. B., Haydon, D. A.: Biochim. biophys. Acta **274**, 294 (1972).
Hladky, S. B., Rink, T. J.: J. Physiol. **263**, 287 (1976).
Hoffman, J. F., Laris, P. C.: J. Physiol. **239**, 519 (1974).
Hoffman, J. F., Lassen, U. V.: Proc. Internat. Union Physiol. Sci. **9**, 253 (abstract 746) (1971) Publ.: German Physiological Society, 1971.
Hoffmann, E. K., Simonsen, L. O., Sjøholm, C.: (in preparation) (1978)

HUNTER, M.J.: J. Physiol. **218**, 49p (1971).
JANZ, G.J., IVES, D.J.G.: Ann. N.Y. Acad. Sci. **148**, 210 (1968).
JUNG, W.G.: IC op-amp. Indiana: Sams 1974.
KIRKWOOD, J.G., OPPENHEIM, I.: Chemical Thermodynamics. New York: McGraw-Hill 1961.
KRIPKE, B.R., OGDEN, T.E.: Electroenceph. clin. Neurophysiol. **36**, 323 (1974).
KRNJEVIĆ, K., MITCHELL, J.F., SZERB, J.C.: J. Physiol. **165**, 421 (1963).
KROMPHARDT, H.: Europ. J. Biochem. **3**, 377 (1968).
LANTHIER, R., SCHANNE, O.F.: Naturwissenschaften **53**, 430 (1966).
LASSEN, U.V.: In: Oxygen Affinity of Hemoglobin and Red Cell Acid-Base Status (M. Rørth, P. Astrup, Eds). Copenhagen: Munksgaard 1972, p. 291.
LASSEN, U.V.: In: Membrane Transport in Red Cells (J.C. Ellory, V.L. Lew, Eds). New York: Academic Press 1977, p. 137.
LASSEN, U.V., STEN-KNUDSEN, O.: J. Physiol. **195**, 681 (1968).
LASSEN, U.V., NIELSEN, A.-M.T., PAPE, L., SIMONSEN, L.O.: J. Membrane Biol. **6**, 269 (1971).
LASSEN, U.V., PAPE, L., VESTERGAARD-BOGIND, B.: In: Erythrocytes, Thrombocytes, Leukocytes (E. Gerlach, K. Moser, E. Deutsch, W. Williams Eds). Stuttgart: Thieme 1973, p. 33.
LASSEN, U.V., PAPE, L., VESTERGAARD-BOGIND, B.: 5th International Biophysics Congress, Copenhagen (abstract) p. 102, 1975.
LASSEN, U.V., PAPE, L., VESTERGAARD-BOGIND, B.: J. Membrane Biol. **39**, 27 (1978).
LAVALLÉE, M., SZABO, G.: In: Glass Microelectrodes (M. Lavallée, O.F. Schanne, N.C. Hébert, Eds). New York: Wiley 1969.
LAVALLÉE, M., SCHANNE, O.F., HÉBERT, N.C.: Glass Microelectrodes. New York: Wiley 1969.
LETTVIN, J.Y., HOWLAND, B., GESTELAND, R.C.: I.R.E. Trans. med. Electron. **PGME-10**, 26 (1958).
LEV, A.A., ARMSTRONG, W.McD.: In: Current Topics in Membranes and Transport, Vol. 6 (F. Bronner and A. Kleinzeller, Eds). New York-London: Academic Press 1975.
LINDEMANN, B.: Biophys. J. **15**, 1161 (1975).
LING, G., GERARD, R.W.: J. cell comp. Physiol. **34**, 383 (1949).
LOEWENSTEIN, W.R.: Ann. N.Y. Acad. Sci. **137**, 441 (1966).
MOORE, J.W.: In: Biophysics and Physiology of Excitable Membranes (W.J. Adelman, Jr., Ed.). New York: Van Nostrand-Reinhold 1971.
MOORE, J.W., COLE, K.S.: In: Physical Techniques in Biological Research, Vol. 6 (W.L. Nastuk, Ed.). New York: Academic Press 1963, p. 263.
MOORE, W.J.: Physical Chemistry. Englewood Cliffs, N.J.: Prentice-Hall 1972.
MÖLLER, W.J.H.M., VAN OS, G.A.J., OVERBEEK, J.T.G.: Trans. Faraday Soc. **571**, 312 (1961a).
MÖLLER, W.J.H.M., VAN OS, G.A.J., OVERBEEK, J.T.G.: Trans. Faraday. Soc. **571**, 325 (1961b).
NASTUK, W.L.: J. cell. comp. Physiol. **42**, 249 (1953).
NASTUK, W.L., HODGKIN, A.L.: J. cell. comp. Physiol. **35**, 39 (1950).
OVERBEEK, J.T.G.: Prog. Biophys. and biophys. Chem. **6**, 58 (1956).
PALLMANN, H.: Kolloidchem. Beihefte **30**, 334 (1930).
PFISTER, H., PAULY, H.: Biophysik **6**, 94 (1969).
PLAMONDON, R., GAGNÉ, S., POUSSART, D.: Vision Res. **16**, 1355 (1976).
REDMANN, K., KALKOFF, W.: Experientia **24**, 975 (1968).
REUSS, L., FINN, L.A.: Biophys. J. **15**, 71 (1975).
ROBINSON, R.A., STOKES, R.H.: Electrolyte Solutions. London: Butterworths 1959.
SELBY, C.C., BIESELE, J.J., GREY, C.E.: Ann. N.Y. Acad. Sci. **63**, 748 (1956).
SIMONSEN, L.O., NIELSEN, A.-M.T.: Biochim. biophys. Acta **241**, 522 (1971).
SIMS, P.J., WAGGONER, A.S., WANG, C.-H., HOFFMAN, J.F.: Biochemistry **13**, 3315 (1974).
SMITH, T.C., LEVINSON, C.: J. Membrane Biol. **23**, 349 (1975).
SNELL, F.M.: In: Glass Microelectrodes (M. Lavallée, O.F. Schanne and N.C. Hébert, Eds). New York: Wiley 1969, p. 111.
STONER, L.C., KREGENOW, F.M.: Biophys. J. **16**, 170a (1976).
TASAKI, I., SINGER, I.: Ann. N.Y. Acad. Sci. **148**, 36 (1968).
TASAKI, K., TSUKAHARA, Y., ITO, S., WAYNER, M.J., YU, W.Y.: Physiol. Behav. **3**, 1009 (1968).
TOYOSHIMA, Y., THOMPSON, T.E.: Biochemistry **14**, 1525 (1975).
ZEUTHEN, T.: Acta physiol. scand. **81**, 141 (1971).
ZEUTHEN, T.: J. Membrane Biol. **33**, 281 (1977).

Chapter 6

Chemical Composition of Membranes

D. J. Hanahan

A. Introduction

The tremendous accomplishments of scientists in delineating various metabolic pathways in cells and their current massive frontal attack on membrane structure and behavior have closely paralleled the emergence of data on the chemical and biochemical composition of cells and their membranous components. Obviously, without the availability of data on the chemical nature of cellular components, little progress could have been made in understanding the many physiological processes operative in these complex biological systems.

Over the past several years, an impressive amount of information has accumulated on the chemical and enzymatic composition of the membranes of many cells. With specific reference to well-defined membrane systems subjected to detailed analyses, certainly the mammalian erythrocyte must be regarded as the prime choice. Even a cursory glance at the literature of the past five years will amply confirm this statement. Two major reasons are evident in the selection of the mammalian erythrocyte for membrane study. The first is the ease and reproducibility of sampling and the second is the presence of only one membrane system, *i. e.,* the plasma membrane. Though the cells from several mammalian species have been investigated, particular attention has focused on the human erythrocyte.

The data collected up to now on the composition of these cells have provided useful and intriguing insights into their comparative characteristics. In many instances, this information has allowed reasonable approaches to elucidation of particular physiological processes. On the other hand, the diversity of the lipid components, for example, present in various species of erythrocytes has not provided a definitive clue as to the uniformity of structure or behavior of these cells based solely on these criteria. The same type of conclusion can be reached as regards the protein (and enzyme) composition of erythrocytes. This emphasizes the probability that elucidation of the "membrane structure" of the human erythrocyte, for example, will not necessarily solve the structure of the membrane of the cow, calf, sheep, or any other erythrocyte. Nonetheless, it can give hints and suggestions as to plausible explanations for a specific behavior. In general, however, each membrane system must be regarded as a unique entity in itself.

The purpose of this chapter is to describe unique features of the chemical or biochemical composition of the erythrocyte, to attempt a critical evaluation of the status of our knowledge in this area at this time, and to provide an overall evaluation of the field. While one could repeat the chemical composition of all erythrocytes examined to date, this approach would be of little value here since this information is readily available in excellent form (NELSON, 1972). Rather the approach here will be to devote attention mainly to the composition or characteristics of a few select species of cells and to consider in some detail the information accumulated on the two major components of these membranes, namely lipids and proteins. Again, there will be not attempt to make this effort into a compendium of facts on these membranes, but rather to focus on a select group of components considered to be of paramount importance to the cell. Again, there is no need to make a wide comparative analysis of membranes from a variety of different cell types and species, but it is hoped that the points selected for discussion will be of evident importance in evaluating other cell membranes.

Three species of cells, human, cow, and pig have been selected for discussion here. There was a purpose in these choices since each has some unique variations in ion composition, lipid composition, and/or metabolic behavior, and yet in each instance a major commitment of the cell is its action as an oxygen carrier. Nonetheless, these variations in compositions or biochemical behavior can provide some pertinent insights into what is necessary for the cell membrane to allow ion discrimination, transport variability, carbohydrate utilization, differences in oxygen diffusion rates and behavior, and other facets. The uniqueness of these erythrocytes will also be emphasized.

B. Some General Observations on Erythrocyte Composition

I. General Composition

The general composition of erythrocytes from the three different species is shown in Table 1. These data will emphasize some of the differences between the species with regard not only to inorganic, but also to organic components.

It is readily apparent from the data presented in Table 1 that each cell has some specific characteristics and one could expect differences not only in membrane structure but also in biochemical behavior of each species of cells.

II. Ion Composition

A particularly interesting point centers on ionic composition of these erythrocytes. If one specifically considers the $Na^+ + K^+$ levels of the cells, some interesting contrasts can be developed. It is well established now that all mammalian

Table 1. Some general characteristics of human, cow and pig erythrocytes

	Human	Pig	Cow
Volume of cell, μm^3	85	61	50
Hemoglobin, mg/100 ml	34	35	29
Lipid, mg/100 ml			
Total	507	325	430
Cholesterol	120	81	130
Phospholipid	386	212	295
Glycolipid	1	32	5
Cholesterol/phospholipid, molar ratio	0.89	0.83	0.89
Ion distribution, mg/100 ml			
Na^+	44	21	150
K^+	370	483	97
Mg^{2+}	4.8	14	1.2
Major energy source, substrate	Glucose	Unknown	Glucose
ATP level, μMol/100 ml	130	260	8
2,3-DPG level, μMol/100 ml	470	600	46

erythrocytes are "high-potassium" cells at birth, but shortly thereafter dramatic changes occur in many species. There is also the situation, well documented by TOSTESON (1969), in which a particular gene is responsible for development of LK (low-potassium) dominance in sheep erythrocytes. However, one does find sheep within a flock that have a HK (high-potassium) cell, but this occurs less frequently. An antigenic difference between the two cells may be an important factor in this. However, in the case of the bovine species the situation is not very clear, but nonetheless does involve a dramatic developmental change. This is evident from the data provided in Table 2.

Table 2. Cation levels of the erythrocytes of calves of increasing age[a]

Results expressed as $\mu g\ ml^{-1}$ packed cells			
Age in Weeks	Na^+	K^+	Mg^{2+}
1	5	41	24
2	2	38	25
4	5	29	25
6	7	22	19
8	8	14	16
10	12	9	15
12	17	7	15

[a] HANAHAN, D. J., EKHOLM, J. E., unpublished observations.

One would hope to associate the changes noted in Table 2 with some specific temporal alteration in a membrane component. Inasmuch as a young calf cell contains a Na^+/K^+-stimulated ATPase activity, this would seem to be a logical area for examination. As might be expected, the level of Na^+/K^+ ATPase in these cells, and interestingly, also the Ca^{+2}-stimulated ATPase, does decline with the decrease in K^+ levels and increase in Na^+ content as the animal ages. One could, then, ask whether the lipid composition of the cell changes during this developmental period, and the answer is unequivocally negative. In a detailed study (HANAHAN and EKHOLM, unpubl.) there was no evident difference in the levels of any of the lipid components found in the 5-day-old calf, a 50-day-old calf, and a year-old cow. The only minor exception was a change in the ratio of unsaturated to saturated hydrocarbon units present in the ether phospholipids located in this cell's membrane. However, this hardly constituted a significant alteration. While there was no evident change in lipid composition as described in this manner, it is possible that the orientation or localization of the lipids had changed with age, but it is not possible to determine this point now. It is reasonable to assume that alterations or changes in the protein composition and characteristics can occur in these membranes and reflect the changes noted. This is an obvious area for future investigation.

III. Age-Related Patterns

Two general approaches have been used to define the composition of the intact erythrocyte: one concerned only with the mixed aged population cells normally obtained by venipuncture of a particular subject and the other concerned with a more restricted age group of erythrocytes, which can be obtained by density centrifugation. In the following sections it is hoped that the information provided on the topics will illustrate an important point, *i. e.,* that much can be gathered from analysis or assay of the mixed cell population erythrocytes, but ultimately more understanding of the cell's behavior will be gained from the study of different age groups.

1. Intact Mixed-Age Erythrocytes

In well over 90 percent of the studies reported to date on the mammalian erythrocyte, only mixed-aged populations of cells have been studied, and relatively few attempts have been made to separate cells on the basis of age. Notwithstanding this fact, it has been possible to establish certain age-related differences in a mixed-age erythrocyte sample.

Perhaps the most clearly defined age-related changes in erythrocytes center on the Na^+ and K^+ content. As noted, at birth all mammals are producing cells with a high K^+ and a low Na^+ content. Shortly after birth, several species exhibit a dramatic change in ionic composition, with reversal towards a high-Na^+ cell. This is particularly true of dog, cat, one genetic variant of sheep red cell (LK), and calf. In the calf the molar ratio of K^+ to Na^+ is changing continually during the eleven weeks after birth (HANAHAN, 1973; ISRAEL et al., 1972). Concomitant with this change is a significant decrease in the level of the Na^+/K^+ stimulated ATPase as well as the Ca^{2+} stimulated ATPase. There is also a decided decline in the Mg^{2+} content of the cell (see Table 2). The loss in Ca^{2+} ATPase and the decrease in the Mg^{2+} level of cell may contribute to the instability of the membrane obtained from cow erythrocyte (BURGER et al., 1968). Thus, unless Ca^{2+} or Mg^{2+} was included in the hemolyzing buffer, these investigators found the cow membrane to disintegrate during preparation. On the other hand, the calf erythrocyte membrane is stable during preparation and no divalent cations are required to preserve the membrane structure. It is provocative to consider that Mg^{2+}- as well as the Ca^{2+}-stimulated ATPase contribute to the stability of this cell's membrane structure.

Other evidence of differences in behavior of a mixed-age population of erythrocyte has been reported. Using the calf and the cow as examples again, it can be shown that calf cell responds to incubation in isotonic Tris HCl, pH 7.6, by swelling and ultimately undergoing hemolysis. However, the adult cow cell under comparable circumstances does not show any obvious change.

It is known that erythrocytes from different mammalian species vary considerably in their metabolic (and transport) characteristics. An examination of the respiratory quotients (RQ) of several species will illustrate this fact (Table 3).

Table 3. Glycolytic rates of mammalian erythrocytes[a]

	$\mu M\ ml^{-1}$ cells × h
Pig	0.14
Ox	0.58
Sheep	1.06
Guinea-Pig	1.44
Man	2.72
Rabbit	2.95
Dog	2.99
Rat	5.60

[a] Taken from McMANUS and KIM, in: Metabolism and Membrane Permeability of Erythrocytes and Thrombocytes (ed. by Deutsch, Gerlach, Moser). 1st intern. Symp. Vienna, June 17–20, 1968, Georg Thieme Verlag, Stuttgart, Germany.

Particularly revealing is the wide range of values noted among the different species, the "low" value for the pig erythrocyte being especially noteworthy since the erythrocytes from an adult pig do not glycolyze although they contain all the necessary glycolytic enzymes. It has been established that these cells exhibit limited permeability to glucose (KIM and McMANUS, 1971). Particularly pertinent is the fact that erythrocytes obtained from neonatal pigs do glycolyze and exhibit significant permeability to glucose until shortly after birth (KIM et al., 1972). Thus, one might associate this change with variations in membrane composition. Preliminary evidence (Watts, R. and Luthra, M., pers. comm.) shows that the sphingomyelin to phosphatidyl choline ratio is greater than 1.0 in the neonatal cells, whereas the postnatal cell shows a ratio significantly less than 1.0. Whether this is important in defining glucose permeability remains to be established.

2. Age (density)-Separated Erythrocytes

While a magnificent and impressive amount of data has been published on the composition of erythrocytes under normal and abnormal conditions, these investigations have centered almost exclusively on the use of mixed-age populations of cells. Inasmuch as these cells represent a considerable span in age ranging up to 120 days, it was to be expected that significant differences might be evident in certain biochemical parameters of these various age groups.

The routes to separation of cells into precisely defined age groups have posed considerable problems in the past. One could approach the problem by developing reticulocytosis in an animal by repeated bleeding (phlebotomy) or through the administration of a reagent, e.g. phenylhydrazine, followed by examination of the "young" cells as they developed in the animal. A more producible and less traumatic approach is to separate the isolated cells from a normal animal on the basis of cell density, which bears a close relationship to age. Hence, upon centrifugation, cells exhibiting a high density are found at the bottom of the column of cells, whereas the younger cells, exhibiting a low-density profile, are localized in the upper part of the column.

Many methods for the separation of erythrocytes according to age-dependent density changes have been reported. Until recently, separation by centrifugation of cells in their plasma only has not provided the resolution achieved by the use of artificial gradients such as bovine serum albumin (LEIF and VINOGRAD, 1964; PIOMELLI et al., 1967), but use of these media has shed some doubt on the biochemical data derived from the separated cells. A very precise separation can be achieved by the use of phthalate esters (DANON and MARKOVSKY, 1964). This method also provides only two fractions per centrifugation and presents difficulties inherent in the use of a foreign medium. Other techniques involving artificial media involve rather long preparative procedures, and many are limited as to the volumes which can be handled. Good separations were achieved by O'CONNELL et al. (1965) by means of centrifugation of erythrocytes in plasma combined with a precisely designed and controlled apparatus for removing the fractions.

In a recent publication, MURPHY (1973) demonstrated that successful separation of erythrocytes could be achieved by a simple method of centrifugation in plasma at 30° C in an angle rotor at 39000 × G. As described by MURPHY, this higher temperature decreased the blood viscosity and increased free movement of the cells, and the angle rotor facilitated the circulation of the tube contents. He presented evidence based on hemoglobin values to support his conviction that it was higher-density and lower-density cells that were being examined.

This method has been pursued in some detail in our laboratory, and certain features of this technique for separation of human erythrocytes are presented here. It was our objective to determine whether or not separation of erythrocytes was achieved on the basis of density, and to subject these separated cells to biochemical and hemological evaluation.

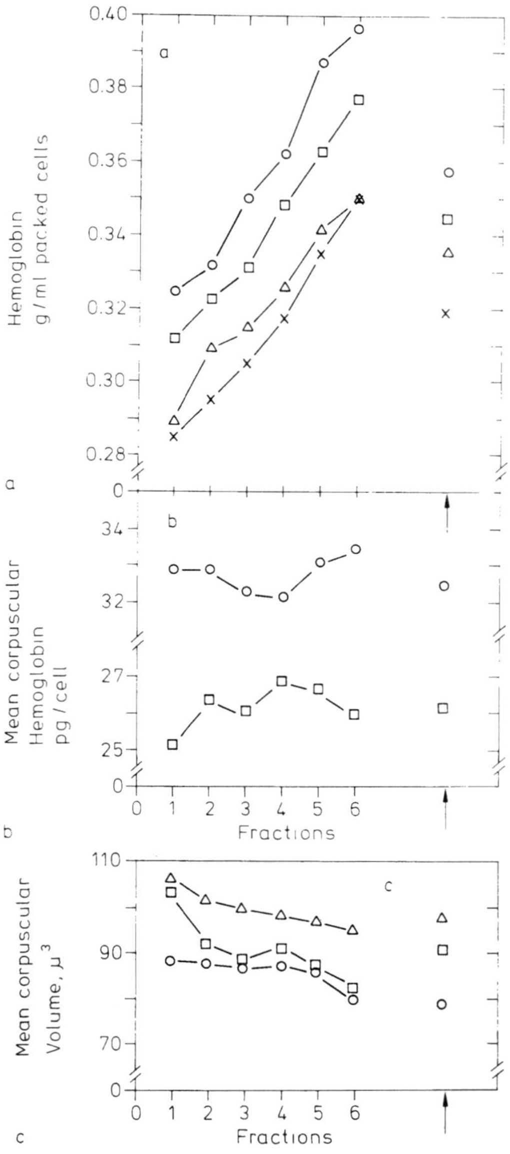

Fig. 1. a–c. Hemoglobin Concentration (a). Mean Corpuscular Hemoglobin (b) and Mean Corpuscular Volume (c) in Density-Separated Human Erythrocytes. The varying density-separated cells are identified as follows: 1, top 10%; 2–5, succeeding 20% layers; and 6, bottom 10%. ↑ refers to unseparated cells (control)

The results of a typical separation experiment are presented in Figure 1. In several separate experiments a similar curve was obtained for distribution of hemoglobin per ml packed cells between varying density of cells. Although these values increase with increasing cell density, calculation of the mean corpuscular hemoglobin (MCH) and mean corpuscular volume (MCV) indicated that MCH was similar in all fractions while MCV decreased with increasing cell density. Hence, calculations of all subsequent data are based on hemoglobin rather than on the volume of packed cells. Figure 2 shows that the cell potassium levels varied inversely with the cell density, but that the sodium content of cells in the different fractions was rather uniform except for the bottom fraction. Cells in the bottom fraction were repeatedly high in sodium.

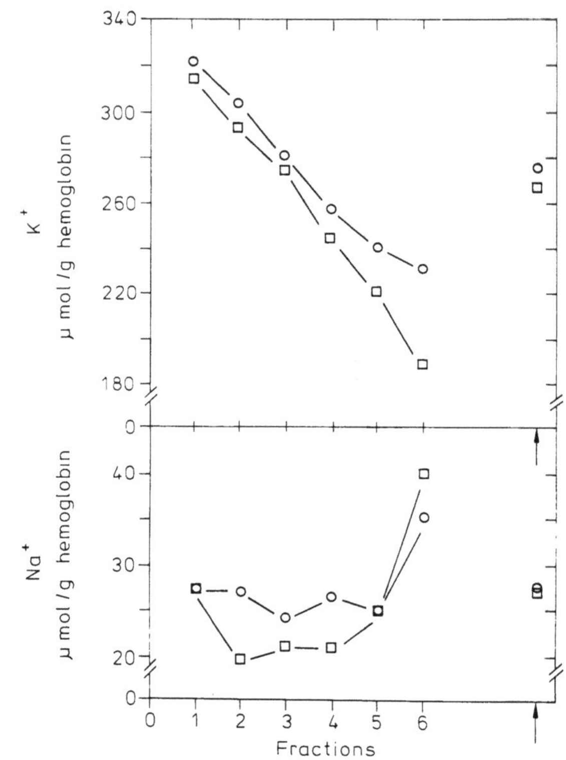

Fig. 2. K^+ and Na^+ Levels in Density-Separated Erythrocytes. See Fig. 1 for identification of fractions

In order to state that the cells were being separated expressly on the basis of density, five equal fractions were recovered from top to bottom of a sample of human erythrocytes after centrifugation by MURRPHY's technique. Each fraction was further separated on phthalate ester mixtures. The results shown in Figure 3 indicate that the average density of the top cells was lower than that of the

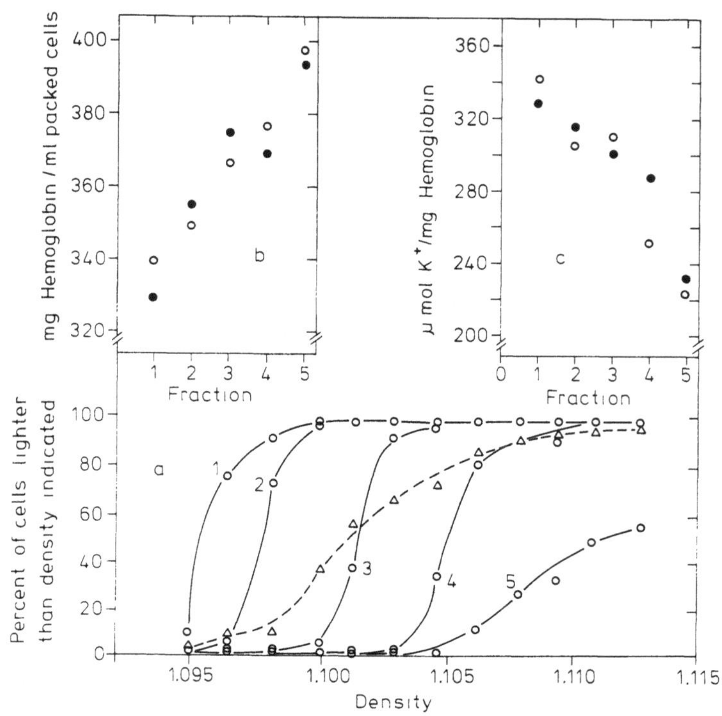

Fig. 3. Density Profile of Density-Separated Erythrocytes. An aliquot from each fraction and total unseparated cells (△----△) were layered on various mixtures of dibutyl phthalate or dimethyl phthalate and centrifuged. The % of cells on top of the phthalate layer was plotted against indicated density (a). In (b) and (c), hemoglobin and K^+ before and after phthalate centrifugation are compared

bottom cells, and that the density of cells increased almost linearly when plotted against the length of the tube. The linear relation between density and length of tube was consistent in each individual, but the slope was found to vary in different individuals. Phthalate itself did not change the density profile, since the hemoglobin per ml packed cells and K^+ per g hemoglobin (Fig. 3) showed the same relationship before and after centrifugation.

Enzymatic activities were examined and the most reproducible relationship between age and density of the cell was exhibited by the acetylcholine esterase, as indicated in Figure 4. The level of this enzymatic activity was consistent and bore a direct relationship to the cell density. While it would have been desirable to have the ATPase activities show a concomitant reproducible decrease from top to bottom cells this was not the case. Though the total ATPase activity showed a reasonably consistent pattern of decreasing levels from top to bottom cells, considerable variability was noted in activity values of the top cells. The ouabain-sensitive ATPase, which is intimately associated with the potassium transport system in erythrocytes, showed variations in pattern, as did the Ca^{2+} stimulated ATPase. Reasons for this variability are not known at this time.

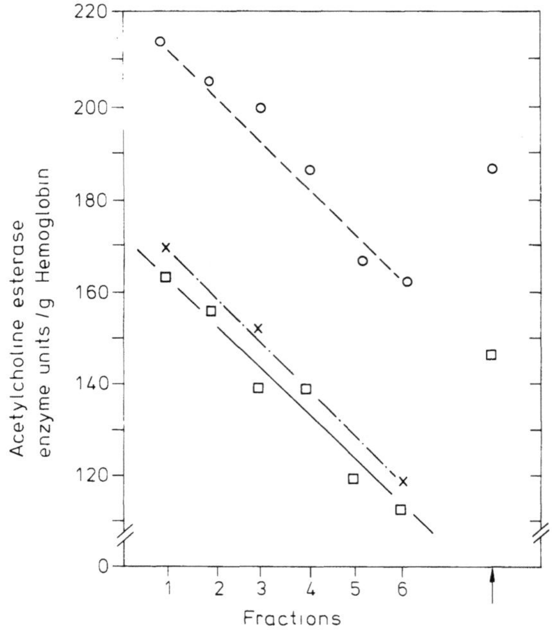

Fig. 4. Acetylcholine Esterase Levels in Density Separated Erythrocytes. See Fig. 1 for identification of fractions

Whereas there was no change in the cholesterol to phospholipid phosphorus (molar) ratio from top to bottom cells, there was a decided decrease in the total amount of each class of lipid (calculated as μmol g^{-1}hemoglobin) as the density (age) of the cells increased. Thus, there is a total loss of lipid (and concomitantly membrane protein) as the erythrocyte ages.

C. Comments on Major Components of the Erythrocyte Membrane

While the use of age-separated cells is strongly recommended for a more definitive evaluation of the erythrocyte membrane structure and behavior, it is recognized that much important information on the composition and characteristics of the erythrocyte can be gained from the use of mixed age populations. Thus, it is appropriate at this time to explore certain features of the lipid and protein composition of the erythrocyte membrane and to evaluate the possible importance of this information in elucidating the structure and function of these cells. Inasmuch as two classes of compounds represent over 95 percent of the cell membrane mass, *i. e.*, lipids and proteins, these are of obvious importance to the cell and will form the basis of the discussion.

I. Lipid: an Appraisal of Composition and Orientation or Localization

All mammalian erythrocytes contain three classes of lipid, *e.g.*, neutral lipid, phospholipids, and glycolipids. The neutral lipids are represented exclusively by cholesterol, whereas the phospholipids and glycolipids are quite heterogeneous in composition. The phospholipids are comprised of structures bearing a *sn*-glycero-3-phosphate backbone and also a sphingosine, or 4-sphingenine moiety. Interestingly, though not containing phosphorus, the glycolipids nevertheless are also comprised primarily of sphingosine or 4-sphingenine derivatives and are often referred to as sphingoglycolipids. Table 4 shows a general level of these three classes of lipids in human, pig, and cow erythrocytes.

Table 4. Lipid distribution in mammalian erythrocytes % of total lipid weight

	Human	Bovine	Porcine
Cholesterol	25	30	28
Phosphatidyl ethanolamine	22	21	26
Phosphatidyl serine	6	8	5
Phosphatidyl inositol	6		<1
Phosphatidyl choline	19	0	16
Sphingomyelin	20	41	16
Gangliosides	< 1	< 1	< 1
Glycolipids	< 1	< 1	12–14

Table 5. Molar ratios of choline to non-choline containing phospholipids in erythrocytes

Human	Bovine	Porcine
1.15	1.20	1.03

A further examination of the phospholipids in these species reveals an interesting pattern. If one is to categorize the phospholipids simply on the basis of whether or not they contain choline, then the data in Table 5 show a most interesting degree of uniformity. This is a consistent and yet unexplained characteristic of these cells as well as of many other cells of mammals. However, it is important to note that while there is uniformity in the ratio of choline- to non-choline-containing phospholipids, it does not mean that the lipids within each class are the same. This is dramatically illustrated in the case of bovine choline-containing phospholipids, which are composed exclusively of sphingomyelin, whereas in the human and the pig there are nearly equal amounts of phosphatidyl choline and sphingomyelin. Examination of the molar ratio of phosphatidyl choline to sphingomyelin will show values of 4:1 for the rat, dog and guinea-pig, 3:2 for human and rabbit, and 1:12 (or higher) for sheep, goat and cow.

1. Some Structural Features of the Erythrocyte Lipids

It is well at this point to outline some of the structural features of the classes of lipids found in the erythrocyte membrane. This approach will provide the focus of ensuing discussions on the unique aspects of these structures and how these may be important to a cell's particular metabolic characteristic. Though it is not important at this point to relate the route to proof to structure of these compounds, it will be of value to focus on some unique characteristics of these structures in a subsequent section:

a) Neutral Lipid

Essentially the only neutral lipid present in mammalian erythrocyte is cholesterol. Its structural formula is depicted below:

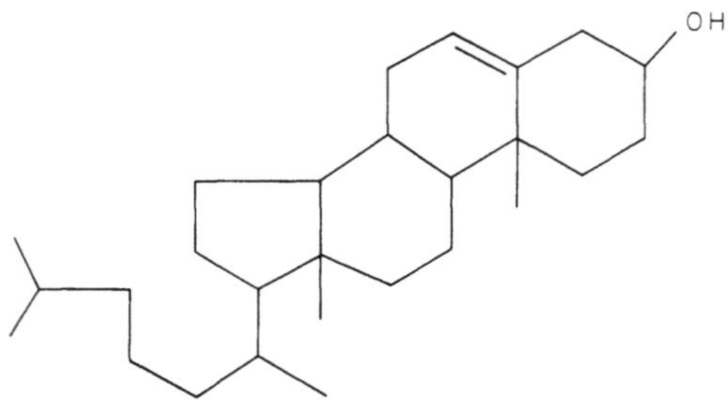

b) Phospholipids

This class of lipids is quite heterogeneous in composition and structure and yet a basic characteristic of their structure is that these compounds contain either an *sn*-glycero-3-phosphate or a sphingosine (4-sphingenine) moiety. The structures of the representative phospholipids in mammalian erythrocytes are shown in the diagram.

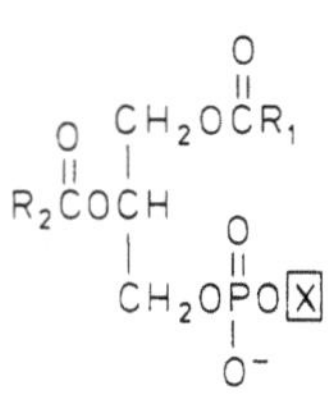

Where [X] = H, inositol, serine ethanolamine, choline

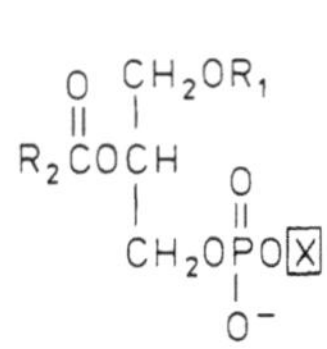

Where [X] = ethanolamine

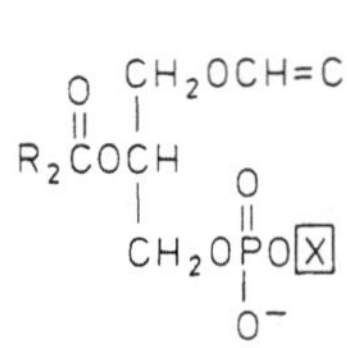

Where [X] = ethanolamine, choline

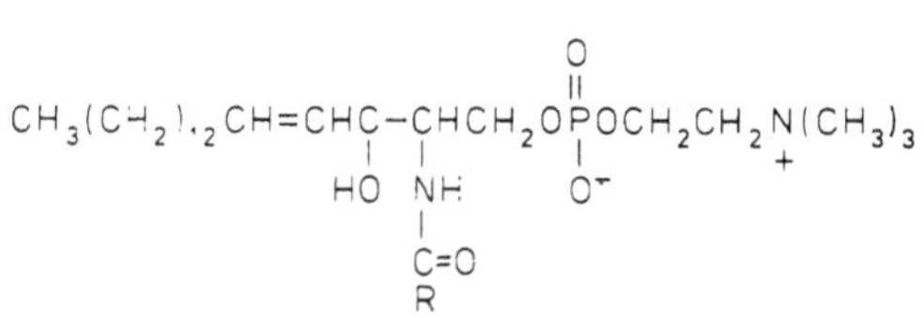

c) *Sphingoglycolipids*

The usual forms of sphingoglycolipids found in most species (*e. g.*, human) are the neutral sphingoglycolipids (containing no sialic acid) and the acidic type, also called gangliosides (containing sialic acid). The ratios of these two classes vary considerably from species to species, with no constant or expected pattern.

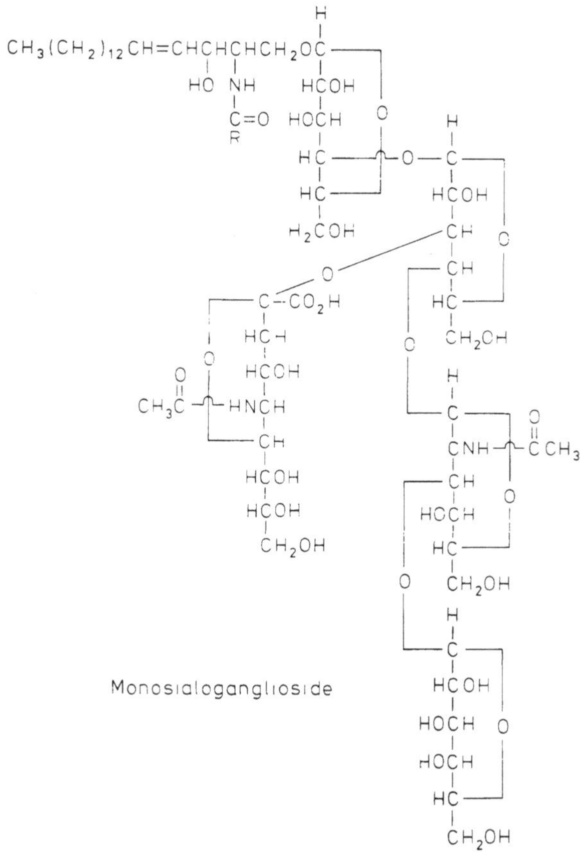

Monosialoganglioside

2. Observations on Types of Phospholipids Present in Human, Cow, and Pig Erythrocyte

While one might expect some uniformity in the distribution of phospholipids within the membranes of erythrocytes, a clue to the fact that this was not the case was noted above in that only sphingomyelin is present (and no phosphatidylcholine) in the cow erythrocyte. Several other distinct patterns emerge on analysis of the phospholipid classes found in these three erythrocytes.

Table 6 shows the types of phospholipids present in the non-choline-containing group, and reveals an interesting pattern illustrating the uniqueness of each cell's lipid pattern.

Table 6. Distribution within non-choline-containing phospholipids of erythrocytes

	Human	Bovine	Porcine
Phosphatidyl Ethanolamine			
Vinyl ether	35	0	0
Glyceryl ether	0	80	0
Diacyl	65	20	100
Phosphatidyl Serine	diacyl type only		
Phosphatidyl Inositol	diacyl type only		

An examination of the types found in the choline-containing phospholipids, as presented in Table 7, again reveals a quite specific pattern for each cell.

Table 7. Distribution within choline containing phospholipids of erythrocytes

	Human	Bovine	Porcine
Phosphatidyl Choline			
Vinyl ether	2		0
Glyceryl ether	0	None	0
Diacyl	98		100
Sphingomyelin	100	100	100

a) Fatty Acid Composition

Though it is unnecessary to present the fatty acid composition of naturally occurring phospholipids and sphingoglycolipids, there are some unusual patterns to the latter compounds that deserve some mention. Usually these facts are not presented in discussions of the composition of membranes. In particular, there is good evidence to support the preferential association of specific types of fatty acids with particular types of lipids, and this is reflected in the data shown in Table 8.

Certain distributional characteristics of these fatty acids are of interest. For example, both phosphatidyl serine and phosphatidyl inositol have high levels of the C_{18} fatty acids, with well over 45 mol percent in most samples. On the other hand, the high content of unsaturated fatty acids in the phosphatidyl ethanolamine of cow erythrocytes reflects the fact that this phospholipid has primarily a saturated ether group at the C-1 position, with the fatty acids found only on the C-2 position (see below). As will be developed below, there is a preferential positioning of unsaturated fatty acids on the C-2 position. One further point centers on the fatty acid composition of the sphingomyelins, where the characteristic feature is a preponderance of 16:0 and 24:0 and 24:1 fatty acids, with minor amounts of the C_{18} series.

Table 8. Major fatty acid distribution in phospholipids of erythrocytes. Data given as % of total fatty acids. Shortened notation for fatty acid: e.g. 16:0, C_{16} with zero olefinic unsaturation

	Human	Cow	Pig
Phosphatidyl Ethanolamine			
16:0	17.0[a]	–	23.4
18:0	10.5	–	11.5
18:1	22.8	57.0[b]	51.5
18:2	7.1	30.0	13.6
20:4	21.7	13.0	–
Phosphatidyl Serine			
16:0	6.8	Not	–
18:0	39.7	available	30.4
18:1	8.2		63.3
18:2	3.0		6.3
20:4	24.1		–
Phosphatidyl Choline			
16:0	34.8	–	32.6
18:0	11.3	–	18.4
18:1	19.6	–	19.8
18:2	23.0	–	21.4
20:4	5.6	–	–
22:1	–	–	8.2
Phosphatidyl Inositol			
16:0	11.3		
18:0	26.0		
18:1	14.7	Data not available	
18:2	3.9		
20:4	22.6		
Sphingomyelin			
16:0	34.1	45.0	36.0
18:0	8.0	3.7	11.2
18:1	4.7	–	2.9
18:2	3.1	–	–
20:4	–	–	–
24:0	18.2	35.0	21.0
24:1	14.7	9.0	14.6

[a] These data do not include the hydrocarbon chains associated with the vinyl ether residues in this fraction.
[b] These data do not include the hydrocarbon chains associated with the saturated ether residues in this fraction.

b) Positioning of Fatty Acids

A more revealing insight into these structures can be gained by examination of the localization of the fatty acids. In particular, the positioning of the fatty acids in phosphatidyl choline (diacyl-*sn*-glycero-3-phosphoryl-choline) of human erythrocytes can be displayed as follows:

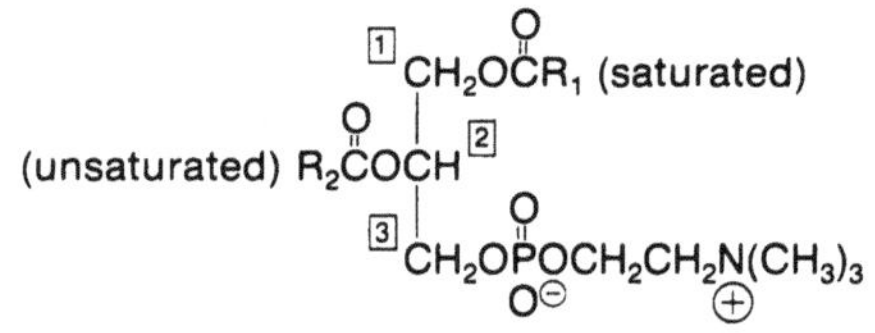

Thus, the C-1 or (1) ester position contains almost exclusively saturated fatty acid residues, while the C-2 or (2) ester position contains unsaturated fatty acid residues. Of considerable interest, the reverse arrangement, *i. e.,* saturated fatty acids only in the C-2 position and unsaturated fatty acids only in the C-1 position, is a rare occurrence. The type of positional asymmetry noted above for phosphatidyl choline is evident also in the major component (~75%) of the "phosphatidyl ethanolamine" fraction of bovine erythrocytes, the ether containing analogue. Its structure can be depicted as follows:

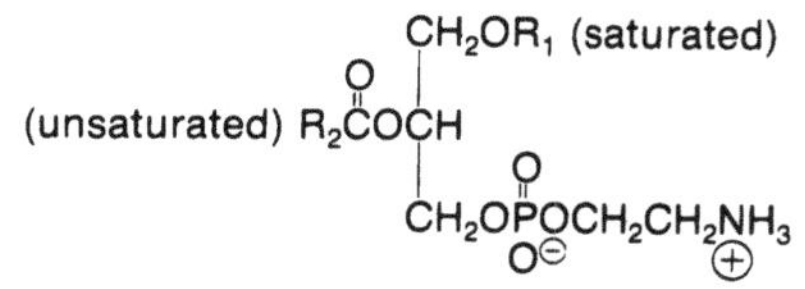

Thus, the hydrocarbon residue, R_1, is comprised of a mixture of primarily saturated chains, with a relatively smaller level of mono-saturated chains. However, the fatty acid at the C-2 ester position contains only unsaturated fatty acid residues (see Table 8). A similar pattern is expressed in the vinyl ether containing phospholipids found in human erythrocytes and other tissues.

A number of approaches can be undertaken to establish the positioning of the fatty acids as illustrated above. Perhaps the most widely used technique has been that of phospholipase A_2 attack on the C-2 ester position of phospholipid. This can be illustrated using phosphatidyl choline as substrate:

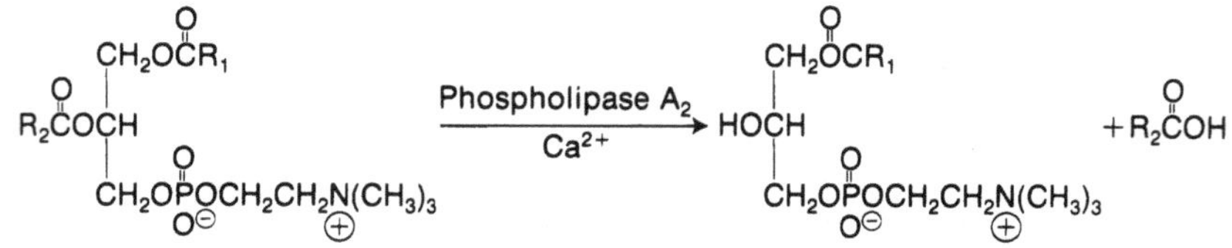

– the last two products, mono-acyl glyceryl phosphorylcholine (lysolecithin) and free fatty acid can be separated easily from each other and from any unreacted substrate, and analyzed by conventional techniques.

c) Importance of Fatty Acid Composition

It is reasonable to question the significance of the above fatty acid patterns and to ask whether they are of any importance to the membrane. One would predict that there should be little difference in the fatty acid composition of these lipids, since presumably there would be a common pool of fatty acyl coenzyme A derivatives for insertion into the lipid backbone structure (*e. g.*, glycerolphosphate, for phosphatidylcholine). However, it is evident that the selection of fatty acids for incorporation is much more complex than anticipated, and that certain specificities must exist at a particular biosynthetic step leading to a specific phospholipid. In addition, there is strong evidence that replacement of the fatty acids can occur after complete synthesis by an exchange mechanism and thus "remodeling" of the molecule for specific purposes can ensue.

The basic point to be stressed here is that each particular fatty acid species of a molecule like phosphatidyl choline may have different metabolic roles and/or metabolic pattern. There is evidence that this may be a correct conclusion, since the concept of "fatty acid pairing" or metabolic heterogeneity of certain phospholipids has been proposed and there is provocative evidence in support of it (Holub, et al., 1971). In addition, there are physicochemical reasons why fatty acids of a certain chain length and with some degree of unsaturation must be present to maintain the "fluidity" of the membrane. Thus, if short-chain fatty acids (*e. g.*, 12:0 and less) were present in phospholipids, it is highly unlikely that they would form stable membranes (at least based on the stability of films of these lipids at an air-water interface). On the other hand, if no unsaturation were present, even with the desired longer-chain saturated (16:0 and greater) fatty acid containing phospholipids, the membrane would become very rigid. Unless the temperature of the membrane could be raised to 40° or above, where these types of fatty acid derivatives essentially "melt" or undergo a phase change, then the cell as we know it now would not survive, due to a dramatic decrease in permeability due to these more rigid films.

Even though only rather vague statements can yet be made regarding the uniqueness of the fatty acid composition and distribution of these complex lipids, an even more puzzling factor is the variety of phospholipids found in these membranes. While little is known about the specific functions of any of these complex lipids, it is possible that each class of phospholipids participates in some quite special metabolic system yet to be determined. At the present time, one can only state that this field is on the threshold of some potentially interesting and important developments, and one of the approaches is to attempt to depict the positioning of these complex lipids in the membranes, *i. e.*, to show whether they are specifically or randomly located. This topic is considered below.

3. Localization of Lipids in Membranes: A Compositional Study of a Different Type

The above general description of the fatty acid composition of membrane lipids did not consider the tantalizing proposition of a specific localization of lipids in membranes. Though a particular orientation or location of lipids within membranes has been a subject of discussion for many years, experimentally it has been most difficult to design or conduct meaningful experiments. The ease of extraction of lipids from the erythrocyte membrane, specifically, and, generally, from most biological tissues, with neutral solvents precluded any specific covalent bonding of lipid to proteins and supported the concept of binding by virtue of electrostatic bonds, hydrogen bonds and/or van der Waals forces. It seemed hardly likely that there would be any orderly or even partially systematic arrangement of lipids within a membrane. However, evidence presented in experiments on the turnover of cholesterol and certain phospholipids in the erythrocyte membrane suggested there was some definitive asymmetry in the membrane. Consequently it is worthwhile to discuss various facets of this topic here.

It is always interesting to speculate on the importance of membrane characteristics as defined by their fatty acid composition. Yet, some caution should be exhibited in attempting such correlations. First, much of our evidence on the "membrane" behavior as related to phospholipids has come from work on *in-vitro* systems in which pure phospholipids are examined at an air-water interface under varying conditions. Furthermore, in the use of certain membrane probes, for example the spin-label types, claims have been made as to the physical state of lipids in a membrane. However, the evidence is clear that these probes disturb the membrane and may not correctly depict the true structure of the lipid environment or position in a membrane. Similarly, the translation of data on behavior of isolated lipids at an air-water interface to the intact cell membrane is indeed most difficult. The importance of protein-lipid interactions, *i. e.,* presumably hydrophobic as well as hydrophilic, in membranes has only recently come into focus, and certainly this could play an important role in determining the permeability or behavior of the membrane. The human erythrocyte, which has been the subject of a multitude of experiments on membrane structure, is an interesting cell since it contains no subcellular membrane structures and except for a brief period in the reticulocyte phase there is no *de-novo* synthesis of lipids, *i. e.* fatty acids, cholesterol, or the complex lipids. However, this is not to say that the components of the erythrocyte membrane do not turn over or change during the life of the cell. It is known that there are certain important exchange reactions involving the cell's lipids and certain transfer reactions operative as such in the cell. Brief mention will be made of these topics, with most attention centering on the question of specific localization of the phospholipids in the human erythrocyte membrane.

a) Some General Observations

Cholesterol Exchange or Replacement. It has been well established that the mammalian erythrocyte can exchange its cholesterol with that of the circulating plasma. This was first reported by HAGERMAN and GOULD (1951). Subsequently, many investigators substantiated and extended these observations (LONDON and SCHWARZ, 1953; GOULD et al., 1955; PORTE and HAVEL, 1961; MURPHY, 1962). Certain conclusions were drawn from the data of these and other investigators. First, the process is rapid and not energy-dependent. Next, the rate of this exchange suggests that the cholesterol is located very close to the cell surface.

Phospholipids. Turnover of phospholipids can occur in the erythrocyte membrane, but it is much slower than that observed for cholesterol. A more detailed outline of the possible mechanisms operative in phospholipid exchange or turnover in the erythrocyte is presented in a recent review by SHOHET (1972) and this topic need not be reconsidered here. In essence, however, experimental studies on the turnover of erythrocyte phospholipids in dogs showed that primarily only phosphatidyl choline was being affected, and in the main by a passive exchange reaction (REED, 1964 and 1968). There is also an active metabolic process by which any lysophosphatidyl choline in the membrane can be acylated *via* fatty acyl coenzyme A to yield a phosphatidylcholine; this is much less readily exchangeable than the phosphatidyl choline, which is in passive equilibrium with the plasma phospholipid (SHOHET, 1972).

The above observations illustrate some differences in the localization of specific lipids within a membrane and lend some support to the concept of a specific ordering of certain types of lipids in these cellular structures.

b) Use of Enzymes as Probes for Location of Lipids in Membranes

Over the past several years, increasing attention has been paid to the use of phospholipases for establishing the location of phospholipids, in particular within the erythrocyte membrane. A number of approaches have been undertaken, in which either intact cells or hemoglobin-free membranes were the primary substrate for the enzymes. In the main, phospholipase A_2 and phospholipase C have been utilized most often but more recently a sphingomyelinase has also become available for use. The actions of these enzymes on some typical substrates are given below.

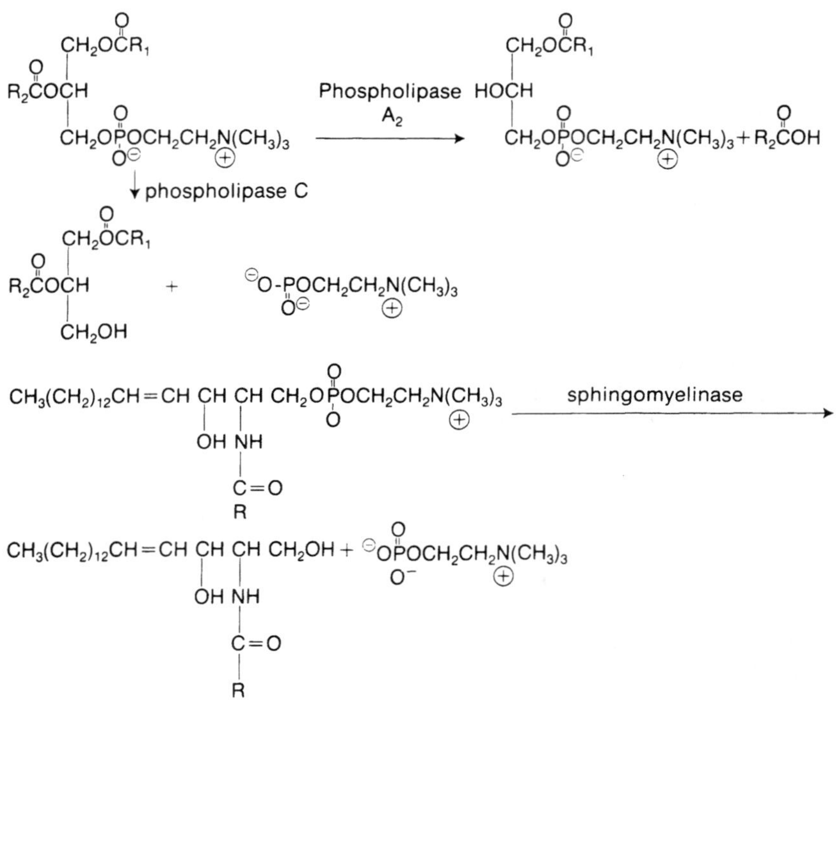

The action of these phospholipases on the intact erythrocyte is very complex and is dependent on a variety of factors ranging from enzyme source to incubation conditions and metabolic state of the cell. This complexity is compounded by the fact that no two investigators have used the same experimental conditions, yet each investigator has repeatedly employed the same enzyme source(s), erythrocyte preparation, pH, buffer, and ionic environment in all experiments. Hence, a comparative evaluation of reports from different laboratories is quite difficult.

Notwithstanding these obstacles, there has emerged a concept that the phospholipids of the mammalian erythrocyte are "asymmetrically" localized in the membrane, with the choline-containing phospholipids at the external surface and the non-choline-containing phospholipids at the internal surface. In the following discussion, an attempt will be made to concisely evaluate the experimental approach and results that led to this asymmetry concept. At the same time, recent results in our laboratories (MARTIN et al., 1975) will show that considerable caution must be exercised in assigning particular locations to specific phospholipids in the erythrocyte membrane. In this discussion, only the mode of action of phospholipase A_2 on the intact erythrocytes will be outlined, since hemoglobin-free membranes, though useful for initial probing experiments, are unsatisfactory for this type of approach due to their highly permeable nature.

4. Summary of Observations on Phospholipase Action on Erythrocytes

As ZWAAL et al., (1973) have very clearly and succinctly summarized the evidence available up to 1973 on phospholipase activity toward erythrocytes, only a brief resumé of the salient features of this report and of subsequent publications on this subject will be presented here.

Essentially three different enzymes, *i. e.*, phospholipase A_2, phospholipase C and sphingomyelinase have been utilized in attempts to establish the location of phospholipids in the erythrocyte membrane. A summary of their behavior is outlined below.

a) Phospholipase A_2

As noted above, this enzyme attacks only the C-2 ester position of phosphoglycerides, such as phosphatidylcholine, phosphatidyl ethanolamine, and phosphatidyl serine, with the release of a free fatty acid and formation of a lysophosphoglyceride.

The phospholipase A_2 in sea snake venom *(Enhydrina schistosa)* and in cobra venom *(Naja naja)* can attack the erythrocyte membrane phospholipids, without causing hemolysis (IBRAHIM and THOMPSON, 1965; GUL and SMITH, 1972). The addition of bovine serum albumin to incubates of human erythrocytes and the *Naja naja* phospholipase causes hemolysis, presumably due to removal of free fatty acids. On the other hand, pancreatic phospholipase A_2 does not attack or hydrolyze the human erythrocyte membrane phospholipids.

According to ZWAAL et al. (1973), *Naja naja* phospholipase A_2 and also bee venom phospholipase A_2 can attack primarily the phosphatidyl choline of the erythrocyte, under their experimental conditions, to the extent of 60–70 percent of the total phosphatidyl choline in the membrane, with little or no degradation of phosphatidyl ethanolamine, phosphatidyl serine or sphingomyelin. There was no obvious hemolysis. Essentially, then, phospholipase A_2 from certain sources and under specific conditions can cause degradation of phosphatidyl choline only. Interestingly, no lysis of these cells occurred, even though the level of lysolecithin produced in the membrane by this enzyme action would cause extensive hemolysis if added extracellularly to a fresh sample of erythrocytes.

b) Phospholipase C

This enzyme, as isolated from *B. cereus*, shows a preferential attack on phosphatidyl choline yielding diglyceride and phosphoryl choline (see above), wherein the enzyme isolated from *Cl. perfringens* exhibits a broader specificity and will attack phosphatidyl choline, phosphatidyl ethanolamine and sphingomyelin.

In general, the *B. cereus* enzyme shows no activity towards the human erythrocyte, nor does it have any action towards the pig erythrocytes, at least under the experimental conditions cited (COLLEY et al., 1973).

c) Sphingomyelinase

This enzyme, which has been isolated in a highly purified form from *S. aureus* (WADSTRÖM and MÖLLBY, 1971) attacks sphingomyelin, with release of phosphoryl choline and N-acyl sphingosine (ceramide) (see above). Of considerable value, this enzyme can attack the sphingomyelin in the human as well as the pig erythrocyte to the extent of 80–85 percent of the total sphingomyelin content without any evident hemolysis or alteration of any other phospholipid therein.

d) Combined Activity of Phospholipase C *and Sphingomyelinase*

COLLEY et al. (1973) and ZWAAL et al. (1973) reported that a combination of these two enzymes will hydrolyze all the phospholipid classes in human erythrocytes and also cause 100 percent hemolysis.

5. Development of the Concept of Asymmetric Location of Phospholipids in Membranes

On the basis of the above experimental observations, ZWAAL et al. (1973) concluded that the choline-containing phospholipids were localized at the exterior surface of the membrane and that the non-choline-containing phospholipids were positioned on the interior surface. Earlier, BRETSCHER (1972) had submitted a similar proposal using an entirely different approach. In effect, BRETSCHER, employing the membrane-impermeable radioactive reagent, formyl-^{35}S-methionyl (sulphone) methyl phosphate, which reacts with free amino groups, investigated the labeling of phosphatidyl ethanolamine in intact (human) erythrocytes and in hemoglobin-free ghosts or membranes. These experiments were conducted in isotonic buffer at pH 7.6. There was little observed labeling of the phosphatidyl ethanolamine in the intact erythrocytes, even after pronase treatment, whereas there was significant labeling of the phosphatidyl ethanolamine in the membranes or ghosts prepared from fresh samples of cells. Interestingly, MADDY (1964) had reported that the fluorescent reagent disodium 4-acetamide-4'-thiocyanostilbene disulfonate labeled the ox erythrocyte surface, but there was little or no labeling of any lipids.

On the basis of these results, BRETSCHER proposed two possible explanations for his own results. One was that the phosphatidyl ethanolamine was equally partitioned between each half of the lipid bilayer of the membrane, but that the phosphatidyl ethanolamine in the external layer was shielded in such a way that it was unreactive to the reagent. The second interpretation was that the phosphatidyl ethanolamine was not equally localized in the bilayer, but was found primarily on the inner surface of the bilayer. He further proposed that the outer layer would be comprised of the choline-containing phospholipids. The second idea appealed more to him than the first one.

6. On the Validity of the Lipid Asymmetry Proposal

As noted, the results of studies to date on phospholipase activity towards erythrocytes, together with the use of non-permeant specific labeling reagents, would favor asymmetry to the phospholipids in the erythrocyte membranes. These conclusions, if valid, would have important implications as to the structure of the erythrocyte membrane.

Certain aspects of the mode of action of these phospholipases are puzzling. In particular, the fact that only certain phospholipases A_2 will hydrolyze phospholipids in the intact erythrocyte is indeed provocative. Furthermore, the possible complex nature of this system is indicated by the observations of GUL and SMITH (1972, 1974) that hemolysis occurs only in the presence of albumin, and the report by GAZITT et al. (1975) that avian erythrocytes are more prone to hemolysis by phospholipases when they are depleted of ATP. It is pertinent to point out that in almost all the studies using phospholipase A_2 as a probe, investigators have employed human erythrocytes, with little attention to other species or the effects of pH and other conditions on the hydrolytic attack. Inasmuch as these phospholipases have assumed considerable importance for structural evaluation of membranes, MARTIN et al. (1975) felt it important to examine the action of phospholipase A_2 from several sources under varying conditions on different erythrocyte species. These results are summarized here and strongly suggest that caution be exercised in the application of these enzymes to elucidate membrane structure in the erythrocyte.

In their investigation, MARTIN et al. (1975) chose to use a basic phospholipase A_2, which is found in the venom of the snake, *Agkistrodon halys blomhofii.* An earlier report by SARKAR and DEVI (1968) that this basic enzyme was directly hemolytic, prompted its use in part of this investigation. The phospholipase A_2

from *Naja naja* venom yielded results similar to those obtained with the basic enzyme. The enzyme from *C. adamanteus* and the acidic enzyme also present in *Agkistrodon* produced only slight hydrolysis and hemolysis.

Other species of erythrocytes, *e.g.*, guinea-pig, monkey, pig, and rat were tested but only those of the guinea-pig behaved comparably to the human cell. Pig, monkey, and rat erythrocytes showed limited hydrolysis of phospholipid and hemolysis in the presence of any of these enzymes.

The data obtained in the investigation by MARTIN et al. showed that the action of the basic phospholipase A_2 on human erythrocytes could be divided into two sequential reactions. The first stage involved the hydrolysis of phosphatidyl choline only to the extent of 70 percent, with the resulting membrane now containing a rather large quantity of lysophosphatidyl choline, and exhibiting stability to hemolysis under the experimental conditions. Specifically the latter were pH 7.4 and low $[Ca^{2+}]$ with no added glucose (see also ZWAAL et al., 1973) and pH 8.0 and high $[Ca^{2+}]$ if glucose was added to the medium. Under these conditions the intracellular ATP remained high. The second stage showed that the above cell membranes containing lyso-phosphatidyl choline were only quasi-stable, since additional phospholipid hydrolysis (phosphatidyl ethanolamine and phosphatidyl serine) and hemolysis could occur under conditions of high pH and $[Ca^{2+}]$ in the absence of glucose. Under these conditions the intracellular ATP levels were low.

The essential facets of this study are summarized below:

Hemolysis Behavior. The extent of hemolysis induced by the basic phospholipase A_2 is dependent on pH and $[Ca^{2+}]$. This can be illustrated by the data in Table 9, where $[Ca^{2+}]$ is held constant at 30 mM and the pH is varied, and in Figure 5, where the effect of varying $[Ca^{2+}]$ at pH 8.0 is noted.

Table 9. The effect of pH on phospholipase A_2-induced hemolysis of human erythrocytes. A 5% hematocrit of freshly prepared cells was incubated at 37° in 20 mM HEPES, 30 mM $CaCl_2$, and NaCl (to make the solution isotonic) with 6.7 μg ml^{-1} of *A. halys blomhofii* basic enzyme for 2 hours. Hemolysis was measured from the absorbance at 540 nm of the supernatant solution after removal of the cells by centrifugation[a]

	% Hemolysis	
pH	with enzyme	without enzyme
6.5	4.5	2.0
7.0	3.5	2.0
7.5	31.0	2.5
8.0	67.5	3.0
8.7	83.5	7.0

[a] Taken from MARTIN et al. (1975)

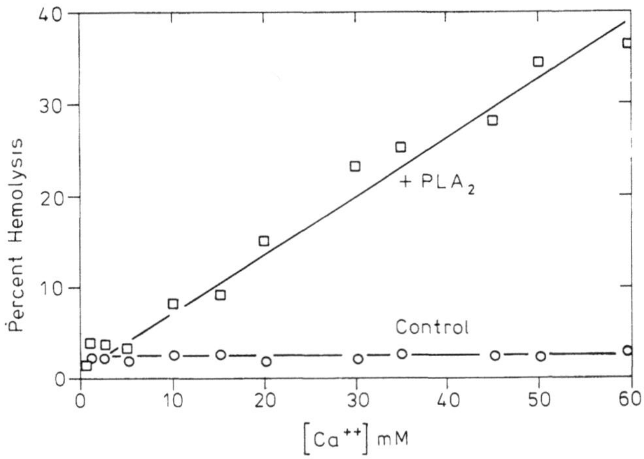

Fig. 5. Influence of Ca^{2+} Concentration on Phospholipase A_2 Induced Hemolysis of Human Erythrocytes

It was observed that preincubation of erythrocytes at 37° C (or storage overnight at 4° C) before the addition of enzyme increases their susceptibility to hemolysis. Hence the possibility that depletion of the erythrocyte energy source, *i.e.*, ATP, could increase the sensitivity of the cells to this enzyme. Conse-

Table 10. Effect of various sugars on the hemolysis of human erythrocytes induced by the basic phospholipase A_2 from *A. halys blomhofii.* A 5% hematocrit of freshly prepared cells was incubated at pH 8.0 and 40 mM Ca^{2+} with 6.7 μg ml^{-1} of enzyme and the sugars indicated below[a]

1. Incubation with 5.3 mM sugars Sugar	$t_{1/2}$ (minutes)
none	97
2-deoxyglucose	95
galactose	104
mannitol	97
sucrose	97
glucose	240
mannose	240

2. Effect of glucose concentration Concentration	$t_{1/2}$ (minutes)
0	90
36 μM	90
62 μM	125
124 μM	190
620 μM	225
5.3 mM	240

[a] Taken from Martin et al. (1975)

quently, a number of carbohydrates were added to incubates of erythrocytes and enzyme and the results are shown in Table 10. These data clearly show that carbohydrates, such as glucose and mannose, which can be metabolized by the erythrocyte, offer considerable protection to the cell against hemolysis by the basic enzyme. On the other hand, carbohydrates such as galactose, 2-deoxy glucose, mannitol and sucrose, which are not metabolized by the cell, afforded no protection. Levels of glucose as low as 124 μM protected the cell against hemolysis by the basic enzyme, and thus the carbohydrate action was not simply a nonspecific effect of polyhydric compounds.

Additional evidence showed that the depletion of ATP from human erythrocytes occurred more rapidly at pH 8.0 and 40 mM Ca^{2+} than at pH 7.4 and 10 mM Ca^{2+} and that the loss of ATP was not markedly affected by the presence of the enzyme at either pH. Essentially then, lowered ATP levels alone in

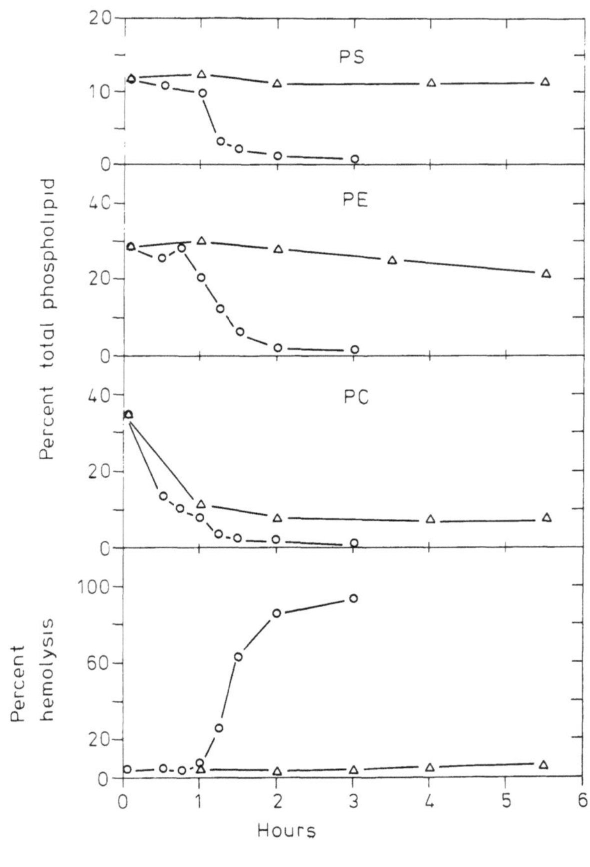

Fig. 6. Influence of Ca^{2+} Concentration and pH on Rate of Hydrolysis of Human Erythrocyte Membrane Phospholipids by Phospholipase A_2 of *Naja Naja* Venom. Isotonic conditions employed throughout. ○——○ 20 mM Hepes buffer, pH 8.0, 40 mM Ca^{2+} and sufficient NaCl to make solution isotonic; △——△ same as above except pH 7.6 and 10 mM Ca^{2+}

the absence of enzyme did not induce hemolysis. However, the fact that phospholipase A_2 induced hemolysis only at pH 8.0 and 40 mM Ca^{2+} in the absence of added glucose showed that this event was related to lowered ATP levels.

Phospholipid Hydrolysis. At pH 7.4 and 10 mM Ca^{2+}, where only limited hemolysis occurs, the use of the basic phospholipase A_2 afforded results similar to those reported by ZWAAL et al. (1973), *viz* about 70 percent hydrolysis of phosphatidylcholine but little or no hydrolysis of phosphatidyl ethanolamine and phosphatidyl serine (Fig. 6). However, when the Ca^{2+} concentration was raised to 40 mM at pH 7.4, or at pH 8.0 with 10 or 40 mM Ca^{2+}, quite different results were noted. In an apparently two-stage reaction, the phosphatidyl choline was first hydrolyzed to the extent of 70 percent, and then the remaining phosphatidyl choline and the phosphatidyl ethanolamine and finally phosphatidyl serine were hydrolyzed. High Ca^{2+} ion concentration and pH increased the second phase of the phosphatidyl choline hydrolysis and the subsequent attack on phosphatidyl ethanolamine and phosphatidyl serine. Hemolysis appeared to occur after the onset of phosphatidyl ethanolamine hydrolysis and phosphatidyl serine hydrolysis was coincident with hemolysis (Fig. 6).

It should be noted that maximal stimulation of hydrolysis is observed at 10 mM Ca^{2+} ion concentration (Fig. 7). These results should be contrasted with the effects of calcium ion concentration on hemolysis as shown in Figure 5. In particular, calcium ion maximally stimulated hydrolysis at concentrations where there was minimal hemolysis. Calcium ion showed saturation characteristics in the case of hydrolysis but not in the case of hemolysis.

It is apparent that the ability of the basic phospholipase A_2 to induce hemolysis and completely hydrolyze the phospholipids of the cell depends not only on pH and calcium ion concentration but also on the energy level of the cell.

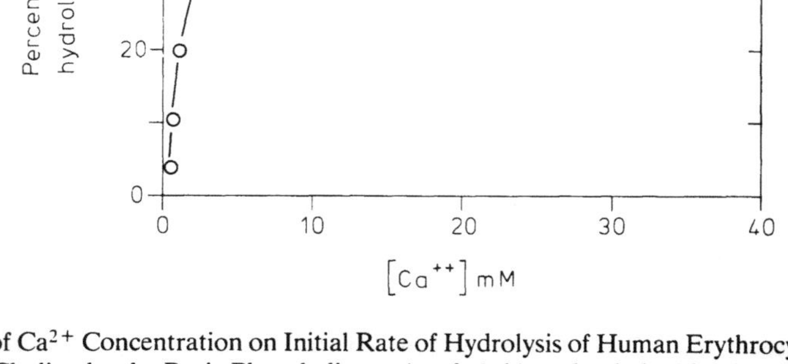

Fig. 7. Effect of Ca^{2+} Concentration on Initial Rate of Hydrolysis of Human Erythrocyte Membrane Phosphatidyl Choline by the Basic Phospholipase A_2 of *Agkistrodon halys blomhofii* venom. Erythrocytes were suspended in media containing 20 mM Hepes buffer, pH 8.0, 6.7 μg ml^{-1} enzyme, indicated concentration of Ca^{2+} and sufficient NaCl to make solution isotonic

Erythrocyte Age. It is known that as the cell ages there are changes in the ATP and cation level (Brok et al., 1966; Astrup, 1971; Hanahan, 1973). Thus it was of value to examine the mode of attack of cells separated by density centrifugation into an older or high-density type and into a lower-density or younger type. As illustrated in Table 11, the older cells are much more susceptible to hemolysis by the enzyme but all cell ages were hemolyzed by the enzyme though at obviously varying rates.

Table 11. Effect of age of human erythrocytes on hemolysis by phospholipase A_2. Cells were fractionated into low, moderate and high density cells, washed three times with isotonic saline, and incubated in 20 mM HEPES pH 8.0, 40 mM Ca^{2+} and sufficient NaCl added to make the solution isotonic. The hematocrit was 5% and the basic enzyme from *A. halys blomhofii* was used at a level of 6.7 μg/ml of incubation mixture[a]

Cells	$t_{1/2}$ (min.)
low density	122
moderate density	100
high density	78

[a] Taken from Martin et al. (1975)

7. Summary Statement

The above experimental results again reflect the complexity of the system under consideration. The observations by Martin et al. (1975) do not eliminate the possibility that various classes of phospholipids are asymmetrically partitioned on the extra- and intracellular sides of the membrane. They do emphasize the susceptibility of phospholipids to hydrolysis by phospholipase A_2 as a function of a number of variables. An understanding of the manner in which these variables, *viz* pH, calcium ion concentration, intracellular ATP concentration, cause membrane phospholipids to become more accessible or susceptible to hydrolysis may provide important clues in the elucidation of the microenvironment of phospholipids in the erythrocyte membrane. Certainly future compositional studies on erythrocyte membrane must be directed toward examination of lipid localization.

II. Protein Composition

Perhaps one of the more formidable problems facing membrane biochemists is a definitive, clear-cut analytical approach to the determination of the amounts and types of specific proteins in a membrane. The situation is complicated by the fact that normally these proteins or polypeptides are insoluble at a "biological" pH of 7–8.0 and any procedure which leads to their solubilization can therefore be assumed to cause some alterations to these compounds. Given that it is possible to solubilize the erythrocyte proteins, the major problem that ensues is a definitive separation of the component polypeptides (which may exist as subunits or as an aggregated species) and subsequent association of these polypeptides with a specific biological function, *e.g.*, enzymatic activity, sugar transport, ion movements, etc. In general, solubilization techniques most often lead to loss of the latter characteristics. Finally, the current widespread practice of using SDS-gel electrophoresis to identify specific types of proteins assumed to be single molecular-weight species must be reconsidered in the light of recent data questioning this technique. This topic will be explored in detail below.

Inasmuch as a number of fine reviews have recently been published on the erythrocyte membrane proteins (MADDY, 1970; JULIANO, 1974; COLEMAN, 1974), it seems inappropriate to restate comparable material here. It is more pertinent to draw attention to some valid questions that have been raised concerning the effectiveness of certain techniques, such as SDS-Page electrophoresis, for establishing the molecular weight and uniqueness of the membrane proteins and polypeptides. Only relatively few publications will be cited here, but they should clearly point out the recurrent problems in this field.

1. General Comments

It would be a major step forward if the composition of the membrane proteins could be established. Some success has been reported in the use of covalent binding reagents for localization of certain proteins in the erythrocyte membrane (CARRAWAY, 1975). It has also provided some input into other structure-function relations of certain proteins. However there are pitfalls in this approach too, since there can be no assurance that a specific chemical reagent or an enzyme-modifying system such as lactoperoxidase will not alter other properties of the membrane; thus the potential for side or adverse reactions cannot be dismissed. It is evident that one must use more than one system or technique for evaluation of the composition of membrane proteins.

An important avenue to understanding the structure (and function) of a membrane is to "solubilize" it and subject this preparation to examination. In a sense, the term solubilization is incorrect here, since it really means dissociation of the membrane into various components. Even so, the concept of solubility is difficult to apply to macromolecules and in fact the "escape clause" has been to consider a component soluble if it is not sedimented after 3 hours at 100,000× g. Nonetheless, this approach has proved to be extremely popular, solubilization with SDS being the technique of choice. There has been a tendency to assume homogeneity of composition within a particular band on these gel patterns, but this has been rendered suspect by some recent findings. These observations are considered in some detail below.

2. Nature of Polypeptide Patterns on SDS-PAGE

Our current knowledge on polypeptide composition of erythrocyte membranes relies heavily on results obtained with SDS-gel electrophoresis. Presumably this technique separates by molecular weight. At least eight bands are normally revealed by Coomassie blue staining of these gels (see I, Fig. 8). Inasmuch as glycoproteins do not stain well with Coomassie blue, they must be stained by the periodic acid-Schiff (PAS) base reagent (see II, Fig. 8). A typical pattern for the human erythrocyte membrane protein solubilized in SDS-gel electrophoresis is given in Figure 8, below:

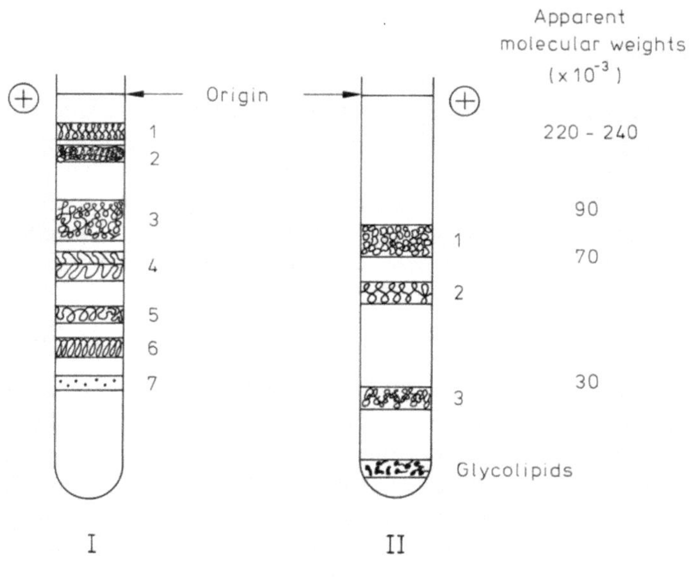

Fig. 8. SDS-Gel Patterns of the Proteins of the Human Erythrocyte Membrane. I represents those proteins stained by Coomassie blue and II the glycoproteins stained by periodic acid – Schiff reagent

The molecular weights proposed for these bands range from ~240,000 for the upper band, to an apparent molecular weight near 40,000.

a) Specific Protein Components Revealed by SDS Gel Electrophoresis

The Coomassie staining components in pattern I can be described as follows:

Bands 1, 2 and 5. These bands are assumed to form a complex, composed of two asymmetrical, high molecular weight proteins (bands 1 and 2 are equivalent to spectrin) and a lower molecular weight component. Bands 1 and 2 appear to be very similar in composition and are easily cross-linked, suggesting a close association. These three bands are presumably easily extractable from the membranes with slightly alkaline solutions of low ionic strength. This is facilitated by the presence of EDTA. Band 5 is usually less well defined; the claim is that it is near actin on molecular size and supposedly can be polymerized at high ionic strength into fibrils which resemble actin and are capable of stimulation of myosin-ATPase, for example.

Band 3. This protein is difficult to isolate because of its hydrophobicity. It contains a relatively large portion of the membrane protein. It may be the anion transport protein or close to it. Essentially the more darkly shaded bands are considered to be in the highest proportions based on the staining technique and the upper three bands corresponding to approximately 55 percent of the membrane protein.

The PAS-staining bands, including glycophorin and two others yielding positive stains, are the glycoproteins of the membranes and include a band composed of glycolipid which migrates very rapidly.

As noted above, current knowledge on the polypeptide composition of membranes rests heavily on results from SDS-PAGE. Recently the assumptions underlying the interpretation of some of these results have been questioned. In addition, experimental data have accumulated indicating that protein aggregates with an apparent molecular weight near 200,000 cannot be dissociated by dodecylsulfate and could be derived from smaller units of 40,000 daltons.

b) Studies on "Spectrin" of the Human Erythrocyte Membrane

In a recent study, FULLER et al. (1974) isolated spectrin from the human erythrocyte membrane by a procedure similar to that described by MARCHESI et al. (1970). They found it difficult to separate this component on a column, so they obtained a doublet of spectrin by slicing an SDS gel and subjected it to qualitative and quantitative N-termini determinations using a dansylation procedure. The molecular weight by SDS gel electrophoresis and sedimentation equilibrium was in the range of 220,000–240,000 daltons. Essentially the experimental results using this dansylation procedure on the spectrin doublet were as follows: a) the doublet had multiple N-terminal aminoacids, implying that the subunit could not be a single polypeptide chain; b) absence of any interchange of covalent cross-bridges (x-glutamyl-lysine or-aspartyl lysine) suggests each unit is probably a mixture of several nonidentical large polypeptide chains. FULLER et al. concluded that previous suggestions that each band of the spectrin doublet was a single polypeptide band chain must be viewed as questionable. Similar observations were published somewhat earlier by KNÜFERMAN et al. (1973), who expressed concern as to the uniqueness of spectrin. The possibility of enzymatic degradation during the EDTA extraction was considered likely, so these investigators directly dansylated SDS-soluble membranes separated on PAGE and identified the DNA aminoacids. A total of five N-termini were indicated.

At this point the message derived from the above experiments was that spectrin contained at least three polypeptide chains, that it was composed of at least two subunits, and that there were strong protein-protein interactions.

c) Observations on the Ox (Bovine) Erythrocyte Polypeptide Heterogeneity

In 1973, MADDY and DUNN suggested that the high molecular weight fractions found in the EDTA extracts (MADDY and KELLY, 1971) of the ox erythrocyte membrane, which other investigators had thought might consist of high molecular weight polypeptides (200,000) as distinct from a stable complex of lower molecular weight proteins (40,000), were dependent entirely on the SDS-gel method. They reported, at that time, that there were multiple N-termini in the 200,000 and 40,000 molecular weight fractions. This was disturbing in that most investigators had erroneously assumed that SDS disrupts all but covalent interactions between aminoacids and hence all proteins are dissociated into their individual polypeptide chains. MADDY and DUNN's studies emphasized that extrapolation of the SDS behavior of soluble proteins to that of insoluble proteins and membranes is a very hazardous approach to be adopted only with great care. In a more recent study, DUNN et al. (1975) reported additional evidence on the N-terminal heterogeneity of the EDTA-extractable (bovine) erythrocyte membrane proteins. Through use of dinitrophenylation and dansylation, some eight different N-termini aminoacids were found in fractions obtained by SDS-gel electrophoresis of these soluble proteins. Coomassie blue staining of the gels revealed far fewer polypeptides than expected from the N-terminal analyses. Examination of the 200,000 molecular weight bands (see peptide I and II, Fig. 8) and the 40,000 molecular weight band (peptide V, Fig. 8) found in the EDTA extract was conducted in the presence and absence of SDS. In the latter case, the molecular weights were estimated by the Ferguson technique, which involves measurement of the influence of acrylamide concentration on mobility on electrophoresis. Essentially they concluded on the basis of some convincing data that the 200,000 and 40,000 molecular weight complexes were heterogeneous with respect to polypeptide chains and that there was a close similarity between the two. In fact DUNN et al. conluded that these bands might consist of the same polypeptide chains in varying stages of aggregation and they were interconvertible. Interestingly. BØG-HANSEN and LJERRUM (1973) noted that human erythrocyte "spectrin" was antigenically heterogeneous, with at least nine different determinants. A number of other laboratories have subsequently confirmed these observations on the human cell and noted N-terminal inhomogeneity in a number of bands separated by SDS gel electrophoresis (KNÜFERMANN et al., 1974; LANGDON, 1974).

The above observations clearly demonstrate the considerable heterogeneity of the membrane protein fractions and that simple SDS gel electrophoresis cannot be used to identify molecular weight characteristics of the membrane proteins from the erythrocyte. Interestingly enough, a recent paper by KNÜFERMANN et al. (1974), in which they described a method for the preparative SDS-PAGE electrophoresis of the major proteins in human erythrocytes, emphasized several points: a) each component in these various bands contains several N-terminal aminoacids, most of which appear constant; b) there are some additional N-terminal aminoacids which vary from one human donor to the next. This particular technique, which they feel would allow for a more definitive investigation of these polypeptides in the membrane, yields up to 75 mg of membrane protein at one time. These authors observed that the high molecular weight spectrin bands (bands 1 and 2) were quite labile. Extraction at room temperature tended to split these proteins into products of lower molecular weight. In contrast, the minor components (2.1 and 2.2 by their numbering system) tended to aggregate yielding components 1 and 2.

3. Summary Statement

It is evident from the above comments on membrane protein identification that this is still a complex, unresolved area of study and many important discoveries can be anticipated in the next few years. After the initial flurry associated with the presumed complete problem-solving abilities of SDS, it seems the field can now settle into, hopefully, a logical approach to analysis and identification of membrane proteins and glycoproteins. Certainly this is a difficult investigative area, since the insolubility of these polypeptides provides an immense barrier to many types of studies and "solubilization" in most instances destroys biological activity of these membrane components. So essentially, the point will be reached where the polypeptides can be dissociated and identified, but then a route must emerge to clarify how they may interact with each other (and with lipids) to produce, for example, a transport protein, or a particular enzymatic activity, and then to further study their localization within the membrane.

Basically, the results of solubilization studies on membrane proteins emphasize the point made earlier, that the use of a single "analytical" technique with such a complicated system is not at all satisfactory, and in fact, no single approach will offer a panacea. The aim of the discussion has been to focus attention on the complexity of the system under study. Membrane biochemistry is one of the most challenging subjects for scientific inquiry and there is no doubt that in the next few years some exciting developments will occur.

References

Astrup, J.: Scand. J. clin. and Lab. Invest. **33**, 231 (1974).

Bøg-Hansen, T. C., Bjerrum, O. J.: Protides Biol. Fluids **21**, 39 (1973).

Bretscher, M. S.: Nature Res. Biol. **236**, 11 (1972).

Brok, F., Ramut, B., Zwang, E., Danon, D.: Israeli J. Med. Sci. **2**, 291 (1966).

Burger, S. P., Fujii, T., Hanahan, D. J.: Biochemistry **7**, 3682 (1968).

Carraway, K. L.: Biochim. biophys. Acta **415**, 379 (1975).

Coleman, R.: Biochim. biophys. Acta **300**, 1 (1973).

Cooley, C. M., Zwaal, R. F. A., Roelofsen, B., Van Deenen, L. L. M.: Biochim. biophys. Acta **307**, 74 (1973).

Danon, D., Markovsky, Y.: J. Lab. clin. Med. **64**, 668 (1964).

Dunn, M. J., McBay, W., Maddy, A. H. Biochim. biophys. Acta **386**, 107 (1975).

Fuller, G. M., Boughter, J. M., Marazzini, M.: Biochemistry **13**, 3036 (1974).

Gazitt, Y., Ohad, I., Loyter, A.: Biochim. biophys. Acta **283**, 65 (1975).

Gould, R. G., LeRoy, G. V., Okita, G. T., Kabara, J. J., Keegan, P., Bergenstal, D. M.: J. Lab. clin. Med. **46**, 372 (1955).

GREEN, J. R., DUNN, M. J., SPOONER, R. L., MADDY, A. H.: Biochim. biophys. Acta **373**, 51 (1974).
GUL, S., SMITH, A. D.: Biochim. biophys. Acta **288**, 237 (1972).
GUL, S., SMITH, A. D.: Biochim. biophys. Acta **367**, 271 (1974).
HAGERMAN, J. S., GOULD, R. G.: Proc. Soc. exp. Biol. **78**, 329 (1951).
HANAHAN, D. J.: Biochim. biophys. Acta **300**, 319 (1973).
HOLUB, B. J., BRECKENRIDGE, W. C., KUKSIS, A.: Lipids **6**, 307 (1971).
IBRAHIM, S. A., THOMPSON, R. H. S.: Biochim. biophys. Acta **99**, 331 (1965).
ISRAEL, Y., MacDONALD, A., BERNSTEIN, J., ROSENMAN, E.: J. gen. Physiol. **59**, 270 (1972).
JULIANO, R. L.: Biochim. biophys. Acta **300**, 341 (1973).
KIM, H. D., McMANUS, T. J.: Biochim. biophys. Acta **230**, 1 (1971).
KIM, H. D., McMANUS, T. J., BARTLETT, G. R.: In: Recent Advances in Membrane and Metabolic Research, 2nd international Symposium, Vienna (E. Gerlach, K. Muser, E. Deutsch and W. Wilmanns, Eds.). Stuttgart: Thieme 1972, p. 146.
KNÜFERMANN, H., BHAKDI, S., SCHMIDT ULLRICH, R., WALLACH, D. F. H.: Biochim. biophys. Acta **330**, 356 (1973).
KNÜFERMANN, H., BHAKI, S., WALLACH, D. F. H.: Biochim. biophys. Acta **389**, 464 (1974).
LANGDON, R. G.: Biochim. biophys. Acta **342**, 213 (1974).
LEIF, R. C., VINOGRAD, J.: Proc. nat. Acad. Sci. (Wash.) **54**, 520 (1964).
LONDON, I. M., SCHWARZ, H.: J. clin. Invest. **32**, 1248 (1953).
McMANUS, T. J., KIM, H. D.: In: Metabolism and Membrane Permeability of Erythrocytes and Thrombocytes, 1st international Symposium, Vienna (E. Deutsch, E. Gerlach and K. Moser, Eds.). Stuttgart: Thieme 1968, p. 43.
MADDY, A. H.: Biochim. biophys. Acta **88**, 390 (1964).
MADDY, A. H.: Sem. in Hemat. **7**, 275 (1970).
MADDY, A. H., DUNN, M. J.: Protides Biol. Fluids, Proc. Colloq. **21**, 21 (1973).
MADDY, A. H., KELLY, P. G.: Biochim. biophys. Acta **241**, 290 (1971).
MARCHESI, S. L., STEERS, E., MARCHESI, V. T.: Biochemistry **9**, 50 (1970).
MARTIN, J. K., LUTHRA, M. G., WELLS, M. A., WATTS, R. P., HANAHAN, D. J.: Biochemistry (1975) **14** 5400
MURPHY, J. R.: J. Lab. clin. Med. **60**, 571 (1962).
MURPHY, J. R.: J. Lab. clin. Med. **82**, 334 (1973).
NELSON, G. J. (Ed.): Blood Lipids and Liproproteins: Quantitation, Composition and Metabolism, New York: Wiley Interscience 1972, p. 317.
O'CONNELL, D. J., CARUSO, C. J., SASS, M. D.: Clin. Chem. **11**, 771 (1965).
PIOMELLI, S., LUNINSKY, G., WASSERMAN, L. R.: J. Lab.Clin. Med. **69**, 659 (1967).
PORTE, D., HAVEL, R. J.: J. Lipid Res. **2**, 357 (1961).
REED, C. F.: J. clin. Invest. **38**, 1032 (1964).
REED, C. F.: J. clin. Invest. **47**, 749 (1968).
SARKAR, N. K., DEVI, A.: In: Venomous Animals and their Venoms, Vol. 1 (W. Bucherl, E. E. Buckley, and V. Deolotev, Eds). New York: Academic Press 1968, p. 167.
SHOHET, S.: New England J. Med. **286**, 577, 638 (1972).
TOSTESON, D. C.: In: Red Cell Membranes, Structure and Function (G. A. Jamieson and T. E. Greenwalt, Eds.). Philadelphia: J. B. Lippincott 1969, p. 291.
WADSTRÖM, T., MÖLLBY, R. (1971) Biochim. biophys. Acta **242**, 288, 308 (1971).
ZWAAL, R. F. A., ROELOFSEN, B., COLLEY, G. M.: Biochim. biophys. Acta **300**, 159 (1973).

Chapter 7

Genetic Determination of Membrane Transport

C. W. SLAYMAN

A. Introduction

During the past twenty years it has become increasingly clear that biological membranes, like other cellular constituents, are under genetic control. Each membrane protein, and each enzyme required for the synthesis of membrane lipids, is now presumed to be coded for by a structural gene, whose nucleotide sequence determines the corresponding amino acid sequence of the protein. In addition, at least in some instances, regulatory genes exist which control the timing and rate of expression of the structural genes. As a result, the investigator interested in membrane transport mechanisms has access to a powerful analytical method. Genetic mutation can cause the specific loss – or even the specific alteration, one amino acid at a time – of a transport system, giving the investigator a chance to study the physiological consequences of the change.

As one might expect from the potential advantages of the method, genetic analysis has been applied to membrane transport processes with increasing frequency over the past few years. In microorganisms, where mutants can readily be selected and mapped, structural and regulatory genes have been identified for a variety of transport systems, and useful information is beginning to emerge about molecular mechanisms of transport. In higher organisms, where it is not possible to select transport mutants at will, work with spontaneously occurring hereditary diseases (for example, HK-LK erythrocytes in sheep and cystinuria and glucose-galactose malabsorption in man) has nonetheless confirmed the general picture that transport systems are under genetic control. And more recently, methods have been developed to induce and select mutants of cultured mammalian cells, making it reasonable to expect that a genetic dissection of transport mechanisms will soon be possible in higher organisms as well.

The purpose of this chapter is to review the present state of knowledge of the genetic determination of transport, both in microorganisms and in higher organisms. No attempt is made at a complete summary; rather, a few examples have been chosen to illustrate the kinds of analysis that are now possible for each group of organisms. More extensive information on the genetic control of transport in microorganisms can be found in recent reviews by LIN (1970), SLAYMAN (1973), BOOS (1974), HALPERN (1974), COX and GIBSON (1974), SIMONI and POSTMA (1975), and OXENDER and QUAY (1975). The genetics of transport in higher organisms has been reviewed by ROSENBERG and SCRIVER (1974) and in several chapters in STANBURY et al. (1972); and current information about the genetics of transport in cultured mammalian cells has been summarized by THOMPSON and BAKER (1973).

B. Microorganisms

I. Isolation of Transport Mutants

The ease with which microbial mutants can be obtained – particularly in species such as *Escherichia coli, Salmonella typhimurium, Neurospora crassa,* and *Saccharomyces cerevisiae* – has led to the extensive genetic analysis of microbial transport systems. A variety of methods have been developed for the isolation of transport mutants: strains can be selected whose growth is resistant to a toxic substrate of the transport system (say, an amino acid or purine analog) or which are unable to grow at low concentrations of the normal substrate. Alternatively, transport-negative strains can be screened directly, using autoradiographic techniques to look for reduced uptake of radioactive substrate (WILSON et al., 1970; WILSON and KUSCH, 1972; BOOS and SARVAS, 1970) or immunological techniques to detect lowered amounts of a particular transport protein (HOGG, 1971). These and other methods have been reviewed recently (SLAYMAN, 1973). In all cases, the precise strategy to be used depends upon the role of the particular transport process in growth. Four situations can be imagined:

1) The substrate of the transport system is not ordinarily required for growth, so mutants lacking the system are readily viable. For example, most microorganisms can use a variety of sugars and complex carbohydrates as sources of carbon and energy, with the result that no individual sugar transport system is essential. In addition, many species can synthesize all of their own amino acids, and consequently can survive without amino acid transport systems under ordinary circumstances.

2) The substrate is required for growth but can be supplied by the experimenter in some other form. In many species, for example, inorganic sulfate can be dispensed with as long as an organic sulfur compound such as cysteine is present; and sulfate transport mutants are isolated as cysteine requirers (DREYFUSS and MONTY, 1963; OHTA et al., 1971). Similarly, phosphate transport mutants can grow in the presence of an alternate phosphorus source such as aminoethylphosphonate (ROSENBERG and LaNAUZE, 1968) or L-α-glycerophosphate (BENNETT and MALAMY, 1970).

3) The substrate is required for growth but two or more transport systems exist; as a result, one system can be lost by mutation and the other will support growth. Such a situation is found in *E. coli,* which has at least two transport systems for potassium. Mutants lacking the high-affinity *(kdp)* system do not grow at very low potassium concentrations (2×10^{-5}M), but can grow at higher concentrations (10^{-4} to 5×10^{-3}M) by using a second *(trk)* transport system (EPSTEIN and DAVIES, 1970; EPSTEIN and KIM, 1971).

4) The substrate is required for growth and cannot be supplied by any other route. This is clearly the least favorable circumstance for the isolation of transport mutants, since one is limited to qualitatively altered mutants and cannot expect to obtain strains lacking the transport system altogether. An example is provided by those microogranisms in which potassium is transported by a single system and, even at high external concentrations, cannot diffuse across the cell membrane rapidly enough to support growth. Here, mutants with qualitative alterations have been described, including Cn_{K6} mutants of *Streptococcus faecalis* which are defective in cation uptake at low pH (HAROLD et al., 1970; HAROLD and PAPINEAU, 1972), Tr_{K8} mutants of *S. faecalis* in which potassium uptake is abnormally sensitive to inhibition by sodium (HAROLD and BAARDA, 1967), *trk*-1 mutants of *Neurospora* with an altered $K_{1/2}$ for potassium (SLAYMAN and TATUM, 1965; SLAYMAN, 1970), and *trk*-2 mutants of *Neurospora* with an abnormally rapid K/K exchange (SLAYMAN, unpublished results). In all of these strains, considerable transport activity remains under the standard growth conditions. Another kind of qualitatively altered mutant that has been useful in the study of essential transport functions is the temperature-sensitive mutant, which transports normally at low temperatures (usually 25°) but not at high temperatures (42°), presumably as a result of a very particular kind of amino acid substitution in the corresponding transport protein. Temperature-sensitive transport mutants have been described in *Escherichia coli* (FOX et al., 1967; EPSTEIN et al., 1970) and *Salmonella* (AMES and LEVER, 1972) and have been useful in identifying molecular components of transport systems, as will be discussed below.

II. Kinds of Genetic Analysis

Once mutants have been obtained with defects in a particular transport system, a variety of questions can be asked. Crossing of the mutants with strains carrying known marker genes allows determination of the chromosomal location of the genes affecting transport. This is a relatively straightforward procedure in the microorganisms commonly used for biochemical genetics: *E. coli* and *Salmonella* have single chromosomes (BACHMANN et al., 1976; SANDERSON, 1976), *Neurospora* has seven chromosomes (RADFORD, 1975), and *Saccharomyces cerevisiae* has 17 chromosomes (SHERMAN and LAWRENCE, 1974), all extensively mapped.

A second kind of genetic analysis, and one that is especially important where there is a possibility of closely linked genes controlling the same overall function, is complementation analysis. Imagine that a group of mutants have been isolated, each lacking the same transport activity, and that the mutations map in the same small chromosomal region. One now wishes to establish whether all of the mutations lie in a single gene, or whether there might be two or three adjacent genes, perhaps coding for multiple transport subunits. The first step is to construct a cell that contains two copies of the genetic region in question. In bacteria, this is done by introducing a small extra fragment of chromosome *via* a transducing phage or an episome, and in the fungi, by allowing two haploid cells (each with a single set of chromosomes) to fuse, forming either a heterokaryon (containing a mixed population of nuclei within a single cytoplasm) or a true diploid cell (in which each nucleus now has two sets of chromosomes). Now imagine that one copy of the genetic region bears one mutation affecting the transport system being studied (call it mutation A) and the other copy of the genetic region bears a second mutation affecting the same transport system (call it B). If the two mutations are in the same gene, the resulting cell will have the genotype

$$\frac{\underline{\quad A \quad}}{\overline{\quad B \quad}}$$

and will be defective in transport, since no normal copy of the gene is present. If, on the other hand, the two mutations are in different genes, the cell will have the genotype

$$\frac{\underline{\quad A \; + \quad}}{\overline{\quad + \; B \quad}}$$

and as long as both mutations are recessive, the cell will have relatively normal transport activity, since the normal copy of gene A and the normal copy of gene B (both designated +) can supply their respective functions. In this manner the number of complementation groups (or "cistrons") affecting a particular transport activity can be determined.

A third and somewhat more complicated kind of genetic analysis is aimed at distinguishing between structural genes, which code for actual transport proteins, and regulatory genes, which determine whether (and at what rate) the structural genes are expressed. In microorganisms, transport systems are well known to be under genetic regulation. One of the earliest cases to be studied, the lactose transport system of *E. coli,* is formed only when cells are grown in the presence of lactose or a related compound (COHEN and RICKENBERG, 1955; RICKENBERG et al., 1956), and the same is true of many other carbohydrate transport systems (reviewed in LIN, 1970, and SLAYMAN, 1973). Amino acid, purine and pyrimidine, and inorganic cation and anion transport systems are also genetically regulated. In *E. coli,* for example, one of the potassium transport systems is repressed during growth at high potassium concentrations (EPSTEIN and KIM, 1971) and one of the magnesium transport systems, by high magnesium (NELSON and KENNEDY, 1972). In *Neurospora,* there are two transport systems for inorganic sulfate, one predominating in conidia and the other in vegetative hyphae (MARZLUF, 1970a, b; ROBERTS and MARZLUF, 1971), and also two transport systems for inorganic phosphate, one repressed by high phosphate (LOWENDORF et al., 1975; LOWENDORF and SLAYMAN, 1975). In each of these cases, certain kinds of mutations in regulatory genes can lead to a loss of transport activity, as can mutations in the corresponding structural genes. Generally the first step in distinguishing between the two kinds of genes is to look for qualitatively altered mutants – those with changes in the kinetic properties or the thermostability of the transport system. Since the most likely way to generate such a change is by aminoacid substitution in the transport protein itself, mutations leading to qualitative alterations are usually tentatively assumed to lie in structural genes. When suitable biochemical methods are available, the conclusion can then be tested rigorously by isolating the protein and looking for evidence of a change in its structure.

III. Examples of the Use of Genetic Analysis

1) *The lactose transport system of Escherichia coli: identification of a transport protein.* The earliest microbial transport system to be studied in detail was the lactose or β-galactoside transport system of *E. coli,* which was shown by COHEN and RICKENBERG (1955) and RICKENBERG et al. (1956) to be induced during growth on lactose and related sugars and to require the product of the *lac*Y gene. Ten years later, FOX and KENNEDY (1965) reported that a specific protein, which they called "M protein," could be detected in the membranes of induced cells by labeling with the sulfhydryl reagent N-ethylmaleimide (NEM) under very special conditions: first with nonradioactive NEM in the presence of thiodigalactoside, a substrate for which the β-galactoside transport system was known to have a high affinity; and then with radioactive NEM, after washing away the thiodigalactoside. The sites labelled in the second step were therefore sulfhydryl groups (later shown to be cysteine residues) that had been specifically protected by bound substrate. Because the number of sites was much higher in induced cells than in noninduced cells, and because the sites were located in the cell membrane, it seemed likely that they formed part of the β-galactoside transport system. Conclusive genetic proof came in 1967, when FOX et al. demonstrated that the specific NEM-binding sites were absent in transport-negative *lac*Y mutants and rapidly inactivated at 42° in temperature-sensitive *lac*Y mutants. From arguments outlined in the preceding section, it could be concluded that the protein containing the binding sites was the product of the *lac*Y gene and a component of the β-galactoside transport system. The M protein has since been extracted from the membrane (JONES and KENNEDY, 1969), and further work has been done on the characteristics of substrate binding by both the isolated protein (KENNEDY, 1970; KENNEDY et al., 1974) and the membrane-bound protein (REEVES et al., 1973; SCHULDINER et al., 1975; RUDNICK et al., 1975; THERISOD et al., 1977).

Because of the wealth of information now available about both the kinetics and the biochemistry of the β-galactoside transport system (reviewed by KEPES, 1971; KENNEDY, 1970), it has become a system frequently used by investigators interested in the molecular mechanisms of transport and of energy coupling. For these kinds of studies, too, the genetic approach is proving useful. In addition to the two kinds of *lac*Y mutants already mentioned (those in which β-galactoside transport is completely missing and those in which it is temperature-sensitive), WILSON and his colleagues have isolated an extremely interesting class of "energy-uncoupled" mutants (WONG et al., 1970; WILSON et al., 1970; WILSON and KUSCH, 1972). These strains contain a normal (or perhaps slightly greater than normal) number of transport sites, as judged by the amount of substrate binding and the rate of facilitated diffusion. Accumulation of lactose and its analogs is reduced drastically, however; and in keeping with the idea that β-galactoside uptake may occur by proton-linked cotransport (as reviewed by HAMILTON, 1975; HAROLD, 1977), very little β-galactoside-stimulated proton influx is observed in the mutants (WEST and WILSON, 1973). Because the mutants leading to energy uncoupling appear to map in the *lac*Y gene (WILSON and KUSCH, 1972) one can assume that they lead to a particular kind of amino acid substitution in the M protein, such that it can still be assembled into the membrane (since the number of binding sites is normal), can still bind and translocate substrate (since the rate of facilitated diffusion is normal), but can no longer interact with the energy source for uphill transport (most likely the electrochemical gradient for protons (RAMOS et al., 1976; RAMOS and KABACK, 1977a, b; FLAGG and WILSON, 1977). Future studies on the details of substrate binding and on the pH- and potential-dependence of translocation in these uncoupled mutants should help to define the transport mechanism in greater detail.

2) *Histidine transport in Salmonella: the use of mutants to sort out multiple transport systems.* Frequently microorganisms – and higher organisms as well – have more than one transport system for a single substrate or class of substrates. In these cases, overlapping substrate affinities can make it extremely difficult to distinguish among the systems on kinetic grounds alone, and mutants can prove helpful. For example, AMES and her co-workers have shown that the bacterium *Salmonella typhimurium* takes up histidine by five separate transport systems: a general aromatic system which transports tryptophan, tyrosine, and phenylalanine as well as histidine (for which it has a $K_{1/2}$ of 10^{-4}M), two specific histidine systems, designated J-P and K-P with $K_{1/2}$'s of 10^{-8}M and 10^{-7}M, and at least two other low-affinity systems which have not yet been characterized in detail (AMES, 1964; AMES and LEVER, 1970). Kinetic analysis by itself could not possibly have described or even enumerated these various systems, and it was only with the help of a series of transport mutants that the kinetic picture became clear. The mutants that have been reported include *aro*P, which lack the general aromatic amino acid transport system (AMES, 1964); *his*P , which is defective in a protein required by both the J-P and K-P systems (AMES and LEVER, 1970); and *his*J, which has lost the J-binding protein (a 26000-molecular weight protein that binds histidine and interacts with the P protein to constitute the J-P mode of transport; LEVER, 1972; AMES and LEVER, 1970). The general aromatic amino acid transport system can be well studied in *his*P mutants; and the J-P and K-P systems, by comparison of the wild-type strain with *his*P and *his*P mutants.

For the J-P system particularly, just as in the case of lactose transport in *E. coli,* the isolation of qualitatively altered mutants has made it possible to pursue the underlying mechanism of transport. Temperature-sensitive *his*J mutants, in which both the J protein and transport via the J-P system show parallel abnormalities, have provided proof for a role of the binding protein in transport

(AMES and LEVER, 1972); and more recently, an interesting new *his*J has been described in which the J protein binds histidine with normal affinity but functions very poorly in transport (KUSTU and AMES, 1974). Further study will be needed to establish the precise nature of the molecular change in this mutant.

3) *Galactose transport in E. coli: identification of a protein which plays a role both in transport and in chemotaxis.* One might imagine, because of the range of intermolecular interactions present in a complex structure such as a membrane, that a single protein is sometimes involved in more than one membrane function. In fact, an example has already been given in the preceding section, where the P protein of *Salmonella* was shown to participate in two distinct histidine transport systems, J-P and K-P. An even more intriguing example was brought to light in 1971 with the discovery that the galactose-binding protein of *E. coli,* a 35000-molecular weight protein readily released from the periplasmic space (the space between the plasma membrane and the cell wall), plays a role both in the β-methylgalactoside transport system and in chemotaxis towards galactose (KALCKAR, 1971; HAZELBAUER and ADLER, 1971). The relationship between transport and chemotaxis, although unexpected until recently, may prove to be quite common: bacterial binding proteins for ribose (HAZELBAUER and ADLER, 1971; AKSAMIT and KOSHLAND, 1972; ADLER et al., 1973) and maltose (HAZELBAUER and ADLER, 1971; ADLER et al., 1973), and the membrane-bound phosphotransferase enzymes II for a variety of sugars (ADLER and EPSTEIN, 1974), have now been shown to play similar dual roles.

The evidence pointing to this relationship in the case of galactose has been summarized by BOOS (1974). In the first place, when galactose and a family of related compounds were compared, nearly identical specificities were observed for the binding of sugar to the purified galactose binding protein (BOOS, 1969; BOOS et al., 1972; PARNES and BOOS, 1973), the transport of sugar *via* the β-methylgalactoside system (PARNES and BOOS, 1973), and the chemotactic response (HAZELBAUER and ADLER, 1971). Even more conclusively, *mgl*B mutants defective in the galactose binding protein were found to be simultaneously defective in transport activity and in chemotaxis. One particularly clearcut example, which serves also to identify the structural gene for the binding protein, was a mutant in which the isolated protein had a 7000-fold reduced binding affinity accompanying a demonstrable change in primary structure (determined by fingerprinting); this same strain was transport-negative and lacked chemotaxis towards galactose and related sugars. When a revertant was obtained by selecting for the ability to transport galactose, the binding protein and chemotaxis also returned nearly to normal (BOOS, 1972 and 1974). This set of results points clearly to an essential role of the binding protein in both transport and chemotaxis.

Transport and chemotaxis are not, however, completely overlapping functions. Mutants have been isolated which map in two genes (*mgl* A and C) closely linked to, but distinct from, the structural gene for the galactose binding protein, and which are defective in transport but not in chemotaxis (ORDAL and ADLER, 1974a and b). Perhaps these two genes code for subunits analogous to the P protein of the *Salmonella* histidine transport systems (see preceding section). Conversely, other mutants (*che*) have been described which map in a cluster of eight genes distant from *mgl* and which are normal in binding protein and β-methylgalactoside transport but defective in chemotaxis (SILVERMAN and SIMON, 1977). Thus, the binding protein can be viewed as providing a recognition site common to transport and chemotaxis, each of which requires additional protein components of its own. This conclusion would have been difficult to substantiate without the aid of *mgl* mutants.

C. Higher Organisms

By contrast with the many flourishing examples of genetic analysis of transport in microorganisms, work with higher organisms has progressed more slowly. Inherited transport defects of mammals have been known for some time, but these organisms do not easily lend themselves to genetic study. Mutants cannot be selected at will, but instead must be recognized from symptoms which are sometimes nonspecific and remote (for example, altered concentrations of a particular compound in the blood or urine); linkage and complementation analyses with whole animals are not often possible; for many tissues, detailed kinetic data are difficult to obtain; and the biochemical isolation of transport components is considerably harder than in microorganisms. Nevertheless, certain kinds of useful data can be gathered: dominance and recessiveness can be determined and can give information about the nature of the mutational change; the tissue specificity of the transport defect can be explored; mutants can be used to distinguish among multiple, overlapping transport systems; and in those few instances where the biochemical analysis of transport has progressed far enough, the presence of a mutationally altered protein can ultimately serve to confirm the role of the protein in transport. Two of the best-studied examples of inherited transport defects in higher organisms are cystinuria in man and HK/LK erythrocytes in sheep and goats, which will now be discussed in turn.

I. Cystinuria

Cystinuria has been known clinically since the early 1800's, and its familial nature was well established by the time of Garrod's famous Croonian lectures on "inborn errors of metabolism" in 1908. It was not recognized as a transport disorder until 1951, however, when DENT and ROSE – influenced by the fact that cystinuria leads to the abnormally high excretion of four structurally related amino acids (cystine, lysine, arginine, ornithine) – proposed that the primary defect lies at the level of a common reabsorptive mechanism in the renal tubules.

Several interesting points have emerged from the genetic study of cystinuria. *First,* it is inherited as a recessive (HARRIS et al., 1955a and b), as one would expect for a mutation leading to the lack of a transport component or to synthesis of a defective component. (In the heterozygous diploid, one normal copy of the gene is still present and can produce enough of the transport component to allow normal, or virtually normal, function.) *Second,* at least three mutant alleles of the gene exist, which can be distinguished on the basis of their patterns of intestinal transport and urinary excretion (types I, II, and III cystinuria) and which presumably arose from three independent mutational events (ROSENBERG, 1966; ROSENBERG et al., 1966a and b). *Third,* cystinuria shows considerable tissue specificity. It is expressed in the renal tubules and in the intestinal mucosa as well (MILNE et al., 1961; ASATOOR et al., 1962; THIER et al., 1964 and 1965; McCARTHY et al., 1964), a combination frequently found with transport disorders. But it is not expressed in leukocytes and cultured fibroblasts from the same patients (BECKER and GREEN, 1958; ROSENBERG and DOWNING, 1965; GROTH and ROSENBERG, 1972), indicating that these cells contain genetically and biochemically distinct amino acid transport systems.

A fourth line of investigation is promising, but has so far been less productive. On the basis of clearance studies, there appear to be three transport systems for cystine and the dibasic amino acids in the kidney: one specific for cystine, one specific for lysine, arginine, and ornithine, and one (mentioned above and deficient in cystinuria) shared by all four amino acids (DENT and ROSE, 1951). Oral loading experiments point clearly to the presence of the shared system in the intestinal mucosa (MILNE et al., 1961; ASATOOR et al., 1962; ROSENBERG et al., 1965); whether the specific systems are also present remains to be established. As a disease inactivating one of the three postulated transport systems, cystinuria should help to simplify the kinetic analysis of transport. Further help could come from two apparently complementary disorders: "isolated" cystinuria, which may be a specific defect in cystine transport in the kidney (BRODEHL et al., 1967), and dibasic aminoaciduria (or protein intolerance), which affects the transport of lysine, arginine, and ornithine in the kidney and possibly in the intestine as well (PERHEENTUPA and VISAKORPI, 1965; KEKOMAEKI et al., 1967; WHELAN and SCRIVER, 1968; OYANAGI et al., 1970). The kinetic comparison of transport in renal and intestinal tissue from normal subjects and patients with these diseases could, in principle, establish the number of transport systems that are present and their properties. Several such studies have been carried out in the case of cystinuria (FOX et al., 1964; THIER et al., 1964 and 1965; McCARTHY et al., 1964) but have been only qualitatively successful, probably because amino acid movements in tissue slices are too complex to allow quantitative resolution of transport events at the brush border. Thus, a complete kinetic picture may have to await the development of micro-methods for fractionating brush border vesicles from small amounts of biopsy tissue, or even better, the isolation of a cell line in which renal and intestinal transport systems are expressed.

II. HK/LK Erythrocytes

Kinetic measurements are much easier with red cells, and here genetic analysis has progressed somewhat further. In 1955, EVANS and his co-workers described a locus (later named *Ka/ka*) that controls the cation composition of sheep red cells, with *Ka* dominant over *ka* (EVANS and KING, 1955; EVANS et al., 1956;

reviewed by LAUF, 1975). Thus, red cells from both *Ka/Ka* homozygotes and *Ka/ka* heterozygotes contain a low concentration of potassium (about 15 mmol/liter cells) and a high concentration of sodium (85 mmol/liter cells) and are called LK cells, while red cells from *ka/ka* homozygotes have the opposite cation composition (90 mmol potassium and 10 mmol sodium/liter cells) and are called HK cells. As one might expect, this difference in composition arises from an underlying difference in cation fluxes across the red cell membrane. TOSTESON and HOFFMAN showed in 1960 that active potassium influx (mediated by the ouabain-sensitive Na^+-K^+-ATPase) is about 8-fold slower, and passive permeability to potassium about 4-fold greater, in LK cells than in HK cells. More recently, DUNHAM and HOFFMAN (1971) have concluded from ouabain-binding experiments that LK cells contain fewer pump sites than HK cells, and HOFFMAN and TOSTESON (1971) have described a major alteration in the kinetic properties of LK pumps, which makes them more sensitive to inhibition by intracellular potassium. Both the quantitative reduction in pump sites and the qualitative change in the remaining pumps contribute to the decreased active potassium influx observed in LK cells. However, in spite of the fact that the physiological differences between HK and LK cells are now clear, the fundamental question of the identity of the *Ka/ka* gene product and its role in transport has not yet been solved. The following genetic clues exist:

1) As stated above, *Ka* is dominant over *ka*. Unlike recessiveness, which simply implies the lack of a functional gene product, dominance requires a more complicated explanation. There are several alternatives: (a) *Ka/ka* might be a regulatory gene, controlling the expression of one or more structural genes whose products are necessary for a stable, functional transport system. (The products need not be actual subunits of the Na^+-K^+-ATPase but could be neighboring proteins in the cell surface or even enzymes needed to synthesize glyco-proteins or lipids. Dominant mutations of regulatory genes are well known in proteins or lipids. Dominant mutations of regulatory genes are well known in bacteria; in the lactose operon of *E. coli*, for example, suitably altered repressor protein [the product of the *lac*I gene] can no longer be removed from the operator by inducer and the genes are not expressed [GILBERT and MUELLER-HILL, 1970]). (b) Alternatively, *Ka/ka* might be a structural gene coding directly for a membrane component – again, in principle, either a subunit of the Na^+-K^+-ATPase or a neighboring molecule in the membrane. The dominance of *Ka* could arise if the membrane component were synthesized in excess and if, in a heterozygote, the abnormal form (*Ka*) competed more effectively than the normal form (*ka*) for insertion into the membrane. Even without such competition, dominance would result if several *Ka/ka* molecules were required per pump site, and if the presence of one or more abnormal molecules led to abnormal function. (For example, suppose that each pump site contained two *Ka/ka* molecules. In a heterozygote, if *Ka* and *ka* were made in equal amounts and if they were incorporated into the membrane at random, then one quarter of the pump sites would contain two *ka* molecules and would be normal, while the remaining pump sites would contain either one or two *Ka* molecules and would be abnormal. If each site contained four molecules, only one-sixteenth of the pump sites would be normal [4 *ka*], and the rest would be abnormal.) One

obvious consequence of a multiple-subunit model of this kind is that, in cells from a heterozygote, the pump sites would be heteregeneous in composition, and might also be expected to display some heterogeneity in physiological function. Specifically, some "mixed" pump sites (those containing more *ka* molecules than *Ka* molecules) might retain partial activity. Although published data indicate that red cells from heterozygous *Ka/ka* animals pump potassium only slightly faster than red cells from homozygous *Ka/ka* animals (DUNHAM and HOFFMAN, 1971), a careful analysis of the kinetic properties of the pump (for example, its sensitivity to intracellular potassium) and the time-courses and extent of ouabain binding (JOINER and LAUF, 1975) might prove worthwhile.

2) A second genetic clue comes from the fact that the LK defect is expressed only in mature red cells from adult animals. In genetically LK lambs, and in adult LK animals that have undergone massive hemorrhage so that large numbers of new cells are entering the circulation, the red cells contain a high concentration of potassium (TOSTESON and MOULTON, 1959; BLECKNER, 1961; BLUNT and EVANS, 1963 and 1965; LEE et al., 1966). Thus, at a particular point in the development of the red cells, the *Ka* allele brings about a change in both active and passive cation fluxes across the cell membrane. Furthermore, the change is at least partially reversible; treatment of LK cells with anti-L antiserum (raised in HK sheep against LK red cells) causes a dramatic increase in active K^+ influx and a simultaneous decrease in passive K^+ movements (ELLORY and TUCKER, 1969 and 1970; LAUF et al., 1970; ELLORY et al., 1972; LAUF, 1975). Both the delayed expression of the LK defect and its reversibility by antiserum serve to limit the number of possible models for the function of the *Ka/ka* gene. In particular, the reversibility argues against a regulatory mutation that simply prevents the synthesis of a required molecule, and fits better with the idea of a structural-gene mutation leading to a defective product that can be "repaired" by interacting with antibody (see, for example, MELCHERS and MESSER, 1973). The delayed expression of the defect suggests either that the activity of the molecule in question decays slowly with time, or that the molecule is inserted into the red cell surface relatively late during development. Recently LODISH has shown that the rabbit reticulocyte – the immediate precursor of the red cell – synthesizes only two major and two or three minor membrane proteins; other membrane proteins are no longer made at this stage (LODISH, 1973; LODISH and SMALL, 1975). A similar study of sheep reticulocyte membrane proteins would be of interest, and might serve to identify the *Ka/ka* gene product.

Clearly much remains to be done, but at the present state of knowledge the HK/LK system illustrates both the usefulness and the limitations of a genetic approach to transport in whole organisms. On the one hand, the study of this particular genetic defect in cation transport has helped in the understanding of the normal processes of volume regulation (TOSTESON and HOFFMAN, 1960) and the dependence of transport upon intracellular cations (HOFFMAN and TOSTESON, 1971). But certain kinds of genetic information cannot be obtained: there is no way at present to determine whether *ka* is a single mutation or two or more closely linked mutations, nor can other mutants with related defects in cation transport be obtained.

D. Cultured Somatic Cells

Many of the obstacles that have hindered the genetic study of transport in higher organisms can potentially be overcome by the use of cultured cells. Clones can be isolated at will with stable, heritable alterations in transport properties, and large batches of homogeneous cells can readily be grown under controlled conditions for influx measurements and for the biochemical study of transport proteins. In several key respects, therefore, somatic cells share the experimental advantages of microorganisms.

There is one important genetic limitation, however. At present, no way is known to cross two somatic cells with different characteristics (say, one normal in a particular transport process and the other abnormal) and to observe the segregation of the characteristics in succeeding generations. Nor are there methods analogous to those used with bacteria for transferring DNA from one cell to another by conjugation, transduction, or transformation, measuring recombination frequencies, and constructing linkage maps. The lack of segregation and recombination data, which have classically played a crucial role in defining "genes" and "mutations," has in the past led to arguments concerning the nature of somatic cell variants. Are they true mutants, bearing heritable changes in the base sequence of their DNA? Or could they be the result of "epigenetic" events, which lead to a change in a differentiated trait without an underlying mutational change in DNA? Much of this debate has subsided during the past few years as particular somatic cell variants have been shown to possess many of the properties expected of true mutants (THOMPSON and BAKER, 1973):

1) Very often the variation is observed to be stable and, even in the absence of the conditions used to select it, to be transmitted through successive cell generations. For example, many cases have been described in which clones, selected for drug resistance and then grown for many months in medium lacking the drug, are still resistant when transferred into medium containing the drug.

2) The frequency with which the variation arises can be increased by chemical or physical treatments known to be mutagenic in microorganisms, and depends upon gene dosage. CHASIN has shown that enzyme-deficient variants arise less frequently in a diploid cell line (presumed to carry two functional copies of the gene in question) than in a heterozygote derived from that line (and presumed to carry only one functional gene copy) (CHASIN, 1974), and even less frequently in a tetraploid line (with four copies) (CHASIN, 1973).

3) The variation can often be traced to the loss of – or even more convincingly, the structural change of – a particular protein. This criterion has been met in an increasing number of cases, where heat-labile or immunologically cross-reacting proteins have been demonstrated and are assumed to have arisen by mutation in the corresponding structural gene (ALBRECHT et al., 1972; CHAN et al., 1972; BEAUDET et al., 1973; SHARP et al., 1973; THOMPSON et al., 1973; CHASIN et al., 1974; WAHL et al., 1975).

4) By forming a hybrid between variant cells from one species and normal cells from another species, and correlating the presence or absence of the variant characteristic with the cytological presence or absence of individual chro-

mosomes, it is often possible to map the variation to a single chromosome (RUDDLE, 1973).

None of these findings is completely unambiguous but, taken together, they serve as strong presumptive evidence that variation can arise from true gene mutation in somatic cells. The development of improved methods of genetic analysis, and at the same time the closer examination of apparently altered proteins by amino acid sequencing, will eventually provide a better test of this notion. In the meantime, the term "mutant" is widely used in somatic cell genetics, with the understanding that its applicability in any particular case must be judged by the criteria listed above, and that some instances of cell variation may prove to be nonmutational.

I. Methods for Selecting Transport Mutants

The range of techniques currently available for cell culture, mutagenesis, and mutant selection has been summarized by TOOZE (1973) and THOMPSON and BAKER (1973), and only a few points of special relevance to the selection of transport mutants will be considered here. As with microorganisms, the most successful approach has been to select cells whose growth is resistant to appropriate drugs or analogues. The list of transport-defective strains obtained in this way is growing steadily, and now includes:

ouabain-resistant cells, suspected (and in several cases demonstrated) to have an altered Na^+-K^+ transport system, and isolated in a wide variety of cell lines (Ehrlich ascites cells [MAYHEW, 1972], Chinese hamster ovary or CHO cells [TILL et al., 1973, BAKER et al., 1974, DAVIES and PARRY, 1974, PITRA and GRAM, 1975], mouse L cells [TILL et al., 1973, BAKER et al., 1974], human diploid fibroblasts [MANKOVITZ et al., 1974], rat myoblasts [LUZZATI, 1974] HeLa cells [ROSENBERG, 1975], and mouse and human lymphocytes [ADELBERG et al., 1975]);

cells resistant to 5-fluorotryptophan or excess phenylalanine and defective in amino acid transport (ENGLESBERG et al; 1976; TAUB and ENGLESBERG, 1976);

cells resistant to azaguanine and defective in purine transport (heteroploid human D98 cells [SZYBALSKI et al., 1962], CHO cells [HARRIS and WHITMORE, 1974]);

cells resistant to bromodeoxyuridine and defective in thymidine transport (haploid frog line ICR 2A [FREED and MEZGER-FREED, 1973]);

cells resistant to actinomycin D (mouse lymphocytes [KESSEL and BOSMANN, 1970; BOSMANN, 1971; BIEDLER and RIEHM, 1970]) or colchicine (CHO cells [TILL et al., 1973]), and believed to have more general changes in permeability.

Two other selection procedures exist which have been less widely used but should, in principle, yield transport mutants. In the "tritium suicide" method, cells are tested with a mutagen and then exposed for a brief period of time to a tritiated compound of extremely high specific activity. The cells are then rinsed, frozen, and stored, and at intervals are tested for viability. The expectation is that most of the cells will take up a significant amount of ^{3}H-labeled compound during the exposure period and will be killed by radioactive decay, while transport-defective mutants will survive. This method has been used successfully in microorganisms and also twice in cultured mammalian cell lines (Chinese hamster fibroblasts for the isolation of thymidine transport mutants; BRESLOW and GOLDSBY, 1969; mouse lymphocytes for the isolation of amino acid transport mutants; FINKELSTEIN et al., 1977).

Alternatively, and again by extrapolation from microorganisms, it should be possible to use a "negative selection" method to look for mutants that are unable to grow at low concentrations of a particular ion, amino acid, or other required substance. Such cells might have undergone a change in the affinity of a transport system for the substance in question, or they might have lost a high-affinity transport system while retaining a low-affinity system. Negative selection methods are described by THOMPSON and BAKER (1973); so far they have been used to select auxotrophic mutants of cultured somatic cells but not transport mutants.

Unfortunately, none of the selection procedures described above is completely specific for transport mutants. Cells can become resistant to drugs or analogs at later steps in metabolism; and in both the "tritium suicide" and "negative selection" methods, any cells which are not growing (or are growing slowly) during the critical step in the procedure may survive, even though they have normal transport systems. It is therefore necessary to test presumptive mutants directly to see whether they have lost (or undergone a change in) the transport process being studied. For cells that grow in monolayer cultures, flux measurements are usually made on cells attached to small coverslips (FOSTER and PARDEE, 1969; ISSELBACHER, 1972) or to scintillation vials (VAUGHAN and COOK, 1972); the cells can then be exposed to radioactive substrate for the desired period of time, rinsed, and counted directly. Cells that grow in suspension culture are handled by the usual methods of centrifugation or filtration.

II. The Kinds of Genetic Information that can be Obtained

Once clear-cut transport mutants have been obtained, they can be studied genetically in a number of ways. Dominance or recessiveness can be determined by fusing mutant cells with cells carrying the normal allele; pairs of recessive mutants can be fused to test for complementation; and the mutant genes can be mapped in human-rodent hybrids (RUDDLE, 1973).

III. Ouabain-Resistant Mutants

The best-studied class of somatic cell transport mutants consists of strains whose growth is resistant to ouabain. Because of the highly specific action of ouabain and other cardiac glycosides on the Na^+-K^+ transport system of animal cells, there was reason to expect that ouabain-resistant mutants would carry alterations in this system, either reducing the ability of the Na^+-K^+-ATPase to bind drug or altering the response of the system to drug when bound; and therefore resistant strains have been actively sought in several laboratories.

Mutants of both rodent cells (CHO cells, mouse L cells, mouse lymphocytes, rat myoblasts) and human cells (HeLa, human lymphocytes, human diploid fibroblasts) have now been isolated (for references, see p. 251). Normal rodent cells are not very sensitive to ouabain, and characteristically require about 10^{-3}M ouabain to decrease the growth rate by 95 percent (ADELBERG et al., 1975) or the plating efficiency by a factor of 10^4 to 10^7 (TILL et al., 1973). Nevertheless, it has been possible to select clones that are significantly more resistant, growing normally at ouabain concentrations as high as 3×10^{-3}M (ADELBERG et al., 1975) and having the minimum toxic dose in plating experiments increased 3- to 50-fold (TILL et al., 1973). More dramatic effects are seen with human cells, which are normally inhibited by extremely low ouabain concentrations (on the order of 10^{-7}M). Here, resistant clones have been isolated which require a 10- to 180-fold increase in ouabain concentration to give equivalent inhibition (MANKOVITZ et al., 1974).

The argument that the ouabain-resistant cells are true mutants rests on the fact that they fulfill most of the criteria listed on page 250: (1) They breed true, remaining resistant to ouabain even when cultured in the absence of the drug for long periods of time (TILL et al., 1973; BAKER et al., 1974). (2) In wild-type populations of cells, ouabain resistance arises randomly at a very low frequency (in the order of 10^{-8} per cell per generation), as would be expected for a spontaneous mutational event (BAKER et al., 1974). The frequency is greatly increased by known mutagens such as ethyl methane sulfonate (EMS) (TILL et al., 1973; BAKER et al., 1974) and ultraviolet light (S. Molnar and M. Rauth, cited in BAKER et al., 1974). (3) Although the altered gene product has yet to be identified directly, it seems almost certain to be either a subunit of the Na^+-K^+ transport system or some neighboring molecule in the membrane. In the several cases that have been tested, ^{42}K uptake in whole cells and Na^+-K^+-ATPase activity in crude homogenates show an increased resistance to ouabain that parallels the increased resistance seen in measurements of growth rate and viability (TILL et al., 1973; BAKER et al., 1974; ADELBERG et al., 1975).

Further information has come from studies of dominance and recessiveness, which shed indirect light on the mechanisms by which resistance can arise. So far, in both rodent and human cell lines, ouabain resistance has usually appeared co-dominant: that is, when a resistant cell is fused with a normal, sensitive cell, and the resulting hybrid is tested over a range of ouabain concentrations, its growth shows an intermediate degree of resistance (TILL et al., 1973; BAKER et al., 1974). This is the response one would expect if the plasma membrane of the hybrid cell contains a mixed population of Na^+-K^+ transport sites, some resistant and others sensitive to ouabain. One exception to this general picture has been reported, however: a ouabain-resistant strain of mouse lymphocytes that appears completely recessive when hybridized with normal cells (ADELBERG et al., 1975). Further work will be required to determine whether this particular strain carries a different kind of mutational lesion from the ordinary co-dominant ouabain-resistant strains.

A long-range genetic goal will be to count, and eventually to map, the genes that can mutate to give ouabain resistance; and a related biochemical goal will be to identify the corresponding gene products. Along these lines, an obvious next step will be to purify the Na^+-K^+-ATPase from normal and mutant cells, solubilize it, and subject it to electrophoresis on polyacrylamide gels in order to separate the component polypeptides. Most preparations of this kind from animal tissues yield two subunits, a large polypeptide of molecular weight 89000–135000 and a smaller glycopeptide of molecular weight 35000–57000; and it is the larger one that binds ouabain and ATP and is phosphorylated during the reaction cycle (reviewed in GLYNN and KARLISH, 1975). Presumably in at least some cases of ouabain resistance, the large subunit will fail to bind ouabain *in vitro,* as a result of a mutation in the corresponding structural gene. But other kinds of mutants can also be imagined: those with alterations in neighboring membrane components rather than in the large subunit directly, and those in which binding still occurs but no longer inhibits pump activity. In all cases, careful physiological study should reveal the extent to which a molecular change leading to ouabain resistance also affects other kinetic properties of the Na^+-K^+ transport system. Such secondary effects seem likely in view of the known interactions between the binding of other ligands (K^+, Na^+, P_i, ATP, Mg^{++}) and the binding of glycoside (reviewed in GLYNN and KARLISH, 1975).

E. Conclusions

To the physiologist interested in complex transport processes, the power of the genetic approach lies in the fact that mutations can produce a wide spectrum of highly specific changes in membrane proteins and lipids. Wherever mutants can be identified and studied, the possibility exists to trace physiological abnormalities in transport back to the molecular level, and thereby to gather information about structure-function relationships in the membrane. The purpose of this chapter has been to show that although these are long-range goals, they are realizable ones, both in microorganisms and in higher organisms.

References

ADELBERG, E. A., CALLAHAN, T., SLAYMAN, C. W., HOFFMAN, J. F.: J. gen. Physiol. **66**; 17a (1975).
ADLER, J., EPSTEIN, W.: Proc. nat. Acad. Sci. (Wash.) **71**, 2895–2899 (1974).
ADLER, J., HAZELBAUER, G. L., DAHL, M. M.: J. Bacteriol. **115**, 824–847 (1973).
AKSAMIT, R., KOSHLAND, D. E., Jr: Biochem. biophys. Res. Commun. **48**, 1348–1353 (1972).
ALBRECHT, A. M., BIEDLER, J. L., HUTCHINSON, D. J.: Cancer Res. **32**, 1539–1546 (1972).
AMES, G. F.: Arch. Biochem. Biophys. **104**, 1–18 (1964).
AMES, G. F., LEVER, J.: Proc. nat. Acad. Sci. (Wash.) **66**, 1096–1103 (1970).
AMES, G. F., LEVER, J.: J. biol. Chem. **247**, 4309–4316 (1972).
ASATOOR, A. M., LACEY, B. W., LONDON, D. R., MILNE, M. D.: Clin. Sci. **23**, 285–304 (1962).
BACHMANN, B. J., LOW, K. B., TAYLOR, A. L.: Bacteriol. Rev. **40**, 116–167 (1976).

BAKER, R. M., BRUNETTE, D. M., MANKOWITZ, R., THOMPSON, L. H., WHITMORE, G. F., SIMINOVITCH, L., TILL, J. E.: Cell **1**, 9–21 (1974).
BEAUDET, A. L., ROUFA, D. J., CASKEY, C. T.: Proc. nat. Acad. Sci. (Wash) **70**, 320–324 (1973).
BECKER, F. F., GREEN, H.: Proc. Soc. exp. Biol. **99**, 694–696 (1958).
BENNETT, R. L., MALAMY, M. H.: Biochem. biophys. Res. Commun. **40**, 496–503 (1970).
BIEDLER, J. L., RIEHM, H.: Cancer Res. **30**, 1174–1184 (1970).
BLECHNER, J. N.: Amer. J. Physiol. **201**, 85–88 (1961).
BLUNT, M. H., EVANS, J. V. B.: Nature **200**, 1215–1216 (1963).
BLUNT, M. H., EVANS, J. V. B.: Amer. J. Physiol. **209**, 978–985 (1965).
BOOS, W.: Europ. J. Biochem. **10**, 66–73 (1969).
BOOS, W.: J. biol. Chem. **247**, 5414–5424 (1972).
BOOS, W.: In: Current Topics in Membranes and Transport, Vol. 5 (F. Bronner, A. Kleinzeller, Eds). New York: Academic Press 1974, pp. 51–136.
BOOS, W., GORDON, A. S., HALL, R. E., PRICE, H. D.: J. biol. Chem. **247**, 917–924 (1972).
BOOS, W., SARVAS, M. O.: Europ. J. Biochem. **13**, 526–533 (1970).
BOSMANN, H. B.: Nature **233**, 566–569 (1971).
BRESLOW, R. E., GOLDSBY, R. A.: Exp. Cell Res. **55**, 339–346 (1969).
BRODEHL, J., GELLISSEN, K., KOWALEWSKI, S.: Klin. Wschr. **45**, 38–40 (1967).
CHAN, V. L., WHITMORE, G. E., SIMINOVITCH, L.: Proc. nat. Acad. Sci. (Wash.) **69**, 3119–3123 (1972).
CHASIN, L. A.: J. Cell Physiol. **82**, 299–308 (1973).
CHASIN, L. A.: Cell **2**, 37–41 (1974).
CHASIN, L. A., FELDMAN, A., KONSTAM, M., URLAUB, G.: Proc. nat. Acad. Sci. (Wash.) **71**, 718–722 (1974).
COHEN, G. N., RICKENBERG, H. V.: C. R. Acad. Sci. **240**, 466 (1955).
COX, G. B., GIBSON, F.: Biochem. biophys. Acta **346**, 1–25 (1974).
DAVIES, P. J., PARRY, J.: Genet. Res. **24**, 311–314 (1974).
DENT, C. E., ROSE, G. A.: Quart. J. Med. **20**, 205–219 (1951).
DREYFUSS, J., MONTY, K. J.: J. biol. Chem. **238**, 1019–1024 (1963).
DUNHAM, P. B., HOFFMAN, J. F.: J. gen. Physiol. **58**, 94–116 (1971).
ELLORY, J. C., SACHS, J. R., DUNHAM, P. B., HOFFMAN, J. F.: Biomembranes **3**, 237–245 (1972).
ELLORY, J. C., TUCKER, E. M.: Nature **222**, 477–478 (1969).
ELLORY, J. C., TUCKER, E. M.: In: Permeability and Function of Biological Membranes (L. Bolis, A. Katchalsky, R. D. Keynes, W. R. Lowenstein, B. A. Pethica, Eds). Amsterdam: North-Holland 1970, pp. 120–127.
ENGLESBERG, E., BASS, R., HEISER, W.: Somatic Cell Genet. **2**, 411–428 (1976).
EPSTEIN, W., DAVIES, M.: J. Bacteriol. **101**, 836–843 (1970).
EPSTEIN, W., JEWETT, S., FOX, C. F.: J. Bacteriol. **104**, 793–797 (1970).
EPSTEIN, W., KIM, B. S.: J. Bacteriol. **108**, 639–644 (1971).
EVANS, J. V., KING, J. W. B.: Nature **176**, 171 (1955).
EVANS, J. V., KING, J. W. B., COHEN, B. L., HARRIS, H., WARREN, F. L.: Nature **178**, 849–850 (1956).
FINKELSTEIN, M. C., SLAYMAN, C. W., ADELBERG, E. A.: Proc. Nat. Acad. Sci. (Wash.) **74**, 4549–4551 (1977).
FLAGG, J. L., WILSON, T. H.: J. Membrane Biol. **31**, 233–255 (1977).
FOSTER, D. O., PARDEE, A. B.: J. biol. Chem. **244**, 2675–2681 (1969).
FOX, C. F., CARTER, J. R., KENNEDY, E. P.: Proc. nat. Acad. Sci. (Wash.) **57**, 698–705 (1967).
FOX, C. F., KENNEDY, E. P.: Proc. nat. Acad. Sci. **54**, 891–899 (1965).
FOX, M., THIER, S., ROSENBERG, L. E., KISER, W., SEGAL, S.: New Engl. J. Med. **270**, 556–561 (1964).
FREED, J. J., MEZGER-FREED, L.: J. Cell Physiol. **82**, 199–212 (1973).
GILBERT, W., MUELLER-HILL, B.: In: The Lactose Operon (J. R. Beckwith, D. Zipser, Eds). New York: Cold Spring Harbor Laboratory 1970, pp. 93–109.
GLYNN, I. M., KARLISH, S. J. D.: Ann. Rev. Physiol. **37**, 13–55 (1975).
GROTH, V., ROSENBERG, L. E.: J. clin. Invest. **51**, 2130–2142 (1972).
HALPERN, Y. S.: Ann. Rev. Genetics **8**, 103–133 (1974).
HAMILTON, W. A.: In: Advances in Microbial Physiology, Vol. 12 (A. H. ROSE, D. W. TEMPEST, Eds.) New York: Academic Press 1975, pp. 1–53.

HAROLD, F. M.: In: Current Tropics in Bioenergetics, Vol. 6 (D. R. SANADI, Ed.) New York: Academic Press 1977, pp. 83–145.
HAROLD, F. M., BAARDA, J. R., PAVLASOVA, E.: J. Bacteriol. **101**, 152–159 (1970).
HAROLD, F. M., PAPINEAU, D.: J. Membrane Biol. **8**, 45–62 (1972).
HARRIS, H., MITTWOCH, U., ROBSON, E. B., WARREN, F. L.: Ann. Human Genet. **19**, 195–208 (1955a).
HARRIS, H., MITTWOCH, U., ROBSON, E. B., WARREN, F. L.: Ann. Human Genet. **20**, 57–91 (1955b).
HARRIS, J. F., WHITMORE, G. F.: J. Cell Physiol. **83**, 43–51 (1974).
HAZELBAUER, G. L., ADLER, J.: Nature New Biol. **230**, 101–104 (1971).
HOFFMAN, P. G., TOSTESON, D. C.: J. gen. Physiol. **58**, 438–466 (1971).
HOGG, R. W.: J. Bacteriol. **105**, 604–608 (1971).
ISSELBACHER, K. J.: Proc. nat. Acad. Sci. (Wash.) **69**, 585–589 (1972).
JOINER, C. H., LAUF, P. K.: J. Membrane Biol. **21**, 99–112 (1975).
JONES, T. H. D., KENNEDY, E. P.: J. biol. Chem. **244**, 5981–5987 (1969).
KALCKAR, H. M.: Science **174**, 557–565 (1971).
KEKOMAEKI, M., VISAKORPI, J. K., PERHEENTUPA, J., SAXEN, L.: Acta paed. scand. **56**, 617–630 (1967).
KENNEDY, E. P.: In: The Lactose Operon (J. R. Beckwith, D. Zipser, Eds). New York: Cold Spring Harbor Laboratory 1970, pp. 49–82.
KENNEDY, E. P., RUMLEY, M. K., ARMSTRONG, J. B.: J. biol. Chem. **249**, 33–37 (1974).
KEPES, A.: In: Current Topics in Membranes and Transport Vol. 1 (F. Bronner, A. Kleinzeller, Eds). New York: Academic Press 1971, pp. 101–134.
KESSEL, D., BOSMANN, H. B.: Cancer Res. **30**, 2695–2701 (1970).
KUSTU, S. G., AMES, G. F.: J. biol. Chem. **249**, 6976–6983 (1974).
LAUF, P. K.: Biochem. biophys. Acta **415**, 173–229 (1975).
LAUF, P. K., RASMUSEN, B. A., HOFFMAN, P. G., DUNHAM, P. B., COOK, P., PARMELEE, M. L., TOSTESON, D. C.: J. Membrane Biol. **3**, 1–13 (1970).
LEE, P., WOO, A., TOSTESON, D. C.: J. gen. Physiol. **50**, 379–390 (1966).
LEYER, J. E.: J. biol. Chem. **247**, 4317–4326 (1972).
LIN, E. C. C.: Ann. Rev. Genetics **4**, 225–261 (1970).
LODISH, H. F.: Proc. nat. Acad. Sci. (Wash.) **70**, 1526–1530 (1973).
LODISH, H. F., SMALL, B.: J. Cell Biol. **65**, 51–64 (1975).
LOWENDORF, H. S., BAZINET, G. F., Jr., SLAYMAN, C. W.: Biochim. biophys. Acta **389**, 541–549 (1975).
LOWENDORF, H. S., SLAYMAN, C. W.: Biochim. biophys. Acta **413**, 95–103 (1975).
LUZZATI, D.: Biochimie **56**, 1567–1569 (1974).
MANKOVITZ, R. M., BUCHWALD, M., BAKER, R. M.: Cell **3**, 221–226 (1974).
MARZLUF, G. A.: J. Bacteriol. **102**, 716–721 (1970a).
MARZLUF, G. A.: Arch. Biochem. Biophys. **138**, 254–263 (1970b).
MAYHEW, E.: J. Cell Physiol. **79**, 441–452 (1972).
MCCARTHY, C. F., BORLAND, J. L., LYNCH, H. J., OWEN, E. E., TYOR, M. P.: J. clin. Invest. **43**, 1518–1524 (1964).
MELCHERS, F., MESSER, W.: Europ. J. Biochem. **35**, 380–385 (1973).
MILNE, M. D., ASATOOR, A. M., EDWARDS, K. D. G., LOUGHRIDGE, L. W.: Gut **2**, 323–337 (1961).
MITCHELL, P.: Symp. Soc. gen. Microbiol. **20**, 121–166 (1970).
NELSON, D. L., KENNEDY, E. P.: Proc. nat. Acad. Sci. (Wash.) **69**, 1091–1093 (1972).
OHTA, N., GALSWORTHY, P. R., PARDEE, A. B.: J. Bacteriol **105**, 1053–1062 (1971).
ORDAL, G. W., ADLER, J.: J. Bacteriol **117**, 509–516 (1974a).
ORDAL, G. W., ADLER, J.: J. Bacteriol. **117**, 517–526 (1974b).
OXENDER, D. L.: Ann. Rev. Biochem. **41**, 777–814 (1972).
OXENDER, D. L., QUAY, S. C.: In: Methods in Membrane Biology (E. D. Korn. Ed.). New York: Plenum 1975.
OYANAGI, K., MIURA, R., YAMANOUCHI, T.: J. Pediat. **77**, 259–266 (1970).
PARNES, J. R., BOOS, W.: J. biol. Chem. **248**, 4436–4445 (1973).
PERHEENTUPA, J., VISAKORPI, J. K.: Lancet 1965 2, 813–816.
PITRA, C., GRAM, I.: Studia biophys. **49**, 177–186 (1975).
RADFORD, A.: C. R. C. Handbook of Biochemistry and Molecular Biology, 3rd ed. (G. D. Fasman, Ed.). Cleveland, Ohio: Chemical Rubber Co. 1976, pp. 739–761.

RAMOS, S., SCHULDINER, S., KABACK, H. R.: Proc. Nat. Acad. Sci. (Wash.) **73**, 1892–1896 (1976).
RAMOS, S., KABACK, H. R.: Biochemistry **16**, 848–854 (1977).
RAMOS, S., KABACK, H. R.: Biochemistry **16**, 854–859 (1977).
REEVES, J. P., SCHECHTER, E., WEIL, R., KABACK, H. R.: Proc. nat. Acad. Sci. (Wash.) **70**, 2722–2726 (1973).
RICKENBERG, H. V., COHEN, G. N., BUTTIN, G., MONOD, J.: Ann. Inst. Pasteur **91**, 829–857 (1956).
ROBERTS, K. R., MARZLUF, G. A.: Arch. Biochem. Biophys. **142**, 651–659 (1971).
ROSENBERG, H., LaNAUZE, J. M.: Biochim. biophys. Acta **156**, 381–388 (1968).
ROSENBERG, H. M.: J. Cell Physiol. **85**, 135–142 (1975).
ROSENBERG, L. E.: Science **154**, 1341–1343 (1966).
ROSENBERG, L. E., DOWNING, S. J.: J. clin. Invest. **44**, 1382–1393 (1965).
ROSENBERG, L. E., DOWNING, S. J., DURANT, J. L., SEGAL, S.: J. clin. Invest. **45**, 365–371 (1966a).
ROSENBERG, L. E., DURANT, J. L., ALBRECHT, I.: Trans. Assoc. Amer. Phys. **79**, 284–296 (1966b).
ROSENBERG, L. E., DURANT, J. L., HOLLAND, J. M.: New Engl. J. Med. **273**, 1239–1245 (1965).
ROSENBERG, L. E., SCRIVER, C. R.: In: Duncan's Diseases of Metabolism, 7th ed. (P. K. Bondy, L. E. Rosenberg, Eds). Philadelphia: Saunders 1974, pp. 465–654.
RUDDLE, F. H.: Nature **242**, 165–169 (1973).
RUDNICK, G., KABACK, H. R., WEIL, R.: J. biol. Chem. **250**, 1371–1375 (1975).
SANDERSON, K. E., HARTMAN, P. E.: Bacteriol. Rev. (in press).
SCHULDINER, S., KERWAR, G. K., KABACK, H. R.: J. biol. Chem. **250**, 1361–1370 (1975).
SHARP, J. D., CAPECCHI, N. E., CAPECCHI, M. R.: Proc. nat. Acad. Sci. (Wash.) **70**, 3145–3149 (1973).
SHERMAN, F., LAWRENCE, C. W.: In: Handbook of Genetics, Vol. 1 (R. C. King, Ed.). New York: Plenum 1974, pp. 359–393.
SILVERMAN, M., SIMON, M.: J. Bacteriol. **130**, 1317–1325 (1977).
SIMONI, R. D., POSTMA, P. W.: Ann. Rev. Biochem. **44**, 523–554 (1975).
SLAYMAN, C. W.: Biochim. biophys. Acta **211**, 502–512 (1970).
SLAYMAN, C. W.: In: Current Topics in Membranes and Transport, Vol. 4 (F. Bronner, A. Kleinzeller, Eds). New York: Academic Press 1973, pp. 1–174.
SLAYMAN, C. W., TATUM, E. L.: Biochim. biophys. Acta **109**, 184–193 (1965).
STANBURY, J. B., WYNGAARDEN, J. B., FREDERICKSON, D. S.: The Metabolic Basis of Inherited Disease, 3rd ed. New York: McGraw-Hill 1972.
SZYBALSKI, W., SZYBALSKA, E. H.: Univ. Mich. Med. Bull., pp. 277–293 (1962).
TAUB, M., ENGLESBERG, E.: Somatic Cell Genetics **2**, 441–452 (1976).
THERISOD, H., LETELLIER, L., WEIL, R., SCHECHTER, E.: Biochemistry **16**, 3772–3780 (1977).
THIER, S. O., FOX, M. S., SEGAL, S., ROSENBERG, L. E.: Science **143**, 482–484 (1964).
THIER, S. O., SEGAL, S., FOX, M., BLAIR, A., ROSENBERG, L. E.: J. clin. Invest. **44**, 442–448 (1965).
THOMPSON, L. H., BAKER, R. M.: In: Methods in Cell Biology, Vol. 6 (M. Prescott, Ed.). New York: Academic Press 1973, pp. 209–281.
THOMPSON, L. H., HAWKINS, J. L., STANNERS, C. P.: Proc. nat. Acad. Sci. (Wash.) **70**, 3094–3098 (1973).
TILL, J. E., BAKER, R. M., BRUNETTE, D. M., LING, V., THOMPSON, L. H., WRIGHT, J. A.: Fed. Proc. **32**, 29–33 (1973).
TOOZE, J. (Ed.): The Molecular Biology of Tumour Viruses. New York: Cold Spring Harbor Laboratory 1973.
TOSTESON, D. C., HOFFMAN, J. F.: J. gen. Physiol. **44**, 169–194 (1960).
TOSTESON, D. C., MOULTON, R. H.: Physiologist **2**, 116–117 (1959).
VAUGHAN, G. L., COOK, J. S.: Proc. nat. Acad. Sci. (Wash.) **69**, 2627–2631 (1972).
WAHL, G. M., HUGHES, S. M., CAPECCHI, M. R.: J. Cell Physiol. **85**, 307–320 (1975).
WEST, I. C., WILSON, T. H.: Biochem. biophys. Res. Commun. **50**, 551–558 (1973).
WHELAN, D. T., SCRIVER, C. R.: Pediat. Res. **2**, 523–532 (1968).
WILSON, T. H., KUSCH, M.: Biochem. biophys. Acta **255**, 786–797 (1972).
WILSON, T. H., KUSCH, M., KASHKET, E. R.: Biochem. biophys. Res. Commun. **40**, 1409–1414 (1970).
WONG, P. T. S., KASHKET, E. R., WILSON, T. H.: Proc. nat. Acad. Sci. (Wash.) **65**, 63–69 (1970).

Chapter 8

Mechanisms of Ion Transport and ATP Formation

E. RACKER

> To stay alive, you have to be able to hold out against equilibrium, maintain imbalance, bank against entropy, and you can only transact this business with membranes in our kind of world.
>
> LEWIS THOMAS

Among the many bewildering tasks which the first cell had to face in life were the retention of wanted ions such as K^+ and the exclusion of unwanted ions such as Na^+ and Ca^{++}. Since the aqueous environment of this planet has an abundance of unwanted and a limited supply of wanted ions, the cell was forced to open up an import-export business of major proportions.

Basically there are two classes of compounds that are translocated across membranes: those that are metabolically utilized, such as aminoacids, glucose and P_i, and those that are not, such as K^+. Membranes are therefore endowed with devices that allow the translocation of "substrates" down a concentration gradient. This can take place by facilitated diffusion or by a symport or antiport mechanism. In this class belong the translocations of ADP, ATP and P_i in the mitochondrial membrane, and of glucose in the plasma membrane; these processes can take place without a direct input of energy. In contrast, the task of establishing and maintaining concentration gradients of ions that are not consumed, such as K^+, Na^+ and Ca^{++}, does require energy. This can be supplied by light (in photosynthetic membranes), by oxidation (in mitochondrial membranes), or by metabolic processes such as glycolysis, either *via* a proton gradient or *via* ATP (in ion pumps).

In this chapter I shall restrict the discussion to processes of ion translocation that are linked directly or indirectly to the generation or dissipation of ATP.

A. Translocation of Protons by the Oxidation Chain of Mitochondria and Chloroplasts

I. The Chemiosmotic Mechanism of Mitchell

The basic scheme of the proton translocation processes in mitochondria and chloroplasts was formulated by MITCHELL (1966), as shown in Figure 1. The catalysts of the oxidation chain are organized asymmetrically in the membrane to permit the separation of charges and formation of a membrane potential in

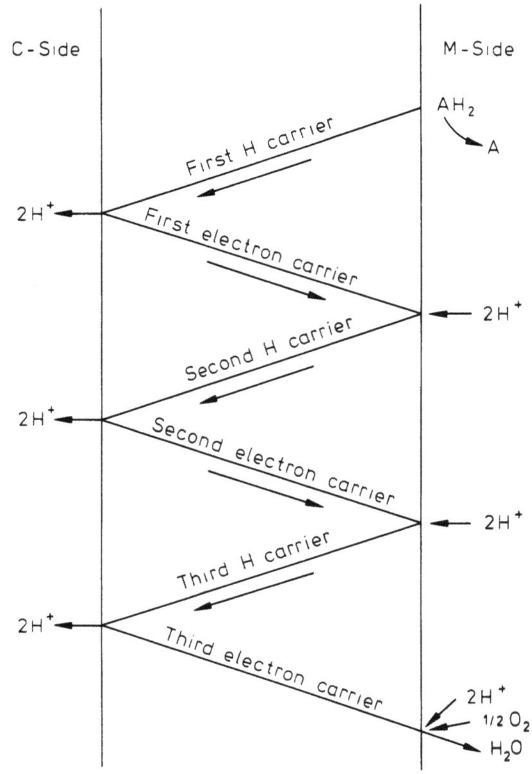

Fig. 1. Stoichiometry of proton translocation in the inner mitochondrial membrane

mitochondria. The hydrogens supplied by the substrates of the Krebs cycle are accepted by a hydrogen carrier on one side of the membrane (matrix side or M-side) and transported to the other side (cytochrome c side or C-side). Here protons are released while the electrons are shuttled back to the M-side. For each electron that arrives at the M-side, one proton is taken up from the water, and the hydrogen thus formed is again carried across the membrane by the second hydrogen carrier to the C-side of the membrane. This process of charge separation is repeated once more until, in the final loop, the electrons reach their final destination in cytochrome a_3 and interact with oxygen and protons to form water. In chloroplasts the hydrogens are supplied by water and the final electron sink is NADP. The $NADPH_2$ thus generated is the reductant of the carbon dioxide fixed in the photosynthetic pentose phosphate cycle.

For each $NADH_2$ that is oxidized in mitochondria six protons are translocated from the inside to the outside; for each succinate that is oxidized, four protons are translocated (MITCHELL, 1966). In submitochondrial particles prepared by sonication of mitochondria the same H^+/O ratios have been established (HINKLE and HORSTMAN, 1971), but here the protons move from the outside to the inside in accordance with the well-established fact that these particles are inside-out (LEE and ERNSTER, 1966; MITCHELL, 1966). The calculations of the H^+/O ratios were based only on the H^+ changes that were transmembranous and therefore abolished by the addition of uncouplers of oxidative phosphorylation which function as proton ionophores. In these experiments, the membrane potential, which restrains the proton flux, was collapsed by addition of valinomycin and K^+. Energy transfer inhibitors such as oligomycin did not inhibit the respiration-driven proton flux; in fact, in submitochondrial particles which lack respiratory control because the membrane is leaky to protons, respiration-linked proton movements can only be detected when oligomycin is added. How oligomycin interacts with the mitochondrial membrane will be discussed later.

II. Asymmetry of Oxidation Chain of Mitochondria and Chloroplasts

The asymmetry of the oxidation chain which is required for the separation of charges was established experimentally (cf. RACKER, 1970). Antibodies against individual oxidoreduction catalysts of the mitochondrial membranes interact with the catalysts at either one or the other side of the membrane. In chloroplasts similar experiments were performed with antibodies against plastocyanin, cytochrome *f* and CF_1 (RACKER et al., 1972). A more quantitative approach was developed by allowing impermeant [^{35}S] diazobenzene sulfonate to interact with either the mitochondria or with inverted submitochondrial particles (SCHNEIDER et al., 1973). The individual catalysts were then purified from these preparations and their radioactivity compared to that of the same protein but treated with [^{35}S] diazobenzene sulfonate after solubilization. Information was thus obtained on the location of the catalysts as well as their relative immersion in the membrane. For example, the radioactivity of cytochrome *c* that was treated while attached to the membrane was about the same as that of cytochrome *c* treated in solution. On the other hand, the radioactivity of cytochrome oxidase exposed to diazobenzene sulfonate on either the C-side or the M-side was only about one-sixth that of the enzyme exposed after solubilization. It was therefore concluded that the enzyme is transmembranous with about two-thirds of the protein not accessible to the negatively charged reagent.

III. The Coenzyme Q Cycle

These experiments have led to tentative assignments of the assembly of the major electron transport components in the mitochondrial and the chloroplast membrane. The identification of the hydrogen carriers has however not as yet been achieved. An ingenious hypothesis was recently proposed by MITCHELL (1975, and pers. comm.) in which a Q cycle functions in the translocations of protons in the cytochrome bc_1 complex. As shown in Figure 2, according to this formulation, QH_2 crosses the membrane from the M-side to the C-side, Q cycles

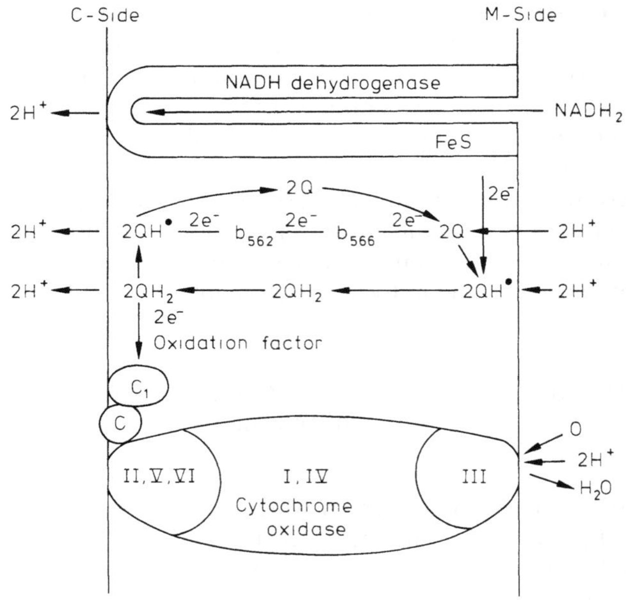

Fig. 2. Hypothetical pathway of protons and electrons in the mitochondrial oxidation chain

back to the M-side. One proton each is given off by FMNH, QH_2 and $QH^•$ on the C-side of the membrane for each electron passing from $NADH_2$ to oxygen; and one proton each is taken up by $QH^•$, Q and $^1/_2$ oxygen on the M-side of the membrane. The electrons are passed from $NADH_2$ *via* the non-heme iron centers to $QH^•$, cytochromes b_{552}, b_{556}, c_1, c, a, a_3 to oxygen. An attractive feature of this cycle is that it includes an electron-lend-lease mechanism. Q accepts an electron from cytochrome b_{566} on the M-side which is later returned by $QH^•$ on the C-side to cytochrome b_{562} and shuttled back across the membrane to cytochrome b_{566}. This mechanism, which increases the efficiency of the overall process, is quite similar to the oxaloacetate lend-lease mechanism operative in the Krebs cycle, which allows the entry of water into the carbon skeleton and more than doubles the number of hydrogens delivered to the respiratory chain during pyruvate oxidation. The assignments for the location of cytochrome b_{562} and b_{566} (WIKSTROM, 1973) and for the subunits of cytochrome oxidase (EYTAN et al., 1975) in Figure 2 are consistent with the available experimental evidence.

Although there is no direct experimental evidence for this hypothesis, it explains the observed stoichiometries of proton translocation and several other hitherto unexplained observations. Moreover, it has already stimulated experiments. One of them, performed recently (TRUMPOWER et al., 1976), showed that the oxidation factor which is required for the reduction of cytochrome c_1 by reduced cytochrome b (NISHIBAYASHI-YAMASHITA et al., 1972), catalyzes the reduction of cytochrome c_1 by QH_2 in line with the proposed Q cycle scheme. Several other experiments, particularly on the electron/proton ratio in this segment of the electron transport chain, should provide evidence in favor or against this mechanism. A similar cycle could be operative in photosynthetic membranes.

The transport of electrons down the thermodynamic ladder of electron acceptors to the final electron sink of oxygen is the driving force that results in the formation of the ΔpH and membrane potential. The return-flux of protons *via* the proton pump of the oligomycin-sensitive ATPase is responsible for the generation of ATP from ADP and P_i.

B. The Translocation of Protons by Bacteriorhodopsin

I. Proton Movements and ATP Formation

It was shown by OESTERHELT and STOECKENIUS (1973) that bacteriorhodopsin, which is localized in *H. halobium* in purple patches of the bacterial envelope, is responsible for the light-induced translocation of protons from the inside to the outside. During this process the intracellular content of ATP increases (DANON and STOECKENIUS, 1974). The sensitivity of the ATP rise to dicyclohexyl-carbodiimide suggests that this process is similar to oxidative phosphorylation and photophosphorylation, all involving an ATP-driven proton pump operating in reverse. The isolated bacteriorhodopsin preparation, consisting of a single polypeptide chain (75%) and phospholipids (25%), was incorporated into liposomes by the cholate dialysis procedure (RACKER and STOECKENIUS, 1974) or by the sonication procedure (RACKER, 1973). These vesicles catalyzed the light-induced translocation of protons from the outside to the inside. When the bacteriorhodopsin pump was reconstituted together with the mitochondrial oligomycin-sensitive ATPase, a light-dependent generation of ATP from ADP and P_i was observed. These experiments demonstrated that proton translocation catalyzed by bacteriorhodopsin, which does not involve oxidoreduction carriers, can substitute for the mitochondrial electron transport chain in the generation of ATP. Thus it seems reasonable to conclude that the major function of the respiratory chain is in fact the translocation of protons, as formulated in the chemiosmotic hypothesis.

II. Mechanism of Proton Translocation

The details of the mechanism of proton translocation by bacteriorhodopsin are still unknown. It appears likely, however, that a channel mechanism is operative, since under conditions that exclude the operation of a mobile carrier, the pump is still operative (RACKER and HINKLE, 1974). Bacteriorhodopsin preparations were incorporated into liposomes made of phospholipids with characteristic transition temperatures. It can be seen from Figure 3 that above the transi-

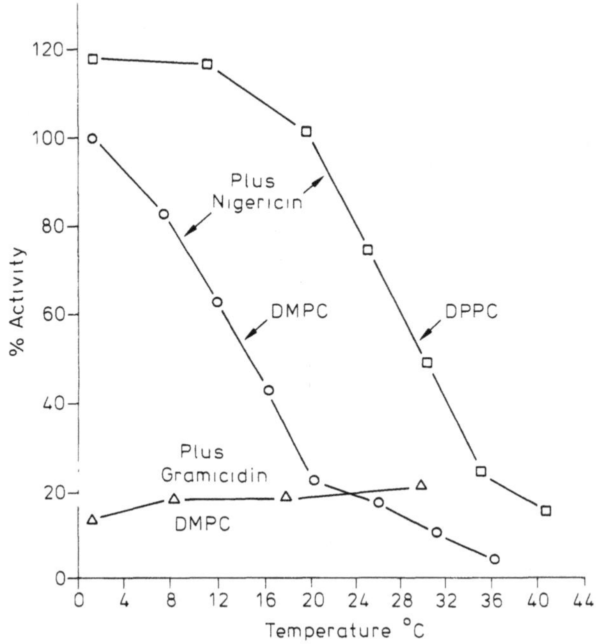

Fig. 3. Effect of temperature on the sensitivity of the bacteriorhodospin proton pump to nigericin and gramicidin. Pump activities of bacteriorhodopsin vesicles with and without nigericin (or gramicidin) tested at various temperatures. Values obtained with nigericin present expressed as % activity of values obtained without nigericin. DMPC – dimyristoyl phosphatidylcholine; DPPC – dipalmitoyl phosphatidylcholine

tion temperature (e. g. 23° C for dimyristoyl phosphatidylcholine) the bacteriorhodopsin pump could not maintain a pH gradient in the presence of a mobile ionophore such as nigericin, which exchanges protons against K^+. When the temperature of the suspension was lowered, pump activity returned, until at about 2° C the total activity (as compared to the control without nigericin) was regained. At this temperature the membrane was frozen and the mobile ionophore had no effect. On the other hand, gramicidin, which functions by a channel mechanism (Krasne et al., 1971), collapsed the proton gradient at all temperatures tested. Dipalmitoyl phosphatidylcholine, which has a transition temperature of about 42° C yielded bacteriorhodopsin liposomes that allowed proton translocation in the presence of nigericin at temperatures that were 20° C higher than those permissive in dimyristoyl phosphatidylcholine vesicles (Fig. 3). Since the transition temperature of dipalmitoyl phosphatidylcholine is above room temperature the preparation of active liposomes required sonication of the phospholipids at 48° C. In the case of gramicidin it was essential to add the ionophore to the vesicles at temperatures at which the membrane was fluid. Once the channel was formed gramicidin was effective in collapsing the ΔpH at all temperatures.

The mode of the propagation of protons through the channel formed by bacteriorhodopsin remains to be elucidated. The light-induced sequences of spectral transformations (STOECKENIUS and LOZIER, 1974; LEWIS et al., 1974) *via* prelumirhodopsin, lumirhodopsin and metarhodopsin resemble in some respects those observed with mammalian rhodopsin. The key event is the light-induced release of a proton from the protonated Schiff base, formed between retinaldehyde and a lysine residue of the protein. How is the release of protons from the Schiff base related to the translocation of protons from one side of the membrane to the other? It is conceivable that the release of the proton from the Schiff base takes place at one side of the membrane and initiates a dominobrigade mechanism of proton movement from the other side of the membrane toward the Schiff base along the polypeptide chain that forms the channel. Such an unidirectional movement may be associated with differences in the pK values of the aminoacid residues that are aligned within the channel from one side of the membrane to the other (Fig. 4). Thus the aminoacid sequence of bacteriorhodopsin now under investigation in several laboratories may be of particular interest in view of these considerations. The well-known spectral changes that accompany the molecular transformations at the active site of mammalian as well as bacteriorhodopsin prior to the release of the proton should help to elucidate the electronic events that are set in motion when a photon interacts with the chromophore.

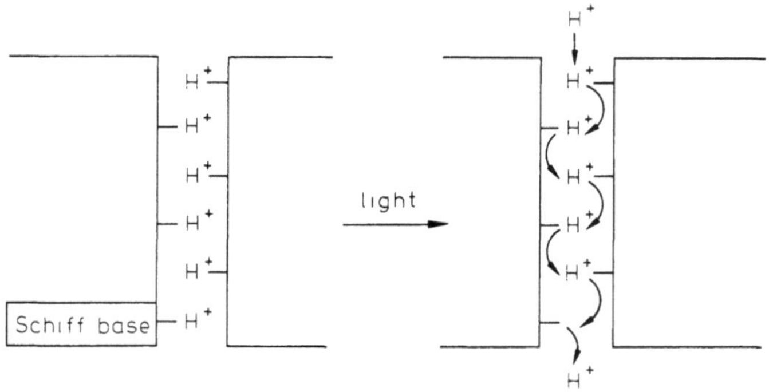

Fig. 4. Hypothetical domino brigade of proton translocation in bacteriorhodopsin pump

C. The Translocation of Protons by the Oligomycin- or Dicyclohexylcarbodiimide-Sensitive ATPase of Mitochondria, Chloroplasts and Bacteria

I. Proton Movements and ATP Formation

In pioneering work, MITCHELL and MOYLE (1968) demonstrated in rat liver mitochondria an ATP-induced proton flux from inside to outside. The experiments suggested that two protons are translocated for each ATP hydrolyzed and that oligomycin inhibits the ATP-dependent translocation of protons (MITCHELL, 1967). THAYER and HINKLE (1973) determined ATP induced proton fluxes in submitochondrial particles. Again H^+/ATP ratios approaching 2 were observed and the process was sensitive to uncouplers of oxidative phosphorylation and to energy transfer inhibitors. Because of the rather high basal ATPase activity exhibited by submitochondrial particles, the measurements of proton translocation were performed at pH 6.25. Thus changes in pH caused by the hydrolysis of ATP were avoided and the translocation of protons could be accurately measured without corrections for the acidification of the medium due to ATP hydrolysis. Since the number of protons translocated per ATP hydrolyzed is of considerable significance for thermodynamic calculations, it must be pointed out that the ratio of 2 is a minimum value. The efficiency of proton translocation is a function of the intactness of the membrane of the submitochondrial particles. The preparations used in the studies of THAYER and HINKLE (1973) catalyzed oxidative phosphorylation with suboptimal P:O ratios suggesting that some damage to the membrane had occurred during preparation rendering it somewhat permeable to protons. In chloroplasts the H^+/ATP ratio has been estimated to be 3 or 4 (RUMBERG and SCHROEDER, 1973; CARMELI et al., 1975; PORTIS and MCCARTY, 1976).

A significant contribution to the recognition of the proton pump as the site of ATP generation was the discovery by JAGENDORF and URIBE (1966) that an artifically imposed proton gradient in chloroplasts gives rise in the dark to the formation of ATP from ADP and P_i. THAYER and HINKLE (1975) showed that in submitochondrial particles the rate of ATP generation by a combination of an imposed ΔpH and membrane potential is more rapid than the rate of ATP generation during oxidation of NADH. This important contribution established the competence of the proton pump as a partner to the respiratory chain in the process of ATP generation linked to electron transport.

II. Properties of the Isolated Oligomycin-Sensitive ATPase Complex

The ATPase complex catalyzes the oligomycin-sensitive hydrolysis of ATP to ADP and P_i. It was isolated from mitochondria and was shown to contain vesicles with the characteristic 90 Å particles of F_1 (KAGAWA and RACKER, 1966). A highly purified ATPase complex was recently obtained in good yield from bovine heart mitochondria (SERRANO et al., 1976). A similar complex was isolated from yeast mitochondria (TZAGOLOFF and MEAGHER (1971) and from thermophilic bacteria (YOSHIDA et al., 1975, SONE et al., 1975). These isolated complexes contain the five subunits of F_1 and three to four additional polypeptide chains that can be seen in scans of acrylamide gels following electrophoresis in the presence of SDS. The best preparations of the mitochondrial complex thus far isolated and still functional in proton translocation (SERRANO et al., 1976) contain a band corresponding to a molecular weight of 30000, one band corresponding to 18000 and two or perhaps more bands in the region between 6000 and 10000. The ATPase complex from thermophilic bacteria contains F_1

(with 5 subunits) and also three to four other polypeptides (19000, 13500 and 6000 to 10000). The rutamycin-sensitive complex from yeast mitochondria contains in addition to the five subunits of F_1 a band corresponding to a molecular weight of 29000 and several additional faster-moving components (TZAGOLOFF and MEAGHER, 1971; TZAGOLOFF and AKAI, 1972). One of them (molecular weight of 7800) is soluble in organic solvents and stains poorly with Comassie blue (TZAGOLOFF et al., 1973). In the heart mitochondrial ATPase complex, in addition to F_1 and phospholipids the functional components required for the oligomycin-sensitive ATP hydrolysis are: OSCP with a molecular weight of 18000 (MacLENNAN and TZAGOLOFF, 1968), F_6 with a molecular weight of about 7000 (FESSENDEN-RADEN, 1972; KANNER et al., 1976) and F_o (KAGAWA and RACKER, 1966), a hydrophobic protein fraction with several polypeptides mentioned above. The most intriguing component of F_o is the proteolipid which interacts with energy transfer inhibitors (CATTELL et al., 1970). In SDS-acrylamide gel electrophoresis it migrates similar to F_6 corresponding to the low molecular region of 6000 to 10000 and is difficult to identify as a distinct band by the Comassie blue staining procedure. It appears to be similar to the hydrophobic component of yeast mitochondria (TZAGOLOFF et al., 1973).

The ATPase complex isolated from bovine heart mitochondria (KAGAWA *et al.*, 1973, SERRANO et al., 1976) and from thermophilic bacteria (YOSHIDA *et al.*, 1975; KAGAWA, pers. comm.) was incorporated into liposomes and shown to function as an ATP-dependent proton pump. The process was abolished by proton ionophores such as 1799 or FCCP (uncouplers of oxidative phosphorylation) as well as by energy transfer inhibitors such as oligomycin or DCCD.

We propose that the proteolipid functions as the proton channel of the ATP driven proton pump. Three arguments justify this proposal. The first is based on the observations that a water-soluble preparation of proteolipid isolated by chloroform-methanol extraction from bovine heart mitochondria according to the procedure of FOLCH-PI and STOFFYN (1972) functions as a proton ionophore (RACKER, 1975a). It collapses the proton gradient generated by bacteriorhodopsin liposomes and it stimulates respiration in reconstituted cytochrome oxidase vesicles with respiratory control. A proteolipid preparation isolated from spinach chloroplasts acted similarly. The second argument is based on the previously mentioned experiments that show that energy transfer inhibitors such as DCCD inhibit proton translocation in mitochondria and in the reconstituted proton pump. Together with the demonstrations (CATTELL et al., 1970; FILLINGAME, 1975) that the site of DCCD action in mitochondria and bacteria is a proteolipid, these considerations strongly point to this component as the key ingredient of the proton channel. The third and most convincing evidence is that a proteolipid isolated from chloroplasts serves in liposomes as a DCCD-sensitive proton translocator (N. NELSON, personal communication).

1. The Water-Soluble ATPase

The composition of the water-soluble ATPase from mitochondria, chloroplasts and bacteria is remarkably similar. There are two major subunits (α and β) with molecular weights between 50000 and 60000, an intermediate-size subunit with a molecular weight of about 35000 (γ), a subunit of about 17000 to 20000 (δ) and one of about 10000 (ε). Some of these numbers vary with the source of F_1. The chloroplast CF_1 subunits are in general somewhat larger than those of F_1. The chloroplast ε subunit (13000 mol wt) is a regulatory subunit which inhibits the ATPase activity (NELSON *et al.*, 1973). The ε subunit of F_1 is less clearly defined and some reported discrepancies in aminoacid composition (KNOWLES and PENEFSKY, 1972; BROOKS and SENIOR, 1972) need clarification. A complicating feature is the presence of an ATPase inhibitor which has a similar molecular weight and is difficult to remove quantitatively from F_1 preparations. In yeast F_1 the δ subunit appears to be much larger than in F_1 from other sources (TZAGOLOFF and MEAGHER, 1971).

Estimates of the number of copies of each subunit in F_1 based on Coomassie blue staining (Senior 1973) suggested three copies of the two larger subunits and one copy of each of the three smaller ones. Such a stoichiometry invariably gives rise to calculations of a molecular weight which is too large when compared to the weight of the protein determined by physical methods. SENIOR (1975) later obtained data that were incompatible with his earlier proposal and suggested that there are two copies of the largest subunit and two copies for the δ subunit. These new suggestions were based on analyses of the location of sulfhydryl groups and of the disulfide bonds. An alternative calculation is based on radioactivity measurements of the polypeptide chains of an ATPase from *E. coli* or *S. typhimurium* grown in the presence of uniformly C-14 labeled protein hydrolysates (BRAGG and HOU, 1975). Once more a composition of α_3, β_3, γ, δ, ε was proposed from these determinations. The same conclusion was drawn from similar experiments with thermophilic bacteria (KAGAWA, pers. comm.). Although these determinations are much more convincing than those based on the intensity of staining with dyes, there are still considerations which suggest that we should use caution in interpretation. The calculated values of the molecular weight of the ATPase based on the summation of the subunits are still higher than those obtained from direct physical measurements of the protein. Moreover, the fit based on the assumption of three copies of each of the two larger subunits is dependent on the accuracy of the molecular weight estimates of the subunits as determined by acrylamide gel electrophoresis. If they are underestimated, *e.g.* by 20 percent, the α_2, β_2 formula also gives a reasonable fit. Actually, the estimations of the sizes of the various subunits of F_1 in the literature vary more than 20 percent (cf. SENIOR, 1973). There is an increasing awareness of the inaccuracies of molecular weight determinations of polypeptide chains based on SDS acrylamide gel electrophoresis.

Although it would be of considerable interest to know the number of polypeptide chains for each subunit, for the construction of a convincing model of F_1, we obviously need more experimental data before arriving at a firm conclusion. It is well established, however, that the catalytic center for ATP hydrolysis resides in the two large subunits. The minor subunits can be removed from CF_1 by digestion with trypsin. The digested protein which contains only α and β subunits has full ATPase activity, but does not bind to the chloroplast membrane and does not serve as a coupling factor (DETERS et al., 1975). The regulatory ε subunit which inhibits the ATPase activity of CF_1 does not inhibit the trypsin-treated enzyme. Very similar observations have been recorded with the ATPase from *E. coli* (SMITH et al., 1975). NBD-chloride which interacts specifically with the β subunit of CF_1 and F_1 (DETERS et al., 1975; FERGUSON et al., 1975) blocks the ATPase activity of the native as well as of the trypsin-treated CF_1. NBD-chloride inhibits the ATPase activity of F_1 by interacting with one tyrosyl residue of the β subunit. Parallel inactivations of ATPase activity and membrane associated energy-linked reactions (*e.g.* enhancement of ANS fluorescence) by NBD-chloride have been recorded (FERGUSON et al., 1975). These experiments point to the involvement of the β subunit as part of the active center of the enzyme. On the other hand, there are experiments pointing to a participation of other subunits also. Antibodies against the γ subunit strongly inhibit photophosphorylation of chloroplasts; a combination of anti α and anti γ antibody inhibits the ATPase activity of CF_1 (DETERS et al., 1975). McCARTY and FAGAN (1974) observed that N-ethylmaleimide specifically interacts in the light with the γ subunit of CF_1 and inhibits photophosphorylation. Although these experiments point to an influence of the γ subunit on the active center, it must be remembered that trypsin-treated CF_1 which contains no γ subunit is fully active as an ATPase.

Experiments on fluorescence energy transfer were used to construct a model of the active center of the ATPase of CF_1 (CANTLEY and HAMMES, 1975). Fluorescent analogs of ATP, *e.g.* 1,N^6-ethenoadenosine imidodiphosphate, were used as donors and NBD-chloride bound to a tryrosine group as acceptors of energy transfer. The distance between the two sites was about 40 Å, with the tight ADP sites located tentatively on the two α subunits and the catalytic sites on the two β subunits.

2. The Oligomycin-Sensitivity Conferral Protein (OSCP)

This protein was discovered as a component of a crude alkaline extract of mitochondria (F_4). It was shown that alkali-treated submitochondrial particles required this factor for oligomycin-sensitive ATP hydrolysis (KAGAWA and RACKER, 1966). It was isolated from F_4 or from submitochondrial particles in pure form by MacLENNAN and TZAGOLOFF (1968) and named OSCP. Although we have accepted this name, we want to point out that it is somewhat misleading since, as will be shown below, more than one protein is required for the conferral of oligomycin-sensitive ATP hydrolysis. Furthermore, OSCP does not interact with oligomycin (BULOS and RACKER, 1968; KNOWLES et al., 1971). OSCP is also a component of the isolated ATPase complex from yeast (TZAGOLOFF and MEAGHER, 1971). The protein has a molecular weight of 18000, with a tendency to form aggregates. It is required for the binding of F_1 to the membrane (MacLENNAN and TZAGOLOFF, 1968; KNOWLES et al., 1971) and is not inhibited by SH-reagents.

It appears that OSCP may have a function in addition to the structural role of binding F_1 to the membrane. We observed that OSCP (also referred to as F_{c1}) inhibited the ATPase activity of a membranous ATPase complex that was deficient in this coupling factor as well as in phospholipids (BULOS and RACKER,

1968). Addition of phospholipids to such a preparation restored the ATPase activity and rendered it sensitive to rutamycin or DCCD. These observations prompted us to suggest that energy transfer inhibitors may act indirectly via the controlling function of OSCP, perhaps by interfering with the interaction between the protein complex and the phospholipid as formulated in the speculative scheme shown in Figure 5.

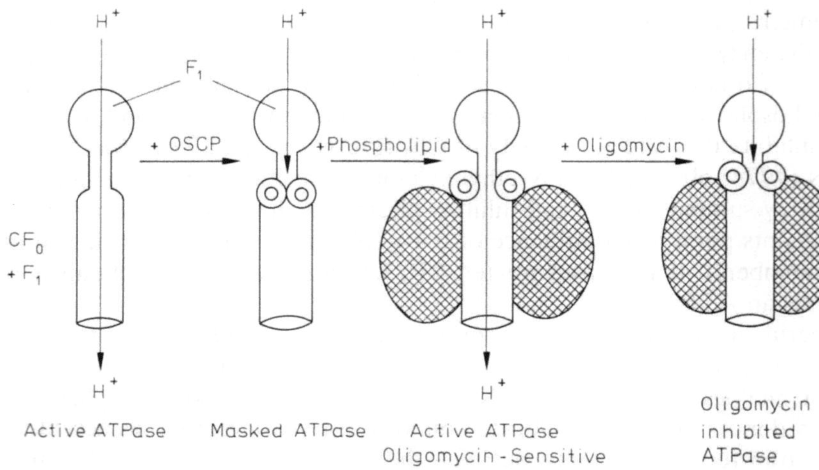

Fig. 5. Mechanism of action of energy transfer inhibitors

3. The Heat-Stable Coupling Factor F_6 (F_{c2})

F_6 is a second protein which is required in particles that had been treated with silicotungstate or with isothianate for the conferral of oligomycin sensitivity of ATP hydrolysis and for the $^{32}P_i$-ATP exchange (Racker et al., 1969; Knowles et al., 1971; Fessenden-Raden, 1972). It was recently isolated in homogeneous form and determined in acrylamide gel electrophoresis in the presence of dodecylsulfate and urea to have a molecular weigth of 7000 (Kanner et al., 1976). Together with OSCP it participates in the tight binding of F_1 to the membrane (Knowles et al., 1971). An indication for a second function in addition to this structural role came from observations of the sensitivity of this factor to trypsin. Whereas conferral of sensitivity of ATP hydrolysis to energy transfer inhibitors was not affected by the exposure of the protein to trypsin, the coupling factor activity as measured by the stimulation of the $^{32}P_i$-ATP exchange in deficient particles was sensitive to trypsin.

4. Coupling Factor 2 (F_2)

This protein stimulates the $^{32}P_i$-ATP exchange in silicotungstate treated submitochondrial particles in the presence of excess F_6 (Racker et al., 1969). It is identical with factor B (SANADI et al., 1968; RACKER et al., 1970). An antibody against factor B inhibits the stimulation of the exchange by F_2. One of the most revealing properties of F_2 is its high sensitivity to sulfhydryl reagents (RACKER and HORSTMAN, 1967; SANADI et al., 1968). The fact that F_2 has never been shown to be required for the oligomycin-sensitivity of ATP hydrolysis and that in several assays oligomycin substitutes for either F_2 or factor B suggests the possibility that this factor may act by decreasing the proton permeability of the membrane during oxidative phosphorylation.

III. Model of the Proton Pump and its Mode of Action

Because of the many uncertainties with respect to the structure and assembly of F_1 and CF_1 as well as of the other coupling factors, any representation of the assembly of the proton pump must be regarded as tentative. The formulation shown in Figure 6 is justified only because it focuses attention on certain features that stimulate the design of further experiments. We have assembled in this scheme of the proton pump: a) F_1, with its five subunits representing the major energy transformation center; b) F_o, composed of proteolipids which it is proposed serve as the transmembranous proton channel; and c) the coupling factors OSCP and F_6. There are currently three hypotheses about the way a proton pump (as shown in Fig. 6) operates to generate ATP from ADP and P_i, these serve for the design of future experiments (Fig. 7).

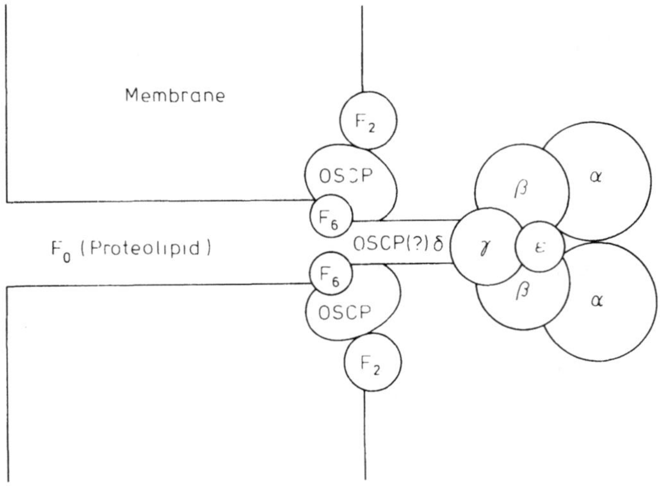

Fig. 6. Hypothetical scheme of the ATP-driven proton pump of mitochondria

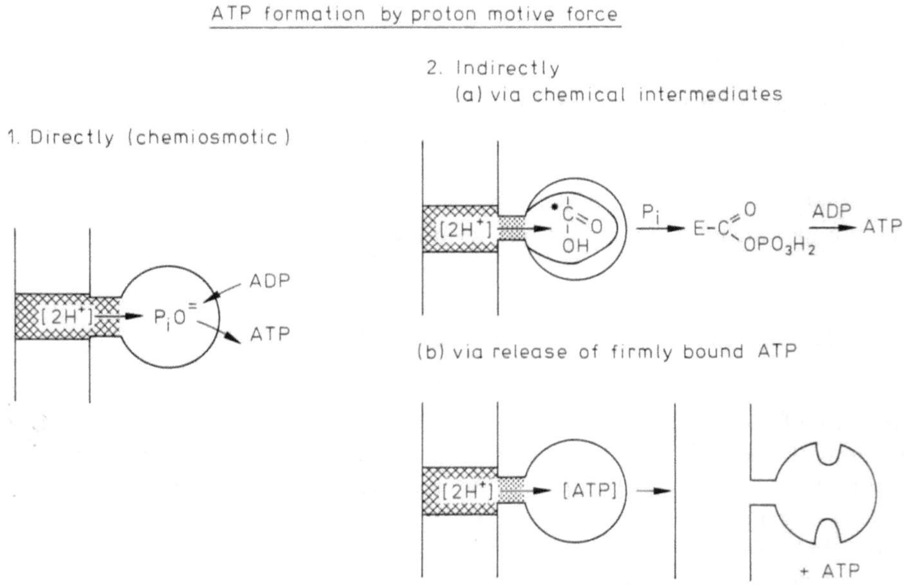

Fig. 7. Three hypotheses on the mechanism of ATP formation by the mitochondrial ATPase complex. Asterisk in 2a indicates activation of the carboxyl group by conformational change of the protein. In 2b the conformational change results in release of bound ATP

1. The Mitchell Hypothesis (Fig. 7-1)

This mechanism invokes a proton flux *via* F_0 and the *direct* interaction of a $P_iO^=$ with protons at the active site followed by an attack by ADP to form ATP (MITCHELL, 1974). The function of the membrane potential is to raise the concentration of H^+ at the M-side of the membrane where F_1 is attached. This formulation has some interesting consequences. It requires that the active site in F_1 is directly exposed to the pH gradient. As already noted, at least part of the active site is in the β subunit, which is located in the 90 Å headpiece of F_1 (Fig. 6). Therefore the connecting stalk, which we believe is not an artifact of staining (RACKER, 1976), should be part of the proton channel. If OSCP or F_6 or both make up the stalk, or if the stalk is an extended subunit of F_1, one might look for proton ionophore activities in these components.

BOGUSLAVSKY et al. (1975) have demonstrated a surface potential with F_1 placed between on octane-water interface. This surface potential was dependent on the presence of a hydrophobic proton acceptor (*e. g.* 2,4 dinitrophenol) in the octane phase. These experiments support the key feature of Mitchell's hypothesis, namely, a direct interaction between the proton flux and inorganic phosphate at the active site of F_1. In any case Mitchell's hypothesis has already proved its value because it has stimulated these interesting experiments.

2. The Phosphoenzyme Intermediate Hypothesis (Fig. 7-2a)

In the case of the mitochondrial proton pump this formulation has no experimental support. It is, however, the most probable mechanism for the Ca^{++} and Na^{+} pumps and will be discussed in detail later. It is proposed here merely because of my faith in the similarity of all ATP-driven ion pumps, but this faith will go no further than its usefulness for the design of experiments. Thus far, all our attempts to demonstrate a phosphoenzyme intermediate with F_1 or CF_1 have failed. However, our experiences with the Ca^{++} and Na^{+}-K^{+}-ATPase have furnished us with excellent excuses for accepting these negative experiments without getting discouraged. According to the phosphoenzyme mechanism the sequence of events should be a) a conformational change in F_1 caused by the ΔpH and/or membrane potential, resulting in an activation of a carboxyl group in an aspartate residue of the protein; b) entry of P_i and formation of an acyl enzyme (which has been demonstrated only in the case of the Ca^{++} and Na^{+}-K^{+}-ATPase) and c) transphosphorylation to ADP to form ATP according to the well-established mechanism of substrate-linked phosphorylation.

3. The Boyer-Slater Hypothesis*

The basic feature of this hypothesis is that ATP is firmly bound by F_1 or CF_1 and that the release of the nucleotide from the enzyme is a major anergy-requiring step which can be brought about by a conformational change in the protein (BOYER, 1974; SLATER, 1974). In a recent communication, BOYER (1975) proposes a specific mechanism by which a proton gradient or a membrane potential induces concormational changes in the enzyme that result in the release of ATP. The basic idea in this scheme includes the migration of a charged group of the protein from one side of the membrane to the other. He points out that a mobile subunit of the protein with several charged groups might account for the movement of two or more protons for each ATP that is released. It is difficult to see how a subunit of F_1 which is attached to the membrane by a stalk could move protons from one side of the membrane to the other. It may be advisable to accept the less specific original formulation until an identification of a moving subunit is possible. The Boyer-Slater hypothesis is attractive because it accounts for the curious fact that mitochondrial, chloroplast and bacterial ATPases contain several moles of firmly bound ATP per mole of enzyme. The terminal phosphoryl group of this bound ATP turns over when the membrane is energized, *e.g.* during illumination of chloroplasts (ROSING et al., 1975). However, the data published thus far do not present kinetic evidence that this turnover is rapid enough to qualify bound ATP as a potential intermediate in the process of photo- or oxidative phosphorylation.

* I should hope that no one will attempt to use an abbreviation of this hypothesis.

D. Translocation of Calcium by the ATPase Complex of Sarcoplasmic Reticulum

I. Properties of the Pump

The sarcoplasmic reticulum is an organelle in the muscle sarcoplasm designed specifically for the accumulation or release of Ca^{++} into and from its inner compartment, processes that are associated with the events of muscular relaxation and contraction. HASSELBACH (1972), MacLENNAN and HOLLAND (1975), and particularly MARTONOSI (1971) have written extensive reviews on the structure and function of the sarcoplasmic reticulum. The present review emphasizes those specific properties that seem to provide clues to the mechanism of Ca^{++} translocation proper.

Electron microscopy of sarcoplasmic reticulum vesicles shows vesicles bound by a single membrane. In negative stains the outer surface is covered by 40 Å particles (IKEMOTO et al., 1968); in freeze-fractured specimens 90 Å particles embedded in the membrane are seen (DEAMER and BASKIN, 1969). Treatment with trypsin removed the 40 Å particles when about 80 percent of the ATPase activity was lost (STEWART and MacLENNAN, 1974), but left the 90 Å particles within the membrane intact. On the basis of these observations, these authors suggest that the transport ATPase consists of a hydrophobic globular protein embedded in the membrane and a hydrophilic extension seen as 40 Å particles in negative stains. Since there are more 40 Å particles than 90 Å globules, JILKA et al. (1975) suggest that several 40 Å particles assemble to connect with each hydrophobic globule in the membrane. There are, however, still uncertainties associated with the globular particles seen in freeze-fractures and their composition is completely unknown. Even size estimates vary considerably (75 to 100 Å). If these globules are dimers or oligomers of the 100000 molecular weight protein, as might be expected from their size, some of the apparent discrepancies in particle distribution disappear. It was pointed out (MacLENNAN and HOLLAND, 1975) that the reported asymmetry in the location of the globules (favoring the external leaflet of the phospholipid bilayer) argues against a transmembranous location of the ATPase. This raises the question, to be discussed later, of how the translocation of ions across the membrane is accomplished. Further information on the morphological assembly of the Ca^{++} pump is therefore very much needed for the evaluation of its mechanism of action.

The protein composition of the isolated sarcoplasmic reticulum vesicles is remarkably simple, consistent with their primary function of removing and releasing Ca^{++}. This simplicity has been a major feature of attraction to biochemists and biophysicists, who have chosen this system for the study of a model membrane. The major protein constituent of sarcoplasmic reticulum is a Mg^{++}- and Ca^{++}-dependent ATPase. In acrylamide gel electrphoresis in sodium dodecylsulfate it is seen as the most predominant band, corresponding to a molecular weight of 100000. In addition there is a proteolipid and several acidic proteins which can bind large amounts of Ca^{++} – some with low, some with high affinity (MacLENNAN and HOLLAND, 1975).

Sarcoplasmic reticulum vesicles isolated from homogenized skeletal muscle by differential centrifugation catalyze a rapid ATP-dependent uptake of Ca^{++}. In the reverse direction the release of Ca^{++} from sarcoplasmic reticulum vesicles is associated with the formation of ATP from ADP and P_i (MAKINOSE and HASSELBACH, 1971; PANET and SELINGER, 1972).

The efficiency of pump operation can be estimated by measuring ATP hydrolysis during Ca^{++} transport. The Ca^{++}/ATP ratio for isolated sarcoplasmic reticulum vesicles is 1.7 and the rate of Ca^{++} translocation is 1.4 μmol per mg protein per min (HASSELBACH, 1972) with some variations around these figures reported from other laboratories (cf. MARTONOSI, 1971). The extent of Ca^{++} uptake is small (0.1 to 0.2 μmol mg protein) unless Ca^{++} precipitating anions such as oxalate are added which precipitate the Ca^{++} within the vesicles and allow up to 8 μmol per mg protein of Ca^{++} to be taken up (MARTONOSI, 1971).

Investigations of the mechanism of pump action with isolated vesicles have been possible because ATP interacts from the outside. The most significant finding is the formation of a phosphoenzyme intermediate in the presence of Ca^{++} and ATP (YAMAMOTO and TONOMURA, 1967 and 1968). Removal of phospholipids eliminates the formation of phosphoenzyme, which is restored by addition of small amounts of phospholipids (KNOWLES and RACKER, 1976), whereas partial removal eliminates ATPase activity before phosphoenzyme formation is completely blocked (MARTONOSI et al., 1971). The phosphoryl group of the phosphoenzyme has been identified as an aspartyl phosphate (DEGANI and BOYER, 1973). Thus far it has only been possible to identify the phosphoryl group after inactivation of the enzyme by denaturation or proteolysis. Attempts to trap the acyl group during action, *e.g.* by hydroxylamine have not been successful. Less direct evidence for a phosphoenzyme intermediate is obtained from studies of the very rapid Ca^{++}- and Mg^{++}-dependent ADP-ATP exchange which is sensitive to low concentrations of mercurials (HASSELBACH and MAKINOSE, 1962).

Of particular interest are observations which show that in the presence of Mg^{++} an acyl-phosphoenzyme intermediate can be formed from P_i even after severe damage of vesicular structures (MASUDA and de MEIS, 1973; KANAZAWA, 1975). These findings, which are similar to those obtained with the Na^+-K^+-ATPase, are consistent with formulations involving two forms of phosphoenzyme intermediates, $E_1{\sim}P$ formed from ATP, and $E_2{\sim}P$ formed from P_i.

Several schemes have been proposed for the molecular events that take place during Ca^{++} translocation in sarcoplasmic reticulum vesicles. Since some of the experiments with reconstituted vesicles have a bearing on these models, I shall discuss them later.

II. Properties of the Ca^{++}-ATPase Complex

1. Latency of the Ca^{++}-ATPase

While the latency of the Na^+-K^+-ATPase is widely recognized (JØRGENSEN, 1975), the latency of the Ca^{++}-ATPase is widely ignored. No one seems to be surprised that the Ca^{++}-ATPase can be purified from sarcoplasmic reticulum about 10-fold, from a specific activity of 3 to one of 30 (MacLENNAN, 1970), yet recent estimates propose (cf. MacLENNAN and HOLLAND, 1975) that the enzyme represents 60 to 70 percent of the total protein present in the vesicles! Accessibility to ATP, perhaps caused by inversion of vesicles or masking of the ATP site during the preparation, may be partly responsible, since 2- to 3-fold activation by deoxycholate without removal of protein is seen (MacLENNAN, 1970). On the other hand, the specific activity of the purified ATPase complex is far greater than could be expected from the rate of ATP hydrolysis during maximal Ca^{++} transport. We therefore postulate that the native enzyme in the membrane is masked either by another protein, as in the case of mitochondrial ATPase or by a phospholipid or by a conformational change of the protein which limits the accessibility of water. This problem, which is relevant to the efficiency of Ca^{++} pumping, needs more generous attention. Recent experiments suggest that Ca^{++} pumped into the vesicles inhibits the ATPase activity and accounts for the low activity of sarcoplasmic reticulum vesicles.

2. Structural Properties of the Ca^{++}-ATPase Complex

In the purified ATPase which contains phospholipids, the same 90 Å globular particles are seen in freeze-fractured preparations as in the sarcoplasmic reticulum vesicles (STEWART and MacLENNAN, 1974). After sonication of the ATPase the 40 Å particles are visible in negative stains. Trypsin treatment eliminates the 40 Å particles as well as ATPase activity. These experiments provide rather convincing evidence that these particles belong to the ATPase complex.

The purified preparation consists of essentially two protein components as revealed by gel electrophoresis in sodium dodecylsulfate. About 90 percent of the protein stain is attached to the 100000 molecular weigth polypeptide chain which has been identified with the ATPase by marking with [γ^{32}] ATP (cf. MacLENNAN and HOLLAND, 1975). A second protein component was identified as a proteolipid (MacLENNAN et al., 1972). It is soluble in chloroform-methanol (2:1) and precipitated by ether. After thin-layer chromatography and fractionation on Sephadex LH-20, the protein contained no phospholipids but one fatty acid per 6000 molecular weight. An estimation of a molecular weight of 6000 was also obtained by acrylamide gel analysis. Based on aminoacid analysis the minimum weight was however calculated to be 12000 and the discrepancy with the electrophoresis pattern was thought to be caused by the fatty acid component of the proteolipid. In view of the possible role of the proteolipid in Ca^{++} translocation (RACKER and EYTAN, 1975) the molecular weight of the proteolipid and its stoichiometric relation to the ATPase molecule becomes rather critical for the formulation of the operation of the pump. We consider it premature to assign a molecular weight to the proteolipid at present, and further investigations of its homogeneity and ability to form aggregates are needed.

3. Catalytic Properties of the Ca^{++}-ATPase Complex

The purified enzyme catalyzes a Mg^{++}- and Ca^{++}-dependent hydrolysis of ATP, an ADP-ATP exchange, and it becomes phosphorylated on addition of ATP (MacLennan, 1970). It also catalyzes a $^{32}P_i$-ATP exchange (Racker, 1972) as well as a net synthesis of ATP from P_i and ADP (Knowles and Racker, 1975a). The hydrolysis of ATP requires about 50 μM Ca^{++} and is strongly inhibited in the presence of excess Ca^{++}. This feature may have a bearing on the mechanism of formation of ATP from ADP and P_i.

The proposed steps in the hydrolysis of ATP involve $E_1{\sim}P$ formation (ADP-sensitive) which is converted to $E_2{\sim}P$ (ADP-insensitive). The latter can also be formed from inorganic phosphate in the presence of Mg^{++} (Kanazawa, 1975; Masuda and de Meis, 1973; Knowles and Racker, 1975a). The phosphoenzyme is rapidly hydrolyzed in the presence of very low concentrations of Ca^{++}.

A purified preparation of the Ca^{++} ATPase can be converted into $E_2{\sim}P$ in the presence of Mg^{++} and P_i provided no Ca^{++} is present (Knowles and Racker, 1975a). Then in a second step both ADP and large amounts of Ca^{++} (10 mM) were added. ATP was formed in amounts almost stoichiometric to the phosphoenzyme. The high concentration of Ca^{++} may have helped to prevent hydrolysis of ATP once it was formed. The ATP was free in solution and available for glucose phosphorylation by hexokinase. Since ATP generation by the same protein can be repeated after reprecipitation of the enzyme and removal of Ca^{++}, it appears that there is no preformed high-energy compound in the enzyme (*e.g.* a thiolester) to account for the reaction. The only remaining alternative concerning the source of the energy for ATP formation is that the binding energy between the enzyme and the divalent cations is used to drive the dehydration reaction. Whether the binding energy of Mg^{++} as well as of Ca^{++} to the protein contribute substantially remains to be resolved by direct calorimetric measurements (now in progress in collaboration with Dr. Y. Kuriki and Dr. R. Biltonen).

III. The Reconstituted Pump and its Mechanism of Action

Reconstitution of an active Ca^{++} pump has been achieved with a highly purified preparation of the Ca^{++} ATPase from sarcoplasmic reticulum and crude or pure phospholipids (Racker, 1972; Warren et al., 1974; Knowles and Racker, 1975b). The early reconstitution experiments yielded vesicles that pumped Ca^{++} with low efficiency. The Ca^{++}/ATP ratios, which serve as a measure of efficiency, were about 0.5 or lower (Racker and Eytan, 1973). Some of the ATPase preparations that were used had a low content of proteolipid. When an appropriate amount of proteolipid was added, the rate of Ca^{++} translocation was increased 3- to 5-fold and the Ca^{++}/ATP ratio was raised to 1.7 (RACKER and EYTAN, 1975). We therefore called the proteolipid a coupling factor in analogy to the coupling factors of mitochondria which raise the P/O ratio and which are required for the reconstitution of the proton pump (SERRANO et al., 1976).

The rate of Ca^{++} translocation in reconstituted vesicles exceeds 0.8 μmol min^{-1} (mg protein),$^{-1}$ which is of the same order of magnitude as the rate observed with the isolated sarcoplasmic reticulum vesicles (cf. HASSELBACH, 1972). However there is a very significant difference in the process of Ca^{++} translocation. The reconstituted vesicles require the presence of an intravesicular Ca^{++} precipitating anion whereas with sarcoplasmic reticulum vesicles oxalate or phosphate can be added externally (RACKER, 1972). This difference in permeability of the reconstituted vesicles permits the design of critical experiments that are pertinent to the mechanism of action of the ion pump.

Two basically different hypotheses of the mode of action of the Ca^{++} pump have been proposed. In the first, Ca^{++} is translocated together with P_i (cf. MARTONOSI, 1971). This is essentially a symport mechanism which proposes the simultaneous translocation of Ca^{++} and of the terminal phosphate of ATP. It requires the rotation of the active site from the outside to the inside. The second formulation is an antiport mechanism (TONOMURA, 1972), in which Mg^{++} ions are moved from the inside to the outside while Ca^{++} moves from the outside to the inside.

Since sarcoplasmic reticulum vesicles are rather permeable to phosphate and oxalate, it is difficult to test formulations which involve movements of such ions. In reconstituted systems the internal and external environment can be controlled more rigorously since the phospholipid bilayers are rather impermeable to ions. Thus it was possible to rule out the deposition of P_i inside the vesicles from external [$^{32}\gamma$] ATP (KNOWLES and RACKER, 1975b) thereby eliminating the first hypothesis. The second hypothesis is also considered unlikely since in the reconstituted system the translocation of Ca^{++} takes place in the absence of intravesicular Mg^{++} and requires only externally added Mg^{++}. In view of the low anion permeability of the vesicles the only ions that can move to insure electroneutrality are K^+ or Na^+ ions which are present inside the vesicles.

A third hypothesis of Ca^{++} translocation was based on observations with the proteolipid (RACKER, 1975b). This component of the pump had to be titrated very carefully in reconstitution experiments, because an excess inhibited Ca^{++} transport. This observation suggested that the proteolipid may act as an ionophore. This was found to be the case. Addition of the proteolipid to pure phospholipid liposomes preloaded with $^{45}Ca^{++}$ caused the release of radioactivity from the vesicles. The simplest explanation for these dual effects is that during the reconstitution procedure which involves prolonged dialysis, the proteolipid can be immobilized by the ATPase and acts as a channel, as illustrated in Figure 8. On the other hand, the proteolipid present in excess during reconstitution or simply added to liposomes is a free agent and serves as a mobile carrier thereby causing uncoupling of the energy-dependent ion transport. In line with

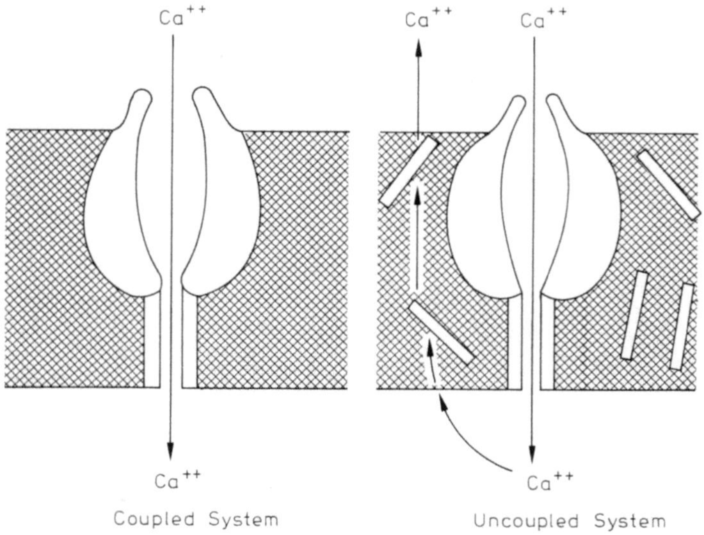

Fig. 8. Hypothetical model for the action of the proteolipid as a coupling factor and as an ionophore

this proposition is the observation that stimulation of ion transport by the proteolipid was not observed when rapid reconstitution methods (either sonication or cholate dilution) were used. The action of the proteolipid from sarcoplasmic reticulum is thus similar to that of the hydrophobic protein fraction in mitochondria. There it was also essential to avoid an excess of hydrophobic protein which caused uncoupling of oxidative phosphorylation in reconstituted vesicles (RACKER and KANDRACH, 1973).

What is the function of the ATPase and what is its relation to the proteolipid? An hypothesis for the function of the ATPase and the proteolipid is shown in Figure 9. The mechanism is based on the experiments on the formation of ATP from P_i and ADP (KNOWLES and RACKER, 1975a). The top line shows the formation of $E_2{\sim}P$ with the phosphoryl group in a hydrophobic pocket of the enzyme. Ca^{++} renders this pocket accessible either to water, resulting in hydrolysis, or to ADP, resulting in ATP formation. In the reverse reaction (bottom line), the conformational changes induced by Ca^{++} permit the phosphorylation of the aspartyl group by ATP. During the conformational changes that take place while P_i is released the Ca^{++} moves into the Ca^{++} channel provided by the proteolipid. Thus the proteolipid channel as outlined here has no bearing on the energy-dependent process. In effect, it only shortens the transmembranous distance. In view of the asymmetric location of the 90 Å particles within the membrane, it should now be easier to formulate the conformational changes of the protein required to translocate Ca^{++}. Of the many hypotheses of how conformational changes in the protein can facilitate ion movements the formulation of JARDETZKY (1966) is appealing, since it does not require rotation of the protein. The lateral and internal conformational changes shown in Figure 9 represent a variant of this model. It could be further adapted to accommodate

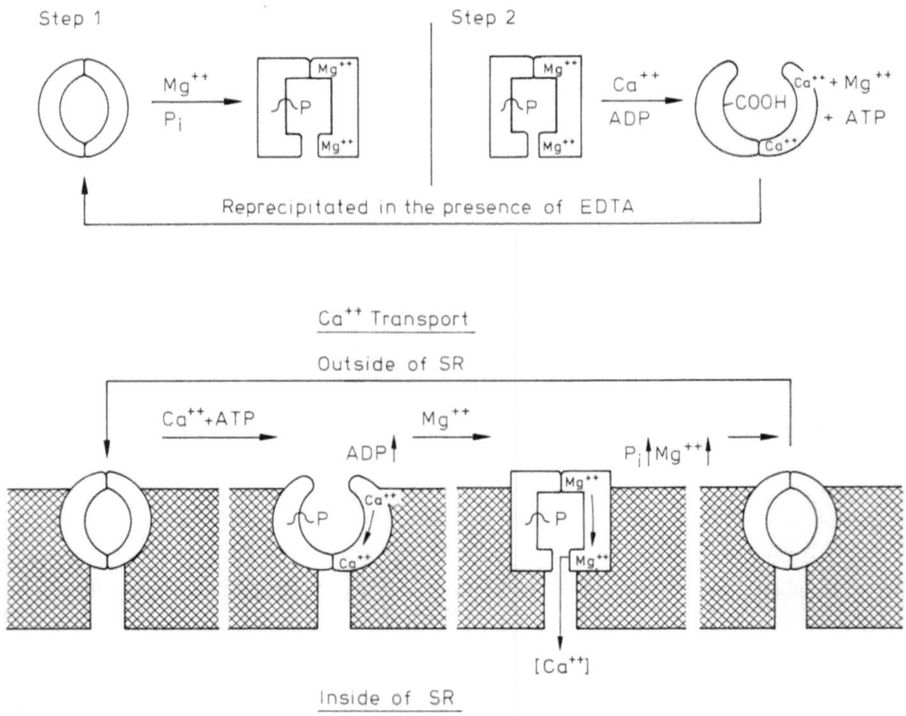

Fig. 9. Mechanism of ATP formation and ion translocation by Ca^{++}-ATPase complex

half-reaction sites as proposed for the Na^{+}-K^{+} pump (STEIN et al., 1973, REPKE et al., 1974, SIEGEL et al., 1974). Although the counter ion has not been established in the case of the Ca^{++} pump, it appears from reconstitution experiments that monovalent cations (either K^{+} or Na^{+}) inside the vesicles allow for the effective operation of the pump (KNOWLES and RACKER, 1975b).

The proteolipid discussed above should not be confused with the "ionophore" described by SHAMOO and MACLENNAN (1974). These authors have shown that a fragment of the ATPase obtained after tryptic digestion of the 100000 molecular weight protein acts as a Ca^{++}-dependent ionophore in black membranes of the Mueller-Rudin type. It is quite likely that this interesting fragment represents the Ca^{++} reactive site of the enzyme, perhaps functioning as the gate of the pump. However, there is no evidence that the intact protein moves as an ionophore across the membrane during ion transport. Nevertheless, the ease of entry of the hydrophobic peptide into the flat black membrane is an interesting phenomenon by itself.

E. Translocation of Sodium and Potassium Ions by the ATPase Complex of the Plasma Membrane

I. Properties of the Pump

The function of the Na^+-K^+ pump which is located in the plasma membrane of animal cells is to export the unwanted Na^+ ions and to import the wanted K^+ ions. This pump plays an important role in the excitability of nerves and muscle, it participates in the salt metabolism of the kidney and in the secretion of salt by specialized glands. The major characteristics of the system, which have been extensively reviewed (cf. SKOU, 1975; GLYNN and LEW, 1969; POST et al., 1969; HOKIN, 1969), are the requirement for ATP, which is regenerated inside the cells, and the sensitivity to ouabain (SCHATZMANN 1953), which interacts from the outside of the cell. This asymmetry of organization is illustrated in Figure 10. We shall refer to the two sides of the membrane as the ATP side (A-side) and the ouabain side (O-side) to avoid possible confusion when we discuss the reconstituted vesicles of the sodium pump which are "inside-out" with the A-side facing the outside and the O-side facing the inside. In both native and reconstituted pumps the A-side of the membrane has sites for interaction with Na^+, Mg^{++}, ATP, ADP and P_i, whereas the O-side contains the sites of interaction with K^+ and ouabain. A simpler notation such as Na^+-side and K^+-side is also avoided because there is ample evidence (SKOU, 1975) that there is competition on both sites between Na^+ and K^+. Moreover, this competition appears to play in important role in the regulation of the pump.

The ratios of $Na^+/K^+/ATP$ have been determined in red blood cells and is estimated to be 3/2/1 (POST et al., 1969). Similar ratios have been reported for nerve cells, though some variabilities have been encountered with changes in the internal Na^+ concentration (cf. SKOU, 1975). There are two aspects of these ratios which are of interest. One is the implied electrogenicity of the pump. Although the contribution of the Na^+ pump to the total electrical potential of the plasma membrane is probably too small to be of direct physiological importance, the electrogenicity of the pump may regulate the activity of the pump and may also be pertinent with respect to the mechanism of pump action. The second aspect of the ratio of interest is that it pertains to the efficiency of the pump.

We have observed that in several tumor cells the efficiency of the Na^+-K^+-ATPase is very low compared to that in red blood cells (SUOLINNA et al., 1975). Measurements of the efficiency had to be performed indirectly according to a procedure first used by WHITTAM and AGER (1965). Lactate production was taken as a measure of ATP hydrolysis (1 ADP required for each lactate formed); uptake of monovalent cations was measured with ^{86}Rb. This procedure tends to overestimate the efficiency of the pump when other ADP utilizing systems, *e.g.* mitochondria, compete with glycolysis. Thus the finding of the Rb^+/lactate ratio of below 0.6 indicates a very inefficiently performing pump.

What is the reason for this inefficiency in tumor cells? It appears that the ATPase activity is partly uncoupled with the active site of the enzyme accessible to water. The evidence for this is mainly based on observations on the effect of certain bioflavonoids, which at low concentrations inhibited the ATPase activity without interfering with the translocation of Rb^+. For example, quercetin inhibits lactate production in a variety of tumor cells and increases the Rb^+/lactate ratio from 0.5 to 2 or even higher (SUOLINNA et al., 1975). An alternative explanation, *e.g.* increased pump activity because of a general increase in the leakiness of the plasma membrane, is unlikely since no direct effect of quercetin on the permeability properties of the plasma membrane could be detected.

A more fundamental exploration of the exact lesion in the ion pump of tumor cells must wait for the elucidation of the control mechanism of pump action. A defect in the organization of the ATPase or the presence of an uncoupler of the Na^+-K^+ pump must be considered.

II. Properties of the Na^+-K^+-ATPase Complex

1. Latency of the ATPase

Extensive reviews have been written on the properties of the Na^+-K^+-ATPase (cf. SKOU, 1975; JØRGENSEN, 1975) and this section is restricted to the discussion of some of the properties that are relevant to other sections of this review.

The richest sources of the enzyme are the outer medulla of the rabbit kidney (JØRGENSEN and SKOU, 1971), the rectal gland of the dog fish (HOKIN et al., 1973) and the electric organ of the electric eel (ALBERS et al., 1963). Many crude preparations of the ATPase are "latent," requiring addition of a detergent such as deoxycholate for activation of enzyme activity. It is obvious from Figure 10 that ATP hydrolysis cannot take place with enzyme preparations in which the A-side is buried. It is likely that at least a large part of the latency is simply of such structural origin. However, in view of the existence of a specific ATPase inhibitor in mitochondria (PULLMAN and MONROY, 1963) and chloroplasts (NELSON et al., 1973), the possible participation of such a regulatory subunit in other ion pumps should be considered also. The great variability in the degree of latency between preparations of Na^+-K^+-ATPases from different sources is very reminiscent of the variability in latency of ATPase activity of mitochondria from different sources, which can often be related to the ease of dissociation of the ATPase inhibitor from F_1.

2. Structural Properties of the Enzyme Complex

Purified preparations of Na^+-K^+-ATPase, similar to the Ca^{++}-ATPase of sarcoplasmic reticulum, contain phospholipids and in negative stains show 40 Å particles arranged in clusters (HOKIN et al., 1973). Thin sections of the preparation reveal flat or cup-shaped pieces of membranes and no closed vesicles

(JØRGENSEN, 1975). This is in line with unsuccessful attempts in several laboratories to detect transport activity in these membrane fragments.

The protein has been highly purified from several sources (JØRGENSEN, 1974; HOKIN, 1974; KYTE 1971). In acrylamide gel electrophoresis these preparations reveal two major bands, one corresponding to a molecular weight of about 100000, the other a sialoglycoprotein with a molecular weight of about 55000. In our hands, highly purified preparations of ATPase from electric eel applied to the gel in quantities of 50 to 100 μg, show several minor bands, including one in the low molecular region where the Ca^{++} pump proteolipid is located.

The larger polypeptide is clearly associated with the ATPase activity. It carries the phosphoryl group of the phosphoenzyme, contains the SH groups of the active site as well as the ouabain reactive site (cf. JØRGENSEN, 1974). In contrast, the evidence for the participation of the smaller polypeptide is less convincing, although a physical association with the larger subunit is suggested by experiments with cross-linking agents (KYTE, 1972). On the other hand, there are reports from two laboratories on the isolation of ATPase complexes that lack the 55000 subunit (NAKAO et al., 1974; SMITH et al., 1974). Moreover, estimates of the molecular ratios of the two polypeptide chains vary considerably even when the enzyme from the same source, isolated by a different procedure, is analyzed (LINDENMAYER et al., 1974). These considerations and the fact that no catalytic activity has as yet been associated with the smaller polypeptide and that the sialic acid can be removed from it without loss of activity (HOKIN et al., 1974), suggest that a definite assignment of this polypeptide as a functional subunit of the ATPase is premature.

3. Catalytic Properties of the Enzyme Complex

The kinetics of the ATPase reactions are very complex. The currently most popular model, based mainly on the work of Albers and Post and their collaborators, involves a minimum of four steps: formation of an ATP-enzyme complex, formation of $E_1{\sim}P$, a phosphorylated intermediate capable of phosphoryl transfer to ADP, formation of $E_2{\sim}P$, a phosphorylated intermediate not reactive with ADP, and finally the hydrolysis of $E_2{\sim}P$ to P_i and free enzyme. Rapid kinetic measurements (TONOMURA and FUKUSHIMA, 1974) indicate a much greater complexity in the sequence of events, but the existence of the ADP-sensitive and -insensitive forms of phosphoenzyme remains well established.

This model is particularly useful in the analysis of the mode of action of inhibitors. I shall mention just a few examples. The inhibition of ATPase activity by NEM and oligomycin was localized at the step $E_1{\sim}P \rightarrow E_2{\sim}P$ because the inhibitors did not interfere with $E_1{\sim}P$ formation by ATP, and in fact increased the rate of the ADP-ATP exchange reation (FAHN et al., 1966). Very interesting data on the interaction of NEM with the enzyme were recorded by TITUS and HART (1974). About two SH groups in the native enzyme react with the inhibitor, and about four SH groups were alkylated at the $E_2{\sim}P$ state and about six at the $E_1{\sim}P$ stage. These data indicate conformational changes in the enzyme, which might be related to the process of ion translocation. As pointed out by the authors, it would be illuminating to perform similar experiments with the enzyme in the native membrane using reconstituted red blood cells and vesicles.

The inhibition of ATPase by ouabain and other cardiac glycosides is more complex. One of the most interesting effects is the marked increase of $E_2{\sim}P$ formation from P_i induced by this inhibitor (ALBERS et al., 1968). This finding as well as the optimal conditions for the binding of ^{3}H-digitoxin, namely either ATP, Na^+ and high Mg^{++}, or Mg^{++} and P_i (SCHWARTZ et al., 1974), suggest that the cardiac glycosides interact with the $E_2{\sim}P$ enzyme and stabilize it in that conformational state. This mode of action is rather different from that of the bioflavonoid quercetin, which inhibits the formation and hydrolysis of $E_2{\sim}P$, but does not interfere with the formation of $E_1{\sim}P$ (KURIKI and RACKER, 1976).

It seems appropriate here to point to some of the difficulties in the analysis of the isolated enzyme, and particularly its relevance for the function of the enzyme in its native conformation in the membrane of intact cells. As mentioned earlier, it is likely that, as in the case of the mitochondrial ATPase, the high catalytic rate of hydrolysis may well be an artifact, perhaps because of removal of a controlling "inhibitor," perhaps because of some structural disorientation. There is therefore not much to be gained by attempting quantitative correlations between the isolated enzyme and the intact cell. Moreover, the physiological intracellular Na^+ and the extracellular K^+ concentrations are far from optimal for ATP hydrolysis, as pointed out by SKOU (1974), and may well represent key control mechanisms in the operation of the pump. In the isolated preparation both the A-side and the O-side are exposed to the same ionic environment, and activations and inhibitions are difficult to separate kinetically. An important discussion comment by HOFFMAN (1974) is relevant in this connection. Interaction of NEM with the inside of red blood cells inhibited the binding of ouabain which was added from the outside, whereas NEM added to the outside had no effect. It seems that many of the kinetic experiments might be extended to studies of resealed ghosts (right-side in) and reconstituted vesicles (inside-out) to permit the spatial isolation of the two sides of the membrane and permutations of substrate and ion concentrations.

TANIGUCHI and POST (1974) were the first to report net ATP formation from P_i and ADP by preparations of Na^+-K^+-ATPase from guinea-pig kidney, and the reactions they describe in detail are in principle similar to those catalyzed by the Ca^{++}-ATPase discussed earlier. Like Ca^{++} in the Ca^{++}-ATPase, Na^+ interferes with $E_2{\sim}P$ accumulation from P_i and Mg^{++}. Very high concentrations of Na^+ are required for maximal ATP yield. By analogy with the Ca^{++}-ATPase, one might suggest that the high Na^+ concentration slows down the hydrolysis of ATP. The thermodynamic questions which were discussed in the section on the Ca^{++}-ATPase apply here as well.

III. The Reconstituted Pump

Na^+-K^+-ATPases from kidney, brain, the rectal gland of the dog fish and the electric eel were incorporated into liposomes (GOLDIN and TONG, 1974; HILDEN et al., 1974; SWEADNER and GOLDIN, 1975; RACKER and FISHER, 1975). The first

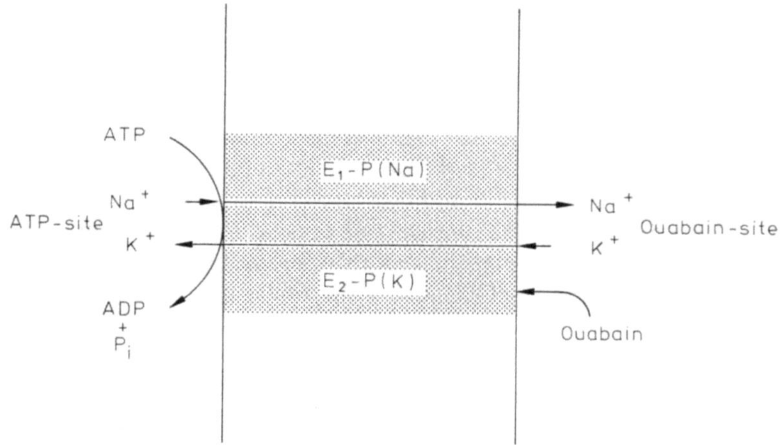

Fig. 10. Asymmetry of Na^+-K^+-ATPase complex assembly in the plasma membrane

three enzymes were reconstituted by the cholate dialysis procedure, the electric eel enzyme by the sonication procedure.

All reconstituted Na^+-K^+ pumps had to be tested with an inverted orientation of the ATPase in the phospholipid bilayer. This is obvious from Figure 10 since ATP cannot penetrate through the phospholipid bilayer without a special device. An ADP/ATP antiport translocator is present in mitochondria and a purified preparation of this carrier has been incorporated into liposomes (SHERTZER and RACKER, 1974 and 1976). Thus it should be feasible now to assay for the presence of a pump reconstituted in the physiological orientation. On the other hand, as pointed out earlier, the availability of inside-out pump vesicles offers a unique opportunity to evaluate the effect on the A-side of ions and inhibitors which cannot always be controlled without ambiguity in experiments with resealed red cell ghosts.

In our hands the sonication method of reconstitution (RACKER, 1973) is far superior to the cholate dialysis procedure (KAGAWA and RACKER, 1971) for the reconstitution of the Na^+-K^+ pump with ATPase from electric eel (RACKER and FISHER, 1975). The observed rates of Na^+ transport were 10 to 20 times more rapid than those reported in the literature (HILDEN et al., 1974) although the incorporation of the protein into the liposomes appears to be greater with the cholate dialysis procedure. The extent of incorporation is estimated from measurements of the ATPase activity in the presence and absence of ouabain. Since neither ouabain nor ATP is permeant, ouabain-sensitive ATPase activity measures the portion of enzyme that has not been incorporated into the liposomes. This conclusion was confirmed by the fact observed in the three different laboratories that Na^+ transport was insensitive to external, but sensitive to internal ouabain. This phenomenon allows us to establish the efficiency of pumping even when the extent of incorporation is very low. In the case of the reconstituted electric eel ATPase the Na^+/ATP ratios were very high, sometimes exceeding a ratio of 3 (RACKER and FISHER, 1975).

The eel ATPase has been reconstituted with phosphatidylethanolamine as the only added phospholipid. This is similar to observations made with the Ca^{++} pump, in which case it is possible to use phosphatidylethanolamine with a protein that was stripped of all endogenous phospholipids (RACKER et al., 1976). Such experiments have not as yet been conducted with the eel enzyme. A second interesting observation is the stimulation of the Na^{+} pump with the reconstituted eel enzyme by the addition of valinomycin under conditions of equimolar concentrations of K^{+} inside and outside the vesicles (RACKER and FISHER, 1975). This suggests that the reconstituted pump is electrogenic.

Finally, the reconstituted ATPase from the rectal gland catalyzes an ATP-dependent Na^{+}/Na^{+} exchange similar to that observed in intact cells (HILDEN et al., 1974).

It can be concluded from the various observations with the reconstituted Na^{+}-K^{+}-ATPase that these artificial inside-out vesicles which catalyze ATP-dependent Na^{+} translocation are useful models that can yield information which would be difficult to obtain with intact cells and which might stimulate new experimental approaches by the physiologist.

F. Concluding Remarks

The physiological functions of the three ATP-driven ion pumps which are discussed in this review differ from each other. The Na^{+} pump of the plasma membrane exchanges intracellular Na^{+} against external K^{+}; the Ca^{++} pump of the sarcoplasmic reticulum sequesters Ca^{++} ions that are not wanted in the sarcoplasma; the proton pumps of mitochondria and chloroplasts generate ATP by utilizing a proton flux emerging from the oxidation chain. They all catalyze a similar reversible reaction, [ions] $\rightleftharpoons$ ATP, and have a number of features in common. The most important one is that they have an ATPase operative as an energy transformer. This transformer is similar in molecular weight (2 to 3×10^5) but contains different numbers of moving parts. The simplest Ca^{++} pump probably contains 2 subunits of 100000 molecular weight. The Na^{+}-K^{+} pump contains the same number, perhaps together with one or two additional subunits of 55000 molecular weight. The proton pump contains four or six subunits of about 50000 molecular weight. The other components of the pumps are connecting parts which allow the flow of ions from one side of the membrane to the other *via* the active center of the transformer. In the proton pump, the transformer is located outside the membrane, connected by a stalk. In the Ca^{++} pump, most of the transformer is intramembranous but an important part containing the active center of the ATPase molecule is extramembranous, visible and accessible to trypsin. In the Na^{+} pump the situation appears similar, with perhaps an even greater portion submerged in the membrane.

In the case of the Na^{+} and Ca^{++} pumps phosphorylated intermediates, derivatives of an aspartyl residue of the ATPase, have been identified, but not in the proton pump. It should be possible in the near future to dissect and map the active center of the ATPase protein, which serves as the energy transformer.

Acknowledgement

This investigation was supported by Grants number Ca-08964 and Ca-14454 awarded by the National Cancer Institute, DHEW, Grant number BC-156 from the American Cancer Society and Grant number BMS-75-17887 from the National Science Foundation.

Abbreviations

1799 — bis-(hexafluoroacetonyl)acetone; FCCP — carbonylcyanide p-trifluoro-methoxyphenylhydrazone; CF_1 — coupling factor 1 from chloroplasts; F_1, F_2, F_6 — coupling factors 1 (ATPase), 2 and 6 respectively; OSCP — oligomycin sensitivity conferring protein; F_o — a membranous preparation from mitochondria conferring oligomycin (or rutamycin) sensitivity to F_1; C-side — the side of the inner mitochondrial membrane which faces the outer mitochondrial membrane and contains cytochrome c; M-side — the side of the inner mitochondrial membrane which faces the matrix; A-side — the side of the plasma membrane where ATP interacts; O-side — the side of the plasma membrane where ouabain interacts; Q — for coenyme Q or ubiquinone; DCCD — N,N'-dicyclohexylcarbodiimide; NBD-chloride — 7-chloro-4-nitrobenzo-2-oxa-1,3-diazole; ANS — 1-anilino-8-naphthalene sulfonate; $E_1{\sim}P$ — pump enzymes phosphorylated by ATP (ADP sensitive); $E_2{\sim}P$ — pump enzymes phosphorylated by P_i (ADP insensitive).

Addendum

Since this review was submitted in December 1975, the direct calorimetric measurements referred to in the text have been completed. We found (Kuriki, Y., Halsey, J., Biltonen, R., Racker, E.: *Biochemistry* **15**, 4956–4961 [1976]) that the interaction between the Na^+-K^+-ATPase and Mg^{2+} or inorganic phosphate results in major heat changes (20 to 50 Kcal mol^{-1} enzyme). Similar heat changes were observed with the Ca^{2+}-ATPase (manuscript in preparation). These findings lend support to the suggestion that the formation of the intermediate $E_2{\sim}P$ and of ATP is driven by the energy of binding between the enzyme and the inorganic ions.

References

Albers, R. W., Fahn, S., Koval, G. J.: Proc. nat. Acad. Sci. (Wash.) **50**, 474 (1963).

Albers, R. W., Koval, G. J., Siegel, G. J.: Mol. Pharmacol. **4**, 324 (1968).

Boguslavsky, L. I., Kondrashin, A. A., Kozlov, I. A., Metelsky, S. T., Skulachev, V. P., Volkov, A. G.: FEBS Letters, **50**, 223 (1975).

Boyer, P. D.: In: Dynamics of Energy-Transducing Membranes (Ernster, Estabrook, E. C. Slater, Eds). Amsterdam: Elsevier 1974, p. 289.

Boyer, P. D.: FEBS Letters, **58**, (1975).
Bragg, P. D., Hou, C.: Arch. Biochem. Biophys. **167**, 311 (1975).
Brooks, J. C., Senior, A. E.: Biochemistry **11**, 4675 (1972).
Bulos, B., Racker, E.: J. biol. Chem. **243**, 3891; 3901 (1968).
Cantley, L. C., Jr., Hammes, G. G.: Biochemistry **14**, 2976 (1975).
Carmeli, C., Lifshitz, Y., Gepshtein, A.: Biochim. biophys. Acta **376**, 249 (1975).
Cattell, K. J., Knight, I. G., Lindop, C. R., Beechey, R. B.: Biochem. J. **125**, 169 (1970).
Danon, A., Stoeckenius, W.: Proc. nat. Acad. Sci. (Wash.) **71**, 1234 (1974).
Deamer, D. W., Baskin, R. J.: J. Cell Biol. **42**, 296 (1969).
Degani, C., Boyer, P. D.: J. biol. Chem. **248**, 8222 (1973).
Deters, D. W., Racker, E., Nelson, N., Nelson, H.: J. biol. Chem. **250**, 1041 (1975).
Eytan, G., Carroll, R. C., Schatz, G., Racker, E.: J. biol. Chem. **250**, 8598 (1975).
Fahn, S., Hurley, M. R., Koval, G. J., Albers, R. W.: J. biol. Chem. **241**, 1890 (1966).
Ferguson, S. J., Lloyd, W. J., Lyons, M. H., Radda, G. K.: Europ. J. Biochem. **54**, 117 (1975).
Fessenden-Raden, J. M.: J. biol. Chem. **247**, 2351 (1972).
Fillingame, R. H.: J. Bacteriol. **124**, 870 (1975).
Folch-Pi, J., Stoffyn, P. J.: Ann. N. Y. Acad. Sci. **195**, 86 (1972).
Glynn, I. M., Lew, V. L.: In: Membrane Proteins (New York Heart Association Symposium). Boston, Mass.: Little, Brown 1969, p. 289.
Goldin, S. M., Tong, S. W.: J. biol. Chem. **249**, 5907 (1974).
Hasselbach, W.: In: Molecular Bioenergetics and Macromolecular Biochemistry. (H. H. Weber, Ed.). Berlin – Heidelberg – New York: Springer 1972, p. 150.
Hasselbach, W., Makinose, M.: Biochem. biophys. Res. Commun. **7**, 132 (1962).
Hilden, S., Rhee, H. M., Hokin, L. E.: J. biol. Chem. **249**, 7432 (1974).
Hinkle, P., Horstman, L. L.: J. biol. Chem. **246**, 6024 (1971).
Hoffman, J. F.: Ann. N. Y. Acad. Sci. **242**, 254 (1974).
Hokin, L. E.: In: Membrane Proteins (New York Heart Association Symposium), Boston, Mass.: Little, Brown 1969, p. 327.
Hokin, L. E., Dahl, J. L., Deupree, J. D., Dixon, J. F., Hackney, J. F., Perdue, F.: J. biol. Chem. **248**, 2593 (1973).
Hokin, L. E.: Ann. N. Y. Acad. Sci. **242**, 12 (1974).
Ikemoto, N., Sreter, F. A., Nakamura, A., Gergely, J.: J. Ultrastruct. Res. **23**, 216 (1968).
Jagendorf, A. T., Uribe, E.: Proc. nat. Acad. Sci. (Wash.) **55**, 170 (1966).
Jardetzky, O.: Nature **211**, 969 (1966).
Jilka, R. L., Martonosi, A. N., Tillack, T. W.: J. biol. Chem. **250**, 7511 (1975).
Jørgensen, P. L.: Ann. N. Y. Acad. Sci. **242**, 36 (1974).
Jørgensen, P. L.: Quart. Rev. Biophys. **7**, 239 (1975).
Jørgensen, P. L., Skou, J. C.: Biochim. biophys. Acta **233**, 366 (1971).
Kagawa, Y., Kandrach, A., Racker, E.: J. biol. Chem. **24S**, 676 (1973).
Kagawa, Y., Racker, E.: J. biol. Chem. **241**, 2461; 2467; 2475 (1966).
Kanazawa, T.: J. biol. Chem. **250**, 113 (1975).
Kanner, B. I., Serrano, R., Kandrach, A. M., Racker, E.: Biochem. biophys. Res. Commun. (1976) **69**, 1050.
Knowles, A. F., Guillory, R. J., Racker, E.: J. biol. Chem. **246**, 2672 (1971).
Knowles, A. F., Penefsky, H. S.: J. biol. Chem. **247**, 6617; 6624 (1972).
Knowles, A. F., Racker, E.: J. biol. Chem. **250**, 1949 (1975a).
Knowles, A. F., Racker, E.: J. biol. Chem. **250**, 3538 (1975b).
Knowles, A. F., Racker, E.: unpublished experiments.
Krasne, S., Eisenman, G., Szabo, G.: Science **174**, 412 (1971).
Kuriki, Y., Racker, E.: Biochem. **15**, 4951 (1976).
Kyte, J.: J. biol. Chem. **246**, 4157 (1971).
Kyte, J.: J. biol. Chem. **247**, 7634 (1972).
Lee, C. P., Ernster, L.: Biochim. biophys. Acta Library **7**, 218 (1966).
Lewis, A., Spoonhower, J., Bogomolni, R., Lozier, R. H., Stoeckenius, W.: Proc. nat. Acad. Sci. (Wash.) **71**, 4462 (1974).
Lindenmayer, G. E., Lane, L. K., Schwartz, A.: Ann. N. Y. Acad. Sci. **242**, 235 (1974).
MacLennan, D. H.: J. biol. Chem. **245**, 4508 (1970).

MacLennan, D. H., Holland, P. C.: Annual Reviews of Biophysics & Bioengineering, **4**, 377 (1975).
MacLennan, D. H., Tzagoloff, A.: Biochemistry **7**, 1603 (1968).
MacLennan, D. H., Yip, C. C., Iles, G. H., Seaman, P.: Cold Spring Harbor Symp. Quant. Biol. **37**, 469 (1972).
Makinose, M., Hasselbach, W.: FEBS Letters **12**, 271 (1971).
Martonosi, A.: Biomembranes **1**, 191 (1971).
Masuda, H., Meis, L.: Biochemistry **12**, 4581 (1973).
McCarty, R. E., Fagan, J.: Biochemistry **12**, 1503 (1973).
Mitchell, P.: Biol. Rev. Cambridge Phil. Soc. **41**, 445 (1966).
Mitchell, P.: Fed. Proc. **26**, 1370 (1967).
Mitchell, P.: FEBS Letters **43**, 189 (1974).
Mitchell, P.: FEBS Letters **56**, 1 (1975).
Mitchell, P., Moyle, J.: Europ. J. Biochem. **4**, 530 (1968).
Nakao, M., Nakao, T., Hara, Y., Nagai, F., Yagasaki, S., Koi, M., Nakagawa, A., Kawai, K.: Ann. N. Y. Acad. Sci. **242**, 24 (1974).
Nelson, N., Deters, D. W., Nelson, H., Racker, E.: J. biol. Chem. **248**, 2049 (1973).
Nishibayashi-Yamashita, H., Cunningham, C., Racker, E.: J. biol. Chem. **247**, 698 (1972).
Oesterhelt, D., Stoeckenius, W.: Proc. nat. Acad. Sci. (Wash.) **70**, 2853 (1973).
Panet, R., Selinger, Z.: Biochim. biophys. Acta **255**, 34 (1972).
Portis, A. R., McCarty, R. E.: J. biol. Chem. (1976) **251**, 1610.
Post, R. L., Kume, S., Tobin, T., Orcutt, B., Sen, A. K.: In: Membrane Proteins (New York Heart Association Symposium) Boston, Mass.: Little, Brown 1969, p. 306.
Post, R. L., Taniguichi, K., Toda, G.: Ann. N. Y. Acad. Sci. **242**, 80 (1974).
Pullman, M. E., Monroy, G. C.: J. biol. Chem. **238**, 3762 (1963).
Racker, E.: In: Essays in Biochemistry Vol. 6 (P. N. Campbell, F. Dickens, Eds). 1970, p. 1.
Racker, E.: J. biol. Chem. **247**, 8198 (1972).
Racker, E.: Biochem. biophys. Res. Commun. **55**, 224 (1973).
Racker, E.: In: Proc. Internatl. Sym. on Electron-Transfer Chains and Oxidative Phosphorylation (Quagliariello, et al., eds.). Fasano, Italy: North Holland Publishing Co., Amsterdam 1975, pp. 401–406.
Racker, E.: Biochem. Soc. Trans. **3**, 785 (1975b).
Racker, E.: A New Look at Mechanisms in Bioenergetics. New York: Academic Press 1976.
Racker, E., Eytan, E.: Biochem. biophys. Res. Commun. **55**, 174 (1973).
Racker, E., Eytan, E.: J. biol. Chem. **250**, 7533 (1975).
Racker, E., Fessenden-Raden, J. M., Kandrach, M. A., Lam, K. W., Sanadi, D. R.: Biochem. biophys. Res. Commun. **41**, 1474 (1970).
Racker, E., Fisher, L. W.: Biochem. biophys. Res. Commun. **67**, 1144 (1975).
Racker, E., Hauska, G. A., Lien, S., Berzborn, R. J., Nelson, N.: In: Second International Congress of Photosynthesis Research, Stresa, 1971. The Hague: Junk 1972, p. 1097.
Racker, E., Hinkle, P. C.: J. Membrane Biology **17**, 181 (1974).
Racker, E., Horstman, L. L.: J. biol. Chem. **242**, 2547 (1967).
Racker, E., Horstman, L. L., Kling, D., Fessenden-Raden, J. M.: J. biol. Chem. **244**, 6668 (1969).
Racker, E., Kandrach, A.: J. biol. Chem. **248**, 5841 (1973).
Racker, E., Knowles, A. F., Eytan, E.: Ann. N. Y. Acad. Sci. **264**, 17 (1975).
Racker, E., Stoeckenius, W.: J. biol. Chem. **249**, 662 (1974).
Repke, K. R. H., Schoen, R., Henke, W., Schoenfeld, W., Streckenbach, B., Dittrich, F.: Ann. N. Y. Acad. Sci. **242**, 203 (1974).
Rosing, J., Harris, D. A., Slater, e. C., Kemp, A., Jr.: J. Supramolec. Structure, **3**, 284 (1975).
Rumberg, B., Schroeder, H.: In: Proceedings 6th International Congress on Photobiology, Bochum, 1972 (O. Schenk, Ed.). 1973, Abstract 036.
Sanadi, D. R., Lam, K. W., Ramakrishna Kurup, C. K.: Proc. nat. Acad. Sci. (Wash.) **61**, 277 (1968).
Schatzmann, H. J.: Helv. physiol. pharmacol. Acta **11**, 346 (1953).
Schneider, D. L., Racker, E.: In: Oxidases and Related Redox System (Proceedings of the Second

International Symposium) (T. E. King, H. W. Mason and M. Morrison, Eds). Baltimore, Md: University Park Press 1973.
SCHWARTZ, A., LINDENMAYER, G. E., ALLEN, J. C., MCCANS, J. L.: Ann. N. Y. Acad. Sci. **242**, 577 (1974).
SENIOR, A. E.: Biochim. biophys. Acta **301**, 249 (1973).
SENIOR, A. E.: Biochemistry **14**, 660 (1975).
SERRANO, R., KANNER, B., RACKER, E.: J. biol. Chem. **251**, 2453 (1976).
SHAMOO, A. E., MACLENNAN, D. H.: Proc. nat. Acad. Sci. (Wash.) **71**, 3522 (1974).
SHERTZER, H. G., RACKER, E.: J. biol. Chem. **249**, 1320 (1974).
SHERTZER, H. G., RACKER, E.: J. biol. Chem. **251**, 2446 (1976).
SIEGEL, G. J., GOODWIN, B. B., HURLEY, M. J.: Ann. N. Y. Acad. Sci. **242**, 220 (1974).
SKOU, J. C.: Quart. Rev. Biophysics **7**, 401 (1975).
SLATER, E. C.: In: Dynamics of Energy-Transducing Membranes (Ernster, Estabrook, E. C. Slater, Eds). Amsterdam: Elsevier 1974, p. 1.
SMITH, J. B., STERNWEISS, P. C., HEPPEL, L. A.: J. Supramolec. Structure, **3**, 248 (1975).
SMITH, T. W., WAGNER, H., Jr., STROSBERG, A. D., YOUNG, M.: Ann. N. Y. Acad. Sci. **242**, 53 (1974).
SONE, N., YOSHIDA, M., HIRATA, H., KAGAWA, Y.: J. biol. Chem. **250**, 7917 (1975).
STEIN, W. D., LIEB, W. R., KARLISH, S. J. D., EILAM, Y.: Proc. nat. Acad. Sci. (Wash.) **70**, 275 (1973).
STEWART, P. S., MACLENNAN, D. H.: J. biol. Chem. **249**, 985 (1974).
STOECKENIUS, W., LOZIER, R. H.: J. Supramolec. Structure **2**, 769 (1974).
SUOLINNA, E-M., BUCHSBAUM, R. N., RACKER, E.: Cancer Research **35**, 1865 (1975).
SWEADNER, K. J., GOLDIN, S. M.: J. biol. Chem. **250**, 4022 (1975).
TANIGUCHI, K., POST, R. L.: Fed. Proc. **33**, 1289 (1974).
THAYER, W. S., HINKLE, P.: J. biol. Chem. **248**, 5395 (1973).
THAYER, W. S., HINKLE, P. C.: J. biol. Chem. **250**, 5330 (1975).
TITUS, E. O., HART, W. M., Jr.: Ann. N. Y. Acad. Sci. **242**, 246 (1974).
TONOMURA, Y.: "Muscle Proteins, Muscle Contraction and Cation Transport," University of Tokyo Press, 1972.
TONOMURA, Y., FUKUSHIMA, Y.: Ann. N. Y. Acad. Sci. **242**, 92 (1974).
TONOMURA, Y., INOUE, A.: In: Energy Transducing Mechanisms (E. Racker, Ed.) MTP International Review of Science, Vol. 3. London: Butterworths, 1975, p. 121.
TRUMPOWER, B., KATKI, A., RACKER, E.: unpublished experiments.
TZAGOLOFF, A., AKAI, A.: J. biol. Chem. **247**, 6517 (1972).
TZAGOLOFF, A., MEAGHER, P.: J. biol. Chem. **246**, 7328 (1971).
TZAGOLOFF, A., RUBIN, M. S., SIERRA, M. F.: Biochim. biophys. Acta **301**, 71 (1973).
WARREN, G. B., TOON, D. A., BIRDSALL, N. J. M., LEE, A. G., METCALFE, J. C.: Proc. nat. Acad. Sci. **71**, 622 (1974).
WHITTAM, R., AGER, M. E.: Biochem. J. **97**, 214 (1965).
WIKSTROM, M. K. F.: Biochim. biophys. Acta **301**, 155 (1973).
YAMAMOTO, T., TONOMURA, Y.: J. Biochem. Tokyo **62**, 558 (1967).
YAMAMOTO, T., TONOMURA, Y.: J. Biochem. Tokyo **64**, 137 (1968).
YOSHIDA, M., SONE, N., HIRATA, H., KAGAWA, Y.: J. biol. Chem. **250**, 7910 (1975).

Chapter 9

Membrane Immunological Reactions and Transport

P. K. Lauf

A. Introduction: The Concept

More than ten years ago we formulated the basic concept of enhancing our knowledge on biological transport processes by immunological approaches. The question was raised as to whether one could prepare by immunization antibody molecules specific for antigenic determinants on transport enzymes. It was assumed that such antibodies would be useful 1) for an analysis of the topographic distribution of the membrane transport units, 2) in the biochemical identification of transport proteins during their purification from whole membranes, and 3) to find out whether antibodies would modify basic physiological functions of transport systems, enabling the membrane physiologist to dissect further and understand the complexities of some biological transport processes.

The first two assumptions are obvious because they are based on the general and successful experience in identifying and purifying many macromolecules by immunochemical techniques. With respect to membrane transport systems, antibodies are now available which, when coupled to electron-dense particles, can be used to localize transport proteins and to visualize their asymmetric transmembranous orientation. These aspects will be discussed in section B of this chapter, in which the antibodies against the Na^+-K^+-ATPase as well as those against the Ca^{++}-ATPase have been considered in detail.

The third assumption, however, is founded on the hypothesis that antigenic dissimilarities between transport systems may reflect evolutionary changes in physiological functions. It was hoped that, by inhibition or activation, the immunological combination between transport-associated antigens and antibodies (antigen-antibody reaction) would shed light on the means by which structural differences influence physiological functions. Genetic mutations of transport systems appear to provide the best models for this purpose, and the genetic association between surface antigens and Na^+-K^+ transport in red cells of sheep, goat and cattle (discussed in section C) is certainly the most unique example. By binding to the outer surface of the red cell membrane, antibodies are capable of modifying the cation affinities at the cytoplasmic side of the Na^+-K^+ pump profoundly, resulting in a quite dramatic increase of the turnover rate of each pump site. The relevance of these studies on ruminant red cells to other systems is best demonstrated by our recent work on antigenic mutants of human red cells (Rh_{null} red cells), and may be borne out again as new genetic variants are found where antigenic changes signify structural alterations of important transport proteins (for example the En(a)-negative erythrocyte).

In certain nucleated cells, a number of immunological reactions exert profound effects on membrane cation permeability. Upon immunological challenge, lymphocytes and tumor cells respond with significant changes in passive and active cation permeabilities (section D). Attempts are being made to understand these early events and their relationships to the morphological and biochemical transformation of these cells, which is required for clonal expansion and immunological defense. Membrane antigens and receptors functionally connected to transport units may serve as transducer systems.

The list of effects of immunological reactions on membrane transport systems could be expanded to include recent work on hormone effects on cation and sugar transport systems. However, the complexity of the phenomena discussed in sections B, C, and D requires the author to exclude these other interesting and somewhat parallel aspects from our discussion.

B. Immunological Reactions and Membrane Transport Proteins

I. Introduction

In recent years membrane physiologists and biochemists have begun to use immunological techniques to study the transmembranous orientation (sidedness) and partial reactions of the Na^+-K^+-ATPase. The purpose of this chapter is not to review the detailed biochemical aspects of the Na^+-K^+-ATPase reactions. (The reader should consult the expert reviews published recently by HOKIN and DAHL (1972), SCHWARTZ et al. (1975), GLYNN and KARLISH (1975), WHITTAM and CHIPPERFIELD (1975), and ALBERS (1976)). Several interesting aspects of the interactions of antibodies with the Na^+-K^+-ATPase, however, will be discussed in detail. First we will consider the enzyme as an antigen, *i. e.* the source and purity of the enzyme used to produce antibodies. Second, the immunochemical nature of the antibodies obtained, and also their locus of binding and their biochemical effects with respect to the sidedness of the membrane will be discussed. The molecular aspects of antibodies relevant to functional changes of membrane cation transport by antigen-antibody reactions have been recently delineated (LAUF, 1975a), and much more detail on the structural and functional diversity of the antibody molecule can be obtained from the monograph by NISONOFF et al. (1975). Third and fourth, an analysis will be given of the modes by which antibody-antigen reactions have been found to alter partial reactions of the enzyme, as well as cation parameters in resealed red cell ghosts, in which the sidedness of the Na^+-K^+-ATPase is best reconstituted. Finally, the immunological cross-reactivity of the anti-Na^+-K^+-ATPase antibodies will conclude the section on antibodies against the Na^+-K^+ transport enzyme. Recent immunological studies on the structural analysis of the Ca^{++}-ATPase of sarcoplasmic reticulum also warrant particular emphasis, since antibodies were prepared against proteolytic fragments of this enzyme and were used to deduce its probable transmembranous assembly.

II. Antibodies Against the Na^+-K^+-ATPase

1. Properties of Antigens and Antibodies

As the techniques to isolate membrane Na^+-K^+-ATPase advanced and our knowledge about the possible subunit structure of this enzyme was furthered, more purified enzyme preparations were used for immunization purposes. Table 1 lists species and tissues from which the Na^+-K^+-ATPase was prepared by eight major groups of investigators; the specific activities of the preparations reflect their purity. It can be seen that human red cell membranes, rat brain, renal outer medulla from pig and dog, and rectal gland from dogfish shark served as major sources of the Na^+-K^+-ATPase used for immunization. Recently, microsomal membranes from electric eel also were used as antigens. The Na^+-K^+-ATPase has not yet been purified from red cell membranes which contain less than 1/3500 enzyme protein by weight. Consequently, the specific activity of the ATPase preparations used for immunization by AVERDUNK et al. (1969) is very low (0.5 or less). In terms of specific activity and low degree of contaminating proteins, the most intensively purified whole (holo- or native) enzyme preparations used for immunization were those of KYTE (1974 and 1976a), MCCANS et al. (1974 and 1975), RHEE and HOKIN (1975), JEAN et al. (1975), JEAN and ALBERS (1976). It is now well established that "native" or "whole" Na^+-K^+-ATPase preparations can be biochemically further partitioned into polypeptide subunits of different molecular weights. Large, unconjugated polypeptides (molecular weight about 100000), now considered to participate in the processes of phosphorylation, ion translocation and ouabain binding (KYTE, 1972; RUOHO and KYTE, 1974; HEGYVARY, 1975), were used as antigens by KYTE (1974, 1976b, c) and JEAN and co-workers (JEAN et al., 1975; JEAN and ALBERS, 1976). The smaller and glycosylated proteins (molecular weight 50000–60000), which are associated with the large chains but to which no clear functional role has been attributed as yet, were recently also used for immunization by RHEE and HOKIN (1975) and JEAN et al. (1975, 1976).

Table 1 reveals that most investigators used rabbits for immunization, except for MCCANS et al. (1975), who also prepared antibodies in sheep. According to whether the Na^+-K^+-ATPase or its subunits are used for immunization, the antibodies obtained were termed anti-Na^+-K^+-ATPase complex antibodies (anti-CA, ASKARI and RAO, 1972), anti-whole or anti-holo enzyme antibodies (JØRGENSEN et al. 1973; KYTE, 1974; MCCANS et al. 1974, 1975; SMITH et al., 1973 and SMITH and WAGNER 1976; RHEE and HOKIN, 1975; JEAN et al. 1975; JEAN and ALBERS, 1976) and anti-glycoprotein or anti-small chain antibodies (RHEE and HOKIN, 1975; JEAN et al. 1975; JEAN and ALBERS, 1976). Immunochemically, most of these antibodies belong to the class of immunoglobulin G

Table 1. Some Properties of the Na^+-K^+-ATPase Antigen and its Antibodies

Na^+-K^+-ATPase Antigen				Na^+-K^+-ATPase Antibodies			References
Species	Tissue	Preparations	Specific activity	Species	Class	Specificity	
Human	Red cell membranes	Whole membranes	0.2–0.5	Rabbit	n. i.	Anti-whole membrane serum	AVERDUNK et al. (1969)
Rat	Brain	Microsomal	124	Rabbit	n. i.	Anti-complex antibody (anti-CA)	ASKARI & RAO (1972)
Pig	Renal outer medulla	Microsomal	15–18	Rabbit	IgG	Anti-whole enzyme	JØRGENSEN et al. (1973)
Dog	Renal outer medulla	Large & small chains Native enzyme	460	Rabbit	IgG	Anti-large chain Anti-native enzyme	KYTE (1974, 1976)
Dog	Renal outer medulla	Native enzyme	1500?	Rabitt	IgG	Anti-native enzyme Anti-catalytic site Anti-digitalis receptor	McCANS et al. (1974)
Dogfish shark	Rectal gland	Native enzyme glycopeptide	n. i.	Rabbit	IgG IgM	Anti-native enzyme Anti-glycopeptide	RHEE & HOKIN (1975)
Electric eel	–	Native enzyme Large chains Small chains	n. i.	Rabbit	n. i.	Anti-native enzyme Anti-large chain Anti-small chain	JEAN et al. (1975, 1976)
Dog	Renal outer medulla	Native enzyme	280	Rabbit	IgG	Anti-native enzyme	SMITH & WAGNER (1976)

n. i. = not indicated by authors

proteins (IgG, molecular weight 160000), which have two sites to combine with and affect the Na^+-K^+-ATPase. However, bivalency is not required for combination with the antigens, since McCANS et al. (1974) and SMITH and WAGNER (1976) demonstrated binding and action of monovalent (Fab) fragments obtained from native antibody by enzymatic cleavage. For detailed information on the techniques of preparing monovalent antibody fragments by the methods of PORTER (1959) or NISONOFF et al. (1960) consult NISONOFF et al. (1975).

2. Sidedness of Binding and Immunological Effects on the Na^+-K^+-ATPase and its Partial Reactions

a) Sidedness of Binding

The point of binding and action of almost all anti-whole enzyme antibodies prepared is at the cytoplasmic aspect of the Na^+-K^+-ATPase (ASKARI and RAO, 1972; KYTE, 1974; MCCANS et al. 1974; SMITH et al., 1973 and SMITH and WAGNER, 1976). The binding site for these antibodies is probably on the cytoplasmic portion of the large chain polypeptide, since ferritin conjugated anti-whole enzyme as well as anti-large chain antibodies preferentially labeled the inner membrane side and not the outer surface, which was identified by adsorption of virus particles specific for sialoglycoproteins (KYTE, 1974). The binding of these anti-large chain antibodies is specific for the large chain subunit of the enzyme: when conjugated to ferritin (KYTE, 1976a), they can be used to visualize the membrane localization of the Na^+-K^+-ATPase in kidney tubules (KYTE, 1976b and c). The enzyme was tightly packed at the basolateral membrane of proximal and distal tubules of the canine kidney which participate in the active sodium reabsorption process. KYTE (1976b and c) was also able to demonstrate some Na^+-K^+-ATPase molecules at the luminal membrane of these epithelial cells.

Anti-holo enzyme antibodies, however, may also contain antibodies against the glycosylated protein subunit of the NA^+-K^+-ATPase (KYTE, 1974, 1976a). In fact, RHEE and HOKIN (1975) and JEAN et al. (1975; JEAN and ALBERS, 1976) prepared antibodies against the glycopeptides associated with the enzyme. It is likely that these antibodies can bind at the inside or at the outside of the enzyme complex and thus to the nonconjugated, internal as well as to the glycosylated, N-terminal outer portion of these glycoproteins. KYTE (1974) demonstrated binding of these antibodies at the outer surface membrane of vesicles.

b) Immunological Effects on the Na^+-K^+-ATPase Activity

Immunization with such complex molecules as the Na^+-K^+-ATPase is expected to produce antibodies with differing specificities and affinities, since the immune system of the responding animal is capable of recognizing a multitude of antigenic determinants on the injected complex. Furthermore, as the time after immunological challenge progresses, antibodies of lower molecular weight (*i. e.* IgG proteins) and of greater affinity for a particular antigenic determinant will appear, while the titer of low-affinity antibodies may diminish. This is an important point with respect to physiological effects of such antibodies, since in some systems antibodies of lower affinity affected the physiological function tested, while antibodies with high affinity did not. Sometimes it is even difficult to reproduce in another rabbit a previously obtained antibody which exhibited a particular effect. Obviously, the physicochemical state of the antigen used as well as the genetic background of the animals chosen for immunization are decisive factors. It is a general rule that the evolutionary distance between the animal from which the antigen is prepared and the animal used for immunization determines the specificity of the antibodies produced. Thus in the light of these complexities, it is not surprising that there is a considerable discrepancy in the effects of anti-Na^+-K^+-ATPase antibodies observed by various investigators.

Most of the anti-holo enzyme antibodies reported possessed a strong inhibitory action on the total Na^+-K^+-ATPase activity, apparently when bound at the cytoplasmic aspect of the enzyme. Thus AVERDUNK et al. (1969), ASKARI and RAO (1972), JØRGENSEN et al. (1973), GLYNN et al. (1974), MCCANS et al. (1974), SMITH and WAGNER (1976), RHEE and HOKIN (1975) and JEAN et al. (1975) demonstrated 50–100 percent inhibition of the catalytic activity of the Na^+-K^+-ATPase. It is likely that, in the majority of these reports, the antibody prepared against the holo enzyme is of anti-large chain specificity, since JEAN et al. (1975; JEAN and ALBERS, 1976) found a similar enzyme inhibition by antibodies prepared against the large polypeptides. It is also possible that anti-glycoprotein antibodies are responsible for the inhibitory effect. Using different enzyme sources for preparation of the glycoprotein moiety, RHEE and HOKIN (1975) observed 30 percent and JEAN et al. (1975; JEAN and ALBERS, 1976) 80 percent inhibition of the enzyme activity. Finally, it cannot be ruled out at present that antibodies may be formed against minor contaminants (peptides, lipids) which are closely associated with the transport enzyme and may mediate antibody effects. Thus KYTE (1974) observed a minor inhibitory anti-Na^+-K^+-ATPase antibody fraction that could be removed by prior absorption with lipids.

In contrast to these inhibitory antibodies, KYTE's (1974) major antibody against the Na^+-K^+-ATPase did not affect its catalytic activity. It is this antibody that KYTE (1976b and c) used in his elegant experiments to demonstrate the transmembranous orientation of the Na^+-K^+-ATPase as well as its distribution on kidney tubules. Since the Na^+-K^+-ATPase was fully active although bound to the antibody, KYTE (1974) inferred that the catalytic activity of and cation translocation by the enzyme does not require large protein movements within the lipid bilayer.

The sidedness of the action of anti-holo enzyme antibodies was also demonstrated by SMITH and WAGNER (1976) where inhibition of red cell membrane Na^+-K^+-ATPase was only seen in inside-out red cell membrane ghosts. This finding is in agreement with the inhibition of cation fluxes in resealed ghosts, when the antibody was incorporated into the cell prior to resealing (see section B/II/3).

c) Effects of Partial Reactions of the Na^+-K^+-ATPase

In the light of the various conformer states proposed for the entire reaction sequence of the Na^+-K^+-ATPase it is of interest to know which of the partial reaction steps of the enzyme is altered by the antigen-antibody reaction. Here again, considerably different results have been reported. ASKARI and RAO (1972), and ASKARI (1974) found a blockade of the Na^+-dependent E-P formation as well as of the ADP/ATP exchange reactions, while no such effects were observed by JØRGENSEN et al. (1973). The last-named authors also did not observe any effect on ouabain binding, which is consistent with their other findings that the Na^+-dependent E-P formation is not altered by the antibody. The K^+-dependent para-nitrophenyl phosphatase (pNPPase) was found to be inconsistently affected (ASKARI, 1974; GLYNN et al., 1974). It is not known whether the K^+-dependent pNPPase is affected by an antibody different from that acting on the Na^+-dependent reaction steps. If it is, then the antibody might bind to a site of the Na^+-K^+-ATPase more distant from its cytoplasmic site. This would imply involvement of more 'outer membrane aspects' in the immunological reactions since the K^+-dependent partial reactions reflect the interaction of the enzyme with external K^+ ions.

Such an effect on more outward-oriented parts of the Na^+-K^+-ATPase must be considered for the anti-digitalis receptor antibody (anti-DR) of MCCANS et al. (1974), which inhibits the binding of ouabain to the external aspect of the transport enzyme. Alternatively, of course, anti-DR by binding from the inside may lock the enzyme into a conformation not conducive to ouabain binding. Indeed, recent experiments by JEAN and ALBERS (1976) suggest that antibodies may induce (or prevent) conformational changes of the enzyme by modifying cation binding sites.

3. Immunological Alteration of Cation Fluxes in Resealed Ghosts

The best evidence for the cytoplasmic action of some of the anti-Na^+-K^+-ATPase or pump antibodies stems from their dramatic effect on cation fluxes in red cells, when incorporated into resealed red cell ghosts. Four out of five ouabain-sensitive reaction modes, which are thought to be representative of the Na^+-K^+ pump, are affected: the normo-vectorial Na^+-K^+ pump, K^+/K^+ exchange, Na^+/Na^+ exchange and the uncoupled Na^+ efflux (JØRGENSEN et al., 1973; GLYNN et al., 1974; SMITH and WAGNER, 1976). The fact that all of the partial reactions of the Na^+-K^+ exchange system except one are affected by the immunological reaction diminishes to some extent the usefulness of antibodies as molecular tools to dissect the individual components of the Na^+-K^+ pump. As for the effect of anti-DR on ouabain binding, it is not impossible that the anti-Na^+ pump antibody severely affects the cation binding sites at the cytoplasmic membrane side, as proposed by JEAN and ALBERS (1976).

4. Species and Organ Specificities of Immunological Reactions Involving the Na^+-K^+-ATPase

Evidence for additional heterogeneity of antibodies prepared against the Na^+-K^+-ATPase stems from their ability to cross-react with enzyme preparations from various organs of the donor species as well as with enzymes of different species. The rabbit anti-rat brain ATPase immunoglobulin of ASKARI and RAO (1972) and ASKARI (1974) also affected the membrane enzyme of rat kidney, heart and red cells. The same antiserum also reacted toward other heterologous systems, inhibiting enzymes of brain tissue from guinea-pig, dog, and cow but not affecting the kidney enzymes of these species. The rabbit antibody against the pig kidney Na^+-K^+-ATPase of JØRGENSEN et al. (1973) inhibited rabbit kidney enzyme (*i. e.* even in the autologous system), ox brain enzyme and also the Na^+ fluxes in human red cell ghosts when present at the cytoplasmic side. These results reflect the presence in antisera of several immunologically identical and different antibody specificities. Different antibody specificities may be the result of minute structural differences between enzymes from various organs within homologous species and between the same tissue of heterologous species. MCCANS et al. (1975) clearly demonstrated antibodies which did not bind to or inhibit the dog brain enzyme but inhibited the dog kidney Na^+-K^+-ATPase. Furthermore, there are differences in the immunological response of different animals to the injection of the same purified Na^+-K^+-ATPase. The inhibition of dog, rat and ox enzyme by sheep antisera was comparable to that obtained with rabbit antisera; however, rabbit antiserum had a greater effect on the catalytic rate of the brain enzyme than sheep antiserum (MCCANS et al., 1975). The anti-canine enzyme antibody of SMITH and WAGNER (1976) affected dog heart, calf brain and human red cell membrane enzymes. In contrast to these complex cross-reactivity patterns, rabbit anti-eel Na^+-K^+-ATPase did not cross-react with Na^+-K^+-ATPase from mammalian tissues, implying a structural and class-specific difference between enzymes derived from fish and mammals. Although more work is needed, it appears that anti-Na^+-K^+-ATPase antibodies may be convenient tools to study evolutionary aspects of the Na^+-K^+ pump. Elsewhere, we have reviewed for example the specificities of antibodies raised against bacterial membrane ATPase which seem to parallel taxonomical boundaries (LAUF, 1978).

III. Antibodies Against Ca^{++}-ATPase of Sarcoplasmic Reticulum

Antibodies raised against Ca^{++}-ATPase from sarcoplasmic reticulum bind but do not inhibit the catalytic activity of the enzyme (MARTINOSI and FORTIER, 1974; STEWART and MACLENNAN, 1975; STEWART et al., 1976) and thus resemble the anti-Na^+-K^+-ATPase antibody of KYTE (1974). These, as well as antibodies against proteolytic fragments of the enzyme, turned out to be very useful tools for study of the transmembranous orientation of the Ca^{++}-ATPase. The purified enzyme consists of a major polypeptide of molecular weight 102000 and a proteolipid of molecular weight 6000–12000, and requires phospholipid for its full activity (MACLENNAN and HOLLAND, 1976). Upon tryptic digestion, the large polypeptide is cleaved into two fragments of 45000 and 55000 daltons (STEWART et al., 1976). The 55000 dalton moiety, exposed at the outer membrane surface, can be further split into two polypeptides of molecular weight 20000 and 30000. Antibodies against the 55000 dalton fragment agglutinated vesicles of sarcoplasmic reticulum membranes, while antibodies against the internal 45000 dalton fragment had no such effect. Antibodies against the 20000 dalton peptide failed to react with these vesicles. It was concluded that the 30000 dalton fragment, which is part of the 55000 dalton unit, must be located most peripherally at the outer surface of the sarcoplasmic reticulum membrane (STEWART and MACLENNAN, 1975; STEWART et al., 1976) while the 20000 dalton fragment, also part of the 55000 dalton moiety, is the structural but buried link between the outer (30000 dalton) and cytoplasmic (45000 dalton) parts of the Ca^{++}-ATPase.

IV. Conclusion

The diversity of antibodies against the Na^+-K^+-ATPase and Na^+-K^+ pump reflects the structural and functional complexities of the partial reactions of this system. Apparently, some antibodies interfere with all Na^+-dependent steps and thus block formation of the phosphorylated intermediates, while others affect the catalytic activity at some later step in the reaction sequence, maybe when the enzyme changes from the Na^+-dependent to the K^+-sensitive conformation. There are also antibodies which do not affect the Na^+-K^+-ATPase activity but bind to certain aspects of the pump and thus become useful membrane topographical tools. The diversity of these antibodies can in part be explained by the macromolecular complexity of the Na^+-K^+-ATPase molecule and the degree of purity of the enzyme which is injected into the animal for immunization. The immunochemical approach to the structure and transmembranous orientation of the Ca^{++}-ATPase points to experimental ways which will probably be taken in the work on the Na^+-K^+-ATPase. Moreover, the use of antibodies as taxonomical tools to study the evolutionary relationship between Na^+-K^+-ATPase preparations from widely different tissues is an interesting and quite promising aspect for future work.

C. Immunological Reactions at the Outer Membrane Surface and Cation Transport in Erythrocytes

I. Introduction

The outer surface of the erythrocyte membrane possesses a large number of antigenic determinants capable of binding antibodies and various plant agglutinins. For many years these membrane antigens have been considered in terms of blood group genetic markers important for intraspecies blood transfusion: severe immunological reactions occur when antigenically incompatible blood is injected to an individual who has natural isoantibodies (clonal selection theory of BURNET, 1959) or antibodies from previous immunological challenges with the particular blood group (immune antibodies). These immunological reactions occur *in vivo* and can be studied *in vitro* by cellular agglutination (hemagglutination), or immune cytolysis (immune hemolysis) when the ensuing antigen-antibody reaction activates the hemolytically active complement system. In general, hemagglutination has not been reported to alter membrane cation permeability. In contrast, immune hemolysis severely affects the passive membrane permeability by generating membrane lesions, biochemically still ill-defined, through which ionic gradients dissipate rapidly, resulting in colloid osmotic immune hemolysis. The physiological basis of this extraordinary process of cellular destruction has been treated in a recent review (LAUF, 1978) and therefore will not be further discussed in this chapter.

The purpose of the present section is to illuminate an entirely different aspect of the role of membrane antigens: the subtle changes induced by a number of immunological reactions which have profound implications with respect to physiological functions of the cell membrane. The underlying concept of this approach rests on the assumption that evolutionary changes in membrane transport systems may also be manifested in the concomitant appearance of antigenically new or different membrane constituents. For example, mutation of a particular antigenic structure of a membrane may significantly alter another membrane component, changing its physiological function. Alternatively, a genetic mutation of a functional membrane component may result in the detection of a new antigen by the immune system. Physiologically, the latter example is most interesting, since it implies that antigenic determinants may reflect structural (genetic) changes of transport systems and that the antibodies elicited by these determinants may be used to alter membrane functions experimentally. It is this concept which we first adopted when we set out to investigate whether immunological reactions may be useful to probe deeper into the physiological basis of cation transport polymorphism in sheep red cells (LAUF, 1969) and which later we applied to prove that the Rhesus antigen, suspected to be associated with the Na^+-K^+-ATPase system, indeed may not be part of this transport system (LAUF and JOINER, 1976; LAUF, 1977). It will become apparent from this presentation that the success of studying membrane transport systems by immunological reactions depends on the fortunate fact that the cation transport mutations observed in erythrocytes of sheep, goats and cattle are genetically associated with particular antigens. These red cells therefore constitute unique models in the mammalian kingdom, since they permit direct study of the effect of immunological reactions on membrane cation transport and promise to help identify isolated membrane transport units by immunological tools. The only parallel to this work is that reported in various bacterial and fungal mutants where antigen-antibody reactions were used to activate membrane functions by conformational modulations (POLLOCK et al., 1966; ROTMAN and CELADA, 1968; CELADA et al., 1970; MESSER and MELCHERS, 1970; ROTH and ROTMAN, 1975).

II. Sheep Red Cells

1. Cation Transport, Genetics and Immunological Parameters

a) Cellular Cations and Genetics

Erythrocytes of individual sheep maintain either high potassium, low sodium (HK) or low potassium, high sodium (LK) steady-state levels (ABDERHALDEN, 1898; KERR, 1937; EVANS, 1954 and 1957). Since the gene responsible for the LK trait is claimed to be dominant, the heterozygous animals also have LK-type red cells (EVANS and KING, 1955). Table 2 depicts the cellular potassium $[K^+]_c$ and sodium $[Na^+]_c$ concentrations of 21 male and 21 female pure-bred Dorset sheep, recently analyzed by our laboratory. It is evident that the $[K^+]_c$ levels are somewhat higher in heterozygous than in the homozygous LK red cells, an observation consistent with the findings of EVANS (1957) and EAGLETON et al. (1970), and thus questioning the complete dominance of the LK gene (EAGLETON et al., 1970). The alleles responsible for the HK and LK properties (see Table 2) are termed ka^h and Ka^L, respectively (symbols for the locus responsible for high and low K^+ [kalium] levels; see RASMUSEN and HALL, 1966a) or Ke^h and Ke^L (symbols after KERR, 1937, for recessive high-[h] and dominant low-[L] potassium cells; see TUCKER, 1975).

Table 2. The Genetic Correlation between Membrane Surface Antigens and Cation Polymorphism in Red Blood Cells of 42 Pure-Bred Dorset Sheep

Cation geno/ phenotypes	Immunological Terminology		Genetic Terminology		Cellular cation concentrations mmol/l cells (± S.D.)	
n=Number, F=Female, M=Male	Presently used	Used until 1969***	Antigens (1974/ 1975**)	Cations (1966*/ 1975**)	$[K^+]_c$	$[Na^+]_c$
HK; M(n=5) F(n=6)	MM or MM′	MM or MM′	$M^aM^a(M^c)$	ka^hka^h or Ke^hKe^h	88.6 ± 11.0 100.0 ± 3.2	14.1 ± 2.2 16.6 ± 2.8
LK; Heteroz. M(n=12) F(n=9)	LM or LMM′	mM or mMM′	$M^bM^a(M^c)$	Ka^Lka^h or Ke^LKe^h	20.4 ± 3.9 19.8 ± 1.9	84.4 ± S.3 81.4 ± 8.2
LK; Homozyg. M(n=4) F(n=6)	LL	MM	M^bM^b	Ka^LKa^L or Ke^LKe^L	14.9 ± 4.2 18.3 ± 2.4	84.3 ± 9.3 83.7 ± 2.3

* RASMUSEN and HALL, 1966; ** RASMUSEN et al. 1974; TUCKER, 1975; *** ELLORY and TUCKER, 1969.

b) Membrane Antigens and Genetics

RASMUSEN and HALL (1966a) discovered that immunization of LK sheep with HK red cells produced an antibody reacting only with HK or heterozygous LK red cells. The antigen was termed M and, because no reagent was available at that time for LK red cells, HK cells were genetically defined as MM, heterozygous LK cells as Mm and homozygous LK cells as mm (RASMUSEN and HALL, 1966a; TUCKER, 1968; see also Table 2). Anti-m, the antibody exclusively reacting with LK red cells, was discovered three years later by injecting LK red cells to HK sheep (RASMUSEN, 1969; ELLORY and TUCKER, 1969a) and shortly thereafter it was agreed to use MM, ML and LL in the immunologic terminology rather than MM, Mm and mm. This terminology will also be used throughout this chapter, although recently an additional genetic terminology has been introduced which defines the HK/M allele as M^a and the LK/L allele as M^b (RASMUSEN et al., 1974a; TUCKER, 1975). The detection of an M antigen slightly distinct from but always together with the M antigen has led to the proposal of a second M allele, M^{ac} (TUCKER, 1975). The recent finding of M' (M^c) without M in LK heterozygotes (M^{bc} cells; see TUCKER, 1975) indicates a third M allele, M^c, which may be related to RASMUSEN's (1960) M^x allele. The genetics of the HK/M and LK/L system in sheep, recently reviewed by TUCKER (1971, 1975) and LAUF (1975a, 1978) are obviously complex; however, its details become important in the interpretation of the number of Na^+-K^+ pumps as well as M and L antigens and their quantitative relationship to each other.

Table 3 considers the membrane immunological and cation transport properties in more detail. Recent measurements of the number of M and L antibody molecules bound to these cells revealed about 3000 and 1600 M antigenic sites in MM and LM sheep red cells, respectively, while there are some 1500 and 1200 L antigens per homozygous and heterozygous LK red cell, respectively (LAUF and SUN, 1976). These numbers are based on the assumption that both

Table 3. Membrane Immunological and Cation Transport Properties of High- and Low-Potassium Sheep Erythrocytes

Parameter	HK	LK	LK	Units
Antigens	MM	LM	LL	–
M-Antigenic sites	3000	1600	None	Molec. Anti-M/cell
L-Antigenic sites	None	1200	1500	Molec. Anti-L/cell
Total K^+ Influx	0.72	0.28	0.32	mM/l cells × h
K^+ Pump Influx	0.63	0.16	0.16	mM/l cells × h
$K_{1/2}$ of ${}^iM_K^P$	2.6	2.7	2.7	mM K_o^+/l
K^+ Leak influx	0.09	0.12	0.16	mM/l cells × h
${}^ik_K^L$	0.013	0.024	0.032	h^{-1}
Na^+-K^+-ATPase	41	n. t.	11	nM P_i/mg Prot. × h
Mg^{++}-ATPase	11	n. t.	38	nM P_i/mg Prot. × h
Na^+-K^+ pumps	120	50	50–70	Molec. ^{3}H-ouab./cell
Ouabain-binding rate	Fast	Slow	Slow	Molec./cell × min
Turnover	1700	1100	1100	K^+ ions/site × min

n. t. = not tested

antibodies bind with only one valency to their antigenic sites. The average K_D values for the M and L antibody binding were 10^{-9} and 10^{-10} M/l, respectively (LAUF and SUN, 1976). We have observed a considerable heterogeneity index for M antibody binding. A more complex and functionally interesting heterogeneity of the L antigenic sites may exist and will be dealt with in section C/II/2c.

Both the M and the L antibodies are able to hemolyze their respective target cells in the presence of complem nt (LAUF and TOSTESON, 1969); LAUF and DESSENT, 1972 and 1973). For some as yet undefined reason LK cells show a slow hemolytic response to anti-L and complement (LAUF and DESSENT, 1973). The weakest hemolytic reaction was found with LM cells exposed to anti-L and complement, supporting the finding that the number of L antigens is somewhat different between LL and LM sheep red cells.

c) Active and Passive Cation Transport

Table 3 also summarizes the transport physiological parameters which were first described by TOSTESON and HOFFMAN (1960) and TOSTESON (1963) for sheep red cells. Since a thorough evaluation of the general kinetic parameters of the Na^+-K^+ pump will be given by others elsewhere in this series, only the properties of the system characteristic for sheep red cell membranes will be briefly discussed, in order to illuminate later the effect of immunological reactions on these cells. The Table does not contain information on cation exchange diffusion (high Na^+/Na^+ exchange diffusion, see TOSTESON and HOFFMAN, 1960; MOTAIS and SOLA, 1973) or anion distribution (net and carrier mediated transport, see TOSTESON et al. 1972; GUNN et al., 1973) because these processes are not known to be affected by the immunological reactions described below. The data in Table 3 for active and passive Na^+-K^+ transport and the membrane ATPase are from the author's laboratory, and the number of Na^+-K^+ pump sites and thus the presumable turnover number of cations/pump site are based on the recent work of JOINER and LAUF (1975).

In accordance with TOSTESON and HOFFMAN (1960) the values for active K^+ influx (K^+ pump influx, ${}^iM_K^P$) are several times higher in HK cells than in LK cells. Low-potassium sheep red cells maintain their LK steady-state levels by a much lower K^+ pump influx and a greater passive K^+ permeability, yielding a ratio of ${}^iM_K^P/{}^iM_K^L$ (β-parameter of TOSTESON and HOFFMAN, 1960) which is about one order of magnitude smaller than in HK erythrocytes. Because in HK cells, P_{Na} is slightly greater and P_K much smaller than in LK cells, the α-parameter of TOSTESON and HOFFMAN (1960), defining the ratio of ${}^ik_{Na}^L/{}^ik_K^L$, is much closer to unity than in LK cells. The most important implication of TOSTESON's and HOFFMAN's finding is that in their low-$[K^+]_o$ high-$[Na^+]_o$ environment of the plasma K^+, pump and leak fluxes are balanced, to guarantee precise volume control of both cell types. However, LK cells lose this compensation when incubated in high-K^+ media, because the higher P_K causes K^+ entry and uncompensated cellular swelling (TOSTESON, 1964 and 1966a).

The kinetics of the Na^+-K^+ pump in the two types of sheep red cells have been worked out by HOFFMAN and TOSTESON (1971), and are entirely different with respect to internal cation affinities. In LK cells, K^+ ions are potent inhibitors at the Na^+ loading site, and this inhibition was only partially relieved when $[K^+]_c$ was reduced to levels below 1 mM by Na^+ replacement. On the other hand, the activation by K_o^+ ions in the presence of Na_o^+ yielded similar $K_{1/2}$ values (see Table 3), and in both cells activation of Na^+-K^+ pump transport was independent of the concentration of the transported ions in the medium. It was concluded that active translocation of Na^+ and K^+ ions is a simultaneous process in both cells (HOFFMAN and TOSTESON, 1971) which was also found in human red cells by others. However, the pump is kinetically very different in LK cells as compared to HK cells.

d) Ouabain Binding

The differences between the Na^+-K^+ pump systems of HK and LK cells, however, are not entirely qualitative. As Table 3 shows, HK cells have about 120 and LK cells some 50 pumps/cell as measured by ^3H ouabain binding (JOINER and LAUF, 1975). These numbers permit an estimation of the turnover of K^+ ions/site/minute of less than 2000 for each cell type. The binding values of JOINER and LAUF (1975) are considerably different from those of DUNHAM and HOFFMAN (1971a), which were 48 for HK and 7 for LK cells, and almost double those of ELLORY and TUCKER (1970b) which were 71 for HK and 37 for LK cells, although ELLORY and TUCKER (1969c) had reported earlier 8.4 ouabain-binding sites/LK cell. The discrepancy between these reports is not immediately obvious. JOINER and LAUF (1975) clearly ruled out nonspecific ouabain binding at the ouabain concentrations used (10^{-7} to 10^{-8} M/l), since a linear correlation between K^+ pump inhibition and the number of ouabain molecules bound was established. Extrapolation to 100 percent K^+ pump inhibition yielded values for ouabain binding which were not different from those obtained by kinetic binding experiments and which were unaffected by extracellular K^+ and Cs^+ ions. More important with respect to the qualitative differences between HK and LK pumps, JOINER and LAUF (1975) showed that the association rate of ouabain with the HK pump is faster than with the LK pump. These findings have to be interpreted in terms of the internal cation activation kinetics for these pumps reported by HOFFMAN and TOSTESON (1971). Intracellular K^+ ions thus not only inhibit the turnover of the LK pump but also affect its transmembranous orientation in such a way that the rate of ouabain binding is considerably reduced. The work of SACHS et al. (1974b) on LK goat red cells is consistent with this interpretation: the rate at which inhibition of the pump by ouabain developed in LK cells with 0.8 mM $[K^+]_c$ was much greater than in LK cells with 23 mM $[K^+]_c$. Thus the turnover of the pump and the rate of ouabain binding appear to be directly proportional in sheep and goat red cells.

e) Na^+-K^+-ATPase

The data on the Na^+-K^+-ATPase of the two types of sheep red cell membranes exhibit similar complexities to those discussed for the Na^+-K^+ pump fluxes, and their interpretation and comparison to the cation flux measurements are difficult, since broken membrane preparations have lost the asymmetry of intact red cells, and since measurements have been done at normal as well as at very low ATP concentrations (at which Na^+-K^+ pump fluxes cannot be measured). TOSTESON'S (1963) first analysis of the Na^+-K^+-ATPase activity showed that the specific activity of the enzyme is about fourfold higher in membranes from HK than in membranes from LK red cells. Table 3 contains our measurements (LAUF et al., 1972) which are within the order of magnitude reported by TOSTESON (1963). The K_D for ATP is in the range of 10^{-5} to 10^{-6} M (TOSTESON, 1963). As the ATP concentration is reduced from physiological (10^{-3} M) to very low concentrations (10^{-6} M and less), the Na^+-K^+-ATPase activity diminishes and only the Na^+-dependent enzyme activity appears with its pronounced sensitivity to K^+ ions (presumably at the Na^+ loading site). Thus WHITTINGTON and BLOSTEIN (1971) found that the HK/LK activity ratios were 13 for the overall Na^+-K^+-ATPase, 10 for the Na^+-ATPase and 2.7 for the ADP/ATP exchange reaction. Their finding that K^+ ions are much more potent inhibitors of the Na^+-ATPase in LK membranes is consistent with the work on K^+ pump fluxes reported by HOFFMAN and TOSTESON (1971). Other qualitative differences between the two enzymes were also found in studies using oligomycin (BLOSTEIN and WHITTINGTON, 1973). In enzyme preparations of high specific activity oligomycin inhibits the K^+-dependent dephosphorylation step, and at low ATP concentrations there is little effect of oligomycin on the steady state level of the Na^+-dependent phosphorylated intermediate (Na-E · P_i). Thus hydrolysis into Na-E_i and P_i is little altered by oligomycin in HK enzyme preparations. In LK membrane preparations, however, oligomycin causes a 400 percent accumulation of Na-E · P_i accompanied by a 400 percent stimulation of the Na^+-dependent ATPase activity. BLOSTEIN and WHITTINGTON (1973) concluded that oligomycin must increase the Na^+-activated step of the LK enzyme. These authors also showed that in LK membranes oligomycin effectively counteracts the inhibitory effect of K^+ ions on the Na^+-ATPase activity. Together, the work of this group is consistent with the cation flux experiments reported above and supports the general hypothesis of a profound qualitative difference between the HK and LK active cation transport systems.

Other membrane parameters similar or different in the two cell types have been reviewed elsewhere (LAUF, 1975a).

2. The Effect of Antibodies on Cation Transport

a) Modification of Cation Pump and Leak Fluxes

Tables 2 and 3 presented the genetic association between sheep erythrocyte cation composition and membrane surface antigens. However, if the ML membrane antigens are indeed expressions of structurally and kinetically different Na^+-K^+ pumps, the finding that there are some 20 times more M and L antigenic sites per cell than there are pumps is unexpected. Thus more evidence is required to demonstrate that the ML antigens are functionally, if not structurally, related to the Na^+-K^+ pump and ATPase system in these cells. It was shown by this laboratory that ouabain did not affect M antibody binding to HK red cells (LAUF and TOSTESON, 1969), supporting TUCKER's hypothesis (1968) that not the M but the m (L) allele controls the cation permeability. However, ELLORY and TUCKER (1969a) reported that anti-L dramatically stimulated active Na^+-K^+ transport in LK sheep red cells, a finding simultaneously confirmed by our laboratory (LAUF et al., 1969 and 1970). Table 4 summarizes these events by expressing the effects of anti-L in factorial in rements. In the following discussion it will be seen that anti-L does not simpry confer HK properties upon LK cells as originally argued (ELLORY and TUCKER, 1969a).

Table 4. Effect of L-Antigen/Antibody Reaction on Na^+-K^+ Transport Parameters in LK Sheep Red Cells Expressed as Factorial Changes

Parameters	LL Cells	LM Cells
Na^+-K^+-ATPase Activity	4–8	n.t.
K^+ Pump influx ($^iM_K^p$)	4–8	4–6
$K_{1/2}$ of $^iM_K^{p\,max}$	1.0	1.0
Inhibition of $^iM_K^p$ by $[K^+]_c$	Reduced	n.t.
K^+ Leak influx	0.5	n.t.
Number of Na^+-K^+ pumps	1.0	1.0
Ouabain-binding rate	Increased	Increased
Turnover of K^+ pump	4–8	4–6

n.t. = not tested

The stimulation of K^+ pump influx by anti-L is 6- to 8-fold in LL and somewhat less in LM red cells (LAUF, 1974) and it has not yet been decided whether the slightly reduced number of L antigens in LM cells causes this difference. Using LL red cells, LAUF et al. (1970) clearly demonstrated that the antibody exerts its main action by reducing the inhibitory action of K^+ ions at the cytoplasmic aspec of the Na^+-K^+ pump. In effect these findings mean that the affinity of the cytoplasmic site of the Na^+-K^+ pump for Na^+ relative to K^+ has been increased, a conclusion also reached by GLYNN and ELLORY (1972) for the same cell system and supported by their experiments with LK goat red cells (ELLORY et al., 1972; see also below). This important result should have cast doubt on the early reports of ELLORY and TUCKER (1970b) and LAUF et al. (1970) that anti-L also increased the number of Na^+-K^+ pumps as measured by ^3H-ouabain binding, and its correlation with the degree of K^+ pump inhibition then indeed revealed that the hypothesis of a major increase and thus unmasking of dormant Na^+-K^+ pumps by anti-L was untenable (JOINER and LAUF, 1975). Binding of ^3H-ouabain at saturation was indistinguishable in LK red cells in the presence or absence of anti-L. In addition, anti-L did not change the linear correlation between the molecules of ouabain bound per cell and the degree of K^+-pump inhibition found for LK cells in the absence of antibody (JOINER and LAUF, 1975). However, the association rate of the cardiac glycoside with the LK membrane was increased to levels found in HK cells, irrespective of the LK genotype (JOINER and LAUF, 1975). These findings are consistent with the earlier data of LAUF et al. (1970), based on HOFFMAN and TOSTESON's experiments (1971), that the antibody alters the internal cation affinity of the pump. Thus by enhancing the affinity for Na_c^+ relative to K_c^+, anti-L induces a conformational change in the LK pump leading to a greater outward orientation of the ouabain binding site.

Anti-L thus far has been discussed in terms of its effect on the Na^+-K^+ pump. Initial reports from our laboratory (LAUF et al., 1970) on the effect of anti-L on $^iM_K^L$, the leak influx of potassium, were in contrast to those of DUNHAM (1975a), who reported a marked reduction in the $^iM_K^L$ parameter using a different source of unabsorbed L antiserum. DUNHAM (1975b and 1976) came to the conclusion that there may be two L antibodies, one affecting the Na^+-K^+ pump and the other the K^+ leak influx component. Having changed to a new batch of unabsorbed anti-L serum, we now consistently find a 20–50 percent reduction of K^+ leak influx in anti-L exposed LK sheep red cells (unpublished data). At present, we do not have any evidence to support DUNHAM's (1975b, 1976) claim that the second anti-L, affecting only the K^+ leak component of the overall Na^+-K^+ transport, is related to the hemolytic L antibody which we have differentiated on other grounds (see below, and also LAUF et al., 1971) from the K^+ pump stimulating L antibody. Nevertheless, the observation that the leak fluxes are also affected by anti-L revives an earlier concept of TOSTESON and HOFFMAN (1960), reiterated by DUNHAM and HOFFMAN (1971a), that K^+ pump and leak fluxes may be genetically coupled. Although this interesting finding may increase the complexity of the L antibody effect on the Na^+-K^+ transport system in LK sheep red cells, at present it does not bear on the conclusion we have reached with respect to the mode of activation of the Na^+-K^+ pump by anti-L.

b) Activation of the Na^+-K^+-ATPase

Stimulation of the LK membrane Na^+-K^+-ATPase by anti-L is consistent with the effect of the antibody on active cation fluxes in intact LK red cells (ELLORY and TUCKER, 1969a; LAUF et al., 1969 and 1970; LAUF and TOSTESON, 1972). The low baseline activity of the LK membrane enzyme, however, makes detailed kinetic measurements at 10^{-3} or 10^{-4} M ATP virtually impossible. More successful, but not without certain ambiguities, were the experiments on the Na^+-ATPase by BLOSTEIN et al. (1971) and BLOSTEIN and WHITTINGTON (1973), who used low ATP concentrations. In this group's studies, anti-L markedly increased the velocity of the Na^+-dependent ATP hydrolysis by 3–5 times (BLOSTEIN et al., 1971). Surprisingly, anti-L did not change the inhibition by K^+ ions characteristic for the LK enzyme, which was interpreted in terms of generation of 'new sites,' a conclusion now obviously not in accord with all studies on intact red cells. Although this complexity is not yet resolved, further work by BLOSTEIN and WHITTINGTON (1973) established that anti-L may exert a distinct effect on the Na^+-ATPase reaction in LK membranes which may be analogous to the effect of oligomycin. Pretreatment with oligomycin, which increases Na-E $\cdot$ P_i and activates the Na^+-ATPase, and thus probably shifts the equilibrium of its intermediates from E_1P to E_1 and to the Na^+-sensitive E_0 form of the enzyme, reduced the effect of anti-L when added subsequently, indicating that the action of the antibody may involve conformational changes associated with an increased affinity for Na^+ ions.

c) *Correlation between Antigenic Sites and Na^+-K^+ Pumps*

As pointed out earlier, there is a lack of a one-to-one correlation between the number of M and L antigens and the number of Na^+-K^+ pumps in HK and LK cells. On genetic grounds, it is not unexpected that the number of M antigens does not agree with that of the Na^+-K^+ pumps. If the L antigen were in fact a constituent of the LK pump, a ratio of L antigens/pumps closer to unity should be expected.

The key to understanding this dilemma may reside in the information we have gained from our experiment with the proteolytic enzyme trypsin (LAUF et al., 1971 and 1972) as well as from absorption experiments using LK goat red cells (ELLORY and TUCKER, 1970a). When LK cells were pretreated with trypsin, the dramatic stimulation of K^+ pump influx by anti-L was completely abolished. The effect could be prevented by preincubation with anti-L prior to the trypsin exposure (LAUF et al., 1971), suggesting specific action of the enzyme at a site close to the L antigenic determinant. The addition of ovomucoid, a trypsin inhibitor, also prevented the action of trypsin (LAUF et al., 1977). Table 5 summarizes these findings. The trypsin digestion of the outer cell membrane, however, did not abolish the hemolytic action of anti-L in the presence of complement. The inactivation of the anti-L effect shows a remarkable dependence on the concentration of the enzyme used and the depression of K^+ leak influx by anti-L appears not to be altered by the enzyme (LAUF et al., 1977). Since there was no substantial diminution of anti-L mediated complement hemolysis of trypsinized LK red cells (LAUF et al., 1971), it was concluded that there may be two distinct L antigenic sites and probably two L antibody specificities. One antigen may be associated with the LK pump (L_p site) and the other, independent of the pump, may mediate the complement-dependent lysis by anti-L (L_{ly} site) (LAUF et al., 1971).

Table 5. Abolition by Trypsin of the Anti-L_p Induced K^+ Pump Stimulation in LL (LK) Sheep Red Cells

Experimental Condition	${}^iM_K^P$ (mmol · l^{-1} cells × hr^{-1})	${}^iM_K^L$
Control LK cells	0.144	0.136
Control LK cells + anti-L	1.245	0.070
Trypsinized LK cells	0.142	0.145
Trypsinized LK cells + anti-L	0.167	0.083
Control HK cells	0.636	0.061
Trypsinized HK cells	0.672	0.083

Trypsinization: 2.5 mg enzyme ml^{-1} cell suspension
60 minutes incubation, 37° C.

The absorption experiments of ELLORY and TUCKER (1970a) led to a similar conclusion. Low-potassium goat red cells bound only the L antibody stimulating the sheep K^+ pump influx but did not adsorb the hemolytically active L antibody, clearly implying a dual specificity of anti-L. It is therefore likely that only a small fraction of the L sites are structurally associated with and indicative of a molecular alteration of the LK pump, while the majority of the L sites are functionally unrelated (LAUF and SUN, 1976). This hypothesis is not invalidated by our experiments attempting to correlate the number of L antibody molecules bound for maximal K^+ pump stimulation: correcting for nonspecific binding, at least 500–600 anti-L molecules must be bound before maximum stimulation of the LK pump occurs. There are of course some serious difficulties involved in the proper evaluation of a low number of surface antigens associated with a very low number of Na^+-K^+ pumps. First, we have not separated the number of L_p and L_{ly} sites on the membrane levels; second, we do not know exactly whether all of the anti-L molecules bound are attached bivalently to the allegedly pump-associated antigens; and third, we have no evidence that all L antibody molecules bound are also functionally active after the iodination procedure (LAUF and SUN, 1976).

It is then evident from the preceding discussion that the correlation of M and L antigens with the cation pumps is considerably complex at the membrane molecular level. LAUF and SUN (1976) have discussed the mode by which the M and L antigens may be associated with the Na^+-K^+ pumps. There are several possibilities. For example, it may be that M and L antigens are not both structurally related to the pumps, which genetically is unlikely (see also the work on LK goat red cells by KROPP and SACHS, 1974). Alternatively, some of the ML antigens may not be connected to the Na^+-K^+ pump, while others are structurally part of the Na^+-K^+ transport system. It is likely that at least a fraction of the L antigens (L_p sites) are part of the Na^+-K^+ pump in LK cells serving as antigenic determinants which mediate the effect of anti-L. Even if this is so, one must postulate several determinants per pump, *i. e.* some sort of antigenic clustering. Indirect evidence for clustering of M sites stems from our experiment correlating the degree of hemolysis with the number of M antibody molecules bound (LAUF and SUN, 1976). Complete hemolysis of HK red cells ensues when only some 100 M antibody molecules are bound per cell. On the basis of the observation by HUMPHREY and DOURMASHKIN (1969), this implies that some of these anti-M IgG molecules must be packed in doublets to activate C1, the first component of complement (in other systems at least 800 molecules of IgG antibody/red cell are necessary to provide one IgG doublet capable of activating the complement cascade). In LK red cells, however, the sites may be more randomly distributed since more than 100 L antibody molecules/cell were required for complete lysis by complement (LAUF and SUN, 1976). Although this is indirect evidence for singlet distribution of L sites (some of which may be L_p sites), clustering of L sites may also occur in LK cells. In ATP depleted LK cells, which are more susceptible to immune hemolysis by anti-L and complement (LAUF and DESSENT, 1973) fewer anti-L IgG molecules/cell are required to cause complete hemolysis, although the total number of L antibody molecules bound/cell does not change (LAUF and SUN, 1976). From the standpoint of molecular mutation it is certainly important to separate the L_p from the L_{ly} sites in order to establish how many L antigenic determinants per pump are involved when anti-L modifies active cation transport in LK sheep red cells.

3. Properties of the ML Surface Antigens and Antibodies

a) Antigens

The preceding section raises the question as to the chemical nature of the ML surface antigens and their antibodies. From section B of this chapter it will be recalled that highly purified Na^+-K^+-ATPase from various membranes consists of large, non-conjugated and smaller glycosylated polypeptide chains (KYTE, 1972). From earlier work of DICK et al. (1969) it is known that sulfhydryl (SH) groups are important for the function of sheep red cell Na^+-K^+-ATPase. If at least some of the M and L antigens are structurally associated with the enzyme, one would expect their presence in the large polypeptide chain, particularly for the L antigen. Accordingly, the biochemical properties of these antigens should resemble those of the Na^+-K^+-ATPase with respect of the large chains. However, it is also possible that the ML determinants are on the glycopeptides which may exert a regulatory (inhibitory) function in turn affected by anti-L and not by anti-M. In the last case, L and M antigens should behave like glycoproteins when extracted in appropriate solvents. Thus far, all our experimental evidence speaks against a glycoprotein nature of the M and L antigens, since extraction with n-butanol (LAUF and TOSTESON, 1969) or phenol (SHRAGER et al., 1972) did not yield antigenically active glycoproteins. Rather, the biochemical properties of these antigens are similar to those of complex, relatively insoluble membrane proteins. The M as well as the L activities were always found in the membrane pellets insoluble in polar solvents. As in the Na^+-K^+-ATPase (TOSTESON, 1966b), SH groups and intra- or intermolecular disulfide (S-S) bonds are necessary for full antigenic expression. These groups may be located deeper in the plasma membrane since their reduction and/or alkylation was only possible in the presence of protein dissociating reagents such as 6 M guanidine HCl (SHRAGER et al., 1972).

In addition, the apparently proteinaceous M substance requires membrane phospholipids and cholesterol for full activity (SHRAGER et al., 1972). The M-activity is independent of the source of lipids, as we found that lipids from LK membranes were as effective as lipids from HK membranes in restoring full M antigen activity. We have some preliminary evidence that the L antigen also requires lipid for its activity. However, work on this antigen is much more difficult, since first, there is even less L substance per cell than M substance, and second, the L antigen is even more labile to biochemical manipulation than the M substance (LAUF, 1974).

Nevertheless, the findings of our laboratory are consistent with a model in which proteinaceous components carry the antigenic determinants at the cell surface, and that part of these molecules, containing functionally important SH and S-S groups, may extend into and interact with the lipid bilayer. Future work must establish the molecular relationship of these proteins to the Na^+-K^+ pump system. It might be added here that trypsin, which is known to hydrolyze glycopeptides from red cell surfaces, may just cleave a crucial peptide bond which is not part of glycopeptides (LAUF et al., 1971). The fact that the glycoprotein released by the enzyme from LK red cells did not exhibit L antigen activity, however, does not prove that glycoproteins are not involved in the formation of the L determinant. It is possible that polypeptides that are conformationally altered but contain L antigen no longer bind the L antibody.

b) Antibodies

Among several immune sera tested, there is greater variability in the class specificity of anti-M than of anti-L. The first anti-M serum used to study its interaction with HK membranes contained most of the M antibody in the heavy molecular weight IgM (macroglobulin) protein (LAUF and TOSTESON, 1969). Subsequent studies used different lots of anti-M, which was always of IgG nature (bivalent immunoglobulin G, molecular weight 160000). The reason for this discrepancy is not obvious, and may be due to slight differences in the immunization schedule as well as the time of harvesting of the antisera. In contrast, the L antibody was always found among the IgG protein, and here only in the electrophoretically fast IgG_1 fraction. This is an interesting finding in view of a general immunological experience with synthetic antigens: positively charged molecules tend to elicit antibody molecules with an overall charge more negative than antibodies produced to negatively charged antigens (SELA and MOZES 1966). In terms of possible biochemical properties of the L substance, this finding may lead to the suspicion that the L antigen is not part of a highly negatively charged glycopeptide, a hypothesis consistent with the trypsin experiments discussed above.

There is some evidence that anti-L requires bivalency to fully stimulate the Na^+-K^+ pump in LK red cells (SYNDER et al., 1971). Papain hydrolysis of anti-L IgG_1, yielding monovalent anti-L, inactivated the dramatic action of anti-L on K^+ pump influx while pepsin digestion producing bivalent fragments failed to show any effect on the stimulatory action of anti-L. Using a different procedure to prepare monovalent anti-L, which consisted of preparing the monovalent fragments of anti-L by opening the inter-heavy chain S-S bonds of a bivalent antibody fragment abtained by pepsin hydrolysis, ELLORY et al. (1973) found only a 50 percent reduction of the original L antibody activity. Nevertheless, these authors also conclude that anti-L IgG probably is doubly attached when eliciting its full effect, even in ML cells. In terms of the molecular mechanism by which anti-L modifies the LK pump, these findings are consistent with at least two L antigenic determinants per pump to which the bivalent L antibody molecule must bind in order to produce maximum stimulation, or two subunits that must be approximated.

4. Developmental Aspects of Transport and Antigens

a) Red Cells of Newborn Sheep

Red cells of all newborn lambs exhibit an HK steady-state composition regardless of the cation phenotype predicted by progeny (Tosteson and Moulton, 1959; Blechner, 1961). Within a period of 60 days after birth the cellular cation composition in the genotypically LK sheep change to the final LK steady-state level observed in LK red cells of adult sheep (Tosteson, 1966b). Almost simultaneously, a progressive fall of the $^{i}M_K^P/^{i}M_K^L$ ratio occurs (Tosteson, 1966b), which is paralleled by a concomitant fall in the Na^+-dependent ATPase activity (Blostein et al., 1974). Dunham and Hoffman (1971b) demonstrated that the number of ouabain-binding sites and therefore Na^+-K^+ pumps remains constant in HK cells throughout the first 19 days of life, while that of genotypically LK lambs continuously decreased from 33 sites/cell to an average of about 14/adult red cell, perhaps involving a conversion of pump into leak sites. Tosteson's (1966b) early interpretation of this remarkable HK-LK transition phenomenon was that either fetal HK cells are gradually replaced by adult LK cells or that a gradual change occurs in the membrane properties of the initial cells, thus implying an interconversion of HK into LK cells.

In order to discern between the two possibilities, *i.e.* replacement versus membrane changes, it is useful: a) to look at the kinetic properties of the Na^+-K^+ pumps during development, b) to correlate the cation and ion transport changes with the reactivity pattern of the M/L antigens, and c) to study transport and immunological reactions in young and old red cells during the newborn period.

Using a newly introduced device to measure cell volumes (Kachel et al., 1970), we have recently determined cell volume, cations and surface antigens in newborn sheep of LK genotype (Valet et al., 1978). The data obtained strongly support the concept of Tosteson (1966b) that the newborn HK-LK transition is caused by cellular replacement. When the LK genotype lamb is born it has almost exclusively large (36 μm^3) HK red cells, which are gradually replaced by two subsequent cell populations, the first one with cells 28 μm^3 in volume and the final one made up of LK adult red cells each with a volume of about 30 μm^3. It is not yet clear whether the second, intermediate cell population is HK or LK in type. The data suggest that these cells may have an intermediate type of LK steady-state levels or are being converted to LK red cells on the basis of membrane changes (proposed as an alternative possibility by Tosteson, 1966b). Recent K^+ flux and ^3H-ouabain-binding experiments from our laboratory (Lauf et al., 1978) are consistent with the observation by Valet et al. (1978): red cells of 2-day-old LK lambs have high-K^+ pump fluxes and numbers of ouabain-binding sites characteristic of HK cells. As the lamb matures beyond 40 days, LK characteristics are dominant, such as LK steady-state cation levels, low-K^+ pump influx and smaller numbers of pump sites, and a typical response of the Na^+-K^+ pump to the stimulation by anti-L. Testing for the presence of the M antigen by absorption and hemolysis in the presence of complement showed that the antigen slowly appeared in heterozygous LK (LM) cells as the initial large cell population vanished, a finding consistent with that of VALET et al. (1978) that neither the M nor the L antigens are expressed on the first cell population after birth.

BLOSTEIN et al. (1974) observed that over 4 weeks the K^+ inhibition kinetics of the Na^+-ATPase in membranes from newborn LK genotype cells were characteristic of on HK enzyme even though the actual Na^+-ATPase decreased to the low levels found in mature LK red cells. Since the reduction in Na^+-ATPase was paralleled by the loss of fetal hemoglobin, the authors concluded that HK type fetal red cells are *replaced* by adult cells with LK membrane properties. Similar studies on the kinetics of K^+ pump fluxes have not been done as yet.

TUCKER (1968) showed that at birth the M antigens of HK red cells are either not present or only weakly developed. Similarly, the L antigen activity, tested by hemolytic assay, was very low in red cells of LK lambs of few days age (TUCKER and ELLORY, 1970). However, TUCKER and ELLORY (1970) found that the appearance of the L antigen was similar in old and young red cells of newborn sheep and coincided with a change in cation composition in both cell types. Persistance of fetal hemoglobin was longest (50 days) in the old cell fraction of the newborn blood. By differential hemolysis of the young LK red cells from 14- to 21-day-old lambs, TUCKER and ELLORY (1970) found that the fetal-type red cells were HK erythrocytes which had either weak or no L and M antigen activity, and concluded that the gradual HK-LK changeover is caused by appearance of new red cells with $[K^+]_c$ steady-state levels and antigens characteristic of adult LK red cells. ELLORY and TUCKER (1969c) studied K^+ uptake in newborn LK genotype red cells in the presence and absence of anti-L. As the red cell $[K^+]_c$ fell, stimulation by anti-L gradually increased. Together these data do not support the second alternative given by TOSTESON (1966b), that fetal red cells gradually mature into LK cells, but rather are consistent with TOSTESON's first proposal of cellular replacement.

Although much more quantitative work must be done in order to understand the complex population kinetics of the HK-LK cell transition in newborn lambs, the data from all laboratories provide quite a consistent picture of this process. Apparently, the LK genotype lamb is born with an HK red cell population that resembles mature HK red cells with respect to Na^+-K^+ transport and Na^+-K^+-ATPase but is different antigenically since the M antigen is weak or absent. Gradually, this population is replaced by a short-lived second one, and finally by a third population, the last being of LK nature and complete in its ML antigenic set. The mode of the functional linkup between the L substance and the Na^+-K^+ pump remains to be elucidated.

b) Stress-Induced Erythrocyte Regeneration

As all evidence supports the concept that cellular replacement takes place in the newborn, the experimental data for cellular regeneration after massive bleeding of adult sheep tend to be less consistent with this hypothesis. BLUNT and EVANS (1963 and 1965) first reported that red cells entering the circulation upon massive hemorrhage of LK sheep contained higher K_c^+ levels than those circulating before the blood loss. Subsequently, LEE et al. (1966) confirmed these findings, reporting that the young red cell population, when separated by density gradient centrifugation, possessed a strophanthidin-sensitive K^+ pump influx which, on day 6 after hemorrhage, was four to five times greater than prior to the experiment. The K^+ leak flux of these new cells was doubled. The young cells, monitored by ^{59}Fe labeling, had a normal volume, and it was assumed that the greater K^+ pump activity was due to a larger number of active transport sites per cell or a faster turnover rate per site. These authors concluded that the LK membrane properties developed rather late in the maturation process (LEE et al., 1966). Confirming their findings, ELLORY and TUCKER (1969b) showed that the young cells in anemic LK sheep have 186 ouabain binding sites per cell, relating the higher K^+ pump influx observed by LEE et al. (1966) to a higher number of pumps per cell. Furthermore, ELLORY and TUCKER (1969b) found that the L antigen was expressed on these young cells. However, it is not clear whether the K^+ pump increment can also be ascribed to a higher number of pumps per surface area, since ELLORY et al. (1970) found that the new cells entering the circulation were larger and had a greater surface area. In addition, data on the kinetic parameters of the Na^+-K^+ pump in these cells are lacking.

A study in this direction was made by BLOSTEIN et al. (1974) and DUNHAM and BLOSTEIN (1975). By changing K_c^+, it was found that increased $[K^+]_c$ stimulated the Na^+-K^+ pumps in reticulocytes of both anemic HK and LK sheep (see also discussion of K^+ stimulatory site in HK goat red cells, section III/1/b), while K_c^+ was a strong inhibitor of $^iM_K^P$ in *mature* LK red cells, as earlier found by LAUF et al. (1970). In spite of these HK characteristics of the LK reticulocyte, anti-L also stimulated K^+ pump influx in these cells (DUNHAM and BLOSTEIN, 1975), supporting ELLORY's and TUCKER's finding that the L antigen is present on reticulocytes of anemic LK sheep. At present it is not entirely clear why, in contrast to the K^+ pump flux data, the Na^+-ATPase of reticulocytes from anemic sheep exhibit LK kinetics with respect to K^+ ions (BLOSTEIN et al., 1974). It is therefore difficult to draw any conclusion regarding the exact process of maturation of the cation transport parameters, since it cannot be decided at present whether during this process the L antigen is involved in changing an HK-type into an LK-type pump system (*i. e.* reduction in number of pumps and turnover/site) or a very active LK-type pump into one of lower activity (*i. e.* depression of turnover/site only).

Preliminary experiments carried out at the Max Planck Institute in Munich indicate that in anemic sheep the erythropoiesis follows entirely different patterns than in the newborn animal (VALET and LAUF, unpublished data). Shortly after phlebotomy, larger cells (reticulocytes) appear in the circulation (see also ELLORY and TUCKER, 1969c), and no population trichotomy was found such as is observed in the newborn lamb. It is likely that the newly emerging cells may be dissimilar to true HK cells since their volume is greater and thus their $[K^+]_c$ probably lower than previously assumed. If these findings can be verified, then the immature reticulocyte of anemic LK sheep appears to be predetermined to lose its K_c^+, becoming a mature LK cell by 'membrane differentiation'. This could mean that hematopoiesis is characterized in the newborn lamb by population replacement and in the adult by premature release of one and the same cell population which defines the adult LK sheep.

III. Cation Transport Polymorphism and Antigenic Parameters in Red Cells of Ruminants Other than Sheep

1. Goat Red Cells

a) Cations and Antigens

RASMUSEN and HALL (1966b) analyzed the K_c^+ levels of 84 goats and, confirming earlier work of EVANS and PHILLIPSON (1957), classified these cells as either HK- or LK-type, the latter including HK cells with some intermediate K_c^+ levels. In comparison to the low-K_c^+ steady-state levels in LK sheep red cells, LK goat red cells have K_c^+ levels ranging from about 25 to 60 mM l^{-1}cells (cutoff point between LK and HK type is around 60–65 mM l^{-1}cells; RASMUSEN and HALL, 1966b). In general, similar findings have been made by ELLORY and TUCKER (1970a). The concentration of K_c^+ in HK goat red cells varied from above 60 to 100 mM l^{-1}cells (RASMUSEN and HALL, 1966b; ELLORY and TUCKER, 1970a). Since the dividing line between LK and HK types lies around 60 mM l^{-1}cells, and there is considerable variation among the $[K^+]_c$ of animals, it is likely that the LK gene is less dominant in goats than in sheep. Thus the goat red cell system may be placed between the strictly dimorphic sheep and the more polymorphic cattle red cell systems for which codominance of the LK (Ke^L) and HK (Ke^H) genes has been considered (see below).

When ovine anti-M and anti-L were used, no close correlation was found between HK-type red cells and the presence of the M antigen (RASMUSEN and HALL, 1966b); furthermore, LK goat red cells could not be hemolyzed by anti-L in the presence of complement (ELLORY and TUCKER, 1970a). The serological (hemolytic) activity of anti-L could not be adsorbed onto LK goat red cells, since anti-L absorbed by these cells still hemolyzed LK sheep red cells in the presence of complement (ELLORY and TUCKER, 1970a). This interesting finding supports the concept outlined in section C/II/2/c that ovine anti-L contains at least two L antibodies, one of which has serological reactivity. The other L antibody (anti-L_p) is directed against the LK pump-associated L antigen: it stimulates the K^+ pump influx and Na^+-K^+-ATPase also in LK goat red cells and this serologically silent L antibody can be eluted in physiologically active form from LK goat cells (ELLORY and TUCKER, 1970a). Furthermore, sheep anti-sera with higher anti-L_p activity may be serologically weak and vice versa (ELLORY and TUCKER, 1970a). On the basis of this antigen and antibody diversity, one would expect that the number of L antigens in LK goat red cells is much lower than that reported for LK sheep red cells (LAUF and SUN, 1976), a proposition borne out by the finding of KROPP and SACHS (1974) that LK goat red cells appear to have only some 60 L antigenic sites per cell reacting with sheep anti-L.

b) Cation Transport and its Modification by Antibody

Active K^+ pump influx in HK goat red cells is about twice as high as in LK goat cells (ELLORY and TUCKER, 1970a). Similarly, the Na^+-K^+-ATPase of HK cells is some three times greater in membranes from HK than in membranes from LK cells (ELLORY and TUCKER, 1970a). The major difference between goat HK and LK pumps appears to reside in their kinetic properties rather than in any quantitative variance, since the rate of inhibition by ouabain and thus probably its binding rate are greater in HK than in LK red cells (SACHS et al., 1974b), while the number of ^{3}H-ouabain-binding sites is similar in the two cell types. SACHS et al. (1974b) reported 57 ouabain binding sites for HK and LK cells (measured in the presence of 5 mM $[K^+]_o$), and they surmised that a considerable nonhomogeneity of the pumps may exist with respect to their interaction with ouabain. Since the Na^+-K^+ pump rate is greater in HK than in LK cells, the findings of SACHS et al. (1974b) imply that, with similar or even identical numbers of pumps, the turnover number for HK pumps must be higher than for LK pumps. As in the sheep system, the kinetic differences between HK and LK pumps of goat cells reside in their different response to K^+ ions at the Na^+ loading site of the pump. At high intracellular $[Na^+]_c$ (using the PCMBS method of GARRAHAN and REGA, 1967) low concentrations of K_c^+ ions stimulate, and higher K_c^+ concentrations inhibit the HK pump, while the LK pump is increasingly inhibited at higher $[K^+]_c$. When $[Na^+]_c$ is lowered in HK cells, K_c^+ ions fail to stimulate and, as in LK cells, only inhibit their pumps (SACHS et al., 1974a). Thus it was concluded that in LK goat red cells, K_c^+ effectively competes with Na_c^+ ions at the internal aspect of the Na^+K^+ pump. In HK cells this K^+ competition becomes evident only when $[Na^+]_c$ is considerably reduced. In HK goat red cells, therefore, the affinity of the Na^+ loading site is lower for K_c^+ than for Na_c^+, and in LK goat red cells the affinity for K_c^+ is higher than that for Na_c^+. The observation that low $[K^+]_c$ acts as stimulator of the Na^+-K^+ pump of HK goat red cells with high $[Na^+]_c$ was interpreted in terms of a distinct stimulatory K^+ site (SACHS et al., 1974a). The profiles of the internal K^+ inhibition kinetics show additional differences between HK and LK goat red cells (*e.g.* pump heterogeneity), which are of secondary importance in regard to our discussion (see SACHS et al., 1974a). Consistent with the K^+ influx kinetics, GLYNN and ELLORY (1972) showed that K^+ ions are also powerful inhibitors of the Na^+-K^+-ATPase in ghosts of LK goat red cells.

Ovine anti-L stimulates the Na^+-K^+ pump of LK goat red cells to 1.5- to 8-fold activity and the Na^+-K^+-ATPase to at least twofold (ELLORY and TUCKER, 1970a). This dramatic effect of anti-L is mainly of a kinetic nature and does not apply to passive K^+ influxes in these cells. At very low cellular $[K^+]_c$ anti-L increases the apparent affinity of the pump for Na_c^+ by decreasing the affinity for K_c^+ ions. At about normal, low $[K^+]_c$, however, the effect of anti-L in reducing K^+ pump inhibition by K_c^+ is much less pronounced (SACHS et al., 1974a). Aside from the inhibitory action, K_c^+ is stimulatory for the HK pump at low $[K^+]_c$ and high $[Na^+]_c$, and is also required for the anti-L-mediated stimulation of the LK pump, since at very low $[K^+]_c$ practically no increment of K^+ pump influx is caused by anti-L (SACHS et al., 1974a). It has been argued that there is also a K^+ stimulatory site in LK cells, but that the action of K_c^+ on this site cannot be seen since K_c^+ ions are masking this effect by their potent inhibitory effect at the Na^+ loading sites of the pump. GLYNN and ELLORY (1972) showed that the inhibitory action of K^+ ions on the Na^+-K^+-ATPase of LK goat membranes is also reduced by anti-L, and a recent report clearly demonstrated that anti-L did not alter V_{max}

of the enzyme but decreased the K/Na affinity ratio from 6 to 2 (CAVIERES and ELLORY, 1974).

Thus, as in the sheep red cell system, anti-L primarily alters the kinetic properties of the pump in LK goat red cells by exerting a specific conformational change at the cytoplasmic aspect of the pump. This change consists of an increase in the affinity of the pump or Na^+-K^+-ATPase for Na_c^+ ions and may involve occupation of a stimulatory site by K_c^+ ions, both required for a pump conformer conducive to a higher turnover. Consistent with this conclusion is the finding of SACHS et al. (1974b) that anti-L caused an increase in the rate of ouabain binding. To cause this conformational change, as stoichiometric (1 : 1) binding of anti-L to the LK pump appears to be required, since KROPP and SACHS (1974) reported that maximal K^+ pump stimulation was attained when about 60 L antibody molecules were bound to LK goat red cells (with about 55 pump sites). This close correlation, of course, suggests but does not prove that the L antigen (L_p site) in LK goat red cells is structurally part of the pump. In sheep red cells, however, no such close relationship has been found as yet, and the reasons for this discrepancy have been discussed earlier.

2. Cattle Red Cells

In contrast to the sheep red cell system and somewhat more comparable to the goat cells, cattle red cells do not exhibit a clear-cut dimorphism of cellular cation composition. In general, most cattle red cells have low- (about 20 mM l^{-1}cells) or intermediate-(about 50 mM l^{-1}cells) K_c^+ steady-state levels (RASMUSEN et al., 1974b; CHRISTINAZ and SCHATZMANN, 1972). Animals with high $[K^+]_c$ (about 70 mM l^{-1}cells) are rare, and it has been argued that selection pressure for animals with higher milk production was against HK types (RASMUSEN et al., 1974b). On the basis of progeny studies, RASMUSEN et al. (1974b) hypothesized that in cattle the red cell K^+ levels are controlled by two codominant alleles, Ke^L and Ke^H. The boundaries between the three cation types, however, are not sharp, and HK cattle red cells with as much as 100 mM $[K^+]_c l^{-1}$ cells have been reported (RASMUSEN et al., 1974b) in some breeds. Independent of particular breeds, red cells from all calves are HK type (more than 80 mM $[K^+]_c l^{-1}$ cells) and, as in sheep, the transition to the final LK state occurs within 4–5 months after birth (RASMUSEN et al., 1974b). In a detailed study, ISRAEL et al. (1972) showed that the time required for half change ($t_{1/2}$) of both Na_c^+ and K_c^+ ions is 35–37 days, and is thus somewhat longer than observed in newborn sheep. The fall of $[K^+]_c$ and rise of $[Na^+]_c$ was paralleled by a slightly preceding loss of K^+ pump and Na^+-K^+-ATPase activity, while the ouabain-insensitive K^+ leak influx increased (ISRAEL et al., 1972). Interestingly, the early cells of newborn calves containing fetal hemoglobin remained HK in type, while blood obtained from animals close to $t_{1/2}$ also contained young cells of HK type, but now with adult hemoglobin (ISRAEL et al., 1972). In the light of our findings in newborn sheep, the observation by ISRAEL et al. (1972) may indeed be consistent with the replacement hypothesis.

The different cation steady-state levels of cattle red cells are maintained by appropriate Na^+-K^+ pump and Na^+-K^+-ATPase activities. Thus ELLORY and

TUCKER (1970c) found that in HK cattle red cells the ouabain-sensitive K^+ pump flux is about twice as high as in LK cattle cells. The Na^+-K^+-ATPase activity was found to parallel the K_c^+ steady-state levels (SCHATZMANN, 1974), but ELLORY (1974) pointed out that it is not the absolute amount of Na^+-K^+-ATPase that is different but the relative affinities for K^+ ions at the internal site. Indeed, K^+ inhibition of the ouabain-sensitive ATPase was proportional to the K_c^+ levels of the cells from which the membranes were prepared (ELLORY and CARLETON, 1974). Thus, in contrast to the numerical and qualitative differences found between HK and LK sheep red cell pumps, HK *and* LK cattle red cells may have similar numbers of Na^+-K^+ pumps with kinetically similar (but varying) response to K^+ ions. Another possibility, pointed out by ELLORY and CARLETON (1974), is that cattle red cells of HK type possess HK and LK pumps.

The L as well as the M antigens have in fact been detected on cattle cells with sheep antibodies. However, their association with the cation type of cattle red cells is not as readily evident as in sheep red cells. First, M antigen-positive and L-negative cattle red cells may, unlike MM sheep red cells, have low-K_c^+ steady-state levels, and there may be no significant difference in $[K^+]_c$ between M-positive and L-positive cattle red cells. Second, not every ovine anti-L stimulates the Na^+-K^+ pump in LK cattle red cells (RASMUSEN, pers. comm.). Third, ovine anti-L, which hemolyzes some cattle red cells, does not lyse all LK cattle red cells. For example, anti-L hemolyzes (in the presence of complement) only cattle red cells positive for cattle blood group S (RASMUSEN et al., 1974b). However, RASMUSEN et al. (1974b) attribute, at most, a minor relationship of the S factor to red cell $[K^+]_c$. A similar inference may be drawn for the established correlation (or homology) between blood group S_2 and U_2, detected by cattle iso-antisera, and the M antigen on cattle red cells cross-reacting with ovine anti-M (RASMUSEN and HALL, 1966b).

The foregoing discussion has revealed that cation and antigen polymorphism appears to be at least as complex in cattle red cells as in sheep and goat red cells. Yet if HK and LK cattle red cells have similar numbers of Na^+-K^+ pumps with varying internal K^+ affinities, and provided the L antigen is considered as an expression of a given LK conformer status of the Na^+-K^+ pump, it may be possible to follow immunologically the dynamics of the conformer attainment in cattle red cells of different internal K^+ concentrations. Such an experimental approach, however, may not be feasible at present, since ovine anti-L, although reacting with some cattle cells, may not have the narrow immunological specificity which a bovine anti-L would offer to monitor fine conformational changes of the Na^+-K^+ pump through the L substance. Attempts to produce bovine iso-immune anti-L sera have failed (ELLORY and CARLETON, 1974).

IV. Human Red Cells

1. The Rhesus Antigen Complex and Cation Transport

The Rhesus antigen complex, which is clinically important in blood incompatibility reactions, appears to belong to integral membrane constituents of human red cells since its activity resides in a non-glycoprotein and lipid-dependent membrane protein moiety. Removal of lipids by n-butanol (POULIK and LAUF, 1965; REGA et al., 1967) inactivates the antigen activity, which can be restored by recombination with the extracted phospholipid (GREEN, 1967 and 1972). There is evidence that functional SH groups or S-S bonds are required for full Rh antigen activity since reduction and alkylation of Rh(D)-positive membranes led to inactivation (LAUF and POULIK, 1968). Denaturation of membrane properties in 4 M urea also causes destruction of the Rh antigen (LAUF and POULIK, 1968). These biochemical characteristics compare in some way with those observed for the Na^+-K^+-ATPase of red cell membranes (SH group and lipid requirements, inactivation in urea of guanidine/HCl) and suggested to some workers that the Rh antigen complex may be structurally similar or identical to the molecules comprising the Na^+-K^+ pump (GIBLET, 1969).

A test for this hypothesis of a correlation between the Rh antigen complex (primarily the D antigen, which is the principal determinant of the Rh [CDEce] complex) would involve a similar experiment as reported for the sheep, goat and cattle red cells. Anti-D, however, did not alter cation transport in Rh(D) red cells (LAUF and JOINER, 1976). Similarly, there was no effect of anti-D on the Na^+-K^+-ATPase (LORUSSO et al. 1977). Another way to test the hypothesis above is to study the cation transport system in Rh_{null} red cells which, by gene depletion or gene regulation, completely lack the Rh antigen complex (not to be confused with Rh-negative red cells, which do not possess the major D determinant). For other details on the Rh_{null} cell system consult LAUF (1977 and 1978). We recently analyzed the Na^+-K^+ transport parameters of the Rh_{null} red cells of the deletion type (SEIDL et al., 1972), and found that Rh_{null} red cells have a greater K^+ leak influx which appears to be almost completely compensated by a higher Na^+-K^+ pump activity (LAUF and JOINER, 1976). Interestingly, the higher Na^+-K^+ pump fluxes were entirely due to a greater number of Na^+-K^+ pumps as measured by ^3H ouabain binding, and not to a change in the turnover number, since the rate of ouabain binding was indistinguishable from that in controls. In addition, the Na^+-K^+-ATPase activity was increased in proportion to the K^+ pump flux increment (LAUF and JOINER, 1976). Since the number of Rh(D) antigens/cell is known to be greater than 10000 (HUGHES-JONES et al., 1963; MASOUREDIS et al., 1967; NICOLSON et al., 1971) and the number of Na^+-K^+ pumps was even higher in Rh_{null} red cells (670/cell) than in Rh(D) cells (450/cell) a close structural and functional identity between Rh antigen complex and Na^+-K^+ pumps is unlikely (LAUF and JOINER, 1976).

2. The En(a)-Negative Red Cell as Physiological Model

Human red cells lacking the common En(a) antigen (En[a-] or En cells) have an approximately 60 percent lower sialic acid content and a reduced electrophoretic mobility (FURUHJELM et al., 1969). The membrane molecular basis of this defect has now been resolved by TANNER and ANSTEE (1976). Their findings are of considerable interest to the membrane physiologist attempting to correlate surface antigens with transport processes. TANNER and ANSTEE (1976) found that En(a-) cells lack the major glycoprotein and apparently have an abnormal band-3 protein with a higher apparent molecular weight than that from control En(a) membranes. This difference in the band-3 protein is probably due to an increased carbohydrate content rather than to a change in the protein composition. Since the band-3 protein has been shown to be involved in anion exchange transport (ROTHSTEIN et al., 1976), water transport (BROWN et al., 1975) and possibly active Na^+-K^+ transport (KNAUF et al., 1974a, and b), the En(a-) cell with its altered band-3 protein offers an interesting genetic mutant for the physiologist. Although no physiological studies have been done on this antigenically deficient cell, it may be anticipated that the absence of the membrane glycoprotein (glycophorin of MARCHESI et al., 1973) may not alter active and passive Na^+-K^+ transport, since no other abnormality has been reported (except a slightly higher osmotic fragility; DARNBOROUGH et al., 1969). In addition, anti-En(a) did not modify the Na^+-K^+ transport system in normal En(a) red cells (LAUF, 1978).

V. Conclusion

The previous section described in detail how specific antigen-antibody reactions were applied to a unique genetic model system in which kinetically different cation pumps are associated with distinct membrane surface antigens. At the outset of our work on cation transport and membrane antigens in sheep red cells, it was evident that, although studying the 'simple' red cell, we had chosen an experimental model system of extraordinary complexity. The system required equally complex and sophisticated techniques to resolve the mystery of the dramatic stimulation of the LK cation pump by anti-L. A host of membrane immunological and physiological parameters were defined: the effect of antibodies on partial aspects of Na^+-K^+ pump fluxes and the Na^+-K^+-ATPase; the quantitative and chemical correlation between antigens and pumps; and the emerging association between antigens and transport in the differentiating and maturing erythrocyte. Experimentally, these studies involved the following measurements: cation fluxes across the red cell membrane and determination of the ATP hydrolysis by the Na^+-K^+-ATPase; an analysis of the relative affinities of the pump for the ions translocated across the membrane and for pump inhibitors binding at the outer membrane surface; the immunochemical demonstration of the specificity and valency of antibody binding and the quantitation of heterogeneous surface antigens; and the biochemical definition of complex lipid-dependent membrane antigens. Thus, an integration of membrane immunology with membrane transport physiology evolved as the prime characteristic of our experimental strategy, which is not limited to the esoteric study of these ruminant model red cells but rather is quite applicable to the future analysis of related genetic mutants in red cells of other species. In studying for the first time the Na^+-K^+ transport properties of the rare human Rh_{null} red cell variant, we have illustrated the validity of our immunophysiological approach. In the final section (D) of this chapter it will be shown that again a combination of immunology with membrane physiology may be most appropriate to define the modes by which membrane immunological reactions alter the cation permeability in lymphocytes and tumor cells.

D. Membrane Immunological Reactions and Cation Transport in Lymphocytes and Other Cells

I. Lymphocytes

1. Introduction

Immunological reactions have been shown to modify active and passive transport of cations and other solutes in lymphocytes. In some instances, these membrane transport changes occur minutes after the immune reaction has taken place and, ever since they have been discovered, debates have been going on as to whether or not alterations of membrane transport are the first and crucial events triggering a chain of subsequent cellular processes. These processes precede and/or accompany the biochemical and morphological metamorphosis of primitive, undifferentiated lymphocytes into highly specialized and immunologically competent cells. All of these observations have been made in *in vitro* experiments where purified lymphocytes are exposed to a variety of immunological ligands which may or may not be important *in vivo* but which certainly may simulate processes that occur *in vivo*. These studies therefore inherently tend to underestimate the greater functional and morphological heterogeneity of these cells *in vivo*. Prior to our discussion of the membrane transport changes, it is imperative to review briefly the cellular, functional and receptor heterogeneities known to exist in lymphocytes. This review draws on several thorough and excellent references, which the reader should consult for more detailed information (GREAVES et al., 1974; EDELMAN, 1974; ADA and EY, 1975; SHARON and LIS, 1975; GERSHORN, 1974; LAWTON and COOPER, 1974; WIGZELL, 1974; EDELMAN, 1976).

2. Cellular Differentiation and Membrane Surface Receptors of Lymphocytes

Lymphocytes populate spleen, thymus, and a large number of other reticuloendothelial organs and tissues (tonsils, peripheral lymph nodes, Peyer's plaques of the intestinal walls), and circulate between lymph and blood vessels. Immunological research has established that lymphocytes differentiate from an omnipotential precursor cell into circulating lymphoid stem cells, which subsequently undergo an important specialization into cells capable of *cellular* and *humoral* immune defenses. In birds, this specialization occurs in the thymus and in the 'bursa Fabricia,' and in mammals in the thymus and most probably in the bone marrow. Thus one distinguishes thymus-derived or T lymphocytes from B lymphocytes originating in the bursa or bone marrow. Following their differentiation, T and B cells reappear in the circulation, reseed in spleen and other reticuloendothelial organs, and are then ready for their function in the immunological defense.

Thymus-derived lymphocytes carry out primarily the so-called cellular or tissue immune response, which physiologically is poorly understood and involves cell-cell interactions between T cells and the challenging 'foreign' cell, often resulting in 'killing' of the target cell by the T-cell (killer cell) (BERKE and AMOS, 1973). This process of cell killing seems to require the presence of certain receptors on the target cell (antigenic determinants, cell-bound immunoglobulins, etc) as well as bivalent cations (GOLSTEIN, 1975) and is very temperature-sensitive (MARTZ et al., 1974). Physiologically, the mechanism of cell killing resembles colloid osmotic immune cytolysis (HENNEY, 1974) as known for erythrocytes hemolyzed by antibody and complement (MAYER, 1973; LAUF, 1975b), but a further rigorous biophysical analysis of this phenomenon is necessary. For a detailed analysis of the physiological basis of colloid osmotic immune cytolysis the reader is referred to a recent review (LAUF, 1978).

Upon challenge by soluble antigens, B cells, circulating or resting in the reticuloendothelial system, transform into plasma cells which have a characteristic morphology (plasma-cell nucleus and large basophilic cytoplasm) and produce large quantities of immunoglobulins. These molecules have been previously discussed (Sections B and C) in terms of their high specificity for antigenic determinants on membrane proteins, polysaccharides and lipids (consult also NISONOFF et al., 1975; LAUF, 1975a).

In addition to individual T and B cell responses, T and B cells often interact in a complex manner, although the biochemical events, such as increase in protein and nucleic acid synthesis, accompanying cellular activation and transformation are common to both cell types. The basis of the T and B cell differentiation has been worked out *in vitro* by experiments employing a large variety of immunological ligands which are specifically recognized by T and B cell membrane receptors. The observation that many of these ligands not only bind but also induce the differentiation and peculiar activities of T and B cells is most relevant in terms of the usefulness of these molecular probes to study the early alteration of membrane transport processes. Table 6 defines T and B cells in terms of some receptors detected on these cells by antibodies, lectins and particular cellular reaction modes. All these immunological tools can be equated to *mitogens* once they initiate cellular transformation and mitosis.

It is evident that T and B cells can be identified on the basis of their capacity to bind certain ligands. For example, the θ and H_2 alloantigens occur only on mouse T cells while immunoglobulin receptor molecules and lipopolysaccharide- or polyanion binding sites (DIAMANTSTEIN et al., 1974) are restricted to B cells. Other ligands do not discriminate between the two cell types with regard to binding (all lectins), but do not always induce cellular transformation (nonaggregated phytohemagglutinin [PHA], Concanavalin A [Con A], and lentil lectins bind to T and B cells but activate only T cells). Anti-immunoglobulin (IgG)

Table 6. Distinction between T and B Lymphocytes by Means of Binding and Effects of Some Immunological and Other Ligands

Ligands	T cells		B cells		Ligand Used (+) in Transport Studies
	Binding	Stimulation	Binding	Stimulation	
Plant agglutinins (lectins)					
Phytohemagglutinin (PHA)	+	+	+	−	+
Concanavalin A (Con A)	+	+	+	−	+
Lentil lectins	+	+	+	−	−
Pokeweed mitogens (PWM)	+	+	+	+	−
Antibodies					
Anti-IgG sera	−	−	+	+	+
Anti-lymphocyte sera (ALS)	+	(±)	+	+	+
Anti-Θ antigen sera	+	+	−	−	−
Anti-H2 (mouse) sera	+	+	−	−	−
Cellular "ligands"					
Sheep red cells	+	+	−	−	−
Graft tissue	+	+	−	−	−
Chemicals and enzymes					
$NaIO_4$	+	+	+	+	−
Lipopolysaccharides	−	−	+	+	−
Polyanions	−	−	+	+	−
Glycosidases	+	+	+	+	−
Proteases	+	+	+	+	−

(+) = binding or stimulation
(−) = no binding, no stimulation

sera and lectins such as PHA or Con A have been shown to modify cation transport in peripheral lymphocytes which constitute a mixture of T and B cells (Quastel and Kaplan, 1970a; Averdunk, 1972 and 1976; Averdunk and Lauf, 1975; Segel et al., 1975, 1976a and b). Thus active and passive cation transport are enhanced by these molecules mainly in T cells, and activities are therefore generally underestimated when using peripheral lymphocyte preparations containing both cell types (Table 6, last column). No membrane cation transport studies have been performed for the majority of the conditions listed in Table 6. There are also receptors for a number of other biologically active molecules such as complement factors C1q,4 and 3b (Dierich and Reisfeld, 1975; Sobel and Bokisch, 1975). In addition, chemical treatment of lymphocytes results in similar biochemical and morphological changes seen with lectins and other immunological ligands: early uptake of amino acids, an increase in protein, and nucleic acid synthesis have been produced by oxidation of cell surface carbohydrates with sodium periodate (Novogrodosky and Katchalski, 1972; Novogrodosky, 1975; Parker et al., 1974; Dixon et al., 1975), but ion permeabilities have not been studied under these conditions. However, it is most likely that the passive cation permeability is altered, resulting in an activation of the Na^+-K^+ pump mechanism. Similarly, treatment of lymphocytes with carbohydrate attacking enzymes causes blast cell transformation (NOVOGRODOSKY and KATCHALSKI, 1973 a and b).

3. Cellular and Membrane Morphological and Biochemical Changes Induced by Immunological Reactions in Lymphocytes

When lymphocytes are incubated *in vitro* with any of the immunological ligands discussed above, characteristic changes in cell morphology and membrane structure are observed. Morphologically, these include the cellular enlargement already mentioned, in particular an expansion of cytoplasm and change in the appearance of the nucleus, indicating increased protein and nucleic acid synthesis. The membrane changes have been defined as 'patch' and 'cap' formation (reviewed by GREAVES et al., 1974) and are very sensitive to experimental conditions (temperature) and the molecular form in which the ligands are presented to these cells. Membrane patches and caps can be recognized with the aid of fluorescein-labeled ligands. Patch formation occurs at 37° and 4° C; it appears to be independent of energy and is not affected by metabolic inhibitors. For certain ligands, patch formation is considered to be a cross-linking of membrane receptors. Cap formation, however, happens only at 37° C: fluorescein-labeled ligands or even whole cells (sheep red cells form rosettes on the T cell surface; BACH, 1973) conglomerate in a cap toward the 'uropod' of the cell (ROSENTHAL and ROSENSTREICH, 1974). It is interesting that this process of cap formation is not restricted to lymphocytes, since it can also be induced in fibroblasts with anti-H_2 sera as well as in basophils with anti-immunoglobulin E antibodies, as shown by others.

The membrane morphological changes depend on type, nature and concentration of the ligands. This has been studied in particular with Con A by the group of EDELMAN (EDELMAN, 1974) and others (BERLIN et al., 1974), because the molecular structure of Con A is best defined (EDELMAN et al., 1972). This lectin has also been extensively used for monitoring cation transport changes (AVERDUNK and LAUF, 1975; AVERDUNK, 1976). At concentrations above 10–20 $\mu g\ ml^{-1}$, Con A inhibits capping by "restriction of receptor mobility" at 37° C (YAHARA and EDELMAN, 1972); this restriction can be abolished by colchicine and vinblastin treatment of the Con A-exposed lymphocytes (YAHARA and EDELMAN, 1973). However, if lymphocytes are exposed to Con A (even in excess concentrations above 10–20 $\mu g\ ml^{-1}$) at low temperatures, washed, and then incubated at 37° C, cap formation is possible (EDELMAN, 1974). EDELMAN and co-workers (1974) have constructed a model to account for these observations. Apparently the valency of the tetrameric Con A molecule is important, since dimeric succinylated Con A, although it binds and forms patches, is unable to induce the temperature-dependent cap formation and dosage-dependent unimodality (GUNTHER et al., 1973) unless anti-Con A antibodies are added to restore the higher valency by cross-linking. A similar observation has been made recently with monovalent Con A, which binds to the same receptor sites (WANDS et al., 1976). The model of EDELMAN (1974) proposes that 37° C capping of Con A-bound receptors involves their dislodging from a colchicine-labile microtubular system beneath the plasma membrane, implying that the Con A receptors are part of a glycoprotein extending through the lipid bilayer of the lymphocyte membrane and anchoring the receptor molecule to these submembraneous structures. At lower temperature or with colchicine, the receptor molecules are in a non-anchored state, *i. e.* they may coalesce to caps when the temperature is raised.

The purpose of the preceding discussion was to demonstrate the extraordinary complexity of the lymphocyte/immune-ligand interaction, which must be kept in mind when interpreting membrane permeability changes caused by these ligands (see section D/4). The unimodality and temperature dependence of lymphocyte membrane changes are not restricted to Con A alone; it has been shown that PHA and other mitogens behave similarly. Moreover, transport of aminoacids and nucleosides is maximally increased at lectin concentrations similar to those found to be optimal for capping (AVERDUNK, 1972). However, it should be pointed out that stimulation of transport of these solutes does not require capping, since it can be induced by Con A, PHA, and lentil lectins in lymphocytes without activation of these transport processes and, furthermore, Con A increases amino acid transport and nucleoside uptake even under conditions of restricted receptor mobility (WANG et al., 1975; EDELMAN, 1976).

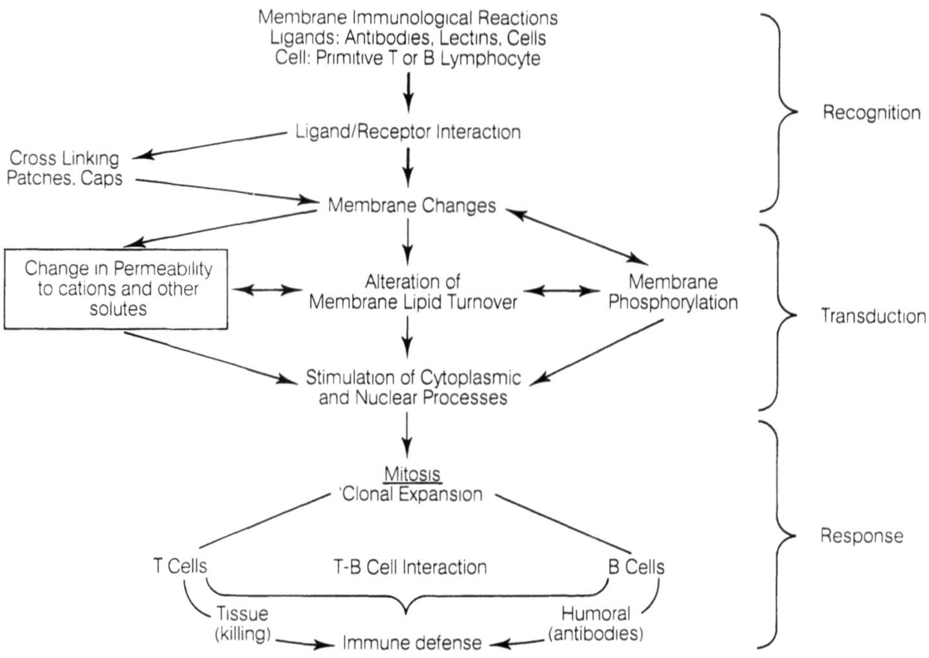

Fig. 1

'Unimodal' activation of Na^+-dependent aminoacid transport by PHA and Con A has been studied extensively by VAN DEN BERG and BETEL (1971, 1973a, 1973b and 1974; see also VAN DEN BERG, 1974), using 2-amino isobutyric acid (AIB), glycine, and asparagine. The Na^+-independent transport of other amino acids (leucine, aspartic acid) was unaffected. The effect of these mitogens was seen as early as 5 minutes after their addition to lymphocytes (VAN DEN BERG and BETEL, 1973a) and only V_{max} and not K_m of the carrier system was altered (VAN DEN BERG and BETEL, 1973b). However, no rigorous analysis of the uphill AIB transport stimulation has been made in terms of Na^+ requirements on both sides of the membrane and flux ratios. Similar to the activation of the Na^+-dependent amino acid transport, the uptake of labeled nucleosides is also stimulated by mitogens reaching maximum uptake some 60–72 hours following the addition of the stimulants to lymphocyte cultures, and the optimum concentration for PHA was found to be around 10 μg ml^{-1} (VAN DEN BERG, 1974). As shown in diagram form in Figure 1, the ultimate effect of immunological reactions on lymphocytes culminates in an increase in protein and nucleic acid synthesis (AHERN and KAY, 1975) required for the "clonal selection and expansion" (EDELMAN, 1974) of these cells into immunologically competent T and B cells. This transformation is accompanied (or necessitates) a significant alteration of membrane lipids consisting of increased phospholipid turnover and a two- to threefold stimulation of acyl transferase activity (RESCH and FERBER, 1972; RESCH et al., 1972; FERBER and RESCH, 1973; FERBER et al., 1974). It is likely that these fundamental changes in membrane lipids influence membrane fluidity (FERBER et al., 1974), an alteration to be considered in terms of membrane cation transport changes.

4. Modification of Monovalent Cation Transport

a) General Aspect of the Effect of Immunological Reactions

The discovery that immunological surface reactions alter the lymphocyte membrane permeability for monovalent cations is rather recent (QUASTEL and KAPLAN, 1970a; AVERDUNK, 1972). The main conclusion from this early work, that various immunological mitogens affect primarily the energy-dependent, active Na^+-K^+ transport in lymphocytes, was supported by two very recent papers on the effect of these substances on the Na^+-K^+-ATPase and ouabain binding (AVERDUNK and LAUF, 1975; QUASTEL and KAPLAN, 1975). However, work from LICHTMAN's group (SEGEL et al., 1975, 1976a, b and c; SEGEL and LICHTMAN, 1975) has shown that treatment of lymphocytes with similar mitogens caused an early significant and selective change in passive K^+ permeability and thus contradicts the hypothesis that it is the Na^+-K^+ pump system which is primarily affected by these reactions. It is important to resolve this controversy since the first concept implies that an early activation on normovectorial Na^+-K^+ pump transport may result in at least temporarily higher $[K^+]_c$ conducive to an enhanced protein synthesis, while the second concept, of a primarily altered passive K^+ permeability, invalidates the association between higher cellular $[K^+]_c$ and enhanced protein synthesis. The first hypothesis is influenced by

LUBIN's reports that $[K^+]_c$ appears to be a determinant of protein synthesis in bacterial and mammalian cells (LUBIN, 1964a and b, 1967; LEDBETTER and LUBIN, 1977).

b) Cation Transport in the Absence of Immunological Reactions

Prior to a more detailed discussion of the effect of immunological reactions on monovalent cation transport, it is useful to survey the cation transport parameters of normal untreated lymphocytes which have been obtained largely by LICHTMAN's group (LICHTMAN and WEED, 1969; LICHTMAN et al., 1972; SEGEL et al., 1975). In looking at the compilation of these data in Table 7 one is impres-

Table 7. Some Properties and Cation Transport Parameters of Lymphocytes

Group	Authors (Year)	Lymphocytes		Preparation	Purity of Preparation
		Species	Source		
I	QUASTEL & KAPLAN (1970a, 1975)	Human	Peripheral	Dextran sedimentation	92–98%
II	AVERDUNK (1972)	Human	Peripheral	Ficoll-Hypaque	>95%
	AVERDUNK and LAUF (1975)	Sheep	Lymphnode	Gradient-centrifugation	
	AVERDUNK (1976)	Mouse	Thymus Spleen		
	AVERDUNK et al. (1976a and b)	Sheep	Peripheral		
III	LICHTMAN & WEED (1969)	Human	Peripheral	Polyvinylpyrrolidone, glass wool	95–98%
	LICHTMAN et al. (1972)	Human	Chronic Lymphatic Leukemia	Polyvinylpyrrolidone, glass wool	>93%
	SEGEL et al. (1975, 1976a–c)	Rat	Thymus	Ficoll gradient	>95%
		Rat	Splenic	Ficoll gradient	>95%
		Mouse	Lymphoblasts Culture		n. g.

Group	Lymphocytes	Cell volume (μm^3)	Cell water (%)	Cell cations 10^{-15} M/cell $[K^+]$	$[Na^+]$	K^+ Influx (10^{-15} moles/cell/h) Total	Ouabain-sensitive	Ouabain-insensitive
I	Human	n. g.	n. g.	n. g.	n. g.	7.2	2.9	n. g.
II	Human	n. g.	n. g.	n. g.	n. g.	8.1	5.6	2.5
	Sheep	n. g.	n. g.	n. g.	n. g.	c. e.	c. e.	c. e.
III	Human	269	79	28	7.9	n. g.	n. g.	n. g.
	Human CLL	203–231	78.5	24.1	5.7	4.72	3.52	1.2
	Rat thymus	136–151	78.5	16.7	3.9	8.29	7.14	1.15
	Rat spleen	n. g.	n. g.	17.8	2.1	n. g.	n. g.	n. g.
	Mouse Lymphoblasts	n. g.	n. g.	170*	20*	0.56**	0.39**	0.17**

Table 7. (Cont.)

Group	Lymphocytes	K^+ efflux	Ouabain binding sites cell	K^+ ions pumped/cell/min	Microsomal Na^+-K^+-ATPase $\frac{M\ P_i}{mg\ protein \cdot h}$	Remarks
I	Human	$t_{1/2}$ = 17 h	1.25×10^5	2.9×10^7	n.g.	Cells washed
II	Human	"reduced"	3×10^4	5.6×10^7	0.14–0.94	Cells spun through phthalate-oil mixture
	Sheep	n.g.	n.g.	n.g.	0.11 1.6**	
III	Human	2.2	n.g.	n.g.	0.36	Cells washed
	Human CLL	n.g.	n.g.	3.5×10^7	n.g.	Cells washed
	Rat thymic	5.9*	n.g.	7.1×10^7	n.g.	Cells washed
	Mouse Lymphoblasts	1.7**	n.g.	n.g.	n.g.	Cells washed

n.g. = not given by authors. c.e. = cannot be estimated from data available
** nmol cm^{-2} × min, volume not given * mmol l^{-1} cells
*** Purified plasma membranes (AVERDUNK et al., 1976a and b).

sed by a host of factors affecting and interfering with cation determinations in these cells. As pointed out by SEGEL et al. (1975), washing of lymphocytes in different media may cause quite significant and variable cation and isotope losses from these cells, affecting the establishment of the steady-state cation concentrations considerably. These effects may be observed with lymphocytes from any species and source (Table 7). Lymphocytes from man, rat, mouse and sheep have been studied, and their degree of purification ranges from 92–98 percent, the only major contaminants being red cells and macrophages. Most of the lymphocyte preparations obtained are a mixture of B and T cells. (Lymphoblasts were studied in the mouse.) Although there is a considerable difference in cell volume between the lymphocytes of various sources, the cell water content and cation concentrations are relatively similar. All types of lymphocytes tested were high-potassium, low-sodium cells (Table 7) with a steady-state ratio of about 25 for $[K^+]_c/[K^+]_o$ and of about 5 for $[Na^+]_o/[Na^+]_c$.

Measurements of K^+ transport in human peripheral lymphocytes were first reported by QUASTEL and KAPLAN (1970a) and AVERDUNK (1972). These early reports cannot be easily interpreted in terms of active or passive K^+ influxes, since either the ouabain concentration used (1.43×10^{-7} M) was too low to inhibit within the short preincubation time (2–5 minutes) all Na^+-K^+ pumps (QUASTEL and KAPLAN, 1970a; QUASTEL et al., 1970) or the data were expressed only in counts per minute without an indication of extracellular K^+ (which makes it impossible to calculate the K^+ influx from the specific activity of ^{42}K; AVERDUNK, 1972). The use of low ouabain concentrations is partly due to the original observation that this and other glycosides specifically inhibit protein and DNA synthesis at low concentrations (QUASTEL and KAPLAN, 1970b; QUASTEL et al. 1971; KAY, 1972). The total K^+ influx of 2.9×10^{-15} mol/cell/h (QUASTEL and KAPLAN, 1970a), however, appears to be within the order of magnitude found later by AVERDUNK and LAUF (1975), based on a lymphocyte volume of 255 μm^3, and by the group of LICHTMAN for various types of lymphocytes (LICHTMAN et al., 1972). The K^+ influx was originally expressed by the latter group in terms of flux per surface area (mol cm^{-2} min^{-1}), assuming that the lymphocyte is a sphere with a smooth surface (which may be a very rough assumption). For comparative reasons, these values have been converted in Table 7 into K^+ influx/cell, taking into account the different cell volumes and surface areas of the various lymphocytes studied. It can be seen that the total K^+ influx amounts to an hourly K^+ exchange of 20–50 percent of steady-state $[K^+]_c$, a finding compatible with the observations by all authors of rapid establishment of isotopic equilibrium. This fact was ignored in the early studies, and a lack of early experimental time points makes it difficult to analyze some of these data in terms of initial ^{42}K uptake velocities.

Distinction between ouabain-sensitive and -insensitive, *i. e.* K^+ pump (${}^iM_K^P$) and leak (${}^iM_K^L$) fluxes, was made by AVERDUNK and LAUF (1975) and LICHTMAN et al. (1972). Table 7 shows that ${}^iM_K^L$ amounts to about 70 percent of the total K^+ influx in human peripheral (normal or chronic lymphocyte leukemia) cells and mouse lymphoblasts (CUFF and LICHTMAN, 1975), while 86 percent of the total K^+ influx is ${}^iM_K^P$ in rat erythrocytes (LICHTMAN et al., 1972), the balance being the ouabain-insensitive ${}^iM_K^L$ Table 7 also depicts some data on K^+ efflux which were obtained either in the presence or absence of ouabain. All these investigations, however, lack a rigorous analysis to relate quantitatively K^+ influxes and effluxes in terms of volume control. Expressing the data given for ${}^iM_K^P$ as turnover of K^+ ions per cell per minute reveals a reasonable agreement among the three groups: the turnover numbers/cell/minute range from about $3–6 \times 10^7$ in human lymphocytes to about 7×10^7 in rat thymocytes. Comparison of these values with turnover numbers for human red cells (LAUF and JOINER, 1976) reveals that K^+ pump transport in human peripheral lymphocytes is about 35–56 times greater than in red cells. Taking into account the calculated surface areas (red cell = 130 μm^2 and lymphocyte $\simeq$ 183 μm^2), K^+ pump influx is about 3×10^5 ions/μm^2/minute and thus still some 30 times greater than in red cells (1.0×10^4 ions/μm^2/minute).

On the basis of these computations one would expect a higher pump site density or a higher cation turnover/site in lymphocytes than in human erythrocytes. Ouabain binding studies in lymphocytes by AVERDUNK and LAUF (1975) and QUASTEL and KAPLAN (1975) indeed indicate that the pump density (160/μm^2 and 690/μm^2, respectively) is about 32 and 133 times greater than that found in human red cells (4–5/μm^2, LAUF and JOINER, 1976). From these data a turnover of K^+ ions/site/minute can be computed, which is about 1800 in the studies of AVERDUNK and LAUF (1975) and only 232 in those of QUASTEL and KAPLAN (1975) (who report a turnover of 600 in their study). The discrepancy between these two reports may be explained by assuming that QUASTEL and KAPLAN (1975) underestimated the K^+ pump influx (which is likely since their calculations of red cell K^+ influx of 0.1 nmoles/10^6 cells/h deviates by a factor of 3 from our data; see LAUF and JOINER, 1976) and/or overestimated the number of ouabain binding sites. Conversely, there may be an underestimation of the number of ouabain binding sites in the studies of AVERDUNK and LAUF (1975). The last-named authors have discussed in detail the problems involved in studying ouabain binding to lymphocytes, which have been also pointed out by COOK et al. (1976), using HeLa cell lines. For example, there are considerable difficulties in dealing with membrane turnover and intracellular accumulation of ^3H-ouabain (COOK et al., 1976). In addition, none of the groups has clearly resolved the problem of 'nonspecific' ouabain binding since, unlike in red cells (LAUF and JOINER, 1976) no rigorous correlation between K^+ pump inhibition and the number of ouabain moleculles bound per cell has been attempted. At any rate, given the K^+ pump flux data, the number of pumps/μm^2 surface area, as evaluated by binding of ^3H-ouabain, is most certainly much higher than in red cells and compares much better with the similar number reported for various other cell lines (BAKER and WILLIS, 1970 and 1972; VAUGHAN and COOK, 1972; COOK et al., 1975). It is also most likely that the turnover of K^+ ions/site/minute may be closer to the value computed by AVERDUNK and LAUF (1975), since this number agrees better with turnover numbers tabulated by SCHWARTZ et al. (1975) for the Na^+-K^+ pump and Na^+-K^+-ATPase of other membranes. It is not helpful in this context to utilize the few data obtained for the ouabain-sensitive Na^+-K^+-ATPase, since the specific activity of this enzyme is only about 2–8 percent of the total ATPase (LICHTMAN and WEED, 1969) and the few determinations done were based on microsomal membrane preparations of lymphocytes (LICHTMAN and WEED, 1969; AVERDUNK and LAUF, 1975). Thus the effect of immunological reactions on lymphocytes must be considered in light of the complexities just discussed.

c) Cation Transport Changes Induced by Immunological Reactions

As discussed in section D/2–3, the mitogenic stimulation of lymphocytes by various immunological reagents is determined by the cell type (T or B cells), the properties of the reagents (mitogenic or nonmitogenic lectins), and moreover, by a 'unimodal' response of these cells to increasing concentrations of these substances. Furthermore, studies of QUASTEL et al. (1971) and SEGEL et al. (1975) have demonstrated that it makes a difference whether or not lymphocytes are washed in tracer flux experiments (this aspect will be taken up again below). Nevertheless, QUASTEL and KAPLAN (1970a) were the first to demonstrate that PHA increased total K^+ influx into lymphocytes by 1–6 times without a change in the K^+ efflux rate. Table 8 compares the general observations made on the effect of immunological reactions in lymphocytes by the three laboratories. AVERDUNK (1972), confirming these findings, reported that K^+ efflux was reduced when lymphocytes were exposed to PHA. Both groups studied the dependence of K^+ influx on the intracellular $[K^+]$ and found that PHA increased V_{max} and not K_m. In neither study, however, was any correction made for the ouabain-insensitive K^+ influx, *i. e.* these early experiments did not distinguish between active and passive K^+ influxes. If indeed the mitogens induced an increase of V_{max} of the Na^+-K^+ pump system, a higher number of ^3H-ouabain-binding sites should be expected.

In agreement with their hypothesis (KAPLAN and QUASTEL, 1975) QUASTEL and KAPLAN (1975) reported that PHA increased the number of ouabain binding sites from 1.25 to 2.3×10^5/cell, and concluded that there is no evidence for

Table 8. Some Effects of Immunological Reactions on Monovalent Cation Transport in Lymphocytes

Group	Mitogen or Antibody Tested		Lymphocytes	General effects on cation transport systems observed		
(see Table 7)	Type	Concentration	Tested	Cation fluxes	Ouabain binding	Na^+-K^+-ATPase
I	Phytohemagglutinin (PHA)	200 μg ml^{-1}	Human	1. 1.6-fold increase in total K^+ influx 2. Activation of ouabain-sensitive K^+ influx via V_{max} 3. No effect on K^+ efflux	1.8-fold increase in ouabain-binding sites without change in affinity	not tested
II	PHA	8 μg ml^{-1}	Human Sheep	1. 2- to 3-fold stimulation of Na^+-K^+ pump influx 2. Early changes in passive (leak) K^+ and Na^+ fluxes 3. Strong unimodal dependence on mitogen concentration	In presence of sodium azide no increase in the number of ouabain-binding sites	2–3 fold stimulation by PHA, ALS, anti-IgG and 4 fold by Con A
	Antilymphocyte serum (ALS)	0.1 μg ml^{-1}	Human			
	Concanavalin A	8 μg ml^{-1}	Human			
	Sheep anti-human IgG	0.1 μg ml^{-1}	Human			
	Rabbit anti-sheep IgG	40 μg ml^{-1}	Sheep			
	Human IgG	986 μg ml^{-1}	Sheep			
III	PHA	30 μg ml^{-1}	Human Rat	1. Increased K^+ permeability 2. No change in K^+ content/cell with PHA	not studied	not studied

synthesis and activation of new Na^+-K^+ pumps (KAPLAN and QUASTEL, 1975). However, some uncertainties remain unresolved in this study: no attempt has been made to correlate the degree of K^+ pump inhibition directly with the number of ouabain-binding sites, which invalidates to some extent the turnover number of 600 K^+ ions/site/minute given by these authors. In addition, the number of ouabain binding sites is higher than that given in an earlier report ($2-4 \times 10^4$/cell, QUASTEL et al., 1974) and identical with that found by AVERDUNK and LAUF (1975) under conditions where membrane turnover was reduced by the presence of sodium azide.

AVERDUNK (1972) and AVERDUNK and LAUF (1975) extended these tranport experiments, studying the effect of a variety of other immunologically active molecules on Na^+-K^+ transport, membrane ATPase and ^{3}H-ouabain binding. Table 9 shows that in addition to PHA, Con A, ALS (anti-lymphocyte serum) and anti-IgG were used. It was found that all of these immune reagents gave maximal stimulation of K^+ influx or total ATPase activity at critical concentrations (*i. e.* the dose-response curve exhibited a unimodal profile, as was demonstrated by VAN DEN BERG (1974) for amino acid transport and protein synthesis, and by WANG et al. (1975) and others for the Con A-induced activation of ^{3}H thymidine uptake). Table 9 shows that, in particular, the active K^+ influx and the ouabain-sensitive Na^+-K^+-ATPase were activated while passive K^+ influx and the ouabain-insensitive ATPase activity seem to be less affected. (However, see discussion on passive cation permeability, below). The degree of Na^+-K^+ pump activation was a function of the immunological stimulants used, yielding a sequence of potency as follows: Con A > PHA > ALS > anti-IgG. Con A and PHA are plant agglutinins binding to manno- and glucopyranosides of the membrane glycocalyx. The interaction of horse anti-lymphocyte serum (ALS), or sheep anti-human immunoglobulin (anti-IgG) with lymphocytes constitutes a true immunological reaction, since ALS recognizes integral membrane antigens and anti-IgG the membrane-bound immunoglobulins on B cells. It was also shown that active Na^+ efflux was increased by these reagents (AVERDUNK and LAUF, 1975) with a K^+/Na^+ coupling ratio of about unity. Thus the findings of

Table 9. Action of Some Mitogens on K^+ Fluxes and Microsomal ATPase Activities in Human Lymphocytes*

Mitogen	Mitogen concentration	$^iM_K^{T}$**	$^iM_K^{P}$**	$^iM_K^{L}$**	Mitogen concentration	Fractional ATPase activities***		
						Total	Ouabain sensitive	Ouabain insensitive
Control	–	8.1	5.6	2.5	Control	1.0	1.0	1.0
PHA	16 μg ml^{-1}	16.6	13.1	3.5	8 μg/ml	2.0	2.4	1.7
Con A	16 μg ml^{-1}	19.3	15.6	3.7	4 μg/ml	2.6	3.9	1.7
ALS	0.1 (v/v)	14.0	11.1	2.9	0.2 (v/v)	2.9	3.1	1.8
Anti-IgG	0.1 (v/v)	11.1	8.2	2.9	0.2 (v/v)	2.6	3.6	1.9

* From: AVERDUNK and LAUF (1975)
** in 10^{-15} moles/cell/h
*** The control activities equal unity.

AVERDUNK and LAUF (1975), that active Na^+-K^+ transport and ATPase are stimulated by various immunological substances, supported the concept of KAPLAN and QUASTEL (1975) that the Na^+-K^+ transport is a prime target in these immunological reactions. Only one report (NOVOGRODOSKY, 1972) has shown that Con A did not affect the Na^+-K^+-ATPase but rather the Mg^{++}-dependent ATPase. However, in this case only rat lymphocytes were tested. It is known that rat tissue is notoriously resistant to the effect fo ouabain, probably because of high dissociation rates of the ouabain-receptor complex. The temporal effect of these reactions is very early, since these studies detected an activation of cation transport as early as 3 minutes after addition of these reagents to lymphocytes (AVERDUNK, 1972).

In their quest to explain the mechanism of active Na^+-K^+ transport stimula tion, AVERDUNK and LAUF (1975) also measured ouabain binding to lymphocytes in the presence and absence of these mitogens. In contrast to QUASTEL and KAPLAN (1975), they found that in the presence of sodium azide at 37° C the rate of ouabain binding was altered and not the maximum number of ouabain-binding sites, which was found to be 3×10^4/cell under these conditions. As pointed out, this number agrees better with the original lower number cited by QUASTEL et al. (1974) and less well with those reported for other cell types (see also BOARDMAN et al., 1972; LAMB et al., 1973). Accordingly, the effect of these immunological reagents is to raise the turnover number of cations per site, a conclusion contradicting the earlier interpretation of an increase only in V_{max} and not in K_m (QUASTEL and KAPLAN, 1970a; AVERDUNK, 1972). It is, however, possible that, as in the sheep red cell system (LAUF, 1975a), the major change occurs at the inner aspect of the Na^+-K^+ pump by alteration of the internal cation affinity, and thus cannot be measured by studying the external K^+ activation kinetics. When ouabain binding was measured in the absence of sodium azide, higher numbers of ouabain-binding sites per cell were found, and a further small increment when mitogens were added (AVERDUNK and LAUF, 1975). These numbers are in closer agreement with those recently reported by QUASTEL and KAPLAN (1975). Furthermore, a study of ouabain binding at 0° C yielded much lower numbers of ouabain-binding sites, and prolonged incubation of lymphocytes in ouabain at 37° C resulted in cytoplasmic accumulation of the cardiac glycoside, which was also observed by COOK et al. (1976) in a study of HeLa cells. It appears, then, that it is difficult to resolve whether the observed Na^+-K^+ pump stimulation is due to unmasking of 'cryptic' (KAPLAN and QUASTEL, 1975) sites or to a change of cation affinity and turnover of each individual site. Better experimental conditions must be sought to measure the number of pumps under conditions where no pump inactivation occurs other than by the cardiac glycoside, and, most importantly, a direct correlation between ouabain binding and Na^+-K^+ pump inhibition must be established.

Nevertheless, the data accumulated by QUASTEL and KAPLAN (1975) and AVERDUNK and LAUF (1975) are consistent with KAPLAN's and QUASTEL's hypothesis (1975) that mitogens and other immunological reagents affect the active Na^+-K^+ transport system in these cells. The mechanism of this stimulation as well as the quantitative correlation between the number of lectin or antibody molecules and the number of activated transport sites awaits further studies.

Furthermore, the nature of the molecules binding these ligands and their relationship to the Na^+-K^+-ATPase moieties must be established. Most likely, glycoproteins which carry the sugar determinants for lectins are involved. Membrane-bound immunoglobulins, against which anti-IgG is directed, are by definition also glycoproteins. Lymphocyte antigens (transplantation antigens?) reacting with ALS probably also contain carbohydrates. The observation that periodate oxidation stimulates mitogenesis, which may also be accompanied by changes in the Na^+-K^+ transport system, raises the question as to the specificity of these reactions. We are still far from understanding why alterations of the carbohydrate shell of lymphocytes are accompanied by rather rapid changes in membrane 'fluidity' or phospholipid turnover (FERBER et al., 1974).

Regardless of the final outcome of the biochemical nature of the membrane receptor, a major unresolved question is whether their prime function is to transduce membrane surface events via the Na^+-K^+ pump system to the cytoplasm and nuclear apparatus and to induce protein and nucleic acid synthesis. This principal question raises another, namely whether the lymphocyte cation transport system, once activated by immunological reactions, is capable of establishing higher $[K^+]_c$ and whether the latter is required for an enhanced protein synthesis. This last proposition has been accentuated by LUBIN, who first demonstrated in *Escherichia coli* mutants (strain B 207) that K^+ ions had a pronounced effect on the transfer of amino acids form charged s-RNA to polypeptides (LUBIN, 1964a and b, 1967). LUBIN (1964b) postulated that the polyribosomal complex formation between the initial s-RNA and an amino acid was highly K^+-sensitive while the peptide chain elongation was K^+-insensitive. Potassium depletion of *E. coli* or *Bacillus subtilis* led to cessation of protein, but not of RNA synthesis. Moreover, LUBIN (1967) showed that when sarcoma S 180 cells were incubated in low-K^+ media and in the presence of amphotericin B, the decrease in the rate of protein and DNA synthesis was proportional to the fall in $[K^+]_c$. Similar observations were made with L cells (LUBIN, 1967). In a recent study on diploid human fibroblasts (LEDBETTER and LUBIN, 1977), blockade of the Na^+-K^+ transport system by 10^{-7} M ouabain resulted in inhibition of protein and DNA synthesis but not of RNA synthesis, which was attributed to lowering $[K^+]_c$ steady-state levels to below 75–80 percent of initial uninhibited cellular concentrations. These changes, accompanied morphologically by greater refractility and lateral aggregation, were reversible when the cells were further incubated in ouabain-free media within the first two days after the first exposure to ouabain. Similar observations have been made by others in L cells, rat sarcoma cells and baby hamster kidney cells. The summary of LUBIN's work is that it proves that protein synthesis requires the normal K^+ steady-state level, but it does not prove that increased $[K^+]_c$ in mammalian cells, brought about by activation of the Na^+-K^+ pump, enhances protein and DNA synthesis, as the earlier work on lymphocytes activated by PHA and other immunological reactions suggested.

Changes in cation composition and concentrations may occur when lymphocytes switch from the resting G_1 phase to the mitotic S phase after being challenged by immunological reactions. Studying mouse leukemic lymphoblasts (L 5178Y cells), YOUNG and ROTHSTEIN (1967) demonstrated that during exponential growth cell volume, K^+ and Na^+ content of the randomly dividing L 5178 Y cells followed the growth curve and that unidirectional K^+ and Na^+ fluxes were constant. However, from the end of the G_1 phase throughout the S phase and into the G_2 phase, a remarkable and complex change of the cation permeability was observed. At the time of division $[Na^+]_c$ fell, followed by a rise when K_c^+ loss occured. Measurements of unidirectional K^+ influx revealed that no change occurred during the cell division phase but that 2 hours after cell division, at which K_c^+ was lowest and Na_c^+ was higher, the rate of K^+ influx for each cell was almost doubled, probably eliminating the previous K_c^+ deficit. At this point K^+ efflux returned to the original lower value permitting net accumulation of K_c^+ (YOUNG and ROTHSTEIN, 1967). Although the enhanced K^+ influx was not directly attributed to an increased K^+ pump influx, it is possible that the drop of intracellular $[K^+]$ and subsequent rise in $[Na^+]$ resulted in an activation of the Na^+-K^+ pump as it is known from cation activation studies in red cells. However, BANERJEE et al. (1976) showed that increased Rb^+ transport during the stationary phase of growing L 5178 Y cells may be accompanied by synthesis of new pump sites. COOK et al. (1975) have recently entertained the idea that the Na^+-K^+ pump rate is stimulated by intracellular Na^+, as has been shown for red cells. When HeLa cells were exposed briefly to nontoxic ouabain concentrations, a dramatic loss of K_c^+ and gain in Na_c^+ occurred, which was reversed during subsequent further incubation of the cells in ouabain-free media. Similarly, the K^+ content dropped remarkably after 2.5 hours incubation at 0° C or in 0.2 mM K^+ at 37° C (an event which could be prevented by high-K^+ media), while the cellular Na^+ content rose. When cells exposed to cold were rewarmed to 37° C, Rb^+ influx more than doubled without a change in the number of ouabain binding sites and therefore Na^+-K^+ pumps. The turnover of cations/site/minute increased from 2400 to 5400. Thus the mode of adaptation of the Na^+-K^+ pump system does not involve net synthesis of pumps but an adjustment, by changing their turnover numbers as intracellular K^+ and Na^+ ion concentrations change, which is quite in line with experiments on red cells. The remarkable constancy of the number of Na^+-K^+ pumps/surface area of HeLa cells is a function of their slow internalization balanced by a cycloheximide-sensitive reinsertion of newly synthesized sites (COOK et al., 1976). It is possible that the concept, dicussed above, of compensatory Na^+-K^+ pump changes cannot be applied to all cell line studies (see also ROZENGURT and HEPPEL, 1975).

In analogy to the phenomena just discussed it may be questionable whether immunological reactions at the lymphocyte surface affect directly and primarily the Na^+-K^+ pump system, causing higher K_c^+ levels and thus enhanced protein synthesis without changing P_K or P_{Na} of the membrane. Furthermore, a primary increase of $[K^+]_c$ above control values may not activate but rather depress K^+ pumping activity, which is in line with COOK's work on HeLa cells and the experiments on red cells (HOFFMAN and TOSTESON, 1971). And, finally, if the active Na^+-K^+ transport system indeed is activated by immunological reactions, as the experiments of QUASTEL and KAPLAN (1970a) and LAUF and AVERDUNK (1975) suggest, this activation may not involve a change in the number of Na^+-K^+ pumps as shown by QUASTEL and KAPLAN (1975), but rather an augmentation of their turnover. This is borne out from the unchanged number of ouabain-binding sites when lymphocytes are exposed to Con A (LAUF and AVERDUNK, 1975). This activation of the Na^+-K^+ pump may then be considered secondary to primary, passive cation permeability changes induced by immunological reactions in the lymphocyte membrane. Thus, in generalizing from YOUNG and ROTHSTEIN's work (1967), immunological reactions may affect and augment certain phases within the individual generation cycle (*e.g.* changes similar to those during cellular synchronization). This hypothesis is not invalidated by our finding that the immunological reactions stimulated the Na^+-K^+-ATPase of microsomal membrane preparations, since these membranes can only be stimulated in vesicular form (AVERDUNK, pers. comm.), again suggesting ionic effects.

It appears then uncertain that the augmented ouabain-sensitive K^+ influx is the primary target of the immunological reactions in lymphocytes, a suspicion nourished by the recent observations of SEGEL et al. (1975, 1976a, b and c; SEGEL and LICHTMAN, 1975) on the selective effect of PHA on the K^+ permeability in these cells. As indicated in Table 8, these authors found that PHA-treated lymphocytes lost some 25 percent of $[K^+]_c$ when washed in Hepes-buffered choline chloride following exposure to the mitogen. The rate constant for K^+ efflux was significantly higher than in control cells: 27 percent in rat thymic and 78 percent in human peripheral lymphocytes. When, however, the PHA-exposed lymphocytes were not washed, the K_c^+ steady-state level remained unchanged over the entire incubation period of 24 hours and in the presence of PHA (SEGEL et al., 1976c). These findings were interpreted in terms of a mitogen-increased P_K of the lymphocyte membrane and were found compatible with the earlier studies of QUASTEL and KAPLAN (1970), which showed that "accelerated K^+ uptake is an early, dramatic consequence of PHA stimulation" (SEGEL et al., 1976c). The authors conclude that "the increased K^+ turnover may be a reflection of changes in membrane permeability after lectin treatment and not necessarily a causal event in blastogenesis" (SEGEL et al., 1976c). It is not yet clear why washing the PHA-activated lymphocytes causes a net loss of K_c^+ from about 22.2 to 4.3×10^{-15} mol/cell while the same lymphocytes, kept in artificial low-K^+ medium, maintain their $[K^+]_c/[Na^+]_o$ ratio of about 22. Possibly, active K^+ pump influx compensates for the increased P_K. However, it was recently reported by AVERDUNK (1976) that passive cation permeability changes must occur very soon after the addition of mitogens. His study demonstrated that Con A treatment of mouse thymic (T cells) and splenic (T and B cells) lymphocytes increased K^+ (and Mg^{++}) efflux by 20 percent and Na^+ (and Ca^{++}) influx by 15 percent. These last-mentioned observations cast some doubt on the selective K^+ permeability changes seen by SEGEL et al. (1975, 1976a–c) in PHA-treated lymphocytes, and they are more compatible with the opposite cation movements observed by YOUNG and ROTHSTEIN (1967). However, the total cation content was decreased in AVERDUNK's study (1976), implying either some volume loss of the cell or maintenance of volume by other solutes. These findings do not contradict the earlier report of stimulation of active K^+ influx in lymphocytes by immunological reactions, since transient changes of $[K^+]_c$ and $[Na^+]_c$ may lead to activation of the Na^+-K^+ pump by increasing its turnover. Another aspect of AVERDUNK's work deserves emphasis. Con A changed the passive monovalent and divalent cation permeability in T and B cells although it is known that Con A binding causes mitogenic transformation primarily in T cells (see section D/2–3). These findings suggest that in T cells other events must occur to stimulate protein synthesis, or that cation permeability and protein synthesis are linked in T cells, but not in B cells.

The purpose of this section was to guide the reader through the problematic aspects of cation permeability changes produced in lymphocytes by immunological reactions. At the outset of this discussion we focused more on the alteration of active Na^+-K^+ transport, but very recent information casts doubt on the role of the Na^+-K^+ pump as a primary target in these reactions. It is more likely that immunological reactions produce first an alteration in passive membrane cation permeability, changes which are then followed by a response of the Na^+-K^+ pump system similar to that found in other cells when intracellular K^+ and Na^+ ion concentrations vary. Attractive as a link between increased K^+ transport and increased protein synthesis would be, there is no solid basis to assert this connection, since the K_c^+ levels were never reported to be substantially elevated. Due to the complexity of the system, it is impossible at present to identify the immunologically induced permeability changes in lymphocytes as causative events of subsequent cellular transformation. Nevertheless, alteration of membrane permeability may precede cellular biochemical changes by temporarily affecting steps crucial for such transformational changes. This is followed by an 'overshoot' reaction of the active cation transport system to reestablish the 'normal' cellular cation content and concentration, and by a permanent elevation of biosynthetic processes of the lymphocyte required for clonal expansion of T and B cells subsequently participating in tissue or humoral immune defense. The schema in Figure 2 illustrates the possible (solid lines) and at present impossible (dashed lines) pathways by which immunological reactions alter cation permeability and protein synthesis in lymphocytes.

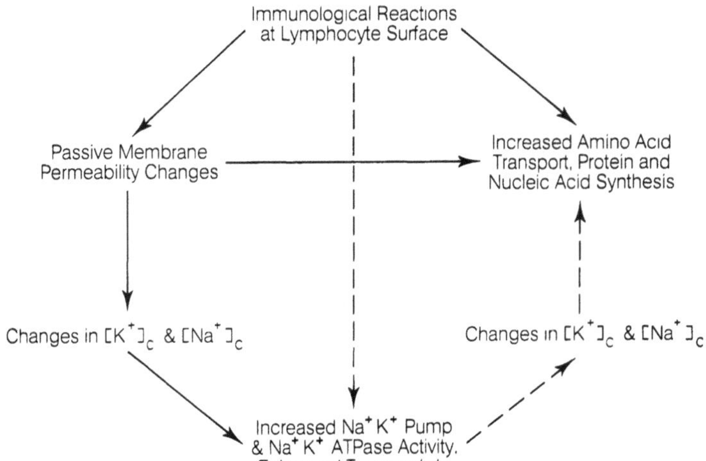

Fig. 2

5. Requirement of Bivalent Cations for Lymphocyte Stimulation by Immunological Reactions

Because of the important role calcium ions play in membrane biological processes (e. g. muscle contractility, exocrine secretion, immunoglobulin E-mediated histamine release, Ca^{++}-induced K^+ efflux in human red cells, erythrocyte membrane shape changes, etc.), a few workers have attempted to relate lymphocyte transformation by immunological reactions to early changes of the membrane permeability to calcium ions. ALLWOOD et al. (1971) found that ^{45}Ca uptake is significantly enhanced 38 minutes after the addition of PHA to human peripheral lymphocytes. WHITNEY and SUTHERLAND (1973a) showed saturation kinetics of calcium uptake in unstimulated lymphocytes at about 1 mM $[Ca^{++}]_o$. When PHA was added to these cells, the uptake of ^{45}Ca was increased, and the authors have ascribed this effect to a slightly enhanced V_{max}, and assuming a carrier type transport, to an increased affinity for calcium ions since K_m dropped from 0.74 to 0.30×10^{-3} M. In a further study, WHITNEY and SUTHERLAND (1973b) concluded that Ca^{++} ions are not essential for the initial interaction of PHA with the lymphocyte membrane but are required for increased amino acid (AIB) transport in PHA-exposed cells. It should be pointed out, however, that the physiological basis of calcium transport in lymphocytes has never been rigorously established.

Recently, calcium ionophores have been used to support the hypothesis that calcium ions play a major role in the initial steps leading to lymphocyte transformation by immunological reactions. For example, the ionophore A23187 has been shown to stimulate calcium-dependent methyl-glucose uptake in rat thymocytes (REEVES, 1975). Using this ionophore, RAFF et al. (1975) demonstrated that calcium induced DNA synthesis and blast cell transformation in T lymphocytes in a similar way to PHA, Con A and pokeweed mitogens. From this analogous behavior RAFF et al. (1975) propose that, by binding to two transmembranous glycoproteins, Con A or other mitogens may affect Ca^{++} channels in the lymphocyte membrane. This effect was seen as early as 45 seconds after addition of the mitogens, reaching a maximum after about 1 minute and was not observed with B cell-specific ligands (lipopolysaccharides). Further quantitative studies on Ca^{++} transport in normal and immunologically stimulated lymphocytes are required to decide whether the initial events of lymphocyte transformation indeed can be related to changes in membrane permeability to calcium and/or to monovalent cations, as discussed in section I/4.

II. Tumor Cells

1. Introduction

Unlike lymphocytes, tumor cells have been studied in great detail with respect to their membrane permeability properties, particularly in the mouse Ehrlich ascites tumor. In part, this is due to early and medically relevant attempts to relate the vigorous growth of cancer cells to alterations in membrane solute transport, and in part to the fact that the diverse picture of T and B cells in immunological defense has only recently begun to unfold. Experimentally, the Ehrlich ascites tumor is easy to produce because of its extraordinary short generation cycle (OEHLERT et al., 1962 and 1963), and the harvested cells are almost free of contaminating erythrocytes, white cells, and macrophages. The ionic properties and the volume control of Ehrlich ascites cells have been described in detail by HEMPLING (1972). In short, the volume of these cells is some eight times larger than that of lymphocytes discussed in section D/I/4a, and accordingly their K^+ and Na^+ ion content is 2.5×10^{-13} and 3×10^{-14} mol/cell, respectively (HEMPLING, 1972). The entire K^+ content/cell is turned over within one hour between one intracellular compartment and the extracellular space (HEMPLING, 1972), while Na^+ ion transport can be partitioned into a slow and a fast component, and thus into a three-compartmental exchange (AULL and HEMPLING, 1963), which has been recently confirmed in experiments involving 2, 4, 6-tri-nitro-benzene sulfonic acid (SMITH and ADAMS, 1976). For the Na^+-K^+ pump system only the external activation kinetics have been studied, yielding a $K_{1/2}$ of about 2 mM $[K^+]_o$ (GROBECKER et al., 1963). The water permeability of the Ehrlich cell membrane is about twofold that in red cells (HEMPLING, 1959) and chloride ion permeability was found in the order of 10^{-7} cm s^{-1} (HEMPLING and KROMPHARDT, 1965). From the chloride distribution ratio and from direct electrode impalement studies a membrane potential of -11 to -12 mV has been estimated (AULL, 1967; SMITH and LEVINSON, 1975). The amino acid transport system has been defined in terms of the Na^+-dependent A and ASC systems and the Na^+-independent L system by CHRISTENSEN (see review by CHRISTENSEN et al., 1973), and their energetics are discussed by HEINZ (1973).

2. Passive Permeability Changes Induced by Lectins

Although the physiological basis of ion transport and volume regulation was studied in Ehrlich cells as early as 1959 (HEMPLING), investigations on the effect of immunological reactions on cation permeability of these cells are only very recent. Thus far the action of three particular ligands, Con A, soybean agglutinin (SBA) and *Ricinus communis* agglutinin (RCA) have been analyzed. When 50 μg Con A ml^{-1} were added to Ehrlich ascites cells, a rapid fall of cellular K^+ content and rise of Na^+ content was observed (AULL and NACHBAR, 1974), which appeared to be much faster than and different from the well-known, cold-induced K^+/Na^+ exchange in these cells (HEMPLING, 1958). In 10–20 minutes after the exposure of these cells to Con A, K^+ loss decreased and Na^+ extrusion began, both of which could be prevented by 10^{-3} M ouabain. Correctly, AULL and NACHBAR (1974) interpreted the second reverse process as mediated by the Na^+-K^+ pump. In contrast, SBA (which is not mitogenic in T lymphocytes) caused much less marked initial changes of cation permeability than Con A. Since both ligands agglutinate these cells, it was concluded that agglutination *per se* cannot be responsible for this effect. INOUE et al. (1975) confirmed the profound effect that Con A exhibited on passive K^+ permeability in these cells. In addition, they showed that succinylated, dimeric Con A, which was incapable of agglutinating these cells, also did not induce changes in K^+ content. Apparently, Con A binding to some 6×10^6 sites/cell (NACHBAR et al., 1976) must be tetrameric to cause the membrane changes, an observation analogous to that made in lymphocytes (see section D/I/3). In a follow-up study, AULL et al. (1977) determined the K^+ flux parameters as well as the Na^+-K^+-ATPase activity in Ehrlich ascites cells exposed to Con A and RCA (10^7 binding sites/cell; see also NACHBAR et al., 1976). Within 4–7 minutes after addition of Con A and RCA, K^+ efflux increased by 300 and 174 percent, respectively, while K^+ influx remained relatively constant. As the time of exposure to the two mitogens progressed, K^+ efflux diminished and the net K^+ influx again became slightly positive. Again, 10^{-3} M ouabain prevented reaccumulation of K_c^+. Since Con A did not stimulate K^+ influx or activate the Na^+-K^+-ATPase, the authors concluded that ascites tumor cells respond in a different manner to Con A than lymphocytes as shown by AVERDUNK and LAUF (1975). AULL et al. (1977) speculate that glycoproteins (MW=75000, NACHBAR et al., 1976) acting as receptors for these lectins may mediate the membrane transport changes, which do not occur below 15° C (INOUE et al., 1975). As discussed for the effect of immunological reactions on lymphocytes, it is tempting to consider the early effect of Con A and RCA on passive cation permeability as relatively similar in both lymphocytes and tumor cells, yet the latter cells may not respond with a increased turnover of cations per site as we proposed for the former in section D/I/4c.

III. Conclusion

The preceding sections attempted to illuminate the effects of immunologically active molecules on cation permeability in lymphocytes and tumor cells, which reveal several phenomena common to alterations in both cell types. Although *immunologically* more conditions have been found to alter lymphocyte cation permeability (*e. g.* Con A, PHA, LPS, anti-IgG, mixed leucocyte cultures) similar modulations of cation permeability have been seen in tumor cells with Con A and RCA. *Biochemically,* the target molecules for these ligands are most probably glycoproteins and only a reaction at the outer membrane surface is required to induce the membrane events. It is at present unknown how cross-linking of these proteins relates structurally and functionally to the transmembranous ion pathways. *Physiologically,* it is not impossible that these molecules participate in some way in the formation or gating of the channels through which cations pass the membrane. In this context the finding of TOSTESON et al. (1973), which demonstrated marked conductance changes when lipid bilayer membranes containing red blood cell glycophorin were exposed to Con A is of interest. The early action of these reagents on passive cation permeability is common to both lymphocytes and tumor cells. However, there is ample evidence in lymphocytes that, at least secondarily, activation of the Na^+-K^+ pump system occurs, probably by kinetic modulation as has been postulated for compensatory changes in other nucleated cells in the absence of immunological reactions. It is not yet clear whether these changes in passive and active cation transport properties can be related to major biochemical events, such as protein and DNA synthesis, which are increased during lymphocyte transformation; nor do we really understand the significance of the induction by Con A and RCA of early profound permeability changes in Ehrlich ascites tumor cells. Nevertheless, it is not impossible that membrane permeability changes are indeed triggering events for cellular processes, a hypothesis not inconsistent with the general finding that cardiac glycosides reversibly inhibit uridine and thymidine incorporation, protein synthesis, cell respiration and mitosis in all these cells, and, in addition, transformation of lymphocytes into blast cells. Clearly, further studies are required to establish the link between membrane permeability and cellular biosynthetic processes.

E. Summary and Prospectus

New scientific developments are often the result of interdisciplinary approaches. As a modern example, this chapter has attempted to show that immunological reactions enable us to probe further into the physiology of membrane transport. In particular, we have discussed: 1. The application of specific antibodies to an analysis of the sidedness and membrane distribution of the Na^+-K^+-ATPase. 2. Antibodies as inhibitors of partial reactions of the Na^+-K^+-ATPase and pump. 3. Antibody-induced conformational modification of genetically and probably structural features of this enzyme (in analogy to the recent work on the Ca^{++}-in relation to an absence of certain genetically determined human red cell antigens. 5. Changes in active and passive cation permeability induced in lymphocytes and tumor cells by membrane immunological reactions.

The possibilities of analyzing structural features of partial reaction steps of the Na^+-K^+-ATPase or pump by means of antibodies are by far not exhausted. It is almost certain that anti-Na^+-K^+-ATPase antibodies will belong to the standard technical repertoire of the membrane physiologist to dissect further the complex structural features of this enzyme (in analogy to the recent work on the Ca^{++} ATPase). Furthermore, there is no doubt that antibodies can be used to recognize evolutionary differences between transport enzymes long before these differences will be brought to light by tedious biochemical analysis.

The application of immunological reactions to cation transport physiology in sheep, goat and cattle red cells has shown that increased turnover rates constitute one particular mode by which antibodies modify transport. Moreover, the effect appears to be a transmembranous alteration of the Na^+-K^+ pump conformation, since binding of the antibody to the outside of the cell profoundly alters cation affinities at the cytoplasmic membrane side. Trypsin completely abolishes the antibody-mediated pump modification and it is likely that the proteolytic cleavage site contains crucial chemical structures through which the antibody affects the pump.

On reflection, the reader may ask why immunization with purified Na^+-K^+-ATPase evokes antibodies primarily directed against and affecting the cytoplasmic, Na_c^+-dependent aspects of the enzyme. The cytoplasmic membrane site is ordinarily not exposed (but rather buried) in living organisms. According to BURNET (1959), it is possible that the immune system does not recognize the cytoplasmic side of the injected enzyme as "non-self" or "foreign" and thus does not produce antibodies. However, one should not overlook the cationic composition of the environment into which the Na^+-K^+-ATPase is injected. Subcutaneous and even muscular interstitial space has a high $[Na^+]_c$ and low $[K^+]_c$. One intriguing possibility is that the immune system recognizes the enzyme in its Na^+-(Mg^{++})E conformation (not as Na-E · ATP, since the extracellular ATP levels are too low). Thus functionally active antibodies are made against the Na^+-sensitive form of the enzyme. This hypothesis would be consistent with the finding that it is difficult to obtain antibodies against the K^+-NPPase and the outer aspects of the enzyme. However, the fact that HK sheep immunologically distinguish an antigenically different LK pump may be due to real structural differences which are exposed on the outer membrane of intact LK red cells used for immunization.

It is possible that immunological reactions in lymphocytes also increase the turnover numbers of cation pumps, perhaps in response to passive permeability changes (such as in tumor cells). Since both passive and active cation transport are altered at the beginning of the morphological and biochemical transformation of these cells, the surface antigen determinants and receptors in their combination with transport systems must serve as transducer units. The fact that a large variety of immunological ligands cause alteration of membrane cation permeabilities is important. The undifferentiated lymphocyte may constitute an ideal nucleated model cell for the study, via immunological methods, of the transmission of information across the plasma membrane and its ion transport constituents to the cytoplasm.

Acknowledgement

This work was supported by USPHS grant HL 12157 and a Career Development Award USPHS K4-GM 50,194. I thank Dr. C. H. Joiner and D. G. Shoemaker for critical reading of the manuscript and Mrs. Rachel Hougom for typing its final version.

References

ABDERHALDEN, E.: Hoppe-Seylers Z. physiol. Chem. **25**, 65 (1898).

ADA, G. L., EY, P. L.: In: The Antigens, Vol. 3 (M. Sela, Ed.). New York: Academic Press 1975, p. 189.

AHERN, T., KAY, J. E.: Exp. Cell Res. **92**, 513 (1975).

ALBERS, R. W.: In: The Enzymes of Biological Membranes, Vol. 3 (A. Martonosi, Ed.). New York: Plenum Press, 1976, p. 283.

ALLWOOD, G., ASHERSON, G. L., DAVEY, M. J., GOODFORD, P. J.: Immunology **21**, 509 (1971).
ASKARI, A.: Ann. N. Y. Acad. Sci. **242**, 322 (1974).
ASKARI, A., RAO, S. N.: Biochem. biophys. Res. Commun. **49**, 1323 (1972).
AULL, F.: J. cell. comp. Physiol. **69**, 21 (1967).
AULL, F., HEMPLING, H. G.: Amer. J. Physiol. **204**, 789 (1963).
AULL, F., NACHBAR, M. S.: J. cell. Physiol. **83**, 243 (1974).
AULL, F., NACHBAR, M. S., OPPENHEIM, J. D.: J. cell. Physiol. **90**, 9 (1977).
AVERDUNK, R.: Hoppe-Seylers Z. physiol. Chem. **353**, 79 (1972).
AVERDUNK, R.: Biochem. biophys. Res. Commun. **70**, 101 (1976).
AVERDUNK, R., GUNTHER, F., DORN, F., ZIMMERMANN, U.: Z. Naturforsch. **24b**, 693 (1969).
AVERDUNK, R., LAUF, P. K.: Exp. Cell Res. **93**, 331 (1975).
AVERDUNK, R., MULLER, J., WENZEL, B.: Hoppe-Seylers physiol. Chem. **357**, 673 (1976a).
AVERDUNK, R., MULLER, J., WENZEL, B.: J. Clin. Chem. Clin. Biochem. **14**, 339 (1976b).
BACH, J. F.: In: Contemporary Topics in Immunobiology, Vol. 2 (A. J. S. Davies and R. L. Carter, Eds). New York-London: Plenum Press 1973, p. 189.
BAKER, P. F., WILLIS, J. S.: Nature **226**, 521 (1970).
BAKER, P. F., WILLIS, J. S.: J. Physiol. **224**, 441 (1972).
BANERJEE, S. P., HAIKIMI, J., BOSMANN, H. B.: Biochim. biophys. Acta **433**, 200 (1976).
BERKE, G., AMOS, D. B.: Transplant Rev. **17**, 71 (1973).
BERLIN, R. D., OLIVER, J. M., UKENA, T. E., YIN, H. H.: Nature **247**, 45 (1974).
BLECHNER, J. N.: Amer. J. Physiol. **201**, 85 (1961).
BLOSTEIN, R., WHITTINGTON, E. S.: J. biol. Chem. **248**, 1772 (1973).
BLOSTEIN, R., LAUF, P. K., TOSTESON, D. C.: Biochim. biophys. Acta **249**, 623 (1971).
BLOSTEIN, R., WHITTINGTON, E. S.: KUEBLER, E. S.: Ann. N. Y. Acad. Sci. **242**, 305 (1974).
BLUNT, M. H., EVANS, J. V.: Nature **200**, 1215 (1963).
BLUNT, M. H., EVANS, J. V.: Amer. J. Physiol. **209**, 978 (1965).
BOARDMAN, L. J., LAMB, J. F., McCALL, D.: J. Physiol. **255**, 619 (1972).
BROWN, P. A., FEINSTEIN, M. B., SHA'AFI, R. I.: Nature **254**, 523 (1975).
BURNET, M. F.: The Clonal Selection Theory of Acquired Immunity. Nashville, Tenn.: Vanderbilt University Press 1959.
CAVIERES, J. D., ELLORY, J. C.: J. Physiol. **245**, 93P (1974).
CELADA, F., STROM, R., BODLUND, K.: In: The Lactose Operon (J. R. Beckwith and D. Zipser, Eds). Cold Spring Harbor: Monographs 1970, p. 291.
CHRISTENSEN, H. N., DECESPEDES, C., HANDLOGTEN, M. E., RONQUIST, G.: Biochim. biophys. Acta **300**, 477 (1973).
CHRISTINAZ, P., SCHATZMANN, H. J.: J. Physiol. **224**, 391 (1972).
COOK, J. S., VAUGHAN, G. L., PROCTOR, W. R., BRAKE, E. T.: J. cell. Physiol. **86**, 59 (1975).
COOK, J. S., WILL, P. C., PROCTOR, W. R., BRAKE, E. T.: In: Biogenesis and Turnover of Membrane Macromolecules (J. S. Cook, Ed.). New York: Raven Press 1976, p. 15.
CUFF, M. M., LICHTMAN, M. A.: J. cell. Physiol. **85**, 217 (1975).
DARNBOROUGH, J., DUNSFORD, I., WALLACE, D. A.: Vox Sang. (Basel) **17**, 241 (1969).
DIAMANTSTEIN, T., BLITSTEIN-WILLINGER, E., SCHULZ, G.: Nature **250**, 596 (1974).
DICK, D. A. T., DICK, E. G., TOSTESON, D. C.: J. gen. Physiol. **54**, 123 (1969).
DIERICH, M. P., REISFELD, R. A.: J. Immunol. **114**, 1676 (1975).
DIXON, J. F. P., O'BRIEN, R. L., PARKER, J. W.: Exp. Cell Res. **96**, 383 (1975).
DUNHAM, P. B.: Fed. Proc. **34**, 237 (1975a).
DUNHAM, P. B.: J. gen. Physiol. **66**, 13a (1975b).
DUNHAM, P. B.: Biochim. biophys. Acta **443**, 219 (1976).
DUNHAM, P. B., HOFFMAN, J. F.: J. gen. Physiol. **58**, 94 (1971a).
DUNHAM, P. B., HOFFMAN, J. F.: Biochim. biophys. Acta **241**, 399 (1971b).
DUNHAM, P. B., BLOSTEIN, R.: Biophys. J. **15**, 211a (1975).
EAGLETON, G. E., HALL, J. G., RUSSEL, W. S.: Anim. Blood Group Genet. **1**, 135 (1970).
EDELMAN, G. M.: In: Cellular Selection and Regulation of Immune Response (G. M. Edelman, Ed.) New York: Raven Press 1974, p. 1.
EDELMAN, G. M., CUNNINGHAM, B. A., REEKE, G. N., BECKER, J. W., WAXDAL, M. J., WANG, J. L.: Proc. nat. Acad. Sci. (Wash.) **69**, 2580 (1972).
EDELMAN, G. M.: Science **192**, 218 (1976).

ELLORY, J. C.: Nature **249**, 864 (1974).
ELLORY, J. C., CARLETON, S.: Biochim. biophys. Acta **363**, 397 (1974).
ELLORY, J. C., O'DONNEL, J. M., TUCKER, E. M.: J. Physiol. **210**: 111P (1970).
ELLORY, J. C., TUCKER, E. M.: Nature **222**, 477 (1969a).
ELLORY, J. C., TUCKER, E. M.: J. Physiol. **204**, 101P (1969b).
ELLORY, J. C., TUCKER, E. M.: J. Physiol. **208**: 18P (1969c).
ELLORY, J. C., TUCKER, E. M.: Biochim. biophys. Acta **219**, 160 (1970a).
ELLORY, J. C., TUCKER, E. M.: In: Permeability and Functions of biological Membranes (L. Bolis, A. Katchalsky, W. R. Loewenstein, and B. A. Pethica, Ed.) North Holland, 1970b, p. 120.
ELLORY, J. C., TUCKER, E. M.: J. Agric. Sci. Cambridge **74**, 595 (1970c).
ELLORY, J. C., SACHS, J. R., DUNHAM, P. B., HOFFMAN, J. F.: In: Passive Permeability of Cell Membranes (F. Kreuzer and J. F. G. Slegers, Eds). Biomembranes **3**, 237 (1972).
ELLORY, J. C., FEINSTEIN, A., HERBERT, J.: Immunochemistry **10**, 785 (1973).
EVANS, J. V.: Nature **174**, 931 (1954).
EVANS, J. V.: J. Physiol. **136**, 41 (1957).
EVANS, J. V., KING, J. W. B.: Nature **176**, 171 (1955).
EVANS, J. V., PHILLIPSON, A. T.: J. Physiol. **139**, 87 (1957).
FERBER, E., RESCH, K.: Biochim. biophys. Acta **296**, 335 (1973).
FERBER, E., REILLY, C. E., BARCINSKY, M. A., CUKIERMAN, S.: In: Lymphocyte Recognition and Effector Mechanisms. (K. Lindahl-Kiessling and D. Osoba, Eds), New York-London: Academic Press 1974, p. 529.
FURUHJELM, U., MYLLYA, G., NEVANLINNA, H. R., NORDLING, S., PIRKOLA, A., GAVIN, J., GOOCH, A., SANGER, R., TIPPET, P.: Vox Sang. (Basel) **17**, 256 (1969).
GARRAHAN, P. J., REGA, A. F.: J. Physiol. **193**, 459 (1967).
GERSHORN, R. K.: In: Contemporary Topics in Immunobiology, Vol. 3 (M. D. Cooper and N. L. Warner, Eds). New York: Plenum Press 1974, p. 1.
GIBLET, E.: In: Genetic Markers in Human Blood (E. Giblet, Ed.). Oxford: Blackwell Scientific Publishers 1969, p. 268.
GLYNN, L. M., ELLORY, J. C.: In: Role of Membranes in Secretory Processes (L. Bolis, R. D. Keynes, and W. Wilbrandt, Eds). New York: North Holland/American Elsevier 1972, p. 224.
GLYNN, I. M., KARLISH, S. J. D.: Ann. Rev. Physiol. **37**, 13 (1975).
GLYNN, I. M., KARLISH, S. J. D., CAVIERES, J. D., ELLORY, J. C., LEW, V. L., JORGENSEN, P. L.: Ann. New York Acad. Sci. **242**, 357 (1974).
GOLSTEIN, P.: In: Membrane Receptors of Lymphocytes (M. Seligmann, J. L. Preud'Homme and F. M. Kourilsky, Eds). New York: North-Holland/American Elsevier 1975, p. 399.
GREAVES, M. F., OWEN, J. J. T., RAFF, M. C.: T and B Lymphocytes: Origins, Properties and Roles in Immune Responses. Amsterdam: Excerpta Medica, New York: American Elsevier 1974.
GREEN, F. A.: Immunochemistry **4**, 247 (1967).
GREEN, F. A.: J. biol. Chem. **247**, 881 (1972).
GROBECKER, H., KROMPHARDT, H., MARIANI, H., HEINZ, E.: Biochem. Z. **337**, 462 (1963).
GUNN, R. B., DALMARK, M., TOSTESON, D. C., WIETH, J. O.: J. gen. Physiol. **61**, 185 (1973).
GUNTHER, G. R., WANG, J. L., YAHARA, I., CUNNINGHAM, B. A., EDELMAN, G. M.: Proc. nat. Acad. Sci. (Wash.) **70**, 1012 (1973).
HEGYVARY, C.: Mol. Pharmacol. **11**, 588 (1975).
HEINZ, E.: Biophysik **9**, 291 (1973).
HEMPLING, H. G.: J. gen. Physiol. **41**, 565 (1958).
HEMPLING, H. G.: J. gen. Physiol. **44**, 365 (1959).
HEMPLING, H. G.: J. cell. comp. Physiol. **60**, 181 (1962).
HEMPLING, H. G.: In: Transport and Accumulation in Biological Systems, 3rd ed. (E. J. Harris, Ed.). London; Butterworth for University, Baltimore, Md, Park Press (1972), p. 271.
HEMPLING, H. G., KROMPHARDT, H.: Proc. Soc. exp. Biol. **24**, 709 (1965).
HENNEY, C. S.: Nature **249**, 456 (1974).
HOFFMAN, J. F.: In: Membrane Proteins. Boston: Little, Brown and Co. 1969, p. 343.
HOFFMAN, P. G., TOSTESON, D. C.: J. gen. Physiol. **58**, 438 (1971).
HOKIN, L. E., DAHL, J. L.: In: Metabolic Pathways, Vol. 6 (L. E. Hokin, Ed.). New York: Academic Press 1972, p. 269.

HUGHES-JONES, N. C., GARDNER, B., TELFORD, R.: Biochem. J. **88**, 435 (1963).
HUMPHREY, J. H., DOURMASHKIN, R. R.: Advanc. Immunol. **11**, 75 (1969).
INOUE, M., UTSUMI, K., SENO, S.: Nature **255**, 556 (1975).
ISRAEL, Y., MacDONALD, A., BERNSTEIN, J., ROSENMAN, E.: J. gen. Physiol. **59**, 270 (1972).
JEAN, D. H., ALBERS, R. W.: Fed. Proc. **35**, 1663 (1976).
JEAN, D. H., ALBERS, R. W., KOVAL, G. J.: J. biol. Chem. **250**, 1035 (1975).
JOINER, C. H., LAUF, P. K.: J. Memb. Biol. **21**, 99 (1975).
LAUF, P. K., SHOEMAKER, D. G., JOINER, C. H.: Biochim. Biophys. Acta **507**, 544 (1978).
LORUSSO, D. J., BINETTE, J. P., and GREEN, F. A.: In: Human Blood Groups (J. F. Mohn, Ed.) Basel: Karger 1977 p. 226.
JØRGENSEN, P. L., HANSEN, O., GLYNN, I. M., CAVIERES, J. D.: Biochim. biophys. Acta **291**, 795 (1973).
JUNG, G., ROTHSTEIN, A.: J. gen. Physiol. **50**, 917 (1967).
KACHEL, V., METZGER, H., RUHEUSTROTER-BAUER, G. Z.: Exp. Med. **153**, 331 (1970).
KAPLAN, J. G., QUASTEL, M. R.: In: Immune Recognition (A. S. Rosenthal, Ed.). New York: Academic Press 1975, p. 391.
KAY, J. E.: Exp. Cell Res. **71**, 245 (1972).
KEPNER, G., TOSTESON, D. C.: Biochim. biophys. Acta **266**, 471 (1972).
KERR, S. E.: J. biol. Chem. **117**, 227 (1937).
KNAUF, P. A., PROVERBIO, F., HOFFMAN, J. F.: J. gen. Physiol. **63**, 305 (1974a).
KNAUF, P. A., PROVERBIO, F., HOFFMAN, J. F.: J. gen. Physiol. **63**, 324 (1974b).
KROPP, D. L., SACHS, J. R.: Nature **252**, 244 (1974).
KYTE, J.: J. biol. Chem. **247**, 7642 (1972).
KYTE, J.: J. biol. Chem. **249**, 3652 (1974).
KYTE, J.: In: The Enzymes of Biological Membranes, Vol. 1 (A. Martonosi, Ed.). New York: Plenum Press 1976a, p. 213.
KYTE, J.: J. Cell Biol. **68**, 287 (1976b).
KYTE, J.: J. Cell Biol. **68**, 304 (1976c).
LAMB, J. F., BOARDMAN, L. J., NEWTON, J. P., AITON, J. F.: Nature **242**, 115 (1973).
LAUF, P. K., POULIK, M. D.: Brit. J. Haematol. **15**, 191 (1968).
LAUF, P. K.: Fed. Proc. **28**, 319 (1969).
LAUF, P. K.: Hematologia **6**, 259 (1972).
LAUF, P. K.: Ann. N. Y. Acad. Sci. **242**, 324 (1974).
LAUF, P. K.: Biochim. biophys. Acta **415**, 173 (1975a).
LAUF, P. K.: J. exp. Med. **142**, 974 (1975b).
LAUF, P. K.: In: Human Blood Groups (J. F. Mohn, Ed.). Karger. Basel: 1977 p. 383.
LAUF, P. K.: In: The Physiological Basis of the Disorders of Biomembranes (T. E. Andreoli, D. Fanestil, and J. F. Hoffman, Eds). New York: Plenum Press 1978.
LAUF, P. K., DESSENT, M. P.: In: Metabolism and Membrane Permeability (E. Gerlach, K. Moser, and W. Wilman, Eds). Stuttgart: Thieme 1972, p. 112.
LAUF, P. K., DESSENT, M. P.: Immunol. Commun. **2**, 193 (1973).
LAUF, P. K., JOINER, C. H.: Blood **48**, 457 (1976)
LAUF, P. K., SUN, W. W.: J. Memb. Biol. **28**, 351 (1976)
LAUF, P. K., TOSTESON, D. C.: J. Memb. Biol. **1**, 177 (1969).
LAUF, P. K., TOSTESON, D. C.: Biomembranes **3**, 229 (1972).
LAUF, P. K., RASMUSEN, B. A., HOFFMAN, P. G., DUNHAM, P. B., PARMELEE, M. L., COOK, P., TOSTESON, D. C.: In: 3rd Int. Biophys. Cong. of the Int. Union for Pure and Applied Biophys, Cambridge, Mass., Abstract, p. 71 (1969).
LAUF, P. K., RASMUSEN, B. A., HOFFMAN, P. G., DUNHAM, P. B., COOK, P., PARMELEE, M. L., TOSTESON, D. C.: J. Memb. Biol. **3**, 1 (1970).
LAUF, P. K., PARMELEE, M. L., SNYDER, J. J., TOSTESON, D. C.: J. Memb. Biol. **4**, 52 (1971).
LAUF, P. K., PARMELEE, M. L., TOSTESON, D. C.: In: 6th International Symposium on Structure and Function of Erythrocytes. Berlin: Akademie Verlag 1972, p. 639.
LAUF, P. K., STIEHL, B. J., JOINER, C. H.: Biophys. J. **17**, 262a (1977).
LAUF, P. K., SHOEMAKER, D. G., JOINER, C. H.: Biochim. Biophys. Acta **507**, 544 (1978).
LAWTON, A. R., COOPER, M. D.: In: Contemporary Topics in Immunobiology, Vol. 3 (M. D. Cooper and N. L. Warner, Eds). New York: Plenum Press 1974, p. 193.

LEDBETTER, M. L. S., LUBIN, M.: Exp. Cell Res. **105**, 223 (1977).
LEE, P., WOO, A., TOSTESON, D. C.: J. gen. Physiol. **50**, 379 (1966).
LICHTMAN, M. A., WEED, R. I.: Blood **34**, 645 (1969).
LICHTMAN, M. A., JACKSON, A. H., PECK, W. A.: J. cell. Physiol. **80**, 383 (1972).
LUBIN, M.: Fed. Proc. **23**, 994 (1964a).
LUBIN, M.: In: Symposium on Cellular Function of Membrane Transport (J. F. Hoffman, Ed.). New Jersey: Prentice-Hall 1964b, p. 193.
LUBIN, M.: Nature **213**, 451 (1967).
McCANS, J. L., LANE, L. K., LINDENMAYER, G. E., BUTLER, V. P., SCHWARTZ, A.: Proc. nat. Acad. Sci. (Wash.) **71**, 2449 (1974).
McCANS, J. L., LINDENMAYER, G. E., PITTS, B. J. R., RAY, M. V., RAYNOR, B. D., BUTLER, V. P., SCHWARTZ, A.: J. biol. Chem. **250**, 7257 (1975).
MacLENNAN, I. C. M.: In: Contemporary Topics in Immunobiology, Vol. 2 (A. J. S. Davies and R. L. Carter, Eds). New York-London: Plenum Press 1973, p. 175.
MacLENNAN, D. H., HOLLAND, P. C.: Ann. Rev. Biophys. Bioengr. **4**, 377 (1976).
MARCHESI, V. Y., JACKSON, R. L., SEGREST, J. P., KAHANE, I.: Fed. Proc. **32**, 1833 (1973).
MARTINOSI, A., FORTIER, F.: Biochem. biophys. Res. Commun. **60**, 382 (1974).
MARTZ, E., BURAKOFF, S. J., BENACERRAF, B.: Proc. Nat. Acad. Sci. (Wash.) **71**, 177 (1974).
MASOUREDIS, S. P., DUPUY, M. E., ELLIOT, M.: J. clin. Invest. **46**, 681 (1967).
MAYER, M. M.: Proc. nat. Acad. Sci. (USA) **69**, 2954 (1973).
MESSER, W., MELCHERS, F.: In: The Lactose Operon (J. R. Beckwith and D. Zipser, Eds), Cold Spring Harbor Monographs, p. 305 (1970).
MOTAIS, R., SOLA, F.: J. Physiol. **233**, 423 (1973).
NACHBAR, M. S., OPPENHEIM, J. D., AULL, F.: Biochim. biophys. Acta **419**, 512 (1976).
NICOLSON, G. L., MASOUREDIS, S. P., SINGER, S. J.: Proc. nat. Acad. Sci. (Wash.) **68**, 1416 (1971).
NISONOFF, A., WISSLER, F. C., LIPMAN, L. N., WOERNLEY, D. L.: Arch. Biochem. Biophys. **89**, 230 (1960).
NISONOFF, A., HOPPER, J. E., SPRING, S. B.: The Antibody Molecule, New York: Academic Press 1975.
NOVOGRODOSKY, A.: Biochim. biophys. Acta **266**, 343 (1972).
NOVOGRODOSKY, A.: J. Immunol. **114**, 1089 (1975).
NOVOGRODOSKY, A., KATCHALSKI, E.: Proc. nat. Acad. Sci. (Wash.) **69**, 3207 (1972).
NOVOGRODOSKY, A., KATCHALSKI, E.: Proc. nat. Acad. Sci. (Wash.) **70**, 1824 (1973a).
NOVOGRODOSKY, A., KATCHALSKI, E.: Proc. nat. Acad. Sci. (Wash.) **70**, 2515 (1973b).
OEHLERT, W., LAUF, P., SEEMAYER, N.: Naturwissenschaften **49**, 137 (1962).
OEHLERT, W., SEEMAYER, N., LAUF, P. K.: Beitr. path. Anat. **63**, 127 (1963).
PARKER, J. W., O'BRIEN, R. L., STEINER, J., PAOLILLI, P.: Exp. Cell Res. **83**, 220 (1974).
POLLOCK, M. R., FLEMING, J., PETRIE, S.: In: Antibodies to Biologically Active Molecules (B. Cinader, Ed.). New York: Pergamon Press 1966.
PORTER, R. R.: Biochem. J. **73**, 119 (1959).
POULIK, M. D., LAUF, P. K.: Nature **208**, 874 (1965).
QUASTEL, M. R., KAPLAN, J. G.: Exp. Cell Res. **63**, 230 (1970a).
QUASTEL, M. R., KAPLAN, J. G.: Exp. Cell Res. **62**, 407 (1970b).
QUASTEL, M. R., KAPLAN, J. G.: Exp. Cell Res. **94**, 351 (1975).
QUASTEL, M. R., DOW, D. S., KAPLAN, J. G.: In: Proc. 5th Leucocyte Culture Conference (J. Harris, Ed.). New York: Academic Press 1970, p. 97.
QUASTEL, M. R., WRIGHT, P., KAPLAN, J. G.: In: Proc. 6th Leucocyte Culture Conference (M. R. (M. R. Schwarz, Ed.). New York: Academic Press 1971, p. 185.
QUASTEL, M. R., MILTHORP, P., KAPLAN, J. G., VOGELFANGER, I. J.: In: Lymphocyte Recognition and Effector Mechanism, Proc. 8th Leucocyte Culture Conference (K. Lindahl-Kiessling and D. Osoba, Eds). New York: Adademic Press 1974, p. 493.
RAFF, M. C., FREEDMAN, M., GOMPERTS, B.: In: Membrane Receptors of Lymphocytes (M. Seligman, J. L. Preud'Homme and F. M. Lourilsky, Eds). Amsterdam-Oxford: North-Holland, New York: American Elsevier 1975, p. 393.
RASMUSEN, B. A.: Genetics **45**, 1595 (1960).
RASMUSEN, B. A.: Genetics **61**, 49s (1969).
RASMUSEN, B. A., HALL, J. G.: Science **151**, 1551 (1966a).

RASMUSEN, B. A., HALL, J. G.: In: Polymorphismes biochimiques des animaux, 10th Europ. Congress on Animal Blood Groups and Biochemical Polymorphism, p. 453 (1966b).
RASMUSEN, B. A., HALL, J. G., HAYTER, S., WIENER, G.: Anim. Prod. **18**, 141 (1974a).
RASMUSEN, B. A., TUCKER, E. M., ELLORY, J. C., SPOONER, R. L.: Anim. Blood Grps. Biochem. Genet. **5**, 95 (1974b).
REEVES, J. P.: J. biol. Chem. **250**, 9428 (1975).
REGA, A. F., WEED, R. I., REED, C. F., BERG, G., ROTHSTEIN, A.: Biochim. biophys. Acta **147**, 297 (1967).
RESCH, K., FERBER, E.: Europ. J. Biochem. **27**, 153 (1972).
RESCH, K., GELFLAND, E. W., HANSEN, K., FERBER, E.: Eur. J. Immunol. **2**, 599 (1972).
RHEE, H. M., HOKIN, L. E.: Biochem. biophys. Res. Comm. **63**, 1139 (1975).
ROSENTHAL, A. S., ROSENSTREICH, D. L.: In: Biomembranes, Vol. 5 (L. A. Manson, Ed.). New York-London: Plenum Press 1974, p. 1.
ROTH, R. A., ROTMAN, B.: J. biol. Chem. **250**, 7759 (1975).
ROTHSTEIN, A., CABANTCHIK, Z. I., KNAUF, P.: Fed. Proc. **35**, 3 (1976).
ROTMAN, M. B., CELADA, F.: Proc. nat. Acad. Sci. **60**, 660 (1968).
ROZENGURT, E., HEPPEL, L. A.: Proc. nat. Acad. Sci. (Wash.) **72**, 4492 (1975).
RUOHO, A., KYTE, J.: Prot. nat. Acad. Sci. (Wash.) **71**, 2352 (1974).
SACHS, J. R., ELLORY, J. C., KROPP, D. L., DUNHAM, P. B., HOFFMAN, J. F.: J. gen. Physiol. **63**, 389 (1974a).
SACHS, J. R., DUNHAM, P. B., KROPP, D. L., ELLORY, J. C., HOFFMAN, J. F.: J. gen. Physiol. **64**, 536 (1974b).
SCHATZMANN, H. J.: Nature **248**, 58 (1974).
SCHWARTZ, A., LINDENMAYER, G. E., ALLEN, J. C.: Pharmacol. Rev. **27**, 3 (1975).
SEGEL, G. B., LICHTMAN, M. A.: Blood **46**, abstract 1034 (1975).
SEGEL, B. G., HOLLANDER, M. M., GORDON, B. R., KLEMPERER, M. R., LICHTMAN, M. A.: J. cell. Physiol. **86**, 327 (1975).
SEGEL, G. B., LICHTMAN, M. A., HOLLANDER, M. M., GORDON, B. R., KLEMPERER, M. R.: J. cell. Physiol. **88**, 43 (1976a).
SEGEL, G. B., HOLLANDER, M. M., GORDON, B. R., KLEMPERER, M. R., LICHTMAN, M. A.: J. cell. Physiol. **86**, 325 (1976b).
SEGEL, G. B., GORDON, B. R., LICHTMAN, M. A., HOLLANDER, M. M., KLEMPERER, M. R.: J. cell. Physiol. **87**, 337 (1976c).
SEIDL, S., SPIELMANN, W., MARTIN, H.: Vox Sang. (Basel) **23**, 182 (1972).
SELA, M., MOZES, E.: Proc. nat. Acad. Sci. (Wash.) **55**, 445 (1966).
SHARON, N., LIS, H.: In: Plasma Membranes, Vol. 3 (E. D. Korn, Ed.). New York: Plenum Press 1975, p. 147.
SHRAGER, P., TOSTESON, D. C., LAUF, P. K.: Biochim. biophys. Acta **290**, 186 (1972).
SMITH, T. C., LEVINSON, C.: J. Memb. Biol. **23**, 349 (1975).
SMITH, T. C., ADAMS, R.: J. cell. Physiol. **87**, 53 (1976).
SMITH, T. W., WAGNER, H.: J. Memb. Biol. **25**, 341 (1976).
SMITH, T. S., WAGNER, H., YOUNG, M., KYTE, J.: J. Clin. Invest. **52**, 78a (1973).
SNYDER, J. J., RASMUSEN, B. A., LAUF, P. K.: J. Immunol. **107**, 772 (1971).
SOBEL, A. T., BOKISCH, V. A.: In: Membrane Receptors of Lymphocytes, (M. Seligmann, J. L. Preud'Homme and F. M. Kourilsky, Eds). New York: North-Holland, American Elsevier 1975, p. 151.
STEWART, P. S., MacLENNAN, D. H.: Ann. N. Y. Acad. Sci. **264**, 326 (1975).
STEWART, P. S., MacLENNAN, D. H., SHAMOO, A. E.: J. biol. Chem. **251**, 721 (1976).
TANNER, M. J. A., ANSTEE, D. J.: Biochem. J. **153**, 271 (1976).
TOSTESON, D. C.: Fed. Proc. **22**, 19 (1963).
TOSTESON, D. C.: In: Cellular Functions of Membrane Transport Processes (J. F. Hoffman, Ed.). Englewood Cliffs, N. J.: Prentice-Hall 1964, p. 3.
TOSTESON, D. C.: In: The Myocardial Cell, Structure, Function and Modification by Cardiac Drugs (S. A. Briller and H. L. Conn, Eds). Philadelphia: University of Pennsylvania Press 1966a, p. 111.
TOSTESON, D. C.: Ann. N. Y. Acad. Sci. **137**, 577 (1966b).
TOSTESON, D. C., MOULTON, R. H.: Physiologist **2**, 116 (1959).

TOSTESON, D. C., HOFFMAN, J. F.: J. gen. Physiol. **44**, 169 (1960).
TOSTESON, D. C., GUNN, R. B., WIETH, J. O.: In: Erythrocytes, Thrombocytes and Leucocytes, Vol. 2. (E. Gerlach, K. Moser, E. Deutsch, and W. Willmanns, Eds). Stuttgart: Thieme 1972, p. 62.
TOSTESON, M. T., LAU, F., TOSTESON, D. C.: Nature **243**, 112 (1973).
TUCKER, E.: J. Physiol. **198**, 33p (1968).
TUCKER, E. M.: Biol. Rev. **46**, 341 (1971).
TUCKER, E.: In: The Blood of Sheep. Composition and Function (M. H. Blunt, Ed.). New York-Heidelberg-Berlin: Springer 1975, p. 133.
TUCKER, E. M., ELLORY, J. C.: Anim. Blood Grps. Biochem. Genet. **1**, 101 (1970).
VALET, G., FRANZ, G., LAUF, P. K.: J. Cell Physiol. **94**, 215 (1978).
VAN DEN BERG, K. J.: The Role of Amino Acids in the Mitogenic Activation of Lymphocytes; Thesis, University of Leiden. Rijswijk, Netherlands: Radiobiological Institute of the Organization for Health Research 1974.
VAN DEN BERG, K. J., BETEL, I.: Exp. Cell Res. **66**, 257 (1971).
VAN DEN BERG, K. J., BETEL, I.: FEBS Letters **29**, 149 (1973a).
VAN DEN BERG, K. J., BETEL, I.: Exp. Cell Res. **76**, 63 (1973b).
VAN DEN BERG, K. J., BETEL, I.: Exp. Cell Res. **84**, 412 (1974).
VAUGHAN, G. L., COOK, J. S.: Proc. nat. Acad. Sci. (Wash.) **69**, 2627 (1972).
WANDS, J. R., PODOLSKY, D. K., ISSELBACHER, K. J.: Proc. nat. Acad. Sci. (Wash.) **73**, 2118 (1976).
WANG, J. L., MCCLAIN, D. A., EDELMAN, G. M.: Proc. nat. Acad. Sci. (Wash.) **72**, 1917 (1975).
WHITNEY, R. B., SUTHERLAND, R. M.: Biochim. biophys. Acta **298**, 790 (1973a).
WHITNEY, R. B., SUTHERLAND, R. M.: In: Proc. 7th Leucocyte Culture Conference (F. Dgaillard, Ed.). New York: Academic Press 1973b, p. 63.
WHITTAM, R., CHIPPERFIELD, A. R.: Biochim. biophys. Acta **415**, 149 (1975).
WHITTINGTON, E. S., BLOSTEIN, R. J.: Biol. Chem. **246**, 3518 (1971).
WIGZELL, H.: In: Contemporary Topics in Immunobiology, Vol. 3 (M. D. Cooper and N. L. Warner, Eds). New York: Plenum Press 1974, p. 77.
YAHARA, I., EDELMAN, G. M.: Proc. nat. Acad. Sci. (Wash.) **69**, 608 (1972).
YAHARA, I., EDELMAN, G. M.: Exp. Cell Res. **81**, 143 (1973).
YOUNG, C., ROTHSTEIN, A.: J. gen. Physiol. **50**, 917 (1967).

Chapter 10

Membrane Receptors, Cyclic Nucleotides, and Transport

S. A. RUDOLPH and R. J. LEFKOWITZ

A. Introduction

The concept of a receptive mechanism for drugs and hormones is generally attributed to LANGLEY (1905) and was developed to explain the great specificity of drug action, with particular reference to the actions of alkaloids such as curare and nicotine.

The term "receptor" refers to that component of a responsive target cell with which a drug or hormone initially interacts. It implies two distinct yet complementary functions. The first is that of discrimination, which is accomplished by specific binding. The particular spectrum of drug action at any receptor is a manifestation of the binding specificity of that receptor. This binding specificity is in turn presumably a reflection of certain complementary aspects of structure between the receptors and the biologically active drug molecules with which it interacts.

The second function of receptors is that of activation of a biological process, often through alteration of various enzymatic activities. Binding of a ligand to a receptor may lead to conformational alterations in that receptor, which in turn, through molecular perturbations as yet largely not understood, lead to changes in ion transport or other physiological processes. Drugs which bind to receptors and cause such biological effects are referred to as *agonists*. Other drugs may bind to the same receptors without altering the receptors and without perturbing a physiological process. Such compounds may act as *competitive antagonists*.

Until about seven or eight years ago, receptors remained hypothetical structures the properties of which were inferred from the observation of physiological effects often many steps removed (in biochemical terms) from the receptors. The receptors were in fact defined in terms of the potency series of various drugs for activating or inhibiting such physiological effects. More recently, receptors have been investigated directly by studying the binding of radioactively labeled hormones, drugs, and antagonists to the actual receptor structures in a variety of cellular and subcellular preparations. This approach has led to rapid progress in elucidating the molecular properties of a wide variety of receptors for polypeptide hormones (ROTH, 1973), neurotransmitters, such as acetyl choline (KARLIN, 1974) and catecholamines (LEFKOWITZ et al., 1976; LEFKOWITZ, 1975 and 1976), and steroid hormones (BAULIEU et al., 1971). On the basis of such studies it has become clear that there are several major classes of receptors. One group is localized in the cell membranes. Of these, many appear to be linked to the enzyme adenylate cyclase, such as the beta-adrenergic receptors for catecholamines as well as the receptors for a variety of polypeptide hormones such as parathormone, vasopressin, ACTH, glucagon, etc. Not all membrane-bound receptors are linked to this enzyme. For example, the nicotinic cholinergic receptors do not seem to be related to the adenylate cyclase-cAMP system.

Another class of receptors, those for the steroid hormones such as aldosterone or progesterone, are intracellular cytoplasmic macromolecules. Finally, the receptors for the thyroid hormones appear to be located in the nuclei of sensitive cells (OPPENHEIMER et al., 1972).

As noted above, direct radioligand binding techniques have greatly advanced our understanding of the molecular properties of receptors. In several cases the receptors have been purified partially (DUFAU and CATT, 1975) or to homogeneity (ONG and BRADY, 1974).

Both membrane bound and soluble receptors may mediate the effects of drugs and hormones on certain transport processes. Of the membrane-bound receptors there are examples of adenylate cyclase-coupled and adenylate cyclase-independent receptors related to transport. In the first group would be beta-adrenergic receptors, parathormone and vasopressin receptors. In the second group would be the nicotinic cholinergic receptor. An example of soluble receptors which may mediate effects on transport are the renal receptors for aldosterone.

Several reviews of receptor identification by radioligand binding have appeared recently (ROTH, 1973; KARLIN, 1974; LEFKOWITZ et al., 1976; LEFKOWITZ, 1975 and 1976; BAULIEU et al., 1971), and hence this area will not be reviewed in detail here. Similarly, no attempt will be made to cover comprehensively and systematically the data dealing with characterization of all receptors that are linked to effects on ion transport. Rather, one or two model systems will be covered in some detail. The emphasis will be on general principles, both physiologic and investigative. Beta-adrenergic effects of catecholamines on sodium and potassium fluxes in avian and amphibian erythrocytes will be especially stressed. These simple red blood cell systems have proved invaluable for dissection and study of the individual steps from receptor binding to alterations in ionic fluxes.

B. Beta-Adrenergic-Receptor Binding in Avian and Amphibian Erythrocytes

Recently several groups have developed methodology for studying beta-adrenergic receptors directly. In each case a high-affinity, high-specific-activity radiolabeled beta-adrenergic antagonist was used. The agents used are (−)[^{3}H]dihydroalprenolol (LEFKOWITZ et al., 1974; MUKHERJEE et al., 1975 and 1976), (±)[^{125}I]hydroxybenzylpindolol (AURBACH et al., 1974; BROWN et al., 1976) and (±)[^{3}H]propranolol (LEVITZKI et al., 1974; ATLAS et al., 1974). Each of these agents was initially used to study the beta-adrenergic receptors in membrane fractions derived from avian or amphibian erythrocytes. These cells, unlike human erythrocytes, possess beta-adrenergic receptors which appear to be coupled to adenylate cyclase. As will be described in detail below, this receptor-enzyme system mediates the effects of catecholamines on ion fluxes in these cells.

The ease with which homogeneous membranes can be prepared from these cells, and the high catecholamine sensitivity of their adenylate cyclase has made them a popular source material for studies of beta-adrenergic receptors and adenylate cyclase. The characteristics of radioligand binding to such erythrocyte membrane preparations are identical to those which would be expected of binding to the physiologic beta-adrenergic receptors. In all cases the binding is rapid and reversible. Binding is saturable. There are about 600–900 receptors per turkey erythrocyte (LEVITZKI et al., 1974) and about 1200 per frog erythrocyte (MUKHERJEE et al., 1975). For comparison there are approximately 2000 receptors on a human leukocyte (WILLIAMS et al., 1976).

Binding to sites in the erythrocyte membranes displays the specificity and stereospecificity expected of true beta-adrenergic receptor binding. An example is provided in Figures 1 A and B. The potency series of agonists for stimulation

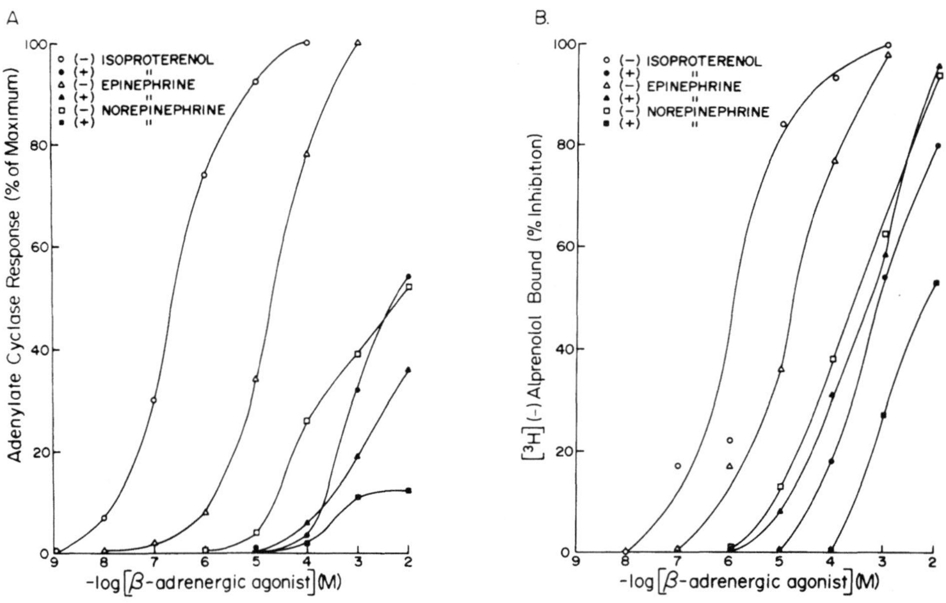

Fig. 1. (A) Stimulation of frog erythrocyte membrane adenylate cyclase by stereoisomers of beta-adrenergic agonists. Maximum response refers to the amount of cyclic AMP generated in the presence of a maximal stimulatory concentration of (−)isoproterenol (0.1 mM). This was generally a 5- to 10-fold stimulation of basal enzyme acticity. (From MUKHERJEE et al., 1975). (B) Inhibition of (−)[^{3}H]dihydroalprenolol binding to frog erythrocyte membranes by stereoisomers of beta-adrenergic agonists. 100% inhibition of binding refers to complete blockade of specific binding, *i. e.* inhibition of binding equivalent to that observed with 10 μM (±)propranolol. (From MUKHERJEE et al., 1975)

of frog erythrocyte membrane adenylate cyclase is isoproterenol > epinephrine > norepinephrine, which is typical of a beta-adrenergic receptor. Moreover, there is marked stereospecificity apparent in that the (−)isomer of each agent is considerably more potent than the (+)isomer in stimulating the enzyme.

Figure 1B demonstrates the ability of each of these agents to compete with (−)[^{3}H]dihydroalprenolol for binding to sites in the frog erythrocyte membranes. The potency series is essentially identical to that found for stimulation of adenylate cyclase. The same marked stereospecificity was also noted. When a much larger group of agents was tested for their ability to interact with the erythrocyte membrane adenylate cyclase and the (−)[^{3}H]dihydroalprenolol binding sites, comparable results were obtained (MUKHERJEE et al., 1976). Thus, in all cases studied, activity in the enzyme and binding assays were directly parallel. Data such as these form the basis for the contention that the sites identified with these techniques are in fact equivalent to the adenylate-cyclase coupled beta-adrenergic receptors.

In turkey erythrocyte membranes similar results have been obtained with the ligands (±)[^{3}H]propranolol (LEVITZKI et al., 1975) and (±)[^{125}I]hydroxybenzylpindolol (BROWN et al., 1976). However, whereas the beta-receptors of frog erythrocytes are beta$_2$-adrenergic receptors (MUKHERJEE et al., 1975) (isoproterenol> epinephrine> norepinephrine), those of turkey erythrocytes are beta$_1$-receptors (isoproterenol> epinephrine = norepinephrine) (AURBACH et al., 1974).

The characteristics of these receptors are currently under study. They appear to be proteins and possibly lipoproteins (LIMBIRD and LEFKOWITZ, 1976). The beta-adrenergic receptors of frog erythrocyte membranes have been solubilized with the plant glycoside digitonin and an approximate molecular weight determined (CARON and LEFKOWITZ, 1976a and b).

As noted above, the beta-adrenergic receptor – adenylate cyclase complex appears to mediate the effects of catecholamines on ion fluxes in avian and amphibian erythrocytes. It should be realized, however, that effects of catecholamines on ion fluxes in these cells were known long before the relationship of cAMP to these processes was appreciated.

C. Beta-Adrenergic-Mediated Transport Processes in Avian and Amphibian Erythrocytes

It was first shown by ORSKOV (1956) that epinephrine and norepinephrine have specific effects on cation transport in nucleated erythrocytes. Using whole blood from both frogs and pigeons, he found that a substantial uptake of plasma potassium could be induced by these catecholamines. In the case of the pigeon erythrocyte, this amounted to as much as 10 mMol K^+ l^{-1} cells. The uptake process was essentially complete after 60 min of incubation with catecholamine and could be half-maximally stimulated with about 5×10^{-9}M norepinephrine. Significant effects were observed with concentrations as low as 3×10^{-10}M.

Epinephrine was also effective in causing potassium uptake, but higher concentrations were required for a given effect. Up to 18 mMol K^+ l^{-1} cells were absorbed by the frog erythrocyte, but with a very different time course; the absorption process required 18 hours for completion. At least part of this difference can be attributed to the fact that the pigeon erythrocyte experiments were performed at 43° C, while those with the frog erythrocytes were done at 22° C. It was also observed that the sensitivity of the potassium uptake process to catecholamines was lower in the frog, and the relative potencies of epinephrine and norepinephrine were reversed. Thus, in the frog, half-maximal responses were achieved with about 3×10^{-7}M epinephrine, while more than 3×10^{-6}M norepinephrine was required. This implies some difference at the receptor level between the two species of erythrocytes, which has been confirmed by more recent studies on adenylate cyclase activity and receptor binding, as discussed earlier.

Following the discovery that a variety of beta-adrenergic effects were mediated by cyclic AMP, SUTHERLAND and his co-workers investigated the role of catecholamines in avian erythrocytes (KLAINER et al., 1962; DAVOREN and SUTHERLAND, 1963a and b; OYE and SUTHERLAND, 1966). Using both pigeon and turkey red cells, they found that epinephrine, norepinephrine, and isoproterenol were all effective in raising cyclic AMP levels in intact cell preparations. Moreover, these catecholamines stimulated adenylate cyclase activity in plasma membrane fractions prepared from avian erythrocytes and this effect could be blocked by beta-adrenergic antagonists. It thus appeared that the potassium absorption phenomenon observed by Orskov was a beta-adrenergic effect which might be mediated by cyclic AMP.

The effects of norepinephrine and dibutyryl cyclic AMP on cation transport in duck erythrocytes have been studied in detail by RIDDICK et al. (1971). They found that the beta-adrenergic receptor of the duck erythrocyte controls both the sodium and the potassium permeability of these cells, and that under appropriate conditions adrenergic stimulation leads to a net uptake of potassium and sodium and an increase in cell volume. These results will now be discussed.

When freshly drawn duck erythrocytes were incubated in their own plasma, they maintained their water and electrolyte composition for at least 90 minutes. If, however, propranolol was added to the plasma or the cells were washed and suspended in an osmotically equivalent medium free of catecholamines, they lost approximately 8 mMol potassium l^{-1} cells and 0.8 mMol sodium l^{-1} cells, and cell water decreased by approximately 2 percent. This shrinkage of the cells could be prevented if the resuspension medium contained 10^{-6}M norepinephrine. It thus appears that the duck erythrocyte can exist in two states of volume and ionic composition; these were termed the upper steady state (USS) and the lower steady state (LSS) by RIDDICK et al. In addition, the presence of a beta-adrenergic agonist (which was endogenous to the normal duck plasma) was required for maintenance of the cells in the USS. Further experiments showed that the transition between the two states was reversible. Thus, LSS cells could absorb potassium, sodium and water and return to the USS if norepinephrine was added to the medium. In addition, the transition to the USS required elevated extracellular potassium. Maximal effects were observed with potassium concentrations above 10 mM, while there was no net uptake of potassium or increase in cell volume with potassium concentrations below 2.5 mM. These results are summarized in Table 1. The transition from the LSS to the USS also required the presence of sodium in the medium; norepinephrine was without effect if sodium was replaced by equiosmolar amounts of either choline or magnesium. Interestingly, although elevated extracellular potassium was required for the transition to the USS, it was not required for maintenance of the USS.

Table 1. Changes in cation and water content of LSS duck erythrocytes due to norepinephrine and db − cyclic AMP

		Δ Cation Content, mmol l^{-1} cells		
$(K)_o$	Additions	Δ K	Δ Na	Δ % cell H_2O (w/w)
2.5mM	none	0	+0.3	+0.2
2.5mM	10^{-6}M norepinephrine	+1	0	+0.1
15 mM	none	0	0	−0.1
15 mM	10^{-6}M norepinephrine	+20	+1.1	+3.3
15 mM	10^{-6}M norepinephrine + 10^{-4}M propranolol	0	0	−0.1
15 mM	10^{-2}M db-cyclicAMP	+17.5	+1.2	+3.6

Adapted from Riddick et al. (1971)

The effects of norepinephrine on LSS cells were quantitatively mimicked by the addition of 10 mM dibutyryl cyclic AMP to the medium, suggesting that cyclic AMP is indeed the intracellular messenger for beta-adrenergic effects on ion transport in these cells. However, resuspension of USS cells in medium containing 10 mM dibutyryl cyclic AMP maintained the cells in the USS in only three of six experiments; this finding was not explained.

These results raise the question of the physiological significance of an adrenergically controlled mechanism for potassium uptake. It is not clear whether the significant effect is the increase in cell volume, the reduction in plasma potassium, or both. The volume increase could affect deformability, shape and plasma volume. However, the potassium uptake may well be the more important effect, since at their normal hematocrit (40–50%) the cells have an enormous capacity for potassium. The red cells would thus act as a buffer (under adrenergic control) for circulating potassium levels, maintaining the concentration at about 2.5 mM, below which they no longer absorb potassium. It is interesting to note that in most mammals, mature erythrocytes do not appear to have a beta-adrenergic receptor and their transport properties are not affected by catecholamines or cyclic AMP. The mammalian kidney, however, is far more sophisticated than the avian one and may thus have assumed any role of the red cell in electrolyte metabolism during the course of evolution.

The specific changes in unidirectional cation fluxes which account for the transition from the LSS to the USS in duck erythrocytes have been studied by Kregenow (1973). Measurements of both sodium and potassium influx and efflux showed that norepinephrine caused a general increase in membrane permeability to these cations. Thus, during the first 15 minutes after the addition of catecholamine to LSS cells, both sodium and potassium influx were increased about six-fold. This effect was observed under conditions where the increase in cell volume occurred ($K^+ = 17$ mM) or under conditions where it did not occur ($K^+ = 2.5$ mM), although the effect on sodium influx was somewhat diminished in the low-potassium medium. There was, however, a striking dependence of the time course of the norepinephrine effect on potassium concentration. Both sodium and potassium influx were unchanged for 60 minutes in the presence of norepinephrine in medium containing 2.5 mM K^+; in medium containing 17 mM K^+, however, there was a progressive decline in sodium and potassium influxes, so that 60 minutes after the addition of catecholamine, the sodium influx was almost down to its control value and potassium influx was only two-fold greater than in untreated control cells. These data are shown in Table 2. It thus appears that a mechanism exists for relieving the effect of norepinephrine on cation permeability and that this mechanism is dependent upon some feedback from the increase in cell volume that occurs in the presence of high extracellular potassium. Under conditions where there is no net uptake of cations or water (2.5 mM K^+), the effect of norepinephrine on cation fluxes is undiminished for at least 60 minutes.

Table 2. The effect of norepinephrine on potassium influx as a function of time in LSS duck erythrocytes

	Potassium influx, mmol l^{-1} cells	
	$(K)_o = 2.6$mM	$(K)_o = 17$mM
control	2	2.5
$+10^{-6}$M norepinephrine		
0–15 min	17	18
15–30 min	16	16
30–45 min	16	7
45–60 min	17	5

Adapted from KREGENOW (1973)

The mechanism for rendering cation fluxes insensitive to catecholamines in USS cells does not involve desensitization of the adenylate cyclase. Although this has not been shown directly in duck red cells, it has been demonstrated in turkey erythrocytes. Turkey erythrocytes undergo similar changes in cation permeability and volume to duck erythrocytes when exposed to beta-adrenergic stimulation (GARDNER et al., 1973 and 1975a; RUDOLPH et al., 1977). Cyclic AMP levels continue to increase for at least 3 hours following catecholamine addition, although the USS is reached within 1 hour. Thus, the transitory response of the cation fluxes that is required to bring the cells from the LSS to the USS is not a reflection of a transitory response at the adenylate cyclase level, but involves some intervention at a step distal to the generation of cyclic AMP. In fact, continued activation of the adenylate cyclase is necessary for maintenance of the USS, although it is not clear what mechanism is responsible for this effect.

The turkey erythrocyte has been extensively studied with respect to catecholamine-dependent changes in bidirectional cation fluxes (GARDNER et al., 1973, 1974b and 1975a; RUDOLPH et al., 1977). As with the duck erythrocyte, there were striking increases in sodium and potassium influx caused by beta-adrenergic stimulation. Concentrations of 10^{-9}M – 10^{-8}M(−)isoproterenol were required for half-maximal stimulation of all these fluxes. Catecholamines were also found to raise cyclic AMP levels in turkey erythrocytes, and this effect was half-maximal with 10^{-6}M(−)isoproterenol (Figs 2 and 3). This discrepancy between the hormone concentrations required for physiological effects and for adenylate cyclase activation is widely observed (*e. g.* for the inotropic effects of epinephrine on cardiac muscle [MAYER, 1972], the induction of glycogenolysis by glucagon in the liver [EXTON and PARK, 1971], and the activation of lipolysis by epinephrine and glucagon in the adipocyte [ROBISON et al., 1971]). This phenomenon does not imply a dissociation between the cyclic AMP system and the hormonal effects, since cyclic AMP or its derivatives are effective in eliciting the physiological response in the absence of the hormones. If adenylate cyclase activation is a direct reflection of hormone receptor occupancy, as seems to be the case from the binding data discussed earlier, then only a small fraction of the total receptors need to be occupied for full physiological activity. Thus, most of

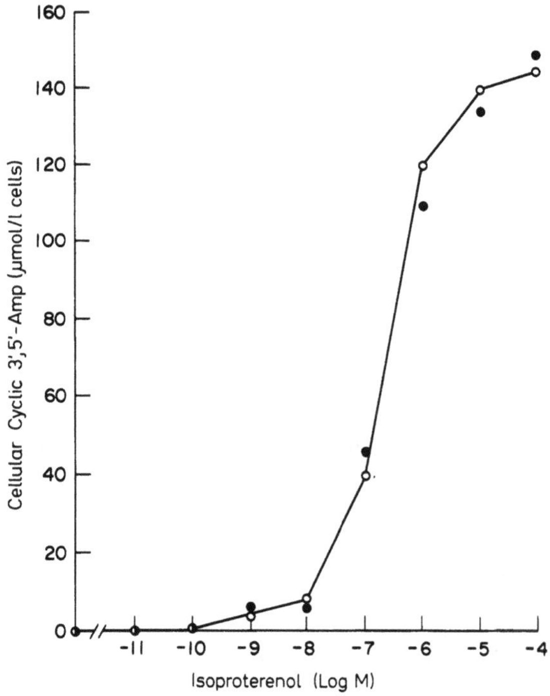

Fig. 2. Effect of (±)isoproterenol on cellular accumulation of cyclic AMP in turkey erythrocytes. Cells were incubated for 60 min in the presence of various concentrations of (±)isoproterenol with (•) or without (0) theophylline (10^{-3}M). Erythrocytes were then processed for determination of cellular cyclic AMP. Results are means of four experiments. (From GARDNER et al., 1973)

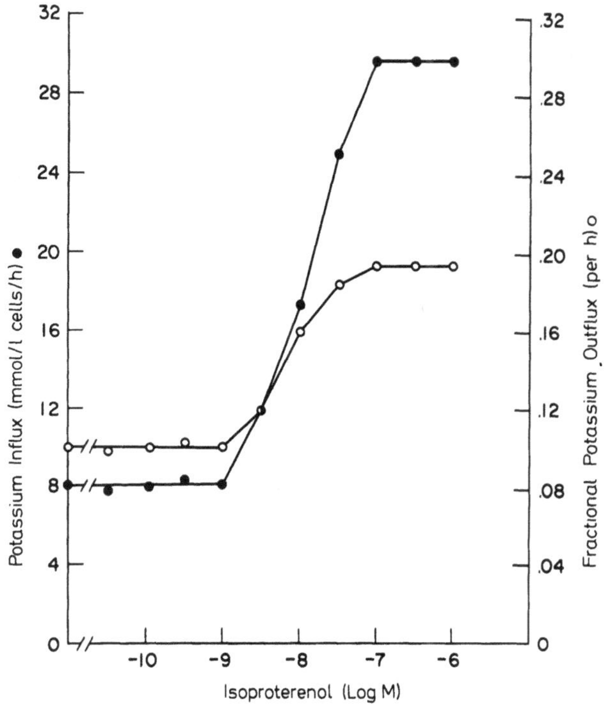

Fig. 3. Effect of (±)isoproterenol on potassium influx and fractional potassium outflux in turkey erythrocytes. Results are means of three experiments. (From GARDNER et al., 1973)

the increase in cyclic AMP levels observed with maximal hormone concentrations is extraneous, only very small increases being necessary to effect physiological changes. In the turkey erythrocyte, 10^{-8}M (−)isoproterenol causes a doubling of cyclic AMP levels and a sixfold increase in potassium influx; further increases in isoproterenol concentration are without effect on potassium influx, but can increase the cyclic AMP level by more than 100-fold.

The question naturally arises as to how increases in intracellular cyclic AMP can cause changes in membrane function. Following the discovery of cyclic AMP-dependent protein kinases in a wide variety of tissues, KUO and GREENGARD (1969) postulated that all of the effects of cyclic AMP might be mediated through protein phosphorylation and that the specificity of cyclic AMP in different tissues resides in the endogenous substrate proteins available for phosphorylation. This mechanism has been worked out in some detail for cyclic AMP control of three processes involving soluble enzymes: glycogenolysis, gluconeogenesis and lipolysis (for review, see STEINBERG and HUTTENEN, 1972; STEINBERG et al., 1975). Much less is known about cyclic AMP effects on membranes and membrane permeability, although cyclic AMP-dependent protein kinases and endogenous substrates have been demonstrated in membrane preparations from many tissues (MALKINSON et al., 1975; RUBIN and ROSEN, 1975). Thus, although the protein kinase hypothesis has not been definitively established, it provides a good starting point for investigating the mechanism of action of cyclic AMP. In addition, no other enzyme has been isolated which is directly activated by cyclic AMP.

Following this line of reasoning, RUDOLPH and GREENGARD (1974, 1977) have studied the effects of catecholamines on protein phosphorylation in the intact turkey erythrocyte. The object of these experiments was to establish whether catecholamine-induced changes in cyclic AMP levels could cause changes in membrane protein phosphorylation and whether these changes correlated with effects on cation fluxes. Suspensions of turkey red cells which had been pre-loaded with ^{32}P were exposed to various catecholamines or cyclic AMP derivatives; the whole cells were then solubilized with sodium dodecyl sulfate, and the proteins were separated by polyacrylamide gel electrophoresis. Analysis of these gels revealed that beta-adrenergic catecholamines were effective in stimulating ^{32}P incorporation into a membrane protein of molecular weight 240000. A typical experimental result is shown in Fig. 4. Table 3 shows that only under

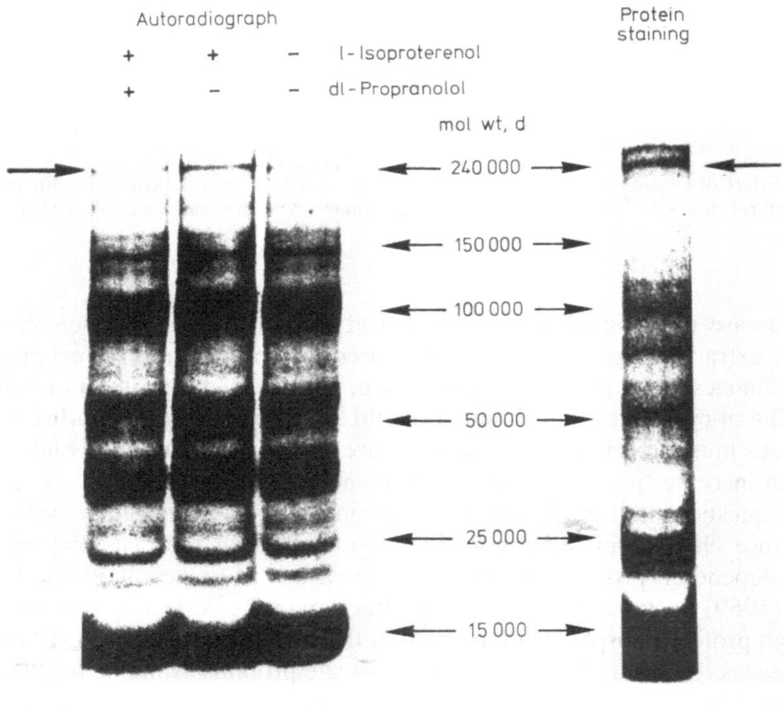

Fig. 4. Effect of (−)isoproterenol (3×10^{-7}M) and (±)propranolol (2×10^{-6}M) on the net incorporation of ^{32}P into proteins of the intact turkey erythrocyte. Erythrocytes were incubated in the presence of the test agents for 10 min. Incorporation of ^{32}P into the 240000 molecular weight protein band in the experiment illustrated was 1,200 and 2,500 cpm in the absence and presence of (−)isoproterenol, respectively. (From RUDOLPH and GREENGARD, 1974).

Table 3. Effect of various substances on net incorporation of ^{32}P into 240000 molecular weight protein, on cAMP levels, and on sodium uptake in intact turkey erythrocyte

Additions	^{32}P Incorporation into 240000 mol wt protein	Intracellular cAMP	Sodium uptake
	% control	μ moles/liter cells	% control
None	100	N. D.[a]	100
(−)Isoproterenol 10^{-6}M	236 ± 9	26.2 ± 1.9	311 ± 8
(−)Epinephrine, 10^{-6}M	198 ± 20	19.5 ± 1.0	312 ± 7
(−)Norepinephrine, 10^{-6}M	198 ± 14	22.3 ± 0.8	323 ± 6
Dopamine, 10^{-6}M	110 ± 11	N. D.[a]	104 ± 1
(−)Isoproterenol, 10^{-6}M, + (±)Propranolol, 2×10^{-5}M	114 ± 6	N. D.[a]	106 ± 2
(−)Isoproterenol, 10^{-6}M + Phentalamine, 10^{-4} M	211 ± 5	23.6 ± 0.8	296 ± 8
cAMP, 2×10^{-3}M	213 ± 8		302 ± 8
N^6-Monobutyryl cAMP, 2×10^{-3}M	238 ± 8		313 ± 14
N^6, $O^{2'}$-Dibutyryl cGMP, 2×10^{-3}M	105 ± 10		86 ± 3

[a] N. D., not detectable (< 0.1 μ mol l^{-1} cells).

In each experiment, net ^{32}P incorporation into the 240000 molecular weight protein during a 10-min incubation was calculated as percentage of that found in erythrocytes incubated for 10 min in the absence of any addition; data represent the means ± S. E. for three or more experiments. In the absence of additions, the incorporation of ^{32}P was 0.29 to 0.39 μCi per mg of 240000 molecular weight protein in this series of experiments. Intracellular cAMP accumulated during a 10-min incubation at 37° C was measured; data represent the means ± S. E. for four or more determinations. Sodium uptake was calculated from intracellular ^{22}Na accumulated during the interval between 8 and 12 min of incubation; data represent the means ± S. E. of triplicate determinations. In the absence of additions, sodium uptake was 2.84 ± 0.2 mmol/l cells/hour. (Adapted from RUDOLPH and GREENGARD, 1974)

conditions where there was an increase in cyclic AMP level was there an increase in sodium influx and protein phosphorylation. The increase in ^{32}P incorporation into this protein occurred without any apparent lag after catecholamine addition and was maximal within 30 minutes, after which it was stable for at least two hours. This time course for the increase in phosphorylation is similar to that observed by GARDNER et al. (1973) for the catecholamine-induced increase in sodium influx. In addition, the dependence of both protein phosphorylation and cation fluxes on catecholamine concentration was similar; approximately 10^{-8}M(−)isoproterenol was half-maximal for each effect. It was further found that three other membrane proteins had increased ^{32}P incorporation, but that these phosphorylation effects were characterized by a 30-minute induction period following catecholamine addition and thus did not correlate with rapid changes in membrane function. They might, however, have some connection with maintenance of the USS in these cells.

In summary, nucleated erythrocytes are capable of controlling their cation content and volume through specific changes in unidirectional sodium and potassium fluxes. Further, this mechanism is under beta-adrenergic control, with cyclic AMP as an intracellular mediator. Intracellular cyclic AMP appears to be necessary both to initiate the increased fluxes necessary to cause the transition from the LSS to the USS and to maintain the USS after the fluxes have returned to basal values. Cyclic AMP may act through the activation of protein kinase, which causes increases in phosphorylation of specific membrane proteins; the state of phosphorylation of these proteins could then determine membrane permeability and the activity of active transport systems responsible for the observed flux changes.

This material concerning the beta-adrenergic receptor-mediated regulation of cation transport in erythrocytes has been presented in some detail because, in biochemical terms it is one of the most thoroughly documented examples of receptor-mediated effects on transport. Thus there is specific information available concerning the receptor binding step, the activation of adenylate cyclase, the effect of cyclic AMP on protein kinases, the phosphorylation of a specific endogenous membrane protein and the effects of cyclic AMP on transport processes.

D. The Amphibian Bladder

A somewhat more complex tissue in which the relationship between hormone receptors and transport has been studied is the amphibian urinary bladder. The inner or mucosal surface of the amphibian bladder consists in part of epithelial cells whose function is the transport of sodium and water across the bladder to the serosal surface. A peptide known as antidiuretic hormone (ADH, or vasopressin) stimulates the rate of sodium and water transport across these cells. Much evidence has been accumulated to show that these effects of ADH are mediated by cyclic AMP; ADH raises intracellular cyclic AMP levels by binding to a cell surface receptor and activating adenylate cyclase. Transport of sodium from the lumen of the bladder is a two-step process. The rate-limiting step is thought to be the passive diffusion of sodium from the luminal fluid across the apical (luminal) membrane of the mucosal epithelial cells; the sodium is then pumped out at the serosal side of the cell by a Na^{+}-K^{+} dependent ATPase (FERGUSON and PRICE, 1972). The chain of events by which the binding of ADH to its receptor influences sodium and water transport will now be discussed. It was first observed in 1961 by ORLOFF and HANDLER (1961 and 1962) that the effects of ADH on the toad bladder could be mimicked by the addition of cyclic AMP or theophylline to the bathing medium. Coupled with the finding that physiologically effective doses of ADH caused substantial increases in the cyclic AMP content of this tissue (HANDLER et al., 1965), it seems clear that cyclic AMP is indeed the intracellular mediator of ADH action. It has also been shown that membranes prepared from both the amphibian bladder and the mammalian kidney possess highly specific binding sites for ADH and its analogues and that the affinity of this binding correlates with the activation of a membrane-bound adenylate cyclase (JARD et al., 1975). It is interesting to note that when ADH is introduced into the lumen of the bladder, it has no effect on either cyclic AMP levels or transport; these effects are observed only when ADH is applied serosally (LEAF et al., 1958). This observation is consistent with the fact that it is the serosal side of the mucosal epithelial cells that is accessible to circulating ADH in the intact animal. Knowing the site of the ADH receptor, it is then of interest to know the end-point site of action of cyclic AMP in the cell. This problem was solved by a series of elegant experiments by CIVAN and FRAZIER (1968). In these experiments, mucosal epithelial cells of the toad bladder were impaled with a micropipette and the DC resistance between the luminal fluid and the pipette was measured along with the resistance across the entire bladder; this was done in both the absence and the presence of ADH. Because of the increased sodium transport caused by ADH, there was a drop in resistance across the bladder and it was found that this change could be completely accounted for by a decrease in the resistance between the luminal fluid and the pipette tip; *i. e.,* by an increase in the sodium permeability of the apical membrane rather than an increase in active transport at the serosal side. Thus, one of the physiological effects of ADH is manifested on the side of the cell opposite to the one containing the ADH receptor and presumably, the ADH sensitive adenylate cyclase. It is not known whether the cyclic AMP formed in response to ADH must diffuse across the cell to affect the sodium permeability of the apical membrane or whether it affects some intracellular cyclic AMP receptor which then alters membrane permeability. Working on the hypothesis that cyclic AMP might exert its effects through changes in protein phosphorylation, GREENGARD and his co-workers DE LORENZO et al., 1973), have studied ^{32}P incorporation into intact toad bladders. They found that ADH caused a decrease in ^{32}P incorporation into a protein of molecular weight 50000 (by polyacrylamide gel electrophoresis in the presence of SDS). The time course of this decrease in ^{32}P incorporation was somewhat more rapid than that for the increase in sodium permeability (measured by potential difference across the bladder) caused by ADH. Also, N^6-monobutyryl cyclic AMP, which has the same effect as ADH on sodium permeability, caused a similar decrease in ^{32}P incorporation into the 50000 MW protein. In further studies, DE LORENZO et al. have found that several agents which cause changes in the sodium permeability of the toad bladder also cause dephosphorylation of the 50000 MW protein and of a 42000 MW protein as well. This correlation with protein phosphorylation did not extend to changes induced by these agents in water transport. DE LORENZO and GREENGARD (1973) have also shown that in homogenates of toad bladder, cyclic AMP reduces phosphorylation of a 50000 MW protein (by gamma-^{32}P-ATP) by apparent activation of a phosphoprotein phosphatase. Although the subcellular localization of the phosphoprotein affected by ADH has not been established, the possibility that cyclic AMP affects sodium permeability in the toad bladder by changes in protein phosphorylation seems strong.

The mineralocorticoid, aldosterone, which regulates sodium permeability in the toad bladder via a mechanism involving protein synthesis, has also been shown to enhance the extent of protein dephosphorylation induced by cyclic AMP (LIU and GREENGARD, 1974). If, in fact, dephosphorylation plays a role in regulating sodium permeability in this tissue, it may be that ADH receptors and steroid receptors share some common biochemical steps in achieving their end-point effects.

E. Cholera Enterotoxin

Cholera enterotoxin (CE) is a protein (MW=84000) secreted by the Vibrio Cholerae organism. It is this protein which causes the striking reversal of intestinal water and electrolyte transport characteristic of the disease. The effects of CE on intestinal transport appear to be a result of its ability to elevate cyclic AMP levels by irreversibly activating adenylate cyclase. Since the discovery and purification of CE, it has been shown to activate adenylate cyclase in a wide variety of tissues and cell culture lines, and in some cases to render adenylate cyclase more susceptible to hormone stimulation. These effects require an induction period of 30–120 min and occur only when intact cells are treated with CE, although the effects on adenylate cyclase persist after the cells are broken (for review, see FINKELSTEIN, 1974). Recent work (GILL and KING, 1975) has shown that the induction period may be necessary for the transport of an active subunit of the toxin into the cell, since this subunit can activate adenylate cyclase rapidly in broken cell preparations. NAD is also required for activation in broken cells (GILL, 1975).

The effects of CE on cyclic AMP levels (RUDOLPH et al., 1977; FIELD, 1974) and on ion and water transport (RUDOLPH et al., 1977) in turkey erythrocytes have been studied in detail. While CE raises basal cyclic AMP levels in these cells by partially activating adenylate cyclase it also lowers by more than tenfold the concentration of epinephrine required for half-maximal stimulation of adenylate cyclase. These changes are reflected in the transport properties of the cell; CE causes partial activation of the catecholamine-dependent cation flux mechanism and causes this mechanism to respond to lower concentrations of catecholamines. These results suggest that the clinical manifestations of cholera may be a result not only of adenylate cyclase activation, but possibly a result of transport mechanisms responding to endogenous levels of hormones that might otherwise be ineffective.

In an effort to find out whether the modulation by CE of catecholamine effects on transport in turkey erythrocytes might be caused by an increase in either the number of beta-adrenergic receptors or their affinity for catecholamines, the binding of $(-)[^3H]$dihydroalprenolol to membranes from control and toxin treated cells has been studied (RUDOLPH et al., 1977). In freshly prepared membranes, there were about 1200 receptors per cell and the specific binding of $(-)[^3H]$dihydroalprenolol was half-maximal at about 2–3 nM; with sub-maximal concentrations, about 85 percent of the alprenolol binding was displaceable with 10^{-5}M $(-)$isoproterenol. Identical results were obtained with both control and toxin-treated membranes, indicating that the effect of CE is not at the level of the hormone-receptor interaction. However, if the membranes were aged prior to the binding assay, specific binding sites for alprenolol decreased less rapidly in the membranes from the toxin-treated cells than in the membranes from untreated cells. Thus, after 24 hours, only about 100 receptors per cell remained detectable in control membranes, but there were 500 per cell in toxin-treated membranes. It thus appears that CE alters the environment of the receptor in the membrane, affording it some protection from denaturation.

F. The Superior Cervical Ganglion

Since a variety of neurotransmitters mediate post-synaptic potentials, the receptors for these substances must be involved in the initial events that lead to changes in the permeability of the post-synaptic membrane. These effects have been worked out in some detail in the mammalian superior cervical ganglion (SCG) (recently reviewed in GREENGARD and KEBABIAN, 1974). In this tissue, the post-ganglionic neurons are innervated by pre-ganglionic cholinergic fibers in three different ways. First, there is direct innervation at a nicotinic cholinergic terminal. There is also direct innervation at a muscarinic cholinergic terminal, and, also through a muscarinic cholinergic fiber, there is innervation by a dopaminergic interneuron. The types of postsynaptic events mediated by these three receptors have been worked out through a series of pharmacological and electrophysiological experiments which have been reviewed in detail elsewhere. Stimulation of the nicotinic cholinergic receptors gives rise to a fast excitatory post-synaptic potential (f-EPSP), as would be expected for this type of receptor. Stimulation of the muscarinic cholinergic receptors gives rise to a slow excitatory post-synaptic potential (s-EPSP) and stimulation at the dopaminergic sites leads to a slow inhibitory post-synaptic potential (s-IPSP). Inhibition of phosphodiesterase activity in the post-ganglionic neuron by theophylline greatly potentiates the size of both the s-EPSP and the s-IPSP, but not the f-EPSP, suggesting that the muscarinic and dopaminergic effects may be mediated by cyclic nucleotides. Indeed, dopamine has been found to raise cyclic AMP levels in slices of the SCG, and acetylcholine raises levels of cyclic GMP. Also, application of cyclic AMP to the post-ganglionic neuron causes hyperpolarization, while application of cyclic GMP causes depolarization. Thus, the susceptibility of the post-ganglionic neuron to firing by stimulation of its nicotinic cholinergic receptors is modulated bi-directionally by a muscarinic cholinergic receptor and a dopaminergic receptor, which in turn exert their effects through cyclic GMP and cyclic AMP respectively.

G. Nicotinic Cholinergic Receptors

Nicotinic cholinergic receptors modulate the effects of acetylcholine at the neuromuscular junction, at certain synapses in the peripheral nervous system, and in the electroplax of the electric eel. While these effects involve changes in ion transport across the post-synaptic membrane, they are very different from the transport changes mediated by cyclic AMP that have been discussed for the nucleated erythrocyte and the amphibian bladder. The nicotinic receptors themselves appear to be very closely coupled or identical to the ion channels that they control (for review see RANG, 1974); this is in contrast to the effects of catecholamines and peptide hormones on transport, where a second messenger is involved and the actual permeability changes may occur at sites in the cell distant from the receptor. The tight coupling between nicotinic receptors and the ion channels they control allows their effects to be extremely rapid; in fact they appear to be diffusion-limited, rather than dependent on any enzymatic or metabolic processes. Thus, the depolarization of the post-synaptic membrane at a nicotinic synapse occurs within milliseconds, whereas the receptor-mediated effects on ion fluxes in the nucleated erythrocyte or amphibian bladder develop over a time course of several minutes or more.

The nicotinic cholinergic receptors have been studied directly by radioligand binding techniques. Such studies have utilized as ligands a variety of agents, including tritiated and radioiodinated snake venom toxins, which bind with high affinity and specificity to the nicotinic cholinergic receptors. This area of research has been the subject of excellent recent reviews (KARLIN, 1974). These receptors have been extensively purified by affinity chromatography and other techniques (KARLIN, 1974).

H. The Heart

The positive inotropic effects of catecholamines appear to be mediated through beta-adrenergic receptors coupled to adenylate cyclase. Although the end-point effects of stimulation of these receptors do not involve bulk transport of water or electrolytes, as in the nucleated erythrocytes or the amphibian bladder, the mechanism by which these effects are brought about involves changes in the transport properties of certain membranes. The increase in calcium concentration which initiates contraction is a result of increases in calcium transport across both the plasma membrane (extracellular calcium) and the sarcoplasmic reticulum (intracellular calcium). It is by regulating the supply of calcium to the contractile apparatus that epinephrine exerts its inotropic effects (for reviews, see SOBEL and MAYER [1973] and BERRIDGE [1975]). There is evidence that this process is at least in part mediated by cyclic AMP-dependent phosphorylation of specific proteins in the sarcoplasmic reticulum.

It has been shown that addition of cyclic AMP and purified cyclic AMP-dependent protein kinase caused a more than two-fold increase in the initial rate of calcium uptake by canine cardiac microsomes (primarily sarcoplasmic reticulum) (KIRCHBERGER et al., 1972; TADA et al., 1974). Further experiments have indicated a positive correlation between the extent of phosphorylation of these membranes and their calcium uptake properties (KIRCHBERGER et al., 1974). The specific membrane components whose phosphorylation may regulate calcium uptake have not been identified; however, most of the phosphate incorporated into cardiac microsomes in the presence of cyclic AMP and added protein kinase is associated with a protein of molecular weight 22000, which has been separated on polyacrylamide gels in the presence of SDS (TADA et al., 1975). Further identification of this protein may reveal the mechanism by which epinephrine, via cyclic AMP, regulates calcium transport and cardiac contractility.

J. General Comments and Conclusions

A large body of evidence exists relating hormone receptor interactions to transport across biological membranes. In a number of systems of varying complexity, hormone-dependent transport processes have been measured and described in great detail. However, the molecular mechanisms of these processes remain for the most part obscure. The discovery of cyclic AMP and its central role as a second messenger in mediating the effects of catecholamines and peptide hormones has provided great insight into hormonal control of transport. However, recognizing the role of cyclic AMP has not provided any final answers, since it is clearly not the final step. It would therefore appear that elucidation of the mechanism of action of cyclic AMP is one of the important problems to be solved in the coming years. The primary handle on this problem at present is the protein kinase hypothesis; that is, that the effects of cyclic AMP in eucaryotic cells are expressed through protein kinase and the specific proteins phosphorylated by this enzyme. Cyclic AMP-dependent protein kinases appear to be very widely distributed, and have been purified and studied in many tissues. However, very little is known about the endogenous substrates for these enzymes, particularly those that might be involved in membrane function. Thus, experiments aimed at demonstrating correlation between hormonally induced protein phosphorylation in intact cells and hormonally induced changes in transport should prove fruitful, although these experiments are technically difficult. The next step, which would involve identification of these phosphoproteins and an understanding of the molecular effects of phosphorylation, will add greatly to our understanding of receptor-mediated transport. The possibility that cyclic AMP receptors other than protein kinase exist in eucaryotic cells must also be considered, although to date no such receptors have been demonstrated. An interesting approach to this problem involves the use of photo-affinity analogs of cyclic AMP, which have recently been developed. With these compounds it is possible to covalently label cyclic AMP-binding sites with ^{3}H, ^{14}C or ^{32}P with great specificity. This approach will undoubtedly yield interesting results on the nature and identity of intracellular receptors for cyclic AMP.

References

ATLAS, D., STEER, M. L., LEVITZKI, A.: Proc. nat. Acad. Sci. (Wash.) **71**, 4246 (1974).

AURBACH, G. D., FEDAK, F. A., WOODARD, C. J., PALMER, J. S., HAUSER, D., TROXLER, F.: Science **186**, 1223 (1974).

BAULIEU, E. E., ALBERGA, A., JUNG, I., LEBEAU, M. C., MERCIER-BODARD, C., MILGROM, E., RAYNAUD, J. P., RAYNAUD-JAMMET, C., ROCHEFORT, H., TRUONG, H., ROBEL, H.: Rec. Prog. Horm. Res. **27**, 351 (1971).

BERRIDGE, M. J.: In: Advances in Cyclic Nucleotide Research, Vol. 6, P. Greengard and G. A. Robison (Eds). New York: Raven Press 1975.

BROWN, E. M., RODBARD, D., FEDAK, S. A., WOODARD, C. J., AURBACH, G. D.: J. biol. Chem. **251**, 1239 (1976).

CARON, M. G., LEFKOWITZ, R. J.: Biochem. biophys. Res. Commun. **68**, 315 (1976a).

CARON, M. G., LEFKOWITZ, R. J.: J. biol. Chem. **251**, 2374 (1976b).

CIVAN, M. M., FRAZIER, H. S.: J. gen. Physiol. **51**, 589 (1968).

DAVOREN, P. R., SUTHERLAND, E. W.: J. biol. Chem. **238**, 3009 (1963a).

DAVOREN, P. R., SUTHERLAND, E. W.: J. biol. Chem. **238**, 3016 (1963b).

DE LORENZO, R. J., GREENGARD, P.: Proc. nat. Acad. Sci. (Wash.) **70**, 1831 (1973).

DE LORENZO, R. J., WALTON, K. G., CURRAN, P. F., GREENGARD, P.: Proc. nat. Acad. Sci. (Wash.) **70**, 880 (1973).

DUFAU, M., CATT, K. J.: J. biol. Chem. **250**, 4822 (1975).

EXTON, J. H., PARK, C. R.: Unpublished data shown in "Cyclic AMP" (G. A. Robison, R. W. Butcher and E. W. Sutherland, Eds). New York, London: Academic 1971.

FERGUSON, D. R., PRICE, R. H.: In: Advances in Cyclic Nucleotide Research, Vol. I. (P. Greengard, R. Paoletti and G. A. Robison, Eds). New York: Raven Press 1972, pp. 113–119.

FIELD, M.: Proc. nat. Acad. Sci. (Wash.) **71**, 3299 (1974).

FINKELSTEIN, R. A.: Crit. Rev. Microbiol. **2**, 553 (1973).

GARDNER, J. D., KLAEVEMAN, H. L., BILEZIKIAN, J. P., AURBACH, G. D.: J. biol. Chem. **248**, 5590 (1973).

GARDNER, J. D., KLAEVEMAN, H. L., BILEZIKIAN, J. P., AURBACH, G. D.: Endocrinology **95**, 499 (1974a).

GARDNER, J. D., KLAEVEMAN, H. L., BILEZIKIAN, J. P., AURBACH, G. D.: J. biol. Chem. **249**, 516 (1974b).

GARDNER, J. D., MENSH, R. S., KIINO, D. R., AURBACH, G. D.: J. biol. Chem. **250**, 1155 (1975a).

GARDNER, J. D., KIINO, D. R., JOW, N., AURBACH, G. D.: J. biol. Chem. **250**, 1164 (1975b).

GARDNER, J. D., JOW, N., KIINO, D. R.: J. biol. Chem. **250**, 1176 (1975c).

GILL, D. M.: Proc. nat. Acad. Sci. (Wash.) **72**, 2064 (1975).

GILL, D. M., KING, C. A.: J. biol. Chem. **250**, 6424 (1975).

GREENGARD, P., KEBABIAN, J. W.: Fed. Proc. **33**, 1059 (1974).

HANDLER, J. S., BUTCHER, R. W., SUTHERLAND, E. W., ORLOFF, J.: J. biol. Chem. **246**, 4524 (1965).

JARD, S., CHRISTIAN, R., BARTH, T., RAJERISON, R., BOCKAERT, J.: In: Advances in Cyclic Nucleotide Research, Vol. 5 (G. I. Drummond, P. Greengard and G. A. Robison, Eds). New York: Raven Press 1975.

KARLIN, A.: Life Sci. **14**, 1385 (1974).

KIRCHBERGER, M. A., TADA, M., REPKE, D. I., KATZ, A. M.: J. Mol. cell. Cardiol. **4**, 673 (1972).

KIRCHBERGER, M. A., TADA, M., KATZ, A. M.: J. biol. Chem. **249**, 6166 (1974).

KLAINER, L. M., CHI, Y.-M., FREIDBERG, S. L., RALL, T. W., SUTHERLAND, E. W.: J. biol. Chem. **237**, 1239 (1962).

KREGENOW, F. M.: J. gen. Physiol. **61**, 509 (1973).

KUO, J. F., GREENGARD, P.: Proc. nat. Acad. Sci. (Wash.) **64**, 1349 (1969).

LANGLEY, J. N.: J. Physiol. **33**, 374 (1905).

LEAF, A., ANDERSON, J., PAGE, L. B.: J. gen. Physiol. **41**, 657 (1958).

LEFKOWITZ, R. J.: Biochem. Pharmacol. **24**, 1651 (1975).

LEFKOWITZ, R. J.: Life Sci. **18**, 461 (1976).

LEFKOWITZ, R. J., MUKHERJEE, C., COVERSTONE, M., CARON, M. G.: Biochem. biophys. Res. Commun. **60**, 703 (1974).

LEFKOWITZ, R. J., LIMBIRD, L. E., MUKHERJEE, C., CARON, M. G.: Biochim. Biophys. Acta **457**, 1 (1976).
LEVITZKI, A., ATLAS, D., STEER, M. L.: Proc. nat. Acad. Sci. (Wash.) **71**, 2773 (1974).
LEVITZKI, A., SEVILIA, N., ATLAS, D., STEER, M. L.: J. Mol. Biol. **97**, 35 (1975).
LIMBIRD, L. E., LEFKOWITZ, R. J.: Molecular Pharmacology **12**, 559 (1976).
LIN, A., Y.-C., GREENGARD, P.: Proc. nat. Acad. Sci. (Wash.) **71**, 3869 (1974).
MALKINSON, A. M., KRUEGER, B. K., RUDOLPH, S. A., CASNELLIE, J. E., HALEY, B. E., GREENGARD, P.: Metabolism **24**, 331 (1975).
MAYER, S. E.: J. Pharm. exp. Ther. **181**, 116 (1972).
MUKHERJEE, C., CARON, M. G., COVERSTONE, M., LEFKOWITZ, R. J.: J. biol. Chem. **250**, 4869 (1975).
MUKHERJEE, C., CARON, M. G., MULLIKIN, D., LEFKOWITZ, R. J.: Molecular Pharmacol. **12**, 16 (1976).
ONG, D. E., BRADY, R. N.: Biochemistry **13**, 2822 (1974).
OPPENHEIMER, J. H., KOERNER, D., SCHWARTZ, H. L., SURKS, M. I.: J. clin. Endocr. and Metab. **35**, 330 (1972).
ORLOFF, J., HANDLER, J. S.: Biochem. biophys. Res. Commun. **5**, 63 (1961).
ORLOFF, J., HANDLER, J. S.: J. clin. Invest. **41**, 702 (1962).
ORSKOV, S. L.: Acta physiol. scand. **37**, 299 (1956).
ØYE, I., SUTHERLAND, E. W.: Biochim. biophys. Acta **127**, 347 (1966).
RANG, H.: Quart Rev. Biophys. **7**, 283 (1974).
RIDDICK, D. H., KREGENOW, F. M., ORLOFF, J.: J. gen. Physiol. **57**, 752 (1971).
ROBISON, G. A., BUTCHER, R. W., SUTHERLAND, E. W. (Eds): Cyclic AMP. New York, London: Academic Press 1971.
ROTH, J.: Metabolism **22**, 1059 (1973).
ROY, C., BOCKAERT, J., RAJERISON, R., JARD, S.: Fed. Eur. Biochem. Soc. Lett. **30**, 329 (1973).
RUBIN, C. S., ROSEN, O. M.: Annual Rev. Biochem. **44**, 831 (1975).
RUDOLPH, S. A., GREENGARD, P.: J. biol. Chem. **249**, 5684 (1974).
RUDOLPH, S. A., SHAEFER, D. E., GREENGARD, P.: (1977) J. biol. Chem., **252**, 7132.
RUDOLPH, S. A., BEAM, K. G., GREENGARD, P.: In: Membrane Transport Processes, Vol. I (J. F. Hoffman, ed.) New York, Raven Press, 1977, pp. 107–123.
SOBEL, H., MAYER, S. E.: Circulat. Res. **32**, 407 (1973).
STEINBERG, D., HUTTENEN, J. K.: In: Advances in Cyclic Nucleotide Research, Vol. 1 (P. Greengard and G. A. Robison, Eds). New York: Raven Press 1972, pp. 47–62.
STEINBERG, D., MAYER, S. E., KHOO, J. C., MILLER, E. A., MILLER, R. E., FREDHOLM, B., EICHNER, R.: In: Advances in Cyclic Nucleotide Research, Vol. 5 (G. I. Drummond, P. Greengard and G. A. Robison, Eds). New York: Raven Press 1975.
TADA, M., KIRCHBERGER, M. A., REPKE, D. I., KATZ, A. M.: J. biol. Chem. **249**, 6174 (1974).
TADA, M., KIRCHBERGER, M. A., KATZ, A. M.: J. biol. Chem. **250**, 2640 (1975).
WILLIAMS, L. T., SNYDERMAN, R., LEFKOWITZ, R. J.: J. clin. Invest. **57**, 149 (1976).

Chapter 11

Permeability Properties of Unmodified Lipid Bilayer Membranes

O. S. ANDERSEN

A. Introduction

Biological membranes are dynamic, mosaic structures where different transport functions usually occur simultaneously. The molecular mechanisms responsible for the transmembrane movement of solutes, ions or nonelectrolytes, are therefore only understood in part, due to the difficulties in isolating and identifying the properties of particular transport systems – whether *in situ* or *in vitro*. Much of the available information has been obtained in a few particularly simple biological systems. Most of the conceptual knowledge, however, has been obtained in studies on artificial model membranes.

There is strong evidence that the lipid moiety in the membranes of all cells, and most subcellular organelles, is organized as bimolecular leaflets, in which the hydrocarbon tails are sequestered away from the aqueous phases to form the central core of the membranes (STEIM et al., 1969; STOECKENIUS and ENGELMAN, 1969; WILKINS et al., 1971; SINGER, 1971; BRANTON and DEAMER, 1972). Artificial bimolecular lipid membranes are therefore very useful and relevant models for biological membranes (MUELLER et al., 1962; BANGHAM, 1968; MUELLER and RUDIN, 1969; FINKELSTEIN, 1972; ANDERSEN, 1978a)[1].

This chapter will attempt to elucidate some of the presently understood biophysical principles underlying the permeability characteristics of thin membranes, and to review the movement of ions and nonelectrolytes through unmodified lipid bilayers. Specific information about the permeability changes induced by ion carriers and uncouplers of oxidative phosphorylation, which are discussed by STARK (see Chapter 12), the permeability changes induced by channel formers, which are discussed by HALL (see Chapter 13), and the photoelectric effects observed in BLM doped with photosensitive pigments (*e.g.*, BERNS, 1976; HONG, 1976; TIEN, 1974, 1976) will therefore be omitted.

Planar bimolecular lipid membranes of the MUELLER-RUDIN (MUELLER et al., 1963) or Montal-Mueller type (MONTAL and MUELLER, 1972) are particularly suited to kinetic studies of ion movement across membranes. These membranes constitute very thin, almost impermeable structures, of relatively well-known composition, which separate two bulk aqueous phases. One can thus control,

[1] The first measurements on artificial lipid bilayers as models for biological membranes were apparently those of DEAN et al. (1940).

and vary at will, the chemical composition of the aqueous phases, as well as the electrical potential difference between the two bulk aqueous phases. The properties of these membranes have been extensively studied and reviewed frequently (BANGHAM, 1968; MUELLER and RUDIN, 1969; HAYDON, 1970a; LIBERMAN, 1970; HAYDON and HLADKY, 1972; LÄUGER, 1972; EISENMAN et al., 1973; EISENMAN, 1973; TIEN, 1974; HLADKY et al., 1974; EISENMAN, 1975; MCLAUGHLIN and EISENBERG, 1975; de LEVIE, 1976; MONTAL, 1976). The articles by HAYDON and HLADKY (1972), LÄUGER and NEUMCKE (1973), and de LEVIE (1978) are particularly relevant as they provide valuable information about the nonmediated movement of hydrophobic ions through artificial lipid bilayer membranes.

Hydrophobic ions are very important tools in the investigation of the biophysics of ion (*i. e.* solute) movement across these membranes, because it is possible to study the *kinetics* of ion *movement* directly, without any of the complications that can arise from chemical reactions between the ion and a membrane-bound carrier or the simultaneous movement of the loaded and unloaded carrier trough the membrane. Hydrophobic ions are thus well suited for experimental tests of models for the behavior of and interactions among charged molecules located at the membrane-solution interfaces or moving through the hydrocarbon interior of bimolecular lipid membranes. These ions have therefore proven useful as probes with which one can study changes in the electrostatic potential within lipid bilayers, as well as changes in membrane structure.

B. Lipid Bilayer Membranes

The properties of a lipid bilayer are, as a first approximation, those of a thin slab, 2.5–5 nm, of wet hydrocarbon[2]. The specific conductance of membranes formed in 0.1 M salt (NaCl, KCl) is 10^{-9} to 10^{-8} Scm^{-2}, corresponding to a bulk resistivity of 10^{14}–10^{15} Ω cm. The cation vs. anion selectivity or intercation specificity of these unmodified membranes is usually poor (ANDREOLI et al., 1966; LEV et al., 1966)

I. Capacitance

The specific capacitance of lipid bilayer membranes can be measured with a precision of about 0.3 percent (*e.g.* WHITE, 1977). Measurements of specific capacitance thus yield important, albeit indirect, information about the properties and composition of a lipid bilayer. The specific capacitance varies from

[2] It is assumed that the reader is familiar with the basic techniques for making artificial lipid membranes, and the essentials of the measurement techniques. TIEN (1974), FETTIPLACE et al. (1975) and articles in FLEISCHER and PACKER (1975, Section IV) may be consulted for details.

about 0.3 μFcm^{-2} to about 0.8 μFcm^{-2}, depending upon the fatty acid composition of the lipids, and the type of hydrocarbon solvent.

The thickness, *d*, of the hydrophobic core of the membrane can be calculated from:

$$C_g = \frac{\varepsilon_o \cdot \varepsilon_r^m}{d} \tag{1}$$

where C_g is the specific geometric capacitance of the bilayer (corrected for the electrolyte-bilayer dispersion (WHITE and BLESSUM, 1975), the polarization capacitance (EVERITT and HAYDON, 1968; WHITE, 1973), and if necessary, the dispersion due to the polar headgroup region (HANAI et al., 1965; COSTER and SMITH, 1974; SANDBLOM et al., 1975) ε_o is the capacitivity of free space, and ε_r^m is the dielectric constant of the membrane interior, similar to that for bulk hydrocarbons, 2.0–2.1 (REQUENA and HAYDON, 1975b; HUANG and LEVITT, 1977). From the known molecular area per lipid molecule in a lipid bilayer (about 0.6 nm^2 for phospholipids (FETTIPLACE et al., 1971; LEVINE and WILKINS, 1971) and about 0.4 nm^2 for monoglycerol esters (FETTIPLACE et al., 1971; REQUENA and HAYDON, 1975b)), the length of the acyl chains (*i. e.*, the molecular volume), and the membrane thickness[3] one can calculate the volume-fraction of hydrocarbon solvent of the membrane interior. The most relevant values are those for membranes made from monoglycerides (ANDREWS et al., 1970; FETTIPLACE et al., 1971; REQUENA and HAYDON, 1975b; WHITE, 1975, 1976, 1977), as these systems appear to be in thermodynamic equilibrium with the torus and aqueous phases (WHITE, 1977). The volume fraction varies from about 0.5 for n-decane to about 0.2 for n-hexadecane (ANDREWS et al., 1970; FETTIPLACE et al., 1971; WHITE, 1975), and can be extrapolated to be close to zero for heicosane (WHITE, 1977). For phospholipid membranes, the mole fraction of the hydrocarbon solvent is less, about 0.3 for n-decane and about 0 for n-hexadecane (FETTIPLACE et al., 1971; REQUENA and HAYDON, 1975b). The specific capacitance of these latter membranes does, however, vary as a function of time (FETTIPLACE et al., 1971; BENZ and JANKO, 1976) and of stirring of the aqueous phases (ANDERSEN and FUCHS, 1975). Phospholipid bilayers are therefore not necessarily in thermodynamic equilibrium with their surroundings, and the quantitative significance of the calculated volume fractions of hydrocarbon is uncertain.

The capacitance of nominally hydrocarbon-free bilayers of the Montal-Mueller type[4] is higher than the capacitance of the hydrocarbon-containing Mueller-Rudin membranes (BENZ et al., 1975; FETTIPLACE et al., 1975). The estimated thickness of Montal-Mueller membranes is, in fact, very close to the thickness of phospholipid lamellae as determined by X-ray diffraction, as emphasized by REQUENA and HAYDON (1975b).

[3] The term membrane thickness will often be used synonymously with the thickness of the hydrocarbon core of the membrane.

[4] WHITE et al. (1976) have shown that it is very unlikely that planar bilayers can be completely solvent-free. A similar conclusion was reached by BENZ et al. (1975).

The capacitance of a solvent-containing membrane is a function of the applied potential (BABAKOV et al., 1966; LÄUGER et al., 1967; ROSEN and SUTTON, 1968; WHITE, 1970; ANDREWS et al., 1970). This capacitance change is presumably due to solvent being squeezed out of the bilayer into a small number of lenses (BENZ et al., 1975; REQUENA et al., 1975). The average membrane thickness will thus decrease. This potential-dependent thickness change has been used with advantage by BAMBERG and LÄUGER (1973) to create conductance transients in bilayers doped with gramicidin A[5]. In general, one must minimize the effect of such thickness changes when studying the voltage-dependent kinetics of charge translocation through lipid bilayers. Very useful information has been presented by BENZ (BENZ et al., 1975; BENZ and JANKO, 1976). At low potentials, < 100 mV, the thickness changes are usually small (ANDREWS et al., 1970; but see also WHITE and THOMPSON, 1973). The membrane area may, however, change as a function of potential even at these low potentials, as the contact angle between the bilayer and the torus will vary according to the integrated Lippman equation (REQUENA and HAYDON, 1975a).

The capacitance transients have been studied in some detail by SARGENT (1974, 1975). The dielectric behavior of membranes made from oxidized cholesterol/n-decane or from dioleoylphosphatidylcholine/n-decane showed several characteristic time constants. The interpretation of the relaxation pattern is tentative, but one can observe distinct changes in the relaxation pattern due to valinomycin with Na^+ as the only cation present (SARGENT, 1974). This approach is thus a potentially very useful method to study the properties of lipid bilayers, especially the molecular packing and dynamics of membrane constituents.

II. Composition

The composition of solvent-containing membranes has been determined directly by chemical techniques (HENN and THOMPSON, 1968; PAGANO et al., 1972; BUNCE and HIDER, 1974), as well as from a combination of measurements of surface tension at the bulk hydrocarbon interface, the contact angle between the bilayer and the Gibbs-Plateau border, and the specific capacitance (COOK et al., 1968; ANDREWS et al., 1970; FETTIPLACE et al., 1971; see also Section B.I).

The chemical composition of a bilayer will always differ from that of the membrane-forming solutions, as the amount of hydrocarbon solvents is much less in the membrane than in the bulk solution (see Section B.I). The relative composition of the bilayer with respect to phospholipids and cholesterol will also differ (COOK et al., 1968; BUNCE and HIDER, 1974), the mole fraction of cholesterol being less in the bilayer. The volume fraction of hydrocarbon is much less in cholesterol-containing bilayers than in cholesterol-free bilayers

[5] The quantitative contributions of these membrane thickness changes in the genesis of gramicidin A conductance transients is unclear, however, as BAMBERG and BENZ (1976) observed qualitatively similar transients in the much less compressible Montal-Mueller membranes.

(HAYDON, 1970a). In glycerolmonooleate membranes the molar ratio of cholesterol to glycerolmonooleate is about half of the molar ratio in the bulk solution (PAGANO et al., 1972). The molar ratio of glycerolmonooleate to glycerolmonostearate in a lipid bilayer, on the other hand, is similar to that of the bulk solution (PAGANO et al., 1972).

C. Transport Model and the Potential Energy Barrier

To a first approximation a lipid bilayer can be regarded as a thin slab of hydrocarbon having a thickness, *d*, which is 2.5–5 nm and a dielectric constant, ε_r^m, which is approximately 2. The membrane separates two aqueous phases with a dielectric constant, $\varepsilon_r^{H_2O}$, which is approximately 80. Solute movement from one bulk aqueous phase to the other will, in general, be determined by the aqueous phases adjacent to the membrane as well as by the membrane itself. The flux at any point will be determined by the *mobility* of the solute particles, the *concentration* of the particles, and the *driving forces* acting upon the particles (see STEN-KNUDSEN, Chapter 2).

In a symmetrical system at equilibrium one can express the solute concentration, $C(x)$ at any point, as

$$C(x) = C_{aq} \cdot \exp\left(\frac{-\Delta W(x)}{kT}\right), \tag{2}$$

where C_{aq} is the solute concentration in the well-mixed bulk phases and $\Delta W(x)$ is the potential energy for the solute at x, relative to the bulk aqueous phases, k is Boltzmann's constant and T is the temperature in Kelvin. In a nonequilibrium system (*i. e.*, finite net fluxes) the concentration profile can be calculated from the flux equations integrated with proper boundary conditions (STEN-KNUDSEN, Chapter 2). An analysis of membrane transport phenomena will thus depend upon a realistic model of the membrane, as well as information about the various contributions to $\Delta W(x)$.

I. The Transport Model

The elementary steps involved in the movement of solutes across unmodified lipid bilayers (KETTERER et al., 1971) are illustrated in Figure 1. The solute, X, is present at constant concentrations, C_I and C_{II}, in the left and right bulk aqueous phases. The bulk phases are assumed to be well-mixed and are separated from the membrane-solution interfaces, at $x = 0$ and $x = d$, by "unstirred" layers. The flux of solutes across a lipid bilayer can thus be limited by translocation through the membrane interior, by the rate of adsorption[6] or desorption at the

[6] The words "absorption" and "adsorption" will both be used in the text. ABSORPTION is used to emphasize that solutes which permeate through a bilayer are dissolved in the membrane interior

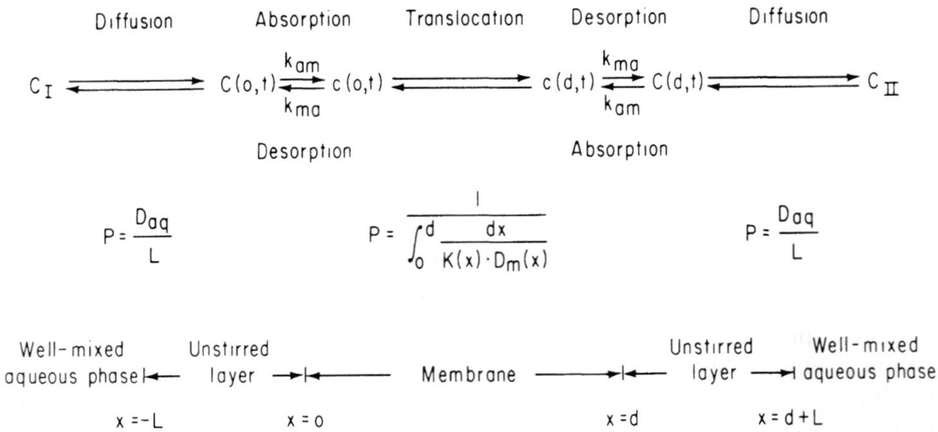

Fig. 1. Model for movement of nonelectrolytes and hydrophobic ions across bimolecular lipid membranes. Concentrations in the aqueous phase denoted by *C*, in the membrane phase by *c*. The diffusion coefficient in the aqueous phase is D_{aq}, in the membrane phase it is D_m, k_{ma} and k_{am} denote rate constants for desorption and adsorption, respectively. *K(x)* is the partition coefficient for the solute, which may vary as a function of *x*. *P* denotes the permeability coefficient. Subscripts $_{I}$ and $_{II}$ refer to the left and right sides of membrane, respectively. Fluxes are from left to right

membrane-solution interfaces, or by a combination of convection and diffusion through the aqueous unstirred layers.

It is generally assumed that the absorption coefficient, *K*, of solutes into the membrane is independent of the aqueous concentration of the solutes, and that the net fluxes vary linearly with the concentration differences for the permeant solutes, $C_{II}-C_{I}$. The correctness of these assumptions must be established experimentally (see Fig. 8). The experimentally observable quantities are: (1) the zero current potential as a function of the bulk aqueous concentration of permeable (and impermeable) ions; and (2) the flux as a function of time, membrane potential, and composition of the bulk aqueous phases with respect to both permeable and impermeable solutes.

A particularly simple and important case occurs when the membrane is permeable to one charged species only, and no current is flowing across the membrane. The electrostatic potential difference between the two bulk aqueous

(Footnote 6, cont.)
to some extent. The quantity of solute will be in concentration units. *Adsorption* is used to quantitate the amount of solute per unit membrane area, *N*. $N_{I}(t) = \int_{-r}^{d/2} c(x,t)dx$, where *r* is the radius of the solute. $N(t)$ should be proportional to the solute concentration, $c(t)$, in a potential minimum near the membrane-solution interfaces. The calculation of the proportionality factor will depend upon detailed information about the shape of the potential energy minimum.

phases, V_m, must then equal the equilibrium (Nernst) potential for the permeable ion:

$$V_m = V_{II} - V_I = -\frac{kT}{z \cdot e} \cdot \ln\left\{\frac{C_{II}}{C_I}\right\}, \tag{3}$$

where z is the valence of the permeant ion, and e is the elementary charge[7].

A quantitative description of net fluxes demands a quantitative description of transport through the aqueous unstirred layers, transport through the membrane itself, and the accumulation or depletion of solutes absorbed into the membrane-solution boundary regions.

II. Unstirred Layers

The aqueous phases are usually assumed to be divided into well-mixed phases and a completely unstirred layer of thickness L, between the bulk phases and the membrane-solution interface (the Nernstian assumption; NERNST, 1904). This separation is clearly fictional, but it emphasizes that there is always a layer of fluid near the membrane that remains (almost) stagnant, despite vigorous stirring of the bulk solutions. Transport perpendicular to the membrane will thus occur by diffusion alone near the membrane-solution interfaces, while transport occurs by both diffusion and convection further from the membrane (see VETTER, 1967; Chapter 2.B, for a more detailed discussion). The thickness of the idealized unstirred layers is always less than the distance over which the aqueous concentration of the solute changes, and is very much less than the distance over which the convection flow rate varies (VIELSTICH, 1953).

The thickness of the unstirred layers, L, is an operational parameter that allows one to express the stationary flux, $J(\infty)$, of a solute as:

$$J(\infty) = -\frac{D_{aq}(C(0,\infty) - C_I)}{L} = -\frac{D_{aq}(C_{II} - C(d,\infty))}{L} = -D_{aq} \cdot \left(\frac{dC}{dx}\right)_{x=0}. \tag{4}$$

The value for L is 50 to 100 μm in the aqueous phases (HOLZ and FINKELSTEIN, 1970; ANDREOLI and TROUTMAN, 1971; HAYDON and HLADKY, 1972), and will depend upon the size and geometry of the membrane, as well as the rate of stirring (DAINTY, 1963; EVERITT et al., 1969; GREEN and OTORI, 1970; HAYDON and HLADKY, 1972; see also Section E.II.1).

[7] Eq. 3 is only valid when $C_I = C(0,t)$ and $C_{II} = C(d,t)$. If the ion participates in chemical reactions, *e.g.* $X^- + H^+ \rightleftarrows XH$, then the flux of XH through the membrane can produce appreciable changes in $C(0, t)$ and $C(d,t)$. The membrane potential will then deviate from the predictions of Eq. (3) even though X^- is the only charged species that can cross the membrane interior (Le BLANC, 1971).

1. Stationary Fluxes

A quantitative analysis of the stationary flux is straightforward but lengthy (HAYDON and HLADKY, 1972). If, however, one neglects the resistance due to the interfacial adsorption and desorption reactions and assumes that the solute is uncharged and distributed uniformly throughout the membrane interior, then the following result is easily obtained (HELFFERICH, 1962; Section 8.3):

$$J(\infty) = \frac{-D_m \cdot K \cdot (C_{II} - C_I)}{d} \cdot \frac{1}{1 + 2 \cdot K \cdot L \cdot D_m/(D_{aq} \cdot d)} \,. \tag{5}$$

The first term describes the ideal behavior of the membrane itself, while the second factor accounts for the effects of unstirred layers. If $2 \cdot K \cdot L \cdot D_m/(D_{aq} \cdot d) << 1$, the existence of unstirred layers can be disregarded. If $2 \cdot K \cdot L \cdot D_m/(D_{aq} \cdot d) >> 1$, then the flux is totally unstirred layer-limited:

$$J(\infty) = -\frac{D_{aq} \cdot (C_{II} - C_I)}{2 \cdot L} \,. \tag{6}$$

The inequality $2 \cdot K \cdot D_m \cdot L << D_{aq} \cdot d$ is a strong limitation: the thickness of the unstirred layers may be 10^5 times the thickness of the membrane, while D_m may be 10 to 1000 times smaller than D_{aq} (see Section D.II). It is thus necessary for K to be less than 10^{-5}–10^{-3} to ensure that the stationary flux is limited by the membrane interior. The effect of unstirred layers is minimized in osmotic flow experiments, due to the "natural convection" observed in this case (DAINTY, 1963; EVERITT and HAYDON, 1969). H_2O permeability coefficients obtained by tracer diffusion measurements, P_D, will thus be less than permeability coefficients determined from osmotic flow measurements, P_f (see Section D.III).

2. Transient Fluxes

Transient transport phenomena are usually observed as current changes (in voltage-clamp experiments), or potential changes (in current-clamp or charge-pulse experiments) due to the movement of ions (or ion-carrier complexes) through the membrane. For example, at time $t = 0$, a potential, V_m, is applied across the membrane. The ions will move through the membrane, and the ion concentration within the membrane will shift from its equilibrium profile to some new profile which will be a function of V_m and time. The time course of the concentration changes will be given by the continuity equation (STEN-KNUDSEN, Chapter 2):

$$\frac{\partial J}{\partial x} = -\frac{\partial C}{\partial t} \,, \tag{7}$$

where $J(x, t)$ denotes the flux and $C(x, t)$ denotes the concentration of X in the membrane and adjoining aqueous phases, as functions of x and t. It is assumed that at $t = 0$, the aqueous phases are symmetrical and that the concentrations are constant up to the membrane, $C(-r, 0) = C(d + r, 0) = C_{\mathrm{I}} = C_{\mathrm{II}}$, where r is the radius of the ion.

Equation (7) can now be integrated from $x = -r$ to $x = d/2$, to give:

$$J\left(\frac{d}{2}, t\right) - J(-r, t) = -\int_{-r}^{d/2} \frac{\partial c}{\partial t}\, dx \tag{8a}$$

or

$$J\left(\frac{d}{2}, t\right) = -\frac{dN_{\mathrm{I}}}{dt} + J(-r, t). \tag{8b}$$

That is, the net flux of X through the membrane, $J(d/2, t)$, is equal to the net flux into the membrane, $J(-r, t)$, plus the rate of decrease in the number of adsorbed ions, $-(dN_{\mathrm{I}}/dt) = -\int_{-r}^{d/2} (\partial c/\partial t)\, dx$ (KETTERER et al., 1971; HAYDON and HLADKY, 1972; ANDERSEN and FUCHS, 1975; SZABO, 1976). One can further integrate Eq. (8a) with respect to time and obtain:

$$\int_{0}^{t} J\left(\frac{d}{2}, \eta\right) d\eta = \int_{0}^{t} J(-r, \eta)\, d\eta - \int_{-r}^{d/2} [c(x, t) - c(x, 0)]\, dx \tag{9}$$

where η is an integration variable. That is, the integrated flux through the membrane, $\int_{0}^{t} J\left(\frac{d}{2}, \eta\right) d\eta$, is equal to the integrated flux of ions into the membrane, $\int_{0}^{t} J(-r, \eta)\, d\eta$, minus the change in the number of ions adsorbed into one half of the membrane: $\int [c(x, t) - c(x, 0)]\, dx = N_{\mathrm{I}}(t) - N_{\mathrm{I}}(0)$.

It is possible to obtain an upper estimate for the integrated flux into the membrane if the flux is assumed to be totally diffusion-limited (HAYDON and HLADKY, 1972; ANDERSEN and FUCHS, 1975). For short times, $t < t_{\mathrm{o}} = \dfrac{l^2}{\pi^2 \cdot D_{\mathrm{aq}}}$ (VETTER, 1967; Fig. 80, see also STEN-KNUDSEN, Chapter 2), one can regard the unstirred layers as semifinite media and obtain the following result (CRANK, 1956, Section 3.3):

$$\int_{0}^{t} J(-r, \eta)\, d\eta \leq 2 \cdot C_{\mathrm{I}} \cdot \sqrt{\frac{D \cdot t}{\pi}}. \tag{10}$$

This limit applies for diffusion-limited entry of solutes into the membrane *and* for diffusion-limited exit of solutes. Equation (10) provides a useful guide for the interpretation of transient flux experiments (ANDERSEN and FUCHS, 1975): if $J(d/2, 0) >> 0$, $J(d/2, t_{\mathrm{o}}) \approx 0$, and $2C_{\mathrm{I}} \cdot (Dt/\pi)^{1/2} << \int_{0}^{t} J(d/2, \eta)\, d\eta$, then the observed charge movement is effectively confined to the membrane, irrespective of whether the membrane-solution interfaces are barriers for solute movement or not. So many ions are absorbed into the membrane-solution boundary regions (in a potential energy minimum) that the flux across the membrane (between the two energy minima) driven by the electrical field transiently is several orders of magnitude larger than the diffusion controlled flux in the aqueous phases, up to the membrane or away from the membrane (Figs. 8 and 9; HAYDON and HLADKY, 1972, Fig. 3).

3. The Membrane-Solution Interface

It has been established that the membrane-solution interface can be the rate-limiting step for solute translocation, which involves a chemical reaction between the solute and a membrane-bound component (STARK and BENZ, 1970; STARK et al., 1972; HLADKY, 1975; NEUMCKE and BAMBERG, 1975; KRASNE and EISENMAN, 1976). It is not clear, however, whether the membrane-solution interface can be rate-limiting for simple solute movement. It has been stated, *a priori,* that interfaces cannot be rate-limiting for simple physical transport phenomena (HELFFERICH, 1962). Many of the experimental demonstrations of interfacial barriers may indeed be explained by diffusion polarization in one or both macroscopic phases (SCHOLTENS and BIJSTERBOSCH, 1976); but it is known that the outer region of lipid bilayers can act as a barrier for solute movement (see Section D.II.3). One can regard this as an example of an interfacial barrier or not, depending upon how one *defines* the interface. This latter point is most strikingly illustrated by experiments on thick membranes (TOSTESON et al., 1968; ANDREOLI and TOSTESON, 1971). A plot of resistance vs. thickness of unmodified membranes will not pass through the origin, but will show a finite "interfacial" resistance, which is presumably due to two lipid monolayers (one at each membrane-solution interface) – the "interfacial" properties of a macroscopic membrane are thus similar to the bulk properties of a lipid bilayer membrane. Larger permeating solutes will, of course, produce a local disruption of the membrane-solution interface when they cross it. It is possible, for example, that hydrogen bonds between polar groups of adjacent lipid molecules will be broken[8]. This will be a barrier for solute movement, but it may not be rate-limiting, and it will only be an interfacial barrier if one *defines* it to be so.

4. Chemical Reaction in Unstirred Layers

It has been assumed in the previous discussion that the permeant species does not participate in chemical reactions in the unstirred layers. Such reactions are of importance for the permeation of weak acids or bases through membranes,

[8] The very slow increase in membrane conductance induced by 5,6-dichloro-2-trifluoreomethyl-benzimidazole (DTFB) (FOSTER and Mc LAUGHLIN [1974]) may be due to such a mechanism. This, however, is at variance with the very high permeability coefficient for the neutral form of DTFB estimated by COHEN et al. (1977). An alternative possibility is that the torus acts as a sink

however, as neutral species generally are much more permeable than charged species of similar size and structure. The flux of the neutral solute may thus be limited by the aqueous unstirred layers, in which case it is important that one can replenish the neutral species in the unstirred layers by chemical reactions, *e.g.*, $X^- + H^+ \rightleftarrows HX$. If the ratio $(X^-)/(XH)$ is large, and the aqueous phases are well buffered such that the pH remains constant throughout the unstirred layers (the ratio $[X^-]/[XH]$ is approximately constant), and if the "relaxation length"[9] of the chemical reaction is small compared to the thickness of the unstirred layers (Le BLANC, 1971; GUTKNECHT et al., 1972), then one can in fact determine the membrane permeability coefficient for the neutral species (Le BLANC, 1971; GUTKNECHT and TOSTESON, 1973; BORISOVA et al., 1974; COHEN et al. 1977). It is thus possible to measure permeability coefficients ranging from 0.7 cm s^{-1} for salicylic acid in lecithin/n-decane membranes (GUTKNECHT and TOSTESON, 1973) to 11 cm s^{-1} for carbonylcyanide *m*-chlorophenylhydrazane (CCCP) in lecithin-cholesterol (molar ratio = 1:2)/n-decane membranes (Le BLANC, 1971). These values are about three orders of magnitude higher than the permeability coefficients of the unstirred layers for XH! The flux of X through the unstirred layers is carried by X^-, but it is carried through the membrane by XH.

III. Potential Energy of Ions Within Lipid Bilayers

A lipid bilayer membrane is essentially impermeable to small inorganic ions such as Na^+, K^+ and Cl^-, while it is permeable to large hydrophobic ions such as tetraphenylborate or tetraphenylarsonium. The barrier properties of a lipid bilayer are primarily determined by the low *concentration* of ions within the membrane (FINKELSTEIN and CASS, 1968; PARSEGIAN, 1969). This low concentration of ions is due to the large electrostatic charging energy (work) necessary to remove an ion from the aqueous phase, with its high dielectric constant and electrolyte content, and insert it into the hydrocarbon phase of the membrane interior with its low dielectric constant.

The potential energy of an ion in the bilayer will, in addition, be determined by many other factors, both electrostatic (Section C. V) and chemical (Section C. VI). The electrostatic charging energy is, however, the quantitatively most important contribution to $\Delta W(x)$.

1. The Born Energy

If one considers the membrane as a macroscopic phase, one can approximate the charging energy by the Born energy, $W_B(r)$:

for the material, as COHEN et al. observed that the membrane conductance does depend upon the rate of stirring in the aqueous phases. This result is similar to that observed by HLADKY (1973) for valinomycin-induced conductance changes in lipid bilayers (see also Section E.II.2.a, β).

[9] The "relaxation length" is the mean square displacement of a particle during one time constant of the chemical reaction.

$$W_B(r) = \frac{z^2 \cdot e^2}{8 \cdot \pi \cdot \varepsilon_o \cdot r} \cdot \left(\frac{1}{\varepsilon_r^{m}} - \frac{1}{\varepsilon_r^{H_2O}} \right) \tag{11}$$

where r is the ion radius. The calculation of the Born energy, and the limitations of this calculation, are discussed in detail by BOCKRIS and REDDY (1970, Chap. 2).

If one ignores other contributions to $\Delta W(x)$ one can express the partition coefficient, K, for the ion into the membrane by

$$K = \frac{c}{C_{aq}} = \exp\left(- \frac{W_B(r)}{kT} \right). \tag{12}$$

$W_B(r)$ is inversely proportional to the radius of the ion. If $|z| = 1$, and $r = 0.2$ nm, the $W_B(r) = 68\ kT$ (169 kcal/mol) or $K = 2 \times 10^{-30}$. If, on the other hand, $r = 0.4$ nm, then $W_B(r) = 34\ kT$ or $K = 1.5 \times 10^{-5}$. A doubling in ion radius has increased the partition coefficient by a factor of 10^{15}! This model is useful, as it explains why a thin lipid membrane is an almost perfect barrier for small ions (for a criticism and more detailed calculations, see MACDONALD, 1976). This model furthermore illustrates why membrane permeable ions are monovalent: If $r = 0.4$ nm, and $[z] = 2$ then $W_B(r) = 136\ kT$, or $K = 4 \times 19^{-60}$. The partition coefficient has decreased by a factor of 10^{45}! Divalent ions need special transport mechanisms to facilitate their movement. It is not realistic, however, to regard a membrane 3 to 5 nm thick as a macroscopic phase, as the electrostatic field around the ion can induce large polarization charges at the membrane-solution interfaces. These polarization charges will establish an electrostatic field within the membrane, which is experienced by the ion and will change its potential energy relative to $W_B(r)$.

2. The "Image" Force

The electrostatic field induced by an ion within the membrane is usually calculated by the *method of images* (FEYNMAN, 1964, Chap. 6; BLEANEY and BLEANEY, 1965, Chap. 2). This method, or any other method, demands specific, idealized geometric assumptions concerning the membrane-solution interfaces, as well as assumptions about the dielectric properties of the membrane and the aqueous phases. It is generally assumed that the interfaces are mathematically smooth plane-parallel surfaces (NEUMCKE and LÄUGER, 1969; PARSEGIAN, 1969; HAYDON and HLADKY, 1972; ANDERSEN and FUCHS, 1975; BRADSHAW and ROBERTSON, 1975), or concentric smooth spheres (KOZAK, 1975). The calculation is simple if one neglects the presence of two interfaces and regards the membrane as a semi-infinite medium, the aqueous phases as perfect conductors, and the ion as a nonpolarizable sphere. The force, $F(x)$, acting upon the ion (drawing it towards the aqueous phase) is:

$$F(x) = - \frac{z^2 \cdot e^2}{4 \cdot \pi \cdot \varepsilon_r^{m} \cdot \varepsilon_o \cdot (x)^2} \tag{13}$$

where x is the distance between the center of the charge and the interface. This force, the *image-force,* is mathematically similar to that due to a charge of equal magnitude but opposite sign to that of the ion in the membrane, located in the aqueous phases as a mirror image[10]. The minus sign in Eq. (13) indicates that the force is directed towards the interface (from the low dielectric constant medium to the conducting medium). This position of the charge is fictional. In a conducting medium the charge is related to the actual aquedes countercharges to the ion. Equation (13) is therefore only one (the simplest) of many mathematically equivalent solutions to the electrostatic boundary value problem encountered when one wants to calculate the field around an ion near the interface between two media with different dielectric properties.

When an ion within the membrane is near one of the membrane-solution interfaces, Eq. (13) is still useful. It becomes a progressively poorer approximation however, as the ion moves towards the center of the membrane. When the ion is near one interface the ion and its image will act as a dipole, the resultant field of the other interface will be small and the polarization of this interface will be minimal. When the ion is near the middle of the membrane, however, there will be appreciable polarization of both interfaces. The relative magnitude of the polarization charges at the two interfaces will vary inversely with the distance between the ion and the interfaces – their sum will, of course, remain equal to but opposite to the charge on the ion. One must therefore satisfy the appropriate boundary conditions at both membrane-solution interfaces (see EMERSLEBEN, 1927; or NEUMCKE and LÄUGER, 1969; for details). The variation in the electrostatic field within the membrane, $E_p(x)$, will be due to the polarization charges induced directly by the ion, as well as the mutual polarization of the two interfaces themselves (EMERSLEBEN, 1927; NEUMCKE and LÄUGER, 1969; ANDERSEN and FUCHS, 1975). If the absolute electrostatic potential at a point somewhere in the membrane were known, $E_p(x)$ could be integrated through the membrane, and the electrostatic potential profile, $V_p(x)$, experienced by an ion as it traverses the membrane (see Fig. 2a) could thus be described. The image-force potential energy, $\Delta W_i(x)$, is then obtained as the sum of the Born energy and the electrostatic potential energy due to the image charges (NEUMCKE and LÄUGER, 1969; HAYDON and HLADKY, 1972; ANDERSEN and FUCHS, 1975):

$$\Delta W_i(x) = W_B(r) + z \cdot e \cdot V_p(x) \tag{14}$$

(see Fig. 2b).

[10] The aqueous phases are neither perfect dielectric media nor perfect conductors, which makes any model calculations problematical. One can use the approximation of BELL et al. (1966) to allow for the fact that the aqueous solutions are not perfect conductors (HAYDON and HLADKY, 1972). These corrections may, however, exceed the difference between exact results calculated assuming the aqueous phases to be either perfect conducting media or ideal dielectric media with $\varepsilon_r^{aq} = 78.3$ (ANDERSEN, unpubl.; see also BARLOW and MACDONALD, 1967). It is further known that the classic image law is an approximation, which disregards molecular details and is only valid for large separations between the charge and the interface (BARDEEN, 1940; SACHS and DEXTER, 1950). Eq. (13) is thus a fairly rough approximation, but more refined models are unlikely to provide more realistic descriptions (ANDERSEN and FUCHS, 1975), especially near the interfaces, where the major differences between the various models become accentuated (ANDERSEN, unpubl.).

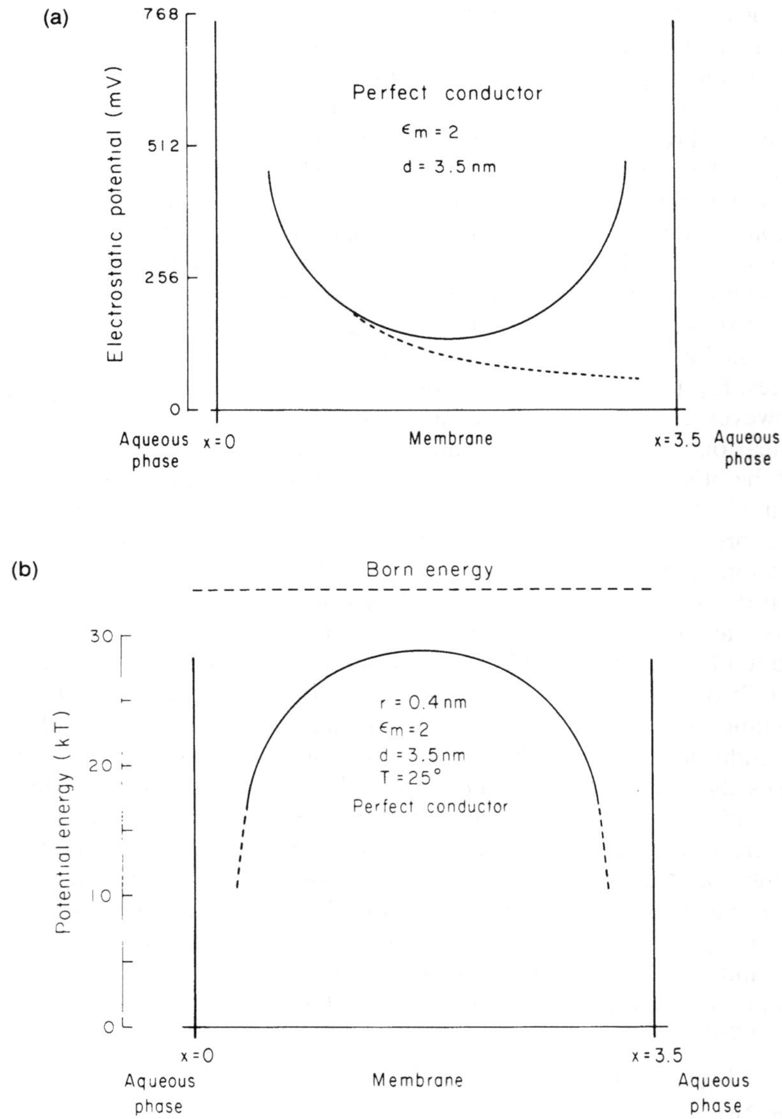

Fig. 2. (a) Electrostatic potential profile due to image-forces of an ion moving through the membrane. The stipled curve denotes the potential profile if only one interface is present, and image force is described by Eq. (13). The solid curve describes potential profile within a thin membrane with $d = 3.5$ nm. The calculation is only performed for $0.4\ \text{nm} < x < 3.1$ nm; (b) the image-force potential energy barrier for an ion with $r = 0.4$ nm. Note how the electrostatic potential in a bilayer *lowers* the energy below $W_B(r)$

This electrostatic model will not provide information about the absolute height of the potential energy barrier within the membrane (ANDERSEN and FUCHS, 1975), but it will provide information about the *difference* in peak potential energy (in the middle of the membrane), between two membranes, $\Delta W(d_1, d_2)$, as a function of membrane thickness, d_1 and d_2 respectively (PARSEGIAN, 1975):

$$\Delta W(d_1, d_2) = \frac{z^2 \cdot e^2 \cdot \ln(2)}{4 \cdot \pi \cdot \varepsilon_0 \cdot \varepsilon_r^m} \cdot \left(\frac{1}{d_1} - \frac{1}{d_2}\right) \tag{15}$$

If the major barrier for ion movement is the middle of the membrane (i. e., no significant contribution from interfacial barriers) then one can estimate the conductance, G_1, of a membrane with thickness d_1, if one knows the conductance, G_2, of an otherwise identical membrane with thickness d_2:

$$G_1 = G_2 \cdot \exp\left(\frac{\Delta W(d_1, d_2)}{kT}\right) \tag{16}$$

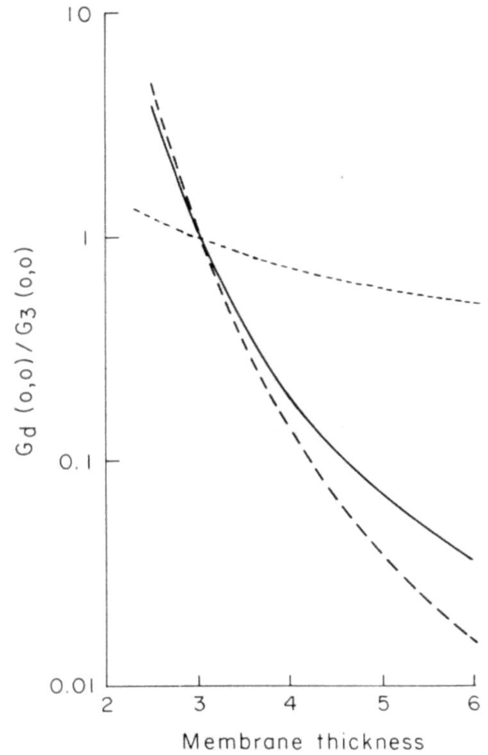

Fig. 3. Membrane conductance as a function of membrane thickness, d. The instantaneous small-signal conductance, $G_d(0, 0)$, is normalized by the conductance at $d = 3$m. $\varepsilon_r^m = 2$, $z = 1$, $T = 25°$C. (——): predictions of Eq. (16), (– –): predictions of the complete integrated image-force energy barrier, (----): predictions if $G \propto d^{-1}$

Figure 3 illustrates how the membrane conductance varies as a function of d. Even small changes in thickness can cause large changes in membrane conductance.

3. Diffusion or Distortion?

The image-forces are very large, and will attract the ion towards the nearest aqueous phase. They may, however, also deform the membrane-solution interface and pull the aqueous phase in towards the ion in the membrane interior (PARSEGIAN, 1969, 1975). The force acting upon the membrane will be equal to but opposite the force acting upon the ion. The magnitude of this force is of the same order of magnitude as the electrical fields observed during dielectric breakdown of lipid bilayers (PARSEGIAN, 1975). The mechanical model one chooses to describe ion movement through lipid bilayer membranes will thus depend upon whether the transit time for a single ion, τ_c:

$$\tau_c = \frac{d^2}{2 \cdot D_m} \tag{17}$$

is small compared to the time it takes to create a significant distortion (dimple) of the interface. τ_c is about 0.5–5 μs, depending upon what assumptions one makes about the viscosity of the membrane interior (see Section D.II). The distortion time is not known. Studies on the capacitance relaxations of lipid bilayers (WOBSCHALL, 1972; BENZ et al., 1975; BENZ and JANKO, 1976) indicate, however, that lipid bilayers are very resiliant in the μs time scale. Most of the usually observed thickness changes as a function of potential occur in the millisecond to second time scales, and are, at least in part, due to solvent extrusion from the bilayer (REQUENA et al., 1975). Scattered observations of the time-course of dielectric breakdown of lipid bilayer membranes (*e.g.*, BRUNER, 1975, Fig. 1; ANDERSEN, unpubl.) indicate that this process is slow (many μs). Smaller ions should thus be able to cross the membrane interior by a diffusion-like mechanism. Larger ions, which are less mobile and inherently more disruptive, however, may cross the membrane by a process that involves transient instabilities in membrane structure. The consequences of such membrane distortions are, at least, three-fold: Firstly, any deformation of the membrane-solution interfaces will make *a priori* calculations of the magnitude of the image-force even more dubious than indicated in Section C.III.2 (see footnote 10), both the *height* and the *shape* of the potential energy barrier may thus be quite different from the predictions of the simple image-force calculations; secondly, the mobility profile of an ion may be affected by these deformations; thirdly, the relative importance of any membrane distortions should depend upon both the geometry and the size of a particular current carrier. One should therefore exert considerable caution when comparing the conductance behavior observed with structurally dissimilar ions to probe changes in membrane structure.

IV. Potential Energy of Dipolar Molecules Within Lipid Bilayers

The electrostatic field around neutral dipolar molecules will polarize the medium in which they are imbedded, similar to what was observed for ions. One can therefore calculate the electrostatic work necessary to move a dipolar molecule from the aqueous phase into a lipid bilayer by means of a procedure similar to that used in Section C.III.1 and C.III.2. Dipoles are mathematically less tractable than ions, but BELL (1931 a and b) obtained solutions for the ideal dipole molecule – a rigid nonpolarizable sphere with a point dipole in the center. The model provided a reasonable prediction of the distribution of dipolar molecules between a gas phase and bulk liquids as a function of the dielectric constant of the fluids.

Similar calculations, as well as image-force calculations, for the distribution of dipolar molecules between a bulk aqueous phase and a hydrocarbon phase (ANDERSEN, unpubl.) lead to the result that hydrogen-bonding to water is more important than these electrostatic interactions in determining the low solubility of polar solutes in the membrane interior. The partition coefficient will be very high at the membrane-solution interface, where the solute can hydrogen-bond to H_2O. The magnitude of the partition coefficient will then fall off very rapidly as a function of distance from the interface, and should be unaffected by any contributions from image forces less than 1 nm from the interface. These electrostatic interactions, between the solute and the aqueous phases, are therefore unlikely to affect solute movement through the membrane interior, but they may play an important role for solute movement through the outer parts of the bilayer (see Section D. IV), and may thus be important in determining the permeability pattern for polar non-electrolytes (SIMON, 1977).

V. Interfacial Potentials

The major determinant of the barrier properties of lipid bilayers is the electrostatic potential energy barrier (see Section C.III). The absolute height of the barrier will be determined by the potential difference between the membrane interior and the bulk aqueous phases (see Fig. 4), as the conductance of the membrane, G, will be proportional to the ion concentration in the middle of the membrane, c:

$$G \sim D_m \cdot c \sim D_m \cdot C_{aq} \cdot \exp\left(\frac{-z \cdot e \cdot \Delta V}{kT}\right) \tag{18}$$

where ΔV is the electrostatic potential in the middle of the membrane with respect to the bulk aqueous phases, explicitly disregarding any contributions from polarization charges (see Section B.III.).

This potential difference can arise from (a) fixed or adsorbed charges at the surface of the membrane, which will produce an aqueous diffuse double-layer potential (ψ in Fig. 4), (b) an oriented layer of fixed or adsorbed dipoles in the membrane-solution interface, which will produce a dipole potential at the outer region of the membrane (Φ in Fig. 4), and (c) fixed or absorbed charges in the low dielectric constant membrane interior, which will produce large potential changes within the membrane, in addition to the aqueous diffuse double layer potentials[11].

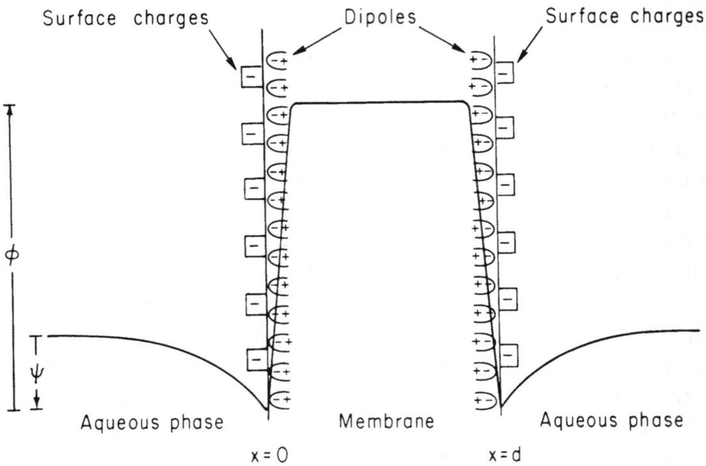

Fig. 4. Schematic representation of electrostatic profile across a lipid bilayer having fixed surface charges and interfacial dipoles. ψ is aqueous diffuse double-layer potential, generally negative due to the scarcity of positively charged lipids (WHITE, 1973). Φ is dipole potential. It is assumed that no potential changes occur in the membrane interior

The dipole potential difference, and thus the electrostatic potential difference, between the bulk phases and the membrane interior is *in principle* not measurable (PLANCK, 1890b; GUGGENHEIM, 1929, 1930; see also PARSONS, 1954 or BOCKRIS and REDDY, 1970), because there is no unambiguous way to separate the electrochemical potential of an ion into its chemical and electrical components. It may, however, be possible to *estimate* the magnitude of the electrical potential difference between two dissimilar media by the use of some extrathermodynamic, physical model (GUGGENHEIM, 1930; PARSONS, 1954). Any value for the potential difference between the membrane interior and the bulk aqueous phases (Le BLANC, 1970; ANDERSEN and FUCHS, 1975) must be interpreted in the light of the assumptions used to *calculate* it.

[11] It is generally assumed that *changes* in the diffuse double layer potentials or dipole potentials will be reflected fully in changes in ΔV. This can only be correct, however, for thin membranes (thickness less than the Debye length within the membrane). In thick membranes the potential changes will be screened by compensating ions in diffuse double layers in the membrane.

1. Diffuse Double-Layer Potentials

Charges at the membrane-solution interface may arise from either the polar head group of phospholipids (*e. g.*, phosphatidic acid, phosphatidyl glycerol, or phosphatidyl serine) or from the adsorption of charged compounds (*e. g.*, weak acid uncouplers of oxidative phosphorylation (McLaughlin, 1972) or local anesthetics (McLaughlin, 1975)). The simplest description of the potential changes due to such charges is the Gouy-Chapman theory of the diffuse double layer, as modified by Stern (1924) to account for the effects of charge adsorption. Bockris and Reddy (1970) and Aveyard and Haydon (1973) give thorough, but readable, introductions to the subject; Grahame (1947), Parsons (1954), Haydon (1964, Barlow (1970) and Bockris and Reddy (1970 provide more sophisticated presentations. McLaughlin (1977) should be consulted for an authoritative discussion of aqueous double-layer potentials as applied to lipid bilayers and biological membranes.

The Gouy-Chapman theory neglects the discreteness of both mobile and adsorbed charges, and deals with smeared (averaged) surface charge densities and volume changes. The distribution of ions in the aqueous phases will be determined by the thermal energy of the ions and the electrostatic repulsion of coions (ions having the same charge as the membrane-bound charges) from the membrane-solution interface and the corresponding attraction of counterions (ions having the opposite charge of the membrane-bound charges). The surface potential, ψ, will be a function of the surface charge density, σ^{o}, and the composition of the aqueous phases. For uni-univalent electrolytes the result is

$$\psi = \frac{2kT}{e} \cdot \text{arc sinh} \left[\frac{\sigma^{o}}{A\sqrt{C}}\right] \tag{19}$$

where

$$A = \sqrt{8 \cdot N \cdot k \cdot T \cdot \varepsilon_{o} \cdot \varepsilon_{r}^{aq}} \tag{20}$$

where N is Avogadro's number, and ε_{r}^{aq} is assumed to be equal to its bulk value.

The results of experimental tests of the Gouy-Chapman theory on lipid bilayers have been in rather remarkable agreement with the theoretical predictions (see McLaughlin, 1977, for a comprehensive review). The agreement is not as good, however, in other, rather similar systems such as the oil/water interface (Haydon, 1962; Levine et al., 1963; Carroll and Haydon, 1975). The applicability of the Gouy-Chapman equation depends, therefore, upon some specific properties of the lipid bilayer membrane, possibly the interposition of a polar head group region between the aqeuous phase proper and the hydrocarbon membrane interior. More detailed models of the aqueous diffuse double layer which explicitly incorporate the existence of the polar head group region, the discrete nature of the surface charges (Cole, 1969; Brown, 1974; Nelson & McQuarrie, 1975), etc. will, however, demand more, and essentially unavailable, experimental information to determine the model parameters. (Haydon, 1964, should be consulted for a critical analysis of the deficiencies of the Gouy-

Chapman theory.) Despite these theoretical shortcomings, experience shows that the Gouy-Chapman theory, at worst, is a very simple approximation, capable of predicting the magnitude of surface potentials to within 20 percent of those observed (McLaughlin, 1977). It should be recognized, however, that parameters calculated on the basis of the Gouy-Chapman theory, for example adsorption coefficients, become model-dependent and may suffer from systematic discrepancies (*e.g.*, Stigter, 1975).

2. Dipole Potentials

At physiological ionic strength the dipole potential is quantitatively much larger than the aqueous surface potentials (McLaughlin, 1977). The polarity of the dipole potential makes the membrane interior positive with respect to the aqueous phases. The magnitude of this potential can be estimated from surface potential measurements at the air-water interface (Bangham and Papahadjopoulos, 1966; Standish and Pethica, 1967; Papahadjopoulos, 1968; Paltauf et al., 1971; see also Hladky and Haydon, 1973; Hladky, 1974; Haydon, 1975). The magnitude of the dipole potential can also be estimated directly in lipid bilayers (Le Blanc, 1970; Andersen and Fuchs, 1975).

Surface potentials of lipid monolayers at the air-water interface do not measure the absolute potential difference between the bulk aqueous phases and the hydrocarbon phase in the monolayer (as this potential is unmeasurable), but rather the difference in potential, $\Delta(\Delta V)$, that occurs when a lipid monolayer is spread on a clean surface of an aqueous solution:

$$\Delta(\Delta V) = \Delta V_{\mathrm{m}} - \Delta V_{\mathrm{a/w}}, \tag{21}$$

where $\Delta V_{\mathrm{a/w}}$ is the potential of the clean interface, and ΔV_{m} is the potential measured when the monolayer is spread. This convention makes surface potential measurements independent of reference electrodes, etc. One can also express $\Delta(\Delta V)$ as

$$\Delta(\Delta V) = V_{\mathrm{a/hc}} + V_{\mathrm{hc/w}} - V_{\mathrm{a/w}}, \tag{22}$$

where $V_{\mathrm{a/hc}}$ is the dipole potential at the hydrocarbon-air interface, $V_{\mathrm{hc/w}}$ is the dipole potential (for a net neutral monolayer) at the membrane-water interface, while $V_{\mathrm{a/w}}$ is the dipole potential of the air-aqueous solution interface. This latter potential has been estimated to be -40 ± 100 mV (Nedermeijer-Denessen and De Ligny, 1974). The value of $V_{\mathrm{a/hc}}$ is unknown and although it may have large values in some systems (Adam, 1941; Gaines, 1966), it is often assumed to be zero. The surface potential of hydrocarbon-free bacterial phosphatidylethanolamine monolayers is about +480 mV (Hladky, 1974; Andersen, unpubl.). The potential of the interior of lipid bilayers with respect to the aqueous phases can thus be calculated to be between +340 to +540 mV – assuming $V_{\mathrm{a/hc}} = 0$, which is in reasonable, but not good, agreement with the value of +310 mV estimated in phosphatidylethanolamine bilayers (Andersen and FUCHS, 1975). The agreement becomes better, however, if one measures the surface potentials of hexadecane-containing monolayers where the surface potentials are only +390 mV (ANDERSEN, unpubl.; similar results were obtained with decane-containing films). These results therefore suggest that the interior of a lipid bilayer does indeed have a large positive potential relative to the aqueous phases. This potential will endow the membranes with a strong intrinsic selectivity for anions over cations, and changes in the potential will lead to changes in anion and cation conductances (see Eq. (18)).

The dipole potential is generally assumed to be associated with the bonds of the glycerol backbone (STANDISH and PETHICA, 1967; HAYDON and HLADKY, 1972) and there exists some direct evidence that carbonyl oxygens of the two fatty acids contribute to the dipole potential as the surface potentials of monolayers from ether phospholipids are about 100–200 mV less than those for ester phospholipids (PALTAUF et al., 1971). There is evidence that the carbonyl oxygen of the β or "2" fatty acid of phosphatidylethanolamine faces out towards the aqueous phase (HITCHCOCK et al., 1974; SEELIG and SEELIG, 1975), and that it can hydrogen-bond with H_2O (SCHMIDT et al., 1977). The dipole moment of a $\rangle C{=}O$ group is 7.7×10^{-30} coulombmeter (2.3 Debye, SMYTH, 1955 Table 4.1) while the molecular area per phospholipid molecule in a bilayer is about 0.6 nm^2 (FETTIPLACE et al., 1971; LEVINE and WILKINS, 1971). The dipole potential due to the $\rangle C{=}O$ groups oriented perpendicular to the plane of the membrane can thus be about 1.5 V!, the membrane phase being positive. In practice, one observes only a small fraction of such calculated dipole potentials (ADAM, 1941; GAINES, 1966), so the actual contribution will be much less than the above estimate. Water molecules, for example those hydrogen-bonded to the carbonyl oxygens, can also contribute to the dipole potential. If only 2 percent of the water molecules adsorbed at the hydrocarbon-water interface were oriented perpendicular to the interface, a potential of about 100 mV would be generated (HAYDON and HLADKY, 1972). It is therefore possible to have very large interfacial dipole potentials indeed; the magnitude of the potentials may, however, be very sensitive to the structure of the surface phase. Most, but not all, dipole potentials reported for lipid monolayers at the air-water interface are positive[12]. This may be due to the common structure (fatty acid – glycerol ester bonds) of most surface active lipids (HAYDON, 1975). This common structure does, however, make an unambiguous assignment of dipole potential contributions to various bond moments exceedingly difficult.

VI. Hydrophobic Interactions

Hydrocarbons and other nonpolar molecules will induce a significant ordering of water molecules in their immediate vicinity (*e. g.*, TANFORD, 1973). Transfer of hydrocarbon-like molecules from the aqueous phases into the membrane

[12] The exceptions may, however, be explained by the presence of a large negative contribution from $\Delta V_{a/hc}$ (BERNETT et al., 1964).

interior will thus be entropically favored (SIMON et al., 1977a). This contribution, ΔW_h, to the potential energy of an ion within the membrane may be large. For tetraphenylmethane, a neutral molecule of similar size and structure to the hydrophobic anion tetraphenylborate, one can, for example, estimate ΔW_h to be $-15\ kT$ (-37 kJ/mole^{-1}) (ANDERSEN and FUCHS, 1975). This estimate assumes, however, that the solvent properties of the hydrocarbon membrane interior are similar to those of bulk hydrocarbon liquids. This may be a rather dubious assumption particularly for larger solutes (see Section D.I.1). Estimates of ΔW_h should be used with caution.

The hydrophobic interactions are essentially surface interactions (REYNOLDS et al., 1974; STONE, 1975). For a spherical molecule their magnitude can thus be estimated as:

$$\Delta W_h = -4\pi \cdot r^2 \cdot \gamma, \tag{23}$$

where γ is the surface tension of the hydrocarbon-water interface, γ is 35 mNm^{-1} for the benzene-water interface (AVEYARD and HAYDON, 1973, Table 3.1), and r is 0.42 mm for tetraphenylborate. The estimated ΔW_h is $-18\ kT$, in reasonable agreement with the value estimated for tetraphenylmethane.

The hydrophobic effects are due to short-range forces between adjacent molecules. The maximal change in the potential energy, $\Delta W_h(x)$, will thus occur near the membrane-solution interfaces. The distribution coefficient for hydrocarbons in lipid bilayers is not constant throughout the membrane interior, however, but appears to increase towards the middle of the membrane (ANDREWS et al., 1970; WHITE, 1976, 1977). The quantitative variation is not known at present. Any spatial variation in the short-range interactions between a solute and the membrane matrix will formally be incorporated into ΔW_h.

VII. The Potential Energy Barrier

The sum of the image-force potential energy, $\Delta W_i\ (x)$, and the hydrophobic potential energy, $\Delta W_h\ (x)$, gives a potential energy barrier in the membrane interior, which separates two potential energy minima near, but not necessarily at, the membrane-solution interfaces (KETTERER et al., 1971; ANDERSEN and FUCHS, 1975; BRUNER, 1975). The *shape* of the barrier in the middle of the membrane is determined by ΔW_i and ΔW_h. The absolute *height* of the barrier in the middle of the membrane will be determined by ΔW_i and ΔW_h as well as by the electrostatic potential difference between the bulk aqueous phases and the membrane interior, ΔV, due to surface charges and oriented dipoles, and by any chemical contribution (such as hydrogen bonding to H_2O) to the potential energy of the solute, $\Delta W_{chem}\ (x)$:

$$\Delta W(x) = \Delta W_i(x) + \Delta W_h(x) + z \cdot e \cdot \Delta V(x) + \Delta W_{chem}\,(x). \tag{24}$$

The potential energy, $\Delta W(x)$, of an ion traversing a symmetrical membrane is shown in Figure 5. $\Delta W(x)$ will begin to change when the surface of the ion

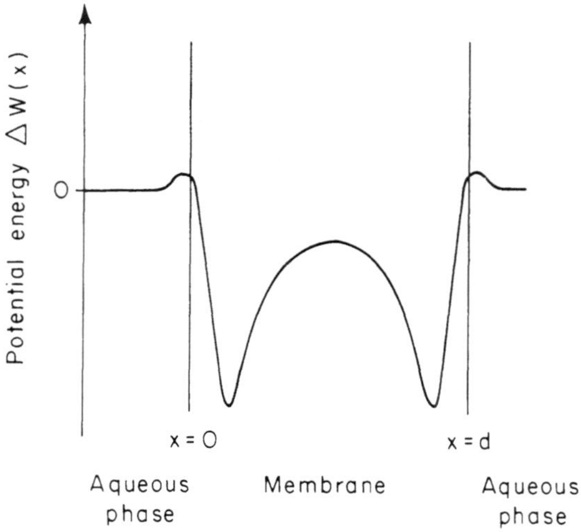

Fig. 5. Schematic representation of the potential energy barrier for ion movement through a lipid bilayer. The position of any potential energy minima with respect to the interfaces will vary according to size of and charge on the ion as well as its chemical structure. A qualitatively similar variation in $\Delta W(x)$ will also be observed for dipolar nonelectrolytes

approaches the polar head group region (the diffuse double-layer potential is assumed to be zero). The sum of the terms contributing to $\Delta W(x)$ has a deep and narrow minimum near the two interfaces. The potential energy barrier is due to the electrostatic work associated with bringing an ion into the low dielectric constant membrane interior. The existence of the potential energy minima near, but not necessarily at, the membrane-solution interfaces is characteristic of large hydrophobic ions and ion-carrier complexes, and for dipolar solutes. The potential energy of an ion (or dipole) within the membrane will also be affected by an applied potential, V_m. This, however, is not a part of the potential energy barrier.

The permeability of a lipid bilayer is determined by the height and shape of the potential energy barrier for the solute in question. The detailed kinetics and voltage-dependence of ion movement through the membrane interior will thus be determined by the combination of the image-force and hydrophobic potential energy barrier, as these are the only terms that change significantly throughout the entire membrane. There is some evidence that the mobility profile also varies significantly through the bilayer (see Section D.II.3), which could also contribute to any observed voltage-dependence. The absolute magnitude of the conductance (permeability) will, of course, be determined by all the contributions to the potential energy barrier, as well as the mobility.

Any of the terms contributing to the potential energy barrier, as well as the mobility profile and the rate constants, for absorption and desorption, may be of importance when the solute crosses the membrane-solution interface.

D. Permeability to Neutral Solutes

As a first approximation, studies on the permeability properties of lipid bilayer membranes to small solutes can be interpreted by the so-called *solubility-diffusion* concept. According to this concept, the translocation of a solute across a membrane proceeds by partitioning into the hydrocarbon-like membrane interior, diffusion through the membrane, and then entry into the opposite aqueous phase. The relative permeability coefficients for a series of neutral solutes should thus be ranked as their distribution coefficients between the aqueous phases and a suitable hydrocarbon-like bulk phase according to Overton's rules (see WARTIOVARA and COLLANDER, 1960, for a review). It is implicit in the solubility-diffusion concept, but not necessary for its validity, that the properties of the membrane interior can be mimicked by a suitable bulk phase, *e. g.*, olive oil, ether, octanol, or hexadecane. LIEB and STEIN (1969, 1971), for example, have proposed that solute permeation across lipid bilayer membranes resembles permeation of a soft, nonporous polymer, rather than solubility and diffusion through a bulk hydrocarbon phase. This concept is not an alternative to the solubility-diffusion mechanism, but rather an extension of the simple solubility-diffusion concept, to account for the greater order that exists among the fatty acid chains in a lipid bilayer, compared with the hydrocarbon chains in a bulk liquid hydrocarbon[13]. This ordering of the hydrocarbon chains *may* cause a striking dependence of the solute mobility upon the molecular volume (cross-sectional area) of the solute.

It is a truism that the interior of a lipid bilayer is not a bulk hydrocarbon phase. It is indeed likely that both distribution coefficients and diffusion coefficients for solutes into a membrane may differ significantly from the corresponding parameters in bulk solvents. It should be realized, however, that one can predict both the magnitude of the permeability coefficient for H_2O (HANAI and HAYDON, 1966; VREEMAN, 1966; FINKELSTEIN and CASS, 1968) and the temperature-dependence of the permeability coefficient (PRICE and THOMPSON, 1969; REDWOOD and HAYDON, 1969), to within a factor of 2 to 3 (see Table 2). This agreement may be fortuitous and due to cancellation of errors (SIMON, 1977). The simple solubility-diffusion concept has, nevertheless, been a very valuable guideline in the analysis of permeability phenomena.

[13] LIEB and STEIN (1969, 1971) proposed their theory as an alternative to the so-called "lipoid-sieve" concept of COLLANDER and BARLUND (see WARTIOVARA and COLLANDER, 1960). This latter hypothesis was developed to reconcile the *increase* in permeability coefficient occurring in homologous series of hydrophobic solutes upon addition of methyl or methylene groups with the *decrease* in permeability coefficient observed with small hydrophilic molecules of increasing molecular weight. The plasma membrane was regarded as a hydrocarbon-like matrix, through which the hydrophobic molecules permeated *via* the solubility-diffusion mechanism and interspersed aqueous channels that selected nonelectrolytes according to size. The validity of the lipoid-sieve concept has been vindicated for the ADH-induced increase in H_2O permeability of various epithelia (FINKELSTEIN, 1976b; AL-ZAHID et al., 1977), where there is strong evidence that H_2O moves through channel-like structures.

I. Partition Coefficients

1. Nonpolar Solutes

The capacitance measurements of ANDREWS et al. (1970), FETTIPLACE et al. (1971), REQUENA and HAYDON (1975b), and WHITE (1976, 1977) have unambiguously demonstrated that the solvent properties of the hydrocarbon core of a lipid bilayer differ from that of bulk hydrocarbon solvents, as the estimated volume-fraction of hydrocarbon solvent in the bilayer changes dramatically as the chain-length of the hydrocarbon solvent varies. Direct measurements of the distribution of n-hexane between water (0.1 M NaCl) and phosphatidylcholine liposomes (SIMON et al., 1977a) have likewise shown that the hydrocarbon core of a lipid bilayer is rather different from a bulk hydrocarbon solvent[14], the major effect being a large negative entropy of transfer from bulk hydrocarbon to the liposome. Similarly, MILLER et al. (1977) have shown that solutes as small as methane have (2–4 times) lower partition coefficients in phospholipid bilayers compared to bulk hydrocarbons, which was attributed to different entropies of transfer. These results appear to depend upon the ordered structure of the lipid bilayer interior, as WISHNIA (1963) found similar low partition coefficients, and a negative entropy of transfer of small hydrocarbon molecules from bulk hydrocarbon solvents to sodium dodecyl sulfate micelles.

The solvent properties of a lipid bilayer are dependent on the chemical structure of the acyl chains of the lipids (WHITE, 1977). The partition coefficient for hexane between water and liposomes, for example, is twice as high into egg phosphatidylcholine liposomes as into dioleoylphosphatidylcholine liposomes (SIMON et al., 1977a).

The effect of cholesterol on the solvent properties of lipid bilayers has been studied in few instances. SIMON et al. (1977a) report that the distribution coefficient for hexane into egg lecithin liposomes decreases by a factor of 2.5 at 25° C (molar ratio of phospholipid to cholesterol = 1:1). For dioleoyl-lecithin the corresponding decrease was 1.8. This change is similar to that inferred from composition studies in planar bilayers (HAYDON, 1970a). These effects are consistent with the notion that the presence of cholesterol "tightens" lipid bilayers. Cholesterol may both decrease the partition coefficient into the bilayer and increase the viscosity of the bilayer. The effect of cholesterol upon partition coefficients into phospholipid bilayers may depend on the molecular volume of the solute. Miller et al. (1977) found, for example, quite small differences between the partition coefficients of hydrocarbon gases into phosphatidylcholine: phosphatidic acid and phosphatidylcholine: phosphatidic acid: cholesterol bilayers.

It is implicit in the elegant work of WHITE (1977) that the solvent properties of monoglyceride bilayers will differ from those of phospholipid bilayers, as the molecular area per acyl-chain is 0.4 nm^2 in the former compared to 0.3 nm^2 in

[14] The large branched hydrocarbon, squalane, does not appear to interact (distribute into?) the hydrocarbon core of phospholipid liposomes (SIMON et al., 1977b).

the latter (see Section B.I). The *potential volume* available for a solute (WHITE, 1977) should therefore be decreased. It is, in fact, also found that the volume-fraction of hydrocarbon solvents is less in phospholipid bilayers than in monoglyceride bilayers (see Section B.I). It is therefore possible that a glycerolmonooleate bilayer will be rather different from a phospholipid bilayer with respect to both enthalpy and entropy of transfer from bulk hydrocarbon to the bilayer interior. The available data does not allow a discussion of this important problem.

2. Polar Solutes

The partition coefficient of the moderately polar molecule di-tert-butylnitroxide (DTNB) has been estimated from electron-spin-resonance (ESR) spectra by DIX et al. (1974) in dipalmitoylphosphatidylcholine liposomes and by TOMKIEWICZ and CORKER (1975) in egg phosphatidylcholine liposomes. The temperature dependence of the partition coefficient into dipalmitoylphosphatidylcholine liposomes is very similar to that observed for the quite polar n-butyramide into dimyristoylphosphatidylcholine liposomes (KATZ and DIAMOND, 1974b). This may mean that both solutes probe the same environment, possibly near the membrane-solution interfaces. The large dipole moment of the nitroxide group, about 1×10^{-29} coulomb-meter (*e.g.* GRIFFITH et al., 1974), would make this a possibility (see Section C.IV). TOMKIEWICZ and CORKER (1975) did, in fact, suggest that a fraction of the nitroxide is located in the membrane near the interfaces.

The partition coefficients for a number of polar solutes between dimyristoylphosphatidylcholine and water have been measured by KATZ and DIAMOND (1974b). The partition coefficients were determined for the sucrose-excluding space associated with the liposomes. The partition coefficients thus measure the amount of solute in the hydrocarbon core of the bilayer, plus the amount adsorbed to the interface, plus any solute in the so-called nonsolvent water (the number of water molecules associated with the phospholipid head groups that effectively exclude sucrose). Le NEVEU et al. (1977) have found that the sucrose-excluding space (about 13 H_2O/head group; KATZ and DIAMOND, 1974a; Le NEVEU et al., 1977) is a relative, and not an absolute space. It is thus likely that the partition coefficient of KATZ and DIAMOND actually does reflect the solvent properties of the polar headgroup region as well as those of the membrane interior. The actual partition coefficient into the membrane interior (*e.g.* middle third of the hydrocarbon region) may thus be (much) less than the average partition coefficients measured in the liposomes. The measured partition coefficient for n-buturamide between hexadecane and water, for example, is 4×10^{-4} (FINKELSTEIN, 1976a), while the partition coefficient measured by KATZ and DIAMOND (1974b) is 5×10^{-1}. This discrepancy is similar to that observed by LANGE et al. (1974) for valeramide and isovaleramide when comparing the distribution into bulk hydrocarbon and phosphatidylcholine vesicles. The discrepancy is even larger for urea (see Section D.IV).

II. Mobility

1. Indirect Measurements

Estimates of the mobility of a solute in a lipid bilayer have usually been indirect. The most common method is to estimate some average "microviscosity" of the membrane interior, based upon measurements of the fluorescence anisotropy of a probe molecule in the membrane in conjunction with a calibration curve obtained in a bulk model liquid (SHINITZSKY et al., 1971; COGAN et al. 1973). One can then calculate the mobility of the solute from Walden's rule (see Section D.II.1.b).

a) Microviscosity

The interpretation of the fluorescence anisotropy spectra will, unless special precautions are taken (*e.g.*, COGAN et al., 1973), depend upon a mono-exponential decay of the time-resolved fluorescence anisotropy, which is not observed in general (CHEN et al., 1977; KAWATO et al., 1977; VEATCH and STRYER, 1977). The problem becomes particularly significant in cholesterol-containing liposomes where the rate of in-plane rotations is less than that of out-of-plane rotations of perylene (COGAN et al., 1973) or the decay of fluorescence anisotropy observed with diphenylhexatriene becomes strikingly nonexponential (VEATCH and STRYER, 1977). A further complication in the interpretation of these measurements is that the estimated microviscosity of phospholipid liposomes may vary by more than one order of magnitude, depending upon the probe molecule and calibration curve chosen (HARE and LUSSON, 1977). The available data for microviscosity, which is about 1.2 P in egg phosphatidylcholine liposomes at 25° C and about 10 times higher in cholesterol-containing liposomes (COGAN et al., 1973), should therefore only be regarded as order-of-magnitude estimates of an *average* microviscosity.

b) Walden's Rule

The mobility of a solute within the membrane is usually calculated from Walden's rule: $u \cdot \eta =$ constant, where u is mobility and η is viscosity. If, for example, the aqueous mobility of a solute, u_{H_2O}, and the microviscosity of the membrane, η_m are known, one should in principle be able to calculate the mobility in the membrane, u_m, as $u_m = (u_{H_2O} \cdot \eta_{H_2O})/\eta_m$ where η_{H_2O} is the viscosity of water, about 1.0 cP (WEAST, 1972).

Walden's rule is empirical, but related to the Stokes-Einstein equation (BOCKRIS and REDDY, 1970, Chap. 4; ERDEY-GRUZ, 1974, Chap. 4). It describes the movement of a solute in a continuum, or in a medium where the solvent particles are much smaller than the solute particles. It is thus of dubious value for the movement of small particles through a lipid bilayer (FINKELSTEIN, 1977). Table 1 summarizes information on the diffusion coefficient of H_2O in different bulk liquids with different viscosities. The macroscopic viscosity appears to be irrele-

Table 1. Comparison of Viscosity and Diffusion Coefficient for H_2O in Various Bulk Solvents. $T = 25°$ C

Solvent	Viscosity (cP)	Diffusion Coefficient ($cm^2\ s^{-1}$)	$\frac{D_{H2O} \cdot \eta_{H2O}}{\eta}$ ($cm^2\ s^{-1}$)
H_2O	0.9	2.6×10^{-5}	2.6×10^{-5}
Hexadecane	3.1	4.2×10^{-5}	7.5×10^{-6}
Squalane	≈ 27	1.7×10^{-5}	$\approx 9 \times 10^{-7}$

The viscosity of H_2O is from WEAST (1972), the viscosity of hexadecane is from GRAY (1972), the viscosity of squalane is calculated from the value of 6.08 Engler at 20° C (STECHER et al., 1968) and the values of 20.5 and 4.2 centistokes at 37.7° and 98.9° C respectively (SAX and STROSS, 1957). D_{H_2O} is from WANG (1965), the values for hexadecane and squalane are from SCHATZBERG (1965).

vant for the description of the movement of very small solutes in a solvent where the solvent molecules are large relative to the solute molecules. The microviscosity may be similarly irrelevant for predicting solute mobilities, unless the probe molecule has similar physical dimensions to those of the solute molecules.

2. Direct Estimates

VANDERKOOI and CALLIS (1974) have measured the translational diffusion coefficient of pyrene in egg phosphatidylcholine vesicles in studies on the diffusion-controlled excimer formation. The diffusion coefficient thus estimated was about $3 \times 10^{-8} cm^2 s^{-1}$ at 20°C. For dimyristoylphosphatidylcholine vesicles at 30° C the diffusion coefficient was about $3 \times 10^{-8}\ cm^2 s^{-1}$ and it fell by a factor of two in cholesterol-containing vesicles. These values are average diffusion coefficients of the bilayer, and may well vary considerably through the membrane (see Section 3).

The diffusion coefficient for O_2 has been estimated by the quenching of pyrene fluorescence (FISCHKOFF and VANDERKOOI, 1975) or by analysis of ESR spectra (PLACHY et al., 1977). In dimyristoylphosphatidylcholine the diffusion coefficient is about three orders of magnitude larger than for pyrene. These results reinforce the limited applicability of Walden's rule in predicting transport behavior.

3. Variation through the Membrane

The variation in microviscosity perpendicular to the plane of a bilayer may be estimated from an analysis of nuclear magnetic resonance (NMR) (SEELIG and SEELIG, 1974; GODICI and LANDSBERGER, 1974; LEE et al., 1976; STOCKTON et al., 1976) or ESR spectra (HUBBELL and MCCONNEL, 1971; SEELIG and NIEDERBERGER, 1974). Both the NMR and the ESR studies provide important information, as they reflect different aspects of the physical situation in a bilayer. The NMR-studies (deuterium or ^{13}C) reflect the structure of the unperturbed bilayer, while the ESR-studies detect the response of the bilayer to small physical stress (SEELIG and NIEDERBERGER, 1974), as when a medium-sized solute molecule moves through the bilayer. As emphasized by SEELIG and SEELIG (1974) fluidity then refers to the *rate* of motion, but not to ordering – which is the parameter usually measured. There is no simple way to relate changes in order parameter to changes in fluidity – and such a correlation may even be qualitatively inappropriate (SEELIG and SEELIG, 1974). It is, however, generally assumed that the interior of lipid bilayers, where the order parameter is low, is more fluid than the outer region, where the order parameter is high. It is also of interest that the order parameter measured by deuterium NMR is significantly higher than that measured by ESR (SEELIG and NIEDERBERGER, 1974). A quantitative estimate of the microviscosity gradient has been attempted by LEE et al (1976), based upon ^{13}C spin-lattice relaxation times. These authors estimate that the "microviscosity" is about 70 times less at the terminal methyl group than at the C_2 in dimyristoyl phosphatidylcholine liposomes. The quantitation should, however, be viewed with some caution, as the calculation is sensitive to the particular model used.

The studies of DIX et al. (1974) and TOMKIEWICZ and CORKER (1975) do, however, provide evidence for a slow exchange (on the ESR timescale) of DTBN between an aqueous solution and the membrane interior. The estimated transition time is in the order of 10^{-7} seconds (DIX et al., 1974), which is about one order of magnitude slower than the time estimated from the apparent diffusion coefficient of DTBN in the interior of the vesicles. These data thus provide additional evidence for a low mobility near the membrane-solution interface.

III. Permeability

1. H_2O

The permeability coefficient has been determined both by tracer diffusion measurements, P_D, and by osmotic flow measurements, P_f. It is a general finding that $P_f > P_D$ (HUANG and THOMPSON, 1966; HANAI et al., 1966; CASS and FINKELSTEIN, 1967; EVERITT and HAYDON, 1969; EVERITT et al., 1969). The experiments of CASS and FINKELSTEIN (1967) and those of REDWOOD (EVERITT and HAYDON, 1969; EVERITT et al., 1969), conclusively demonstrated that this finding is due to aqueous unstirred layers (see also VREEMAN, 1966). There is no evidence for small aqueous pores in unmodified lipid membranes.

The H_2O permeability coefficient is a function of lipid composition (FINKELSTEIN and CASS, 1967; GRAZIANI and LIVNE, 1972; FINKELSTEIN, 1976a) and of cholesterol content. The permeability coefficient will in general decrease with increasing mole fraction of cholesterol in the membrane-forming solution (FINKELSTEIN and CASS, 1967; PRICE and THOMPSON, 1969; GRAZIANI and LIVNE, 1972). Table 2 summarizes information on the osmotic water permeability of

Table 2. Osmotic H_2O permeability coefficients, P_f, and activation energy for permeation, E_a. Bimolecular membranes formed from egg phosphatidylcholine in n-decane, egg phosphatidylcholine and cholesterol in n-decane, or oxidized cholesterol in octane. All data were obtained at 36° C, or normalized to this temperature

Lipid	P_f cm s^{-1}	E_a kJ mol^{-1}	References
Egg phosphatidylcholine	$4.2 - 5.2 \times 10^{-3}$	54	a, e
Plant phosphatidylcholine	6.1×10^{-3}	45	e
Hydrogenated egg phosphatidylcholine	1.7×10^{-3}		a
Dipalmitoyl-phosphatidylcholine	3.0×10^{-3}	58	e
Egg phosphatidylcholine: cholesterol			
(molar ratio 1:1)	$2.3 - 6.2 \times 10^{-3}$	55–56	a, c, e
(molar ratio 1:2)	$2.0 - 6.2 \times 10^{-3}$	55–61	a, c, d
(molar ratio 1:4)	9.0×10^{-4}		a
(molar ratio 1:8)	7.5×10^{-4}		a
Oxidized cholesterol	1.2×10^{-3}	28	b
Egg phosphatidylcholine vesicles	7.1×10^{-3}	35–36	g, h
Calculated on the basis of the solubility-diffusion model, d=4nm:			
Hexadecane	9.2×10^{-3}	47	f
Squalane	4.2×10^{-3}	52	f

References: a Finkelstein and Cass (1967); b Tien and Ting (1968); c Price and Thompson (1969); d Redwood and Haydon (1969); e Graziani and Livne (1972); f Schatzberg (1963, 1965); g Reeves and Dowben (1970); h Cohen (1975)

lipid bilayers made from phosphatidylcholine/n-decane, alone or in various mixtures with cholesterol or oxidized cholesterol, and phosphatidylcholine liposomes. The predicted H_2O permeability coefficients, calculated on the assumption that the membrane interior can be regarded as a thin slab of bulk hydrocarbon are also listed. The properties of lecithin and lecithin: cholesterol bilayers are in reasonable agreement with those of a thin slab of hydrocarbon, while those of liposomes differ somewhat. This difference may be due to the presence of hydrocarbon solvent in the planar bilayers. The agreement is not perfect, however, as Graziani and Livne (1972) find a *decrease* in P_f with increasing temperature in monogalactosyldiglyceride: cholesterol/n-decane membranes, while the cholesterol-free membranes had a "normal" behavior. The interpretation of the activation energies is not unambiguous, however, (Haydon, 1969), as the temperature dependence of P_f reflects the combined variation of membrane composition, membrane thickness, partition coefficient, and mobility.

It is, of course, possible that the agreement observed between P_f of lipid bilayers and that predicted by the data of Schatzberg is atypical, or fortuitous (Simon, 1977). Finkelstein (1976a) did in fact conclude that P_{DH_2O} was about 3 to 10 times *higher* than that predicted from a series of somewhat larger solutes, while that of formamide was three times higher. This discrepancy can easily be explained by assuming that H_2O (and possibly also formamide) can move through a lipid bilayer along "kinks" (mobile structural defects formed by neighboring gauche ± rotations; TRÄUBLE, 1971; SEELIG and SEELIG, 1974) in the hydrocarbon interior. The diffusion coefficients of such kinks is calculated to be about 10^{-5} cm^2s^{-1} (TRÄUBLE, 1971), a value close to the diffusion coefficient of O_2 in phosphatidylcholine vesicles (Section D.II.2). The concentration of H_2O near the membrane-solution interface may be higher than the concentration in the bulk membrane (SIMON, 1977; see also Section C.V), and thus compensate for the decreased mobility expected in this region. The H_2O permeability would therefore reflect the properties of the middle two-thirds or so of the membrane (Fig. 6), a region that should be better modelled by a bulk hydrocarbon (see also FINKELSTEIN, 1977).

2. Organic Solutes

Measurements of organic nonelectrolyte permeability in lipid bilayers have always been diffusion measurements. Such data should be evaluated critically, as they may be affected by the presence of aqueous unstirred layers (see Section C.II). These experiments are of importance, however, as they can provide much additional information about the interior of a lipid bilayer. A major question is whether the interior of lipid bilayer membranes really can be modelled by a bulk hydrocarbon liquid (as would seem to be the case for H_2O), or whether it behaves more like a soft polymer, as proposed by LIEB and STEIN (1969, 1971).

Small alcohols and amides form a consistent pattern: their permeability coefficients can be calculated on the basis of the distribution coefficients in decane if one assumes that the membrane interior (phosphatidylcholine/n-decane bilayers) has a viscosity similar to that of bulk hydrocarbons (FINKELSTEIN, 1976a). For larger solutes significant deviations appear, and the "fluidity" of the membrane interior appears to be a function of solute size (COHEN and BANGHAM, 1972; WOLOSIN and GINSBURG, 1975)[15]. FINKELSTEIN (1976a) found a similar result for codeine. The properties that can be inferred about the membrane interior may thus depend upon the probe molecules used.

The permeability coefficients measured for urea, and thiourea in egg phosphatidylcholine n-decane membranes (VREEMAN, 1966; LIPPE, 1969; GALLUCCI et al., 1971; GALLUCCI et al., 1975; FINKELSTEIN, 1976), or in phosphatidylcholine/n-tetradecane-(chloroform-methanol) membranes (POZNANSKY et al., 1976), differ significantly from the pattern set by other solutes (GALLUCCI et al., 1971; FINKELSTEIN, 1976a; POZNANSKY et al., 1976; WOLOSIN, et al., 1977). The permeability coefficients are 10 to 30 times less than those predicted and ob-

[15] The permeability coefficients reported by WOLOSIN and GINSBURG (1975) are, however, spuriously low, and may in part be unstirred layer limited. These authors employed an experimental approach similar to that of GUTKNECHT and TOSTESON (1973), but used *unbuffered aqueous solutions* which invalidates the analysis presented in Section C.II.4. WOLOSIN and GINSBURG (1975) did try to correct for unstirred layers, but it is not clear that they were successful as their permeability coefficient for acetic acid, for example, was 2.4×10^{-4} cm sec^{-1}, 20 times less than the value determined in similar membranes, but with well buffered aqueous solutions, 5×10^{-3} cm/sec (J. GUTKNECHT, personal communication). A similar discrepancy has been observed for butyric acid (E. OHRBACH and A. FINKELSTEIN, personal communication).

served for solutes of similar size, based upon the assumption that the interior of the membrane is bulk hydrocarbon (FINKELSTEIN, 1976a). The molecular basis for this deviation is obscure, but may reside in the translocation of urea across the outer (polar head group?) region of the membrane. WOLOSIN et al. (1977) have pointed out that the enthalpy and entropy of partitioning into phosphatidylcholine liposomes (KATZ and DIAMOND, 1974b) are abnormally low.

The measured permeability coefficients depend upon the presence or absence of cholesterol (BEAN et al., 1968; GALLUCCI et al., 1975; FINKELSTEIN, 1976a). The permeability coefficients in cholesterol-containing membranes are generally less than those in cholesterol-free membranes (BEAN et al., 1968; GALLUCCI et al., 1975; FINKELSTEIN, 1976a). The permeability coefficients are less in membranes based upon sphingomyelin than in membranes based upon egg phosphatidylcholine (FINKELSTEIN, 1976a), which may reflect either the presence of longer and more saturated hydrocarbon tails (FINKELSTEIN, 1976a) or the hydrogen-bonding between adjacent molecules suggested by BROCKERHOFF (1974).

The adsorption of polar molecules, *e.g.* phloretin, the aglycone of phlorizin, on the other hand can increase the permeability of bilayers for acetamide (ANDERSEN et al., 1976) and for other amides (POZNANSKY et al., 1976). This may reflect a nonspecific permeability increase of cholesterol-containing membranes (ANDERSEN et al., 1976).

IV. The Rate-Limiting Barrier for Solute Movement

Permeability coefficients will be determined by whatever part of a membrane represents the largest barrier for solute movement. This barrier may be due to a low concentration, to a low mobility, or to both. It is generally accepted that the outer region of a bilayer is less "fluid" than the membrane interior (see Section D.II.3). The concentration of hydrophobic solutes is possibly also higher in the membrane interior than in the interfacial regions (DIAMOND and KATZ, 1974; SIMON et al., 1977a). The outer regions of the bilayer are thus likely to be the major barriers for permeation of these hydrophobic solutes. The concentration of hydrophilic solutes will probably be smallest in the membrane interior (DIAMOND and KATZ, 1974) and may vary by many orders of magnitude through the membrane (see Section C.IV). The partition coefficient for urea into dimyristoylphosphatidylcholine vesicles is about 10^5 times higher than into hexadecane (KATZ and DIAMOND, 1974b; FINKELSTEIN, 1976). This ratio may overestimate the urea concentration at the interfaces relative to that in the membrane interior, but is unlikely to be off by many orders of magnitude. The largest estimated fluidity ratio between the central and outer region of a lipid bilayer is 70 (LEE et al., 1976). The major barrier to the movement of hydrophilic solutes will thus be the membrane interior. For neither hydrophobic nor hydrophilic solutes will measurements of the average partition coefficient into liposomes determine the solute concentration at the major permeability barrier (DIAMOND and KATZ, 1974; see also Fig. 6). For example, if one uses the permeability coefficient of

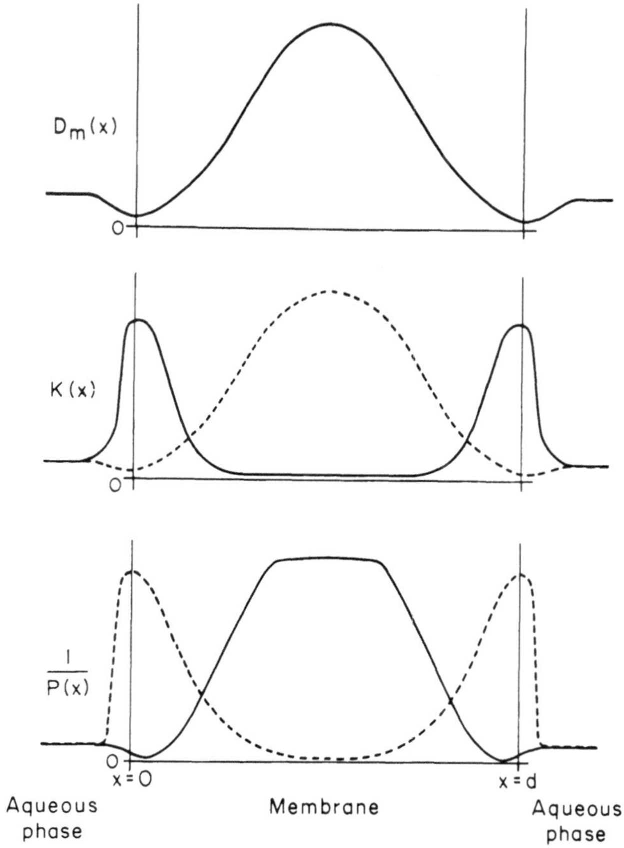

Fig. 6. Schematic illustration of parameters of importance for nonelectrolyte permeation through lipid bilayers *Top:* Variation of diffusion coefficient D_m through a lipid bilayer. *Middle:* Variation in the partition coefficient for polar solute (solid line) and nonpolar solute (broken line). *Bottom:* Resistance to solute movement through the bilayer for polar solute (solid line) and nonpolar solute (broken line)

urea in egg phosphatidylcholine/n-decane bilayers, 4×10^{-6} cm s^{-1} (FINKELSTEIN, 1976a), and the partition coefficient determined by KATZ and DIAMOND, (1974b), 0.2, one will calculate a diffusion coefficient for urea in a lipid bilayer of about 10^{-11} cm^2 s^{-1}, more than three orders of magnitude lower than the value determined for large hydrophobic molecules like pyrene (VANDERKOOI and CALLIS, 1974), about four orders of magnitude less than the value one can estimate from the data for DTBN (DIX et al., 1974), and about six orders of magnitude less than the value determined for O_2 (FISCHKOFF and VANDERKOOI, 1975).

A determination of the average solvent properties of a lipid bilayer will thus be misleading for the interpretation of permeability phenomena. The use of polar bulk solvents such as n-octanol, amylalcohol, etc., as models for the solvent properties of lipid bilayers should similarly be critically assessed. The available permeability data of COHEN and BANGHAM (1972) and WOLOSIN and GINS-

BURG (1975)[15] are qualitatively consistent with the concept that the membrane interior behaves like a soft polymer. It may, however, be misleading to quantitate the relation between molecular volume and mobility within the membrane in the light of the present limited information about the solvent properties of lipid bilayers, especially with regard to their spatial variation through the membrane.

Pertinent information can possibly be obtained by ESR spectroscopy, as the spectra depend upon the polarity of the medium surrounding the nitroxide (HUBBEL and McCONNELL, 1971; GRIFFITH et al., 1974). GRIFFITH et al. (1974) have proposed on this basis that there exists a polarity profile in the outer region of the bilayer, possibly due to the presence of H_2O in the outer region hydrocarbon core. This interpretation is in agreement with the predictions of Section C.IV, but not unambiguous, as the nitroxide probes are fairly polar. The position of the nitroxide group with respect to the interface may thus be somewhat different from that deduced from its structural position on the fatty acid chains of the phospholipids. It is even possible that it may make contact with the aqueous phase (CADENHEAD et al., 1975).

E. Ion Permeability

Figure 7 illustrates some of the substances that produce conductance changes characteristic of simple hydrophobic (lipophilic) ions. Only the negative ions have been studied in detail.

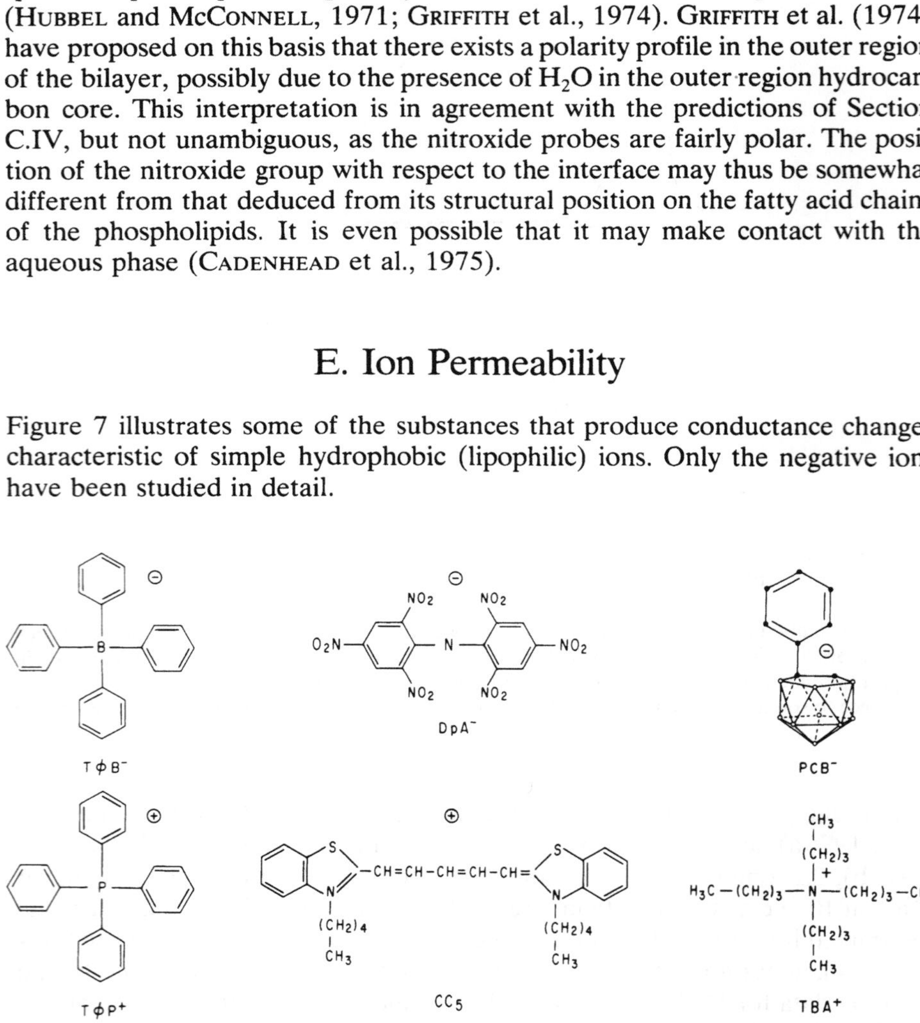

Fig. 7. Chemical structure of some membrane-permeable ions: tetraphenylborate (TØB$^-$); dipicrylaminate (DpA$^-$); phenyldicarbaundecaborane (PCB$^-$); tetraphenylphosphonium (TØP$^+$); 3,3'-dipentyl-S-dicarbocyanine (CC5); tetrabutylammonium (TBA$^+$). Three-dimensional structures determined for the tetrahedral TØ$^-$ compounds: for TØB$^-$ (HOFFMAN and WEISS, 1974; for TØP$^+$ (CORBRIDGE, 1974; for tetraphenyl arsonium (TØAs$^+$), (not illustrated) (MOONEY, 1940). The X-ray structure of the rather planar DpA$^-$ has been determined by GUPTA and DUTTA (1975)

I. Tracer Flux Measurements

The high electrical resistance of lipid bilayers suggests that the permeability coefficients for small inorganic anions and cations should be almost unmeasurably low. PAGANO and THOMPSON (1968) found P_{Na^+} to be about 2×10^{-8} cms^{-1} at 30°C, in good agreement with the value predicted from electrical measurements (the specific conductance of the membranes used was rather high: 1.3–5.3 $\times 10^{-6}$ Scm^{-2}). P_{Cl}, on the other hand, was about 4.5×10^{-7} cms^{-1} at 30°C, about two orders of magnitude higher than predicted from electrical measurements. The Cl^- permeation was therefore an electroneutral process. The unidirectional flux was a function of the Cl^- concentration on the "cold" side of the membrane, and could be abolished by substituting NO_3^- for Cl^-.

A similar discrepancy between tracer fluxes and electrical fluxes was observed for another halide, Br^-, by GUTKNECHT et al. (1972). In this case the discrepancy could by explained by movement of the much more permeable Br_2 through the membrane and an isotopic exchange reaction in the unstirred layers adjacent to the bilayer, as the flux was decreased by a factor of 10 upon addition of 1×10^{-3}M $Na_2S_2O_3$ to the aqueous phases.

TOYOSHIMA and THOMPSON (1975a) found in bilayers formed from the non-oxidizable diphytanoylphosphatidylcholine/n-decane that P_{Cl^-} was 6.8×10^{-8} cms^{-1}. The electrical conductance was about 1×10^{-9} Scm^{-2}, so that the predicted P_{Cl^-} is more than three orders of magnitude less than that observed. $Na_2S_2O_3$ had no effect upon the measured chloride fluxes. The flux could, however, be almost completely eliminated by the addition of I^- to the aqueous phases. This might indicate a role of the choline groups, as I^- binds to choline groups with higher affinity than Cl^- (De GEISO et al., 1954).

A separate study on Cl^- movement across small homogenous vesicles formed from diphytanoylphosphatidycholine yielded similar results (TOYOSHIMA and THOMPSON, 1975b). The activation energy for translocation was about 80 kcal-mol^{-1}, indicating a very substantial potential energy barrier. The flux could be separated into a rather slow net flux of the (neutral) chloride species and quite a rapid isotopic exchange of Cl^-. The net flux (and the isotopic exchange) decreased with increasing pH. The flux was proportional to $\sqrt{[H^+]}$ and depended upon the buffer capacity of the aqueous solutions.

The observed translocation rate corresponded quite well with the rate of transmembrane exchange (flip-flop) of spin-labeled phospholipids (KORNBERG and MCCONNELL, 1971). It was therefore proposed that the observed Cl^- flux was due to the transmembrane movement of a neutral phosphatidylcholine-H^+-Cl^- complex. The difference between the net flux and the exchange flux in the vesicles is presumably due to the pH gradients produced between the very small intravesicular volume and the extravesicular medium. More recent estimates of the rate of transbilayer exchange of nonspin-labeled phospholipids are, however, several orders of magnitude *less* than that found by KORNBERG and Mc CONNELL for spin-labeled lipids (ROSEMAN et al., 1975; ROTHMAN and DAWIDOWICZ, 1975; THOMPSON, 1977). The observed P_{Cl^-} can therefore not be attributed to phospholipid flip-flop. ROBERTSON and THOMPSON (1977) have recently put forward an interesting hypothesis to account for these observations. It is suggested that Cl^- in the membrane interior moves as HCl, while the transfer across the interfaces is catalyzed by the phosphatidylcholine molecules: Cl^- binds to the choline group and H^+ titrates the phosphate group. This neutral species can "sink" into the hydrocarbon part of the bilayer, where the Cl^- and H^+ will combine to form HCl and cross the membrane to the other side, where they leave the bilayer. This model emphasizes the dynamic structure of a lipid bilayer, and it appears to be energetically feasible, but it remains to be demonstrated that it predicts the observed kinetics. The observation that $P_{Cl} \propto \sqrt{[H^+]}$ is difficult to account for, and it is not clear how the obligatory Cl^--Cl exchange in planar bilayers is accomplished.

II. Anion Permeability

The behavior of large hydrophobic anions is well understood at low aqueous concentrations, where the model of KETTERER et al. (1971) (see also Figs. 1, 5 and 9) predicts, and experiments confirm, that the conductance and the number of absorbed ions increase linearly with concentration, that the ions move between two interfacial potential energy wells with a single time constant when a potential is applied across the membrane, and that the potential required to move a given fraction of absorbed ions from one potential energy well to the other is independent of the adsorbed charge density. At higher concentrations, however, serious anomalies are observed. These will be discussed in Section E.IV.

1. Stationary Conductance Changes

The large interfacial dipole potential (inside positive) affects the partitioning of ions into the hydrocarbon interior of a lipid bilayer (LIBERMAN and TOPALY, 1969; Le BLANC, 1970; ANDERSEN and FUCHS, 1975). The stationary conductance of a lipid bilayer in the presence of large hydrophobic anions, *e.g.* DpA^-, PCB^- and $TØB^-$, will be determined by the aqueous unstirred layers. LIBERMAN and TOPALY (1969) did, for example, observe *quantitatively* identical conductance vs. concentration characteristics for $TØB^-$ and PCB^- at aqueous concentrations less than 10^{-4} M, where the conductance increased linearly with concentration (see Fig. 8). Le BLANC (1969) again found conductance-vs.-concentration characteristics for $TØB^-$ that were quantitatively identical with those observed by LIBERMAN and TOPALY (1969). The zero-current potential as a function of the bulk aqueous concentration of $TØB^-$ were described by the Nernst equation (Eq. 3) for 10^{-8} M $\leqq$ $[TØB^-]$ and pH > 6 (LIBERMAN and TOPALY, 1968a; Le BLANC, 1969). At pH $\leqq$ 6 there is evidence of an additional process, which may be the movement of TØBH across the membrane. The pK of TØBH is less than −1, if the neutral compound exists at all (COOPER and POWELL, 1962). At a pH of 6 or above it should be permissible to disregard any such process.

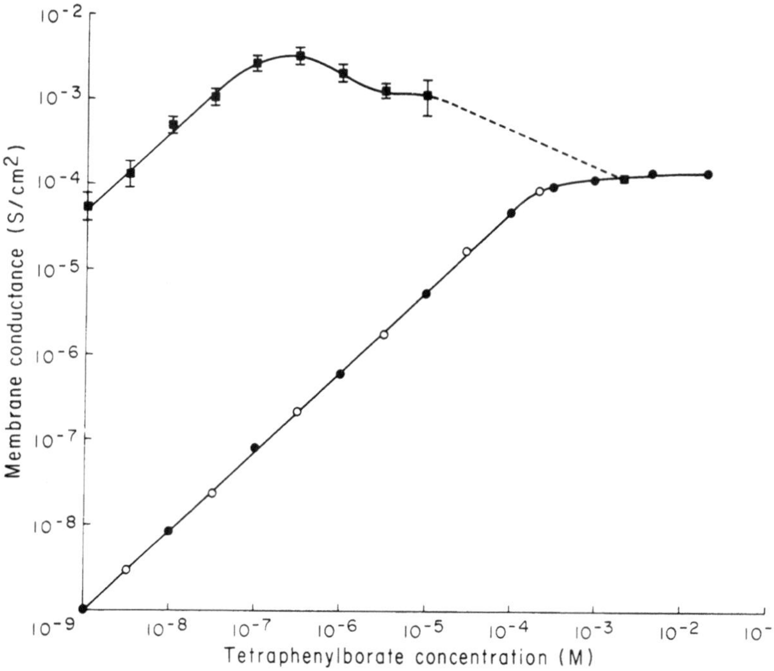

Fig. 8. Conductance-vs.-concentration characteristics of $TØB^-$ (filled symbols) and PCB^- (open symbols). The lower curve (circles) shows stationary conductance data from LIBERMAN and TOPALY (1969), obtained on ox brain phospholipid membranes. The upper curve (squares) is the instantaneous conductance in phosphatidylethanolamine membranes. The data for $[TØB] \leqslant 10^{-5}$M are from ANDERSEN et al. (1978). The conductance at 2×10^{-3}M was determined about 20 ms after application of potential (ANDERSEN, unpubl.)

Le BLANC concluded that the rate-limiting step for stationary currents was indeed the aqueous unstirred layers[16]. NEUMCKE (1970) confirmed this conclusion, on the basis of a more detailed analysis of Le BLANC's data. Further independent evidence that the stationary currents due to $TØB^-$ are unstirred layer-limited was provided by HLADKY (quoted in HAYDON and HLADKY, 1972). He found that the stationary current depended upon stirring: L was about 180 μm in unstirred systems, while it was about 60 μm in well-stirred systems. The actual values decreased with decreasing membrane area. The time-course of the diffusion polarization-dominated current transient was also studied. (HAYDON and HLADKY, 1972; should be consulted for details). The time course of these transients was independent of membrane area, and the thickness of the unstirred layers thus calculated was 350–500 μm in unstirred systems and about 70 μm in well-stirred systems.

[16] It was further concluded that the current-voltage characteristics displaying a "negative resistance" reported by LIBERMAN and TOPALY (1969) could be attributed to aqueous diffusion polarization.

2. Translocation through the Membrane Interior

Stationary conductance measurements are of little use in studies of anion transport through the membrane interior. It was therefore a major step forward when LÄUGER and co-workers (KETTERER et al., 1971; LÄUGER, 1972; STARK et al., 1972) introduced the use of "relaxation" methods to study the kinetics of ion translocation through the interior of lipid bilayers. The term "relaxation" describes a retarded process of self-adjustment of a perturbed molecular system to its (new) thermodynamic equilibrium state (EIGEN and De MAYER, 1963). The macroscopic relaxation behavior is connected to the microscopic molecular phenomena, *i. e.,* the return toward equilibrium from spontaneous fluctuations and the relaxation due to external perturbations follow the same time-course, according to the fluctuation-dissipation theorem (REIF, 1965, Section 15.8).

Relaxation measurements on lipid bilayers are: (1) voltage-clamp measurements where one applies a rectangular potential difference across the membrane and measures the magnitude and time-course of the resulting current transients; (2) a. c.-impedance measurements, where one applies a sinusoidal membrane potential variation around a steady potential, $\bar{V}_m$, usually 0 mV, and measures the magnitude and phase shift of the stationary membrane current signal as a function of frequency; (3) charge-pulse measurements, where one charges the membrane capacitance very rapidly (< 1 μs) and observes the decay of the membrane potential during open-circuit conditions. The information obtained with these three techniques is equivalent, at least for small perturbations from equilibrium.

Experimental studies on the potential dependence of charge translocation through the membrane interior have, almost exclusively, been voltage-clamp studies (see Fig. 9). The absorbed charges, *e. g.* $TØB^-$, are located in the two potential energy minima (Fig. 5). They will thus be concentrated in two very narrow strips near the membrane-solution interfaces. When the membrane potential is zero, the concentration of $TØB^-$ will be identical in both potential energy wells, as depicted by the solid line in Figure 9a. At $t = 0$, the membrane potential is changed from this initial value to a new value, V_m, and a current will flow in the external circuitry (Fig. 9b). One observes an initial fast current transient, which charges the membrane capacitance, C_m. This is followed by a much slower current transient, which results from ion translocation within the membrane (the on-response). The current will initially be high but will fall with time due to the depletion of ions on one (the negative) side of the membrane and the corresponding accumulation on the right. If the membrane-solution interfaces were perfect barriers the current would decline to zero and the concentration profile would be described by the broken line in Fig. 9a. At the end of the transient, when the membrane potential is returned to zero, the capacitance will be discharged by an initial fast transient of opposite polarity to that seen at $t = 0$. The ions move back through the membrane to establish the original distribution. This charge movement (the off-response) is registered as a current transient of opposite polarity to that observed during the on-response.

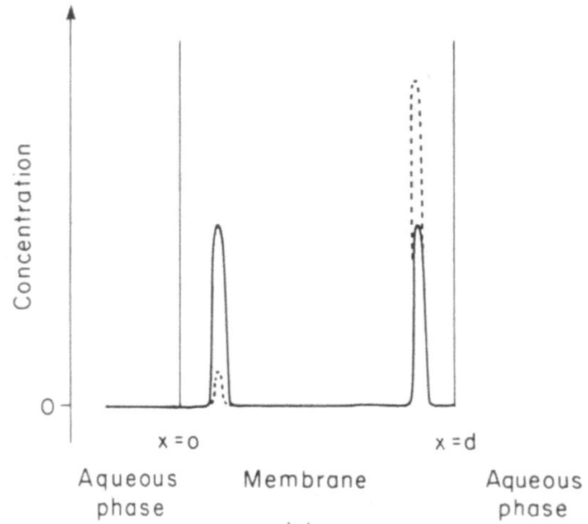

(a)

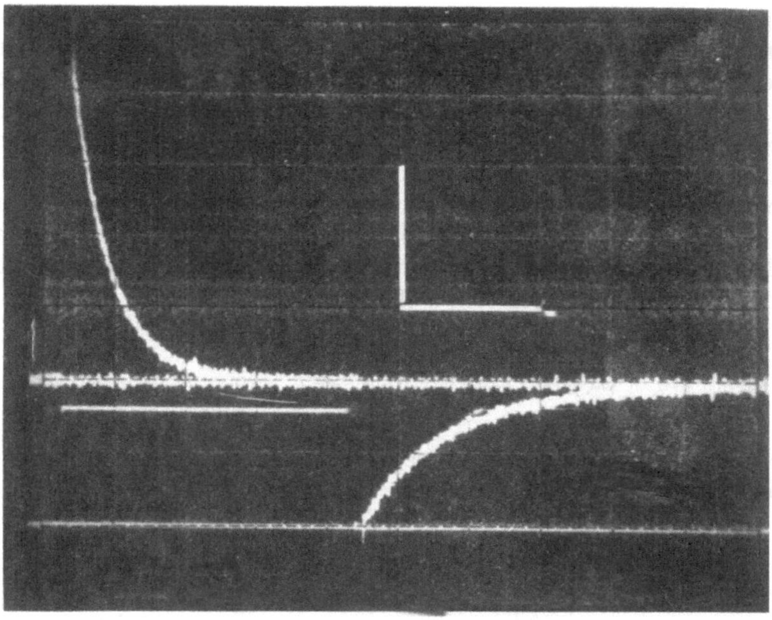

(b)

Fig. 9. (a) Schematic concentration profile for a hydrophobic anion in lipid bilayer. The solid line is the profile in a symmetrical membrane at equilibrium with $V_m = 0$. The broken line shows the concentration profile when a potential of about 80 mV is applied across the membrane and ion redistribution between the two potential energy minima is complete, at equilibrium; (b) Oscillogram of a TØB$^-$- induced current transient in a phosphatidylethanolamine/n-decane membrane. The top trace shows current, the top line shows potential. Stimulus 80 mV, duration 2.5 ms. The aqueous phases contain 1.0 M NaCl + 0.1 M phosphate buffer, pH 7.3, and 10^{-8}M Na TØB. $T = 24°$ C. Calibration bars: vertical 100 mV and 2×10^{-6} A cm^{-2}, horizontal 1.0 ms

a) The Transport Model

α. The Model of KETTERER et al.

DpA^-- and $TØB^-$-induced current relaxations in dioleoylphosphatidylcholine/n-decane were studied by KETTERER et al. (1971). The membrane conductance was a linear function of $[X^-]$ at low concentrations, while significant deviations were observed at higher concentrations (see Section E.IV, and similar data in Fig. 8). The current transients due to low concentrations of DpA^- were exponential functions of time for many time constants, and the stationary current was negligible:

$$I_c(V,t) = I_c(V,0) \cdot \exp\left(-\frac{t}{\tau(V)}\right), \tag{25}$$

where $I_c(V,t)$ is the *measured* current in the external circuitry (Acm^{-2}) as a function of time, $I_c(V,0)$ is the instantaneous current obtained by extrapolating the slow transient back to $t = 0$ (KETTERER et al., 1971; ANDERSEN and FUCHS, 1975), and $\tau(V)$ is the time constant of the relaxation, which is a function of potential. Similar exponential current transients have been reported for $TØB^-$ (KETTERER et al., 1971; ANDERSEN and FUCHS, 1975; SZABO, 1976). The observed current transients are thus similar to those observed for two-compartment kinetics (STEN-KNUDSEN, Chapter 2), and the current transients were interpreted as due to a potential-dependent redistribution of the absorbed ions between two potential energy minima, one at each membrane-solution interface. No direct evidence for such a polarization within the membrane was presented.

The data were interpreted in terms of a Eyring rate theory model (ZWOLINSKY et al., 1949; PARLIN and EYRING, 1954), with the assumption that the translocation through the membrane occurred as a single jump. The potential dependence of the instantaneous current is then predicted to be:

$$I(V,0) = G(0,0) \cdot \frac{2e}{kT} \cdot \sinh\left(\frac{eV}{2kT}\right), \tag{26}$$

where $G(0,0)$ is the instantaneous specific membrane conductance in the limit of zero applied potential. The potential dependence of $\tau(V)$

$$\tau(V) = \frac{\tau(0)}{\cosh\left(\frac{eV}{2kT}\right)} \tag{27}$$

where $\tau(0)$ is the relaxation time constant in the limit of zero applied potential and

$$\tau(0) = \frac{1}{2 \cdot k_i}, \tag{28}$$

where k_i is the rate constant for translocation through the membrane interior in the absence of an applied potential. The observed current-voltage characteristics and time constant-voltage characteristics for TØB$^-$ were described reasonably well for $V_m < 130$ mV. For DpA$^-$ significant deviations between predicted and observed current-voltage characteristics were present at $V_m > 100$ mV.

β. Impedance Measurements

Evidence that the observed current relaxations are indeed due to a potential-dependent ion redistribution between two potential energy minima near the membrane-solution interfaces rather than due to conductance changes was obtained in small-signal a. c. impedance measurements (GRIGOREV et al., 1972; De LEVIE et al., 1974b; De LEVIE and VUKADIN, 1975; SZABO, 1977).

In a symmetrical system with $\bar{V}_m = 0$, the two membrane-solution interfaces are equivalent and the analysis is very much simplified (De LEVIE et al., 1974a and b; De LEVIE and ABBEY, 1976; De LEVIE, 1978). The model in Figure 1 leads to the equivalent circuit illustrated in Figure 10 (KETTERER et al., 1971; De LEVIE et al., 1974a; De LEVIE, 1978). This equivalent circuit is valid for small perturbations, where the changes in current are a linear function of V_m[17]. The slower current transients seen in Fig. 9b due to the translocation of absorbed ions through the membrane are represented by the lower branch. These current transients are equivalent to charging a capacitance, the adsorption capacitance C_a, through a resistor R_m representing the rate of translocation through the membrane. The even slower current transients associated with the absorption/desorption reactions and with the aqueous diffusion polarization are represented by the phase-transfer resistance, R_{pt}, and the Warburg impedance, W. GRAHAME (1952) may be consulted for a discussion of the Warburg impedance

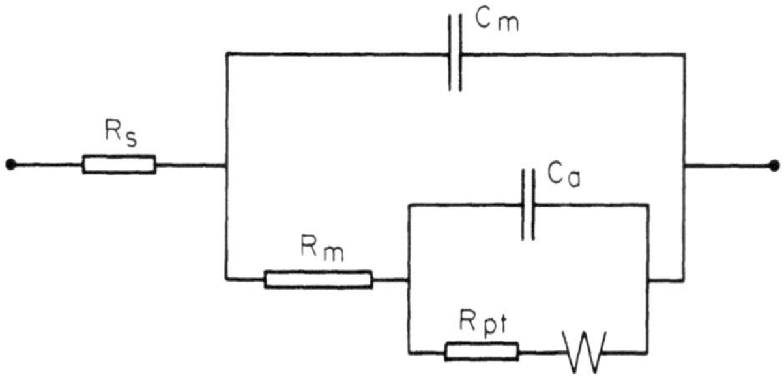

Fig. 10. Equivalent circuit for intramembrane translocation of hydrophobic anions through lipid bilayers. R_s is the series resistance to membrane, C_m is the geometric membrane capacitance. See text for further details

[17] The current transients at high potentials, where the current-voltage characteristics are non-linear, can also be represented by an equivalent circuit as in Fig. 10. The physical significance of the parameters R_m and C_a is quite different, however, as they are no longer calculated on the basis of a specific model (ANDERSEN, 1978b).

and its significance. Many apparently different equivalent circuits reduce to the circuit shown in Figure 10 (see ZOBEL, 1923; De LEVIE, 1978; and compare Fig. 10 with Fig. 1 of MARKIN et al., 1971). This equivalent circuit is thus an *irreducible* representation of the linear electrical properties of a lipid bilayer in the presence of DpA^- or $TØB^-$[18]. The magnitude of the adsorption capacitance is related to the magnitude of the adsorbed charge density (De LEVIE, 1974; ANDERSEN and FUCHS, 1975; De LEVIE, 1978; ANDERSEN et al., 1978b). In the limit of very low adsorbed charge density, the model of KETTERER et al. predicts:

$$C_a = \frac{e}{2kT} \cdot \beta \cdot \Delta q_{c,\,max.} \tag{29}$$

where β is the fraction of the applied potential actually effective in the intramembrane charge translocation, and $\Delta q_{c,\,max}$ is the maximal amount of charge transfer which can be measured in the external circuitry (ANDERSEN et al., 1978b, see also Section E.II.2.a.γ).

The experimental impedance spectrum for $TØB^-$ in phosphatidylethanolamine/n-decane membranes (De LEVIE et al., 1974b) demonstrates the same processes as observed in voltage-clamp experiments (ANDERSEN and FUCHS, 1975). It was further concluded that the interfaces are negligible barriers compared with the aqueous unstirred layers. The experimental data for DpA^- are also described by the equivalent circuit in Figure 10. In this case, however, there was evidence for a phase transfer resistance separate from the Warburg impedance. It thus appears that the interfacial adsorption/desorption reactions can play an important role under some circumstances. Even more striking evidence for interfacial barriers has recently been obtained by GRIGOREV and YERMISHKIN (1976) for the weak acids perfluoropinacal and auratine.

γ. Voltage Clamp Measurements

Direct evidence that the observed current transients are due to the redistribution of ions between the two potential energy minima was obtained by ANDERSEN and FUCHS (1975) for $TØB^-$ and by BRUNER (1975) for DpA^-. It was shown that the current-transient during the on response is associated with a current transient of opposite polarity during the off-response (see Fig. 9b). The $TØB^-$-induced current transients during both the on-response and the off-response were exponential for four time constants (ANDERSEN and FUCHS, 1975), and continued with a time-course consistent with aqueous diffusion polarization. The amount of charge movement measured in the external circuitry, Δq_c, can therefore be approximated as

$$\Delta q_c(V) = I_c(V, 0) \cdot \tau(V). \tag{30}$$

[18] The equivalent circuit in Fig. 10 is more general than the model used to derive it (de LEVIE et al., 1974b; De LEVIE, 1978). The a.c. impedance measurements are therefore only consistent with the model proposed by KETTERER et al., (1971), but they do not exclude other models. The model proposed by De LEVIE et al. (1974a and b) is, in fact, rather different from the model proposed by KETTERER et al. (1971).

The magnitude of the charge moved during the on-response is equal (but opposite in sign) to the magnitude of the charge moved during the off-response (ANDERSEN and FUCHS, 1975), see also Fig. 11. The time-course of the on-response and the off-response will, of course, be different due to the potential-dependent kinetics of charge translocation through the membrane. The time constant of the off-response is independent of the potential applied during the on-response as this transient represents diffusional redistribution of the ions between the two potential energy wells. Less than 1 percent of the intramembrane charge translocation can be explained by diffusional entry of $TØB^-$ from the aqueous phase according to Eq. 10. The magnitude of the measured charge translocation reaches an upper limit with increasing membrane potential (see Fig. 11). One can thus deplete one potential energy well (almost) totally of $TØB^-$, and there is no evidence for any alternative conduction mechanisms in bilayers doped with $TØB^-$. One may conclude that the observed $TØB^-$-induced current transients indeed do represent charge redistribution between the two potential energy wells, as suggested by KETTERER et al. (1971). The membrane solution interfaces and the adjoining aqueous unstirred layers act as almost ideal barriers for the movement of $TØB^-$, at least during the first 10 ms or so after the application of a potential across the membrane. Artificial lipid membranes are thus well suited to the investigation of both diffusion processes, during the off-response, and electrodiffusion processes, during the on-response, of simple hydrophobic ions through the membrane interior.

At the end of a current transient, $I_c(t)$ is much less than the initial current for both the on-response and the off-response. One can then approximate the concentration distribution of $TØB^-$ between the potential energy wells by means of the Boltzman relation (ANDERSEN and FUCHS, 1975):

$$\frac{c_{II}}{c_I} = \exp\left(\frac{-\Delta W}{kT}\right), \tag{31}$$

where c_I and c_{II} are the $TØB^-$ concentrations in the two potential energy wells, and ΔW is the potential energy difference between ions in the two wells. If there are *no interactions* among the absorbed ions, the charge movement as a function of concentration will be described by:

$$\Delta q_c(V) = \Delta q_{c,\,max} \cdot \tanh\left(\frac{e \cdot \beta \cdot V}{2kT}\right), \tag{32}$$

where $\Delta q_{c,max}$ is the limiting value of $\Delta q_c(V)$ when $V \rightarrow \infty$, and $\beta \cdot V$ is the fraction of the applied potential actually effective in translocating the ions through the membrane (ANDERSEN and FUCHS (1975). A plot of $\Delta q_c(V)$ vs. V (see Fig. 11) should thus determine β, provided there are no interactions among the absorbed ions. The value of β thus calculated is 0.75–0.80 in bacterial phosphatidylethanolamine/n-decane membranes with 0.1 M NaCl in the aqueous phases (ANDERSEN and FUCHS, 1975; MELNIK et al., 1977) and about 0.85–0.90 in 1.0 M NaCl (MELNIK et al., 1977; ANDERSEN et al., 1978b). This estimate of β is independent of any particular *kinetic* model for intramembrane

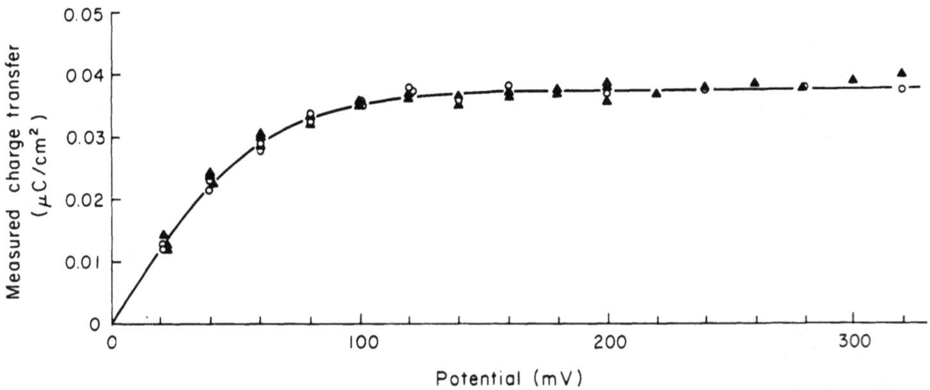

Fig. 11. Charge translocation, $\Delta q_c(V)$, as function of applied potential. ▲ represents the magnitude of the measured charge movement during the on-response, ○ represents the magnitude of measured charge movement during off-response. The data is from 5 independent experimental runs on single phosphatidylethanolamine/n-decane membrane. The composition of the aqueous phases as in Fig. 9b. $\Delta q_{c,\,max} = 3.8 \times 10^{-8}$ C cm^{-2}. The solid curve is drawn according to Eq. (32) with $\beta = 0.86$

charge translocation. It is, however, very dependent upon which *electrostatic* model one uses to quantitate any interactions among adsorbed charges. See Section E.IV for further discussion.

Alternatively one can write:

$$G_c(V, 0) \cdot \tau(0) = \frac{\Delta q_c}{V} = \Delta q_{c,\,max} \cdot \frac{\tanh\left(\frac{e \cdot \beta \cdot V}{2kT}\right)}{V} \tag{33}$$

where

$$G_c(V, 0) = \frac{I_c(V, 0)}{V} \tag{34}$$

is the instantaneous conductance at $V_m = V$. It follows that one can estimate the magnitude of the adsorbed charge density by an analysis of the small signal behavior. This estimate, $\Delta q_{c,\,ss}$, can be obtained as:

$$\Delta q_{c,\,ss} = \beta \cdot \Delta q_{c,\,max} = \frac{2kT}{e} \cdot G_c(0, 0) \cdot \tau(0) \tag{35}$$

(Szabo, 1976). Neither the small-signal analysis, nor the $\Delta q_c(V)$ vs. V analysis determines the adsorbed charge density, q^o, when $\beta < 1$. The deviations are less for the extended analysis, however, which determines

$$\Delta q_{c,\,max} = \beta \cdot q^o = \beta \cdot N \cdot |z| \cdot e \tag{36}$$

(ANDERSEN et al., 1978b), than for the small-signal analysis, which determines

$$\Delta q_{c,ss} = \beta \cdot \Delta q_{c,\max} = \beta^2 \cdot q^{o}. \tag{37}$$

The ratio of $\Delta q_{c,ss}/\Delta q_{c,max}$ should thus allow determination of β (see Table 5, p. 433).

The results of BRUNER (1975) with DpA^- are qualitatively similar to those of ANDERSEN and FUCHS (1975) for $TØB^-$. The current transients were not, however, described well by a single exponential, and more importantly, the magnitude of the charge during the on-response was greater than that of the charge translocation during the off-response (for V_m 50 mV). This discrepancy is unexpected and unexplained, as the a. c. impedance measurements of De LEVIE and VUKADIN (1975) indicate that DpA^- is even more confined to the membrane phase than $TØB^-$. The results indicate the existence of at least two separate contributions to the measured current – one being the DpA^- itself moving as a simple ion through the membrane interior. The other could be a DpA^--mediated movement of K^+ or H^+, through the membrane, as DpA^- is a weak acid and a known chelator of alkali metal ions (GABORIAUD, 1966; MOTOMIZA et al., 1969). This is in accordance with the low diffusion potential and the stationary current-voltage characteristics reported by MUELLER and RUDIN (1969, Fig. 5 A). The current-voltage characteristics reported by these investigators at potentials < 100 mV showed evidence of saturation similar to that observed by Le BLANC (1969), and quantitatively of the right magnitude. At higher potentials the current-voltage characteristics became hyperbolic, indicating a new current pathway due to the DpA^--mediated movement of other ions, probably K^+ and H^+, as this process was not observed with Na^+ as the sole cation in the aqueous phase (MUELLER, pers. comm.). The experimental evidence of BRUNER (1975) provides strong support for the validity of the basic transport model in Figure 1. DpA^- does, however, exhibit additional complicating behavior which deserves further study.

δ. Charge Pulse Measurements

Charge-pulse experiments (BENZ et al., 1976) provide additional evidence that the model of KETTERER et al. (1971) is correct. If the absorbed charges are effectively confined to the membrane, the voltage transients following the initial charging of the membrane capacitance will decline from an initial value, $V_m(0)$:

$$V_m(0) = \frac{q_c}{C_m}, \tag{38}$$

where q_c is the charge injected into the aqueous phases charging the membrane capacitance, to a final value, $V_m(\infty)$:

$$V_m(\infty) = \frac{q_c - \Delta q_c}{C_m} \tag{39}$$

where Δq_c is the amount of charge translocated as registered in the external circuitry. The value of Δq_c is determined by $V_m(\infty)$ through the BOLTZMANN relation:

$$\Delta q_c = \Delta q_{c,\,max} \cdot \tanh\left(\frac{e \cdot \beta \cdot V_m(\infty)}{2kT}\right); \tag{40}$$

this is an obvious extension of Eq. (32), and is dependent upon the same assumptions. For $V_m(0) << \frac{kt}{e}$

one can express the time constant, τ_1, of the exponential voltage-transient from $V_m(0)$ to $V_m(\infty)$ as:

$$\tau_1 = \frac{2\,\frac{kT}{e} \cdot C_m \cdot \tau(0)}{2\,\frac{kT}{e} \cdot C_m + \beta \cdot \Delta q_{c,\,max}} \tag{41}$$

where $\tau(0)$ is defined by Eq. (28).
The observed voltage transients show that $TØB^-$ and DpA^- are reasonably well confined to dioleoyl phosphatidylcholine n-decane membranes (BENZ et al., 1976). The potential declines from $V_m(0)$ to a pseudostationary value $V_m(\infty)$, followed by a decline 200 to 2000 times slower to the final value of $V_m = 0$. The first transient represents the redistribution of membrane bound ions. An analysis of this transient will determine the value of both $\beta \cdot \Delta q_{c,\,max}$ and $\tau(0)$ from Eqs. (38)–(41).

The slow transients represent net movement of charge across the membrane, from one aqueous phase to the other. The time course of these transients is, however, quite fast. The time constant is 2.5–4 seconds for $TØB,^-$ and 0.2 s for DpA^-, which is one to two orders of magnitude less than estimated for transients due to aqueous diffusion polarization (HAYDON and HLADKY, 1972; BENZ et al., 1976). These slow transients must, therefore, represent a second current path dependent upon the presence of the hydrophobic ion but due to the movement of some other ions, possibly similar to that found by BRUNER (1975). The relative magnitude of this second process however, is much less than in BRUNER's experiment at high potentials.

ε. Noise Analysis

The kinetics of intramembrane charge translocation is well understood, but the molecular details are poorly resolved. These problems will demand other experimental approaches, such as optical studies of the amount of charge absorption (WULF et al., 1976) or noise analysis (KOLB and LÄUGER, 1977). The noise analysis will determine the rate constant for charge translocation through the membrane at thermal equilibrium ($V_m = 0$) as well as the amount of charge absorbed into the membrane (KOLB and LÄUGER, 1977). The spectral intensity

$S_I(f)$ of the current fluctuations, per unit membrane area, as a function of frequency, f, is:

$$S_I(f) = \frac{(2\cdot\pi\cdot f)^2 \cdot \tau(0) \cdot \beta \cdot e \cdot 2 \cdot \Delta q_{c,\,max}}{1 + (2\cdot\pi\cdot f\cdot \tau(0))^2} \tag{42}$$

A plot of $S_I(f)$ vs f will thus determine both τ (0) and $\beta \cdot \Delta q_{c,max}$. KOLB and LÄUGER found excellent agreement between the value of τ (0) for DpA^- in dioleoyl phosphatidylcholine/n-decane membranes with the value determined by BENZ et al. (1976), but the values of $\Delta q_{c,\,max}$ differed significantly for $(DpA^-) > 3 \times 10^{-8}$M.

Noise analysis offers the additional possibility to determine the translocation rate across the membrane-solution interfaces, as well as the more hypothetical possibility of determining the single ion translocation time, τ_c, through the membrane (STEVENS, 1972). This latter measurement is extremely difficult due to the high frequency range, which must be covered experimentally. The former problem was analyzed with the assumption that at all times there was partition equilibrium between DpA^- in the membrane and DpA^- on the adjacent aqueous phases (no interfacial barrier). Under these circumstances the aqueous phases will only contribute to the very low frequencies of $S_I(f)$. A similar analysis for the more difficult case, where one relaxes the assumption of partition equilibrium between the membrane and the adjacent aqueous phases, has not been reported. Such an analysis may conceivably permit the low frequency deviations between the observed spectra and Eq. (42) to be explained in terms of the interfacial translocation barrier.

b) Kinetics of Charge Translocation

The kinetics of the voltage-dependent translocation of hydrophobic ions through the interior of lipid membranes have been studied in detail by KETTERER et al. (1971) ($TØB^-$ and DpA^-), and by ANDERSEN and FUCHS (1975) ($TØB^-$). The results were analyzed in terms of specific models for the potential energy barrier for intramembrane ion translocation. KETTERER et al. (1971) proposed a single-jump rate theory model which provided an adequate prediction of their results for $TØB^-$ when $V_m < 140$ mV. The data of ANDERSEN and FUCHS (1975) is also reasonably well described by this model when $V_m < 100$ mV; at potentials >100 mV significant deviations appeared, however. These deviations could be rationalized by a model which takes the detailed shape of the potential energy barrier into account.

Within experimental error, it is possible to predict any observed current-voltage characteristics by an infinite number of potential energy barrier shapes (ANDERSEN and FUCHS, 1975; GINSBURG and NOBLE, 1976). The detailed shape of any proposed barrier is thus somewhat irrelevant. It is, however, of importance if one can *predict* the observed current-voltage characteristics on the basis of a (simple) physical model for the *potential energy* profile of an ion within the membrane, and a reasonable *mechanical model* for solute movement within the membrane. Agreement between the predictions of the simple model and the experimental results provide some confidence that the underlying physical picture is a reasonable approximation to reality[19]. The detailed model, however, will have to be modified each time one expands the range of membrane potentials covered.

Theoretical models of ion translocation through lipid bilayers can be constructed on the basis of the transition state theory (ZWOLINSKY et al., 1949; PARLIN and EYRING, 1954) often denoted as the Eyring rate theory, or on the basis of the SMOLUCHOWSKI equation describing the movement of particles in a field of force (SMOLUCHOWSKI, 1916; KRAMERS, 1940; CHANDRASEKHAR, 1943; see also STEN-KNUDSEN, Chapter 2):

$$\frac{\partial c}{\partial t} = \frac{\partial}{\partial x}\left\{u(x)\cdot\left\{kT\cdot\frac{\partial c}{\partial x} + c(x,t)\cdot\left[z\cdot e\cdot\frac{\partial \varphi}{\partial x} + \frac{\partial \Delta W}{\partial x}\right]\right\}\right\}, \quad (43)$$

where $u(x)$ is the mechanical mobility of the ion, $\varphi(x)$ is the variation in electrostatic potential due to an externally applied field and the mobile charges within the membrane. Eq. (43) is valid as long as the thermal energy of the ions is much larger than their kinetic energy (KRAMERS, 1940; CHANDRASEKHAR, 1943), a condition that is satisfied for all reasonable mobilities and masses of ions in lipid bilayer membranes. One can regard the transport as pseudostationary provided the single particle transit time, τ_c, (see Section C. III. 3), is much less than the observed relaxation time of the current transient, $\tau(V)$ (PLANCK, 1890a; ANDERSEN and FUCHS, 1975) or that the potential energy barrier is high compared with kT (CHANDRASEKHAR, 1943). τ_c is an upper estimate on the time constant changing the concentration profile in the membrane interior (ANDERSEN and FUCHS, 1975). Transients with $\tau(V) > 50\ \mu s$ should thus be pseudostationary. One can then follow PLANCK (1890a) and integrate Eq. (43) *via* the continuity equation, Eq. (7), to obtain the following result (ANDERSEN and FUCHS, 1975):

$$I_c(V,t) = \beta\cdot I(V,t) = \frac{\beta\cdot z\cdot e\cdot\left\{c_{II}\cdot\exp\left(\frac{z\cdot\beta\cdot eV}{2\,kT}\right) - c_I\cdot\exp\left(\frac{-z\cdot\beta\cdot eV}{2\,kT}\right)\right\}}{\exp\left(\frac{-z\cdot e\cdot V}{2\,kT}\right)\cdot\int_{\delta}^{d-\delta}\frac{1}{u(x)}\cdot\exp\left(\Delta W_1(x) + z\cdot e\cdot\varphi(x)\right)dx} \quad (44)$$

where δ denotes the distance of the energy minima from the interfaces, $\Delta W_1(x) = \Delta W(x) - \Delta W(\delta)$, and $I_c(V,t)$ is the measured current (CIANI, 1976; ANDERSEN et al., 1978b). The denominator in Eq. (44) contains all information about the interactions between an ion in the membrane and its surroundings – including the aqueous phases. The potential dependence of the integral is thus a very important parameter in analyzing a transport system, but one can only relate this potential dependence to the shape of the current-voltage characteristics if one has determined β. The reader is referred to HAYDON an HLADKY

[19] The reader should consult PARSEGIAN (1975) for a less optimistic point of view, see also Section C.III.3.

(1972), LÄUGER and NEUMCKE (1973), HALL et al. (1973), CIANI et al. (1973), HLADKY (1974), ANDERSEN and FUCHS (1975), and GINSBURG and NOBLE (1976) for detailed discussions of Eq. (44) and its consequences for intramembrane charge translocation.

It is apparent from Figure 2a that the force moving an ion through a bilayer will differ from the force due to the applied potential, and will *vary with position through the membrane.* This force is due to the interactions between the ion and the aqueous phases, and is dependent upon the thinness of the membrane and the low dielectric constant. Ions in thick membranes will not interact with the exterior solutions, and the current-voltage characteristics are linear in symmetrical solutions (see STEN-KNUDSEN, Chapter 2). In lipid bilayers, on the other hand, the current-voltage characteristics are nonlinear. The nonlinearity is due to a potential-dependent shift in the concentration profile of ions within the membrane. Figure 12 illustrates how the relative concentration varies through the membrane as a function of V_m, as well as the corresponding current-voltage characteristics. For comparison, Figure 12b also illustrates the current-voltage characteristics predicted by the model of KETTERER et al. (1971). The articles by KAUFFMAN and MEAD (1970), HAYDON and HLADKY (1972), HALL et al. (1973), ANDERSEN and FUCHS (1975) and GINSBURG and NOBLE (1976) should be consulted for further information on the relations that exist between the variation of $\Delta W(x)$ through the membrane and the shape of the theoretically predicted or experimentally observed current-voltage characteristics.

Equation (44) is derived on the basis of an essentially macroscopic model for solute translocation through lipid bilayers. Alternative models have been developed on the basis of the transition state theory (CHIZMADZHEV et al., 1971a and b; KETTERER et al., 1971; MARKIN et al., 1971) to describe both the stationary and transient conductance behavior of lipid bilayers in the presence of hydrophobic ions. KETTERER et al. (1971) should be consulted for a very clear presentation of the theory and its experimental consequences (see also Fig. 12b). The mathematically simplest of these models are the single-jump models, in which the initial charging transient is neglected and it is assumed that the translocation proceeds as a jump across a single activation energy barrier. This assumption, however, is very strong, as the rate constant, k_i, and the diffusion coefficient, D_m, are related through $D_m = k_i \cdot d^2$ (ZWOLINSKI et al., 1949; LÄUGER, 1972). The measured value of k_i for DpA^- and $TØB^-$ in dioleoylphosphatidylcholine/n-decane membranes is 380 s^{-1} and 9 s^{-1} respectively (KETTERER et al., 1971). The predicted values for D_m are thus about 8×10^{-11} cm^2 s^{-1} and 2×10^{-12} cm^2 s^{-1} if one assumes $d = 4.5$ nm. These values are three to four orders of magnitude less than diffusion coefficients estimated by more direct techniques using molecules of similar size as DpA^- and $TØB^-$ (see Section D. III). The time constant of the current transients therefore reflects primarily the much lower ion concentration in the membrane interior compared to the concentration in the potential energy wells in the interfaces (KETTERER et al. 1971; LÄUGER and NEUMCKE, 1973; ANDERSEN and FUCHS, 1975; SZABO, 1976). The mechanical movement of the ions is presumably dominated by a diffusion-like process through many small jumps (LÄUGER and NEUMCKE, 1973; ANDERSEN and FUCHS, 1975). The transition state theory and the electrodiffusion forma-

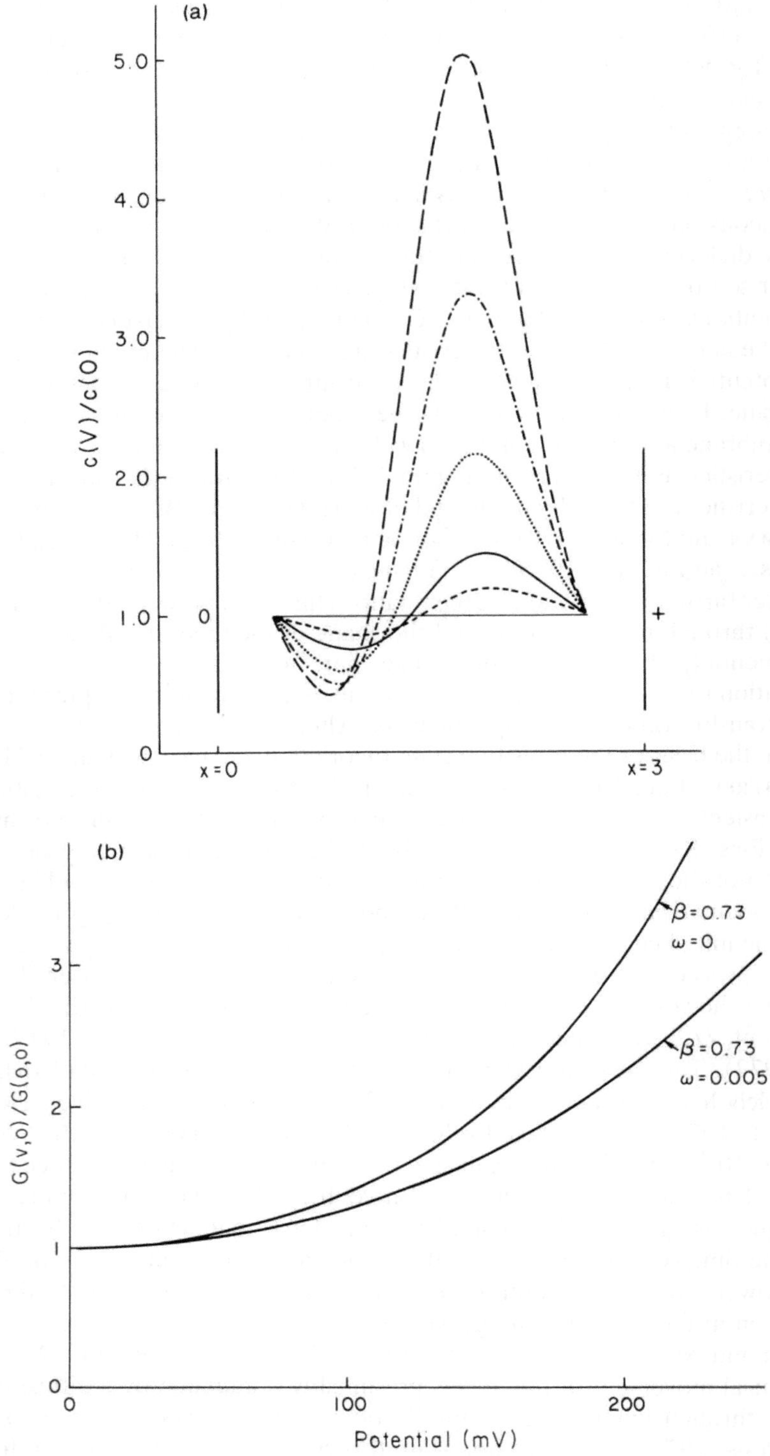
(a)
5.0
4.0
3.0
2.0
1.0
0
c(V)/c(0)
0
+
x=0
x=3
(b)
3
2
1
0
G(v,o)/G(o,o)
β=0.73
ω=0
β=0.73
ω=0.005
0
100
200
Potential (mV)

lism will then predict essentially identical results provided the particle exhibits more than 6 individual jumps (CIANI, 1965; ANDERSEN and FUCHS, 1975). Furthermore, the time resolution of most experiments is such that one measures a space and time average of the individual particle motion, and loses most of the information which might allow a discrimination between the two models (see also AITYAN et al., 1976a–c). It is unlikely, however, that there are preferred stable potential energy minima for ion movement through a fluid bilayer. The advantage and reality of assuming a fixed number of jumps is therefore dubious.

c) Temperature-Dependence

The temperature-dependence of the adsorption coefficient, K_{ads}, and relaxation time constant, $\tau(0)$, has been studied by BRUNER (1975), BENZ et al. (1975) and KOLB and LÄUGER (1977) in phosphatidylcholine/n-decane membranes (see Table 3). The temperature-dependence of the membrane conductance is the sum of the temperature-dependence of K_{ads} and $1/\tau(0)$. There is no obvious explanation for the discrepancy between ΔH_{ads} as determined by BRUNER (1975) and KOLB and LÄUGER (1977), although the aqueous support electrolyte was different in the two cases.

A value of ΔH_{ads} less than zero is similar to the findings for transfer of aromatic hydrocarbons from water to bulk hydrocarbons (TANFORD, 1973), or for transfer of hexane from water to lecithin liposomes (SIMON et al., 1977a), but *opposite* to the findings of KATZ and DIAMOND (1974b) on the transfer of polar nonelectrolytes from water to phosphatidylcholine liposomes. The available data are therefore consistent with the notion that the potential energy minimum for DpA^- is located very deep in the polar head group region or in the outer region of the hydrocarbon core of the membrane. The large change in ΔH_{ads} with changes in the fatty acid chain length observed by KOLB and LÄUGER is also consistent with this interpretation. If the potential energy minimum for DpA^- were in the outer part of the polar headgroup region one would expect ΔH_{ads} to be relatively insensitive to the fatty acid chain length.

d) Ion Translocation as a Function of Membrane Composition

The composition of lipid bilayers will be reflected in the adsorption coefficient, K_{ads}, the translocation rate constant, $\tau(0)$, and the membrane conductance,

◁ Fig. 12. (a) Concentration profiles within a lipid bilayer as a function of applied potential, assuming that potential energy barrier is the image-force barrier (see Fig. 2). The parameters used were $d =$ 3nm, $\varepsilon_r^m = 2.0$, $r = \delta = 0.4$ nm, $z = +1$. Concentrations have been normalized by the concentration at $V = 0$. The concentration profile is calculated from Eq. (50) of ANDERSEN and FUCHS (1975) for $t = 0$, *i. e.*, after the new concentration profile has been established but before the concentrations in the two potential energy wells have changed. The lines represent V_m = 25 mV---; 50 mV——; 100 mV . . .; 150 mV-·-·-; and 200 mV– – –; (b) Predicted current-voltage characteristics in lipid bilayers, $\beta = 0.73$. The lower curve is calculated from Eq. (44), with same assumptions as in (a). The parameter ω is a "shape factor" that accounts for the potential dependence of the denominator (ANDERSEN and FUCHS, 1975). Upper curve, $\omega = 0$, is the prediction of the single-jump rate theory with $\beta = 0.73$

Table 3. Translocation of DpA^- as a function of membrane thickness and temperature

Fatty acid chain length	Thickness[c] nm	τ (0) ms		K_{ads} cm	ΔH_{ads} kJ mol^{-1}	G_c (0, 0) / DpA^- S cm^{-2} M^{-1}	References
16:1	4.8	0.5		2.6×10^{-2}		9.7×10^4	c
		0.6	−33	1.6×10^{-2}	−15	4.9×10^4	d
18:1	5.0	1.3		2×10^{-2}		3×10^4	a
		1.1		5.0×10^{-2}		7.5×19^4	c
		1.2		1.9×10^{-2}	−19	3.1×10^4	d
20:1	5.2	2.8		6×10^{-2}		3.8×10^4	c
		3.6	−50	3.2×10^{-2}	−39	1.7×10^4	d
22:1	5.7	7.5	−59	2.5×10^{-2}*	0	6×10^4	b
				6×10^{-2}**			b
		6.5		6.1×10^{-2}		1.8×10^4	c
		6.7	−65	5.0×10^{-2}	−49	1.4×10^4	d

Phosphatidylcholine/n-decane membranes, 0.1 M NaCl (a, c, d) or 0.1 M KCl (b); T = 25° C
References: a KETTERER et al. (1971), voltage-clamp; b BRUNER (1975), voltage-clamp, * based on $\Delta q_{c,\ ss}$; ** based on $\Delta q_{c,\ max}$; c BENZ et al. (1976), charge-pulse and voltage clamp; d KOLB and LÄUGER (1977), thermal noise

Table 4. Intramembrane translocation of DpA^- and $TØB^-$ as a function of the lipid composition of the membrane-forming solution

Lipid	τ (0) ms	K_{ads} cm	G_c (0, 0)/[X^-] S cm^{-2} M^{-1}	References
a) DpA^-				
Glycerolmonooleate	0.35	2.8×10^{-3}	1.5×10^4	b
Dioleoylphosphatidylcholine (ester)	1.2	3.7×10^{-2}	$6\ \times10^4$	e
Dioleoylphosphatidylcholine: cholesterol (1:4)	0.22	3.9×10^{-2}	3.3×10^5	e
1-0-oleyl-2-0-palimityl-3-phosphatidyl-choline (ether)	21	3.1×10^{-2}	2.8×10^3	e
Dioleoylphosphatidylethanolamine (ester)	0.2	1.1×10^{-2}	$1\ \times10^4$	e
Di-0-oleylphosphatidylethanolamine (ether)	1.4	1.6×10^{-2}	2.1×10^4	e
Phosphatidylserine (3.0 M NaCl)	1.3	2.3×10^{-2}	3.3×10^4	e
b) $TØB^-$				
Glycerolmonooleate	18	2.6×10^{-3}	2.7×10^2	c
	6.6	2.9×10^{-3}	7.5×10^2	f
Glycerolmonooleate: cholesterol (1:4)	0.6	1.7×10^{-3}	5.3×10^2	c
Dioleoylphosphatidylcholine	55	$3\ \times10^{-2}$	1.0×10^3	a
Egg phosphatidylcholine	30	$4\ \times10^{-2}$	1.3×10^3	d
Egg phosphatidylcholine: cholesterol (1:4)	2.0	$6\ \times10^{-3}$	$8\ \times10^3$	d
Bacterial phosphatidylethanolamine	1.2	3.7×10^{-2}	$5\ \times10^4$	f

Temperature 23–25°.
References: a KETTERER et al. (1971); b BENZ et al. (1976); c SZABO (1976); d ANDERSEN (1978a) e BENZ and LÄUGER (1977); f ANDERSEN et al. (1978b)

$G_c(0,0)$. Each one of these parameters will, of course, be a function of the two other parameters:

$$K_{ads} \sim G_c(0,0) \cdot \tau(0). \tag{45}$$

The variation in K_{ads}, $\tau(0)$, and G_c (0,0) between different membranes may reflect changes in (a) membrane thickness, (b) the electrostatic potential between the bulk aqueous phases and the membrane interior, and (c) short-range interactions between the ions and the membrane matrix. It is impossible to quantitate the relative importance of each of these variables unambiguously – especially in view of the scarcity of data on isosteric positive ions. Tables 3 and 4 summarize information for DpA^- and $TØB^-$.

α. Effect of Membrane Thickness

The change in translocation rate of DpA^- as a function of membrane thickness has been studied by BENZ et al. (1976) and KOLB and LÄUGER (1977), using monoglycerides or phosphatidylcholines with different fatty acid chains lengths (Table 3), and by BENZ and LÄUGER (1977), using glycerolmonooleate or dioleoylphosphatidylcholine and hydrocarbon solvents of different chain lengths. In good agreement with the predictions of Figure 3 (see Section C. III.), these latter experiments showed a change in translocation rate by up to 20-fold when the membrane thickness was varied between 2.5 and 5 nm. The adsorption coefficients were constant within experimental error.

β. Effect of the Interfacial Potential

i. Surface Potentials and Ionic Strength

The adsorption of DpA^- into phosphatidylserine/n-decane membranes has been studied by BENZ und LÄUGER (1977). These membranes carry a net negative charge, and K_{ads} varies as a function of ionic strength (0.03M–3 M NaCl) in accordance with the GOUY-CHAPMAN equation (Eq. 19). Control experiments in net neutral membranes (BENZ et al., 1976) show that the K_{ads} and τ (0) are independent of ionic strength. WANG and BRUNER (1977b) similarly found that K_{ads} for DpA^- into dioleoylphosphatidylcholine/n-decane membranes ([DpA] $\leq 10^{-8}$M or $\Delta q_{c,\,max} \leq 0.2\ \mu C/cm^{-2}$) is independent of the type of electrolyte (NaCl, $BaCl_2$, or $MgSO_4$) and concentration (10^{-2} M to 1 M) used. But K_{ads} for $TØB^-$ is dependent upon the ionic strength. SZABO (1978) finds that K_{ads} in glycerolmonooleate/n-decane membranes increases about threefold when [NaCl] increases from 10^{-2} to 1.0 M. $\tau(0)$, however, is independent of [NaCl].[20]

ii. Dipole Potentials

The interfacial dipole potential of phospholipid bilayers is about 130 mV higher (inside positive) than of glycerolmonooleate membranes (HLADKY and

[20] The low K_{ads} observed by ANDERSEN and FUCHS (1975) in 0.1 M NaCl compared with the value of 3.7×10^{-2} found by ANDERSEN et al. (1978) in 1.0 M NaCl in bacterial phosphatidylethanolamine/n-decane, may also reflect an ionic strength effect on K_{ads}. The adsorption coefficient observed by ANDERSEN and FUCHS (1975), however, is anomalously low, the reason for it is not understood.

HAYDON, 1973; HLADKY, 1974; HAYDON, 1975; ANDERSEN et al., 1978b). One would thus predict that the TØB⁻ conductance of phosphatidylethanolamine membranes should be 150–160 times higher than the conductance of glycerolmonooleate membranes, in qualitative agreement with the observed ratio of 70 (ANDERSEN et al., 1978b)[21]. The different K_{ads} may also reflect the difference in interfacial dipole potentials (BENZ and LÄUGER, 1977; ANDERSEN et al., 1978b). If the difference in absorption coefficients were related only to the difference in interfacial dipole potential, one would conclude that a DpA⁻ or TØB⁻ ion absorbed into a phospholipid bilayer experiences an electrostatic potential some 40 to 70 mV more positive than if they were absorbed into a glycerolmonooleate bilayer.

If the location of the dipoles contributing to the overall potential were known, it would thus be possible to place an *outer limit* on the location of the potential energy minimum for the ions. Such an analysis is, however, at best a very rough approximation. Neither DpA⁻ or TØB⁻ is a point charge, as the area per molecule in the interface is of the same order of magnitude as that for the lipids themselves. The absorbed ions will thus create a major disturbance in that part of the dipole potential that originates in the glycerol-ester bonds. The effective dipole potential experienced by the ion in the interface may be a small fraction of the potential experienced by a point (nonperturbing) charge located in the potential energy minimum.

The large difference in dipole potentials between ether and ester phospholipids (see Section C.V. 2) is not reflected in K_{ads} (BENZ and LÄUGER, 1977). If anything K_{ads} is higher in membranes formed from ether lipids! The translocation rate constant is about 10-fold less in membranes formed from ether lipids than in those formed from ester lipids, so at least some of the dipole potential difference observed at the air-water interface is reflected in the bilayer. These results thus suggest that DpA⁻ absorbs into phospholipid membranes in a region where it experiences part of the difference in dipole potentials between phospholipid glycerolmonooleate membranes (possibly a contribution from the phosphoric acid-glycerol ester bond), but it does not experience the dipole potential due to the ester groups linking the hydrocarbon tail to the glycerol backbone.

An alternative possibility, namely that DpA⁻ (and TØB⁻) absorbed into phospholipid membranes are associated with the ethanolamine or choline moieties, is difficult to reconcile with the comparatively deep position of the adsorption plane of these hydrophobic ions compared with the amphipathic ions sodium dodecyl sulphate (HAYDON and MYERS, 1973) and 2,6 toludinyl-naphtalene sulfonate (MCLAUGHLIN and HARARY, 1976). The adsorption of these latter ions can be described well by a simple theory reminiscent of the Gouy-Chapman theory (MCLAUGHLIN, 1977), while the absorption of DpA⁻ (WANG and BRU-

[21] The difference between $\tau(0)$ (and thus $G(0,0)$) observed by SZABO (1976) and ANDERSEN et al. (1978b) is probably due to a different thickness of the bilayers, as the membranes used by ANDERSEN et al. (1978b) were made from a very concentrated solution of glycerolmonooleate in decane. These membranes have a much higher capacitance (0.47 μF cm^{-2}) than the membranes usually studied, which have a specific capacitance of about 0.39 μF cm^{-2} (BENZ et al., 1977).

NER, 1977a and b) and TØB⁻ (ANDERSEN et al., 1978a and b) produces large boundary potentials not predicted by the simple Gouy-Chapman theory (see Section E.IV). The available data cannot, however, exclude this possibility of specific interactions between DpA^- and $TØB^-$ and the polar headgroups. But it should be noted that the behavior of DpA^- could be affected by hydrogen bonding between the $-NO_2$ groups and H_2O in the headgroup region, or between the $-NO_2$ groups and the polar headgroups of cholesterol, glycerolmonooleate, etc.

iii. Dipole Potential Changes

The adsorption of neutral dipolar molecules such as salicylamide (MCLAUGHLIN, 1973) and the unionized form of phloretin (CASS et al., 1973; ANDERSEN et al., 1976), and 2,4-dichlorophenoxyacetic acid (SMEJTEK and PAULIS, 1978), can produce large and opposite changes in the membrane conductance due to both positive and negative carriers, and thus presumably change the interfacial potential of phospholipid and phospholipid-cholesterol bilayers. Phloretin at 2.5 $\times$ 10^{-4} M, for example, can produce a 2000-fold increase in cation conductance and a corresponding 2000-fold decrease in anion conductance, which indicates a change (decrease) in the interfacial dipole potential by about 200 mV in phosphatidylethanolamine/n-decane membranes (ANDERSEN et al., 1976). Phloretin and 2,4-dichlorophenol noxyacetic acid affect both K_{ads} and τ (0) for $TØB^-$, the greatest effect being on τ (0) (ANDERSEN et al., 1976; MELNIK et al., 1977; SMEJTEK and PAULIS, 1978). The interpretation of these results will, of course, depend upon whether one regards $TØB^-$ as a probe for the mechanism of action of phloretin, or phloretin as a probe for the location and behavior of $TØB^-$. However, the results are in agreement with the notion that the potential energy minimum is located in the low dielectric constant membrane interior (KETTERER et al., 1971; GRIGOREV et al., 1972; ANDERSEN and FUCHS, 1975).

The effect of phloretin on K_{ads} is larger in cholesterol-containing membranes than in cholesterol-free membranes (ANDERSEN et al., 1976; MELNIK et al., 1977; ANDERSEN, 1977). In the former membranes, $TØB^-$ experiences a considerable fraction ($\approx$ 1/3) of the dipole potential change induced by phloretin, in the latter membranes the ratio is less (MELNIK et al., 1977). The addition of phloretin to only one side of a phosphatidylethanolamine/n-decane bilayer will decrease K_{ads} on that side only, and produce rectifying current-voltage characteristics (ANDERSEN, 1977) similar to those observed in other experiments with asymmetric dipole potentials (HANSEN et al., 1977; LA TORRE and HALL, 1977). Concomitantly with the rectification one observes that the maximal relaxation time constant no longer occurs at V_m = 0 mV, but at V_m = ln $(K_{I\ ads}/K_{II\ ads})$ (ANDERSEN, 1977).

γ. Cholesterol

The presence of cholesterol in the membrane-forming solution *increases* the conductance observed with negative current-carriers and *decreases* the conductance observed with positive current carriers. These effects are observed with bilayers formed from glyceroldioleate (SZABO et al., 1972), glycerolmonooleate (SZABO, 1974, 1976; BENZ et al., 1977), and from phosphatidylcholine (HLADKY and HAYDON, 1973; SMEITEK et al., 1976; ANDERSEN, 1978; BENZ and LÄUGER,

1977). These results suggest that the presence of cholesterol in lipid bilayers increases the interfacial dipole potential, as was suggested by SZABO et al. (1972) for glyceroldiooleate membranes.

The most clear cut results are those of SZABO (1974, 1976, 1978), obtained on glycerolmonooleate/n-decane membranes. With both negative and positive current carriers it was shown that the anion concuctance increases with increasing mole fraction of cholesterol to glycerolmonooleate in the membrane-forming solution. An essentially parallel fall in cation conductance was also observed, indicating that the conductances indeed are due to a change in interfacial dipole potential. At very high mole fractions of cholesterol, the conductance changes for anions and cations are no longer symmetrical. The anion conductance increases less than the cation conductance decreases. The anion conductance did in fact go through a maximum, indicating that other factors than the interfacial dipole potential are of importance.

These results can be analyzed by writing the approximation:

$$G(0,0) = \frac{e \cdot F}{kT} \cdot P_n \cdot [X^z] \cdot \exp\left(\frac{-z \cdot e \cdot \Delta V}{kT}\right) \tag{46}$$

(SZABO, 1976; ANDERSEN et al., 1978), where P_n is the dipole potential-independent permeability coefficient:

$$P_n = \frac{1}{\int \frac{1}{U_m(x)} \cdot \exp\left(\Delta W_2(x)/kT\right) dx} \tag{47}$$

where

$$\Delta W_2(x) = \Delta W_i(x) + \Delta W_h(x) + \Delta W_{chem}(x) \tag{48}$$

is the dipole potential-independent potential energy profile through the membrane. The integration limits in Eq. (47) go from $-r$ to $d+r$ for ions that do not absorb strongly into the membrane, and from δ to $d-\delta$ for ions that do absorb strongly. If one assumes that P_n changes proportionally for all current carriers as a function of the cholesterol mole fraction, and that the major barrier for ion movement is the membrane interior, one can separate the measured conductance changes into their relative dipole potential-dependent and dipole potential-independent contributions (SZABO, 1974, 1976, ANDERSEN et al., 1978b). The estimated dipole potential increased by 100–120 mV when going from cholesterol-free bilayers to bilayers with a cholesterol mole fraction of about 0.9. P_n decreased about three-fold. This rather small effect reflects both mobility terms and changes in membrane thickness (see Fig. 3). The estimated thickness of cholesterol-free glycerol-monooleate/n-decane membranes is 5 nm, while the thickness of cholesterol-containing membranes is less (as estimated from the capacitance data of BENZ et al., 1977). One would thus predict that P_n would *increase* if no "viscosity" changes occurred. The mobility of the current carriers in cholesterol-rich membrane may thus be very much less than in the cholesterol-free membranes, a result expected from nonelectrolyte permeability studies (Sections D. II and III).

In contrast to the marked effects of cholesterol on the translocation rate of TØB⁻ (and of trinactin-K^+ complexes [SZABO, 1977]) almost no effects are observed on the adsorption coefficients of TØB⁻ into glycerolmonooleate bilayers (SZABO, 1976). K_{ads} decreases by at most a factor of 2.

The effect of cholesterol on TØB⁻ in phosphatidylcholine membranes is particularly striking, as G (0,0) *increases* while K_{ads} *decreases* (ANDERSEN, 1978a). This can be compared with the more than 200-fold decrease in nonactin-K^+ conductance observed by HLADKY and HAYDON under similar circumstances. Part of the conductance increase observed with TØB⁻ can probably be attributed to changes in membrane thickness (BENZ and LÄUGER, 1977); but this fails to account for the large drop in K^+-nonactin conductance. It thus appears that the presence of cholesterol in phosphatidylcholine bilayers both *increases* the interfacial dipole potential and decreases P_n. It is, however, problematical that the surface potential of phosphatidylcholine monolayers at the air-water interface is more positive than that in monolayers of phosphatidylcholine-cholesterol (molar ratio 1:4) (HLADKY and HAYDON, 1973; ANDERSEN et al., 1976). The reader is referred to ANDERSEN et al. (1976) for a discussion of other discrepancies between surface potential changes in cholesterol-containing monolayers and bilayers. Possibly the neglect of the potential contribution from the hydrocarbon tail-air interface is unwarranted.

The marked decrease in K_{ads} for TØB⁻ with increasing cholesterol content in phosphatidyl-choline membranes (ANDERSEN, 1978a) is similar to the effect reported by SIMON et al., (1977a) for hexane in phosphatidylcholine liposomes. This may indicate that TØB⁻ is located some distance (a few tenths of 1 nm) into the hydrocarbon interior of the bilayer. TØB⁻ may possibly be located deeper into the membrane than DpA⁻, as K_{ads} for this ion is marginally affected by the presence of cholesterol (BENZ and LÄUGER, 1977).

III. Positive Ions

The behavior of positive hydrophobic ions is much less understood than that of their negative counterparts. A number of investigators have used the small signal conductance of positive ions to characterize the properties of lipid bilayer membranes (LIBERMAN and TOPALY, 1969; LE BLANC, 1970; ANDERSEN and FUCHS, 1975; SHCHIPUNOV and BOGUSLAVSKI, 1975; BUGOSLAVSKI et al., 1976; SZABO, 1976). The conductance of TØB⁻ has, for example, been compared to the conductance of its positive counterparts, TØP⁺ and TØAs⁺. It is thus possible to estimate the absolute magnitude of the interfacial dipole potential (LE BLANC, 1970; ANDERSEN and FUCHS, 1975, see Section C.V. 2).

The behavior of TBA⁺ has been investigated in some detail by LÜSCHOW et al. (1975) in membranes formed from oxidized cholesterol without support electrolyte in the aqueous phases. The conductance vs. concentration characteristics were linear, and the current-voltage characteristics were strikingly superlinear.

The results were analyzed by means of an electrodiffusion regime with a flat potential energy barrier. The nonlinearity was ascribed to potential-dependent changes in the concentration of the permeant ions in the aqueous phases adjacent of the membranes, the so-called ion-injection mechanism (WALZ et al., 1969). The charge injected to change the potential across the membrane capacitance will increase the concentration of positive ions at the positive side of the membrane and decrease their concentration at the negative side (EVERITT and HAYDON, 1968; WALZ et al., 1969; LÜSCHOW et al., 1975). This mechanism predicts nonlinear current-voltage characteristics at ionic strengths $<10^{-3}$M. The magnitude of the nonlinearity should be dependent upon the ionic strength, however, and this was not observed. The observed current-voltage characteristics are, on the other hand, consistent with the existence of a potential energy barrier in the membrane interior.

The kinetics of translocation of CC5 through glycerolmonooleate membranes has been investigated by WAGGONER et al. (1977) in an attempt to elucidate the mechanism by which these dyes sense changes in the electrical field across the membrane. The results indicate a rapid translocation between two interfacial potential energy wells, and a slower aqueous diffusion layer-limited process. The stationary conductance is therefore only in part limited by the membrane itself. These results are rather different from those of KRASNE (1977) in phosphatidylethanolamine/n-decane membranes. The current-voltage characteristics in these membranes are far too steep to be explained by the movement of single ions through the membrane. It is therefore suggested that cyanine dyes can form transmembrane channels that have voltage-dependent association-dissociation rates. The different results of WAGGONER et al. and of KRASNE may just represent the effect of the dipole potential on the behavior of charged molecules in the membranes. The conductance due to the translocation of single cations may be 500 to 1000 times higher in glycerolmonooleate membranes than in phosphatidylethanolamine membranes (HLADKY, 1974; ANDERSEN et al., 1978b). The relative contribution of the single ion mechanism will thus be much larger in glycerolmonooleate membranes.

IV. Interactions Among Ions Absorbed into Lipid Membranes

At high aqueous concentrations of hydrophobic anions, membrane conductance (LIBERMAN and TOPALY, 1969; LE BLANC, 1969; KETTERER et al., 1971; ANDERSEN and FUCHS, 1975; BRUNER, 1975; GAVACH and SANDAUX, 1975; BENZ et al., 1976; SZABO, 1976; ANDERSEN et al., 1978) and charge adsorption (GRIGOREV et al., 1972; BRUNER, 1975; BENZ et al., 1976; ANDERSEN et al., 1977; WANG and BRUNER, 1977a and b; ANDERSEN et al., 1978) deviate from the linear behavior observed at lower concentrations. The instantaneous conductance may even go through a maximum (see Fig. 8). Furthermore, one finds that the conductance of a lipid bilayer in the presence of both a negative and a positive current carrier is much larger than the sum of the membrane conductances observed when only one of the current carriers is present (LIBERMAN and TOPALY, 1968a and b,

1969; HINKLE, 1970; MARKIN et al., 1971; DEMIN et al., 1974; ANDERSEN et al., 1978b).

Associated with the deviation of conductance and charge absorption from linearity one also observes that: (a) the relaxation time constant increases with increasing charge absorption (ANDERSEN and FUCHS, 1975; BRUNER, 1975; ANDERSEN et al., 1978a; WULF et al., 1977; ANDERSEN et al., 1978b), and the current transients may even become nonexponential at high absorbed charge densities (ANDERSEN and FUCHS, 1975; ANDERSEN et al., 1978a and b; FELDBERG and DELGADO, 1978); (b) that increasingly large potentials must be applied to move a given fraction of the absorbed charge through the membrane (ANDERSEN et al., 1978a and b); (c) that the absorption of hydrophobic anions is associated with a change in the potential within the membrane in a negative direction (ANDERSEN et al., 1978a and b); and (d) that similar potential changes have been found in phospholipid monolayers at the air-water interface (BABAKOV et al., 1972; ANDERSEN, 1978a; ANDERSEN et al., 1978b).

All of these phenomena indicate that appreciable interactions may occur among ions absorbed into lipid bilayers. An elucidation of the interactions observed in these simple systems, may be of value in understanding the behavior of charged or dipolar molecules in biological membranes. The emphasis will be upon interactions among an ions. The very interesting interactions which occur among anions and cations are less well studied and more problematical to interpret.

1. Space Charge-Limited Conductance

Le BLANC (1969) and LIBERMAN and TOPALY (1968a, 1969) proposed that the interactions were due to changes in the electrostatic potential in the membrane, produced by the absorption of charges into the membrane interior. Electrostatic interactions are, as emphasized by Le BLANC (1969), of longer range than most other interactions among molecules in a membrane. The potential around an ion will vary in proportion to $(\text{distance})^{-1}$ if one neglects the induced polarization charges, and as $(\text{distance})^{-2}$ if one assumes ideal imaging in only one interface. Le BLANC neglected these complications and suggested that the interactions could be quantitated by a model that assumes that the conductance is space charge-limited. This model is of dubious physical significance, however, as emphasized by Le BLANC himself (quoted in HAYDON and HLADKY, 1972), as the estimated physical separation of ions within the membrane, according to this theory, was about 30 nm. This distance is hardly small compared to a membrane thickness of 3–5 nm. Indeed it is known (VON LAUE, 1918; HERRING and NICHOLS, 1949, Appendix 1) that the theory of space charge-limited conductance is only a valid approximation when x is large enough to satisfy

$$\frac{e^2}{4 \cdot \varepsilon_0 \cdot \varepsilon_r \cdot (2x)^2} < e \cdot \frac{d\varphi}{dx} \qquad (49)$$

and the potential energy of an ion due to its interactions with the interface is less than kT. The membrane thickness must be on the order of 100 nm to satisfy Eq. (49). The important point, however, is that the model emphasized the electrostatic interactions among ions within the membrane. The model can also account for the experimentally observed deviations in charge absorption from ideal behavior (LIBERMAN and MARGULIS, 1974), but it will not predict a maximum in conductance, as can be seen from the analyses of NEUMCKE and LÄUGER (1970), de LEVIE et al. (1972), and MARGULIS (1974).

2. Blocking Phenomena

An alternative model based upon the "blocking" theory of BRUNER (1970) was proposed by KETTERER et al. (1971), on the basis of their relaxation measurements on DpA^- and $TØB^-$. According to this model there exists a limited number, N_s, of binding sites for hydrophobic ions at the membrane-solution interface. The probability of an ion crossing the membrane from left to right will depend upon the number of empty spaces on the right $(N_s - N_{II})/N_s$. This model predicts the observed maximum in instantaneous membrane conductance, as well as a not observed maximum in the stationary conductance (BRUNER, 1970; KETTERER et al., 1971; HAYDON and HLADKY, 1972). The time-course of the on-response current transient will be prolonged and in general nonexponential at high adsorbed charge densities[22]. The time-course of the off-response will, however, be exponential at all adsorbed charge densities, and the relaxation time constant will be invariant with adsorbed charge density, which is not observed.

[22] According to the formalism of KETTERER et al. (1971) one can describe the kinetics of intramembrane charge translocation as

$$I = z \cdot e \cdot k_i \cdot \exp\left(\frac{e \cdot z \cdot V}{kT}\right) \cdot N_I \cdot \left(1 - \frac{N_{II}}{N_s}\right) - \exp\left(\frac{-ezV}{kT}\right) \cdot N_{II} \left(1 - \frac{N_I}{N_s}\right)$$

and

$$I = -z \cdot e \cdot \frac{dN_I}{dt} = z \cdot e \cdot \frac{d\Delta N}{dt}$$

where

$$\Delta N = N_I(0) - N_I(t) = -(N_{II}(0) - N_{II}(t))$$

By substitution one arrives at the following equation which describes the time course of ion translocation through the membrane:

$$\frac{d(\Delta N)}{dt} = 2 \cdot k_i \cdot \left\{\sinh\left(\frac{z \cdot e \cdot V}{2\,kT}\right) \cdot \left[N_I(0) \cdot \left(1 - \frac{N_I(0)}{N_s}\right) + \frac{(\Delta N)^2}{N_s}\right] - \cosh\left(\frac{z \cdot e \cdot V}{2\,kT}\right) \cdot \Delta N\right\}$$

In the limit of $V_m = 0$ this equation reduces to a linear differential equation with a time constant given by Eq. (28). For $|V_m| > 0$ it is a non-linear differential equation of the Ricatti-type (FRANK

The limiting charge density for TØB⁻ was estimated using a Langmuir adsorption isotherm. The result was that N_s is about 8×10^{-7} C cm^{-2} or 5×10^{12} sites cm^{-2} (1 ion/20 nm^2); a similar value was obtained for DpA⁻. These values are surprisingly small, less than those determined by ZINGSHEIM and HAYDON (1973) on glycerolmonooleate/n-decane membranes for 1-anilino-8-naphtalene sulphonate (ANS⁻) ($N_s = 3 \times 10^{13}$ cm^{-2}), and by HUANG and CHARLTON (1972) on phosphatidylcholine vesicles for 2,6-toludinyl-naphthalene sulfonate (TNS⁻) ($N_s = 2 \times 10^{13}$ cm^{-2}). These values were, however, determined by fitting to a simple Langmuir absorption model, which disregards electrostatic interactions among the adsorbed ions. This model is, however, of limited validity when the adsorption of charged molecules is being studied (SCATCHARD et al., 1950; EDSALL and WYMAN, 1958; MCLAUGHLIN, 1977). The electrostatic potentials produced by the adsorption of the charges will cause significant interactions to occur long before saturation of the number of sites occurs. If one incorporates the self-limitation on adsorption that arises from diffuse aqueous double-layer potentials calculated according to the Gouy-Chapman theory, values of N_s that are two to six times higher than the above estimates for TNS⁻ and ANS⁻ (HAYNES and STAERK, 1974; MCLAUGHLIN and HARARY, 1976) are obtained. The adsorbed charge densities observed with DpA⁻ and TØB⁻ are not large enough to explain the observed deviations by the appearance of aqueous double-layer potentials (KETTERER et al., 1971; ANDERSEN et al., 1978b). The inter-

and VON MISES, 1935, Chapter VI), and can thus be solved analytically. The resulting expression for the current is:

$$I(V, t) = I(V,0) \cdot \exp(-t/\tau(V)) \cdot \left\{ \frac{(\lambda_1 - \lambda_2)}{(\lambda_1 \cdot \exp(\lambda_2 \cdot t) - \lambda_2 \cdot \exp(\lambda_1 \cdot t))} \right\}$$

where

$$I(V,0) = e \cdot 2 \cdot k_i \cdot \cosh\left(\frac{eV}{2\,kT}\right) \cdot N_I(0) \cdot \left(1 - \frac{N_I(0)}{N_s}\right),$$

τ is expressed by Eqs. (27) and (28),

$$\lambda_1 = -k_i \cdot \cosh\left(\frac{e \cdot V}{2 \cdot k \cdot T}\right) \cdot (1 + \sqrt{1 - z}),$$

$$\lambda_2 = -\, k_i \cdot \cosh\left(\frac{e \cdot V}{2 \cdot k \cdot T}\right) \cdot (1 - \sqrt{1 - z}),$$

and

$$z = 4 \cdot \left(\tanh\left(\frac{e \cdot V}{2 \cdot k \cdot T}\right)\right)^2 \cdot \frac{N_I(0)}{N_s} \cdot \left(1 - \frac{N_I(0)}{N_s}\right)$$

That is, the current transient in the presence of a "block" will be non-exponential and slower than the current transient in the "unblocked" situation.

actions can be due to electrostatic potential changes in the membrane interior, however, if the charges absorb into a region with a low dielectric constant (MARKIN et al., 1971; see also Section 3).

3. The Three-Capacitor Model

There is considerable evidence that hydrophobic anions absorb into the low dielectric constant interior of lipid bilayers. Hydrophobic ions absorbed into a bilayer and their counterions in the aqueous diffuse double layer are separated by a region with a relatively low dielectric constant. It is, for simplicity, assumed that the charge associated with the ions absorbed into the membrane, $z \cdot q^o$, is smeared uniformly over a plane located some distance, δ, into the membrane. The charge associated with counterions in the aqueous phases, $-z \cdot q^o$, is assumed to be smeared over a plane one Debye length, L_o, from the membrane-solution interfaces (see Fig. 13). The four charged planes at $x = -L_o$, $x = \delta$, $x = d-\delta$, and $x = d + L_o$, thus define three dielectric regions: two outer regions and an inner region (see Fig. 13). One can associate a specific capacitance with each region, two outer capacitances, C_o, and an inner capacitance, C_i, (ANDERSEN, et al., [1978b] should be consulted for a detailed definition of these terms.) C_o will be the series combination of the capacitance of the aqueous diffuse double layer (between $x = -L_o$ and $x = 0$) and the capacitance of the

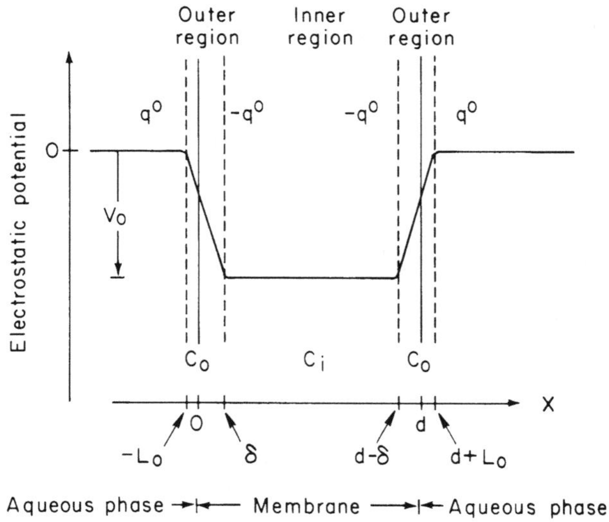

Fig. 13. Schematic representation of the three-capacitor model. The loci of the adsorbed ions are at $x = \delta$ and $x = d - \delta$, respectively. The loci of aqueous countercharges are at $x = -L_D$ and $x = d + L_D$ respectively. The *change* in electrostatic potential due to adsorbed TØB$^-$, in a membrane at equilibrium, is denoted by the solid line (it is for simplicity assumed that the dielectric coefficient of the outer regions is constant). Note that the major change in the boundary potential occurs in the membrane phase, and that the interfacial dipole potentials have been neglected, as the three-capacitor model describes potential *changes*

outer region of the membrane (between $x = 0$ and $x = \delta$). This model is similar to that proposed by STERN (1924) to account for specific adsorption at the electrode-electrolyte solution interface (De LEVIE, 1977; ANDERSEN et al., 1978b).

a) Charge Adsorption

The three-capacitor model was first developed by GRIGOREV, MARKIN and YERMISHKIN and was described in two important papers: MARKIN et al., (1971); GRIGOREV et al., (1972), to account for the potential *changes* that occur across the outer region of the membrane upon adsorption of $T\O B^-$ into bilayers. This potential *change*, the *boundary potential*, V_o, can be defined as

$$V_o = \frac{z \cdot q^o}{C_o}. \tag{50}$$

If, for example, $C_o = 35\ \mu F\ cm^{-2}$ (ANDERSEN et al., 1978b), and $q^o = 1.4\ \mu C\ cm^{-2}$, then $V_o = -40$ mV. The diffuse double-layer potential, ψ on the other hand, will be -6.1 mV when the ionic strength of the aqueous phase is 1.0 M. At high ionic strength the potential change within the membrane is thus much larger than the potential change in the aqueous phases.

The boundary potential will affect the adsorption of hydrophobic ions into the potential energy wells (MARKIN et al., 1971; ANDERSEN et al., 1978a and b). If one combines Henry's Law and the Boltzmann relation the adsorption should vary as:

$$q^o = K \cdot F \cdot [X^-] \cdot \exp\left(\frac{e \cdot V_o}{kT}\right) \tag{51}$$

or

$$q^o = K \cdot F \cdot [X^-] \cdot \exp\left(\frac{-e \cdot q^o}{kT \cdot C_o}\right), \tag{52}$$

where F is Faraday's constant and the valency is assumed to be -1. (For a discussion of adsorption isotherms, and the combination of Henry's law and the Boltzmann relation, see AVEYARD and HAYDON, 1973, Chap. 1 and 2). Experimental tests of charge adsorption as a function of aqueous $T\O B^-$ concentration show that Eq. (52) can describe the data quite well (GRIGOREV et al., 1972; ANDERSEN et al., 1978a and b). This model can also predict the deviations in DpA^- absorption from Henry's law observed by BRUNER (1975) (McLAUGHLIN, 1977, Fig. 20). The agreement between theory and experiment is, in fact, somewhat better for DpA^-[23].

[23] WANG and BRUNER (1977a and b) have suggested that the magnitude of C_o for the adsorption of DpA^- into dioleoylphosphatidylcholine/n-decane bilayers, is entirely due to the capacitance of the aqueous diffuse double layer. That is, they interpret their adsorption data using a combina-

Independent evidence that the absorption of $TØB^-$ into lipid bilayers produces boundary potentials of appropriate sign and magnitude has been provided by measurements of the changes in anion or cation conductance produced by the absorption of $TØB^-$ (ANDERSEN et al., 1978a and b). Surface potential measurements on phospholipid monolayers spread at the air-water interface likewise demonstrate that the absorption of $TØB^-$ produces large changes in the boundary potential of appropriate sign (BABAKOV et al., 1972; ANDERSEN et al., 1978b; see also Fig. 14). Qualitatively similar potential changes presumably occur in lipid bilayers.

b) Charge Translocation

The dependence of Δq_c on V_m is also predicted by the three-capacitor model. The fraction of the applied potential that is effective in translocating ions between the two potential energy wells, b, is defined as

$$b = 1 - \frac{2 \cdot C_m}{C_o} = \frac{C_m}{C_i}. \tag{53}$$

Formally b is equivalent to β (see Section E. II. 2. d. γ). The effective potential translocating charge through the membrane interior, V_i, is however not equal to $b \cdot V_m$ but rather to

$$V_i = b \cdot V_m - \frac{(1-b) \cdot \Delta q_c}{C_m} = b \cdot V_m - 2\Delta V_o, \tag{54}$$

where

$$\Delta V_o = \frac{\Delta q_c}{C_o} \tag{55}$$

is the *change* in boundary potential associated with the translocation of charge through the membrane. Two important consequences arise from the dependence of V_i on Δq_c:

(Footnote 23, cont.)

tion of Eqs. 19, 20 and 51. This interpretation leads them to the conclusion that ε_r^{aq} near the membranes decreases, as the magnitude of the surface potential (and the associated field) increases. This interpretation is certainly consistent with their adsorption data, but the conclusion is in quite severe disagreement with a larger number of studies on the adsorption of amphipathic ions at lipid bilayers where is no evidence for such changes in ε_r^{aq} (for a comprehensive review see MCLAUGHLIN, 1977; see also Section E.II.2.d.β.i, and ANDERSEN et al., 1978b). It is, indeed, possible to predict the shape of the experimental DpA^- adsorption isotherms (and their dependence upon the aqueous electrolyte concentration) quite satisfactorily if one assumes that the boundary potential is the sum of (a) the boundary potential proper (between $x = 0$ and $x = \delta$) and (b) a simple aqueous diffuse double layer potential with a constant ε_r^{aq} (ANDERSEN, unpublished analysis of data supplied by Dr. L. J. BRUNER). The limiting value of C_o at very high ionic strengths is 110 $\mu F\,cm^{-2}$, independent of the electrolyte (NaCl, $BaCl_2$, $MgSO_4$). There is therefore no need to assume that ε_r^{aq} varies as a function of q^o.

Firstly, the charge translocation as a function of potential is no longer described by Eq. (32), which neglects the existence of boundary potentials, but rather by

$$\Delta q_c = \Delta q_{c,\,max} \cdot \tanh\left(\frac{e \cdot V_i}{2kT}\right)$$

$$= \Delta q_{c,\,max} \cdot \tanh\left(\frac{e}{2kT} \cdot \left(b \cdot V_m - \frac{(1-b) \cdot \Delta q_c}{C_m}\right)\right). \tag{56}$$

The two formalisms become equivalent when $\beta = b = 1.0$ or when q^o is so small that $(1-\beta)$. $\beta \cdot q^o \cdot e/(C_m \cdot kT) << 1$ (ANDERSEN et al., 1978b). The correction term in Eq. (56) may be considerable (up to 57 mV, if $\Delta q_{c,\,max} = 1$ μC cm^{-2}, $b = 0.97$, and $C_m = 0.53$ μF cm^{-2}). The shape of the Δq_c vs. V_m relations will thus be profoundly affected by the existence of boundary potentials (see ANDERSEN et al., 1978b; (Figs. 7 and 10)). The potential necessary to move 90 percent of $\Delta q_{c,\,max}$ through the bilayer will increase from about 100 mV at 10^{-7} M TØB$^-$ to about 200 mV at 3×10^{-6} M TØB$^-$ (ANDERSEN et al., 1978a, Table I; ANDERSEN et al., 1978b, Fig. 7).

Secondly, the three-capacitor model predicts that the relaxation time constant for the off-response, $\tau(0)$, will vary as a function of $\Delta q_{c,\,max}$ (FELDBERG and DELGADO, 1978):

$$\tau(0, \Delta q_{c,\,max}) = \frac{\tau(0,0)}{1 + \dfrac{e \cdot \Delta q_{c,\,max} \cdot (1-b)}{2\,kT \cdot C_m}}, \tag{57}$$

where $\tau(0,0)$ is the time constant in the limit of $\Delta q_{c,\,max} = 0$, while $\tau(0, \Delta q_{c,\,max})$ is the time constant observed at finite $\Delta q_{c,\,max}$. An analysis of $\Delta q_{c,\,ss}$, from the small-signal measurements will thus result in systematic deviations from $\Delta q_{c,\,max}$:

$$\Delta q_{c,\,ss} = \beta \cdot \Delta q_{c,\,max} \cdot \frac{1}{1 + \dfrac{e \cdot \Delta q_{c,\,max}(1-b)}{2\,kT \cdot C_m}} \tag{58}$$

Table 5. Comparison of $\Delta q_{c,\,max}$ and $\Delta q_{c,\,ss}$

TØB$^-$ M	$\Delta q_{c,\,max}$ μC cm^{-2}	$\Delta q_{c,\,ss}$ μC cm^{-2}	$\Delta q_{c,\,ss}$ / $\Delta q_{c,\,max}$	Predicted ratio, Eq. 58
1×10^{-8}	$3.5 \pm 0.8 \times 10^{-2}$	$3.1 \pm 0.6 \times 10^{-2}$	0.87	0.93
3×10^{-8}	$9.5 \pm 1.3 \times 10^{-2}$	$8.1 \pm 1.2 \times 10^{-2}$	0.85	0.88
1×10^{-7}	$3.5 \pm 0.6 \times 10^{-1}$	$2.5 \pm 0.4 \times 10^{-2}$	0.71	0.71
3×10^{-7}	$7.6 \pm 0.6 \times 10^{-1}$	$4.5 \pm 0.2 \times 10^{-1}$	0.59	0.53
1×10^{-6}	1.43 ± 0.13	$5.0 \pm 0.5 \times 10^{-1}$	0.35	0.39

Experiments on bacterial phosphatidylethanolamine/n-decane membranes with TØB$_-$. $\Delta q_{c,\,max}$ from ANDERSEN et al. (1978), $\Delta q_{c,\,ss}$ from the same experiments (ANDERSEN, unpubl.). The values are mean ± S. D. on at least 5 different membranes. $b = 0.97$.

Table 5 lists the deviations observed for TØB⁻. The discrepancy between the estimates of $\Delta q_{c,\,max}$ and of $\Delta q_{c,\,ss}$ may be similar to that noted by BRUNER (1975) for DpA⁻ in dierucylphasphatidylcholine/n-decane membranes, see Table 3. WULF et al. (1977) also found very significant deviations between the amount of DpA⁻ adsorption estimated by an optical method, which presumably measures $q°$, and the charge adsorption estimated by a small-signal analysis. The $\Delta q_{c,\,ss}/\Delta q_{c,\,max}$ ratio observed by WULF et al. (1977) can be fitted by Eq. (58) if one assumes $C_o \approx 50$–$70\ \mu F/cm^{-2}$ for DpA⁻ adsorption into dierucylphosphatidylcholine/n-decane membranes.

At very low charge densities, there should be no interactions among the adsorbed species, and β should equal b. This is not observed, although one finds that the value of b is low ($b = 0.93$ at $[TØB^-] = 10^{-8}M$) than at high charge densities ($b = 0.97$ at $[TØB^-] = 10^{-7}M$) (ANDERSEN et al., 1978b, Table I). This difference is not by itself very impressive, but it signifies a change in $(1-b)$ from 0.07 to 0.03, that is a two-fold variation in the estimated magnited of the outer capacitance (see Eq. 53). Such a variation in b may reflect the fact that this parameter is used to quantitate two, possibly quite different, aspects of charge adsorption and intramembrane charge translocation. Firstly, $b \cdot V_m$ is used to denote the fraction of applied potential which falls between the two adsorption planes before any charge translocation has occurred, $2 \cdot C_m \cdot (1-b)^{-1}$ is thus the specific *geometric* capacitance of the outer region of the bilayer. Secondly, C_o is used to quantitate the electrostatic interactions which exist among the adsorbed ions, Eq. (50). According to the three-capacitor model, with its assumption that the adsorbed charge density is *smeared uniformly* over the adsorption plane, one can equate these two expressions for the outer capacitance through Eq. (53). This assumption will clearly be satisfied at very high adsorbed charge densities. It is not clear, however, whether a charge density of $0.35\ \mu C \cdot cm^{-2}$ (corresponding to an average lateral separation of ≈ 7 nm) is "very high" when the ions are adsorbed into a membrane with a thickness of about 4 nm. There is, indeed, some evidence that charges absorbed into the low dielectric constant membrane interior should be regarded as discrete entities (see Section E.IV.4), in which case b and β in general will estimate quite different parameters. β will be related to the *geometric capacitance* of the outer region of the bilayer, it can only be unambiguously estimated at "low" adsorbed charge densities. b, on the other hand, will estimate a weighted average of the geometric capacitance of the outer region and the bilayer and the *effective interaction capacitance* of the outer region, as determined by electrostatic interactions among the adsorbed ions. These two capactiances may differ significantly if discrete charge effects occur, the interaction capacitance will be much larger than the geometric capacitance (GRAHAME, 1958, BARLOW and MACDONALD, 1964).[24]

[24] *If* one assumes that such discrete-charge effects exist, then one can use the data of ANDERSEN et al. (1978b) to estimate the *effective interaction capacitance* to be about $70\ \mu F/cm^2$ (from an analysis of the adsorption isotherm) which should be compared with an estimate of the *geometric capacitance* of the outer region, 7–8 $\mu F/cm^2$ (from an analysis of the surface potential changes at the air/water interface, see ANDERSEN et al. (1978b) footnote 10). If one accepts this analysis one may then estimate β to be 0.85–0.87 ($C_m = 0.53\ \mu F/cm^2$), in reasonable agreement with estimates based upon the $\Delta q_{c,\,ss}/\Delta q_{c,\,max}$ ratio (Table 5) or the simple $\Delta q_c\,(V)$ vs V analysis (Fig. 11), at low adsorbed charge densities.

4. Discrete Charge Effects?

The three-capacitor model can account for many of the interactions observed among hydrophobic ions absorbed into lipid bilayer membranes. It does not, however, account for the maximum observed in the instantaneous conductance (Fig. 8) or for the associated increase in relaxation time constant (ANDERSEN et al., 1978b). Similar conductance maxima have been reported for $TØB^-$ by ANDERSEN and FUCHS (1975) and GAVACH and SANDAUX (1975) and for DpA^- by KETTERER et al. (1971), BRUNER (1975), and GAVACH and SANDAUX (1975). The time constant changes have been reported by ANDERSEN and FUCHS (1975) for $TØB^-$, and by BRUNER (1975), BENZ et al. (1976), and WULF et al. (1976) for DpA^-. These phenomena occur in glycerolmonooleate, phosphatidylethanolamine and phosphatidylcholine membranes, and are seen with chemically quite dissimilar ions. They are thus general, and may contribute important additional information about the interactions among hydrophobic ions absorbed into lipid bilayer membranes.[25]

The time constant changes are qualitatively of the wrong sign compared to those predicted by the three-capacitor model (Eq. 57). This change in $\tau(0)$ is, however, qualitatively consistent with the observations that the potential changes estimated by negative probes are larger than those estimated from the deviations from Henry's law through Eq. (51). The change in $\tau(0)$, and the quantitative discrepancies that exist among the various estimates of V_0, may also be related to the observation that the surface potential change upon absorption of $TØB^-$ in a phospholipid monolayer at the air-water interface varies by more than 2.303 kT/e per decade of $TØB^-$ concentration (ANDERSEN et al., 1978; see also Fig. 14). These results are reminiscent of predictions of the so-called *discrete charge* theories for specific adsorption at the interface between a medium with a high dielectric constant and a medium with a low dielectric constant, where the *measured potential* changes associated with the adsorption of ions into the interface appears to be much larger than the *average potential* experienced by the ions themselves (GRAHAME, 1958; BARLOW and MacDONALD, 1967).

According to the discrete charge theory the electrical field around an ion absorbed into a lipid bilayer will be the sum of the field due to the ion itself plus the field due to the induced polarization charges. The field along the plane of the membrane will therefore fall off faster than if the ion behaved like a monopole in an infinite medium. If, for example, the ion is located a distance, δ^*, from the membrane-solution interface, and the interface provides perfect imaging, then the ion will behave like a dipole with a dipole of $2 \cdot \delta^* \cdot e$. The interaction energy of an array of such ions will thus be that of an array of dipoles and *not* that of an array of monopoles. The net result is that the interaction energy, or the average electrostatic potential experienced by the ions, will be less than calculated for the original array of monopoles.

Any direct comparison between the predictions of a discrete charge model and experimental data obtained in lipid bilayer membranes becomes very com-

[25] HLADKY (1975) and KNOLL and STARK (1975) find that $\tau(0)$ increases with increasing amounts of trinactin-K^+ or valinomycin-K^+ absorbed into glycerolmonooleate membranes.

plex due to mathematical difficulties in describing the system. (BARLOW and MacDONALD [1964, 1967] and MacDONALD and BARLOW [1966] may be consulted for an analysis of a similar problem in the inner Helmholtz layer at the electrode-electrolyte interface.) A much more tractable system is that of charge absorption into a phospholipid monolayer spread at the air-water interface. In this case one needs only to consider the existence of one interface adjacent to the potential energy wells where the ions are absorbed. The potential change measured with an electrode much further away from the plane of adsorption than the average distance between the adsorbed ions is the macropotential, V^∞:

$$V^\infty = \Delta(\Delta V) = \frac{z \cdot q^o}{C_o^*}, \tag{59}$$

where C_o^* is the specific capacitance of the outer region of the bilayer. When the charge associated with the adsorbed ions is not uniformly smeared over the adsorption plane, the adsorption of ions will be determined by a micropotential, φ:

$$q^o = K^* \cdot F \cdot [\text{TØB}^-] \cdot \exp\left(\frac{-z \cdot e \cdot \varphi}{kT}\right), \tag{60}$$

where K* is the adsorption coefficient for TØB^- into a monolayer. The micropotential will, in general, be lower in magnitude than the macropotential. In

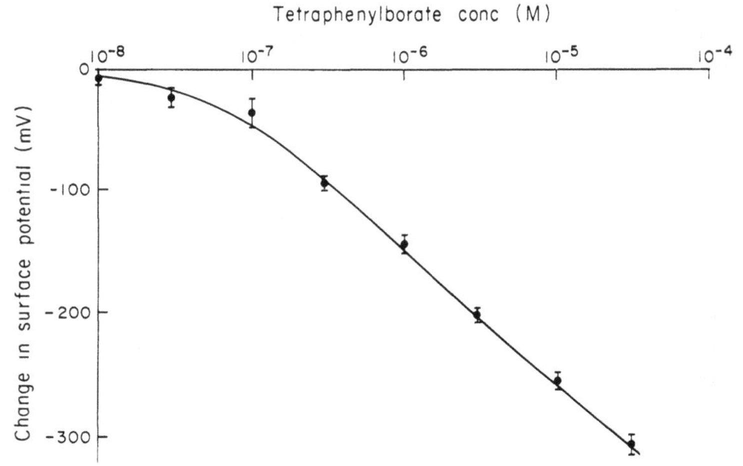

Fig. 14. Changes in surface potential, $\Delta(\Delta V) = V^\infty$ of phosphatidylethanolamine monolayers as a function of $[\text{TØB}^-]$. The potential of the clean interface of air/1.0 M NaCl + 0.1 M Na-phosphate buffer (pH = 7.3) was < −260 mV. The potential of the monolayer-covered interface was about +220 mV. The surface potential of the (hydrocarbon-free) monolayer is therefore +490 mV. Small aliquots of ethanolic Na TØB were added to the aqueous phase, and the surface potential reached a new stationary value after a few minutes. The resulting potential changes were added and plotted above. The points represent mean ± S. D. of 5 experiments. $T = 23°$ C. The solid line is calculated according to Eq. 61 with $\alpha = 0.62\ V^{-1/2}$ and $K^*/C_o^* = 5.9 \times 10^3 V \cdot cm^2\ con^{-1}$

terms of the three-capacitor model, with its assumption that the adsorbed charge is uniformly smeared, both V^{∞} and φ are equal to V_0. The purpose of the three-capacitor model is then to calculate φ in terms of the adsorbed charge density (that is V^{∞}).

A simple model, which encompasses the salient features of the discrete charge concept, has been developed by ANDERSEN et al. (1978b). It is assumed that the monolayer is a uniform slab of hydrocarbon with dielectric constant, ε_r^*, that the aqueous phase is a homogenous conductor, and that the interface is a mathematically planar, sharp, boundary. The thermal movement of the adsorbed ions is neglected, and the ions are assumed to be located in a regular, square or hexagonal, planar lattice at a distance δ^* from the monolayer-solution interface. Each charge will induce a polarization charge (of opposite sign) at the interface, and the electrostatic field variation around each ion in the monolayer will be identical to that of a dipole in a macroscopic medium. The surface potential as a function of TØB⁻ concentration can then be calculated to be (ANDERSEN et al., 1978b):

$$\Delta(\Delta V) = V^{\infty} = \frac{z \cdot K^* \cdot F}{C^*} \cdot [\text{TØB}^-] \cdot \exp\left(\frac{-e \cdot \alpha |V^{\infty}|^{3/2}}{kT}\right), \quad (61)$$

where α is a parameter dependent upon ε_r^* and δ^*. The solid line in Figure 14 is drawn according to Eq. (61) with the value of $\alpha = 0.62\ V^{-1/2}$ and $K^* / C^*_0 = 5.9 \cdot 10^3 \text{V} \cdot \text{cm}^2 \cdot \text{coul}^{-1}$. The fit ist good, but should not be taken to indicate that the discrete charge model is valid. It is important to note, however, that the data *cannot* be fitted by a model that assumes that the micropotential is a constant fraction of the macropotential. It is therefore unlikely that the observed discrepancies between the simple smeared model and the experimental results are due to a change in the pre-existing dipole potential due to the disordering produced by the adsorbed TØB⁻.[26]

The available results are thus consistent with the predictions of the discrete charge concept. It is, in fact, difficult to explain them by any other model. It should, however, be realized that this simple model makes strong (and probably unrealistic) assumptions about (1) the electrostatic interaction between the absorbed ions and their surroundings, (2) the thermal movement of the ions, both in the plane of the membrane and perpendicular to the membrane, and (3) the shape of the adsorption isotherm. In the absence of any alternative explanations, the discrete charge concept is a reasonable working hypothesis.

It is tempting to speculate that similar discrete charge effects may exist in lipid bilayer membranes[27]. It has previously been established (see Section E. II. 2)

[26] Results quantitatively similar to those in Fig. 14 have been observed by HALL and SIMON (pers. comm.). Qualitatively similar results have been reported by HAYDON and HLADKY (quoted in HLADKY, 1975, Appendix II) where the surface potential changes due to the absorption of trinactin-NH_4^+ complexes into glycerolmonooleate monolayers changes by 80 m V (from 27 to 107 mV) for a ten-fold increase in trinactin concentration (from 10^{-4}M to 10^{-3}M).

[27] Other investigators (HAYDON and MYERS, 1973; McLAUGHLIN and HARARY, 1976) have found good agreement between adsorption phenomena of amphipathic molecules at monolayers or lipid vesicles and electrostatic changes in lipid bilayers (see also Mc-LAUGHLIN, 1977).

that the electrostatic interactions between ions within the bilayer can qualitatively account for (a) the shape of the current-voltage characteristics, (b) the dependence of the relaxation time constant on membrane thickness and (c) the existence of potential energy minima for hydrophobic ions near the membrane solution interfaces. Each of these phenomena will, of course, depend upon other, nonelectrostatic, interactions between the ions and the membrane matrix. The available evidence does, however, provide considerable support for the notion that the electrostatic field around an ion within the bilayer is the sum of two separate contributions: (1) the spherical symmetrical field around the ion itself, and (2) the field due to the polarization charges at the membrane-solution interfaces. The magnitude of this second contribution may be significant if the "imaging" in the aqueous phases is good (see BARLOW and MacDONALD, 1967). The electrostatic interactions among ions absorbed into a lipid bilayer may thus be modified by the presence of image-charges. This effect is recognized in the theory of thermionic emission (VON LAUE, 1918; HERRING and NICHOLS, 1949, Appendix 1). According to these authors one can disregard such complications, when the membrane can be regarded as a macroscopic space-charge region. This restriction is probably unreasonably strong. What is important is the relative magnitude of the potential energy of an ion due to interactions with other ions in the membrane (neglecting boundary effects) compared to the potential energy due to interactions with the "images" of these other ions.

The strongest *a-priori* argument against the discrete charge model is that smeared models, *e. g.* the Gouy-Chapman theory of the diffuse double layer, are very successful in predicting the adsorption of amphipathic molecules at the membrane-solution interfaces, or the change in surface potential due to fixed charges as a function of the aqueous concentration of univalent and divalent electrolytes (McLAUGHLIN, 1977). This parallelism may, however, be misleading. The crucial property of the lipid bilayer is its thinness and its low dielectric constant, which makes interactions between the ion within the membrane and the adjacent interfaces much stronger than those usually observed in the diffuse double layer. But more importantly, the distortions of the electrostatic field around an ion in the membrane (a low dielectric constant medium adjacent to a high dielectric constant medium) are qualitatively quite different from the field distortions around an ion, in the aqueous diffuse double layer (a high dielectric-constant medium adjacent to a low dielectric-constant medium) (BROWN, 1974). The field distortion *within the membrane* will *minimize* lateral interactions among the absorbed charges as the major flux will be almost perpendicular to the aqueous phases. The field distortion in *the aqueous phase,* on the other hand, will *maximize* the interactions among neighboring charges. The physical mechanism that will contribute to discrete charge phenomena within the membrane will thus contribute to the adequacy of the smeared model in the aqueous diffuse double layer adjacent to the same membrane.

In summary: The smeared three-capacitor model provides a consistent description of many of the observed interactions among hydrophobic ions absorbed into a lipid bilayer membrane. The assumption that the charges are smeared leads to a drastic simplification of an otherwise intractable problem, and the three-capacitor model does predict the deviation of charge adsorption from

Henry's law, the discrepancy between charge adsorption as measured by small-signal analysis and a complete Δq_c vs V_m analysis (see Table 5) as well as other phenomena. The physical, as opposed to the mathematical, reasonableness of the assumption that charges are smeared is, however, dubious. The possibility that discrete phenomena may exist is of obvious importance for any attempts to study the properties of other polar or charged molecules in thin lipid membranes.

Acknowledgements

I would like to thank L. J. Bruner, S. W. Feldberg, A. Finkelstein, H. Ginsburg, H. A. Kolb, S. G. A. McLaughlin, V. A. Parsegian, S. A. Simon, P. Smejtek, G. Szabo, T. E. Thompson, and W. Veatch for helpful discussions and for providing manuscripts in advance of publication.

It is a pleasure to thank Rochelle Lester and Cheryl Martin for excellent assistance during the writing of this chapter.

This work was, in part, supported by a Senior Investigatorship from the New York Heart Association and an N. I. H. grant GM-21342.

References

ADAM, N. K.: The Physics and Chemistry of Surfaces. London: Oxford University Press (Reprinted by Dover Publications) 1968.
AIT'YAN, S. Kh., MARKIN, V. S., MALEV, V. V.: Biofizika **21**, 253 (1976a).
AIT'YAN, S. Kh., MARKIN, V. S., MALEV, V. V.: Biofizika **21**, 257 (1976b).
AIT'YAN, S. Kh., MARKIN, V. S., MALEV, V. V.: Biofizika **21**, 261 (1976c).
AL-ZAHID, G., SCHAFER, J. A., TROUTMAN, S. L., ANDREOLI, T. E.: J. Membrane Biol. **13**, 103 (1977).
ANDERSEN, O. S.: Macy Conference on Renal Function (in press) (1977) a, b). In: Renal Junction (G. N. Giebisch and E. Purcell, eds.) Post Washington, N. Y.: Independent Publishers Group.
ANDERSEN, O. S.: Abstract of the 27th International Congress of Physiological Sciences (1977b).
ANDERSEN, O. S.: (in preparation, for submission to Biophys. J. (1978).
ANDERSEN, O. S., FUCHS, M.: Biophys. J. **15**, 795 (1975).
ANDERSEN, O. S., FINKELSTEIN, A., KATZ, CASS, A.: J. gen. Physiol. **67**, 749 (1976).
ANDERSEN, O., FELDBERG, S., NAKADOMARI, H., LEVY, S., MCLAUGHLIN, S. In: Ion Transport Across Membranes – the Proceedings of a Joint U. S.-U. S. S. R. Conference (D. C. Tosteson, Yu. A. Ovchinnikow and L. Latorre, Eds). New York: Raven Press 1977, p. 327.
ANDERSEN, O. S., FELDBERG, S., NAKADOMARI, N., LEVY, S., MCLAUGHLIN, S.: Biophys. J. **21**, 35 (1978).
ANDREOLI, T. E., TOSTESON, D. C.: J. gen. Physiol. **57**, 526 (1971).
ANDREOLI, T. E., TROUTMAN, S. L.: J. gen. Physiol. **57**, 464 (1971).
ANDREOLI, T. E., BANGHAM, J. A., TOSTESON, D. C.: J. gen. Physiol. **50**, 1729 (1966).
ANDREWS, D. M., MANEV, E. D., HAYDON, D. A.: Spec. Discuss. Faraday Soc. **1**, 46 (1970).
AVEYARD, R., HAYDON, D. A.: An Introduction to the Principles of Surface Chemistry. London: Cambridge University Press (1973).

BABAKOV, A. V., ERMISHKIN, L. N., LIBERMAN, E. A.: Nature **210**, 953 (1966).
BABAKOV, A. V., MYAGKOV, I. V., SOTNIKOV, P. S., TEREKHOV, P.: Biofizika **17**, 347 (1972).
BAMBERG, E., LÄUGER, P.: J. Membrane Biol. **11**, 177 (1973).
BAMBERG, E., BENZ, R.: Biochim. biophys. Acta **426**, 570 (1976).
BANGHAM, A. D. In: Progress in Biophysics and Molecular Biology (J. A. V. Butler and D. Noble, Eds). Oxford and New York: Pergamon Press 1968, pp. 29–95.
BANGHAM, A. D., PAPAHADJOPOULOS, D.: Biochim. biophys. Acta **216**, 181 (1966).
BARDEEN, J.: Phys. Rev. **58**, 727 (1940).
BARLOW, C. A.: In: Physical Chemistry. An Advanced Treatise, Vol. IXA (H. Eyring, Ed.). New York: Academic Press 1970, pp. 167–246.
BARLOW, C. A., MacDONALD, J. R.: J. chem. Phys. **40**, 1535 (1964).
BARLOW, C. A., MacDONALD, J. R.: In: Advances in Electrochemistry and Electrochemical Engineering, Vol. 6 (P. Delanay, Ed.). New York: Interscience 1967.
BEAN, R. C., SHEPHERD, W. C., CHAN, H.: J. gen. Physiol. **52**, 495 (1968).
BELL, G. M., MINGINS, T., LEVINE, S.: Trans. Faraday Soc. **62**, 949 (1966).
BELL, R. P.: J. chem. Soc. **139**, 1371 (1931a).
BELL, R. P.: Trans. Faraday Soc. **27**, 797 (1931b).
BENZ, R., JANKO, K.: Biochim. biophys. Acta **455**, 721 (1976).
BENZ, R., LÄUGER, P.: Biochim. biophys. Acta **468**, 245 (1977).
BENZ, R., FRÖHLICH, O., LÄUGER, P., MONTAL, M.: Biochim. biophys. Acta **394**, 323 (1975).
BENZ, R., LÄUGER, P., JANKO, K.: Biochim. biophys. Acta **455**, 701 (1976).
BENZ, R., FRÖHLICH, O., LÄUGER, P.: Biochim. biophys. Acta **464**, 465 (1977).
BERNETT, M. K., JARVIS, N. L., ZISMAN, W. A.: J. phys. Chem. **68**, 3520 (1964).
BERNS, D. S.: Photochem. Photobiol. **24**, 117 (1976).
BLEANEY, B. I., BLEANEY, B.: Electricity and Magnetism, 2nd ed. London: University Press 1965.
BOCKRIS, J. O. M., REDDY, A. K. N.: Modern Electrochemistry, Vol. I. London: Macdonald (1970).
BOGUSLAVSKI, L. I., KRYLOV, V. S., SHCHIPUNOV, Yu., A.: Biofizika **21**, 266 (1976).
BORISOVA, M. P., ERMISHKIN, L. M., LIBERMAN, E. A., SILBERSTEIN, A. Y., TROFIMOV, E. M.: J. Membrane Biol. **18**, 243 (1974).
BRADSHAW, R. W., ROBERTSON, C. R.: J. Membrane Biol. **25**, 93 (1975).
BRANTON, D., DEAMER, D. W.: Protoplasmatologia II **E, 1**, 1 (1972).
BROCKERHOFF, H.: Lipids **9**, 645 (1974).
BROWN, R. H.: Prog. Biophys. molec. Biol. **28**, 343 (1974).
BRUNER, L. J.: Biophysik **6**, 241 (1970).
BRUNER, L. J.: J. Membrane Biol. **22**, 125 (1975).
BUNCE, A. S., HIDER, R. C.: Biochim. biophys. Acta **363**, 423 (1974).
CADENHEAD, D. A., KELLNER, B. M. J., MÜLLER-LANDAU, F.: Biochim. biophys. Acta **382**, 253 (1975).
CARROLL, B. J., HAYDON, D. A.: Faraday Trans. **12**, 361 (1975).
CASS, A., FINKELSTEIN, A.: J. gen. Physiol. **50**, 1765 (1967).
CASS, A., ANDERSEN, O. S., KATZ, I., FINKELSTEIN, A.: Biophys. Soc. Abst., 108a (1973).
CHANDRASEKHAR, S.: Rev. mod. Phys. **15**, 1 (1943) (Reprinted in: Selected Papers on Noise and Stochastic Processes [N. Wax, Ed.]. New York: Dover Publications 1954).
CHEN, L. A., DALE, R. E., ROTH, S., BRAND, L.: J. biol. Chem. **252**, 7, 2163 (1977).
CHIZMADZHEV, Yu. A., MARKIN, V. S., KUKLIN, R. N.: Biofizika **16**, 230 (Eng. trans. **16**, 235 [1971a]).
CHIZMADZHEV, Yu. A., MARKIN, V. S., KUKLIN, R. N.: Biofizika **16**, 437 (Eng. Trans. **16**, 451 [1971b]).
CIANI, S.: Biophysik **2**, 368 (1965).
CIANI, S. M., EISENMAN, G., LAPRADE, R., SZABO, G.: In: Membranes, Vol. 2 (G. Eisenman, Ed.). New York: Dekker 1973.
CIANI, S.: J. Membrane Biol. **30**, 45 (1976).
Cogan, U., SHINITZKY, M., WEBER, G., NISHIDA, T.: Biochemistry **12**, 521 (1973).
COHEN, B. E.: J. Membrane Biol. **20**, 205 (1975).
COHEN, B. E., BANGHAM, A. D.: Nature **236**, 173 (1972).
COHEN, F. S., EISENBERG, M., McLAUGHLIN, S.: J. Membrane Biol. **37**, 361 (1977).
COLE, K. S.: Biophys. J. **9**, 465 (1969).

COOK, G.M.W., REDWOOD, W.R., TAYLOR, A.R., HAYDON, D.A.: Kolloid Zeitschrift und Zeitschrift für Polymere **227**, 28 (1968).
COOPER, J.N., POWELL, R.E.: J. Amer. Chem. Soc. **85**, 1590 (1962).
CORBRIDGE, D.E.C.: The Structrual Chemistry of Phosphorus. Amsterdam: Elsevier 1974.
COSTER, H.G.L., SMITH, J.R.: Biochim. biophys. Acta **373**, 151 (1974).
CRANK, J.: Mathematics of Diffusion. London: Oxford University Press 1956.
DAINTY, J.: Botanical Res. **1**, 279 (1963).
DEAN, R.B., CURTIS, H.J., COLE, K.S.: Science **91**, 50 (1940).
de GEISO, R.C., RIEMAN, W., LINDENBAUM, S.: Anal. Chem. **26**, 1840 (1954).
de LEVIE, R.: J. electroanal. Chem. **69**, 265 (1976).
de LEVIE, R.: J. electroanal. Chem. **82**, 361 (1977).
de LEVIE, R.: Adv. Chem. Phys. **37**, 99 (1977b).
de LEVIE, R., ABBEY, K.M.: J. theoret. Biol. **56**, 151 (1976).
de LEVIE, R., VUKADIM, D.: J. electroanal. Chem. **62**, 95 (1975).
de LEVIE, R., SEIDAH, N.G., MOREIRA, H.: J. Membrane Biol. **10**, 171 (1972).
de LEVIE, R., SEIDAH, N.G., MOREIRA, H.: J. Membrane Biol. **16**, 17 (1974a).
de LEVIE, R., SEIDAH, N.G., LARKIN, D.: J. electroanal. Chem. **49**, 153 (1974b).
DEMIN, V.V., BABAKOV, A.V., SHKROB, A.M., OVCHINNIKOV, Yu.A.: Biofizika **19**, 661 (1974).
DIAMOND, J.M., KATZ, Y.: J. Membrane Biol. **17**, 121 (1974).
DIX, J.A., DIAMOND, J.M., KIVELSON, D.: Proc. nat. Acad. Sci. (Wash.) **71**, 474 (1974).
EDSALL, J.T., WYMAN, J.: Biophysical Chemistry, Vol. 1. New York: Academic Press 1958, pp. 591–660.
EIGEN, M., de MAYER, L.: Techniques of Organic Chemistry, Vol. 8 (A. Weissberger, Ed.). New York: Interscience 1963, pp. 895–967.
EISENMAN, G. (Ed.): Membranes, Vol. 2: Lipid Bilayers and Antibiotics. New York: Dekker 1973.
EISENMAN, G. (Ed.): Membranes, Vol. 3: Lipid Bilayers and Biological Membranes: Dynamic Properties. New York: Dekker 1975.
EISENMAN, G., SZABO, G., CIANI, S., MCLAUGHLIN, S., KRASNE, S.: In: Progress in Surface and Membrane Science, Vol. 6. New York: Academic 1973.
EMERSLEBEN, O.: Ann. Physik **82**, 713 (1927).
ERDEY-GRUZ, T.: Transport Phenomena in Aqueous Solutions. New York: Wiley 1974.
EVERITT, C.T., HAYDON, D.A.: J. theoret. Biol. **18**, 371 (1968).
EVERITT, C.T., HAYDON, D.A.: J. theoret. Biol. **22**, 9 (1969).
EVERITT, C.T., REDWOOD, W.R., HAYDON, D.A.: J. theoret. Biol. **22**, 20 (1969).
FELDBERG, S.W., DELGADO, A.B.: Biophys. J. **21**, 71 (1978).
FEYNMAN, R.P., LEIGHTON, R.B., SANDS, M.: The Feynman Lectures on Physics, Vol. 2. Reading: Addison-Wesley 1964.
FETTIPLACE, R., ANDREWS, D.M., HAYDON, D.A.: J. Membrane Biol. **5**, 277 (1971).
FETTIPLACE, R., GORDON, L.G.M., HLADKY, S.B., REQUENA, J., ZINGSHEIM, H.P., HAYDON, D.A.: In: Methods in Membrane Biology (E.D. Korn, Ed.). New York: Plenum 1975, pp. 1–95.
FINKELSTEIN, A.: Arch. int. Med. **129**, 229 (1972).
FINKELSTEIN, A.: J. gen. Physiol. **68**, 127 (1976a).
FINKELSTEIN, A.: J. gen. Physiol. **68**, 137 (1976b).
FINKELSTEIN, A.: J. gen. Physiol. **70**, 125, (1977).
FINKELSTEIN, A., CASS, A.: Nature **216**, 717 (1967).
FINKELSTEIN, A., CASS, A.: J. gen. Physiol. **52**, 145 (1968).
FISCHKOFF, S., VANDERKOOI, J.M.: J. gen. Physiol. **65**, 663 (1975).
FLEISCHER, S., PACKER, L.: Methods in Enzymology, Vol. 32, New York: Academic Press 1975.
FOSTER, M., MCLAUGHLIN, S.: J. Membrane Biol. **17**, 155 (1974).
FRANK, P., MISES, R. v.: In: Die Differential und Integralgleichungen der Mechanik und Physik. New York: Dover Publications (1961).
GABORIAUD, R.: C. R. Acad. Sci. (Paris) **C 263**, 911 (1966).
GAINES, G.L.: Insoluble Monolayers at Liquid-Gas Interfaces. New York: Interscience 1966.
GALLUCCI, E., MICELLI, S., LIPPE, C.: Arch. internat. Physiol. Biochem. **79**, 881 (1971).
GALLUCCI, E., MICELLI, S., LIPPE, C.: Nature **255**, 722 (1975).
GAVACH, C., SANDEAUX, R.: Biochim. biophys. Acta **413**, 33 (1975).
GINSBURG, S., NOBLE, D.: J. Membrane Biol. **29**, 211 (1976).

GODICI, P. E., LANDSBERGER, F. R.: Biochemistry **13**, 362 (1974).
GRAHAME, D. C.: Chem. Rev. **41**, 441 (1947).
GRAHAME, D. C.: J. Electrochem. Soc. **99**, 370C (1952).
GRAHAME, D. C.: Zeitschrift für Elektrochemie **62**, 264 (1958).
GRAY, D. E. (Ed.): American Institute of Physics Handbook. New York: McGraw-Hill 1972.
GRAZIANI, Y., LIVNE, A.: J. Membrane Biol. **7**, 275 (1972).
GREEN, K., OTORI, T.: J. Physiol. **207**, 93 (1970).
GRIFFITH, O. H., DEHLINGER, P. J., VAN, S. P.: J. Membrane Biol. **15**, 159 (1974).
GRIGOREV, P. A., YERMISHKIN, L. N.: Biofizika **21**, 385 (1976).
GRIGOREV, P. A., YERMISHKIN, L. N., MARKIN, V. S.: Biofizika **17**, 788 (1972).
GUGGENHEIM, E. A.: J. phys. Chem. **33**, 842 (1929).
GUGGENHEIM, E. A.: J. phys. Chem. **34**, 1540 (1930).
GUPTA, M. P., DUTTA, B. P.: Acta Crystallogr. B **31**, 1272 (1975).
GUTKNECHT, T., TOSTESON, D. C.: Science **182**, 1258 (1973).
GUTKNECHT, J., BRUNER, L. J., TOSTESON, D. C.: J. gen. Physiol. **59**, 486 (1972).
HALL, J. E., MEAD, C. A., SZABO, G.: J. Membrane Biol. **11**, 75 (1973).
HANAI, T., HAYDON, D. A.: J. theoret. Biol. **11**, 370 (1966).
HANAI, T., HAYDON, D. A., TAYLOR, J.: J. theoret. Biol. **9**, 278 (1965).
HANAI, T., HAYDON, D. A., REDWOOD, W. R.: Ann. N. Y. Acad. Sci. **137**, 731 (1966).
HANSEN, B. D., KARENBROT, J. I., HARRIS, A.: Biophys. J. **17**, 132a (1977).
HARE, F., LUSSAN, C.: Biochim. biophys. Acta **467**, 262 (1977).
HAYDON, D. A.: Kolloid-Zeitschrift und Zeitschrift für Polymere **185**, 148 (1962).
HAYDON, D. A. In: Recent Progress in Surface Science, Vol. 1. (J. F. Danielli, D. Parkhurst and A. C. Riddiford, Eds). New York: Academic Press 1964, pp. 94–158.
HAYDON, D. A.: In: Molecular Basis of Membrane Function. (D. C. Tosteson, Ed.). Englewood Cliffs, N. J.: Prentice-Hall 1969, pp. 111–131.
HAYDON, D. A.: In: Membranes and Ion Transport, Vol. 1. (E. E. Bittar, Ed.). New York: Wiley-Interscience 1970a, pp. 64–92.
HAYDON, D. A.: In: Permeability and Function of Biological Membranes (L. Bolis, A. Katchalsky, R. D. Keynes, W. R. Loewenstein, B. A. Pethica, Eds). Amsterdam: North-Holland 1970.
HAYDON, D. A.: Ann. N. Y. Acad. Sci. **264**, 2–16 (1975).
HAYDON, D. A., HLADKY, S. B.: Quart. Rev. Biophys. **5**, 187 (1972).
HAYDON, D. A., MYERS, V. B.: Biochim. biophys. Acta **307**, 429 (1973).
HAYNES, D. H., STAERK, H.: J. Membrane Biol. **17**, 313 (1974).
HELFFERICH, F.: Ion Exchange. New York: McGraw Hill 1962.
HENN, F. A., THOMPSON, T. E.: J. molec. Biol. **31**, 227 (1968).
HERRING, C., NICHOLS, M. H.: Rev. mod. Phys. **21**, 185 (1949).
HINKLE, P.: Biochem. biophys. Res. Commun. **41**, 1375 (1970).
HITCHCOCK, P. B., MASON, R., THOMAS, K. M., SHIPLEY, G. G.: Proc. nat. Acad. Sci. (Wash.) **71**, 3036 (1974).
HLADKY, S. B.: Biochim. biophys. Acta **307**, 261 (1973).
HLADKY, S. B.: Biochim. biophys. Acta **352**, 71 (1974).
HLADKY, S. B.: Biochim. biophys. Acta **375**, 327 (1975).
HLADKY, S. B., HAYDON, D. A.: Biochim. biophys. Acta **318**, 464 (1973).
HLADKY, S. B., GORDON, L. G. M., HAYDON, D. A.: Ann. Rev. Phys. Chem. **25**, 11 (1974).
HOFFMAN, K., WEISS, E.: J. organometallic Chem. **67**, 221 (1974).
HOLZ, R., FINKELSTEIN, A.: J. gen. Physiol. **56**, 125 (1970).
HONG, F.: Photochem. Photobiol. **24**, 155 (1976).
HUANG, C., CHARLTON, J. P.: Biochemistry **11**, 735 (1972).
HUANG, C., THOMPSON, T. E.: J. molec. Biol. **15**, 539 (1966).
HUANG, W.-T., LEVITT, D. G.: Biophys. J. **17**, 111 (1977).
HUBBELL, W. L., MCCONNELL, H. M.: J. Amer. Chem. Soc. **93**, 2: 314 (1971).
KATZ, Y., DIAMOND, J. M.: J. Membrane Biol. **17**, 87 (1974a).
KATZ, Y., DIAMOND, J. M.: J. Membrane Biol. **17**, 101 (1974b).
KAUFFMAN, J. W., MEAD, C. A.: Biophys. J. **10**, 1084 (1970).
KAWATO, S., KINOSITA, K., Jr, IKEGAMI, A.: Biochemistry **16**, 2319 (1977).
KETTERER, B., NEUMCKE, B., LÄUGER, P.: J. Membrane Biol. **5**, 225 (1971).

Kolb, H. A., Läuger, P.: J. Membrane Biol. **37**, 321 (1977).

Kornberg, R. D., McConnell, H. M.: Biochemistry **10**, 1111 (1971).

Kozak, J. J.: Proc. nat. Acad. Sci. (Wash.) **72**, 683 (1975).

Kramers, H. A.: Physica **7**, 284 (1940).

Krasne, S.: Biophys. J. Abst. **17**, FAM-C11 (1977).

Krasne, S., Eisenman, G.: J. Membrane Biol. **30**, 1 (1976).

Lange, Y., Gary-Bobo, C. M., Solomon, A. K.: Biochim. biophys. Acta **339**, 347 (1974).

Latorre, R., Hall, J. E.: Nature **264**, 361 (1976).

Läuger, P.: Science **178**, 24 (1972).

Läuger, P., Neumcke, B.: In: Membranes, Vol. 2 (G. Eisenman, Ed.). New York: Dekker 1973.

Läuger, P., Lesslauer, W., Marti, E., Richter, J.: Biochim. biophys. Acta **135**, 20 (1967).

Le Blanc, O. H., Jr: Biochim. biophys. Acta **193**, 300 (1969).

Le Blanc, O. H., Jr: Biophys. Soc. Abst., 94a (1970).

Le Blanc, O. H., Jr: J. Membrane Biol. **4**, 227 (1971).

Lee, A. G., Birdsall, N. J. M., Metcalfe, J. C., Warren, G. B., Roberts, G. C. K.: Proc. roy Soc. Lond. B. **193**, 253 (1976).

Le Neveu, D. M., Rand, R. P., Parsegian, V. A., Gingell, D.: Biophys. J. **18**, 209 (1977).

Lev, A. A., Gollib, V. A., Buzhinsky, E. P.: J. evolutionary Biochem. Physiol. **2**, 109 (1966).

Levine, Y. K., Wilkins, M. H. F.: Nature New Biol. **230**, 69 (1971).

Levine, S., Mingins, J., Bell, G. M.: J. phys. Chem. **67**, 2095 (1963).

Liberman, E. A., Margulis, D. M.: Biofizika **19**, 450 (1974).

Liberman, E. A., Topaly, V. P.: Biofizika **13**, 1025 (1968a).

Liberman, E. A., Topaly, V. P.: Biochim. biophys. Acta **163**, 125 (1968b).

Liberman, E. A., Topaly, V. P.: Biofizika **14**, 452 (1969).

Liberman, E. A., Topaly, V. P., Tsofina, L. M.: Biofizika **15**, 69 (1970).

Lieb, W. R., Stein, W. D.: Nature **224**, 240 (1969).

Lieb, W. R., Stein, W. D.: In: Current Topics in Membranes and Transport (F. Bronner and A. Kleinzeller, Eds). New York: Academic Press 1971, pp. 1–39.

Lippe, C.: J. molec. Biol. **39**, 669 (1969).

Lüschow, U. L., Heckmann, K. D., Pring, M.: Biochim. biophys. Acta **389**, 1 (1975).

MacDonald, R. C.: Biochim. biophys. Acta **448**, 193 (1976).

MacDonald, J. R., Barlow, C. A.: J. Electrochem. Soc. **113**, 978 (1966).

McLaughlin, S.: J. Membrane Biol. **9**, 361 (1972).

McLaughlin, S. G. A.: Nature **243**, 234 (1973).

McLaughhlin, S.: (1975). In: Molecular Mechanism of Anesthesia. Ed. B. R. Fink. Progress in Anesthesiology. Vol. 1. New York: Raven Press.

McLaughlin, S.: In: Current Topics Membranes and Transport, Vol. 9 (F. Bronnen and A. Kleinzeller, Eds). New York: Academic Press 1977, pp. 71–144.

McLaughlin, S., Eisenberg, M.: Ann. Rev. Biophys. Bioeng. **4**, 335–366 (1975).

McLaughlin, S., Harary, H.: Biochemistry **15**, 1941 (1976).

Margulus, D. M.: Biofizika **19**, 655 (1974).

Markin, V. S., Grigorev, P. A., Yermishkin, L. N.: Biofizika **16**, 1011 (1971).

Melnik, E., Latorre, R., Hall, J. E., Tosteson, D. C.: J. gen. Physiol. **69**, 243 (1977).

Miller, K. W., Hammond, L., Porter, E. G.: Chem. & Phys. Lipids **20**, 229 (1977).

Montal, M.: Ann. Rev. Biophys. Bioeng. **5**, 119–175 (1976).

Montal, M., Mueller, P.: Proc. nat. Acad. Sci. (Wash.) **69**, 3561 (1972).

Mooney, R. C. L.: J. Amer. Chem. Soc. **62**, 2955 (1940).

Motomiza, S., Toei, K., Iwachido, T.: Bull. Chem. Soc. Japan. **42**, 1006 (1969).

Mueller, P., Rudin, D. O.: In: Current Topics in Bioenergetics, Vol. 3. New York: Academic Press 1969, pp. 157–249.

Mueller, P., Rudin, D. O., Tien, H. T., Wescott, W. C.: Circulation **26**, 1167 (1962).

Mueller, P., Rudin, D. O., Tien, H. Ti., Wescott, W. C.: J. phys. Chem. **67**, 534 (1963).

Nedermeijer-Denessen, H. J. M., de Ligny, C. L.: Electroanal. Chem. **57**, 265 (1974).

Nelson, A. P., McQuarrie, D. A.: J. Theoret. Biol. **50**, 13 (1975).

Nernst, W.: Z. physik. Chem. **47**, 52 (1904).

Neumcke, B.: Biophyzik **7**, 95 (1970).

NEUMCKE, B., BAMBERG, E.: In: Membranes. Vol. 3 (G. Eisenman, Ed.). New York: Dekker 1975, pp. 215–253.
NEUMCKE, B., LÄUGER, P.: Biophys. J. **9**, 1160 (1969).
NEUMCKE, B., LÄUGER, P.: J. Membrane Biol. **3**, 54 (1970).
PAGANO, R., THOMPSON, T. E.: J. molec. Biol. **38**, 41 (1968).
PAGANO, R. E., RUYSSCHAERT, J. M., MILLER, I. R.: J. Membrane Biol. **10**, 11 (1972).
PALTAUF, F., HAUSER, H., PHILLIPS, M. C.: Biochim. biophys. Acta **249**, 539 (1971).
PAPAHADJOPOULOS, D.: Biochim. biophys. Acta **163**, 240 (1968).
PARLIN, R. B., EYRING, H.: In: Ion Transport Across Membranes (H. T. Clarke, Ed.). New York: Academic Press 1954.
PARSEGIAN, A.: Nature **221**, 844 (1969).
PARSEGIAN, A.: Ann. N. Y. Acad. Sci. **264**, 161 (1975).
PARSONS, R.: In: Modern Aspects of Electrochemistry (J. O. M. Bockris, Ed.). London: Butterworths 1954, pp. 103–179.
PLACHY, W. Z., WINDREM, D., DROBNY, G.: Biophys. J. **17**, 85a (abst.) (1977).
PLANCK, M.: Ann. Phys. Chem. **2**, 161 (1890a).
PLANCK, M.: Ann. Phys. Chem. **4**, 561 (1890b).
POZNANSKY, M., TONG, S., WHITE, P. C., MILGRAM, J. M., SOLOMON, A. K.: J. gen. Physiol. **67**, 45 (1976).
PRICE, H. D., THOMPSON, T. E.: J. molec. Biol. **41**, 443 (1969).
REDWOOD, W. R., HAYDON, D. A.: J. theoret. Biol. **22**, 1 (1969).
REEVES, J. P., DOWBEN, R. M.: J. Membrane Biol. **3**, 123 (1970).
REIF, F.: Fundamentals of Statistical and Thermal Physics. New York: McGraw-Hill 1965.
REQUENA, J., HAYDON, D. A.: J. Colloid Interface Sci. **51**, 315 (1975a).
REQUENA, J., HAYDON, D. A.: Proc. roy. Soc. A **347**, 161 (1975b).
REQUENA, J., HAYDON, D. A., HLADKY, S. B.: Biophys. J. **15**, 77 (1975).
REYNOLDS, J. A., GILBERT, D. B., TANFORD, C.: Proc. nat. Acad. Sci. (Wash.) **71**, 2925 (1974).
ROBERTSON, R. N., THOMPSON, T. E.: FEBS Letters **76**, 16 (1977).
ROSEMAN, M., LITMAN, B. J., THOMPSON, T. E.: Biochemistry **14**, 4826 (1975).
ROSEN, D., SUTTON, A. M.: Biochim. biophys. Acta **163**, 226 (1968).
ROTHMAN, J. E., DAWIDOWICZ, E. A.: Biochemistry **14**, 2809 (1975).
SACHS, R. G., DEXTER, D. L.: J. appl. Physics **21**, 1304 (1950).
SANDBLOM, J., HAGGLUND, J., ERICKSSON, N. E.: J. Membrane Biol. **23**, 1 (1975).
SARGENT, D. F.: In: Molecular Aspects of Membrane Phenomena (H. R. Kaback, H. Neurath, G. K. Radda, R. Schwyzer and W. R. Wiley, Eds). New York: Springer 1974, pp. 104–120.
SARGENT, D. F.: J. Membrane Biol. **23**, 227 (1975).
SAX, K. J., STROSS, F. H.: Anal. Chem. **29**, 1700 (1957).
SCATCHARD, G., SCHEINBERG, I. H., ARMSTRONG, S. H.: J. Amer. chem. Soc. **72**, 540 (1950).
SCHATZBERG, P.: J. phys. Chem. **67**, 776 (1963).
SCHATZBERG, P.: J. Polymer Sci. Part C **10**, 87 (1965).
SCHMIDT, C. F., BARENHOLZ, Y., HUANG, C., THOMPSON, T. E.: Biochemistry **16**, 3948 (1977).
SCHOLTENS, B. J. R., BIJSTERBOSCH, B. H.: FEBS Letters **62**, 233 (1976).
SEELIG, J., NIEDERBERGER, W.: Biochemistry **13**, 1585 (1974).
SEELIG, A., SEELIG, J.: Biochemistry **13**, 4839 (1974).
SEELIG, A., SEELIG, J.: Biochim. biophys. Acta **406**, 1 (1975).
SHCHIPUNOV, YU. A., BOGUSLAVSKI, L. I.: Biofizika **20**, 1024 (Eng. Trans. Vol. No. 20 1042 [1975]).
SHINITZKY, M., DIANOUX, A. C., GITLER, C., WEBER, G.: Biochemistry **10**, 2106 (1971).
SIMON, S. A., STONE, W. L., BUSTO-LATORRE, P.: Biochim. biophys. Acta **468**, 378 (1977a).
SIMON, S. A., LIS, L. J., MACDONALD, R. C., KAUFFMAN, J. W.: Biophys. J. **19**, 83 (1977b).
SIMON, S. A.: J. gen. Physiol. **70**, 123 (1977).
SINGER, S. J.: In: Structure and Function of Biological Membranes (L. Rothfield, Ed.). New York: Academic Press 1971, pp. 145–222.
SMEJTEK, P., PAULIS, M.: (in preparation) (1978).
SMEJTEK, P., KWAN, H., PERMAN, W. H.: Biophys. J. **16**, 319 (1976).
SMOLUCHOWSKI, M. V.: Physik. Zschr. **17**, 557 (1916).
SMYTH, C. P.: Dielectric Behavior and Structure. New York: McGraw-Hill 1955.
STANDISH, M. H., PETCHICA, B. A.: Trans. Faraday Soc. **64**, 1113 (1968).

STARK, G., BENZ, R.: J. Membrane Biol. **5**, 133 (1971).

STARK, G., KETTERER, B., BENZ, R., LÄUGER, P.: Biophys. J. **11**, 981 (1972).

STECHER, P. G., WINDHOLZ, M., LEAHY, D. S., BOLTON, D. M., EATON, L. G., (Eds): The Merck Index, 8th ed., New Jersey: Merck 1968.

STEIM, J. M., TOURTELLOTTE, M. E., REINERT, J. C., McELHANEY, R. N., RADAR, R. L.: Proc. nat. Acad. Sci. (Wash.) **63**, 104 (1969).

STERN, O.: Zschr. f. Elektrochem. **30**, 508 (1924).

STEVENS, C. F.: Biophys. J. **12**, 1028 (1972).

STIGTER, D.: J. phys. Chem. **79**, 1015 (1975).

STOCKTON, G. W., POLNASZEK, C. F., TULLOCH, A. P., HASAN, F., SMITH, I. C. P.: Biochemistry **15**, 954 (1976).

STOECKENIUS, W., ENGELMAN, D. M.: J. Cell Biol. **42**, 613 (1969).

STONE, W. L.: J. biol. Chem. **250**, 4368 (1975).

SZABO, G.: Nature **252**, 47 (1974).

SZABO, G.: In: Extreme Environment: Mechanisms of Microbial Adaptation (M. R. Heinrich, Ed.). New York: Academic 1976, pp. 321–348.

SZABO, G.: In: Ninth Rochester International Conference on Environmental Toxicity (R. W. Miller, Ed.). New York: Plenum 1977.

SZABO, G.: (in preparation) (1978).

SZABO, G.: EISENMAN, G., McLAUGHLIN, S. G. A., KRASNE, S.: Ann. N. Y. Acad. Sci. **195**, 273 (1972).

TANFORD, C.: The Hydrophobic Effect: Formation of Micelles and Biological Membranes. New York: Wiley 1973.

THOMSON, T. E.: In: Molecular Specializations and Symmetry in Membrane Functions (A. Solomon, Ed.). Cambridge, Mass.: Harvard University Press. 78–98 (1977).

TIEN, H. T.: Bilayer Lipid Membrane (BLM). New York: Dekker 1974.

TIEN, H. T.: Photochem. Photobiol. **24**, 97 (1976).

TIEN, N. T., TING, H. P.: J. Colloid Interface Sci. **27**, 702 (1968).

TOMKIEWICZ, M., CORKER, G. A.: Biochim. biophys. Acta **406**, 197 (1975).

TOSTESON, D. C., ANDREOLI, T. E., TIEFFENBERG, M., COOK, P.: J. gen. Physiol. **51**, 373 (1968).

TOYOSHIMA, Y., THOMPSON, T. E.: Biochemistry **14**, 1518 (1975a).

TOYOSHIMA, Y., THOMPSON, T. E.: Biochemistry **14**, 1525 (1975b).

TRAUBLE, H.: J. Membrane Biol. **4**, 193 (1971).

VANDERKOOI, J. M., CALLIS, J. B.: Biochemistry **13**, 4000 (1974).

VEATCH, W. R., STRYER, L.: Biophys. J. **17**, W-PM-G1 (Abst.) (1977).

VETTER, K. J.: Electrochemical Kinetics. Theoretical and Experimental Aspects (S. Bruckenstein and B. Howard, Eds). New York: Academic Press 1967.

VIELSTICH, W.: Z. Elektrochem. **57**, 646 (1953).

VON LAUE, M.: Jahrb. d. Radioaktiviät u. Elektronik **15**, 205 (1918).

VREEMAN, H. J.: Proc. roy. Soc. Lond. Ser. B **69**, 542 (1966).

WAGGONER, A., WANG, C. H., TOLLES, R. L.: J. Membrane Biol. **33**, 109 (1977).

WALZ, D., BAMBERG, E., LÄUGER, P.: Biophys. J. **9**, 1150 (1969).

WANG, C.-C., BRUNER, L. J.: Biophys. J. (Abst.) **17**, 131a (1977a).

WANG, C.-C., BRUNER, L. J.: **38**, 311 (1978).

WANG, J. H.: J. phys. Chem. **69**, 4412 (1965).

WARTIOVARA, V., COLLANDER, R.: Protoplasmatologia II, **C. 8. d.**, 1 (1960).

WEAST, R. C. (Ed.): Handbook of Chemistry and Physics. Cleveland, Ohio: The Chemical Rubber Co. 1972.

WHITE, D. A.: In: Form and Function of Phospholipids (G. B. Ansell, J. N. Hawthorne and R. M. C. Dawson, Eds). Amsterdam: Elsevier 1973, pp. 441–482.

WHITE, S. H.: Biophys. J. **10**, 1127 (1970).

WHITE, S. H.: Biochim. biophys. Acta **323**, 343 (1973).

WHITE, S. H.: Biophys. J. **15**, 95 (1975).

WHITE, S. H.: Nature **262**, 421 (1976).

WHITE, S. H.: Ann. N. Y. Acad. Sci. **303**, 243 (1977).

WHITE, S. H., BLESSUM, D. N.: Rev. Sci. Instrum. **46**, 1462 (1975).

WHITE, S. H., THOMPSON, T. E.: Biochim. biophys. Acta **323**, 7 (1973).

White, S. H., Petersen, D. C., Simon, S., Yafuzo, M.: Biophys. J. **16**, 481 (1976).
Wilkins, M. H. F., Blaurock, A. E., Engelman, D. M.: Nature New Biol. **230**, 72 (1971).
Wishnia, A.: J. phys. Chem. **67**, 2079 (1963).
Wobschall, D.: J. Colloid Interface Sci. **40**, 417 (1972).
Wolosin, J. M., Ginsburg, H.: Biochim. biophys. Acta **389**, 20 (1975).
Wolosin, J. M., Ginsburg, H., Lieb, W. R., Stein, W. D.: J. gen. Physiol. **71**, 93 (1978).
Wulf, J., Benz, R., Pohl, W. G.: Biochim. biophys. Acta **465**, 429 (1977).
Zingsheim, H. P., Haydon, D. A.: Biochim. biophys. Acta **298**, 755 (1973).
Zobel, O. J.: Bell System Techn. J. **2**, 1 (1923).
Zwolinski, B. J., Eyring, H., Reese, C. E.: J. phys. Colloid Chem. **53**, 1426 (1949).

Chapter 12

Carrier-Mediated Ion Transport Across Thin Lipid Membranes

G. STARK

A. Introduction

The molecular basis of ion transport across biological membranes is largely unknown at present. Its study is complicated above all by the complex nature of living systems, which allows the application of suitable physical methods only to a small number of especially favorable systems. Even in these cases, interpretation of experimental results is usually hampered by the presence of several other transport systems, even when these can be partially eliminated by the use of special blocking agents. A different approach, designed to circumvent these difficulties, consists in a transfer of the transport system under study to model membranes. Planar lipid membranes, which are formed according to a technique originally developed by MUELLER et al. (1962), seem to be especially appropriate for the study of ion transport, since they easily allow for the application of an electric voltage difference across the membrane and the measurement of the electric current. In addition, as is shown in detail in the previous article, the resistance of an unmodified black film assumes rather high values, typically 10^8 Ωcm^2. Therefore, the incorporation of relatively few, and in some cases even of single "transport entities" into a black film may yield a significant change in the membrane resistance. The isolation of special transport proteins from biological membranes and their active reconstitution into black films have proved to involve extreme difficulties, and the results must be considered as rather modest up to now. The main aim of such reconstitutions, namely the elucidation of molecular mechanisms, has not yet been achieved. Nevertheless, this should be possible in the future, when improved reconstitution techniques will be available. The outstanding advantages of thin lipid membranes for the study of transport mechanisms are clearly apparent in studies performed with a series of model compounds. Special attention has been devoted to the action of small macromolecules, which are produced by certain microorganisms and have been found to induce a highly selective increase in ion permeability of both natural and artificial membranes. The mechanisms through which these substances exert their influence have been intensively studied throughout the last decade, since they might give a clue as to how the native biological transport proteins function.

I. Carriers and Pores

Although our knowledge about the mechanisms of ion transport across biological membranes has been rather incomplete up to now, simple concepts have been developed which allow us to understand how ions could cross the high energy barrier of a hycrocarbon like layer about 100 Å thick. Some of these ideas are schematically summarized in Figure 1. Two fundamental possible means of compensation can be envisaged for the high electrostatic energy difference that an ion experiences when moving from a medium of high dielectric constant such as water to a medium of low dielectric constant such as a lipid phase (PARSEGIAN, 1969). The lipid phase could be interrupted by a molecule of sufficient size, such as an integral membrane protein, spanning the membrane and forming an aqueous pore. If the diameter of such a pore is large enough, the ion concentration inside the pore should be comparable to the ion concentration in water (Fig. 1B). Another possible means of increasing the ion concentration in the membrane is a complex formation with another molecular species, which, due to energetically favorable interactions, has a high partition coefficient inside the membrane and balances the positive electrostatic energy difference of the pure ion between the membrane and water. In contrast to a pore, the translocation of the ion from one membrane interface to the other occurs through a movement of the whole complex. Because of this coupled movement such transport systems are frequently called "carriers." According to the kind of movement one can distinguish "rotational" and "translational" carriers (Fig. 1A, D).

The classification of transport systems into carriers and pores is, of course, very rough. An overwhelming number of other transport mechanisms can be envisaged, which contain features of both and which seem to be more realistic approaches to the interpretation of rather complex transport phenomena of biological transport proteins. Nevertheless, the carrier and pore mechanisms illustrated in Figure 1 can be considered as fundamental limiting possibilities. While a pore rather represents a structural part of the membrane, which appears more or less fixed relative to the moving ion, a carrier undergoes a coupled movement with the ion. An advantage of these definitions consists in the existence of physical methods which – at least in favorable cases – allow the distinction of carriers and pores. The mobility of an ion inside a pore with a large

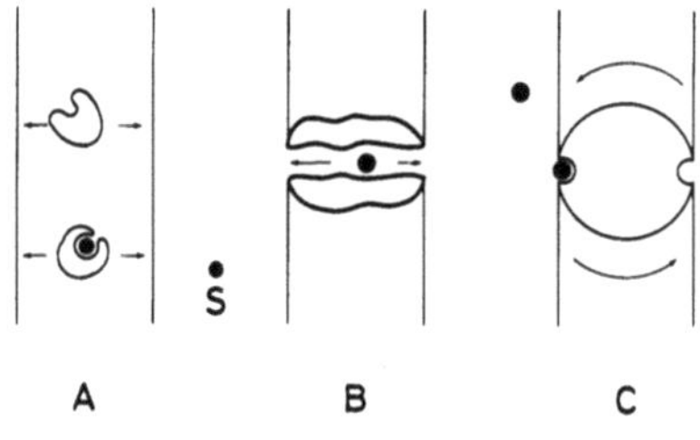

Fig. 1. Schematic presentation of some simple transport channels for a substrate S: (A) translational carrier; (B) pore; (C) rotational carrier

diameter will be comparable to the mobility in water. On the other hand, the mobility of a relatively large carrier-ion complex inside the membrane will be considerably lower, especially through the rather high viscosity of a lipid structure compared to water. Therefore, the "transport capacity" of a pore of sufficient size will by far exceed that of a single carrier molecule (given by the maximum number of ions transported in one direction per second). As will be shown in the next article, the sensitivity of available instrumentation is sufficient to provide evidence for single pores. Due to their smaller transport capacity, the existence of carriers has had to be inferred indirectly up to now.

Throughout the last decade a considerable number of compounds have been found to increase the permeability of thin lipid membranes to monovalent and divalent ions. Most of these so called "ionophores" have been classified as carriers or pores.

II. A Survey of Suggested Ion Carriers

While it is comparatively easy to demonstrate the presence of large pores showing high transport rates, the distinction between narrow pores with low transport rates and translational or rotational carriers is much more difficult. Perhaps the best indirect evidence in favor of a special transport mechanism is a quantitative study on the basis of a clearly formulated model, as will be shown in the next section. Steady-state and time-dependent properties of a given transport system often allow discrimination between different mechanisms. But qualitative considerations may also help to answer some basic questions:

a) One essential prerequisite for a carrier mechanism is the possibility of complex formation between the "transport unit" and the transported ion. In many cases, complex formation can also be studied in macroscopic solutions of different polarity, as has been done for most of the compounds mentioned below. The detection of ion binding is, of course, no unequivocal evidence for a carrier mechanism, since a pore may have one or several distinct binding sites too, which might be active outside the membrane.

b) If the binding species is a molecule with a much smaller diameter than the thickness of membrane, a pore and also a rotational carrier should consist of an aggregate of several molecules. Then, dependence of the number of aggregates, and therefore of the ion permeability and conductance of a membrane on the monomer concentration is expected to be more than linear. Most substances currently believed to form pores show a high-order dependence ($n \simeq 5$–10) of the membrane conductance on carrier concentration. Those thought to function as ion carriers exhibit a linear or square relationship between membrane conductance and carrier concentration ($n \simeq 1$–2). Exceptions from this general rule are, however, known.

c) Another essential property of an ion carrier is its free mobility across the membrane, which should depend strongly on structural properties as well as on the microviscosity of the membrane. A very drastic change of the fluidity of membranes occurs at the phase transition of the lipid molecules, which, for lipids with saturated fatty acid residues, frequently lie within the experimentally accessible temperature region for black films ($0 \leqq t \leqq 70°$ C). While the mobility of an ion carrier would presumably be very sensitive to a stepwise fluidity change, a pore, defined as a structural part of the membrane, might not be influenced by a phase transition of the surrounding lipid molecules. Such different behavior for carriers and pores was reported by KRASNE et al. (1971) for valinomycin and nonactin on the one hand, and gramicidin A on the other. However, one should be careful in generalizing these findings obtained with special transport systems, since a change in the lipid structure could change the structure of the pore and thus influence its transport properties.

In summarizing, it must be concluded that a general and simple method of distinguishing between carriers and pores with low transport rates does not exist at present. It is rather the sum of many individual observations which enables us to introduce such a classification, which in some cases might be subject to modifications in the future. This section contains a review of some fairly well studied transport systems which are believed to act as ion carriers; later sections will be devoted to a description of the physical methods and to an analysis of their results for neutral macrocyclic carriers of alkali ions. All compounds described in this article are believed to act as translational carriers, since their functional unit is much smaller than the thickness of the membrane.

B. Carriers of Hydrogen Ions

There is a series of weak acids, which are known to act as uncouplers of oxidative phosphorylation in mitochondria and which induce a pH-dependent conductance increase in thin lipid membranes (BIELAWSKI et al., 1966; SKULACHEV et al., 1967; HOPFER et al., 1968; LIBERMAN and TOPALY, 1968). Some of these compounds are shown in Figure 2. All of them possess dissociable proton groups

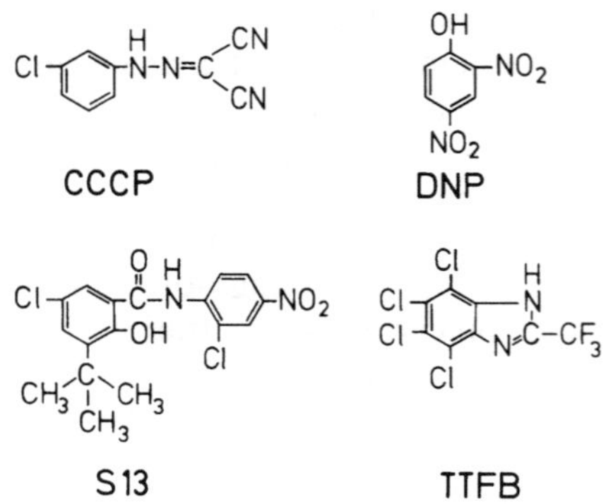

Fig. 2. Structure of some proposed carriers for hydrogen ions: carbonylcyanide-m-chlorophenylhydrazone (CCCP), 3-t-butyl, 5-chloro, 2′-chloro, 4′-nitro-salicylanilid (S 13), 2,4 dinitrophenol (DNP), tetrachloro-2-trifluoromethylbenzimidazole (TTFB)

with pK values in water ranging from 4 (DNP) to 6,8 (S13) (HOPFER et al., 1970). Furthermore, they contain π-orbitals which enable them to delocalize the charge of the ionized form, thus increasing the solubility of an ion in a medium of low dielectric constant (PARSEGIAN, 1969). Their classification as proton translocators across membranes arises mainly from two experimental observations on lipid bilayers:

(a) the application of a pH gradient across the membrane at equal uncoupler concentrations yields an open circuit electrical potential which is close to the theoretical value for a membrane selectively permeable for protons (or hydroxyl ions);

(b) the induced membrane conductance as a function of proton concentration (identical on both sides of the membrane) increases with pH at pH < pK, but shows a well pronounced maximum, which in some cases is around pH = pK.

From (a) it follows that protons (or hydroxyl ions) must cross the membrane in some way. Furthermore, the bilayer must be permeable to a charged species which moves across it driven by a pH gradient in such a way that its final distribution on both sides of the membrane is asymmetrical, thus establishing an electrical potential difference. It does not, however, permit any conclusions about the nature of this charge carrier. As will be shown below, even a permeable anion A^- can generate a proton potential, if the membrane shows a sufficiently high permeability for the neutral weak acid (uncoupler) HA. Let us consider a membrane separating two well-buffered aqueous solutions 1 and 2, with different pH values but with identical total uncoupler concentration c_t. The membrane has a certain permeability for the ionized form A^-, which we assume to be the only charge carrier inside the membrane. The equilibrium constant K_1 of the reaction in water

$$HA \rightleftarrows H^+ + A^- \text{ is given by } K_1 = \frac{(c_{H^+}) \cdot (c_{A^-})}{c_{HA}} \tag{1}$$

Because of the different pH, the anion concentrations c'_{A^-} and c''_{A^-} at the left (′) and right (″) aqueous interfaces of the membrane will be different. Therefore an open circuit potential difference $V = \Psi_1 - \Psi_2$ for the anion A^- will develop, i.e.

$$V = -\frac{RT}{F} \ln \frac{c'_A}{c''_A} \tag{2}$$

For a membrane which is not permeable to the neutral form HA, the interfacial concentrations c'_{A^-} and c''_{A^-} will be equal to the corresponding bulk concentrations $c^1_{A^-}$ and $c^2_{A^-}$ far from the membrane interfaces:

$$c_A^{1,2} = c_A^{\prime,\prime\prime} = \frac{K_1 c_t}{c_H^{1,2} + K_1} \tag{3}$$

Normally, however, the concentration of HA inside the membrane — for energetical reasons — will far exceed the concentration of A^-. If the permeability of the membrane to HA is sufficiently high, the concentrations c'_{HA} and c''_{HA} at the interfaces will be different from c^1_{HA} and c^2_{HA}, since the diffusion time through the unstirred water layers (thickness δ) on both sides of the membrane will exceed the diffusion time through the membrane. The proton concentrations c'_{H^+} and c''_{H^+} are equal to the bulk concentrations $c^1_{H^+}$ and $c^2_{H^+}$, since the solutions are well buffered. Then, if reaction (1) proceeds with sufficient velocity, a difference in c^1_{HA} and c'_{HA} will also generate one in $c^1_{A^-}$ and c'_{A^-}. Combination of Eqs. (1) and (2) gives

$$V = -\frac{RT}{F} \ln \frac{(c'_{HA}) \cdot (c^2_{H^+})}{(c''_{HA}) \cdot (c^1_{H^+})} \tag{4}$$

In the limit of a very high permeability for HA we can write $c'_{HA} = c''_{HA}$. Equations (4) and (2) are then reduced to

$$V = -\frac{RT}{F} \ln \frac{c'_{A^-}}{c''_{A^-}} = \frac{RT}{F} \ln \frac{c^1_{H^+}}{c^2_{H^+}} . \tag{5}$$

This case is illustrated in Figure 3.

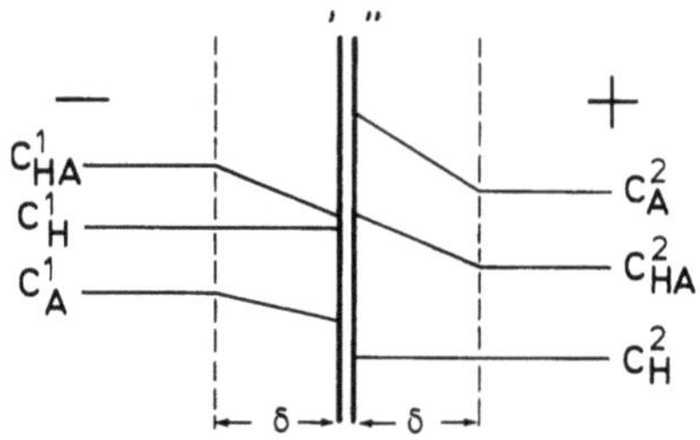

Fig. 3. A membrane permeable to the anion A^-, and very highly permeable to the neutral form HA of a weak acid. The membrane separates two well-buffered solutions of different pH and develops an open circuit potential determined by the relation $c'_A/c''_A = c^1_H/c^2_H$ (see Eq. (5)). δ = thickness of the unstirred layers

Equation (5) indicates that a membrane which is only permeable to one anionic charge carrier may behave like a membrane that is permselective for protons if the neutral acid HA is also highly permeable.

This argument may be extended to anionic complexes, *e.g.* HA_2^-, as FINKELSTEIN (1970) has shown. Such complexes might be formed in water through the reaction

$$HA_2^- \rightleftarrows HA + A^- \text{ with } K_2 = \frac{(c_{HA}) \cdot (c_{A^-})}{c_{HA_2^-}} . \tag{6}$$

In some cases the complexes HA_2^- seem to represent the charge carriers inside the membrane, as has been suggested by LEA and CROGHAN (1969) and by

FINKELSTEIN (1970). If the concentration of HA_2^- in water is much less than that of either HA or A^-, Eqs. (1) and (6), together with $c_t = c_{HA} + c_{A^-}$, yield:

$$c_{HA_2^-} = \frac{K_1}{K_2} c_t^2 \frac{c_{H^+}}{(K_1 + c_{H^+})^2} . \tag{7}$$

Equation (3) predicts a linear dependence of the anion concentration c_{A^-} on the total uncoupler concentration c_t. From Eq. (7), on the other hand, follows a square dependence of $c_{HA_2^-}$ on c_t. The concentration of the species A^- and HA_2^- inside the membrane – at least within a certain concentration range – is proportional to their concentrations in water, *i.e.* related *via* partition coefficients. Another proportionality exists between the conductance λ of a membrane and its charge carrier concentration (for a single charge carrier, neglecting "blocking effects," see below). Therefore one should expect:

(a) $\lambda \simeq c_t$, if A^- is the only charge carrier in the membrane.

(b) $\lambda \simeq c_t^2$, if HA_2^- is the charge carrier.

Indeed, linear and square dependences for different weak acid uncouplers have been found (Fig. 4), which caused several authors to define two classes of uncouplers. Nevertheless, such a classification has to be considered with caution. As shown in Figure 4, different effects may contribute to or even falsify the order of the dependence. In the case of CCCP, for example, a saturation of λ at high concentrations becomes apparent. Such saturation effects have frequently been found – *e.g.* for hydrophobic ions (LeBLANC, 1969; KETTERER et al., 1971) – and have been interpreted as space charge limited currents (LeBLANC, 1969) or more recently, and probably more correctly, on the basis of a limited

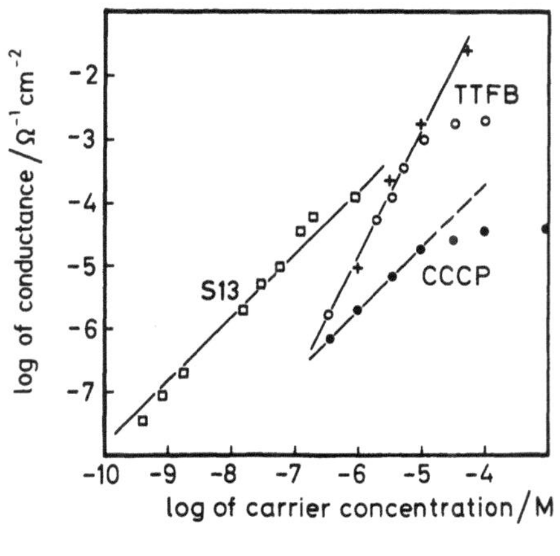

Fig. 4. Dependence of the membrane conductance of the ionophore concentration for S 13 (HOPFER et al., 1970) CCCP (LeBLANC, 1971) and TTFB (BORISOVA et al., 1974) The data for S 13 and CCCP represent stationary measurements, for TTFB stationary (0) and initial conductances (+) (see text). The full lines were drawn with slope 1 (S 13, CCCP) and slope 2 (TTFP). Experimental conditions: (S 13) diglucosyldiglyceride membranes, pH 6.4, 2 mM buffer; (CCCP) egg lecithin/cholesterol membranes, pH 10.2, 50 mM buffer; (TTFB) ox brain lipid membranes, pH 5.6, 20 mM buffer

number of ions which may be incorporated into the two-dimensional lattice of lipid molecules (KETTERER et al., 1971). A more detailed study of this phenomenon has been performed by BRUNER (1975), using the hydrophobic ion dipicrylamine. Another effect which influences the interpretation occurs when the ionic permeabilities across the membrane are very high. As the ionic strength of the aqueous solutions is increased, the applied voltage begins to fall almost completely across the interior of the membrane. In this case the ions inside the membrane are driven both by diffusion and by the electric field, whereas in the unstirred layers adjacent to the membrane only diffusion acts as driving force. This may lead to the further complication that the conductance is not determined by the membrane itself but rather by the unstirred layers. Figure 4 shows that at high TTFB concentrations the stationary conductance saturates towards the value of the unstirred layers. In such cases, the intrinsic conductance of the membrane can be obtained by measurement of the initial conductance by means of pulse experiments. This technique will be described in more detail later in this article. The second-order dependence of TTFB conductance, measured *via* the initial conductance, extends up to very high concentrations, as was first shown by BAMBERG (1971; see also BAMBERG et al., 1975). Finally, a third effect that makes determination of the molecularity of the charge complex inside the membrane so difficult is the possibility of anion adsorption to the membrane interface. MCLAUGHLIN (1972) found that the almost linear relationship between membrane conductance and concentration of the uncoupler DNP is converted into a second-order dependence if the adsorption of DNP is taken into account. The adsorption of anions generates a substantial negative surface potential, which reduces the concentration of anionic complexes at the membrane interface and thus produces some kind of selfhindrance. Apart from DNP, the other uncouplers given in Figure 2 do not seem to change the surface potential by adsorption (MCLAUGHLIN, 1972).

After correction for these effects, two of the uncouplers shown in Figure 2 seem to enhance the conductance of membranes as simple anions (S13 and CCCP). The other two (TTFB and DNP) behave like anionic complexes HA_2^-. If two different uncouplers, HA and HB, are present in the aqueous phases, each of which shows a second-order dependence, their combined effects may indicate the presence of mixed complexes HAB^-. This has been found for DNP and DTFB, an analogue of TTFB (FOSTER and MCLAUGHLIN, 1974).

The interpretation of second-order concentration dependences of the conductance on the basis of HA_2^- complexes as charge carriers in the membrane is also in agreement with the second experimental observation mentioned above. For both TTFB and DNP the membrane conductance as a function of pH shows a maximum at about the pK value of the uncoupler. The concentration of HA_2^--complexes in water and therefore also their concentration inside the membrane assumes a maximum at $c_{H^+} = K_1$, as may be verified from Eq. (7). The members of the other uncoupler class (*e. g.* S13 and CCCP) for which there is a linear dependence between conductance and concentration (in accordance with A^- as charge carrier in the membrane) also exhibit maxima in the pH dependence of conductance, but these are not correlated with the pK value of the acid (LEBLANC, 1971; HOPFER et al., 1970). Rather it can be demonstrated that a carrier-mediated transport may show such maxima depending on the relative velocity of the single transport steps (MARKIN et al., 1969). Alternatively, as has been found by LEBLANC (1971) and NEUMCKE (1971), diffusion polarization in the presence of a chemical reaction in the aqueous phases, which would enhance the amount of permeating charge carriers in the unstirred layers, may also create a maximum in the pH dependence of λ.

In conclusion, all experimental results presented so far can be explained on the basis of a carrier mechanism for hydrogen ions, though different species may carry the current. The question arises as to whether this is the only possible mechanism in agreement with the data, or whether other explanations for hydrogen transport are possible. According to the definition of a carrier given in this article, there must be a coupled movement of the carrier-ion complex, which in the present case would be the neutral form HA. One might, however, think of uncoupler molecules that are adsorbed onto the membrane interface in a more or less immobilized way and exchange a proton across the interior of the membrane. Such a mechanism would explain most of the experiments with TTFB and DNP, and has been suggested by BRUNER (1970). In contrast to a carrier mechanism, it is the hydrogen ion itself that moves from a donor to an acceptor. A fundamental problem, however, consists in the large distance of several tens of Ångstroms a proton would have to move in the free form in a medium of low dielectric constant. This is rather improbable, but could be circumvented by allowing a simultaneous movement of uncoupler molecules adsorbed onto opposite interfaces to a closer distance. Such a modified "proton jump model" would, of course, have features of a carrier mechanism, since a proton would be transported through part of the membrane by one uncoupler molecule before it jumps to the A^- form of another. During the encounter of HA and A^-, the resulting "dimer" might be considered as an HA_2^--complex. Although there is some experimental evidence against the strict model of BRUNER (FOSTER and MCLAUGHLIN, 1974), rigorous exclusion of such a modified version has not yet been achieved (BORISOVA et al., 1974). This model could be called a "modified carrier model." Indeed, the existence of mixed complexes HAB^- between chemically different species, as postulated by FOSTER and MCLAUGHLIN (1974), would be easier to understand in the frame of this concept, since a stable configuration between A and B moving together would not be necessary.

In summary, the action of weak acid uncouplers on thin lipid membranes seems to include the following basic facts:

a) The proton permeability is strongly enhanced due to an almost unrestricted diffusion of the neutral complexes HA through the membrane.
b) The electric conductance is increased. While for some substances the anions (A^-) carry the current, for others the anionic complexes (HA_2^-) are essential. It is unclear at present whether these complexes move across the membrane as charge carriers or whether they are only formed as intermediates in a process which includes a transfer of a proton from HA to A^-.

C. Macrocyclic Carriers

I. Neutral Carriers

It has been found that certain macrocyclic compounds, most of which are metabolites of microorganisms, can increase the cationic permeability of lipid bilayers as well as that of natural membranes. These compounds have been classified as carriers and pores using the criteria mentioned above. Those believed to function as carriers can be subdivided into two further groups according to their state of charge as a function of pH. This section deals with the uncharged compounds which bear no ionizable groups, and Section II will discuss the mode of action of the macrocyclic compounds which have dissociable groups, thus bearing variable charge as a function of pH.

Figure 5 shows the structure of the best known neutral compounds, which, apart from the polyether XXXII, seem to behave largely identically in thin lipid membranes, from a mechanistic point of view. Their most interesting common

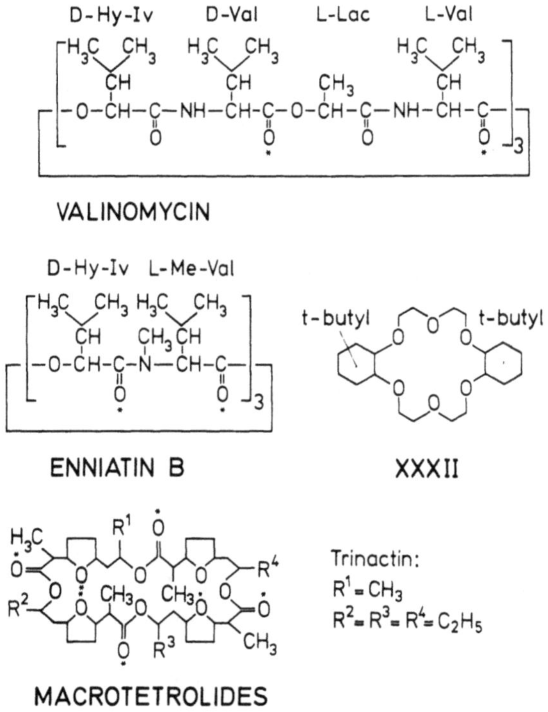

Fig. 5. Structure of some neutral macrocyclic ion carriers. Compound XXXII = bis(t-butyl)dicyclohexyl-18-crown-6. (PEDERSEN and FRENSDORFF, 1972; McLAUGHLIN et al., 1972). Stars indicate oxygen groups interacting with cations during complex formation

property is their ability to form complexes with alkali ions in organic solvents, with a high degree of specificity. In some cases ratios as high as 10^4 can be found in measurements of the stability constant of complex formation for K^+ vs. Na^+. Similar specificities are also observed in the potency of these compounds to promote increases in the conductance of lipid membranes. This stresses the importance of complex formation for the mechanism of ion permeation. Another characteristic common to these compounds is the hydrophobic exterior of their ion complexes, which favors a large partition coefficient in the membrane and counteracts the electrostatic energy difference of the pure ion between membrane and water. These excellent structural prerequisites for a carrier mechanism are illustrated in some more detail for valinomycin.

This molecule (as well as enniatin B) is a cyclodepsipeptide, *i. e.* it consists of an alternating sequence of α-amino- and α-hydroxy acids (BROCKMANN and GEEREN, 1957; SHEMYAKIN et al., 1963). Spectroscopic studies of valinomycin in solvents of different polarity have shown that it assumes a rather compact conformation in nonpolar media, resembling a bracelet (SHEMYAKIN et al., 1969; GRELL et al., 1975). The peculiarities of this conformation are clearly visible in Figures 6 and 7. While the nonpolar side chains of the molecule provide for a hydrophobic exterior, the hydrophilic oxygen carbonyls from the six ester bonds surround a cavity situated in the center of the molecule. The size of this cavity is almost ideally adapted to the diameter of an unhydrated potassium or rubidium

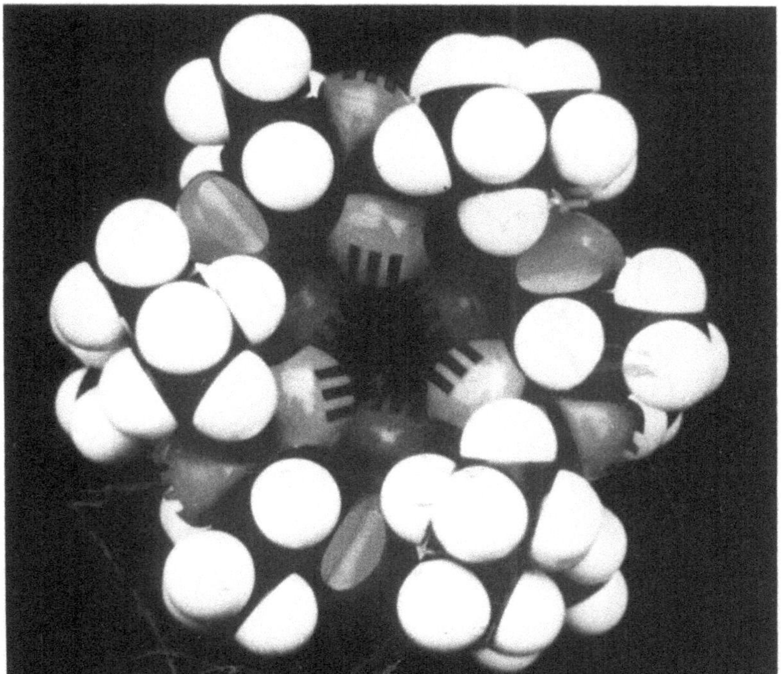

Fig. 6. Molecular model of valinomycin. (Courtesy Dr. B. F. GISIN, Rockefeller University, N. Y.)

ion. These ions may interact with the dipole moments of the inner boundary carbonyls *via* ion-dipole forces in the same way as they interact with their hydration shell in an aqueous medium. Both are electrostatic interactions which, if considered together, permit an understanding of the very high specificity of complex stability for different cations, as KRASNE and EISENMAN have shown (1973). Other groups have considered, in addition, variable conformational energies of valinomycin-ion-complexes for different cations (DIEBLER et al., 1969; SIMON and MORF, 1973; MAYERS and URRY, 1972). The conformation of the complexes are stabilized by up to six hydrogen bonds (depending on the polarity of the medium) which are formed between neighboring amide groups (Fig. 7) (SHEMYAKIN et al., 1969). Enniatin B shares most of the essential structural features of valinomycin, differing in the number of ring atoms (18 versus 36). Its structure is not stabilized by hydrogen bonds. The same is true for trinactin and the other macrotetrolides (nonactin, monactin, dinactin), which contain 4 ester carbonyls and 4 ether carbonyls, all of which act as ligands for cations (SIMON and MORF, 1973; LARDY et al., 1967). In contrast to these natural compounds, which were isolated from certain streptomyces species, the macrocyclic polyether shown in Figure 5 is a synthetic product, which also exhibits excellent qualities of complex formation with alkali ions (PEDERSEN and FRENSDORFF, 1972; CHOCK, 1972).

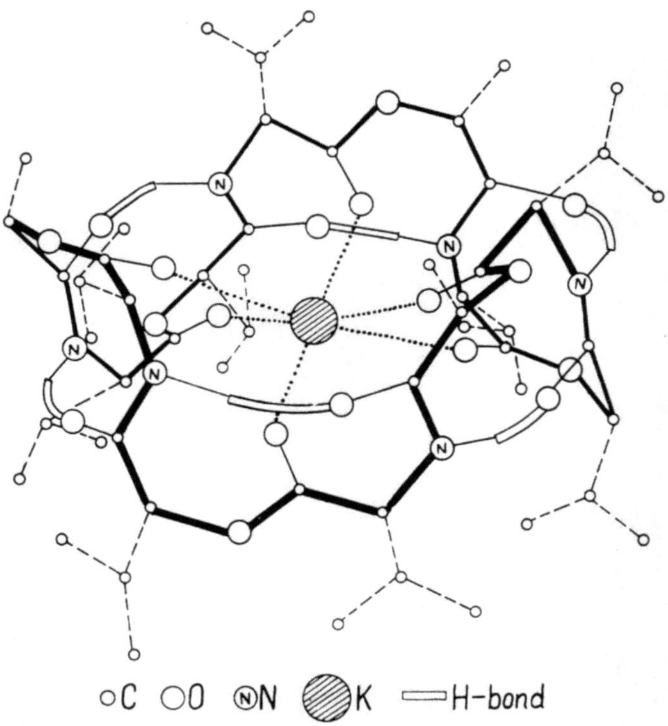

Fig. 7. Conformation of the valinomycin-K^+ complex. (Reproduced with permission from SHEMYAKIN et al., 1969)

The effect of these substances on the ion permeability is not limited to artificial lipid membranes, but seems to be present in all kinds of biological membranes. In fact, it was first shown by MOORE and PRESSMAN (1964) that valinomycin induces alkali ion transport across mitochondrial membranes, which subsequently was also found for the other compounds and in other biological membranes (PRESSMAN et al., 1967; LARDY et al., 1967; HAROLD and BAARDA, 1967). The same has been confirmed for artificial lipid membranes by different groups (MUELLER and RUDIN, 1967; ANDREOLI et al., 1967; LEV and BUZHINSKY, 1967; LIBERMAN and TOPALY, 1968; SZABO et al., 1969; McLAUGHLIN et al., 1972). There seems to be an increasing agreement that the other biological effects produced by the neutral macrocyclic compounds such as their antimicrobial activity (SHEMYAKIN et al., 1969; HAROLD and BAARDA, 1967) or their uncoupling of oxidative phosphorylation appear as a secondary consequence of an increased potassium permeability of the membranes in question. This further emphasizes the importance of the artificial lipid membranes as a model for natural membranes.

The characterization of the compounds shown in Figure 5 as ion carriers stems from a variety of qualitative arguments: both essential features of a translational carrier (Fig. 1 A), namely complex formation and free mobility inside the membrane, have been established. While the importance of the latter has been concluded from the large drop in conductance accompanying a phase change ("freezing") of the membrane (KRASNE et al., 1971), the close relationship between conductance and complex formation is demonstrated in Figure 8 for valinomycin and different monovalent cations. The ability of the considered substances to solubilize ions in media of low dielectric constant as ion-carrier complexes has been found also in macroscopic two phase systems such as water and organic solvents (PRESSMAN et al., 1967; EISENMAN et al., 1969). Furthermore, the presence of valinomycin and related compounds reduces the resist-

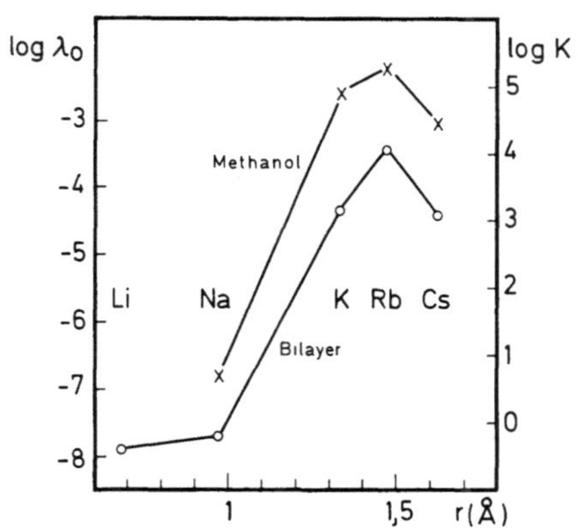

Fig. 8. Ion specificity of complex formation and transport for valinomycin. The membrane conductance λ_o (dioleoyllecithin, 25 °C) was measured in 10^{-2}M solutions of different alkali ions at a valinomycin concentration of 10^{-7} M, and plotted as a function of ion radius r. Stability constant K for complex formation in methanol according to GRELL et al. (1975)

ance of model membranes of macroscopic thickness (TOSTESON et al., 1968; WIPF et al., 1969), in which the formation of transmembrane structures, a prerequisite for the existence of pores, is not possible. Furthermore, in agreement with a simple carrier mechanism, there is a strict proportionality between conductance of a bilayer membrane and ionophore concentration for valinomycin and the macrotetrolides (SZABO et al., 1969; STARK and BENZ, 1971), which was observed over many orders of magnitude (Fig. 9). The same was found for the

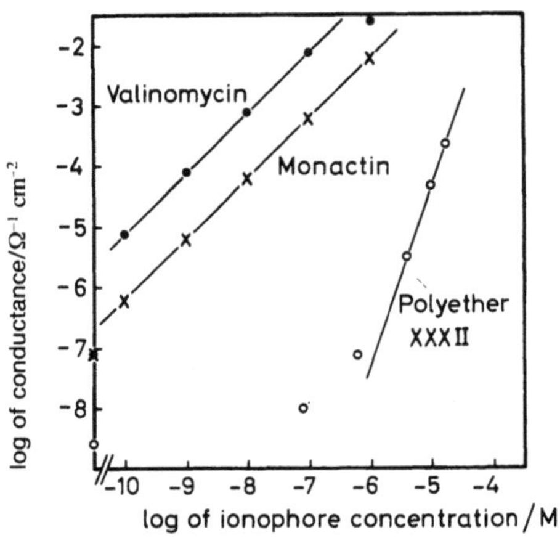

Fig. 9. Membrane conductance as a function of concentration of neutral macrocyclic ionophores. The data for valinomycin and monactin refer to lecithin bilayer membranes at 1 M KCl in the aqueous solutions (STARK and BENZ, 1971). The data for the polyether XXXII were measured with asolectin bilayer membranes at 10^{-3} M KC1 in water (MCLAUGHLIN et al., 1972). The full lines were drawn with slope 1 (valinomycin, monactin) and slope 3 (polyether)

conductance as a function of salt concentration (except at very high salt concentrations). This indicates that a 1:1 complex between one valinomycin (or macrotetrolide) molecule and one cation is the smallest transport unit and seems to exclude transport across pore like aggregates consisting of several valinomycin molecules. A linear relationship between the number of such aggregates and the total valinomycin concentration would be expected only for a very strong interaction between single valinomycin molecules. Indeed cooperative cation binding of valinomycin molecules has been found in valinomycin monolayers at an air-water interface (in contrast to the macrotetrolide trinactin) (KEMP and WENNER, 1973). Within bimolecular lipid membranes however, both antibiotics seem to behave largely identically and no direct evidence for the existence of aggregates has been reported. On the other hand, aggregates with molecularity higher than one seem to be responsible for cation transport induced by synthetic polyethers. MCLAUGHLIN et al. (1972), using polyethers synthesized according to PEDERSEN and FRENSDORFF (1972), found that the conductance increases as the third power of the polyether concentration (see Fig. 9). Therefore, one "polyether-transport unit" seems to consist of the least three "monomers."

Although one could imagine that several polyether molecules "stack up like life savers in a package to form a pore", McLaughlin et al. (1972) presented arguments in favor of a carrier mechanism: The polyether XXXIII is most efficient for large cations like Cs^+ or Rb^+. Cesium, however, is too large to fit easily through the 18-membered polyether ring, which would be necessary for a pore. Finally, the polyether-induced conductance is drastically reduced by freezing the membrane, which seems to stress the importance of free mobility of the transport unit inside the membrane.

The high selectivity of ion binding and transport of most neutral ion carriers has resulted in an increased interest of synthetic chemists in the underlying basic principles which govern structure-function relations. Thus, synthesizing analogues of valinomycin and enniatin, Shemyakin et al. (1969) have studied the influence of the primary structure on their activity. More recently, Gisin and Merrifield (1972) described the synthesis of a cyclopeptide analogue of valinomycin, namely cyclo- (D-val-L-pro-L-val-D-pro)$_3$), which shows excellent binding properties for potassium. Ting-Beall et al. (1974) reported that it also acts as an ion carrier in lipid bilayer membranes.

II. Charged Carriers

While the compounds described in the last section are neutral independent of pH, there is another class of ionophores which has in common the presence of one dissociable carboxyl group, which is responsible for one negative charge above the pK-value of this group of carriers. While the ion complexes of the valinomycin group are positively charged, those of the carboxylic ionophores may be either neutral or positively charged, depending on the valency of the cation and the stoichiometry of the complex. They were first described by Pressman (Pressman, 1968; Pressman et al., 1967), who found that antibiotics such as nigericin or monensin transport alkali ions as well as protons across mitochondrial membranes. Further members of this group have since been reported, which have the outstanding characteristic of also being able to induce transport of divalent cations across membranes. Figure 10 shows the structure of X-537 A, which in addition to monovalent ions like K^+ and Rb^+, binds and transports Ca^{2+} and other divalent cations (Pressman, 1973). Another carboxylic acid ionophore, A23187, which is also obtained from Streptomyces, even favors divalent cations against alkali metal cations and can be used to increase

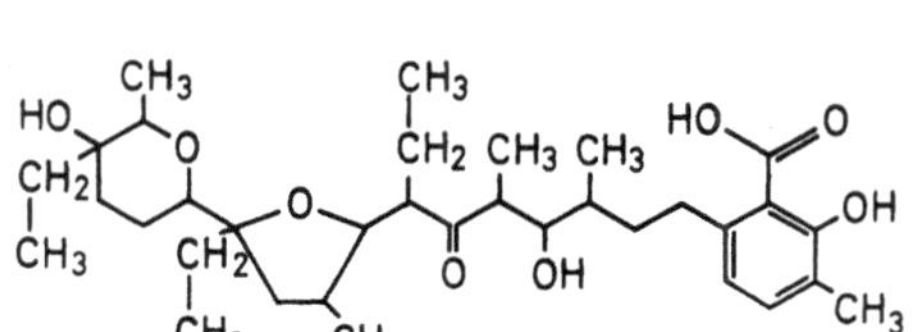

Fig. 10. Structure of X-537 A

selectively the calcium permeability of membranes (REED and LARDY, 1972). Studies performed by CELIS et al. (1974) and SCHADT et al. (1974) have shown that the conductance of black films depends on the square of the X-537A concentration, but linearly on the calcium concentration. They suggested the complex $(CaHA_2)^+$ as permeant species responsible for the conductance, where A is the anionic form of the carrier. A stoichiometry of 2:1 has also been concluded from X-ray analysis of the barium salt (JOHNSON et al., 1970). For the antibiotic A23187 a linear relation between membrane conductance and concentration was reported (CASE et al., 1974), consistent with a $(CaA)^+$-complex as charge carrier. Most authors agree, however, that the total cation flux across the membrane induced by the charged carriers exceeds by far the flux calculated on the basis of electrical conductances. This means that only a small percentage of the total flux is electrogenic. This is equivalent to the conclusion that most complexes are neutral and must have a different stoichiometry than that mentioned above.

The antibiotic X-537A not only facilitates the movement of ions but also induces an increased permeability for biogenic amines like dopamine and norepinephrine (PRESSMAN, 1973). This transport is also electrically silent and rather specific. Its rate depends linearly on the antibiotic concentration. This is a hint at a 1:1 complex functioning as transport unit (SCHADT and HÄUSLER, 1974).

The characterization of these compounds as carriers is largely based on the similarity of their effects with those of the hydrogen carriers and neutral macrocyclic carriers of alkali metal cations (see Sections B and C_I). Like them, they show ion-binding in bulk phases and the capacity to transport ions even through hydrocarbon-like barriers of macroscopic thickness (PRESSMAN, 1973). Even the structure and stoichiometry exhibit some analogies. Like neutral macrocyclic ion carriers, the charged compounds also seem to surround the complexed ion, thus allowing for an interaction between the ion and dipoles of oxygen groups and providing for a hydrophobic exterior of the complex.

Carboxylic ion carriers – again resembling hydrogen and macrocyclic carriers – produce a variety of different biological effects, which at least in part may be due to their ability to influence the permeability of membranes to cations or other biological important metabolites.

D. The Iodide-Iodine System

In the presence of NaCl in the aqueous phases, the conductance of bimolecular lipid membranes is very low, a consequence of the barrier properties of a two-dimensional lipid structure. If Cl^- is replaced by I^-, a very pronounced increase in conductance is observed, as first described by LÄUGER et al. (1967). Subsequently, it was found that this unexpected result requires the presence of small amounts of molecular iodine (LÄUGER et al., 1967; ROSENBERG and JENDRASIAK, 1968; FINKELSTEIN and CASS, 1968). Both ionic and electronic conductance mechanisms have been proposed to explain this striking effect (LÄUGER et al., 1967; ROSENBERG and JENDRASIAK, 1968). FINKELSTEIN and CASS (1968), however, provided convincing evidence that the anion I^- crosses the membrane as a lipid-soluble polyiodide complex. Specifically, they suggested that I_2 acts as a carrier for I^-. Zero-current potentials in the presence of gradients of I^- as well as gradients of I_2 are consistent with this view. Recently, SZABO et al. (1973) reported a square dependence of the conductance on the I_2-concentration and a linear dependence on the I^--concentration in water. Therefore, a I_5^--complex seems to be the predominant charge carrier inside the membrane. Similar results were obtained with Br^- and Br_2 by GUTKNECHT et al. (1972). Combining electrical and tracer experiments, they also showed a very high nonelectrogenic permeability of Br resulting from a high permeability of Br_2 through the membrane and a fast isotopic exchange between Br^- and Br_2 in the unstirred layers adjacent to the membrane.

E. The Carrier-Transport Model

While the last sections were mainly concerned with the nature and stoichiometry of the charge carriers inside the membrane, we will now proceed to ask more detailed questions about the transport mechanism. We will confine the discussion to neutral macrocyclic carriers of cations, such as valinomycin, for two reasons: (a) they show all the essential features of carrier-mediated transport, and (b) they appear as a sufficiently well studied carrier-transport system and allow us to demonstrate the usefulness of artificial lipid membranes as models for the study of transport mechanisms.

One of the first steps in the formulation of models for carrier transport was the determination of the locus where the ion-carrier complexes are formed. Figure 11 shows two possiblities which have to be considered, namely the aqueous phases adjacent to the membrane or the membrane-water interface. The second case (Fig. 11a) corresponds to the "classical" picture of a membrane-bound translational carrier S, which combines with an ion M^+ from the water (see Fig. 1). In the first case (Fig. 11b), a certain solubility of S and of the complex MS^+ in water is absolutely necessary. A distinction between the two

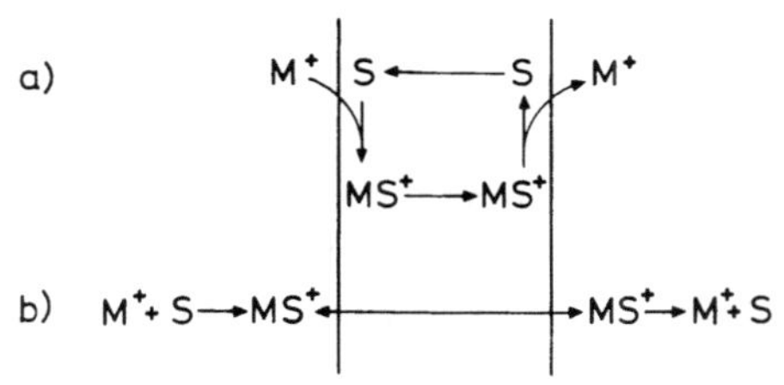

Fig. 11. Formation and dissociation of charge carriers in carrier-mediated transport (a) interface; (b) water

alternatives has been possible for valinomycin and the macrotetrolides, due to the very high conductance increase induced by these compounds. If charge transport through the membrane-water interfaces were to proceed by diffusion of complexes MS^+ formed in water (Fig. 11b), the "diffusion resistance" of the unstirred layers would be higher than the actual observed resistance in the presence of these carriers. This arises from the fact that the charge carriers in the unstirred layers do not experience the electric field, which is present only inside the membrane. Therefore, at high membrane conductances the flux of charge carriers across the unstirred layers is not high enough to account for the flux across the membrane, which is driven by the electric field. In the limiting case of extremely high membrane conductances, the resistance of the system will be given by that of the unstirred layers (see also Fig. 4, data for TTFB). As a consequence, the concentration of charge carriers at opposite membrane interfaces will be different and will also be different from the bulk concentration, *i. e.* diffusion polarization is developing. The presence of these unstirred layer effects is easily verified. After the application of a voltage jump (see below), the current shows a characteristic continuous decrease, ranging from very short times up to minutes (NEUMCKE, 1971). The appearance of diffusion polarization depends directly on the concentration of charge carriers in the aqueous phases. It is the ratio λ/c of membrane conductance to charge carrier concentration that determines whether unstirred layer effects have to be considered or not. From the absence of diffusion polarization in the case of the neutral carriers it had to be concluded that the interfacial reaction mechanism (Fig. 11a) must be present in the cation transport induced by them, since the concentration of metal ions M^+ exceeds the concentration of complexes MS^+ by many orders of magnitude (STARK and BENZ, 1971).

The theory of carrier mediated ion transport across artificial lipid membranes has been developed by using different approaches. Treating the membrane like a thin macroscopic phase, the conductance can be described on the basis of the classical electrodiffusion equation of Nernst and Planck. This has been done by CIANI et al. (1969) and MARKIN et al. (1969), using different boundary conditions. A fundamental difficulty in this kind of approach is that the thickness of a bilayer membrane is only about four times the diameter of most macrocyclic carrier molecules. Therefore, the validity of a macroscopic theory introduced for homogeneous phases is uncertain. Moreover, physical quantities such as diffusion coefficients or mobilities, which depend on the viscosity of a given medium, will not be constant throughout the membrane. Different physical methods have provided evidence for variable microviscosity inside a membrane (comparing the head group region with the interior). In addition, the chemical potential of charged particles inside ultrathin membranes which separate two aqueous phases of high dielectric constant is not constant. Image forces acting near the interface of two media of different dielectric constants give rise to a higher potential energy in the middle of the membrane (NEUMCKE and LÄUGER, 1969). These difficulties may be partly overcome by using a generalized form of the Nernst-Planck equation (CIANI et al., 1973). Even then the theory contains too many unknown functions and is probably not practical. A simplified treatment of carrier-mediated ion transport can be made by taking into account some specific properties of carriers in lipid bilayers. Lipid membranes not only represent barriers to ion transport, but also consist of two interfaces to which compounds showing interfacial activity may adsorb. Many substances believed to act as carriers have been found to be surface-active, *e.g.* they adsorb onto lipid monolayers at an air-water interface or form monolayers by themselves (SHEMYAKIN et al., 1969; KEMP and WENNER, 1973). Convincing evidence for the surface activity of valinomycin in bilayers has been provided by HSU and CHAN (1973), who found in their NMR-studies of dipalmitoyllecithin liposomes that the lines that are influenced by valinomycin are only those which originate from the "interfacial part" of the lipid molecules. It therefore seems reasonable to assume a profile of the potential energy for the carrier molecules inside the membrane, as shown schematically in Figure 12. Its essence is that the carrier molecules are preferentially restricted to the interfaces and have to overcome an energy barrier in order to cross the membrane. This profile also underlies the reaction scheme shown in Figure 13. It describes charge transport through the interface in terms of a simple bimolecular reaction between a metal ion M^+ from the aqueous phase and a free carrier molecule S located at an interfacial energy minimum. Both the interfacial reaction and the diffusion across the membrane are quantitatively described by rate constants. The translocation rate constants k_S and k_{MS} replace a treatment of membrane conductance on the basis of the Nernst-Planck equation. Their physical significance can be illustrated by noting that $1/k_S$ is the mean time for a free carrier molecule S to cross the energy barrier.

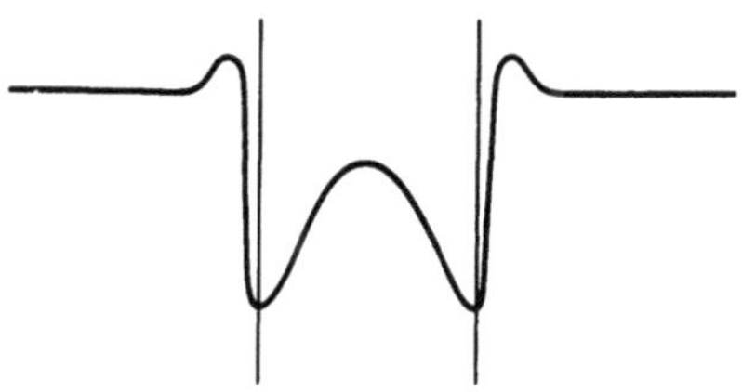

Fig. 12. Simplified energy profile of neutral macrocyclic ion carriers in bilayer membranes

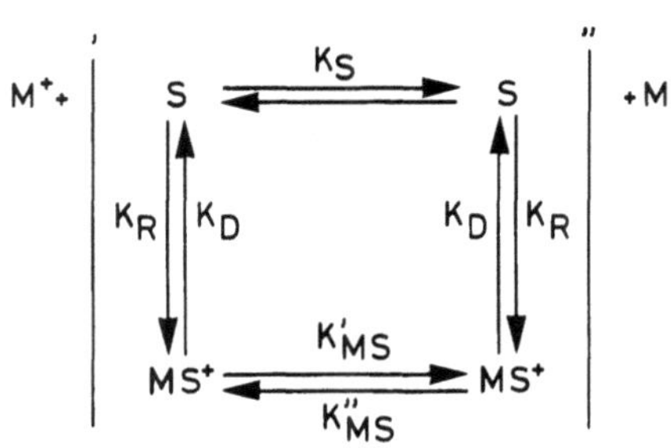

Fig. 13. Simplified reaction scheme of ion transport induced by neutral macrocyclic ion carriers

The analysis of carrier transport on the basis of rate constants allows answers to questions of immediate importance concerning the velocity of single transport steps. Through the determination of these rate constants using non-steady-state experimental techniques (see next section) it is possible to assess whether the charge transfer through the interfaces is the rate-limiting step or whether this step is the transport through the membrane interior. One disadvantage of this kinetic treatment, however, consists in the difficulty to give a precise interpretation of the rate constants in molecular terms. As in the case of chemical reaction mechanisms, the improvement of experimental methods will probably lead to the detection of further "intermediate steps." The description of diffusion in terms of rate constants was introduced by ZWOLINSKY et al. (1949). They considered diffusion as a movement of particles across a series of activation barriers. The height of these barriers can be enhanced or reduced by an electric field, depending on the direction of movement relative to the field. Correspondingly the voltage dependence of the translocation rate constants k'_{MS} and k''_{MS} for the positively charged carrier-ion complex can be expressed as (LÄUGER and STARK, 1970):

$$\begin{aligned} k'_{MS} &= k_{MS}\,e^{-u/2} \\ k''_{MS} &= k_{MS}\,e^{u/2} \end{aligned} \qquad \text{with } u = FV/RT \qquad (8)$$

(F = Faraday constant, V = applied voltage, R = gas constant, T = absolute temperature). Equations (8) are strictly valid only under certain conditions. Firstly, the energy minima must coincide with the interfaces so that the complexes MS^+ on moving from one minimum to the other experience the total voltage applied to the membrane. The rate constants k_R and k_D of the interfacial reaction can then be considered as voltage-independent. Secondly, the energy barrier must have a very steep maximum in the middle of the membrane. Both prerequisites are not strictly met, as has been found by comparison between theory and practice (STARK and BENZ, 1971; HALL et al., 1973; HLADKY, 1974; KNOLL and STARK, 1975). Nevertheless, the simple model shown in Figure 13 can be considered as a good first-order approximation for more complicated phenomena. The next section will show how, by application of suitable stationary and nonstationary electrical methods, a carrier model can be submitted to a critical test.

I. Kinetic Analysis of the Carrier Model

The study of transport phenomena across natural biological membranes, apart from some favorable systems, is largely based on tracer experiments. Planar artificial lipid membranes, however, offer the possibility of electrical measurements, which, apart from greater convenience, greatly expand the information on transport mechanisms.

We consider the electrical conductance of a membrane which incorporates a carrier system according to Figure 13. It depends, in addition to the rate constants determining the velocity of the system, on the total number of carrier molecules per square centimeter of membrane. Usually it is rather difficult to determine concentration- and velocity-quantities separately. One source of information is the nonlinearity of stationary current-voltage characteristics, which does not depend on the carrier concentration and is only a function of the relative magnitude of the single rate constants. If the voltage-dependent translocation of the carrier-ion complex across the interior of the membrane is rate-limiting, then, due to the special properties of the barrier shape, the current rises more steeply than the voltage, *i. e.* a superlinear *J*-*V* curve is observed. If, on the other hand, the dissociation of the complexes at the interface or the translocation of the neutral species S across the membrane is rate-limiting, both of which are assumed to be voltage-independent, a saturating current with increasing voltage should result. In agreement with these ideas, the mathematical analysis of the simple carrier model of Figure 13 shows that the nonlinearity of current-voltage curves, which is conveniently expressed as the voltage-dependent conductance ratio $\lambda/\lambda_o = (J/U) / (J/U)_{U \simeq 0}$, only depends on one parameter A (STARK and BENZ, 1971):

$$A = 2\, k_{MS}/k_D + (k_{MS}/k_D)\,(k_R c_M / k_S), \tag{9}$$

with c_M = concentration of M^+ in water.

Therefore, the concentration dependence of the nonlinearity in favorable cases allows the determination of the two ratios k_{MS}/k_D and k_R/k_S. Figure 14 shows an example, which demonstrates the basic correctness of the ideas thus

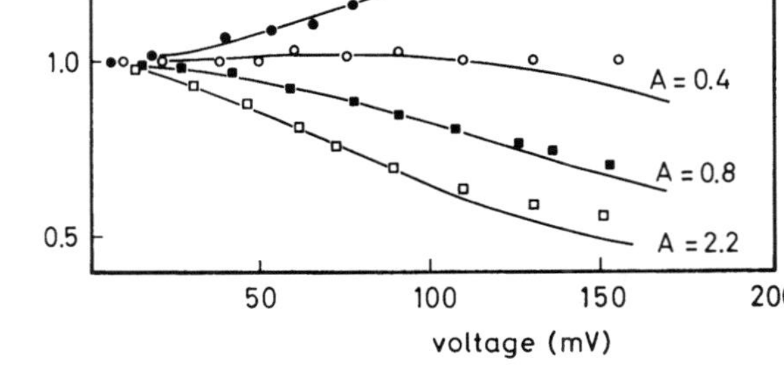

Fig. 14. Conductance ratio λ/λ_o as a function of voltage for trinactin in monoolein membranes at different salt concentrations. □ 1 M NH_4Cl; ■ 10^{-2} M NH_4Cl (+1 M LiCl); ○ 1 M RbCl; ● 10^{-2} M RbCl (+1 M LiCl). Full lines calculated on the basis of Fig. 13, using Eqs. (8). (Reproduced with permission from BENZ and STARK, 1975)

far developed, but also indicates clear deviations at high voltages. These deviations arise from the only approximative treatment of the voltage dependence of the rate constants. While symmetrical systems have been treated up to this point, similar considerations can be applied to asymmetrical membranes or to different aqueous solutions on both sides of the membrane. The rectification behavior observed under such conditions has also been found to be in agreement with predictions from the simple model of Figure 13 (STARK, 1973). However, from steady-state measurements alone only combinations of the rate constants can be determined. Analogous to the kinetics of chemical reactions in homogeneous phases, the determination of the single rate constants of a model requires the measurement of a time-dependent property of the system. Since carriers such as valinomycin are assumed to be very efficient, *i. e.* very fast transport systems, the application of fast relaxation methods seems appropriate. The principle of these methods consist in a sudden change of an external parameter and a measurement of the time dependence of an internal parameter of the system. For experiments with thin lipid films the electric current may serve as the time-dependent property, which is shifted from one steady state (or equilibrium) to another steady state by a stepwise change in the electric field or the temperature. Especially the voltage jump method illustrated in Figure 15 has been successfully used for different kinds of problems. A voltage pulse from a generator is applied to the membrane and the current is followed with an oscilloscope *via* the voltage drop across an external resistor. The current spike accompanying the voltage step arises from the loading of the membrane capacity and limits the time resolution of the method (about 1 μs under favorable conditions). A decrease of the current indicated in the figure is frequently observed with different transport systems and may indicate one of the following time-dependent processes. Firstly, if the unstirred layers influence the conductance, diffusion polarization produces a characteristic decrease of the current, which mirrors the development of an opposite diffusion potential across the membrane (see last section). Secondly, in the case of neutral macrocyclic ion carriers (in the absence of diffusion polarization) a relaxation in the μs region is

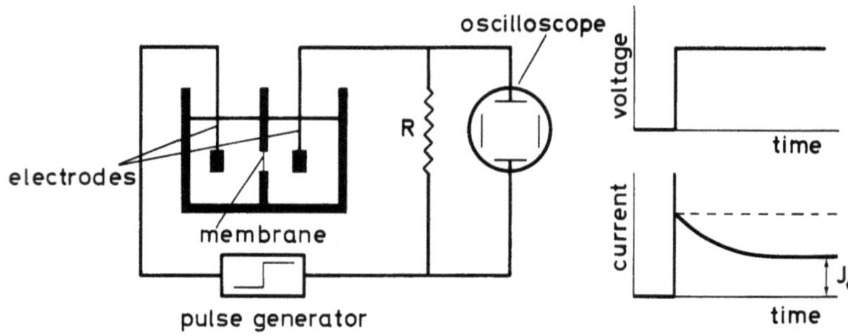

Fig. 15. Principle of voltage jump current relaxation technique with lipid bilayer membranes

often observed. Such a relaxation is expected within the framework of the carrier model of Figure 13. The application of a voltage generates an asymmetrical energy profile, which induces a concomitant shift of the interfacial concentrations of the carrier complexes towards new stationary values. This concentration shift is accompanied by a decrease of the current, which can be described as (STARK et al., 1971):

$$J = J_\infty (1 + \alpha_1 e^{-t/\tau_1} + \alpha_2 e^{-t/\tau_2}), \tag{10}$$

with J_∞ = stationary current reached at long times.

The four parameters τ_1 and τ_2 (relaxation times) and α_1, α_2 (relaxation amplitudes) only depend on the four unknown rate constants of the model (apart from the applied voltage and the ion concentration c_M), which can be determined in this way. If only the slower time can be resolved, the information from the steady-state J-V curves can be used, in addition, to get a full quantitative analysis of the transport model (STARK et al., 1971). Figure 16 shows that in agreement with the predictions of the model, the time dependence of the current in the presence of valinomycin shows two relaxation times (LAPRADE et al., 1975; KNOLL and STARK, 1975). The same has been found for trinactin (unpublished). These results indicate the range of application of the voltage jump method, which is complemented by other relaxation techniques known from chemical kinetics and electrochemistry:

(a) The charge pulse technique, which is related to the voltage jump method. In this case the membrane capacitance is charged rapidly (less than 1 μs). This charge then decays through the membrane's conductive mechanism and this is

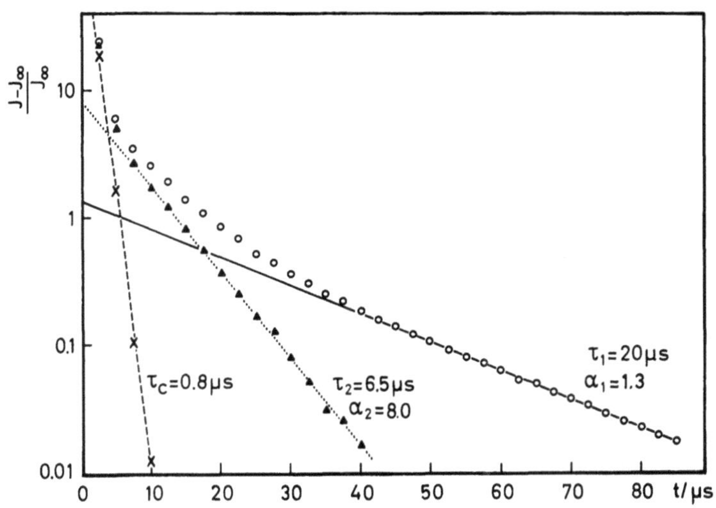

Fig. 16. Typical current relaxation following a voltage jump of 90 mV across a monoolein membrane in the presence of valinomycin and 0.1 M RbCl. The time dependence of the current (o) is composed of two exponentials (Eq. 10). Both relaxation times are well separated from the charging time constant τ_C of the membrane capacity. (Reproduced with permission from KNOLL and STARK, 1975)

measured as a voltage decay which is the result of membrane and boundary layer phenomena only (FELDBERG and KISSEL, 1975). Its potential usefulness consists in an improvement of the time resolution, since the method is not limited by the *RC* time of the "membrane-circuit;"

(b) The temperature jump method uses a stepwise change of the temperature as driving force and takes advantage of the temperature dependence of the conductance. The measurement follows the time course of the current at a fixed voltage. While the voltage jump and the charge pulse method only excite field-dependent processes, the temperature jump method is probably of more general applicability, since most phenomena show a more or less pronounced temperature dependence. A relatively slow version of this method has been used to separate the temperature dependence of the partition coefficient from that of the rate constants of transport in valinomycin induced transport (STARK et al., 1972). A faster method involves generating the temperature jump through absorption of a very intensive flash of light (STARK and KNOLL, unpublished).

II. Valinomycin and Trinactin

The action of valinomycin and trinactin on the cation permeability of lipid membranes has been analyzed along the lines discussed in the last section for different lipids and with increasing sophistication in recent years (STARK et al., 1971; BENZ et al., 1973; GAMBALE et al., 1973; LAPRADE et al., 1975; HLADKY, 1975; BENZ and STARK, 1975; KNOLL and STARK, 1975). Table 1 contains values for the single rate constants as well as for the turnover number *f*. The latter is defined as the maximum number of ions, which may be transported by a single carrier molecule per second and depends on the rate constants as follows (LÄUGER, 1972):

$$f = \left[\frac{1}{k_S} + \frac{1}{k_{MS}} + \frac{2}{k_D}\right]^{-1} \tag{11}$$

Depite some differences in the single rate constants (the conductance in the presence of valinomycin is rather controlled through the surface reaction compared to trinactin), the turnover number is equal for both ion carriers and is more than three orders of magnitude smaller than the number of ions moving through a single channel formed by gramicidin A. This high single-channel conductance can only be explained if an almost unhindered diffusion of ions inside an aqueous pore is assumed. In contrast, the much larger carrier-ion complexes experience the considerable lower fluidity of the membrane interior. The high "transport efficiency" of a gramicidin pore is, however, accompanied by a rather poor ion selectivity compared to valinomycin or trinactin. The association rate constant k_R calculated from Table 1 is several orders of magnitude lower than the one found in methanol (GRELL et al., 1975). This might only indicate a lower ion concentration than in water at the place of complex formation (KNOLL and STARK, 1975). The exact location of the reaction plane at the

Table 1. Rate constants of Rb transport induced by the neutral macrocyclic carriers valinomycin and trinactin. Their turnover number f is compared with the number of ions moved through a single channel of the pore former gramicidin A at 1 M CsCl and 100 mV applied voltage (BAMBERG et al., 1975). All data refer to glycerylmonooleate membranes at 25 °C

	$k_R c_M/s^{-1}$	k_D/s^{-1}	k_{MS}/s^{-1}	k_S/s^{-1}	f/s^{-1}
Valinomycin	$2 \cdot 10^5$	$6 \cdot 10^4$	$2 \cdot 10^5$	$3 \cdot 10^4$	10^4
Trinactin	$1 \cdot 10^5$	$2 \cdot 10^5$	$2 \cdot 10^4$	$5 \cdot 10^4$	10^4
Gramicidin A	–	–	–	–	$(6 \cdot 10^7)$

interface, which also influences the voltage dependence of the interfacial reaction, is a further problem under study at present. The answer to these questions will probably lead to a refinement of the simplified mechanism of Figure 13 and thus to a better physical understanding of carrier-mediated ion transport through membranes.

F. Biological Implications

It seems there is some justification for asking whether and how the studies dealt with in this article offer any answers to questions of biological relevance, since the membranes, as well as most of the substances that have been found to increase their ion permeability, are "artificial." If one regards lipid membranes as a means to a study of the transport mechanisms of biological membranes, one has to admit that this aim has only partly been reached. The active reconstitution of membrane proteins into lipid films has met with considerable difficulties. Nevertheless, the studies throughout the last decade have elucidated the basic principles of ion transport through a lipid barrier. In addition, since their mechanism of action is now better understood, ionophores can be used as probes for the study of transport mechanisms in biological membranes (HAROLD et al., 1974). The use of proton or potassium carriers allows one to make membranes selectively permeable for these ions. In the case of an ion gradient across the membrane, a membrane potential can be generated by adding the corresponding ionophore. Thus, the influence of an electric potential on different membrane processes such as oxidative phosphorylation can now be examined (MITCHELL and MOYLE, 1969). Besides this, the study of ionophore mechanisms in lipid bilayers has led to the development of special techniques and methods, some of which are mentioned in this article. The same methods will also be of great advantage if better techniques for the reconstitution of transport proteins in bilayers become available in the future.

Finally, studies with lipid films have contributed to a better understanding of the principles of ion selectivity (KRASNE and EISENMAN, 1973) and have initiated the synthesis of membrane active peptides (TING-BEALL et al., 1974), which could make it possible in the future to mimic more complex membrane functions, such as active transport, *in vitro*.

References

ANDREOLI, T. E., TIEFFENBERG, M., TOSTESON, D. C.: J. gen. Physiol. **50**, 2527 (1967).

BAMBERG, E.: Elektrochemische Untersuchungen an bimolekularen Lipidmembranen in Gegenwart von Entkopplern der oxidativen Phosphorylierung. Ph. D. Thesis, University of Basel, 1971.

BAMBERG, E., NODA, K., GROSS, E., LÄUGER, P.: Biochim. biophys. Acta **419**, 223 (1976).

BENZ, R., STARK, G.: Biochim. biophys. Acta **382**, 27 (1975).

BENZ, R., STARK, G., JANKO, K., LÄUGER, P.: J. Membrane Biol. **14**, 339 (1973).

BIELAWSKI, J., THOMPSON, T. E., LEHNINGER, A. L.: Biochem. biophys. Res. Commun. **24**, 948 (1966).

BORISOVA, M. P., ERMISHKIN, L. N., LIBERMAN, E. A., SILBERSTEIN, A. Y., TROFINOV, E. M.: J. Membrane Biol. **18**, 243 (1974).

BROCKMANN, H., GEEREN, H.: Liebigs Ann. Chem. **603**, 216 (1957).

BRUNER, L. J.: Biophysik **6**, 241 (1970).

BRUNER, L. J.: J. Membrane Biol. **22**, 125 (1975).

CASE, G. D., VANDERKOOI, J. M., SCARPA, A.: Arch. Biochem. Biophys. **162**, 174 (1974).

CELIS, H., ESTRADA-O., S., MONTAL, M.: J. Membrane Biol. **18**, 187 (1974).

CHOCK, P. B.: Proc. nat. Acad. Sci. (Wash.) **69**, 1939 (1972).

CIANI, S., EISENMAN, G., SZABO, G.: J. Membrane Biol. **1**, 1 (1969).

CIANI, S., LAPRADE, R., EISENMAN, G., SZABO, G.: J. Membrane Biol. **11**, 255 (1973).

DIEBLER, H., EIGEN, M., ILGENFRITZ, G., MAAS, G., WINKLER, R.: Pure appl. Chem. **20**, 93 (1969).

EISENMAN, G., CIANI, S., SZABO, G.: J. Membrane Biol. **1**, 294 (1969).

FELDBERG, S. W., KISSEL, G.: J. Membrane Biol. **20**, 269 (1975).

FINKELSTEIN, A.: Biochim. biophys. Acta **205**, 1 (1970).

FINKELSTEIN, A., CASS, A.: J. gen. Physiol. **52**, 145 (1968).

FOSTER, M., MCLAUGHLIN, S.: J. Membrane Biol. **17**, 155 (1974).

GAMBALE, F., GLIOZZI, A., ROBELLO, M.: Biochim. biophys. Acta **330**, 325 (1973).

GISIN, B. F., MERRIFIELD, R. B.: J. Amer. chem. Soc. **94**, 6165 (1972).

GRELL, E., FUNCK, Th., EGGERS, F.: In: Membranes (G. Eisenman, Ed.), Vol. III, Chap. 1. New York: M. Dekker (1975).

GUTKNECHT, J., BRUNER, L. J., TOSTESON, D. C.: J. gen. Physiol. **59**, 486 (1972).

HALL, J. E., MEAD, C. A., SZABO, G.: J. Membrane Biol. **11**, 75 (1973).

HAROLD, F. M., BAARDA, J. R.: J. Bacteriol. **94**, 53 (1967).

HAROLD, F. M., ALTENDORF, K. H., HIRATA, H.: Ann. N. Y. Acad. Sci. **235**, 149 (1974).

HLADKY, S. B.: J. Membrane Biol. **10**, 67 (1972).

HLADKY, S. B.: Biochim. biophys. Acta **352**, 71 (1974).

HLADKY, S. B.: Biochim. biophys. Acta **375**, 327 (1975).

HOPFER, U., LEHNINGER, A. L., THOMPSON, T. E.: Proc. nat. Acad. Sci. (Wash.) **59**, 484 (1968).

HOPFER, U., LEHNINGER, A. L., LENNARZ, W. J.: J. Membrane Biol. **3**, 142 (1970).

HSU, M., CHAN, I. S.: Biochemistry **12**, 3872 (1973).

JOHNSON, S. M., HERRIN, J., LIU, S. J., PAUL, I. C.: J. Amer. chem. Soc. **92**, 4428 (1970).

KEMP, G., WENNER, C.: Biochim. biophys. Acta **323**, 161 (1973).

KETTERER, B., NEUMCKE, B., LÄUGER, P.: J. Membrane Biol. **5**, 225 (1971).

KNOLL, W., STARK, G.: J. Membrane Biol. **25**, 249 (1975).

KRASNE, S., EISENMAN, G.: In: Membranes (G. Eisenman, Ed.), Vol. II, Chap. 3, V. New York: Dekker p. 277 (1973).

KRASNE, S., EISENMAN, G., SZABO, G.: Science **174**, 412 (1971).

LAPRADE, R., CIANI, S. M., EISENMAN, G., SZABO, G.: In: Membranes (G. Eisenman, Ed.), Vol. III. New York: Dekker p. 127 (1975).

LARDY, H. A., GRAVEN, S. N., ESTRADA-O., S.: Fed. Proc. **26**, 1355 (1967).

LÄUGER, P.: Science **178**, 24 (1972).

LÄUGER, P., STARK, G.: 1970. Biochim. biophys. Acta **211**, 458 (1970).

LÄUGER, P., LESSLAUER, W., MARTI, E., RICHTER, J.: Biochim. biophys. Acta **135**, 20 (1967 a).

LÄUGER, P., RICHTER, J., LESSLAUER, W.: Ber. Bunsenges. Physik. Chem. **71**, 906 (1967 b).

LEA, E. J. A., CROGHAN, P. C.: J. Membrane Biol. **1**, 225 (1969).

LeBlanc, O. H., Jr: Biochim. biophys. Acta **193**, 350 (1969).
LeBlanc, O. H., Jr: J. Membrane Biol. **4**, 227 (1971).
Lev, A. A., Buzhinsky, E. P.: Tsitologiya **9**, 102 (1967).
Liberman, E. A., Topaly, V. P.: Biochim. biophys. Acta **163**, 125 (1968).
McLaughlin, S.: J. Membrane Biol. **9**, 361 (1972).
McLaughlin, S. G. A., Szabo, G., Ciani, S., Eisenman, G.: J. Membrane Biol. **9**, 3 (1972).
Markin, V. S., Krishtalik, L. I., Liberman, Ye. A., Topaly, V. P.: Biofizika **14**, 256 (1969 a).
Markin, V. S., Pastushenko, V. F., Krishtalik, L. I., Liberman, Ye. A., Topaly, V. P.: Biofizika, **14**, 462 (1969 b).
Mayers, D. F., Urry, D. W.: J. Amer. chem. Soc. **94**, 77 (1972).
Mitchell, P.: Biochem. J. **81**, 24 (1961).
Mitchell, P., Moyle, J.: Europ. J. Biochem. **7**, 471 (1969).
Moore, C., Pressman, B. C.: Biochem. biophys. Res. Commun. **15**, 562 (1964).
Mueller, P., Rudin, D. O.: Biochem. biophys. Res. Commun. **26**, 398 (1967).
Mueller, P., Rudin, D. O., Tien, H. T., Wescott, W. C.: Nature **194**, 979 (1962).
Neumcke, B.: Biophysik **7**, 95 (1971 a).
Neumcke, B.: T.-I.-T. J. Life Sci. **1**, 85 (1971 b).
Neumcke, B., Bamberg, E.: In: Membranes (G. Eisenman, Ed.), Vol. III. New York: Dekker, p. 215 (1975).
Neumcke, B., Läuger, P.: Biophys. J. **9**, 1160 (1969).
Ovchinnikov, Y. A., Ivanov, V. T., Shkrob, A. M.: Membrane-active complexons. Amsterdam: Elsevier 1974.
Parsegian, A.: Nature **221**, 844 (1969).
Pedersen, C. J., Frensdorff, H. K.: Angew. Chemie **84**, 16 (1972).
Pohl, G., Stark, G., Trissl, H.-W.: Biochim. biophys. Acta **318**, 478 (1973).
Pressman, B. C.: Fed. Proc. **27**, 1283 (1968).
Pressman, B. C.: Fed. Proc. **32**, 1968 (1973).
Pressman, B. C., Harris, E. J., Jagger, W. S., Johnson, J. H.: Proc. nat. Acad. Sci. (Wash.) **58**, 1949 (1967).
Reed, P. W., Lardy, H. A.: A 23187: J. Biol. Chem. **247**, 6970 (1972).
Rosenberg, B., Jendrasiak, G. L.: Chem. Phys. Lipids **2**, 47 (1968).
Schadt, M., Haeusler, G.: J. Membrane Biol. **18**, 277 (1974).
Shemyakin, M. M., Aldanova, N. A., Vinogradova, E. I., Feigina, M. Yu.: Tetrahedron letters **28**, 1921 (1963).
Shemyakin, M. M., Ovchinnikov, Yu. A., Ivanov, V. T., Antonov, V. K., Vinogradova, E. I., Shkrob, A. M., Malenkov, G. G., Evstratov, A. V., Laine, I. A., Melnik, E. I., Ryabova, I. D.: J. Membrane Biol. **1**, 402 (1969).
Simon, W., Morf, W. E.: In: Membranes (G. Eisenman, Ed.), Vol. II, Chap. 4. New York: Dekker 1973.
Skulachev, V. P., Sharaf, A. A., Liberman, E. A.: Nature **216**, 718 (1967).
Stark, G.: Biochim. biophys. Acta **298**, 323 (1973).
Stark, G., Benz, R.: J. Membrane Biol. **5**, 133 (1971).
Stark, G., Ketterer, B., Benz, R., Läuger, P.: Biophys. J. **11**, 981 (1971).
Stark, G., Benz, R., Pohl, G., Janko, K.: Biochim. biophys. Acta **266**, 603 (1972).
Szabo, G., Eisenman, G., Ciani, S.: J. Membrane Biol. **1**, 346 (1969).
Szabo, S., Eisenman, G., Laprade, R., Ciani, S. M., Krasne, S.: In: Membranes (G. Eisenman, Ed.), Vol. II, Chap. 3. New York: Dekker 1973.
Ting-Beall, H. P., Tosteson, M. T., Gisin, B. F., Tosteson, D. C.: J. gen. Physiol. **63**, 492 (1974).
Tosteson, D. C., Andreoli, T. E., Tieffenberg, M., Cook, P.: J. gen. Physiol. **51**, 373 S (1968).
Wipf, H.-K., Pioda, L. A. R., Stefanac, Z., Simon, W.: Helv. chimica Acta **51**, 377 (1968).
Wipf, H. K., Pache, W., Jordan, P., Zähner, H., Keller-Schierlein, W., Simon, W.: Biochem. biophys. Res. Commun. **36**, 387 (1969).
Zwolinsky, B. J., Eyring, H., Reese, C.: J. Phys. Colloid Chem. **53**, 1426 (1949).

Chapter 13

Channels in Black Lipid Films

J. E. Hall

A. Introduction

This review will deal with both basic conclusions and ways in which pore formers can be studied in detail. The most elementary consideration is how to determine whether conductance is *via* a pore-like mechanism. This will be first in order of business. Next we will discuss the basic properties of the conductance at a macroscopic level such as voltage-current curves, response to a voltage-pulse, and dependence of conductance on pore-former concentration. By "macroscopic" we mean of sufficiently large magnitude that individual conductance events are indistinguishable. We will then consider conductance properties at the pore level in detail and finally discuss how single-channel properties can explain the macroscopic properties of the conductances induced by some pore formers. At this point we will have earned the right to speculate a little on possible molecular mechanisms through which pore formers act.

I have written this review with workers in neurophysiology and membrane transport in mind. In particular I have tried to show how experiments with pore formers in black lipid films compel one to believe that simple molecules interacting with each other, not covalently, but largely through hydrophobic interactions, can mimic many of the essential features of biological membranes. In the black lipid film system[1], it is possible to make a variety of measurements, some of which cannot be performed *in vivo*. For some pore-forming antibiotics, this variety is great enough for a plausible picture of how the molecule acts at the molecular level to emerge from the experimental data. Nonetheless, considerable uncertainty as to the detailed nature of the molecular interactions remains, even for the best understood pore formers.

These pore formers are clearly not the molecules responsible for normal conductances in biological membranes. They do, however, have similar effects on black lipid films and biological membranes, and it is possible that the ways in which pore formers work are similar to the ways in which the normal conduct-

[1] We assume the reader is familiar with the basic techniques of black lipid film formation and the essentials of the measuring apparatus used for making conductance measurements. If not, Hladky and Haydon (1972), Mueller and Rudin (1968) Eisenberg et al. (1973) and Muller and Finkelstein (1972 a) give useful detail on these subjects. In this article, as the necessity arises, we will discuss additional techniques and instrumentation used in specialized experiments on pore formers.

ance mechanisms in biological membranes work. I would ask the reader who works with biological membranes not to think in terms of the obvious discrepancies between the properties of the pore formers discussed and properties of particular mechanisms *in vivo*, but to exercise his imagination and, where possible, to try to see ways in which a particular pore former or its membrane environment could be modified so that it would act more like some actual *in-vivo* mechanism. For example, one should not object that the conductances induced by alamethicin in membranes of certain lipid composition respond to a voltage pulse too slowly for the mechanism of action of alamethicin to be similar to that of the sodium channel in nerve. Instead one should be aware that altered lipid composition speeds up the action of alamethicin without altering other basic features of the conductance. In this way a general rule about the way in which lipid composition affects voltage-dependent conductances may be suggested. Similarly, observation of the conductances induced by pore formers, correlation with antibiotic structure, modification of the pore former, and alteration of membrane composition may enable us to develop a set of general principles of molecular interaction in membranes. While we have not yet reached the point of reaping all the rewards of such a program, we have got the program fairly well underway, have already realized some important results, and can see exciting possibilities ahead.

Several reviews (HAYDON and HLADKY, 1972; MCLAUGHLIN and EISENBERG, 1975) have discussed antibiotics which alter the conductance properties of black lipid films. These articles focus on the properties of the various substances discussed. This article will instead focus on the experimental techniques which have been used to study pore formers and attempt to provide a critical assessment of the rationale behind various courses of experiment.

No single technique provides sufficient data to allow interpretation at a molecular level, even in the simplest cases, but often a combination of techniques can allow construction of a plausible and reasonably detailed molecular picture.

Differing degrees of information are available for different pore formers. For gramicidin we have much information on ion selectivity, the only measurement of an actual membrane concentration of a pore-forming antibiotic, and very nearly complete confirmation that two gramicidin molecules form a pore. For alamethicin we have the dependence of conductance on aqueous concentration of the antibiotic, much statistical information on formation of pores, kinetic data, some data on modified forms of the molecule, many voltage-current curves, and single-channel data. For monazomycin we have much interesting data on inactivation, kinetics, ion selectivity, and voltage-current curves, some interesting "noise" measurements, but only meagre single channel data. For EIM and hemocyanin, we can relate single channel data and high-level conductances, but have almost no information on molecular structure. We can barely demonstrate that the polyene antibiotics amphotericin B and nystatin form pores, but nonetheless by measuring dependence of conductance on antibiotic concentration and steroid requirements for elevated conductance, we can make surprisingly detailed guesses about the molecular structure of the pores they form. For none of the pore formers can we make a molecular model of action that is satisfactory in all respects, but for some we can make some interesting and elegant guesses which suggest, at least, important considerations to be kept in mind when thinking about channels in biological membranes.

In organizing this review, I have followed a strongly experimental approach. In particular I have divided experiments into two categories: "basic" and "advanced." Basic experiments are essentially those experiments that can be performed on most biological membranes with mid-1976 technology. I have also included in the basic category those experiments that enable us to class a substance as a pore former. Advanced experiments are those which necessarily require the use of bilayer technology or which become much more powerful used in bilayers because the system is then well defined. It is these "advanced" experiments that are most useful in evaluating molecular-level models of some of the pore formers.

B. Basic Experiments

We will first consider a set of basic experiments designed to describe a given conductance mechanism at more or less the level at which we understand in squid axon and frog nerve. We will then discuss "advanced" experiments, which can sometimes provide detailed pictures of the mode of action of a given substance. Most of these depend on using the basic black lipid film system and are not now possible in biological preparations.

Convincing explanation in molecular terms of the conductances induced by pore-forming antibiotics depends on our ability to control a wide variety of experimental variables, some of which are accessible neither to manipulation nor to measurement in biological membranes. In particular, the low background conductance of black lipid films makes possible the direct measurement of conductances in the order of 10^{-14} to 10^{-9} mho (depending on the resolution time). Single-channel conductances have been measured in a frog muscle preparation, but not yet in any biological membrane exhibiting a strongly voltage-dependent conductance (NEHER and SAKMANN, 1976 a and b). It seems likely that we will not fully understand many biological conductance mechanisms before we can incorporate the molecules responsible for them into artificial membranes where techniques like those to be discussed can be applied. In addition, such reconstitution studies provide one of the few means of identifying functionally the molecules responsible for the conductance. Studies in which a labeled poison known to block a particular function acts as a marker in the isolation of the molecules responsible for that function are exceedingly interesting and have provided a considerable body of information. They cannot, however, provide us with a functional unit. Only functional reconstitution can do that. We can thus view the results already obtained with substances that fortuitously produce a wide variety of conductance effects in black lipid films in two ways: first as instructive examples of the possible effects of small molecules on membrane conductance, and second as exercises in technique for the future (and necessary) incorporation of biological conductance mechanisms in black lipid films.

I. Demonstration of Conductance by Pore

In this section we will define the terms "carrier", "pore", and "channel" as they are to be understood in this review, and indicate to some extent how usage in the literature differs. By "carrier" we mean a molecule which complexes with an ion or other substrate at the surface of the membrane. The complex then moves across the membrane, translocating the substrate and releasing it at the other side of the membrane. Evidence for the existence of such molecules is discussed by STARK (present Volume). By "pore" we mean a pathway through which an ion or substrate can move in crossing the membrane, and which does not change in structure during the time it takes for an ion or substrate molecule to cross the membrane. By "channel" we will mean in this review a conductance mechanism which experimentally exhibits discrete jumps in conductance. The term "channel" is also used in the literature of neurophysiology to refer to pathways through which various ions move, as the sodium channel or the potassium channel. We will also discuss in this section the degree to which observation of discrete conductance steps (sometimes called single-channel conductances) can be interpreted as conduction by a pore mechanism. Pore mechanisms are also possible even in the absence of the observation of discrete steps in conductance, and experimental means for determining whether this is the case will be discussed as well. The observation that certain substances added to black lipid films produce discrete conductance changes is one made by many authors (to be cited later under specific instances). Such conductance steps generally occur randomly in time, but have specific amplitude patterns characteristic of the particular substance used (to be discussed further in Section C.II). The conductance steps range in size from about 10^{-9} mho for alamethicin to 10^{-11} mho for gramicidin (measured for both in 1 M NaCl). As specification of salt concentration implies, the step conductances usually depend almost linearly on salt concentration, especially if the salt activity remains below about 1 M. These conductance changes must involve some sort of discrete change in state of the membrane components. They might arise from the alteration of the conformation of a hypothetical mobile carrier from a nonconducting to a conducting form. The argument against this is a quantitative one. The conductance of a single carrier is limited either by the rate at which the uncharged carrier can diffuse back across the membrane, or by the rate at which the charged form moves across the membrane. Because of the high energy necessary to move a charge from the high dielectric constant region of the water into the low dielectric constant region of the membrane, movement of the charged form will usually be rate-limiting. We can estimate how often an unimpeded carrier can move across the membrane. A single carrier will not be able to cross the membrane more than D/d^2 times per second on the average, where D is the diffusion coefficient of the carrier and d the thickness of the membrane. The flux of ions through most of the pores we are considering is greater than about 10^8 ions per second. Thus the carrier diffusion coefficient would have to be on the order of 10^{-5} cm^2 s^{-1} to account for the ion fluxes observed. This is about the diffusion constant of small ions (*e. g.* potassium) in water, and it is doubtful whether even the smallest molecule capable of being a carrier would have a diffusion coefficient within an order of magnitude of the required value. A more serious objection, however, is that the charged form of the carrier would have its translocation rate seriously reduced by virtue of its electric charge. In fact the translocation rate for molecules thought to be carriers has been measured and values of about 10^4 per second found (STARK et al, 1971; STARK, this Volume). This is far below the value required for experimental detection. The observation of discrete conductance steps in the order of 10^{-11} mho or greater can thus be taken as very strong evidence that pore formation is responsible for the conductance change. If the step conductance were much less than 10^{-11} mho, however, the conclusion would not be nearly as firmly based, and the possibility that the mechanism of conductance was a mobile carrier would exist. This argument for conductance by pore is thus an essentially quantitative one. The behaviour of the conductance step under experimental manipulation can also provide strong support for the contention that a conductance increase results from pore formation, and will be discussed in detail later.

The method of discrete steps has been successfully applied in a number of cases, and the following substances are considered pore formers because they exhibit discrete conductance steps of sufficiently large magnitude: EIM (BEAN et al., 1969; EHRENSTEIN et al., 1970), gramicidin (HLADKY and HAYDON, 1972), alamethicin (GORDON and HAYDON, 1972; EISENBERG et al., 1973), and hemocyanin (ALVAREZ et al., 1975 a). Figure 1 shows current fluctuations at constant voltage induced by these substances under similar conditions. Note the differences in time scale and the similarities in conductance.

When possible, the method of discrete steps provides the most direct evidence of the existence of a pore, but there are important cases where it cannot be applied. The first limit is experimental resolution. It is possible now to achieve a resolution of current with time such that the product of resolution time and observed current is 10^{-14} to 10^{-15} ampere-seconds. Thus a current of 10^{-12} amps can be resolved in a millisecond but to resolve a current of 10^{-15} amps requires one second. This is not yet a fundamental limit and improvements can be expected. Nevertheless, if the fluctuations in conductance produced by a suspected pore former are either too small or too fast, the method of discrete steps to demonstrate conductance by pore fails, in the first case because steps are unresolvable, and in the second because steps are resolved but attributable to a diffusing carrier. In addition, the pores may be static and never open or close during the course of the experiment, or they may open and close continuously and not have a well defined size. In such cases discrete steps would be undetectable not because they were unresolvable, but because they did not exist. It is thus important to develop other criteria for pore formation when discrete conductance steps cannot be observed.

One such criterion, that of membrane permeability to molecules of varying size, has been applied by physiologists to biological membranes. The method becomes more useful in the black lipid film system, because membrane conductance for a given salt can be measured and correlated with the increase in permeability of any desired neutral molecule in the same membrane. One chooses suitable molecules of graded size and measures their permeability. It is first

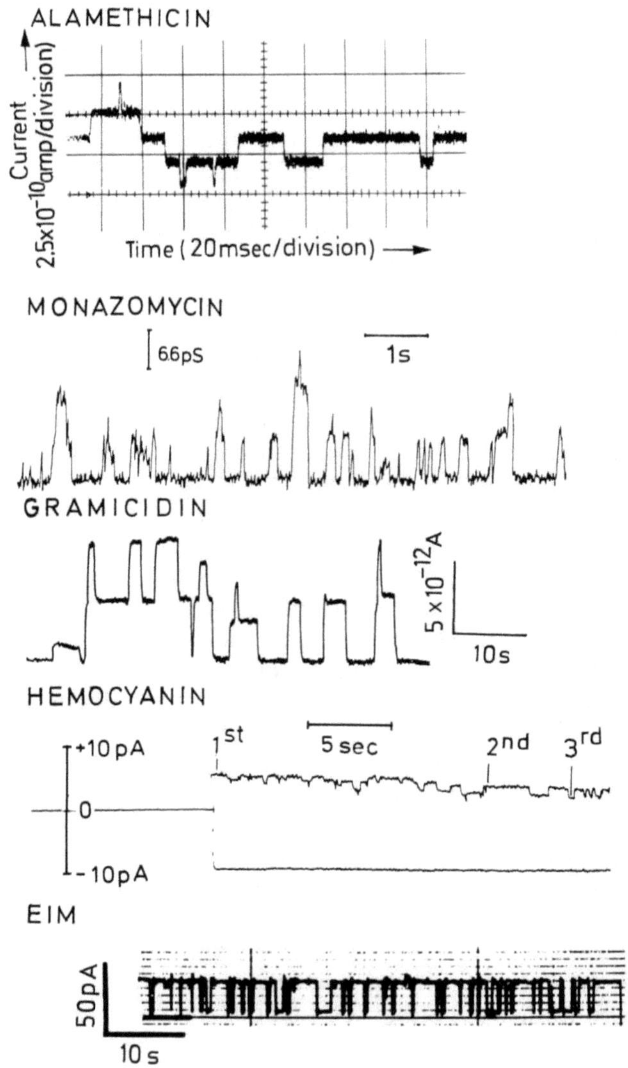

Fig. 1. Current records at fixed voltages for pore formers that exhibit discrete single-channel conductance changes. Note that all these fluctuations have an essentially random component in their behavior. *Alamethicin:* Single channels in 1 M NaCl solution at 145 mV. Small arrow at lower left indicates zero line. (After EISENBERG et al., 1973). PE decane-membranes. *Monazomycin:* Single channels in 4 M CsCl solution at 150 mV (after BAMBERG and JANKO, 1976). L-αdiphytanoyl-PC decane membrane. *Gramicidin:* Single channels in 1 M KCl at 100 mV. (After VEATCH et al., 1975). Glycerol monooleate-alkane membrane. *Hemocyanin:* Two traces: one for positive, one for negative voltage, applied to side of membrane where hemocyanin was added. Positive current trace shows result of application of a + 50 mV pulse. (After LATORRE et al., 1975). Oxidized cholesterol-decane membrane. *EIM:* Single channel in 0.1 M KCl solution at 60 mV. A negative applied voltage would also produce fluctuation in conductance, in contrast to hemocyanin where only one sign of voltage produces fluctuations. (After ALVAREZ et al., 1975b). Brain lipid-decane membranes

necessary to establish that the induced electrical conductance in a given salt solution and the induced permeability to the chosen molecules are proportional. That done, the permeabilities of the chosen molecules, normalized to membranes of standard conductance in a standard salt solution, can be compared. If the permeability of several molecules increases proportionally to the induced conductance of the membrane, the implication that the induced conductance is produced by a pore former is strong. The way in which the relative permeabilities of the chosen molecules vary with molecular size can be used to estimate pore size. If, for example, one measures the permeability and conductance for a series of neutral molecules of increasing size as described, one expects a drop in permeability when the molecular size approaches the size of the presumed pore. We shall call this basic procedure *sieving.*

Clearly sieving is a less direct method for establishing the existence of pores than the observation of discrete conductance steps. It involves the synthesis of several experimental results to be convincing and is not sensitive to the details of pore state. It cannot distinguish, for example, between a pore that is open all the time and one that opens and closes through several states. In particular, it has poor time resolution. Nevertheless the method is often the best available, and it is easy to see that for a static network, the pores of which never open or close, the method of discrete steps cannot in principle be used[2]. The sieving approach can reveal the existence of pores even under these circumstances.

The method of sieving has been successfully applied to artificial membranes with conductances induced by the polyene antibiotics amphotericin-B and nystatin (HOLZ and FINKELSTEIN, 1970; ANDREOLI, 1974). Representative data indicating that these antibiotics form pores appear in Figure 2 and its accompanying table (Table 1). Figure 2 shows a plot of permeability versus electrical conductivity for a membrane doped with nystatin (HOLZ and FINKELSTEIN, 1970). In Figure 2 the permeability is measured for urea and corrected for unstirred layers. Clearly the permeability and conductivity are proportional to each other. Table 1 shows the permeabilities to various small molecules of a membrane in the presence of either nystatin or amphotericin B. As molecular size increases to about a 4 Å Stokes-Einstein radius, the permeability drops two orders of magnitude. This implies that nystatin and amphotericin induce formation of pores with radii on in the order of 4 Å in black lipid films.

Amphotericin and nystatin are included in this review as pore formers essentially on the strength of data exemplified by those shown in Table 1 and Figure 2. There are supporting data which make the case stronger, and these will be discussed subsequently.

[2] If such pores were produced by a substance added to the aqueous phase after formation of the membrane, the pore would clearly have had to form and the formation jump might be detectable. It is possible that the formation of a relatively large pore occurs through a series of very small steps which cannot be detected. Something of this kind may be the case for amphotericin and nystatin, which are pore formers by the sieving criterion but which do not exhibit detectable discrete steps in steady state or during the development of the conductance after addition to the aqueous solutions bathing an already formed membrane.

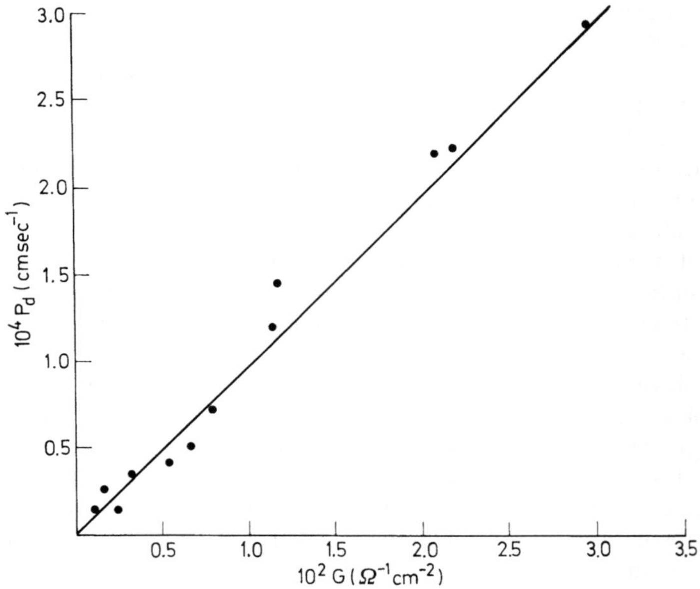

Fig. 2. Permeability to urea versus conductivity in 100 mM NaCl of membrane in the presence of nystatin. Since conductance and permeability are proportional even for different concentrations of nystatin, it is inferred that conductance and permeation occur *via* the same pathway. (From HOLZ and FINKELSTEIN, 1970, with permission of Rockefeller University Press)

Table 1. This table shows permeabilities and reflection coefficients for various solvents in membranes in the presence of nystatin and amphotericin B. The permeabilities are normalized to membranes with a conductance of 100 ohm-cm², it having previously been demonstrated that conductance and permeability are proportional (see Fig. 2). The data in this table show that as size increases, permeability decreases for small nonelectrolytes. Glycerol is very impermeant compared to water, and glucose is almost impermeant. Thus since the Stokes-Einstein radius of glucose is about 4Å, we conclude that amphotericin and nystatin produce pores with radii in the order of 4Å. These data, of course, tell us no details of pore structure. (From HOLZ and FINKELSTEIN, 1970, with permission of the Rockefeller Press)

	Nystatin		Amphotericin B	
	P_d (cm s^{-1})	σ	P_d (cm s^{-1})	σ
Water	12.0×10^{-4}	0	6.0×10^{-4}	0
Urea	0.95×10^{-4}	0.55	0.68×10^{-4}	0.57
Thiourea	0.95×10^{-4}	–	–	–
Ethylene glycol	0.45×10^{-4}	0.67	–	–
Glycerol	0.115×10^{-4}	0.78	0.075×10^{-4}	–
Glucose	–	1.0	–	1.0
Sucrose	–	1.0		1.0
NaCl	$P_f=40\times10^{-4}$		$P_f=18\times10^{-4}$	

Permeability values are normalized to membranes with a resistance of 100 ohm-cm² in 0.1 M NaCl. See Fig. 2 and text.

Undetectable Pores

It is possible to imagine pores which cannot be detected by either of these methods, and for which almost all experiments will yield results which can be interpreted by a carrier mechanism. It is conceivable in fact that nonactin and valinomycin act by forming pores[3] rather than as mobile carriers, and it is only the weight of much accumulated data from a variety of experiments that allows us to believe that this is not the case (cf. STARK, this Volume). This leads to a certain asymmetry in the experimental situation *vis-a-vis* pores and carriers. While there is a single definitive experiment (discrete steps) that allows us to say in some cases, "this is surely a pore," there is as yet no experiment which allows us to say this is surely a "carrier." In addition, one can imagine carrier and pore models which cannot be distinguished by experiment. To further complicate the spectrum of possibilities one can imagine transport mechanisms which look like pores to some and carriers to others (see *e. g.* STARK, this Volume). It is fruitless to engage in essentially semantic arguments over whether or not such mechanisms are pores or carriers, and the only sensible approach is to ask what will experiments reveal in a given case. It may be impossible to distinguish between interpretations of conductance data for a certain class of mechanisms and we should be aware that some biological conductance mechanisms may belong to this class.

II. Basic Conductance Characteristics

The goal of the study of membrane conductance mechanisms is to explain observed conductances in terms of the properties of specific and identified molecules. This implies measurement of both molecular properties and conductance properties, followed by construction of a detailed scheme relating the two. In this section we will discuss measurement of steady-state current-voltage (I–V) curves and dependence of conductance on concentration of pore former. These are important experiments, which define the basic conductance characteristics at a level similar to that at which we understand, say, squid nerve.

1. Steady-State Current-Voltage Curves

The steady-state current-voltage curve produced by a given substance is an important experimental datum, but one which must be measured with care if it is to be useful. Artifacts can arise from two sources: misunderstanding of the experimental technique and failure to allow for the time-dependent properties of the membrane conductance. To avoid the first error it is essential to understand possible sources of error arising from the basic instrumentation. The

[3] It is in fact possible to construct a pore model that gives calculated results analogous to those of some carrier models. Such pores must have intuitively unsatisfying properties, but we cannot say for certain that they do not exist.

minimum time for the system to respond to a change in voltage of the source will be determined by the product of membrane capacitance and the sum of the source impedances, including solution and electrode impedance. If the total source impedance is very small, this time will be short and the voltage across the membrane will be essentially equal to the source voltage. If the source impedance is very large, the value of the source voltage will determine the membrane current. The time for the system to respond to changes in membrane current will then be the membrane time constant, the product of membrane resistance and membrane capacitance. Since membrane resistance is usually much larger than source impedance, the time response in this case is very slow. In addition it is not constant, but varies with membrane conductance. For these reasons *I–V* curves are usually measured under constant[4] voltage or low-impedance conditions. *I–V* curves are often taken point by point, *i. e.* the voltage is fixed and the current measured. This is repeated for a series of voltages. When this technique is employed, it is imperative that the minimum length of time spent waiting for steady state be explicitly noted. This is particularly important for the voltage-dependent pore formers whose kinetic properties vary over an enormous range. Otherwise the value of the data is considerably reduced, and one is more than justified in retaining a lingering doubt as to whether steady state has in fact been reached.

Adequate current-voltage curves have been measured for the pore formers under consideration. Gramicidin, nystatin and amphotericin B have *I–V* curves which show only slight dependence of conductance on voltage. For gramicidin, the dependence of current on voltage is due in part to the voltage dependence of the conductance of a single pore and in part to a slight voltage dependence of the number of pores on voltage. For amphotericin and nystatin, the question cannot be settled definitively, because single-pore conductances cannot be determined, but from comparison with the voltage-dependence of the single channel conductances of other pore formers it appears likely that most of the small observed voltage dependence can be explained by the properties of a single pore and that the average number of pores is constant with voltage. This is consistent with the very fast response of polyene conductances to a voltage pulse.

EIM and hemocyanin show a negative resistance characteristic in most membranes (MUELLER and RUDIN, 1968; LATORRE et al., 1975). Their voltage-current curves are compared in Figure 3. The curves are similar, but analysis of the single channel data indicates that they arise from a different kind of unit event. Increasing voltage in either sense will decrease conductance, but an increase on the side to which the EIM has been added is slightly more effective. In contrast, only a voltage increase positive on the side where hemocyanin was added will decrease the hemocyanin-induced conductance. Both hemocyanin and EIM show maximum conductance at zero field, in contrast to the other pore formers, all of which usually show minimum steady-state conductance at zero voltage.

[4] The term "constant" in this context does not mean "unvarying," but denotes that for a given value of source voltage, either the voltage (constant voltage) or the current (constant current) is prescribed and independent of fluctuations in the impedance of the membrane under test. In both cases the unprescribed parameter may vary considerably.

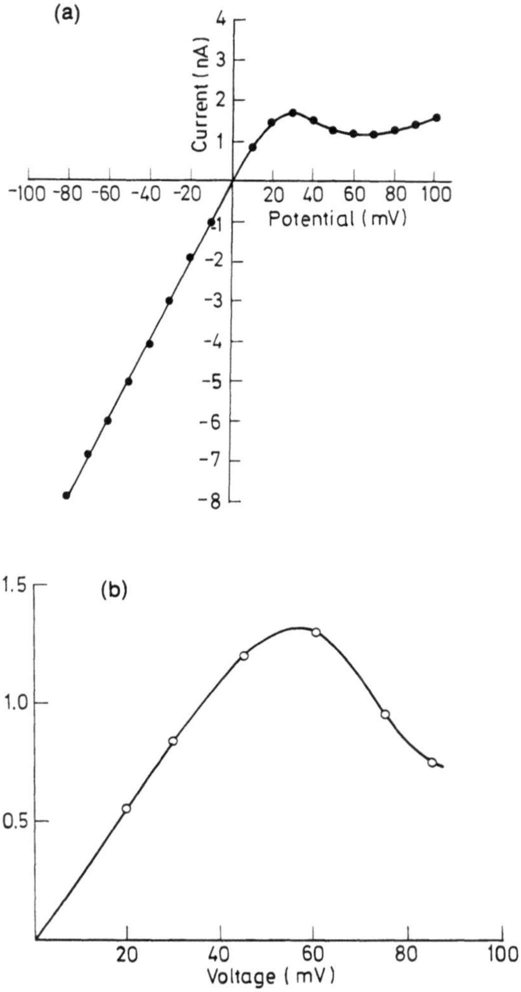

Fig. 3. (a) and (b). Voltage-current curves for EIM- and hemocyanin-doped membranes. In both cases the pore former was added to only one side of membrane. EIM shows turn-off with both signs of voltage (third quadrant not shown), but hemocyanin curve only shows turn-off with positive voltage. (a) steady-state hemocyanin current-voltage curve in 0.1 M KCl solution. (b) steady-state EIM voltage-current curve. (a: from LATORRE et al., 1975; b: from EHRENSTEIN et al., 1970 with permission of Rockefeller Press)

Changing membrane properties can alter the conductance characteristics. The voltage necessary to shut off the conductance induced by EIM can be increased by increasing the cholesterol content of the membrane. Addition of cholesterol to the membrane alters not only the voltage at which the EIM conductance is decreased, but also the kinetics (LATORRE et al., 1976). In terms of the single conductance steps, cholesterol stabilizes an "open" configuration of the pore and increased voltage makes a "closed" configuration more likely.

Alamethicin, its derivatives, and monazomycin all show very strong dependence of conductance on voltage. Conductance induced by these substances increases *e*-fold for a 4 to 12 mV change in potential. Figure 4 shows a typical current-voltage curve for alamethicin. Current is plotted semilogarithmically to emphasize the exponential dependence of conductance on voltage. The exact amount of voltage required to induce an *e*-fold conductance change depends on lipid composition, ionic strength, temperature, and other factors. We will discuss those factors that have systematic effects at the appropriate points.

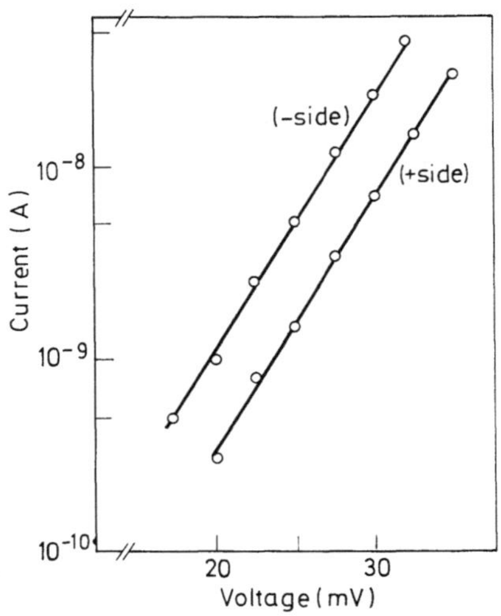

Fig. 4. Logarithm of current versus voltage for a membrane in the presence of alamethicin. Fig. 6 shows similar data for monazomycin and alamethicin as functions of voltage and concentration. Note that both substances induce exponential increase in current with voltage and only 4–5 mV are required to induce an e-fold change in current. Changes in conductance of similar magnitude are seen in excitable biological membranes. (From EISENBERG et al., 1973)

The symmetry of the *I–V* curve for alamethicin depends on whether alamethicin is added to both sides and, if it is added to only one side, on the nature of the lipid used to form the membrane. Generally, for equal alamethicin concentrations on both sides, the *I–V* curve is symmetric. If alamethicin is added to one side only, some membranes show a very nearly symmetric *I–V* curve. Most membranes showing nearly symmetric *I–V* curves are formed from unsaturated lipids. If alamethicin is added to only side of a bacterial phosphatidylethanolamine (PE) or di-isostearoyl phosphatidylcholine membrane, the *I–V* curve is very asymmetric. In fact, for the PE membrane, no increase in conductance is seen with voltage unless the side to which the alamethicin was added is positive (EISENBERG et al., 1973). This implies that alamethicin has either a net positive charge or a large dipole moment, a point to be discussed in more detail later. Monazomycin *I–V* curves show a similar asymmetry, implying a net positive charge or dipole moment, but the effect of lipid on this asymmetry has not been studied.

The I–V curves of alamethicin and monazomycin both exhibit negative resistance in the presence of a salt gradient. This happens because the conductance depends only on the electric field, but the current depends on the conductance and the driving force: this can be expressed by the equation

$$I = g(V)(V - V_{eq}),$$

where I is the current, $g(V)$ the conductance explicitly written as a function of voltage, V the voltage across the membrane and V_{eq} the voltage at which no current flows. Figure 5 shows an I–V curve for alamethicin in the presence of a salt gradient. Similar curves are obtained for monazomycin. Note that the conductance-voltage curves for both alamethicin and monazomycin are of the same form as in symmetric salt solutions, *i. e.* conductance is an exponential function of voltage.

We thus see that pore formers exhibit three types of I–V curves:
1) Nearly linear or only slightly superlinear for gramicidin and the polyenes.

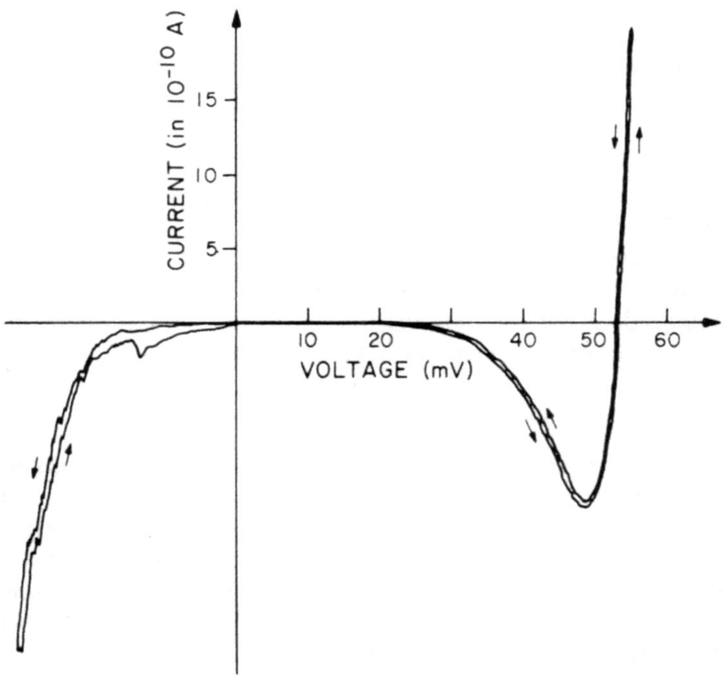

Fig. 5. (a) Voltage-current curve for alamethicin in the presence of a salt gradient. Similar curves are found for monazomycin in the presence of a salt gradient. Front chamber contained 0.5 M KCl, back 0.005 M KCl. Voltage positive when front chamber positive with respect to back. (From EISENBERG et al., 1973)

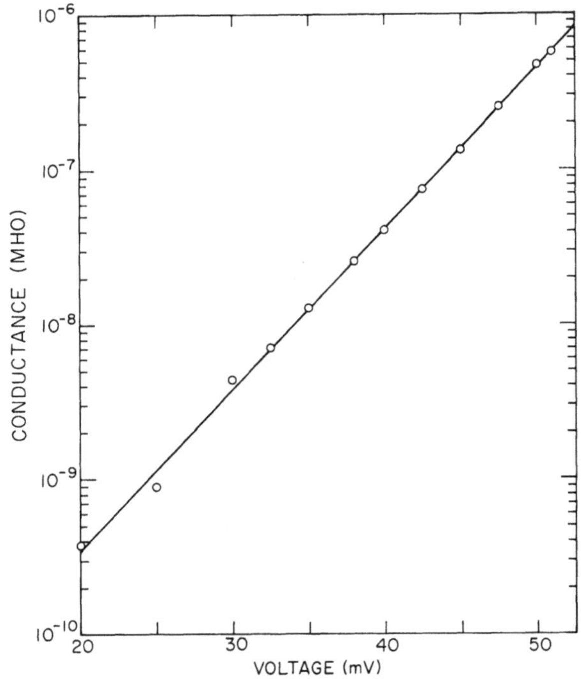

Fig. 5. (b) Conductance-voltage curve for the negative resistance branch of curve in fig. 5a (preceding page); conductance even in this region depends exponentially on voltage just as in the absence of a salt gradient

2) Decrease in conductance with increasing voltage with a negative resistance region (EIM and hemocyanin).
3) Very steep increase in conductance with increasing voltage (alamethicin and monazomycin). The EIM and alamethicin types of *I–V* curve have never been observed except for substances known to be pore formers, but as we have seen, pore formers need not have only these types of *I–V* curve.

2. Conductance and Antibiotic Concentration

One of the striking circumstantial facts about many of the pore formers is that the conductance they induce depends on a high power of pore former concentration (generally powers from 4 to 12 for those pore formers showing strong power dependence). Since this strong power dependence, coupled with other data, leads to an interesting generalization about many of the pore formers, it is very important to understand precisely what we mean by it and the exact nature of the experiments used to determine it.

For a voltage-independent conductance the problem is simplified, and the experiment and interpretation are straightforward: simply add pore former to the aqueous solution, stir, wait for steady state, measure the conductance, and plot the conductance against concentration, on a log-log plot. The slope of the line gives the power dependence of the conductance on concentration. This experiment can be performed with amphotericin and nystatin, but for other pore formers complications arise.

For the voltage-dependent pore formers the effects of voltage must be considered, and for gramicidin the aqueous concentration is difficult to control because of its low solubility and tendency to stick to interfaces. Gramicidin can be detected at concentrations of 10^{-13}M (about 10^7 molecules in a 1-ml cuvette!) and contamination control is very difficult[5]:

The difficulty of maintaining a constant gramicidin concentration has been overcome in an ingenious way by VEATCH et al., (1975). They were able to dansylate gramicidin A and thus render it fluorescent. Dansyl gramicidin A fluoresces strongly only in a hydrocarbon environment and thus fluorescence is sensitive only to gramicidin in the membrane and is not affected by dansyl gramicidin in the aqueous phase. Dansylated gramicidin has a single-channel conductance about half that of the unmodified form. VEATCH et al. showed that the conductance of a membrane doped with dansylated gramicidin was proportional to the square of the fluorescence and thus, since fluorescence is proportional to concentration, to the square of the concentration. This is the only case where it has been possible to measure a number directly proportional to concentration of a substance in the membrane and avoid the problem of partition between the aqueous phase and the membrane. We thus know that the power dependence of conductance on gramicidin in the membrane is two, as first suggested by TOSTESON et al. (1968).

For voltage-dependent pore formers the situation is complex. Since a small change in voltage can produce a large change in conductance, it is important that the effect of voltage be taken into account. Usually this is done by measuring the shift of the conductance-voltage curve with changes in concentration of pore former in the aqueous phase. Increasing pore former concentration generally decreases the voltage at which the conductance reaches a certain value or, alternatively, elevates the conductance observed at a given voltage. Figure 6a shows an example of such an experiment for monazomycin (MULLER and FINKELSTEIN, 1972a). Note that the shape of the conductance-voltage curve (in this case the slope of the semi-log plot is the factor determining slope) is unchanged by an increase in the monazomycin concentration. Similar results have been obtained for alamethicin (MUELLER and RUDIN, 1968; EISENBERG et al., 1973; GORDON and HAYDON, 1975; ROY, 1975), and an example is shown in Figure 6b. Experiments of this kind enable us to determine the power dependence of the conductance on antibiotic concentration at a fixed voltage. This calculation assumes that the conductance-voltage curves can be extrapolated to the same voltage, an assumption born out in the alamethicin and monazomycin systems whenever direct comparison at the same voltages has been possible. From this kind of data we can write that the conductance at constant voltage is given by:

$$G \sim C^n \tag{1}$$

[5] This brings up an interesting point in passing: often the black lipid film is the most sensitive assay for a pore forming antibiotic. Consequently many observed properties may arise from a contaminant undetectable by any means but the bilayer! It is therefore important to use several different sources of antibiotics, which presumably are not contaminated in the same way.

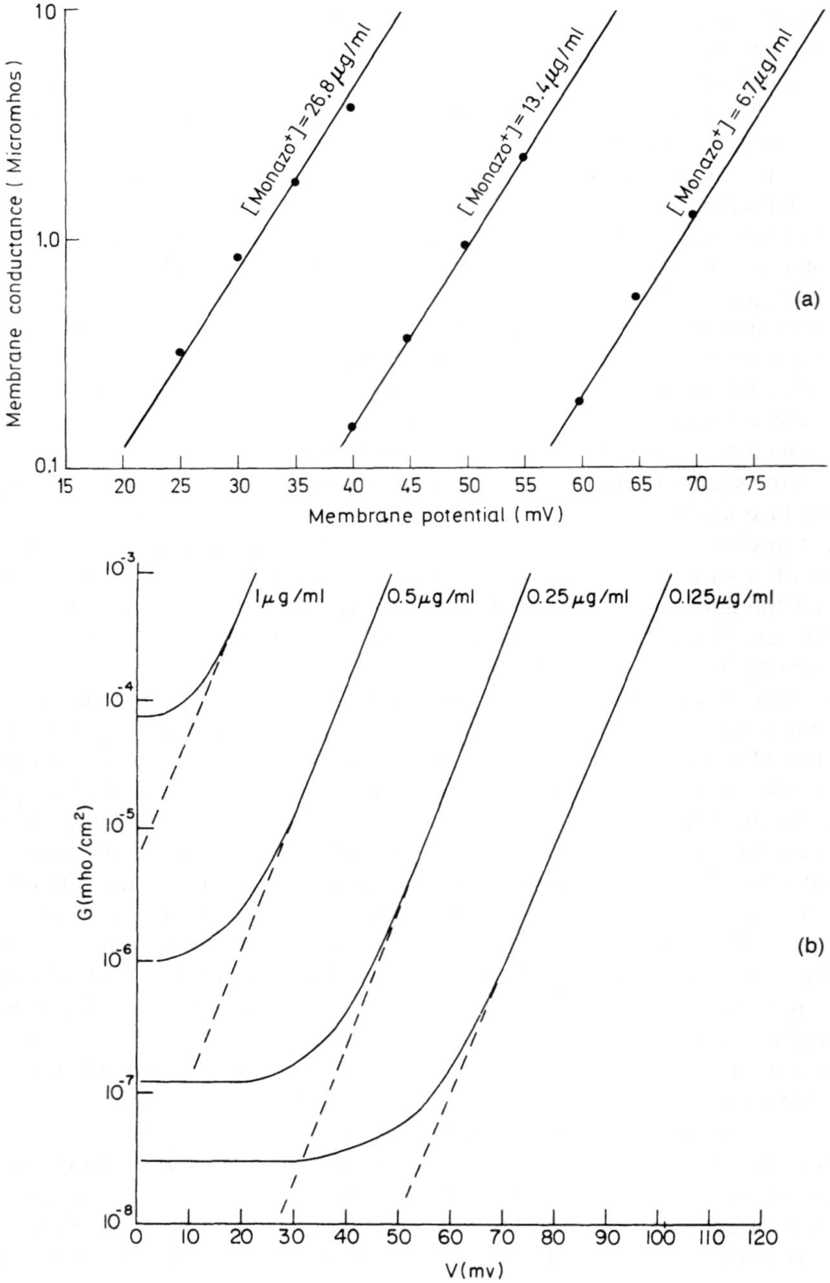

Fig. 6. (a) and (b). Conductance-voltage curves of both alamethicin (b) and monazomycin (a) shift toward lower voltages with increasing antibiotic concentration. Note that the curves remain parallel on logarithmic plot for both pore formers. Form of the voltage-dependence thus is unchanged by changing pore former concentration. For alamethicin the power ranges from 6 to 10, depending on experimental conditions, and for monazomycin the power ranges from 6 to 12. (a: from MULLER and FINKELSTEIN, 1972, with permission of Rockefeller Press; b: from ROY, 1975)

where G is the conductance, C the antibiotic concentration in the aqueous solution and n, the power dependence, for alamethicin the value of n depends somewhat on lipid and measurement techniques. EISENBERG et al. (1973) using PE-decane membranes find $n = 9.2 \pm 1.1$, Gordon and Haydon (1975) find $n = 9.4 \pm 1$. MUELLER and RUDIN (1968) find $n = 6$. For monazomycin, MULLER and FINKELSTEIN (1972a) report $n \simeq 5$. The interpretation of these data is very model-dependent and may not, as one is first tempted to believe, reflect simple properties of the conductance mechanism such as, for example, the number of molecules in a single conducting unit. In conjunction with other data, however, conductance dependence on a concentration can be interpreted with somewhat more confidence, as, for example, in the cases of gramicidin A and its analogues.

Amphotericin B and nystatin exhibit conductances almost independent of voltage. Amphotericin and nystatin also require a steroid for greatest effectiveness. CASS et al. (1970) report that ergosterol-containing membranes are more sensitive to nystatin than cholesterol-containing membranes. ANDREOLI (1974) reports that cholesterol and dihydrocholesterol are effective in promoting the action of amphotericin B and nystatin, while epicholesterol, cholesterol palmitate, dihydrotachysterol, and Δ^5 cholesten-3-one are not. In the presence of steroid, polyene induced conductances increase with a high power of the antibiotic concentration. CASS et al. (1970) find that n ranges from 6 to 12 for different membrane forming solutions, but that n is constant to within 10 percent for any given solution. ANDREOLI and MONAHAN (1968) report an n of 4.5. Both of these results are obtained with equal concentrations of antibiotic on both sides of the membrane. CASS et al. report that if the concentration of amphotericin or nystatin is much higher on one side of the membrane than the other, the power dependence of the conductance on the low concentration (while the concentration on the other side is unchanged) is reduced. They imply that this power is approximately $n/2$, where n is the power dependence for symmetric addition.

Finally the zero-voltage conductance of hemocyanin appears to depend approximately on the fifth power of aqueous concentration (LATORRE and ALVAREZ, pers. comm.). No reliable data on concentration dependence are available for EIM.

3. Kinetics of Conductance Development: Response to a Voltage Pulse

Much of the detailed information we have on the conductance mechanisms in various nerve and muscle cells comes from application of the voltage clamp technique. A feedback amplifier circuit supplies the current necessary to maintain the cell membrane voltage at a value selected by the experimenter. Another circuit measures the current necessary to maintain the voltage at the desired value. The experimenter applies potential steps to the membrane and measures the time response of the current. The data thus obtained are often interpreted to reflect the time dependence of the "opening" or "closing" of various conductance pathways in the membrane. The technique has been enormously useful in neurophysiology and has been applied to the study of conductances in artificial membranes.

Because of the high impedance of artificial membranes relative to available source impedances, it is not usually necessary to use feedback amplifiers to prescribe the voltage, and any low impedance pulse generator with suitable time and amplitude characteristics works very well. A voltage pulse experiment is a low impedance or "constant" voltage measurement.[6]

The response of current to an applied voltage pulse is interesting for gramicidin, alamethicin, monazomycin, EIM and hemocyanin, but not for amphotericin B and nystatin.

Gramicidin

BAMBERG and LÄUGER (1973) have studied the response of gramicidin conductance to a voltage pulse. Figure 7 shows the result of a typical experiment. The voltage pulse induces a rapid initial rise of the current, followed by a slower rise to a final value. The slow conductance increase is exponential with a single time constant. BAMBERG and LÄUGER (1973) find that the time constant measured in this way depends on the square root of the mean conductance of the membrane for a voltage pulse of fixed size. They show that this is consistent with a model in which the conductance depends on the square of the free gramicidin concentration in the membrane, in agreement with the result found by VEATCH et al. (1975). Analysis of the data in terms of this model yields values for the rates of

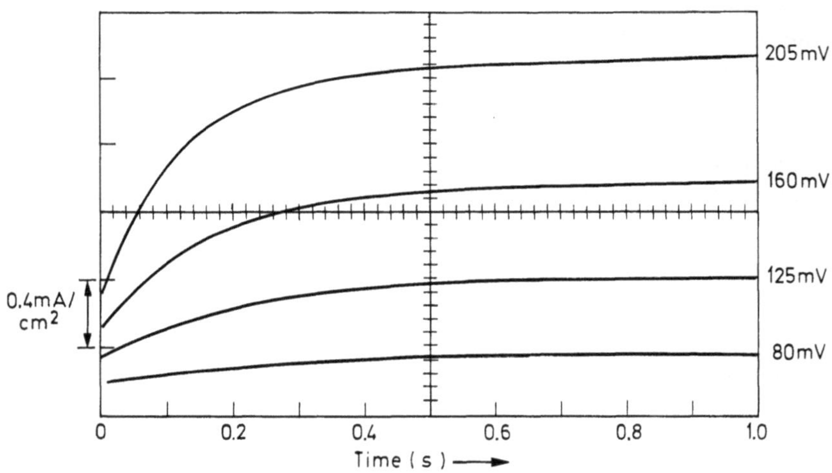

Fig. 7. Kinetic response of gramicidin-induced conductance to voltage pulse. Current rises rapidly to initial value, then increases more slowly with a single time constant to the final value (BAMBERG and LÄUGER, 1973)

[6] It is possible to provide current pulse (high impedance or constant current measurement) and measure the time response of the voltage, but complications of analysis make this mode much less preferable.

formation and disappearance of gramicidin pores. The magnitude of the single-pore conductance must be used in the analysis, to avoid the problem of uncertainties in the concentration of gramicidin in the membrane. There is a small dependence of the rate of pore formation on voltage, but nearly no voltage dependence of the rate of pore disappearance. The dissociation rate is about 2 per second and the formation rate 2×10^{14} cm^2 $mol^{-1}s^{-1}$.

In general, almost all the available experimental results for gramicidin are consistent with the Bamberg-Läuger (1973) dimer model.

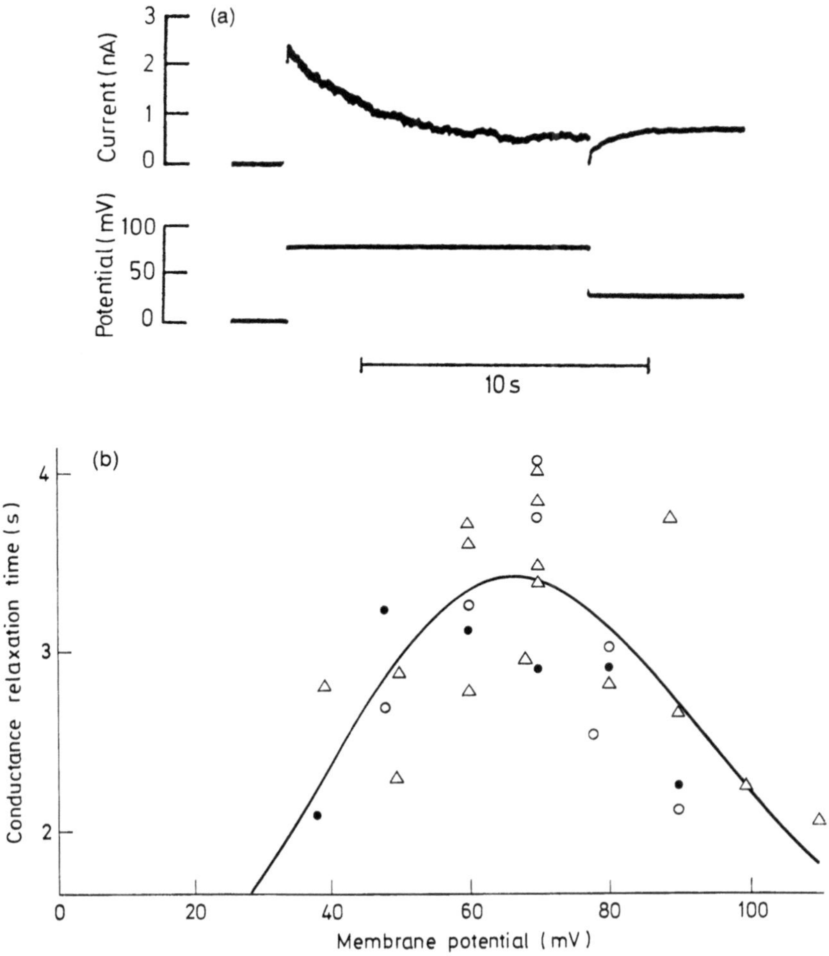

Fig. 8. (a) Response of EIM-induced conductance to a voltage step. Note that the time constant is voltage-dependent and the current decreases with increasing voltage. (From EHRENSTEIN et al., 1974, with permission of Rockefeller Press). (b) Voltage-dependence of time constant for relaxation of EIM-induced conductance to final value after application of voltage pulse. Bell-shaped curve arises because of definite number of EIM channels in membrane and rate of closing channels increases montonically with voltage. (From EHRENSTEIN et al., 1974, with permission of Rockefeller Press)

EIM and Hemocyanin

EIM and hemocyanin both exhibit conductance changes in response to a voltage pulse, but these responses differ strongly from that of gramicidin – the time course of gramicidin conductance is not notably voltage-dependent, while those of EIM and hemocyanin are. We will discuss EIM kinetics first. Figure 8a shows an example of an experiment measuring EIM kinetics (EHRENSTEIN et al., 1974). The voltage is pulsed to different values and the current response observed (a classic voltage clamp measurement).

The decay of the current is exponential in time, and the relaxation time has a bell-shaped dependence on voltage. This result is shown in Figure 8b. We will return to the kinetics of EIM with the single-channel experiments discussed in the next section, but we should note in anticipation that a similar bell-shaped curve is seen in biological membranes exhibiting excitability (see LECAR et al., 1975 for an illuminating discussion of the details of this relationship).

Hemocyanin

The time course of the hemocyanin-induced current response to a voltage pulse is more complicated than that of EIM. The decay does not have a single exponential time constant, but rather has at least three time constants, a fast component with $\tau \sim 200$ μs, a second component with $\tau \sim 3$ s, and a third with $\tau \sim 80$ s. These time constants reflect changes in the conductances of the states of the channel as well as voltage-induced transitions between states (LATORRE et al., 1975).

Monazomycin

The state of our knowledge of monazomycin conductance is transitional at the time of this writing. While single step conductances have been observed, they have not yet been related to the conductances measured at high levels. The kinetic response of monazomycin to voltage pulses is, however, very varied and interesting, suggesting that perhaps the molecular properties of monazomycin allow it to act in several ways depending on the composition of the membrane in which it is studied. MUELLER and RUDIN (1968), MAURO et al. (1972), MULLER and FINKELSTEIN (1972b), and BAUMAN and MUELLER (1974) have measured the kinetic response of monazomycin to a voltage step. The typical behavior is sigmoidal, as shown in Figure 9a (MULLER and FINKELSTEIN, 1972b). A small change in voltage increases both the final current (see Section B.II.1) and the rate at which the final current is achieved. Under other circumstances the monazomycin conductance exhibits inactivation, *i. e.* the current increases immediately on application and subsequently decays to a steady-state value. We will discuss possible explanations of this phenomenon in Section D.

Alamethicin

The alamethicin conductance response to a voltage step is simple compared to that of monazomycin and, as shown in Figure 9b, is not sigmoidal (for exception to this statement, see MAURO et al. [1972]). Several authors have reported on the conductance kinetics of alamethicin, including MUELLER and RUDIN (1968),

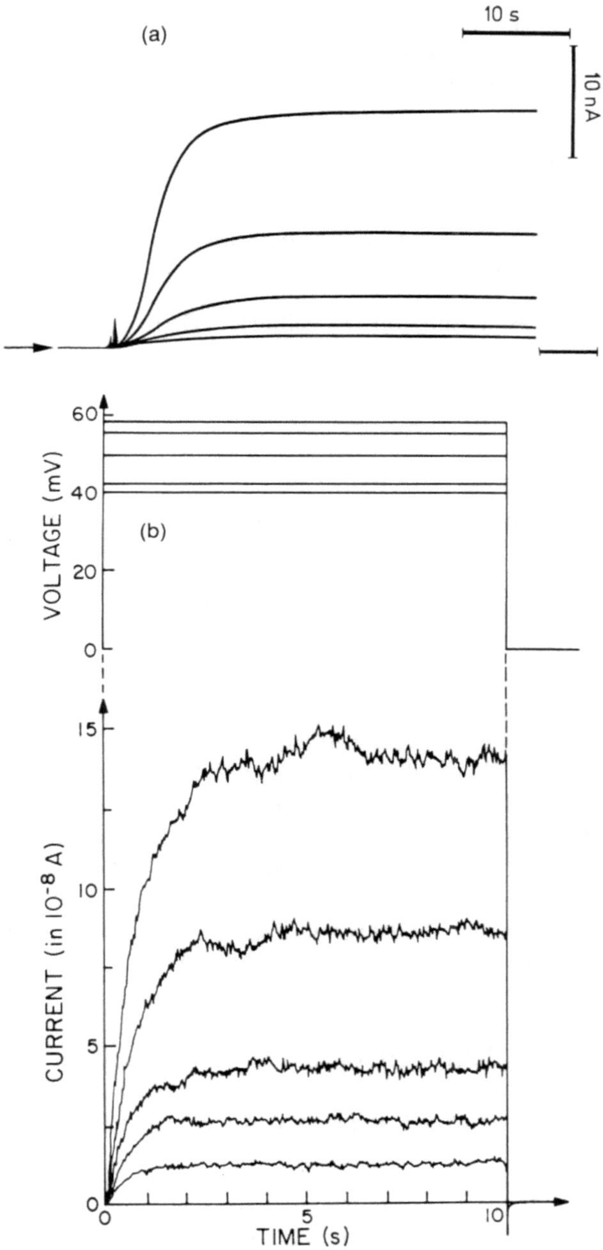

Fig. 9. (a) Kinetic response of monazomycin-induced conductance to voltage pulses of 25, 29, 33, 39 and 41 mV. Note that the initial onset of conductance is delayed somewhat after stimulus and that final value and the rate of conductance increase both depend strongly on voltage. (From MULLER and FINKELSTEIN, 1972, with permission of Rockefeller Press). (b) Kinetic response of alamethicin-induced conductance to voltage pulses of 40, 42, 50, 55 and 58 mV. Note that the delay in turn-on of conductance after application of field is much shorter than for monazomycin. Data like these show that turn-on rate, turn-off rate and steady-state conductance of alamethicin depend exponentially on voltage. (From EISENBERG et al., 1973)

BAUMAN and MUELLER (1974), MAURO et al. (1972), and EISENBERG et al. (1973). EISENBERG et al. report that the alamethicin conductance response to a single pulse obeys the differential equation:

$$\frac{dG}{dt} = \mu(V) - \lambda(V)G, \qquad (2)$$

where G is the conductance, and $\mu(V)$ is the rate of channel formation written explicitly as a function of voltage and $\lambda(V)$ is the rate of channel disappearance per channel formed as a function of voltage. When $\frac{dG}{dt} = 0$ the rates of channel formation and disappearance are equal and

$$G = \frac{\mu(V)}{\lambda(V)}. \qquad (3)$$

$\mu(V)$ and $\lambda(V)$ are both exponentially dependent on voltage. EISENBERG et al. (1973) find

$$\mu(V) = \mu_0 \, e^{V/6.7\,\mathrm{mV}} \text{ and } \lambda(V) = \lambda_0 \, e^{V/9.6\,\mathrm{mV}}. \qquad (4)$$

MUELLER (1975) reports somewhat more complex kinetics based on two pulse experiments. The time constant for disappearance of alamethicin channels is a strong function of ionic strength and can vary from about 1 s^{-1} at 50 mV in 1 M ionic strength salt to 10^3 s^{-1} at 50 mV in 10^{-3}M ionic strength salt (HALL, unpublished work in PE-decane membranes). The specification of voltage in measuring any time constant associated with either alamethicin or monazomycin is vital since a small change in voltage can result in a large change in time constant.

This concludes a summary of basic experiments used to characterize the high-level conductances induced by substances known to be pore formers. It is interesting to note that except for the dependences of conductance on concentration of pore former and the single-step experiments all of these experiments can and for the most part have been performed *in vivo* in many preparations. The experiments in Section C will show how much more we can learn in the fortunate circumstances where single steps in conductance can be observed and studied in detail.

C. Advanced Experiments

I. Introduction

The preceding section shows how conductances induced in black lipid films can be assessed in a preliminary way by a set of crucial basic experiments. These experiments provide a means for determining whether an induced conductance arises by pore formation. They also describe the basic properties of the induced conductance at a phenomenological level: the steady-state *I–V* curve, the dependence on pore former concentration, and the time-dependent response to a voltage step. With the exceptions of the detection of single steps and the variation of pore former concentration, the basic experiments (*I–V* curve, kinetics as a function of voltage) can be done in living systems such as squid nerve, frog node, barnacle muscle and other suitable preparations. Unfortunately these basic experiments do not provide a firm basis for an understanding of the molecular mechanisms. This is so, not because we lack models, even molecular models, which can explain the data, but because we cannot distinguish experimentally between models. What we need is not more theories but more experiments.

In this section we will discuss in detail single channel experiments which directly reflect events at the molecular level. Single channel measurements have been made in a few living systems (NEHER and SACKMAN, 1976a and b), but in general single channel measurements are difficult in living systems. It has, however, been possible to measure fluctuations in conductance in living systems, and several authors have interpreted these data in terms of single channel conductances. Because fluctuation data ("noise analysis") can connect single channel data and high level conductance data we will discuss them in detail in this section (see for example, ANDERSON and STEVENS, 1973; KATZ and MILEDI, 1972; NEHER and SAKMAN, 1976a, b).

A single step conductance increase does not necessarily imply pore formation unless the conductance step is large enough, but it does imply the existence of a stable configuration lasting for the duration of the step increase. If the conductance arises by pore formation, the existence of a sharp step implies that the pore is in a fixed configuration that does not change during the duration of the step. It also implies that the lower conductance is a stable state as well. A rapid transition between conducting levels implies that the configurations through which the system must pass in going from one state to the other must be of very high energy[7]. Experiments measuring relative duration of different steps and transition rates as functions of temperature thus directly report information on the relative energies of the different states.

Note that the existence of sharp and discrete conductance states such as those under discussion is fortuitous. We can easily imagine a pore of somewhat loose and elastic structure, which does not have a discrete conductance, but fluctuates continuously over a range of conductances. Such a pore might be characterized by a distribution describing the probability of finding the pore in a given conductance state. In fact, such pores were discovered as this review was in preparation. A synthetic alamethicin-like compound synthesized according to the structure of MARTIN and WILLIAMS (1975) exhibits just such conductance variations as would be expected from a loose elastic structure (GISIN et al., 1977) capable of existing in many conformations of roughly equal energy.
(The MARTIN and WILLIAMS structure is not cyclic like that originally proposed by PAYNE et al. (1970) and appears to be essentially correct).

[7] Transitions between states of the pore formers discussed in this section are too fast to be resolved by present techniques.

II. Single-Step Experiments

On observing conductance fluctuations of those pore formers that show single steps, one sees immediately that the step events occur at random (cf. Fig. 1). One of the first tasks of the experimenter is to ascertain the probability functions which characterize the random behavior of a given type of conductance. In particular, we must ask whether the events are correlated in some way or are entirely independent.

This investigation will require understanding of what constitutes a "unit event." Alamethicin induced single steps come in bursts of single steps, for example, and while the bursts occur independently, the single steps making up a burst are not independent. For gramicidin, on the other hand, the single steps are quite independent.

1. The Probability Distribution

We can construct a probability distribution of conductance by sampling the conductance for a short time at frequent intervals, measuring the conductance of each sample, and constructing a histogram of the number of samples obtained versus sample amplitude (this procedure was first applied in the study of single channels by EHRENSTEIN et al. 1970). If we perform this procedure for a mem-

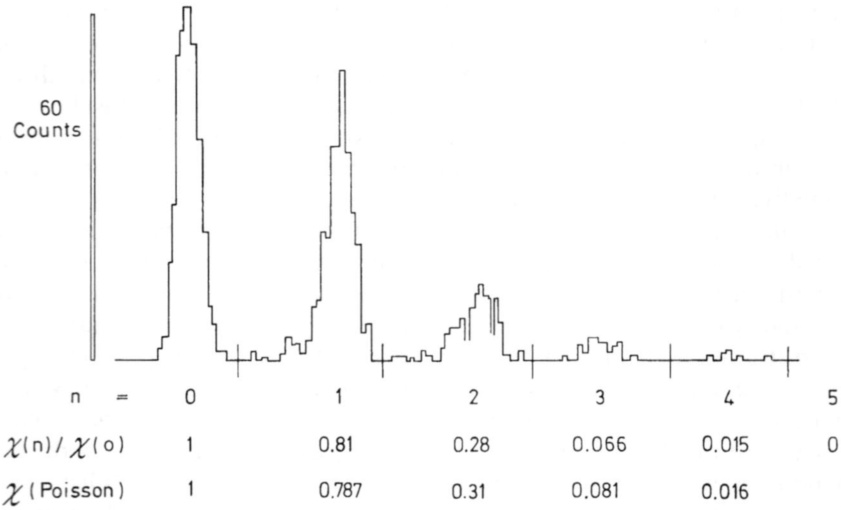

Fig. 10. Histogram of the number of open gramicidin channels versus the probability of that number being open. Probabilities of various numbers of channels being open are Poisson-distributed. This implies that opening of a single gramicidin channel is an independent event. n denotes number of channels open, and χ probability of number of channels indicated being open. Abscissa; experimentally measured current at fixed voltage; the width of the peaks is an indication of experimental resolution. (From HLADKY and HAYDON 1972, with permission of Elsevier)

brane with a small gramicidin induced conductance, the result is a histogram with regularly spaced peaks. Figure 10 is a histogram constructed by HLADKY and HAYDON (1972) for a gramicidin induced conductance. The abscissa shows the number of open pores: $n = 0$ corresponds to bare membrane conductance, $n = 1$ corresponds to one open gramicidin pore, $n = 2$ to two open pores and so on. The area under each peak is proportional to the probability that the conductance arises from the given number of pores. If the probability of a single pore being open at a given time is very small and unaffected by the conditions of other pores, the probability that n pores will be open at any given time will be Poisson-distributed:

$$P(n) = \frac{<n>^n e^{-<n>}}{n!} \tag{5}$$

where $P(n)$ is the probability that N pores are open and $<n>$ is the average number of pores open. From the distribution shown in Figure 10 we can calculate the average number of pores open and then use eq. (5) to calculate the probabilities assuming a Poisson distribution. The average number of pores open is 0.787. Hence the probability that no pores are open is 0.455. Even a casual glance at Figure 10 shows that the area under the first peak is about half the total area under all the peaks, and a quantitative comparison between the ratios of the areas under all the peaks accurately confirms the Poisson formula. This is important because it establishes that the gramicidin pores open and close independently. In other words, the probability of any particular single pore opening or closing is independent of the number of pores open.

Because of the difficulty of establishing a known concentration of gramicidin in the membrane, it is impossible to measure a meaningful rate of formation for the gramicidin pore. However, because the number of open pores is easy to determine, it is possible to measure mean pore lifetimes under a variety of conditions. HAYDON and HLADKY (1972) have shown that mean pore duration depends on membrane thickness. Pores remain open longer in thin membranes than in thick ones. This is shown in Table 2 (after HAYDON and HLADKY, 1972). These lifetimes are not inconsistent with the dissociation constant of 2 s^{-1} found by BAMBERG and LÄUGER (1973).

Table 2. (From HAYDON and HLADKY, 1972, with permission of Elsevier)

Membrane-forming lipid solution	Hydrocarbon thickness (Å)	Mean duration of single channel (s)
Glyceryl monopalmitate + n-hexadecane	26	~60
Glyceryl monooleate + n-hexadecane	31	2.2
Glyceryl monooleate + n-tetradecane	40	1.3
Glyceryl monooleate + n-decane	47	0.4
Polyhydroxystearic acid + glyceryl monooleate + n-decane	~64	~0.03
Glyceryl monooleate + cholesterol + n-decane GMO:CH, 1:1	47	~0.4

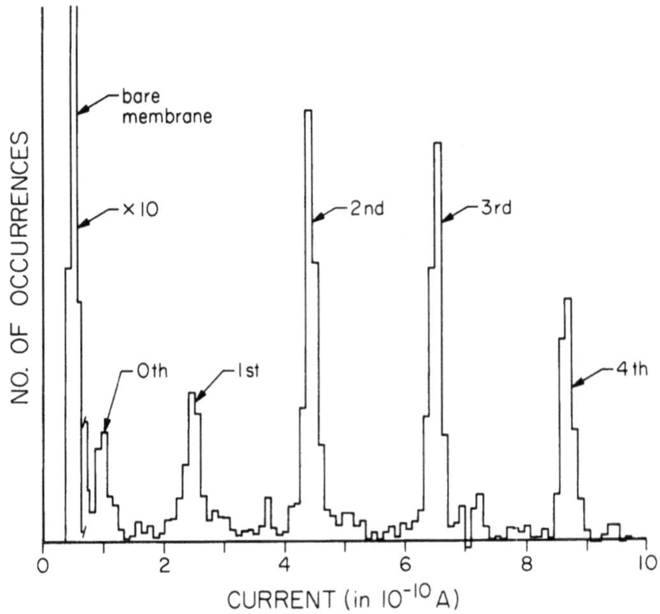

Fig. 11. Histogram of alamethicin-induced conductance versus probability with only one pore complex present. Experimental conditions such that the peaks shown correspond to different possible conductance levels of the pore complex. Note that the bare membrane conductance peak (conductance is at level it would have without alamethicin) has highest amplitude of any peak and does not fit a Poisson distribution including relative heights of the other peaks. (From EISENBERG et al. 1973)

The probability distribution experiment yields a very different result for a membrane with alamethicin induced conductance. A histogram of alamethicin single steps is shown in Figure 11. The largest peak corresponds to the probability of there being no open pores and the other peaks are labeled 0th, 1st, 2nd, 3rd, 4th. We note immediately that the peaks are not equally separated. Thus the increment in conductance of the lowest step is much smaller than the increment of conductance of the highest step. (In fact the increment increases from the lowest to the highest step.) This alone implies that the single steps are not independent, as they are in the case of gramicidin and EIM (cf. Sections C.II.2 and C.II.3). Further, the peak heights do not follow the Poisson formula, an additional argument that the steps do not arise from the opening and closing of identical independent pores. In fact, the distribution, especially the unequal spacing of the peaks in conductance, suggests that the events responsible for different conductance levels are not independent but closely associated in some unknown way.

An additional experiment confirms this view. The probabilities of the different conductance levels (in membranes of certain compositions) do not change relative to each other with increased voltage. But the probability of all the levels does increase with voltage relative to the probability of the bare membrane conductance (zero level) (EISENBERG et al., 1973; GORDON and HAYDON, 1975).

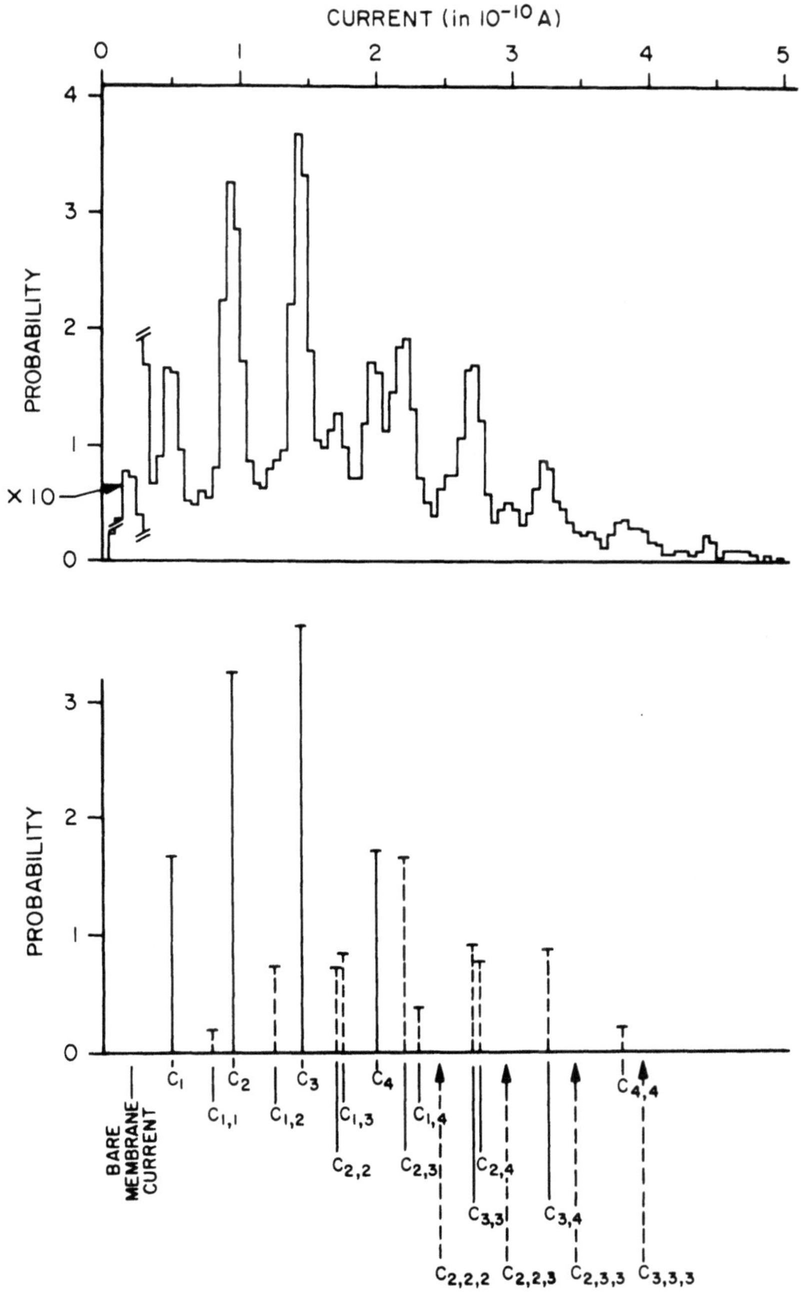

Fig. 12. Histogram of alamethicin-induced conductance versus probability with more than one pore complex present. Upper figure shows experimental data, and lower shows an analysis according to the independent event scheme described in text. C_n denotes normalized probability that pore complex is in n^{th} conducting state. Positions of lines in lower figure are constructed from the known conductances of levels as measured under conditions such that only one pore complex is present. (From EISENBERG et al. 1973)

In addition, if the voltage is increased sufficiently, the distribution changes radically. Figure 12 shows such a distribution. The most immediate difference between Figures 11 and 12 is the difference in the number of peaks. Figure 12 is much more complicated than Figure 11 and contains many more conductance levels. We can understand this apparent complexity if we assume that the alamethicin unit event is not, as for gramicidin, the formation of a pore within a single conductance state. Instead we assume that the unit event is the formation of a pore complex which can have several different conductance levels. Under this assumption the conductance fluctuations analyzed in Figure 11 correspond to one complex fluctuating among its various levels. Those analyzed in Figure 12 correspond to several complexes, all fluctuating simultaneously among their allowed levels. Since we know the conductances and relative probabilities of the levels of a single complex from data such as are shown in Figure 11, we can calculate what the relative probabilities of there being one, two, or more complexes simultaneously are from experimental data like those shown in Figure 12. It turns out that these probabilities are Poisson distributed. Thus complex formation is the independent event, but fluctuation of conductance among the allowed levels of a single complex is not. Table 3 shows an analysis of the data of Figure 12 to determine the relative probabilities of there being zero, one, two, or three complexes. The assumption that complex formation is the independent event enables one to predict correctly the relative amplitude of the zero complex peak corresponding to bare membrane current.

Table 3. Probabilities of there being one, two, three, or zero pore complexes simultaneously. (From the data of Fig. 13). The average number of open pores is taken as 1.44 in the calculation

Number of pore complexes	$\frac{P(n)}{P(o)}$ Experimental	$\frac{P(n)}{P(o)}$ Calculated
0	1	1
1	1.31	1.44
2	1.09	1.04
3	0.61	0.50

The unit events of gramicidin and alamethicin conductance are reasonably well understood, as we have shown above. Amphotericin and nystatin unit events are not accessible to experimental study since no conductance fluctuations have been observed in membranes treated with these substances. Monazomycin single steps have been observed (Muller and Anderson, 1975; Bamberg and Janko, 1976), but because of experimental complexity, the nature of the statistically independent unit event has not been determined. In fact for monazomycin there may not be only a single kind of unit conductance event that can occur but a multiplicity of different possible mechanisms. This view is supported by the complexity of monazomycin kinetics. The fact that conductances induced by gramicidin and alamethicin arise from the occurrence of many essentially identical events of a single type is fortuitous, and there is no reason to assume this will always be the case in either artificial or living systems.

2. EIM and Hemocyanin: The Unit Event Explains High Level Conductance

We also understand the unit events of hemocyanin and EIM. The two substances both show a decrease in conductance with increasing voltage, and it is instructive to discuss them simultaneously. For both, the conductance is maximum at low voltage, and if the voltage is made more positive in the compart-

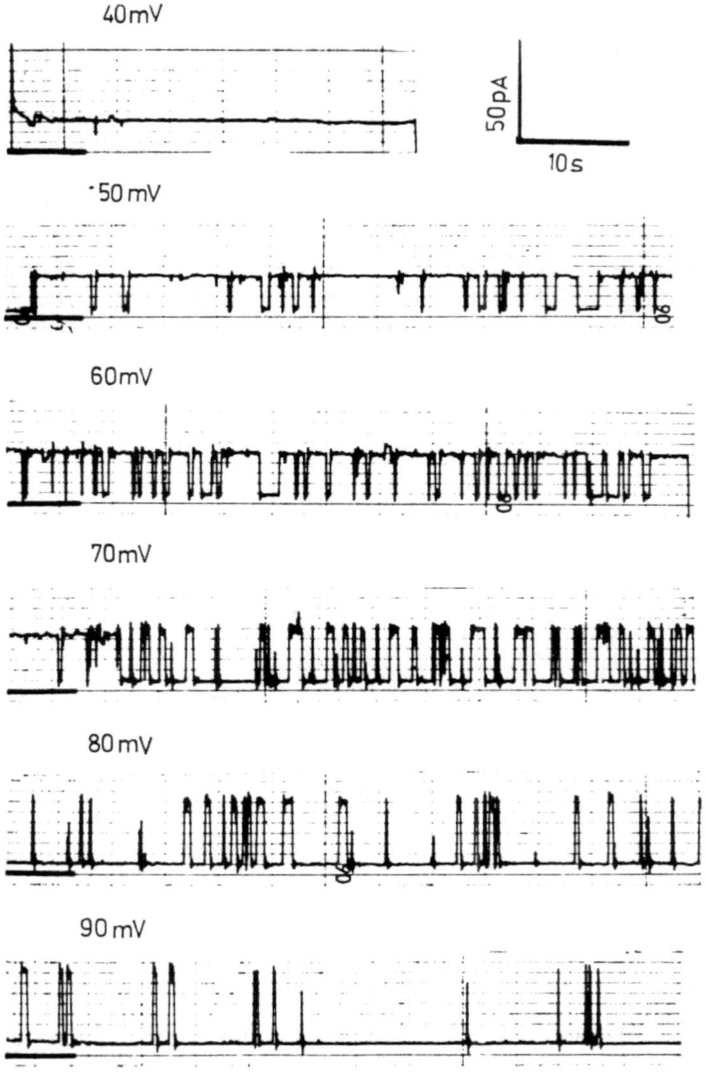

Fig. 13. Single-channel recordings of EIM at different voltages. At low voltages, the channel is almost always in high-conductance state and, as the voltage is increased, the channel spends more and more time in low-conductance state. This particularly instructive series of recordings shows just how voltage affects the relative probabilities of the channel being on or off. (From ALVAREZ et al., 1975b, with permission of Rockefeller Press)

ment to which the substance was added, the conductance decreases (see Fig. 3 for I–V curves). As it turns out this can be understood precisely in terms of the unit event.

That this so was first shown for EIM by EHRENSTEIN et al. (1970). They found that when a very small amount of EIM is added to one of the aqueous phases bathing a membrane, the conductance eventually increases in a single step to some fixed amount (at small voltages). If the voltage is kept small, the conductance remains at this level, but if the voltage is increased, the conductance eventually jumps down to a lower level, where it remains briefly before returning to the original value. The lower conductance level is still higher than the conductance of the unmodified membrane.

EHRENSTEIN et al. showed that increasing the voltage increases the frequency of downward jumps and decreases the frequency of upward jumps. The time-averaged conductance of the single pore thus decreases as the voltage is increased. Furthermore, it does so with exactly the same voltage dependence as the conductance of a membrane containing many pores. Figure 13 (after ALVAREZ et al. 1975b) is a series of single pore recordings of EIM taken at progressively increasing voltages. Note that the conductance at low voltage is almost always in the high level state, but as the voltage is increased it begins to spike downward more and more often until finally the conductance is almost always in the low state.

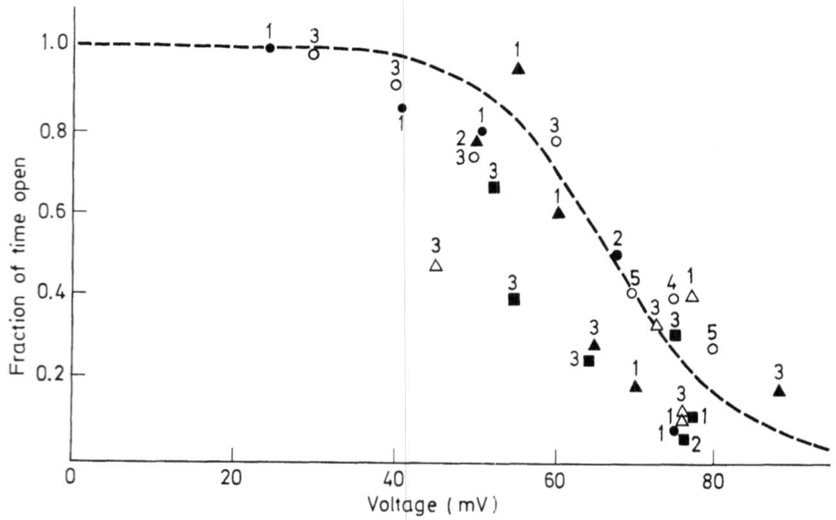

Fig. 14. Comparison of the conductance-voltage curve of a many-channel EIM membrane with the dependence on voltage of the time-averaged probability that a single channel is open. Dotted curve shows conductance of many-channel membrane divided by conductance at zero applied voltage. Points show fraction of time a single channel remains open as function of voltage. Conductance of a many-channel membrane and fraction of time a single channel remains open have same voltage-dependence. Thus, the voltage-dependence of a many-channel membrane is explained by the voltage-dependences of the turn-on and turn-off rate constants for single channel. (From LATORRE et al., 1972, with permission of Rockefeller Press)

The unit event of EIM is thus the voltage dependent opening and closing of a single pore. The statistics of the unit event regulate the conductance of a membrane containing many pores. Figure 14 compares the voltage dependence of the average conductance of a single pore to the unit conductance (LATORRE et al., 1972) and demonstrates that the voltage dependence of the two is the same.

The hemocyacin unit event differs from that of EIM in that it is considerably more complicated. Instead of having only an open state and a closed state, the hemocyanin pore has several intermediate states between the largest conductance state and the smallest[8]. The relative probabilities of these states are voltage dependent, and, as with EIM, the probability of the lower states increases as the voltage is increased. For hemocyanin, however, there is an additional complication. The conductances of the levels are themselves voltage dependent (see Fig. 15). As voltage is increased, level conductance itself decreases. Never-

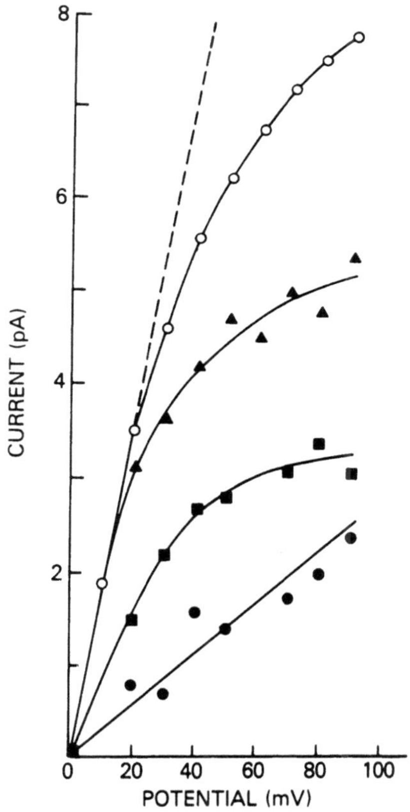

Fig. 15. Voltage-dependence of conductances of the levels of a single hemocyanin channel. (From LATORRE et al., 1975.) Note that the conductance of each level decreases with increasing voltage

[8] EIM also exhibits intermediate levels under certain conditions (Bean et al., 1969) but these events are not yet well studied.

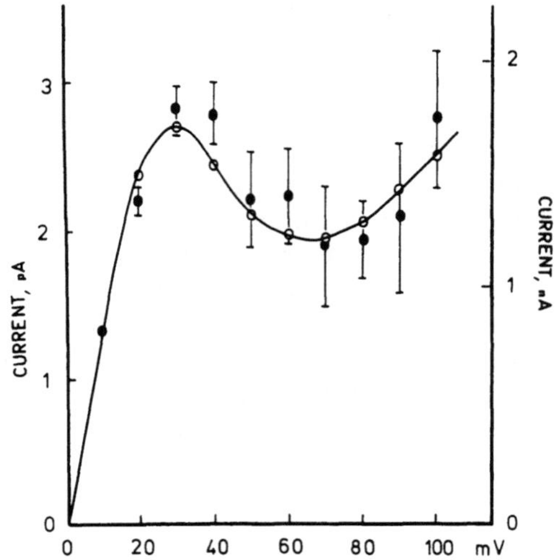

Fig. 16. Voltage-current curve of many-channel hemocyanin-containing membrane (right hand current scale) superimposed on a series of average single channel points (left-hand scale) similar to those shown for EIM in Figs 13 and 14. Average current of each histogram has the same voltage-dependence as many-channel voltage-current curve. This demonstrates that the voltage-dependence of the unit event of hemocyanin-induced channels explains the many-channel voltage-current curve. (From LATORRE et al. 1975)

theless, LATORRE et al. (1975) have shown that the time averaged conductance of a single-channel hemocyanin membrane has the same voltage dependence as the conductance of a many channel membrane. Figure 16 shows a comparison of these two measurements. The smooth trace (left-hand scale) is a many channel voltage current curve and the filled circles show the average single channel conductance (right-hand scale) at corresponding voltages. The figure shows that the shapes of the two curves are the same.

We have seen that the measured properties of the hemocyanin and EIM unit events exactly explain the measured properties of conductance in membranes with many channels. Thus, in describing the unit event (or channel) for EIM and hemocyanin, we have simultaneously demonstrated that the properties of the unit event lead directly to the properties of a membrane containing many channels. We have, however, not yet shown that the properties of the unit event of alamethicin and gramicidin can account for the conductances observed at high levels. We will now attack this problem and in doing so will have to discuss techniques of fluctuation ("noise") analysis. It is here that several interesting and instructive contacts between the study of artificial systems and living systems occur.

3. Noise Measurements and the Unit Conductance

We have seen how the properties of a single channel lead directly to the properties of a many-channel membrane for hemocyanin and EIM and we have established, at least in membranes with few channels, that a defined unit event exists for gramicidin and alamethicin. We have not shown how the properties of the unit events of gramicidin and alamethicin explain conductances of many channel membranes.

For gramicidin this can be done fairly straightforwardly by using the technique of noise analysis. ANDERSON and STEVENS (1973) and KATZ and MILEDI (1972) have used noise analysis to study the acetylcholine sensitive conductance of the frog neuromuscular junction, and FISHMAN (1973) and CONTI et al. (1975b) have applied it to the squid axon. Noise analysis of gramicidin and alamethicin is relatively simple compared to that of the physiological systems, and has the considerable advantage that we can compare the conclusions of noise analysis to the single channel data. In fact we will use the noise data accumulated for gramicidin and alamethicin to establish that the high level conductances of many membranes can be explained by the low-level single channel data. This is just the reverse of the procedure used in the physiological systems, where the noise data were the first clues as to the physiological unit conductance events.

A brief discussion of some of the basics of noise measurements will be useful. For a more complete treatment STEVENS' article (1972) is a good starting place. CONTI and WANKE (1975) have also written an excellent review detailing the applications of noise measurements to biological problems. The reader seriously interested in noise studies might consult PAPOULIS (1965), which will steer her (or him) to the important classic works. Noise analysis involves the construction by experimental methods of one of two statistical functions: the autocorrelation fucntion or the "power" spectrum. Both contain, in principle, identically the same information, since the power spectrum is the Fourier transform of the autocorrelation function. The function constructed then contains two types of information: something about the mean square amplitude of the fluctuations (the variance) and something about the rates at which these fluctuations occur. The unit of the autocorrelation function will be the square of the unit of the quantity measured: for conductance, mho^2; for current, A^2. The unit of the power spectrum will be the same as that of the autocorrelation function per unit band width (either per radian per second or per cycle per second). The autocorrelation function is a function of time and the power spectrum a function of frequency. Although the autocorrelation function and the power spectrum are equivalent, the autocorrelation function is somewhat easier to analyze, and we will conduct the following discussion in terms of the autocorrelation function.

Knowing the autocorrelation function is of course not enough. It must be analyzed. The amplitude of the autocorrelation function at zero time is always equal to the variance of the quantity measured (the variance is the square of the standard deviation). One can usually measure the mean of the quantity simultaneously or has already known it for years. For example, we can measure conductance induced by gramicidin and its variance. If we assume that the conductance arises from independent unit events (which we already know is the case, at least at low conductance levels) we can use the quantities to calculate the amplitude of the unit event.

The average conductance, $\bar{\lambda}$, will be $\bar{n}\,\Lambda$ and the variance of the conductance will be $\sigma_\lambda^2 = \sigma_n^2\,\Lambda^2$, where $\bar{n}$ is the average number of conducting pores, Λ the unit conductance of one pore, and σ_n the variance of the number of open pores. The experimental quantities are $\bar{\lambda}$ and σ_λ^2. We can construct the quantity:

$$\frac{\sigma_\lambda^2}{\bar{\lambda}} = \frac{\sigma_n^2 \Lambda}{\bar{n}} \tag{6}$$

from experimental data, and if the variance of the number of channels equals the mean, this quantity given a value for Λ, the conductance of the single channel.

NEHER and ZINGSHEIM (1974) and KOLB et al. (1975) have measured the autocorrelation function of gramicidin induced noise. Figure 17 shows the autocorrelation function of gramicidin at low conductance levels (Fig. 17a) and high levels (Fig. 17b). From the amplitude of the autocorrelation function at time zero both NEHER and ZINGSHEIM and KOLB et al find (using eq. 6) a single channel conductance value within about 10 percent of the value given by direct single channel measurements. This is good evidence that the single channel events are in fact responsible for the observed conductance at high levels. It also implies that there are no other events taking place that are washed out by high conductance.

Analysis of the time domain of the the autocorrelation function strengthens this conclusion. BAMBERG and LÄUGER (1973) have proposed a model for formation of the gramicidin channel. The model's essential features are that a conducting pore is a dimer of two gramicidin molecules and that the total amount of gramicidin in the membrane remains constant during a measurement. KOLB et al (1975) have calculated the autocorrelation function expected from this model and find that it should decrease exponentially with a single time constant. The time constant is a mix of the lifetimes of a single channel and the rate of formation of a single channel. Making the measurement in membranes with different average conductances yields the necessary information to give both constants. Both NEHER and ZINGSHEIM and KOLB et al. find that the single channel lifetimes measured in this way agree very well with those measured at the single-channel level ($\tau = 0.78 \pm 0.05$ s from autocorrelation and $\tau = 0.76 \pm 0.05$ s from single channel; NEHER and ZINGSHEIM, 1974).

Noise technique thus enables us to state with some certainty that the measured single channel properties of gramicidin are really responsible for the high level conductances. The technique will enable us to draw similar conclusions about alamethicin but an important *caveat* will arise, namely that without an experimental determination of the unit event, interpretation of noise results is somewhat risky.

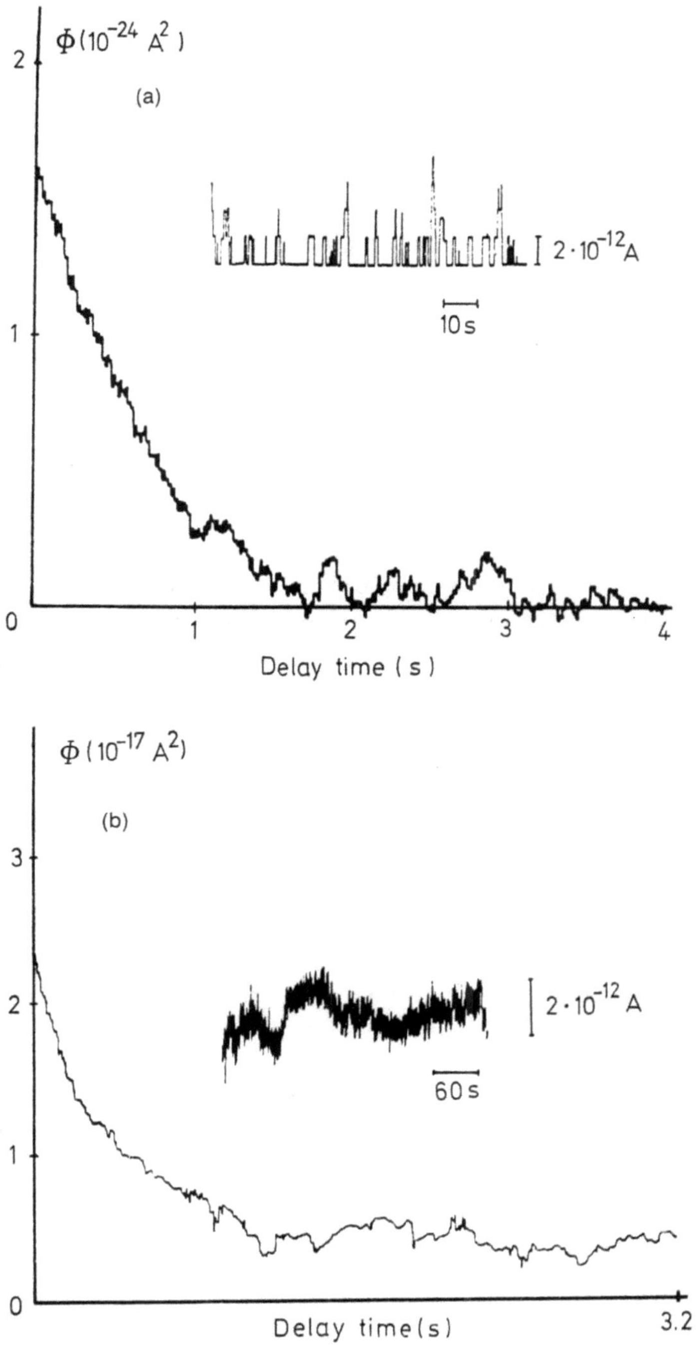

Fig. 17. (a) and (b). Autocorrelation function of gramicidin A-induced conductance for single-channel membrane (a) and many-channel membrane (b). Inset traces show the general appearance of the analyzed current in both cases. Note that time scales are similar in both cases and that the autocorrelation functions for both decay exponentially with time. (NEHER and ZINGSHEIM, 1974)

4. Noise Measurements on Alamethicin

No one has published a detailed noise study for alamethicin induced conductance, but EISENBERG et al. (1973) have measured the variance of the high conductance at several voltages, and MOORE and NEHER (1976) have published an alamethicin power spectrum more or less consistent with their results, but containing more information. These data are sufficient to say something about whether or not the unit event can explain the high level conductance.

We must first calculate what the variance of the conductance should be, given the nature of the unit event of alamethicin conductance. We can then compare the expected result with the experiment, and if the two agree, we can conclude that it is likely that the high level conductance is made up of the simultaneous occurrence of many unit events.

While calculating the entire autocorrelation function for alamethicin on the basis of unit event properties is a formidable task, it is easy to calculate the variance of the conductance (EISENBERG et al., 1973). We know already from the data of Figure 12 that the number of pore complexes present at any given time is Poisson distributed. Hence we can calculate the variance divided by the mean. Because the number of complexes is Poisson distributed, this ratio depends only on the statistical properties of a single complex:

$$\frac{\sigma_G^2}{<G>} = \frac{<\gamma^2>}{<\gamma>} = <\gamma> + \frac{\sigma_\gamma^2}{<\gamma>}, \tag{7}$$

where $< >$ indicates an average, γ is the conductance of a single complex, σ^2 denotes variance of the subscripted quantity and G means conductance measured at a high level. This result should be compared to the corresponding result obtained for gramicidin in Eq. (6). We have used G and γ to denote the alamethicin conductance and unit conductance and λ and Λ for the corresponding gramicidin conductances, following the usual notations found in the literature. Note that the variance of the alamethicin induced conductance divided by its mean is much greater than the average conductance of a pore complex.

Table 4 (EISENBERG et al., 1973) shows the variance of the conductance divided by its mean for average numbers of pore complexes (called "pores" for short) ranging from 57 to 664. The average conductance of a pore complex in 1 M NaCl is 2.5×10^{-9} mho. Note that the experimental value of the variance of the conductance (in 1 M NaCl) divided by its mean is $3.20 \pm 0.27 \times 10^{-9}$ mho. This agrees well with the value of 3.05×10^{-9} mho calculated from the known properties of the unit event.

Suppose, however, we had not known what the form of the unit event was and had assumed a single pore with only one level for alamethicin. We would have estimated its conductance at 3.2×10^{-9} mho, 30 percent higher than the average conductance and 356 percent higher than the conductance of the first level! The only reason the calculated and experimental values agree so well in Table 4 is our prior knowledge of the unit event. If we had not known the form of the unit event, we would not have known what the interpretation of the variance of the conductance divided by its mean should have been.

Table 4. (From EISENBERG et al., 1973)

1.0 M NaCl, 6×10^{-7} g ml^{-1} alamethicin		
Voltage (mV)	$\sigma_G^2/\bar{G}$ (mho)	No. of patches
24	3.25×10^{-9}	57
26	2.16×10^{-9}	109
28	3.48×10^{-9}	164
30	3.71×10^{-9}	264
34	3.38×10^{-9}	664

$\frac{\sigma_G^2}{\bar{G}} = 3.20 \pm 0.27 \times 10^{-9}$ mho (average of column 2)

$\overline{\gamma^2}/\bar{\gamma} = 3.05 \times 10^{-9}$ mho (from single channel data)

0.05 M NaCl, 2×10^{-6} g ml^{-1} alamethicin		
Voltage (mV)	$\sigma_G^2/\bar{G}$ (mho)	No. of patches
67.5	1.00×10^{-10}	10
67.5	3.13×10^{-10}	8.1
70	2.25×10^{-10}	20
70	2.23×10^{-10}	20
72	2.63×10^{-10}	31

$\frac{\sigma_G^2}{\bar{G}} = 2.24 \pm 0.52 \times 10^{-10}$ mho (average of column 2)

$\overline{\gamma^2}/\bar{\gamma} = 2.55 \times 10^{-10}$ mho (from single channel data)

This was the state of affairs in physiological noise measurement before the single channel acetylcholine receptor measurements of NEHER and SAKMAN (1976). ANDERSON and STEVENS (1973) and KATZ and MILEDI (1972) measured the variance of the acetylcholine receptor conductance divided by its mean. KATZ and MILEDI assumed a shot-noise model and Anderson and Stevens a single-step model. As it turns out Anderson and Stevens guessed right. NEHER and SAKMANN (1976a and b) measured single conductance steps in frog neuromuscular junction and found a unit event in agreement with the one estimated by ANDERSON and STEVENS. Note, however, that before NEHER and SAKMANN'S measurement, the noise measurements themselves were incapable of distinguishing between the shot-noise model and the single-step model. This is the situation now for other preparations. We have only the noise measurements and must thus be careful to understand that noise analysis alone cannot provide model independent single channel conductance estimates.

This ambiguity can sometimes be resolved partially by considering the temporal components of the noise. Nonetheless even with all the information contained in a complete power spectrum, single-channel data still may result in a new interpretation. For example, in the case of alamethicin one would expect at first approximation a power spectrum consisting of the sum of two Lorentzians, one corresponding to the formation of a single complex and one to the fluctuations between the levels of a pore. This would be the case if the levels within a pore were uniformly spaced and if the lifetimes of the states of each pore were the same. Both these conditions are approximately true (BOHEIM, 1974; EISENBERG et al., 1973; GORDON and HAYDON, 1976). But even assuming this, the power spectrum of alamethicin (MOORE and NEHER, 1976) still leaves us without any knowledge of the conductance pattern of a single pore complex. On the other hand, prior knowledge of the single channel structure and power spectrum make the macroscopic power spectrum a very powerful tool for investigating whether or not single channels act independently at high conductance levels. In fact, comparison of single channel level results with noise measurements at high conductances can yield results in principle unobtainable by either method alone. Investigation of the conductance induced by monazomycin provides an excellent illustration of how these two types of measurements can interact.

5. Noise Measurements on Monazomycin

The monazomycin induced conductance is much more complicated kinetically than that induced by alamethicin, and it presents a more difficult experimental problem because of the small size of the unit conductance steps (BAMBERG and JANKO, 1976; MULLER and ANDERSON, 1975). In fact it has not yet proven possible to show how the unit conductance induced by monazomycin is related to the macroscopic conductance. Comparison of single channel data and conductance noise induced by monazomycin gives a good idea of how these two different types of measurements support and augment each other. Two groups have measured noise spectra of monazomycin, WANKE and PRESTIPINO (1976) and MOORE and NEHER (1976). Both find a noise spectrum made up of a fast component and a slow component. MOORE and NEHER show that the fast component is not voltage dependent and that the slow component is voltage dependent. MULLER and ANDERSON (1975) and BAMBERG and JANKO (1975) have measured single channel conductances on the order of about 10×10^{-12} mho in 4 M CsCl. Both WANKE and PRESTIPINO and MOORE and NEHER estimate single channel conductances to be in the order of 10^{-12} mho in 1 M KCl, a value consistent with the single channel conductances seen in 4 M CsCl. Thus the two types of data are at first glance consistent.

Even more information is available, however, when the variation in the noise data under various conditions is examined. WANKE and PRESTIPINO found that changing the solution bathing the membranes from KCl to tetraethylammonium chloride greatly reduces the single channel conductance estimated from measurement of the power spectrum, but does not much reduce the average current under the same conditions. This, they argue, indicates that much of the unit conductance may not fluctuate. Thus a single channel conductance in the order of 10^{-12} mho, as deduced from the power spectrum or autocorrelation function, could be consistent with a channel which on the basis of macroscopic measurements apparently allows passage of tetraethylammonium ions and should therefore have a considerably greater conductance.

MOORE and NEHER note that the fast and slow relaxation processes are best studied by different methods. The fast voltage independent process is most evident in fluctuation analysis which looks at statistical deviations from equilibrium. But the slower voltage dependent process is best studied by relaxation methods (essentially the voltage pulse technique described in Section B.III.3). The voltage-dependent step may be identified with a field-dependent insertion after BAUMAN and MUELLER (1974), and the voltage-independent step may represent field-independent aggregation and dissociation. As yet, however, we have no direct evidence that this is the case.

Most importantly, we should note that while the single channel data are consistent with the conductance estimates from noise measurements, they do not indicate the existence of the voltage dependent relaxation at all! On the other hand, noise data themselves (*vide* alamethicin) do not adequately describe the structure of the unit event. A synthesis of both approaches is vital for best understanding of a given conductance mechanism.

6. Compounds with Unknown Unit Events: Summary

Alamethicin, gramicidin, EIM, and hemocyanin all have unit events with experimentally established properties which account for the conductances seen at high levels where single events cannot be resolved. The remaining compounds monazomycin, amphotericin and nystatin do not have experimentally established unit events. Of these, only monazomycin induces a conductance which can be resolved into single steps (BAMBERG and JANKO, 1976; MULLER and ANDERSON, 1975). The monazomycin unit event has not been directly connected with the macroscopic conductance, but noise measurements have shown that at least two processes must take place. Monazomycin unit conductances measured directly and by noise are consistent.

III. Conductance and Ion Selectivity of Unit Channels

1. Introduction

It is becoming increasingly clear that selectivity (and of course conductance) of a given pore for a given ion cannot depend on a single factor. DIAMOND and WRIGHT (1969) have argued that selectivity depends on a mobility factor and a binding factor. HILLE (1975) has proposed a "barrier" model which reflects the essential competition between "binding-like factors and mobility-like factors" to explain the selectivity of the sodium channel in frog nerve. EISENMAN (1962) showed that various selectivity sequences of alkali cations can be derived from consideration of dehydration energy and ion interactions with dipoles of various field strength. LÄUGER (1973) has shown how a rate theory analysis (based on a series of energy barriers) can account for selectivities (the various rates can be estimated in some cases by means of EISENMAN's approach). While by no means intended to be an exhaustive survey, these few remarks indicate that the subject of selectivity is complex and far ranging. We will confine ourselves to those aspects directly relevant to those pore formers that exhibit well studied single channel conductances.

Because of strong electrostatic repulsion, a pore that completely excludes ions of one sign will probably contain only one ion at a time. The relative probability of different ions occupying the pore will thus be an important factor in its selectivity (LÄUGER, 1973). On the other hand, if the pore is relatively large and contains an appreciable amount of water, the electrostatic repulsion between ions of like sign will be screened somewhat (although probably not as much as in bulk water) and the selectivity of the pore will be considerably reduced. There should thus be, to first order, an inverse relationship between single-channel conductance and selectivity: a low conductance pore will discriminate between ions better than a high conductance one. This trend holds in some cases. Gramicidin, with a unit conductance of 2.6×10^{-12} mho in 1 M NaCl is almost ideally cation-selective (MYERS and HAYDON, 1972) but alamethicin with a (level one) conductance of 9×10^{-10} mho in 1 M NaCl is only imperfectly cation selective (EISENBERG et al., 1973).

If only one ion can occupy a pore at one time, the conductance will be proportional to the fraction of time the pore is available to accept ions. This will be determined by the rate at which ions can enter an unoccupied pore and the rate at which ions leave an occupied pore (LÄUGER, 1973). At very low ion concentrations, almost all the pores will be unoccupied and the rate limiting step will be how fast ions can find pores. At very high ion concentrations, essentially all the pores will be occupied and the rate limiting step will be how fast ions can leave the pore. Thus, at low concentrations a voltage-current curve of a single pore will be sensitive to the rate at which ions enter the pore and at high concentrations to the rate at which ions cross the pores.

Alamethicin, EIM and hemocyanin all have more than one conducting state. Changes in selectivity between the various states may, in the light of the above discussion, be useful in clarifying how variations in pore structure result in different conductance levels. We will discuss this point in detail for alamethicin (cf. Section D.III).

Gramicidin, on the other hand, is a much simpler system with a single, well defined conductance level. It therefore provides a paradigm for the study of pore selectivities, and we will accordingly describe the results of measurements of gramicidin selectivity first and then consider the more complex cases.

2. Conductance and Selectivity of the Gramicidin Unit Event

The properties of the gramicidin unit event have been well studied (HLADKY and HAYDON, 1970; MYERS and HAYDON, 1972; NEHER, 1975; EISENMAN et al., 1976). HLADKY and HAYDON found that the single-channel conductance increases to a maximum value in solutions of NaCl, KCl and CsCl as the salt concentration is increased. From their data the concentrations at which the conductance is half of its maximum are 0.24 M for NaCl, 0.28 M for KCl and 0.67 M for CsCl. (Reciprocals are 4.2 M^{-1}, 3.6 M^{-1} and 1.5 M^{-1} respectively.) These data are for glyceryl monooleate-decane membranes and are calculated from the data given by HLADKY and HAYDON according to the formula:

$$G = \frac{G_{max} K C}{1 + K C} \tag{8}$$

where K is the binding constant, C is the concentration and G_{max} is the maximum single-channel conductance. (Concentrations greater than 2 M were not used for CsCl.) Similar data were obtained by NEHER (1975) for Na-acetate and Tl-acetate with $K_{Na} = 6.2\ M^{-1}$ and $K_{Tl} = 12.3\ M^{-1}$ in glycerol monooleate-hexadecane membranes corresponding to a concentration for half-maximal conductance of 0.16 M for Na-acetate and 0.081 M for thallous acetate. NEHER also used LÄUGER's treatment to predict the conductance of mixtures of sodium and thallous ion in sodium acetate solutions. Increasing the ratio of Tl to Na ion in a 1 M acetate solution should result in a monotonic increase in conductance. This apparently does not happen. In fact, a very small addition of Tl ion decreases the single channel conductance; adding still more thallium eventually results in an increase in conductance at about a 3% Tl/Na ratio. The results establish that more is going on than a simple exclusion mechanism, and EISENMAN et al. (1976) discuss how these and other results can be interpreted in terms of a four-site channel.

HLADKY and HAYDON (1972) have also measured the voltage-current curves for gramicidin single channels. At low concentrations of salt, where the rate of ions entering the pore would be expected to be rate limiting, they find saturating voltage-current curves (current reaches a maximum at high voltages). At high salt concentration, where the rate of translocation through the channel would be expected to be rate-limiting, they find superlinear voltage-current curves (current increases faster than linearly with voltage). From the data at high KCl concentrations, the barrier for K^+ entry into the gramicidin pore is about two-tenths of the way down the field (0.2 of the distance across the membrane if constant field can be assumed; cf. LÄUGER, 1973). HAYDON and HLADKY also report that the single-pore conductances decrease in the order $H^+ > NH^+_4 > Rb^+ > K^+ > Na^+ > Li^+$ in 0.5 M chloride solutions. These data are in agreement with bi-ionic potential measurements (MYERS and HAYDON, 1972), but curiously the higher the concentration, the higher the bi-ionic potential and the lower the conductance ratios (MYERS and HAYDON give a detailed comparison for Na^+ and K^+). Competition effects between ions are presumably responsible for these results, as discussed by EISENMAN et al., 1976.

3. Conductance and Selectivity of the Unit Events of EIM and Hemocyanin

The data on single channel conductance for EIM and hemocyanin in various salt solutions are not as extensive as those available for gramicidin. For hemocyanin, data are lacking because it has only recently been shown that hemocyanin is a pore former. In both cases, extension of measurements of open channel conductances to large voltages is difficult because of the tendency of channels to close with increasing voltage.

LATORRE et al. (1972) have measured single EIM channel conductances in 0.1 M chloride salts for various cations. They find the conductance sequence: $Cs^+ > NH_4^+ > K^+ > Na^+ > Li^+$. They also measured bi-ionic potentials as functions of salt concentration ratios. This established that the EIM channel is almost completely cation selective. They found the permeability sequence from single channel measurements to be consistent with the mobility sequence in aqueous solution. BEAN (1971) also measured the conductances of EIM sublevels in sphingomyelin membranes, but only for various concentrations of NaCl. No saturation of any conductance level occurred up to about 1 M.

Because the single step conductance does not saturate with increasing salt concentration, is proportional to ion mobility in water for a given ion, and is relatively large (about 2.6×10^{-9} mho in 1 M NaCl), the EIM pore is probably large and hydrated. Anion exclusion may result from negative charges or oriented dipoles.

Hemocyanin channels have not been much studied, but already they have shown one property exhibited by no other single channel. The conductances of the levels themselves are strongly voltage-dependent. If the voltage is increased sufficiently to turn off the channel, the conductance of the highest level itself decreases before the channel changes state and drops to a lower level (ALVAREZ et al., 1975a; LATORRE et al., 1975). In addition, the conductances of all the levels saturate with both increasing voltage and increasing concentration (LATORRE et al., 1975). Furthermore, at high salt concentration (about 1 M) hemocyanin loses its voltage-dependent conductance.

4. Conductance and Selectivity of the Alamethicin Unit Event Levels

For alamethicin many data on single channel conductances are available. Voltage-current curve data, conductance in various salt solutions as a function of concentration and even some data in aqueous-glycerol solutions. Alamethicin is the least selective of all the pore formers, probably because it forms large pores filled with water.

EISENBERG et al. (1973) found that the selectivity sequence of alamethicin for alkali cations followed the same order as their mobilities in water. In fact they found that the single channel conductances for all levels were proportional to the conductance of the bulk solutions in which the measurements were made. Even for aqueous glycerol solutions, where the solution conductance varies with viscosity of the solvent mixture for a constant molarity of salt, the single channel conductance was proportional to the conductance of the bulk solution. All these facts imply that alamethicin induced conductance arises from relatively large pores containing solvent and ions in nearly the same state as in bulk solution.

GORDON and HAYDON (1975) have investigated alamethicin single channel conductances in solutions of methyl-substituted ammonium chlorides. They find, except for the lowest level, that there is an abrupt drop in the proportionality factor between single channel conductance and bulk solution conductance as methylation is increased from $NHMe_3Cl$ to NMe_4Cl. These results are difficult to reconcile with a single pore which increases in size, and may imply that a pore complex consists of an assembly of similar pores. On the other hand, it is possible that tetramethyl ammonium absorbs specifically to various sites in the pore and thus decreases its own conductance. Experiments to test this latter conjecture have not yet been performed. This is a rather critical point: many data (EISENBERG et al., 1973; BOHEIM, 1974) are consistent with the picture of a single pore which increases in size, but the data of GORDON and HAYDON imply an assembly of pores of equal size. There is an additional puzzle to complicate things still further: the lowest level is anomalous and does not fit either scheme.

IV. Time Course of the Unit Event

We have already discussed in passing the opening and closing rates of EIM and gramicidin channels. In both cases the single channel opening and closing rates are consistent with the macroscopic kinetics of the conductances. Hemocyanin kinetics are also consistent with single channel time dependences, which, because hemocyanin has only recently been shown to be a pore former, have not been extensively studied. Monazomycin presents too severe a technical problem, and adequate single-channel kinetic data are not yet available for it. This leaves only alamethicin for detailed discussion.

We have already discussed the unit of alamethicin and found that essentially two steps are important: the formation of a pore complex and the fluctuation in conductance of the pore complex between allowed conductance levels. In Section C.II we discussed the relative probabilities of the allowed conductance levels of a single pore complex, but have left until now a discussion of the rates of fluctuation between the levels. Clearly these rates and the ways in which they vary with voltage, salt concentration, and membrane composition will be very important in any consideration of the molecular mechanism of alamethicin action. In particular, it is necessary to establish the connection, if any, between single channel kinetic parameters and those of the macroscopic conductance.

A measurement of the rate of transition from one state to another is obtained simply from appropriate analysis of constant voltage conductance data obtained under conditions where discrete conductance changes can be resolved. Having obtained a current record of sufficient duration for good statistics, one can analyze the time spent in each conductance level before an upward or a downward transition occurs and construct a histogram of number of steps observed versus step duration. For a single relaxation time the number of levels with duration longer than a given amount falls off exponentially. The probability of observing a level of duration τ is proportional to $e_0^{-\tau/\tau}$, where τ_0 is the average lifetime of the level. The exponential distribution of level duration is equivalent to the assumption that the probability of the level terminating at any given instant is constant with time[9]. This probability is, incidentally, the sum of the probabilities for upward and downward transitions. For the lowest level a downward transition corresponds to the disappearance of the pore complex, and for the highest level termination of the level can only occur by a downward transition and no upward transition can occur.

The upward transition probabilities per unit time for each level can be measured by constructing histograms of number versus duration for levels terminated only by upward transitions. Downward transition probabilities can be measured by constructing histograms of number versus duration for levels terminated only by downward transitions. For an alamethicin pore complex with five conductance levels there will thus be a set of four upward transition probabilities and four downward transition probabilities, plus the probabilities of pore complex formation and disappearance for a total of ten experimentally accessible parameters.

Clearly the ways in which these parameters vary with voltage, salt concentration, and membrane composition are very important for intelligent evaluation of possible models. Voltage dependent transition probabilities imply that changing from one conductance level to another involves movement of charge, while voltage independent transition probabilities imply that change from one conductance level to another does not involve charge movement parallel to the electric field[10]. The transition probabilities also determine the relative probabilities of the levels. The probability distributions must therefore be consistent with measurements of the transition probabilities. If the probability distribution is voltage dependent, voltage can at most alter all the transition probabilities by the same multiplicative factor. As an illustration, recall that increased voltage lowers EIM induced conductance by increasing the probability of a downward transition and decreasing the probability of an upward transition.

Boheim (1974), Mueller (1975a) and Gordon and Haydon (1976) have measured transition probabilities for alamethicin-induced conductances. Boheim has provided a complete statistical analysis of alamethicin transitions, including the voltage dependence of the transition probabilities in phosphatidyl choline membranes at 3°C. He finds that decreasing the voltage decreases the probability of downward transitions but increases that of upward transitions.

[9] The probability per unit time of level termination is $1/\tau_0$. Thus the probability of termination in any small interval Δt is $\Delta t/\tau_0$. The probability that the level endures without termination is $1 - (\Delta t/\tau_0)$. If we devide an interval of length τ into small intervals of length Δt, the probability that the level does not terminate in time τ will be the product of the probabilities that it does not terminate in any of the small intervals of length Δt. Since there are $\tau/\Delta t$ such intervals, the probability of observing an interval of length τ is $(1 - (\Delta t/\tau_0))^{(\tau/\Delta t)}$. The limit of this quantity as $\Delta t \rightarrow 0$ is $e^{-\tau/\tau_0}$. This connection between the exponential distribution of interval lengths and constant probability of decay per unit time is sufficiently fundamental to be worth pointing out.

[10] Since the applied field is perpendicular to the membrane surface, charge movement parallel to the plane of the membrane is consistent with voltage-independent transition probabilities. Voltage dependence necessarily implies charge movement but the form of the voltage dependence is not in itself adequate to determine the nature of the charge movement. Rotation of dipoles, change in dipole moment, and movement of single charges all or part of the way across the membrane can all lead to identical voltage dependences.

MUELLER found that while increasing voltage always increases the probability of upward transitions, it may either decrease or increase the probability of a downward transition depending on the composition of the membrane. The voltage dependences observed by both for almost all of the transition probabilities are exponential. BOHEIM found that in PC membranes the downward transition rates for all levels except the top and the bottom decreased e-fold for a change of about 50 mV in potential. GORDON and HAYDON measured transition rates between levels for an extensive set of experimental conditions, including varied membrane compositions, different temperature and varied potential. Their results agree in part with those of BOHEIM (1974) and MUELLER (1975a), but provide amplification and clarification of several questions. First, the voltage dependence of the rate constants for upward and downward transitions can depend on lipid composition, as noted by MUELLER. Second, the apparent activation energies for transitions between levels are all about 1.2 kcal mol^{-1} (52 meV). Finally, even at constant alamethicin concentration an increase in the electric field can cause the rate constant for an upward transition of one level to increase and that of another to decrease.

JOHNSTON and HALL (work in progress) have measured lifetimes of various alamethicin levels as a function of temperature from 25° C to 0° C. They find that lifetimes increase with decreasing temperature, in agreement with GORDON and HAYDON. The apparent activation energy is lower at high temperature than at low temperature (about 3 kcal mol^{-1} as against about 40 kcal mol^{-1}). This finding implies the existence of different processes for changing conductance level at low and high temperatures.

EISENBERG et al. (1973), HALL (1975), and GORDON and HAYDON (1975) report that the probability distribution of the levels is nearly unchanged by voltage. These measurements were recorded at 20–25° C and differ from BOHEIM's transition probabilities and probability distributions measured at 3° C, but, in addition to being obtained at different temperature, they were measured for membranes of different composition. BOHEIM finds that the ratio of the probabilities of adjacent levels change exponentially in a way consistent with his measured rate constants. (MUELLER's data also imply voltage dependent probability distributions, but he does not specify the temperature of measurement.) Differences between voltage dependences of rate constants as measured by different authors may be explained by differences in temperature or membrane composition[11]. We will discuss possible implications of these results on time dependences of the various levels in Section D.

V. Alteration of the Molecule and the Membrane

In this section we will briefly treat experiments that involve alteration of the pore forming molecules themselves or of the membrane into which the pore former is inserted. Many different membrane forming solutions have been used

[11] BOHEIM has demonstrated experimentally that his measurement involves a stationary state of the pore conductance, but the same conditions should have held for the measurements of EISENBERG et al. (1973) and GORDON and HAYDON (1975, 1976).

in various experiments, and to give a comprehensive account of all these would take far too much space and result in little more than a list of experimental results with no unifying theme. We will therefore consider only systematically organized experiments designed to answer particular questions regarding the effects of membrane composition.

The number of pore former derivatives available is small. Slightly modified forms of alamethicin, nystatin, and gramicidin (as already mentioned) are available. It may be possible to synthesize new derivatives of these compounds which will prove instructive, and we will discuss this possibility shortly.

1. Effects of Membrane Composition

We can conceive of many possible effects of lipid composition on pore formers. MUELLER (1975a) has suggested that hydrogen bonding may be important under certain circumstances. BAUMANN and MUELLER (1974) argue that membrane viscosity changes may affect the kinetics of pore formation. LEBLANC (1969) suggested that surface dipoles induced by different lipids could have a large effect on ion conductance. Recent results indicate that this suggestion is likely to be correct (HAYDON and MYERS (1975)). SZABO (1974) showed that increasing the cholesterol content of a glycerol monolein membrane increases conductances due to lipophilic anions but decreases conductances due to lipophilic cations. The effect is not small, and there can be as much as a factor of 100 difference in the conductance of a pure glycerol monooleate membrane and a membrane containing glycerol monooleate and cholesterol. These observations are easily understood in terms of LEBLANC's surface dipoles, but because the conductances of anions and cations are differently affected cannot result from changes in viscosity alone. There must be a component of electrostatic interaction, and SZABO's data indicate that the interior of a glycerol monooleate (GMO) membrane containing maximal cholesterol is on the order of 100 mV more positive than a membrane of pure GMO. Increasing the cholesterol content of the membrane forming solution lowers the water permeability of phospholipid membranes by a factor of less than 10 (FINKELSTEIN and CASS, 1968), and this effect almost certainly is a result of an increase in membrane viscosity with increasing cholesterol.

It is thus clear that altering the cholesterol content of a membrane may have two separate effects: an increase in the electrical potential of the interior of the membrane and an increase in membrane viscosity. Both of the changes could conceivably affect the process of pore formation under certain circumstances. We have already noted that sterols greatly enhance the formation of pores by amphotericin B and nystatin. It is not possible, however, to relate this effect to one either of viscosity or of electrical potential with the data currently available.

A tentative separation of viscosity and voltage effects can, however, be made for the voltage dependent pore formers, EIM and alamethicin, and the conclusion reached is in qualitative agreement with results obtained for monazomycin.

BAUMANN and MUELLER (1974) present data showing that the time constant for alamethicin pore disappearance is about 1000 times slower for membranes formed from oxidized cholesterol than for membranes formed from glycerol diolein/diolein phosphate. (Time constants were measured as described in Section B.II.3). They did not, however, take into account the different voltages at which the time constants were measured in the two systems, and did not report the functional dependence of the time constants on voltage. Both of these are very important in determining whether or not the lipid change has any effect *per se* on the membrane. For example, the time constants reported by BAUMANN and MUELLER for the diolein membranes were measured at around 75 mV, while those for the oxidized cholesterol membranes were measured at around 138 mV, a difference in potential of 63 mV. If we use the results of EISENBERG et al. for PE decane membranes that the time constant increases by a factor of e for each 9.6 mV increase in potential, we would expect the time constant to be about 1000 times slower at 138 mV than at 75 mV (exp (63 mV/9.6 mV) = 708). This is in fact about what BAUMANN and MUELLER observe.

The situation is complicated, however, because the voltage dependences of the time constants may be different in the different systems. The point is that we do not have sufficient data to decide whether or not the lipid change is in fact affecting the kinetics of pore formation and disappearance or whether the observed difference arises only because of the different voltages at which the measurements were made. We can say for sure that all of the systems under discussion will reach equilibrium faster at low than at high voltages, and consequently some of the differences reported by BAUMANN and MUELLER must be due to voltage difference and not to variation in lipid composition. This elaborate discussion of possible misinterpretation arising from failure to take into account the voltage dependence of strongly voltage dependent parameters should illustrate the necessity of quantitative treatment in analyzing data. Incidentally, it also shows that such a treatment could be done very simply, in this case just by presenting the voltage dependences of the time constants for all the lipid systems considered and comparing results obtained under equivalent conditions.

ALVAREZ et al. (1975b) have given a more complete analysis, comparing the rates of opening and closing EIM channels in oxidized cholesterol membranes with those in membranes formed from brain lipid extract. By analyzing the kinetics of channel opening and closing as a function of temperature, they calculated an activation energy between the open and closed state of 1.25 eV in oxidized cholesterol membranes and 1.00 eV in brain lipid membranes. (These measurements were made with enough voltage applied so that the energy of the open channel was equal to that of the closed.) They also found that the voltage at which the number of open channels equaled the number of closed channels was greater by 38 mV at 35° C for brain lipids than for oxidized cholesterol lipids. In this case the measured effect clearly results from a change in lipid composition and there is no possibility that another experimental variable may be responsible. BEAN (1973) found that use of different plasticizing agents (tocopherol or different esters) also alters the time constant of conductance change in response to a voltage pulse.

LATORRE et al. (1976) have attempted to explain, by a single mechanism, changes in EIM-alamethicin- and nonactin-induced conductances caused by increases in cholesterol content of phospholipid membranes. They found that increasing the cholesterol content did not much affect alamethicin turn-off kinetics and increased the voltage necessary to turn off half of the EIM channels. In addition, for a constant alamethicin concentration in the aqueous phase the voltage needed to turn on a given number of alamethicin channels increased with increasing cholesterol.

If 10 electronic charges move across the membrane to form one alamethicin channel, measurements of the voltage dependence of the kinetics (Section B.II.3) allow construction of a plot of energy versus distance moved by gating charge in going from a closed to an open alamethicin channel. This plot has two minima, corresponding to the open and closed states, and a maximum corresponding to the activation energy barrier for changing from the closed to the open state. Since the energy to move the gating charge through the low dielectric hydrocarbon is probably the major part of the activation energy barrier, it is reasonable to assume that the maximum is in the center of the membrane. This assumption fixes the positions of the minima, since their separations from the peak are determined by the voltage dependences of the formation rate (for the closed states) and the disappearance rate (for the open state).

LATORRE et al. (1976) argue that the shift in the alamethicin conductance versus voltage curve results from an alteration in relative energy of the open and closed state caused by a cholesterol-dependent change in surface dipole potential. If the decrease in nonactin conductance with increasing cholesterol is entirely attributed to a dipole potential, the magnitude of the dipole potential for a given cholesterol content can be calculated and compared with the shift in voltage necessary to turn on a fixed number of alamethicin channels in a membrane of the same cholesterol composition. LATORRE et al. find that a 100 mV change in dipole potential, as measured by means of nonactin conductance, results in a 42 mV shift in the voltage-current curve of alamethicin. They interpret this to imply that the gating charges in the closed channel position see about half of the cholesterol dipole. The gating charges are about 0.1 of the distance down the field from the surface, according to the kinetic measurements previously discussed. Hence the cholesterol induced dipole, if it started at the surface and produced a constant eletric field, would extend about 0.2 of the distance down the field. Using the geometry of the cholesterol dipole inferred in this way, LATORRE et al. estimate that the shift in voltage where half of the EIM channels are in the open state with cholesterol content should be about 60 mV for a 100-mV change in dipole potential as measured by nonactin conductance. The observed value of 45 mV is in reasonable agreement.

This study thus indicates that a fairly simple idea, alteration of dipole potential by altering lipid composition, can account for the observed changes in conductance induced by three different substances. It seems likely that electrostatic effects of surface dipoles can be important not only for the movement of charged mobile carriers like nonactin across the membrane, but for the movement of the charges associated with the gating function of voltage dependent pore formers[12].

[12] In the membranes used by LATORRE et al., increasing cholesterol decreased the conductance induced by nonactin (a positively charged carrier) but did not much increase the conductance of

The observation of MULLER and FINKELSTEIN (1972) that cholesterol-containing phosphatidyl glycerol (PG) membranes are much less sensitive to monazomycin than PG membranes without cholesterol is consistent with this result[13] (Monazomycin is positively charged.)

2. Alteration of the Pore Forming Molecule

A pore forming molecule may be altered in a known way. The alteration may affect its conductance properties and thus reveal important relations between structure and function. Even if the results obtained do not directly reveal a relation between structure and function, they may serve some other useful purpose. This is well illustrated by studies on modified gramicidin molecules by VEATCH et al. (1975). They were able to synthesize a gramicidin molecule which formed pores of different conductance from those of natural gramicidin. Even though the reasons for this conductance alteration were not understood in terms of the structure of the altered molecule, the experiments were able to make use of the conductance change itself as a diagnostic tool. They added two different types of gramicidin to the same membrane and observed not two separate pore conductances, but three. Two of the observed conductances were the conductances shown by the two types of gramicidin separately, but the third had an intermediate conductance and apparently resulted from a hybrid pore containing one molecule of each type of gramicidin. This suspicion was checked by comparing the relative concentrations of the two kinds of molecule. The probabilities of pure pore formation were proportional to the concentration squared (see Section A.II.2) of the corresponding type of gramicidin molecule, but the probability for formation of the hybrid pore was proportional to twice the product of the concentrations of the two types. This elegantly obtained result implies not only that formation of a gramicidin pore requires two molecules, but also that the structural alterations made did not much alter the ability of the two types of molecules to combine and form a pore. These gramicidin studies of VEATCH and his colleagues are perhaps the most easily interpreted studies in which the pore forming molecule has been altered, but there are several other studies worthy of mention, involving other pore formers.

negatively charged carriers like carbonylcyanide m-cholorophenyl hydrazone (CCCP). It might thus be argued that electrostatic and viscosity effects are of about equal importance. (Anion conductance does not decrease, remember.) If half the change in nonactin conductance were attributed to viscosity, the analysis of LATORRE et al. would lead to the conclusion that the cholesterol induced dipole was very short and located between 0.1 and 0.2 of the way down the field. The EIM and alamethicin data would still be consistent with an essentially electrostatic effect of the dipole potential difference between membranes containing varying amounts of cholesterol.

[13] This observation was made in passing, but this paper contains primarily an elegant analysis of the effects of surface charge on monazomycin induced conductance. MULLER and FINKELSTEIN show in particular that the monazomycin conductance is sensitive to the field across the membrane even when that field cannot be measured by external electrodes, as in the case of asymmetric surface potentials. A complete discussion is not presented here because the clarity of the original paper would suffer from the compression necessary to include it.

CASS et al. (1970) tested various derivatives of small nystatin and amphotericin B, in particular the methylesters of both and N-acetyl nystatin and N-succinyl nystatin. Methylation did not alter activity, but acetylation and succinylation reduced activity by 90 percent. Methylation also altered the rate at which the conductance changed when the antibiotic was removed from the solution. At 25 °C the half-time for conductance decrease was 20 minutes for nystatin and over 2 hours for amphotericin B, but was about 1 minute for the methyl ester derivatives. These results are not particularly enlightening, but they do agree with the results of LAMPEN (1966), which showed that acetylated forms of polyene antibiotics lose biological activity.

Several experiments have been performed with derivatives of alamethicin. LAU and HALL (unpublished results) found that the single channel conductances of methyl ester alamethicin were slightly lower than those of unmodified alamethicin. They also found that in bacterial phosphatidyl ethanolamine membranes the voltage-current curves of methyl ester alamethicin, when added to one side of the membrane, were more symmetric than those of alamethicin under similar circumstances. One possible explanation for this observation is that methyl ester alamethicin can pass through the membrane, not very rapidly, but more rapidly than ordinary alamethicin. The third quadrant branch of the I-V curve would thus result from methyl ester alamethicin which had leaked through the membrane and the asymmetry would arise because some of the material which leaked through the membrane would diffuse out into the water, thus reducing the surface concentration. Since the two branches develop with different time constants, this phenomenon must involve more than simple diffusion, which would imply that both branches would develop with the same time constant.

The reduced asymmetry seen with methyl ester alamethicin in bacterial PE-decane membranes is also seen in membranes of different composition with ordinary alamethicin, but the factors connecting the two phenomena, if they indeed are connected, are not understood.

BOHEIM et al. (1976) have investigated the properties of suzukacillin, an antibiotic produced by a different strain of the soil fungus which makes alamethicin. JUNG et al. (1975) have extensively studied the structure of suzukacillin and find that it has 23–24 aminoacids to alamethicin's 19–20. BOHEIM et al. found that suzukacillin has conductance characteristics very similar to those of alamethicin, including a multi-step unit event. Suzukacillin, however, exhibits inactivation of conductance following a voltage pulse. BOHEIM et al. interpret this result in terms of a statistical model developed earlier by BOHEIM (1974). They attribute the observed inactivation to an alteration in the constants of oligomerization. They were, however, unable to measure these constants directly.

BOHEIM et al. also note that suzukacillin shows a voltage independent conductance which develops slowly with time, and suggest that this conductance may be associated with a slow change in conformations, deduced by JUNG et al. with conducting states, and have proposed a possible structure for the suzukacillin molecule in the pore forming state.

D. Possible Molecular Mechanisms of Pore Formation

The experiments discussed above provide sufficient data, at least for some of the pore formers, to allow discussion of possible molecular mechanisms at a reasonably detailed level. But, while discussion is possible, even quite sophisticated discussion, a number of uncertainties remain. It is particularly important to emphasize that, while it is astonishing and marvelous that simple compounds like alamethicin and monazomycin can mimic certain aspects of the kinetic behaviour of the sodium and potassium channels in nerve membranes, there is no direct evidence whatsoever that the molecular mechanisms are not similar, merely no direct evidence that they are.

On the other hand, one common theme in the mode of action does suggest itself for several of the pore formers, and this idea is very useful in eliciting new experiments and approaches in the study of excitable biological membranes. This is the idea that more than one molecule of the pore forming substance is necessary to form a pore. The possible ways in which molecules can aggregate to form pores are manifold. A sort of "barrel-stave" model has been suggested by BAUMANN and MUELLER (1974), BOHEIM (1974), and LECAR et al. (1975). In this picture the pore former molecules are rod-like and form bundles like the staves of a barrel with their long axes perpendicular to the membrane. This picture is attractive as the basic mode of action for alamethicin, monazomycin, amphotericin B, nystatin, and possibly for EIM and hemocyanin. Aggregation of subunits also appears important in gramicidin induced conductances, but here two (and only two) gramicidin molecules apparently form a pore.

For the polyenes, aggregation of monomers in a barrel-stave like pattern is an attractive picture from several viewpoints. Two pieces of evidence are used: the dependence of the induced conductance on a high power of antibiotic concentration, and the size of the pore, estimated from sieving experiments. These data allow construction of a pore model involving about ten polyene molecules and having a radius consistent with the observed cut-off size for permeable molecules. This pore structure is attractive in terms of space-filling models and even suggests a role for steroid (see HOLZ, 1974; ANDREOLI, 1974). However, several unanswered questions remain. First no data are available on the effect of sterol concentration in the membrane on polyene induced conductance. Second, discrete conductance fluctuations have not yet been observed, in spite of the fact that the conductance of a single pore of the size estimated from sieving studies is well within the values routinely detected for other pore formers. Moreover, as CASS et al. (1970) have pointed out, polyene pores cannot be static but must form and break up continuously, because the conductance disappears in half an hour when the antibiotic is removed from the aqueous solutions. We thus know very little about the dynamics of pore formation and decay of polyene pores, except that they must be somewhat unusual because the expected single channel conductances are not observed. Perhaps the pores which are observed in the sieving experiments form and disappear in a continuous manner and do not have the relatively rigid structure implied by discrete conductance changes.

Gramicidin is probably the best understood and most completely studied of the pore formers, and it appears to be the simplest. Almost certainly two molecules of gramicidin combine to form a pore which must have a relatively fixed structure. Two ways in which this can be done have been proposed. In one (URRY 1971) two gramicidin molecules form identical helices which join together to form a single pore. In the second (VEATCH and BLOUT, 1974) two gramicidin molecules form intertwined parallel and antiparallel helices. Detailed studies of the single channel conductances of gramicidin pores show that competition between ions, formerly considered only in terms of carriers, can also be important in determining the conductances of pores. This is a very important finding, since many physiologically important conductance mechanisms appear more and more pore like. Notable examples are the acetylcholine receptor and the sodium and potassium channels in nerve. Gramicidin, however, shows a conductance which is only slightly voltage dependent, and such mechanisms must therefore be studied in other systems.

EIM was the first voltage dependent pore former to receive serious study, and considerable information is available about it. Unfortunately one crucial piece of information is almost completely lacking, *i. e.* the dependence of conductance on EIM concentration in the membrane. The concentration dependence is important because it provides, as for the polyenes, monazomycin and alamethicin, some estimate of the number of molecules or subunits in one pore. Interpretation of concentration dependence as actually giving the number of molecules in a single pore is very risky, but lack of such makes even a guess, which can be checked for consistency with other measurements impossible. For this reason we cannot speculate very much about the molecular mechanisms by which EIM works. Nonetheless, we can say a little. From the dependence on voltage of the time constants for opening and closing of EIM channels, we can determine a charge-distance product characteristic of the difference in charge distribution between an open and a closed state. Any model we formulate must be consistent with this number. Temperature dependence of the time constants gives an idea of the activation energy between open and closed states, and the magnitude of this energyis in agreement with estimates for the translocation energy of a charge across a region of low dielectric constant (ALVAREZ et al., 1975b). Furthermore, EHRENSTEIN and LECAR (1975) have shown that the EIM channel has two distinct off-states, occurring at slightly different voltages. This taken together with the voltage dependence and activation energy of the time constants is consistent with a model in which two to three charges flip almost all the way across the membrane in going from the open to the closed state. This implies that these gating charges would see a potential barrier similar to that seen by carrier molecules as they move across the membrane. Factors affecting the barrier for carriers should thus similarly affect the movement of the gating charges of EIM. This effect should be manifested not in a change in the conductance of the EIM channel, but in a change in the voltage dependence of conductance properties, determined as we have seen by the time constants of opening and closing of a single channel. Similar effects should be observed for other voltage dependent pore formers, a point we will return to in discussing alamethicin.

The hemocyanin I-V curve is very similar to that of EIM, but we do not really know enough about hemocyanin to speculate very much on how it forms pores.

Studies on hemocyanin are now in progress and we will have to wait and see what they reveal.

Monazomycin and alamethicin are similar in several ways. Both induce strongly voltage dependent conductances and have voltage-dependent kinetics. We can measure concentration dependence, voltage dependence, and single channels for both, although for monazomycin it has not yet proven possible to relate single channel conductances to macroscopic conductances. Several analogues of alamethicin are available, and some attempts have been made to explain the relationship of structure and function, but this cannot yet be done in more than a speculative way. What seems clear is that these molecules interact with lipids and themselves in very complex ways to form aqueous pores across the membrane. The details of the interactions are not yet well known, but for alamethicin at least we can define reasonably well what we do know and what we do not.

We are sure that alamethicin molecules can aggregate to form a pore like structure with several stable conductances, and we know that this structure once formed can fluctuate among different conductance levels. The step that is most voltage dependent is the initial formation of a single pore. We do not yet know how the pore changes its conductance level. BAUMANN and MUELLER (1974) and BOHEIM (1974) have suggested that an alamethicin monomer leaves a formed pore to produce a decrease in conductance, and joins a formed pore to produce an increase. They point out that if the monomers are arranged in the pore as the side of a prism whose cross section is a regular polygon, the area of polygons with different numbers of sides scales closely with the observed conductances of individual steps. Recall that for alamethicin the increment between conductance levels is not uniform (cf. Fig. 1 and Fig. 11). Thus the idea of a cylindrical pore increasing in area with the addition of a monomer and decreasing in area with the loss of a monomer is an attractive one. The voltage-dependent step would be the rotation by the electric field of an alamethicin monomer (or multimer) from the surface into the membrane. Molecules in this "spanned state" (EHRENSTEIN et al., 1975) then would aggregate to form cylindrical pores as described. The lateral diffusion constant of alamethicin in the spanned state is not known, but surface diffusion constants for molecules of similar molecular weight have been estimated as being around $10^{-8} - 10^{-6}$ cm^2 s^{-1} (see WEBB, 1976, for general discussion; KUO and BRUNER, 1976, for valinomycin). These values are sufficient to account for the observed rates of transition between different conducting levels. Some picture of subunits rotated into the membrane from the surface by the electric field is probably correct for alamethicin and monazomycin. A similar sort of mechanism where the field rotates subunits from a high conductance to a low conductance configuration may even be true for EIM. The details of how the subunits fit together and how the conductance of a pore once formed changes are, however, at present very uncertain.

The picture of subunits in the spanned state diffusing in the plane has the rates of aggregation and disaggregation as parameters. BAUMANN and MUELLER (1974) have calculated that certain values of these rate constants can lead to inactivation and other complicated kinetic phenomena which are observed, particularly for monazomycin.

HEYER et al. (1976) have shown, however, that inactivation in the monazomycin system can be explained on an entirely different basis. They show that monazomycin itself can go through monazomycin pores. Consequently when monazomycin is added to only one side of a membrane and a voltage is applied, so as to open pores, the monazomycin concentration on the pore forming side is initially high. This results in a rapid rise of the conductance to a high level. As pores open, monozomycin from the pore forming side rapidly diffuses through its own pores under the influence of the electric field. The monazomycin concentration on the pore forming side thus decreases and the conductance falls. Boheim has observed a similar inactivation of conductance induced by suzukacilin, and analogue of alamethicin. He has explained this result on the basis of a rate constant theory similar to that of BAUMANN and MUELLER.

The rate constants for all of these systems can in principle be measured from the time course of single channel conductances, but no one has yet measured a set of rate constants from single channel measurements and then compared observed kinetic behavior of many channel membranes with that predicted by the measured rate constants.

It is possible that the subunits tend to aggregate on the surface of the membrane before they are rotated into the membrane by the electric field. This picture would imply that single channel conductance changes arise by alteration in configuration of an aggregate of closely associated molecules, and not by association and dissociation of monomers. We are not even sure that the discrete steps of the unit event arise because a single pore changes its cross sectional area (the "barrel-stave" hypothesis). GORDON and HAYDON (1975) have shown that alamethicin single channels show a sharp cut-off in conductance with increasing size of methyl-substituted ammonium chlorides, and they interpret these results as implying that discrete steps of the unit event arise from the opening of closely associated single pores with each open pore, and that closely associated single pores can exhibit the observed conductance pattern of the alamethicin unit event.

We are thus reasonably safe in saying that alamethicin, its analogues, and monazomycin all aggregate to form pores, but we do not know how the field exerts its influence to drive the pore formers into the membrane for alamethicin. For monazomycin, it probably acts on the molecule's intrinsic positive charge, but for alamethicin it is not yet certain that the molecule has its own positive charge, although a recent structure proposed by MARTIN and WILLIAMS (1975) seems to indicate that it might. The binding of cations seems unlikely in view of the work of EISENBERG et al. (1973) and GORDON and HAYDON (1972), who found that cations of different valence produced no detectable change in the voltage-current curve. PRESSMAN (1968) has reported that alamethicin extracts cations into organic solutions, but the experiments were not described in detail and have not been repeated. CHERRY et al. (1972) do report an increase in slope of the I-V curve with increasing cation valence, of a sort consistent with cation binding as the source of the positive charge. GORDON and HAYDON (1975) suggest that the molecule may have a structural dipole moment. We do not yet know the details of the aggregation or even whether it occurs before or after the field has driven the pore forming molecules into the spanned state.

We do not know whether or not single pores are formed in close cooperation, or whether one pore increases in size with the addition of monomers, or even whether or not both processes can occur. Because monazomycin has more complicated kinetics than alamethicin it seems likely that it can interact with itself to form more different kinds of aggregates than can alamethicin, but this is by no means certain. Essentially we have only begun to study the details of the aggregation process and can expect considerable progress in the near future.

There are several experiments, in principle possible, that should be able to answer some of the questions implied above.

First, because monazomycin and alamethicin induce voltage dependent conductances, charge must move in the membrane to turn on the conductance. This movement should manifest itself as a gating current similar to the gating currents observed in nerve membranes (ARMSTRONG and BEZANILLA, 1973; KEYNES and ROJAS, 1973). The nature of the gating current would tell us a great deal. It could tell us how much charge must be moved to open a single channel, and thus enable us to see whether or not subunits aggregate before or after they enter the membrane. If aggregation is after entering the gating current, measurements allow an estimate of how far a subunit diffuses before interacting with another subunit.

Most importantly, the study of pore formers in black lipid membranes has clearly shown that a synthesis of results derived with many different techniques is necessary to build up a reasonable understanding of any type of induced conductance. If this is so for these relatively well defined systems, it is reasonable to assume that a similar approach will be necessary to make sense of conductance mechanisms in biological membranes, where the additional problem of defining and isolating the molecules in the conductance system still remains.

Acknowledgements

I would like to thank *Osvaldo Alvarez, Ramon Latorre* and *Magdalena T. Tosteson* for their helpful comments and critical readings of the manuscript.

References

ALVAREZ, O., DIAZ, E., LATORRE, R.: Biochim. biophys. Acta **389**, 444 (1975a).
ALVAREZ, O., LATORRE, R., VERDUGO, P.: J. gen. Physiol. **65**, 421 (1975b).
ANDERSON, C. R., STEVENS, C. F.: J. Physiol. **235**, 655 (1973).
ANDREOLI, T. E.: Kidney International **4**, 337 (1973).
ANDREOLI, T. E.: Ann. N. Y. Acad. Sci. **235**, 448 (1974).
ANDREOLI, T. E., MONAHAN, M.: J. gen. Physiol. **52**, 300 (1968).
ANDREOLI, T. E., DENNIS, V. W., WEIYL, A. M.: J. gen. Physiol. **53**, 133 (1969).
ARMSTRONG, C. M., BEZANILLA, F.: Nature **242**, 459 (1973).

Bamberg, E., Janko, K.: Biochim. biophys. Acta **426**, 447 (1976).
Bamberg, E., Läuger, P.: J. Membrane Biol. **11**, 177 (1973).
Bauman, G., Mueller, P.: J. supramolec. Struct. **2**, 538 (1974).
Bean, R. C.: J. Membrane Biol. **7**, 15 (1971).
Bean, R. C.: In: Membranes, Vol. 2 (G. Eisenman, Ed.). New York: Dekker 1973.
Bean, R. C., Shepherd, W. C., Eichner, J.: J. gen. Phsiol. **53**, 741 (1969).
Boheim, G.: J. Membrane Biol. **19**, 277 (1974).
Boheim, G., Janko, K., Liebfritz, D., Ooka, T., Konig, W., Jung, G.: Biochim. biophys. Acta **433**, 182 (1976).
Cass, A., Finkelstein, A., Krespi, V.: J. gen. Physiol. **56**, 100 (1970).
Cherry, R. J., Chapman, P., Graham, D. E.: J. Membrane Biol. **7**, 325 (1972).
Conti, F., Wanke, E.: Quart. Rev. Biophys. **8**, 451 (1975).
Conti, F., De Felice, L. J., Wanke, G.: J. Physiol. **248**, 45 (1975).
Diamond, J. M., Wright, E. M.: Ann. Rev. Physiol. **31**, 581 (1969).
Ehrenstein, G., Lecar, H.: Biophys. J. **15**, 167a (1975).
Ehrenstein, G., Lecar, H., Nossal, R.: J. gen. Physiol. **55**, 119 (1970).
Ehrenstein, G., Blumenthal, R., Latorre, R., Lecar, H.: J. gen. Physiol. **63**, 707 (1974).
Eisenberg, M., Hall, J. E., Mead, C. A.: J. Membrane Biol. **14**, 143 (1973).
Eisenman, G.: Biophys. J. **2**, 2: 259 (1962).
Eisenman, G., Sandblom, J., Neher, E.: In: Metal Ligand Interactions in Organic and Biochemistry (9th Jerusalem Symposium) (B. Pullman and N. Goldblum, Eds). Dordrecht: Reidel 1976.
Finkelstein, A., Cass, A.: J. gen. Physiol. **52**, 145 (1968).
Fishman, H. M.: Proc. nat. Acad. Sci. (Wash.) **70**, 876 (1973).
Gisin, B. F., Kobayashi, S., Hall, J. E.: Proc. nat. Acad. Sci. (Wash.) **74**, 115 (1977).
Gordon, L. G. M.: In: Drugs and Transport (B. A. Callingham, Ed.). Baltimore, London, Tokyo: University Park Press 1974, p. 251.
Gordon, L. G. M., Haydon, D. A.: Biochim. biophys. Acta **255**, 1014 (1972).
Gordon, L. G. M., Haydon, D. A.: Proc. Phil. Trans. roy. Soc. B **270**, 433 (1975).
Gordon, L. G. M., Haydon, D. A.: Biochim. biophys. Acta **436**, 541 (1976).
Hall, J. E.: Biophys. J. **15**, 934 (1975).
Haydon, D. A., Hladky, S. B.: Quart. Rev. Biophys. **5**, 187 (1972).
Haydon, D. A., Myers, Valerie B.: Biochim. biophys. Acta **307**, 429 (1975).
Haydon, D. A., Hladky, S. B., Gordon, L. G. M.: In: Mitochondria; Biomembranes. Amsterdam: North-Holland 1972, p. 307.
Heyer, E. J., Muller, R. U., Finkelstein, A.: J. gen. Physiol. **67**, 731 (1976).
Hille, B.: J. gen. Physiol. **66**, 535 (1975).
Hladky, S. B., Haydon, D. A.: Nature **225**, 451 (1970).
Hladky, S. B., Haydon, D. A.: Biochim. biophys. Acta **274**, 294 (1972).
Holz, R. W.: Ann. N. Y. Acad. Sci. **235**, 469 (1974).
Holz, R. W., Finkelstein, A.: J. gen. Physiol. **56**, 125 (1970).
Jung, G., Dubischar, N., Liebfritz, D.: Europ. J. Biochem. **54**, 395 (1975).
Katz, B., Miledi, R.: Nature **226**, 962 (1970).
Katz, B., Miledi, R.: J. Physiol. **224**, 665 (1972).
Keynes, R. D., Rojas, E.: J. Physiol. **233**, 28P (1973).
Kolb, H. A., Läuger, P., Bamberg, E.: J. Membrane Biol. **20**, 133 (1975).
Kuo, K. H., Bruner, L. J.: J. Membrane Biol. **26**, 385 (1976).
Lampen, J. O.: In: Biochemical Studies of Antimicrobial Drugs (16th Symposium for the Society of General Microbiology) (B. A. Heston and P. E. Reynolds, Eds). Cambridge, Mass.: Cambridge University Press 1966, p. 111.
Latorre, R., Ehrenstein, G., Lecar, H.: J. gen. Physiol. **60**, 72 (1972).
Latorre, R., Alvarez, O., Ehrenstein, G., Espinoza, M., Reyes, J.: J. Membrane Biol. **25**, 163 (1975).
Latorre, R., Alvarez, O., Hall, J. E.: Biophys. J. **16**, 80a (1976).
Läuger, P.: Biochim. biophys. Acta **311**, 423 (1973).
LeBlanc, O. H.: Biochim. biophys. Acta **193**, 350 (1969).
Lecar, H., Ehrenstein, G., Latorre, R., Ann. N. Y. Acad. Sci. **264**, 304 (1975).
McLaughlin, S., Eisenberg, M.: Ann. Rev. Biophys. Bioeng. **4**, 335 (1975).

MARTIN, D. R., WILLIAMS, R. J. P.: Biochem. Soc. Trans. **3**, 166 (1975).
MARTIN, D. R., WILLIAMS, R. J. P.: Biochem. J. **153**, 181 (1976).
MAURO, A., NANAVATI, R. P., HEYER, E.: Proc. nat. Acad. Sci. (Wash.) **69**, 374 (1972).
MOORE, L. E., NEHER, E.: J. Membrane Biol. **27**, 347 (1976).
MUELLER, P.: In: MTP International Reviews of Science. Biochemistry Series 1, Vol. 3 (E. Racker, Ed.). London: Butterworths; Baltimore, Md: University Park Press 1975a.
MUELLER, P.: Ann. N. Y. Acad. Sci. **264**, 247 (1975b).
MUELLER, P., RUDIN, D. O.: Nature **217**, 731 (1968).
MULLER, R., ANDERSON, O. S.: In: Fifth International Biophysics Congress. Copenhagen: IUPAC III (1975).
MULLER, R., FINKELSTEIN, A.: J. gen. Physiol. **60**, 263 (1972a).
MULLER, R., FINKELSTEIN, A.: J. gen. Physiol. **60**, 285 (1972b).
MYERS, Valerie B., HAYDON, D. A.: Biochim. biophys. Acta **274**, 313 (1972).
NEHER, E.: Biochim. biophys. Acta **401**, 540 (1975).
NEHER, E., SAKMANN, B.: Biophys. J. **16**, 154a (1976a).
NEHER, E., SAKMANN, B.: Nature **260**, 799 (1976b).
NEHER, E., ZINGSHEIM, H. P.: Pflügers Archiv **351**, 61 (1974).
PAPOULIS, A.: Probability, Random Variables and Stochastic Processes: McGraw Hill, New York 1965
PAYNE, J. W., JAKES, R., HARTLEY, B. S.: Biochem. J. **117**, 757 (1970).
PRESSMAN, B.: Fed. Proc. **27**, 1283 (1968).
ROY, G.: J. Membrane Biol. **24**, 71 (1975).
SMEJTEK, P.: Chem. and Phys. Lipid **13**, 141 (1974).
STARK, G., KETTERER, B., BENZ, R., LÄUGER, P.: Biophys. J. **11**, 981 (1971).
STEVENS, C. F.: Biophys. J. **12**, 1028 (1972).
SZABO, G.: Nature **212**, 47 (1974).
TOSTESON, D. C., ANDREOLI, T. E., TIEFFENBERG, M., COOK, P.: J. gen. Physiol. **51**, 373 (1968).
URRY, D. W.: Proc. nat. Acad. Sci. (Wash.) **68**, 672 (1971).
VEATCH, W. R., BLOUT, E.: Biochem. **13**, 5257 (1974).
VEATCH, W. R., MATHIES, R., EISENBERG, M., STRYER, L.: J. molec. Biol. **99**, 75 (1975).
WANKE, E., PRESTIPINO, G.: Biochim. biophys. Acta **436**, 721 (1976).
WEBB, W. W.: Quart. Rev. Biophys. **9**, 49 (1976).

Subject Index

Absorption 373
Adsorption 374
Affinity of the chemical reaction 145
Agonists 349
Alamethicin, analogues 524
– conductance and concentration 489
– effects of lipids 520, 522
– I–V curves 486, 487
– kinetics for voltage pump 494
– noise measurements 510
– single channels 480
– – –, conductance and selectivity 516
– – –, models 527
– – –, probability distribution 500
Aldosterone 361
Alleles 247
Amino acid transport: 5-fluorotryptophan resistant cells 251
Amphotericin B, analogues 524
–, conductance and concentration 491
–, permeabilities and reflection coefficients 482
–, pore formation model 525
Amplifiers, bias current 176
–, driver shield 177
–, input resistance 176
–, negative capacitance 178
Anions; "blocking" theory 428
–, effect of interfacial potentials 421
–, influence of cholesterol 423
–, kinetics of charge translocation 415
–, tetraphenylboron in bilayers 406
–, three capacitor model 430
–, transport model in bilayers 408, 416
Antidiuretic hormone (bladder) 360
– –, asymmetry of action 361
– –, effects on protein phosphorylation and cation fluxes 361
Antiserum: anti L 249
Autocorrelation function in noise measurements 507

Beta-adrenergic receptor (erythrocytes) 350
– –, agonists binding 351
– –, effects on adenylate cyclase 353
– –, effects on cation transport 352
– –, effects on protein phosphorylation 358
– –, – and relation to cation fluxes 359
Beta-galactoside transport: E. Coli 243
Bilayer conductance and pore former concentration 488
– –, discrete steps 479, 481
– –, voltage dependence 488
Binding protein 245
Born energy 379

Ca^{++}-ATPase antibodies 298
Calcium Pump (sarcoplasmic reticulum) 274
– –, catalytic properties 277
– –, general properties 274
– –, reconstitution liposomes 277
– –, structural properties 276
Carriers 449, 478
–, kinetic analysis of carrier-mediated transport 466
–, macrocyclic (charged) 461
–, macrocyclic (uncharged) 456
– of hydrogen 450
–, rate constants (valinomycin, trinatin) 471
–, transport model 463
Catecholamines (see beta-adrenergic receptor) 350
– and Ca^{++} 364
– in heart muscle 364
Cattle erythrocytes 317
Charge pulse method in carrier transport 469
Charge translocation (bilayers) and membrane composition 419
– –, blocking theory 428
– –, charge-pulse measurements 413
– –, kinetics 415
– –, models 408, 416
– –, noise analysis 414
– –, three capacitor model 430
– –, voltage-clamp measurements 410
Channels 478
–, lifetimes 517
–, noise measurements 506
Chemical potential 145, 162, 164
Chemotaxis 215
Chemiosmotic hypothesis 260, 272
Cholera enterotoxin 362

Colera enterotoxin
– – and catecholamine-induced effects 362
Chromosome 241
Cistron 242
Coefficient of friction 8
Complementation analysis 241
Conductance, anions in bilayers 404
–, unstirred layers 454, 464
Convection 7
– with diffusion 13
– flux 9
Cultured cells 250
– –, somatic 250
Cystinuria 246, 247

Diffusion 6
–, coefficient 10
–, –, H_2O in bulk solvents 396
–, distribution coefficient 17
–, driving force 11
–, equation 14
–, Fick's law 10, 11
–, flux 9, 10, 11, 12
– and convection 53
– –, concentration profile 57
– –, equation of motion 54
– –, flux determination 55
– –, stationary transport through a membrane 55
– –, steady-state concentration profile 55
– –, unidirectional fluxes and flux ratio 56
– and migration 12
– –, Smoluchowski's equation 12, 48
–, macroscopic treatment 13
–, – –, diffusion out of a plate 32
–, – –, distribution coefficient 17
–, – –, equilibrium state 15
–, – –, establishing the stationary concentration profile 37
–, – –, Fick's second law 14
–, – –, permeability coefficient 17
–, – –, quasistationary processes 21
–, – –, solutions of diffusion problems by means of Green's function 29, 30, 31
–, – –, stationary diffusion through two different media 19
–, – –, stationary state 15
–, – –, steady-state in a plate 16
–, – –, time dependent processes 20
–, – –, unidirectional fluxes 26
–, microscopic aspects, Brownian movement 40
–, – –, diffusion coefficient and mobility 51
–, – –, Kramer's Equation 51
–, – –, Einstein-Smoluchowski equation 41, 45
–, – –, Einstein-Stokes' relation 53
–, – –, Einstein's relation 52
–, – –, random walk in one dimension 42
–, – –, Smoluchowski's treatment 41, 48
Diffusion polarization 468
Diffusion potential 80, 91, 172
– –, charging and redistribution time 82
– –, electrical equivalent circuit for the Planck Regime 87
– –, Henderson Regime 84
– –, Planck Regime 85
– –, Planck's general relations 86, 87, 90
Diploid 247
Donnan equilibrium 68
– –, concentration and potential profiles between the phases (Poisson-Boltzmann equation) 73
– –, thermodynamic treatment 68
– potential 64
Dopaminergic receptors (superior cervical ganglion) 363

EIM
–, conductance and selectivity 516
–, effect of lipids 521
–, formation model 526
–, I–V curves 485
–, kinetics for voltage jump 494
–, single channels 480, 503
Electrochemical potential 163
Electrodiffusion 7, 58–112
–, conductance 59
–, constant field 65
–, current density 59
–, diffusion potentials 80
–, Donnan equilibrium 68
–, electrical mobility 59
–, electrochemical potential 62, 63
–, electrodiffusion through membranes 91–110
–, entropy production 157, 167
–, equilibrium potential 66
–, equivalent electrical circuit 94
–, equivalent electrical circuit for the ion-selective membrane 67
–, flux ratio 98
–, Goldman regime 99
–, ion selective membrane 96
–, membrane equilibrium 65
– through membranes electroneutrality 95
–, membranes separating electrolytes having a common ion 98–110
–, molar mechanical mobility 62
–, Nernst equation 65, 66
–, Nernst-Planck electroneutrality condition 64
–, – equation 61
–, nonosmotic equilibrium 65
–, partial conductivity 60
–, Poisson equation 63
–, single salt 91
–, total conductivity 60
Erythrocyte membrane, fatty acids 219
– –, general composition 207

– –, lipid asymmetry 225
– –, – composition 215
– –, localization of fatty acids 220
– –, phospholipase-action 224
– –, phospholipids 218
– –, proteins 233
Erythrocytes (see also by species)
–, cation concentration (bovine) 207
–, cattle 317
–, density separation 210
–, glycolitic rates (mammalian) 209
–, goat 315
–, HK/LK sheep 247
–, lamb 249
–, sheep 302
Erythropoiesis 313
Escherichia coli, transport mutants 240
Ethyl methane sulfonate 253

Fick's second law 14
Flux 6
–, diffusion limited 377
– ratio 98
– – equation 122, 151, 161, 167
– – –, derivation 123
– – –, deviations 132
– – –, nonelectrolytes 130
– – –, nonindependent fluxes 131
– – –, single file diffusion 137
– – – with solvent drag 127
–, stationary 376
–, tracer measurements (bilayers) 403

Galactose binding protein 245
– transport in E. coli 245
Gene, regulators 239
–, structural 239
Genes, cultured somatic cells 250
Genetic analysis 241
– –, complementation 241
– –, cultured somatic cells 250
– –, HK/LK erythrocytes 247
– control 239
– mutation 239
Gibbs-Duhem equation 142
Goat erythrocytes, antibody-modified cation transport 315
– –, cations and antigens 315
Goldman Regime 99
– –, concentration profile and membrane potential 104
– –, ionic conductances and membrane potential 106
– –, ionic currents and diffusion potential 100
– –, membrane current and membrane potential 102
Gramicidin channels, analogues and conductance 523
– –, conductance and selectivity 516
– –, models 525
– –, probability distribution 498
–, dansyl-gramicidin A and single channel conductance 489
–, lifetime 499
–, noise measurements 506
–, single channels 480
Gravity 7

Hemocyanin, conductance 516
–, I–V curves 485
–, kinetics for voltage jump 494
–, single channels 480, 485, 503
Histidine transport in Salmonella 244
HK/LK erythrocytes 247
Human erythrocytes, En(a) antigen 319
– –, Rh-antigen and transport 318

Image-force energy potential 381
Interfacial potentials and hydrophobic ions 421
– –, cholesterol effects 423
– –, diffuse double-layer 387
– –, dipole potentials 388
Isotope flow 141
– –, nonequilibrium thermodynamic approach 156
– –, nonthermodynamic considerations 148
– flows, chemical picture 142
– –, isotope picture 146

Kedem-Essig equation 150

Lactose transport, E. Coli 243
Lipid bilayers, capacitance 371
– –, conductance 370
– –, hydrophobic anions: effect of interfacial potentials 421
– –, –: effect of lipids 419
– –, –: kinetics of charge translocation 415
– –, –: transport models 408, 416
– –, hydrophobic effect 390
– –, interfacial potentials 385
– –, partition coefficients 393
– –, permeability 391
– –, permeability coefficients (nonpolar solutes) 397, 399
– –, permeation barriers 400
– –, potential energy barrier 390
– –, potential energy of ions 379, 380
– –, solute mobility 395
– –, stationary flux 376
– –, transient flux 376
– –, unstirred layers 375
Lymphocytes, biochemical changes 326
–, cation transport 327
–, changes in protein synthesis 337
–, changes in cell and membrane morphology 324
–, changes in transport properties 330
–, differentiation 321
–, divalent cations 338
–, immunological reactions 324
–, ligand binding 322

M protein 243
Macrocyclic carriers (see carriers)
Macrotetrolides (see also carriers) 459, 470
–, rate constants 471
Mechanical mobility 8
Membrane conductance, unstirred layers 454, 464
Membrane potential 171
– –, Amphiuma red cells 197, 199
– –, Ehrlich cells 195
Membrane resistance 92
– –, Amphiuma red cells 197
– –, Ehrlich cells 197
– –, urinary bladder 200
Microelectrode, filling 180
–, junction potential 180
–, resistance 183
–, suspension effect 181
–, tip diameter 179
–, tip potential 180, 183
Microorganisms, transport mutants 240
Migration flux 7
Mitochondria, ATPase 266
–, oligomycin sensitive proteins 269
–, oxidation chain 261
Monazomycin, conductance and concentration 489
–, kinetics of voltage jump 494
–, noise measurements 512
–, single channels 480
Muscarinic cholinergic receptors (superior cervical ganglion) 363
Mutagens 253
Mutants 239, 270
–, cultured somatic cells 250
–, temperature sensitive 241

N-ethyl maleimide 243
Na^+-K^+ ATPase, catalytic properties 283
–, ouabain resistent mutants 254
–, reconstitution in liposomes 284
–, sheep erythrocytes 248
–, structural properties 282
–, antibodies 293
– –, binding (anti-holoenzyme) 295
– –, immunological effects on cation fluxes 297
– –, – on enzymatic activity 295
– –, – on parital reactions 296
– – species and organ specificity 297
Na^+, K^+ transport, ouabain resistant cells 251, 252, 253
Nernst equation 65, 66
Neurospora crassa, transport mutants 240
Nicotinic cholinergic receptors (superior cervical ganglion) 363
– – – and ion channels 364
Noise measurements 506
Nonequilibrium thermodynamics 141
Nystatin, analogues 524
–, conductance and concentration 489
–, permeability and reflection coefficients 482

Onsager's reciprocal relations 157, 158
Ouabain, resistant cells 251, 252, 253 (see also Na^+-K^+ ATPase)

Partition coefficient and Born energy 380
– –, nonpolar solutes 393
– –, polar solutes 394
Passive transport 5
Permeability coefficients, H_2O in bilayers 397
– –, organic solutes in bilayers 399
Piezoelectric transducer 191
Poisson-Boltzmann equation 73
Pores 478
–, effects of lipids 520
–, models 525
–, probability distribution 498
–, voltage-current characteristics 488
Potential energy, barrier 390
– –, dipolar molecules in bilayers 385
– –, hydrophobic effect 390
– –, ions in bilayers 379, 380, 381
– –, macrocyclic ion carrier 465
Protein kinase hypothesis (mechanism of action of c-AMP) 365
Proton translocation, ATP-dependent 264
– –, bacterio rhodopsin 263
– –, models (mitochondria) 271
Purine transport, azaguanine resistant cells 251

Rate constants (see also carriers) 471
– – in carrier transport 466
Receptor 349
Reciprocity relations, Onsager 157, 158
Red cell (see erythrocyte)
Resistant cells, azaguanine 251
– –, bromodeoxyuridine 251
– –, colchicine 251
– –, 5-fluoro tryptophan 251
– –, ouabain 251, 252, 253

Saccharomyces cervisial, transport mutants 240
Salmonella typhimurium, transport mutants 240
Sheep erythrocytes, antigenic sites and pump sites 308
– –, cations and genetics 301
– –, effects of antibodies: cation fluxes 305
– –, HK/LK 247
– –, Na^+-K^+-ATPase 304, 307
– –, properties of antibodies 311
– –, properties of surface antigens 310
– –, transport and antigens in development 312
– –, transport properties, antigens and genetics 302, 304
Sieving 481

Sodium-potassium-activated ATPase 293 (see Na^+-K^+ ATPase)
Superior cervical ganglion 363

Temperature sensitive, transport mutants 241, 243
Thermodynamics, nonequilibrium 141–168
Thymidine transport, bromodeoxyuridine resistant cells 251
Tracer coupling 159, 162
– fluxes, exchange diffusion 120
– –, isotope effects 118
– –, non-steady state 126
– –, unidirectional 118
Transference number 185
Transport coefficients 157, 158, 159
Transport mutants 240
– –, cultured somatic cells 250
– –, energy uncoupled 244
– –, methods of selection 251
– –, micro-organisms 240
– –, temperature sensitive 241
Trinactin 471 (see also carriers)
Tritium suicide 251
Tumor cells 339
– –, Lectin-induced changes 339

Uncouplers, action on bilayers 455
– as ionophores 450
Unstirred layers 375
– –, chemical reactions 378
– –, effects on anion permeability 403, 405
– –, influence on stationary fluxes 376
– – and membrane conductance 454, 464
Ussing's first law 149
Ussing's second law 151

Valinomycin 459, 470 (see also carriers)
–, rate constants 471
Vasopressin (see antidiuretic hormone) 360
Voltage jump method, in carrier transport 468
– – –, in pore-formers studies 492

Walden's Rule 395

Biochemistry of Membrane Transport

FEBS-Symposium No. 42
Editors: G. Semenza, E. Carafoli
1977. 392 figures, 76 tables. XIX, 669 pages
(Proceedings in Life Sciences)
ISBN 3-540-08082-1
This book contains 47 articles, dealing with various aspects of the topic "Biochemistry of Membrane Transport". The area covered spans the molecular architecture of biological membranes and the interaction of membrane lipids and proteins, the chemistry of ionophoric molecules and various theoretical aspects of transport processes, with a detailed discussion of the properties of several natural and reconstituted transport processes. Distinctive for the book is its interdisciplinary character, a quality frequently found in membrane transport research. A further characteristic is the balance between established knowledge, offered in some review-type articles and the numerous contributions presenting the most recent advances in the field.

22. Colloquium der Gesellschaft für Biologische Chemie, 15.-17. April 1971 in Mosbach/Baden

The Dynamic Structure of Cell Membranes

Editors: D.F. Wallach, H. Fischer
1971. 87 figures. IV, 253 pages
ISBN 3-540-05669-6
From the reviews:
"...Drs. F. Wallach and H. Fischer invited eminent biologists to participate at the 22nd Mosbach Colloquium, and the book *The Dynamic Strcture of Cell Membranes* is the outcome of this Colloquium...
The book starts off with a chapter on Molecular Membranology by Dr. F.O. Schmitt. He deals very elegantly with some interesting aspects of membranology such as membrane-gene linkage, membrane behaviour in cell transformation and malignancy. Dr. R. Auerbach's paper on communications between cells is concise but informative. I enjoyed the paper by Dr. U. Hämmerling on cell surface antigens. Dr. S. Hakomori deals with membrane surface changes associated with malignant transformation. He suggests that malignancy is due to incomplete synthesis of carbohydrate chains. Drs. H.D. Klenk's and L.I. Rothfield's papers on structure and assembly of viral and bacterial membranes respectively are of much interest, particularly the aspect of assembly, translocation and migration of membrane components.
In short, this is an excellent book. Cell and molecular biologists will find this a useful reference book for research and also teaching. The book will be a very useful addition to any good library."

Canadian Journal of Genetics and Cytology

Heidelberg Science Library
Volume 18
D.F.H. Wallach

The Plasma Membrane: Dynamic Perspectives, Genetics and Pathology

1972. 27 figures. XI, 186 pages
ISBN 3-540-90047-0
Contents: Genetics. Isolation, Fractionation and Biochemical Properties. Special Methods of Study. Membrane Models and Model Membranes. Medical and Paramedical Aspects. Membrane Effects of Radiation. Transport Defects.

Heidelberger Science Library
Volume 22
M.H. Saier Jr., C.D. Stiles

Molecular Dynamics in Biological Membranes

1975. 47 figures, 8 tables. XIII, 129 pages
ISBN 3-540-90142-6
Contents: Introduction: Cell Structure and Function. – Constituents of Biological Membranes. – Structure of Membranes and Serum Lipoprotein Complexes. – Biological Consequences of Membrane Fluidity and Fusion. – Transmembrane Solute Transport Mechanisms. – Sensory Perception I: Chemoreception. – Sensory Perception II: Transmission Mechanisms. – Hormonal Regulation of Cellular Metabolism. – Cell Recognition. – Role of the Plasma Membrane in Growth Regulation and Neoplasia.

Springer-Verlag
Berlin Heidelberg NewYork

MIX
Papier aus verantwortungsvollen Quellen
Paper from responsible sources
FSC® C105338

If you have any concerns about our products,
you can contact us on
ProductSafety@springernature.com

In case Publisher is established outside the EU,
the EU authorized representative is:
Springer Nature Customer Service Center GmbH
Europaplatz 3, 69115 Heidelberg, Germany

Printed by Libri Plureos GmbH
in Hamburg, Germany